D0985441

2nd edition

physiology

Williams & Wilkins

Managing Editor: Debra Dreger
Project Editors: Martha Gay, Susan Keller
Production: Keith LaSala, Laurie Forsyth
Illustration: Wieslawa B. Langenfeld, Laura Barton, Patricia MacAllen
Composition and layout: June Sangiorgio Mash, Port City Press, Inc.

Library of Congress Cataloging-in-Publication Data

Physiology / [editors] John Bullock, Joseph Boyle III, Michael B.
   Wang. — 2nd ed.
         p.   cm. — (The National medical series for independent study)
      Includes index.
   ISBN 0-683-06258-1 (pbk. : alk. paper)
   1. Human physiology—Outlines, syllabi, etc. I. Bullock, John,
1932-   . II. Boyle, Joseph. III. Wang, Michael B. IV. Series.
   [DNLM: 1. Physiology—examination questions. 2. Physiology-
-outlines.   QT 18 P5784]
   QP41.P492   1990
   612—dc20
   DNLM/DLC
   for Library of Congress   90-5257
                                                                       CIP

ISBN 0-683-06258-1

©1991 Williams & Wilkins

Printed in the United States of America. All rights reserved. Except as permitted under the Copyright Act of 1976, no part of this publication may be reproduced or distributed in any form or by any means or stored in a data base or retrieval system, without the prior written permission of the publisher.

10 9 8 7 6 5 4 3 2

The National Medical Series for Independent Study

2nd edition
physiology

John Bullock, M.S., Ph.D.

Associate Professor of Physiology
New Jersey Medical School
University of Medicine and Dentistry
* of New Jersey*
Newark, New Jersey

Joseph Boyle, III, M.D.

Associate Professor of Physiology
Clinical Assistant Professor of Medicine
New Jersey Medical School
University of Medicine and Dentistry
* of New Jersey*
Newark, New Jersey

Michael B. Wang, Ph.D.

Professor of Physiology
Temple University
* School of Medicine*
Philadelphia, Pennsylvania

NMS

National Medical Series from Williams & Wilkins
Baltimore, Hong Kong, London, Sydney

Harwal Publishing Company, Media, Pennsylvania

Contents

Comprehensive Examination 421

Index 475

Preface

Since its publication in 1984, NMS *Physiology* has garnered considerable response from medical students around the world. Based on these written and oral commentaries, the authors have revised the text to conform better to the needs of students preparing for course examinations and standardized medical or other health science examinations. Accordingly, this new edition of NMS *Physiology* contains a well-defined fund of knowledge the authors consider essential for medical practitioners. Although this book does not focus specifically on pathology, it does provide insight into the pathophysiologic aspects of disease by comparing normal processes in the body with the abnormal.

The outline format has allowed the authors to emphasize the relative importance of the facts presented and to provide students with a source of fundamental information about physiology. Such a core guide to the discipline should prove especially valuable now that medical students are expected to assume more and more responsibility for their own learning.

John Bullock
Joseph Boyle, III
Michael B. Wang

Preface

Acknowledgments

We collectively wish to express our thanks to the professional staff at the Harwal Publishing Company, particularly to Matthew Harris (Medical Editor), Debra Dreger (Managing Editor), and Martha Gay (Project Editor) for their personal support, perseverance, understanding, and expertise. We also recognize the special contribution of Weislawa Langenfeld as medical illustrator. Our greatest measure of gratitude goes to Debra Dreger, who has earned our utmost respect for her energy and commitment to the NMS series.

The authors

To Barbara, who, through her love and willing sacrifice, encouraged me to complete two graduate degrees in physiology and biochemistry. Also to Laura, John, and Katherine—all important well-springs of inspiration. Not to be forgotten are the many medical and dental students together with the foreign medical graduates who have become friends and colleagues over the years.

John Bullock

To Patti, for her love and support over many years. To many colleagues who have made the study of physiology an exciting and stimulating career. And, to the many students who have asked the questions which have enlightened me.

Joseph Boyle, III

My thanks to family, friends, colleagues, and students for their encouragement and support over many years.

Michael B. Wang

To the Reader

Since 1984, the National Medical Series for Independent Study has been helping medical students meet the challenge of education and clinical training. In this climate of burgeoning knowledge and complex clinical issues, a medical career is more demanding than ever. Increasingly, medical training must prepare physicians to seek and synthesize necessary information and to apply that information successfully.

The National Medical Series is designed to provide a logical framework for organizing, learning, reviewing, and applying the conceptual and factual information covered in basic and clinical studies. Each book includes a concise but comprehensive outline of the essential content of a discipline, with up to 50 study questions. The combination of distilled, outlined text and tools for self-evaluation allows easy retrieval and enhanced comprehension of salient information. Each question is accompanied by the correct answer, a paragraph-length explanation, and specific reference to the text where the topic is discussed. Study questions that follow each chapter use current National Board formats to reinforce the chapter content. Study questions appearing at the end of the text in the Comprehensive Exam vary in format depending on the book; the unifying goal of this exam, however, is to challenge the student to synthesize and expand on information presented throughout the book. Wherever possible, Comprehensive Exam questions are presented in the context of a clinical case or scenario intended to simulate real-life application of medical knowledge.

Each book in the National Medical Series is constantly being updated and revised to remain current with the discipline and with subtle changes in educational philosophy. The authors and editors devote considerable time and effort to ensuring that the information required by all medical school curricula is included and presented in the most logical, comprehensible manner. Strict editorial attention to accuracy, organization, and consistency also is maintained. Further shaping of the series occurs in response to biannual discussions held with a panel of medical student advisors drawn from schools throughout the United States. At these meetings, the editorial staff considers the complicated needs of medical students to learn how the National Medical Series can better serve them. In this regard, the staff at Harwal Publishing welcomes all comments and suggestions. Let us hear from you.

1
Neurophysiology
Michael B. Wang

I. CELLULAR HOMEOSTASIS

A. Homeostasis is the process by which an organism maintains the composition of the extracellular fluid (ECF) and intracellular fluid (ICF) in a **steady-state** condition. The composition of the ECF and the ICF differ from each other (Table 1-1; see also Table 4-3A).

1. **Extracellular fluid** consists of the blood plasma and interstitial fluid. The composition of the ECF is maintained by the cardiovascular, pulmonary, renal, gastrointestinal, endocrine, and nervous systems acting in a coordinated fashion.

2. **Intracellular fluid.** The composition of the ICF is maintained by the cell membrane, which mediates the transport of material between the ICF and ECF by diffusion, osmosis, and active transport.

B. Cell membrane. All animal cells are enveloped by a cell membrane that is composed of a variety of lipids and proteins.

1. **Lipids.** The lipids (i.e., primarily phospholipids, cholesterol, and glycolipids) are formed into a bilayer (Figure 1-1) and are amphipathic; that is, they have a hydrophilic polar region at one end of the molecule and a hydrophobic hydrocarbon tail at the other.
   a. The **hydrophilic** ends of lipid molecules, which contain the phosphate group, line up facing the ICF and ECF.
   b. The **hydrophobic** ends of the molecules face each other in the interior of the bilayer.

2. **Proteins.** According to the fluid mosaic model of membrane structure, membrane proteins insert into (and can float within) the bilayer and are anchored by covalent bonds. Some proteins span the entire bilayer; others are contained within the intracellular or extracellular half of the bilayer. Membrane proteins serve several functions (see Figure 1-1).

Table 1-1. Major Ionic Components of the Intracellular and Extracellular Fluids

Ion	Intracellular Concentration		Extracellular Concentration	
	(mOsm/L)	(mEq/L)	(mOsm/L)	(mEq/L)
Na^+	15	15	140	140
K^+	135	135	4	4
Ca^{2+}	10^{-4}	2×10^{-4}	2	4
Mg^{2+}	20	40	1	2
Cl^-	4	4	120	120
HCO_3^-	10	10	24	24
HPO_4^{2-}	10	20	2	4
SO_4^{2-}	2	4	0.5	1
Proteins$^-$, amino acids$^-$, urea, etc.	99	152	0.5	1

Note—The osmolarity of the intracellular fluid (ICF) is the same as that of the extracellular fluid (ECF). Also, the milliequivalents of cations and anions are equal in the ICF as well as in the ECF.

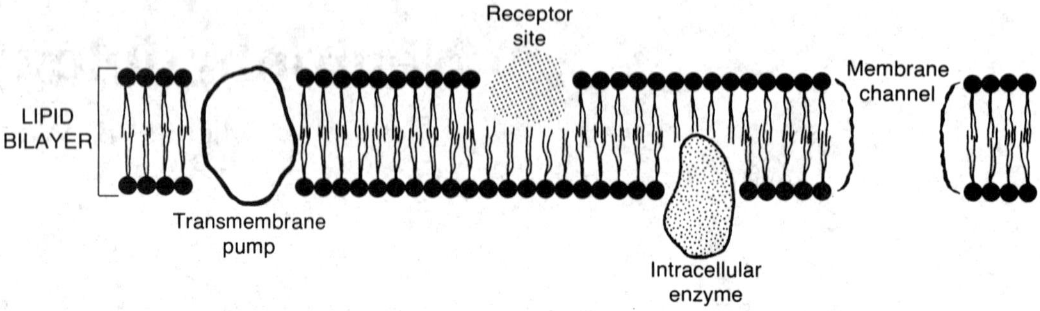

Figure 1-1. Diagram indicating some of the functions performed by proteins within the lipid bilayer of biologic membranes.

    a. Some proteins (**transmembrane proteins**) have active sites on both sides of the bilayer. These proteins form:
        (1) Channels, through which small, water-soluble substances can diffuse
        (2) Carriers, which are used to transport material across the bilayer
    b. Other proteins have an active site on only one side of the membrane.
        (1) Receptor sites for antibodies, hormones, neurotransmitters, and pharmacologic agents usually are contained within proteins located on the **outer surface** of cell membranes.
        (2) Enzymes used to activate or inactivate various metabolic intermediates usually are contained within proteins found on the **inner surface** of cell membranes.

C. Translocation across the cell membrane. Ions, nutrients, and waste products of metabolism are transferred across the cell membrane by simple diffusion and a variety of carrier-mediated processes.

    1. Passive transport processes do not require energy. They are downhill processes.
        a. Simple diffusion occurs because all particles in solution are in constant motion.
            (1) Although a single particle moves in an unpredictable (random) direction, it is more likely that the particle will move from an area where it is highly concentrated to an area where its concentration is low than it is for the particle to move in the opposite direction. Thus, a particle moves down its concentration gradient by diffusion.
            (2) Net movement ceases when the concentration of the particle is equal everywhere within the solution (diffusional equilibrium). Although random movement of the particles does not cease, the concentration remains the same.
            (3) Fick's law of diffusion describes the rate of diffusion through a membrane, which increases as the concentration gradient increases.

$$\text{flux} = -\frac{D \times A}{d}(C_{in} - C_{out})$$

            where flux = the amount of material (mmol) moved per unit of time; D = the diffusion coefficient (cm^2/sec), which is characteristic of the material being moved and the membrane through which it is being moved; A = the area (cm^2) of the membrane; d = the diffusion distance, or thickness of the membrane (cm); and C_{in} and C_{out} = the concentrations of the material (mmol/L or mmol/1000 cm^3) on the inside and outside of the membrane, respectively. The negative sign indicates that the material is moving down its concentration gradient.
            (4) Permeability. Fick's law can be simplified when biologic membranes are considered, because the thickness of the membrane always is about 10^{-6} cm. Dividing D by 10^{-6} yields the permeability coefficient (P) of the membrane, and Fick's law becomes

$$\text{flux} = -P \times A \,(C_{in} - C_{out})$$

            Note that P = D/d and that the units of permeability are cm/sec.
            (5) Diffusion of lipid-soluble materials occurs through the lipid bilayer of the cell membrane. The diffusion coefficient of lipid-soluble materials is proportional to their lipid solubility, which often is expressed as the oil-water partition coefficient.
            (6) Diffusion of water-soluble materials occurs through the aqueous channels formed by transmembrane proteins. The diffusion coefficient of water-soluble materials is

proportional to their molecular size. Particles too large to fit through the aqueous channels (particles with diameters greater than 0.8 nm) cannot cross the cell membrane by simple diffusion.

b. Facilitated diffusion requires a carrier. Some particles (e.g., glucose) are too large to diffuse through membrane channels. However, they can be moved across the membrane by a carrier-mediated process that does not require energy (Figure 1-2A).

   (1) The particle binds to a carrier on one side of the membrane and, while bound to the carrier, is transported through the membrane. The particle dissociates from the carrier when it reaches the other side of the membrane. Because the particle binds to the carrier only if it is highly concentrated, the particle can only move **down** its concentration gradient by this process.

   (2) The rate of diffusion increases as the concentration gradient increases until all of the carrier sites are filled. At this point, the rate of diffusion can no longer increase with increasing particle concentration. This is called **saturation, or Michaelis-Menten, kinetics** (see Figure 1-2B).

2. Active transport processes require energy. They are uphill processes.

   a. Primary active transport. Active transport processes using adenosine triphosphate (ATP) are called **direct energy utilizing, or primary active transport,** processes.

      (1) The most common of these active transport systems is the **sodium-potassium (Na^+-K^+) pump, or Na^+-K^+-ATPase,** which uses the membrane-bound ATPase enzyme as a carrier molecule. The Na^+-K^+ pump is responsible for maintaining the high K^+ and low Na^+ concentrations in the ICF (Figure 1-3). The operation of the pump can be divided into three steps:

         (a) Three ions of Na^+ bind to the carrier on the inside of the cell

         (b) Two ions of K^+ bind to the carrier on the outside of the cell

         (c) The carrier uses the energy obtained from one molecule of ATP to move two ions of K^+ into the cell and three ions of Na^+ out of the cell

         (d) Inhibition of the carrier will occur if the intracellular ATP or Na^+ concentration or the extracellular K^+ concentration is too low. The carrier also can be inhibited by several therapeutic agents, such as **digitalis.**

      (2) Other carriers with a direct energy source are available to transport a variety of ions, such as chloride (Cl^-), calcium (Ca^{2+}), and hydrogen (H^+).

   b. Secondary active transport. Carrier-mediated active transport systems that use the energy stored in the Na^+ concentration gradient are called **indirect energy utilizing, or secondary active transport,** processes. The system can operate only if the extracellular Na^+ concentration is higher than the intracellular Na^+ concentration. Energy is required to establish the

A

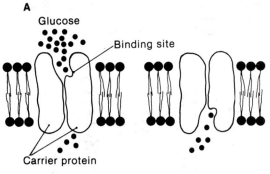

Glucose
Binding site
Carrier protein

B

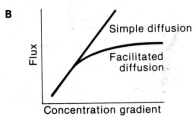

Flux
Simple diffusion
Facilitated diffusion
Concentration gradient

Figure 1-2. (*A*) The transmembrane protein responsible for the facilitated transport of glucose undergoes a conformational change when glucose binds to the transporter on one side of the membrane. The conformational change allows glucose to diffuse across the membrane down its concentration gradient. (*B*) The rate of transport reaches a maximum when all binding sites on the transporter are filled.

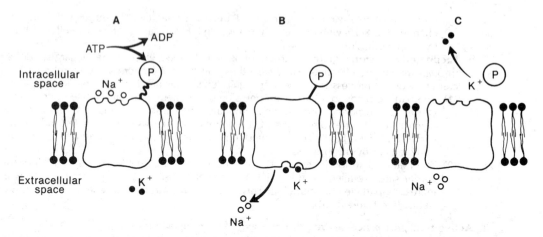

Figure 1-3. (*A*) The energy contained in the high-energy phosphate bond is used to transport three ions of Na$^+$ out of the cell. (*B*) A second conformational change occurs after two K$^+$ ions bind to the carrier, transporting K$^+$ into the cell and (*C*) causing the phosphate to dissociate from the carrier.

Na$^+$ concentration gradient, and it is this energy that is used indirectly in the active transport process.

(1) Transport of glucose and amino acids. Glucose and amino acids are transported against their concentration gradients by a secondary active transport process (Figure 1-4).

   (a) Na$^+$ binds to the carrier on the outside of the cell where the Na$^+$ concentration is high. After Na$^+$ binds to the carrier, the carrier's affinity for glucose increases, enabling it to bind glucose.

   (b) When Na$^+$ and glucose are bound to the carrier, the carrier undergoes a conformational change, during which both glucose and Na$^+$ are exposed to the inside of the cell.

   (c) Because of the low intracellular Na$^+$ concentration, Na$^+$ dissociates from the carrier, reducing the carrier's affinity for glucose. This allows glucose to dissociate from the carrier despite the relatively high intracellular glucose concentration.

   (d) Na$^+$-dependent secondary active transport of glucose and amino acids is used primarily for the transport of these nutrients by epithelial cells of the nephron and the intestine.

(2) Transport of Ca^{2+}. Calcium is transported out of the cell against its concentration gradient by a secondary active transport process.

   (a) Three Na$^+$ ions are transported into the cell for each Ca^{2+} ion transported out of the cell.

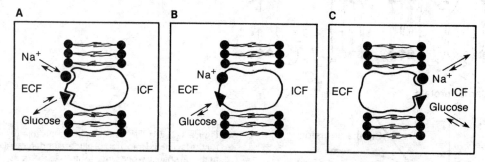

Figure 1-4. Glucose is transported through the membrane by active transport that uses an indirect energy source. (*A*) As indicated by the length of the *arrows*, the affinity of the carrier for glucose is low in the absence of Na$^+$. (*B*) However, after Na$^+$ binds to the carrier, the affinity for glucose is increased. Glucose and Na$^+$ diffuse through the membrane together. (*C*) Once inside the cell, Na$^+$ is removed from the carrier, the affinity for glucose is reduced, and glucose dissociates from the carrier.

(b) The Na^+-Ca^{2+} exchanger is very important for the regulation of intracellular Ca^{2+} and, thus, contractile force in cardiac cells (see Ch 2 II D 2 a).

(3) Other carrier-mediated secondary active transport processes, such as those that pump synaptic transmitter substances into nerve terminals (see III C 1) or that secrete H^+ ions into the proximal tubule (see Ch 5 VI B 1), are vital to normal cellular homeostasis.

(4) Carrier types

(a) Uniporters are carriers that transport a single particle in one direction, such as the facilitated diffusion of glucose.

(b) Symporters transport two particles in the same direction, such as the secondary active transport of glucose.

(c) Antiporters transport molecules in opposite directions, such as the Na^+-Ca^{2+} and Na^+-H^+ exchangers.

3. Filtration and osmosis. In filtration and osmosis, water is forced to flow across a biologic membrane by **hydrostatic pressure.** In the kidney, for example, the capillary blood pressure created by the heart causes plasma water to be filtered out of the glomerular capillaries and into the nephron. In the peritubular capillaries, the osmotic pressure created by the high concentration of plasma proteins causes water to flow from the interstitial space surrounding the nephron into the capillaries.

a. Osmotic pressure is produced when a **semipermeable membrane** (i.e., a membrane permeable only to water) separates two solutions that have different concentrations of particles.

b. Although it is not known how the particle concentration gradient creates an osmotic pressure, the pressure across the membrane can be calculated using the **van't Hoff equation**

$$\pi = \Delta c \times R \times T$$

where π = the osmotic pressure (mm Hg), Δc = the difference in the concentration of particles between the two solutions (mOsm/L), R = the natural gas constant (62 mm Hg $\times$ L/mmol $\times$ ° K), and T = absolute temperature (° K)

c. Osmotic pressure is a **colligative property** of solutions; that is, it is related to the number of particles dissolved in the solution.

(1) Each mmol/L of a particle produces an osmotic pressure of about 19 mm Hg.

(a) For example, a 1 mmol/L glucose solution produces an osmotic pressure of 19 mm Hg, and a 1 mmol/L sodium chloride (NaCl) solution produces an osmotic pressure of 38 mm Hg (because two particles of solute exist for each molecule of NaCl).

(b) Therefore, the osmotic concentration of a substance in solution reflects the number of particles dissolved and not the number of molecules. Thus, a 1 mmol/L glucose solution has an osmotic concentration of 1 mOsm/L, and a 1 mmol/L NaCl solution has an osmotic concentration of 2 mOsm/L.

(2) Osmolarity and osmolality. As a unit of concentration, **osmolarity** is a measure of the moles of solute/L of solution, whereas **osmolality** is a measure of moles of solute/kg of water. If the volume of solute is small (as it is for most biologic solutions), the osmolarity and osmolality of the solution will be approximately equal and, thus, the two terms can be used interchangeably.

d. Change in cell volume due to osmotic pressure differences. Changes in plasma osmolarity cause cells to shrink or swell, as water flows out of or into the cells due to the osmotic pressure difference between the inside and outside of the cell.

(1) Calculations of cell volume change. The final volume of the cell subjected to a change in extracellular osmolarity can be calculated by the expression

$$\pi_i \times V_i = \pi_f \times V_f$$

where i and f refer to the initial and final osmolarities and volumes of the cell.

(a) For example, if a red blood cell with an initial volume of 100 μ^3 and a normal osmolality of 285 mOsm/L is placed in a solution with an osmolality of 325 mOsm/L, its final volume will be

$$285 \times 100 = 325 \times V_f$$
$$V_f = 88 \ \mu^3$$

(b) The volume of water leaving the red cell is so small compared to the volume of the extracellular solution that the osmolarity of the extracellular solution is assumed to remain constant.

(2) Osmotic pressure differences can only cause a steady state change in cellular volume if the particles creating the osmotic pressure difference are impermeable to the membrane. For example, because urea easily flows through cell membranes, a change in the extracellular concentration of urea does not cause a steady state change in cell volume.

(3) The ability of a particle to cause a steady state change in cell volume is referred to as its **tonicity**. An extracellular solution that causes water to flow into the cell, thus making it swell, is called **hypotonic**. An extracellular solution that causes water to flow out of the cell is called **hypertonic**. A solution causing no change in intracellular volume is called **isotonic**.

e. **Flow of water due to osmotic pressure differences.** When a membrane separates two solutions that have different concentrations of nonpermeable particles, the osmotic pressure difference will cause water to flow across the membrane.

(1) The water will flow from the solution with the lower concentration of particles to the solution with the higher concentration of particles (Figure 1-5A).

(a) The flow of water can be calculated using the **osmotic flow equation**

$$\text{flow} = L \times A \, (\pi_1 - \pi_2)$$

where L = the hydraulic conductivity (μL/sec/cm^2/mm Hg) of the membrane, A = the area (cm^2) of the membrane, and π_1 and π_2 = the osmotic pressures on either side of the membrane.

(b) This equation is used to calculate the glomerular filtration rate (see Ch 4 V A 1 c).

(2) The osmotic pressure produced by a concentration difference can be determined experimentally by measuring the amount of hydrostatic pressure that must be applied to the solution with the higher particle concentration (higher osmotic pressure) to prevent water from entering it from the solution with the lower particle concentration (lower osmotic pressure; see Figure 1-5B).

f. **Reflection coefficient.** If the membrane separating two solutions is permeable to the particles in solution, the osmotic pressure created by the concentration difference will be reduced. The relative permeability of the membrane to the solute is expressed by the **reflection coefficient (σ)**.

(1) A reflection coefficient of 1 indicates that the membrane is not at all permeable to the particle. That is, the particle is totally reflected by the membrane.

(2) A reflection coefficient of 0 indicates that the membrane is as permeable to the particle as it is to water.

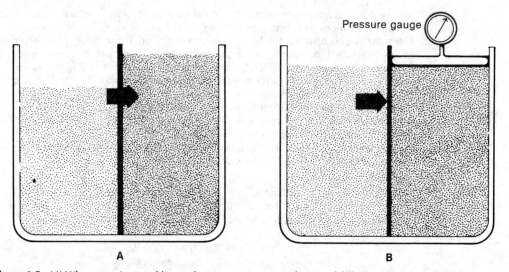

Figure 1-5. (*A*) When a semipermeable membrane separates two solutions of different osmolalities, water flows from the solution with a lower osmotic pressure (concentration) to the solution with a higher osmotic pressure (concentration). Water will flow into the chamber until the hydrostatic pressure created by the increased height of the fluid column equals the osmotic pressure difference between the two chambers. (*B*) The flow of water can also be prevented by applying pressure to the chamber containing the higher solute concentration. The amount of pressure that must be applied to prevent the flow of water is a measure of the osmotic pressure difference between the two chambers.

(3) The reflection coefficient can be calculated using the expression

$$\sigma = 1 - \frac{P_{solute}}{P_{water}}$$

(4) The osmotic flow equation can be used to calculate the osmotic flow of water if it is modified to include the reflection coefficient

$$flow = \sigma \times L \times A \times (\pi_1 - \pi_2)$$

g. Almost all particles dissolved in blood plasma, except proteins, easily cross the capillary and, thus, do not cause water to flow between the capillary and the interstitial fluid. Plasma proteins have an osmolar concentration of about 1.2 mOsm/L and, thus, create an osmotic pressure of about 23 mm Hg.

   (1) The osmotic pressure produced by the plasma proteins, called the **colloid oncotic pressure,** draws water into the capillaries from the interstitial fluid and is counteracted by the hydrostatic pressure of the blood produced by the heart. The movement of water through the capillaries due to hydrostatic or osmotic pressure differences is called **bulk flow**.

   (2) Whether water flows into or out of the capillaries depends on whether the colloid osmotic pressure is greater or less than the hydrostatic pressure of the blood. When water flows through the capillaries, it carries dissolved particles with it. This is called **solvent drag**.

D. Donnan equilibrium

1. **Definition.** When two solutions containing charged particles (ions) are separated by a membrane that is permeable to some of the ions and not to others, a Donnan equilibrium will be established. A Donnan equilibrium is an example of an **electrochemical equilibrium,** because the electrical and chemical energies on either side of the membrane are equal and opposite to each other.

2. **Characteristics of a Donnan equilibrium** (Figure 1-6)
   a. The concentrations of ions on one side of the membrane do not equal those on the other side of the membrane. The tendency for the ions to diffuse down their concentration gradient is balanced by an electrical potential that develops across the membrane.
   b. Within each solution, the concentrations of anions and cations are equal to each other. That is, **electroneutrality** is preserved.
   c. The concentration of diffusible cations is greater in the solution that contains the nonpermeable, negatively charged particles. The concentration of diffusible anions is greater in the solution without the nonpermeable particles.

   (1) The concentrations of the diffusible ions on either side of the membrane are described by the expression

$$[Na^+]_1 \times [Cl^-]_1 = [Na^+]_2 \times [Cl^-]_2$$

   where $[Na^+]$ and $[Cl^-]$ represent the concentrations of diffusible ions in solutions 1 and 2 of the example illustrated in Figure 1-6. The expression is valid for any group of diffusible ions.

   (2) The side of the membrane facing the solution with the nondiffusible anions is negatively charged compared to the other side of the membrane.

   d. The solutions are not in osmotic equilibrium. The total number of particles in the solution with the nonpermeable anions is greater than the total number of particles in the other solution. This creates an osmotic pressure gradient, causing water to flow into the cell. The Na^+-K^+ pump keeps water from flowing into the cells by preventing the establishment of a Donnan equilibrium despite the presence of nondiffusible proteins inside the cells.

   (1) The Na^+-K^+ pump keeps the Na^+ concentration inside the cell low enough to maintain osmotic equilibrium between the inside and outside of the cell.

   (2) In brain ischemia, for example, decreased activity of the Na^+-K^+ pump allows the Na^+ concentration to rise. The increase in intracellular osmolarity causes water to flow into the cell, producing neuronal damage.

II. RESTING AND ACTION POTENTIALS. An electrical potential (voltage) difference exists between the inside and outside of all cells. This is called the **resting membrane potential**. The resting potential is approximately –80 mV in excitable cells (e.g., nerve cells, muscle cells) and approximately –20 to –40 mV in nonexcitable cells (e.g., red blood cells, epithelial cells). When

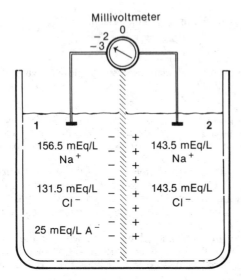

Millivoltmeter

Figure 1-6. Characteristics of an established Donnan equilibrium. Compared to chamber 2 (*2*), chamber 1 (*1*) is negatively charged and has a larger solute concentration, a greater cation concentration, a smaller concentration of permeable anions (*A⁻*).

excitable cells are stimulated, an **action potential** is generated, during which the membrane potential changes from –80 mV to about +45 mV and then returns to the resting potential.

A. Recording resting and action potentials

1. **Intracellular recordings** of membrane potentials are made with glass microelectrodes that have tip diameters of less than 0.5 μ, allowing them to be inserted through the membrane without damaging the cell.

   a. Figure 1-7 shows the recording made as a microelectrode is inserted into a cell at rest. The electrical potential is 0 mV when the microelectrode is outside the cell and drops to –80 mV as soon as the microelectrode passes through the membrane and enters the ICF.

   b. When the cell is stimulated, the microelectrode records the changes in membrane potential (i.e., the action potential; see II C 1).

2. **Extracellular recordings** usually are made with metal electrodes that are placed on or near the nerve or muscle.

   a. Because these electrodes are outside the cell, they **can record only changes in membrane potential** (i.e., action potentials but not resting potentials). Also, they cannot record the exact magnitude or time-course of the action potentials.

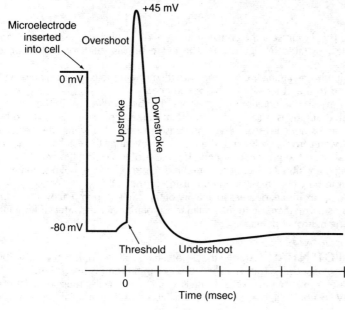

Figure 1-7. When a microelectrode enters a nerve cell, a resting membrane potential is recorded. Stimulation produces an action potential, which also is recorded. The various components of the action potential are indicated on the diagram.

   b. Extracellular recordings are useful in clinical situations when the electrical activity of excitable tissues must be monitored. For example, **electroencephalograms (EEGs)** are used to aid in the diagnosis of brain disease, **electrocardiograms (EKGs)** are used to detect damage to the heart, and **electromyograms** (recordings from skeletal muscle) are used to aid in the diagnosis of neuropathies and myopathies.

3. Patch electrodes are used to record the changes in membrane potential that occur in an isolated patch of membrane. These electrodes can be used to monitor the opening and closing of single channels.

B. The resting membrane potential is the electrical potential difference between the inside and outside of the cell.

   1. Factors that determine the resting potential
   a. The resting potential is produced by the movement of ions across the plasma membrane. The **magnitude of the resting potential** depends on:
      (1) The **concentration gradient** for each ion to which the membrane is permeable
      (2) The **relative permeability** (or **conductance**) of the membrane for each ion
   b. The **resting potential of nerve and muscle** is determined by the concentration gradients and membrane permeabilities (conductances) for Na^+ and K^+. The concentration gradients for these ions are established and maintained by the Na^+-K^+ pump.
      (1) K^+ is much more abundant in the ICF than in the ECF, whereas Na^+ is more abundant in the ECF than in the ICF (see Table 1-1).
      (2) To preserve electroneutrality, the intracellular and extracellular cations must be balanced with anions.
         (a) The primary anions of the ECF are HCO_3^- and Cl^-.
         (b) The primary anions of the ICF are proteins and organic and inorganic acids (e.g., phosphates).
   c. In the resting state, the membrane conductance for K^+ is about 10 times as great as it is for Na^+.
   d. Changing the membrane conductance for an ion will change the membrane potential. Conversely, changing the membrane potential will change the membrane conductance for an ion. This relationship between membrane potential and conductance enables excitable tissues to produce action potentials.

   2. Nernst potential (also called the **diffusion**, or **equilibrium**, **potential**) is used to express the concentration gradient in electrical terms.
   a. The Nernst potential is a measure of the amount of work that can be done by an ion diffusing down its concentration gradient. It is the electrical equivalent of the energy (in mV) in the concentration gradient. The Nernst potential is expressed as

$$E = -\frac{RT}{zF} \ln \frac{C_{in}}{C_{out}}$$

where E = electrical potential (mV), R = natural gas constant, T = absolute temperature (° K), z = volume of the ion, F = Faraday's constant (96,500 coulombs/mol), and C_{in} and C_{out} = the concentration of the ion (mmol/L) inside and outside the cell.
   b. The Nernst equation can be simplified by substituting for the constants (R, T, and F) and converting to common logarithms, yielding

$$E_{ion} = -61 \log \frac{[ion]_{in}}{[ion]_{out}}$$

where E_{ion} = the equilibrium potential (mV) for a particular ion, and $[ion]_{in}$ and $[ion]_{out}$ = the intracellular and extracellular concentrations of that ion. The valence is omitted because it is +1 for both Na^+ and K^+.
   c. Table 1-2 gives Nernst potentials for some important electrolytes. (Note that –61 is divided by –1 to calculate the Nernst potential for Cl^- and HCO_3^- and by +2 to calculate the Nernst potential for Ca^{2+}.)

   3. The resting membrane is in a steady state.
   a. Ionic gradients and membrane potentials are constant in both the steady state and equilibrium conditions. However, in the steady state, free energy is used to maintain the constant membrane potentials and ionic gradients, whereas no free energy is used in equilibrium conditions, such as the Donnan equilibrium (see I D).

Table 1-2. Nernst Potential for Ions Commonly Found in Nerve and Muscle Cells

| Ion | Concentration (mmol/L) | | Nernst Potential (mV) |
	Intracellular	Extracellular	
Na^+	15	140	+58
K^+	135	4	-92
Ca^{2+}	10^{-4}	2	+129
H^+	10^{-4}	40×10^{-6}	-24
Cl^-	4	120	-89
HCO_3^-	10	24	-23

    b. The steady state is illustrated in Figure 1-8A, which shows a typical nerve axon that is permeable to Na^+ and K^+.

        (1) Na^+ leaks into the cell down its concentration gradient; K^+ leaks out of the cell down its concentration gradient.

        (2) The Na^+-K^+ pump keeps the concentration gradients for Na^+ and K^+ from changing by pumping Na^+ out of and K^+ into the cell.

  4. Extracellular K^+ concentration affects the resting membrane potential. Since the permeability of the resting membrane to K^+ is so much higher than it is to Na^+, the value of the resting membrane potential can be approximated by the equilibrium potential for K^+.

    a. Increases in extracellular K^+, which make the equilibrium potential for K^+ more positive, cause the resting membrane potential to become more positive (depolarize).

    b. Decreases in extracellular K^+, which make the equilibrium potential for K^+ more negative, cause the resting membrane potential to become more negative (hyperpolarize).

C. The action potential results from a sequential change in membrane conductance for Na^+ and K^+.

  1. Phases and shape of the action potential (see Figure 1-7)

    a. Threshold. The action potential begins when the membrane is depolarized (made less negative) from its resting potential to its threshold potential by a stimulus.

A

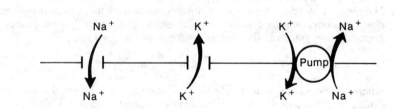

B

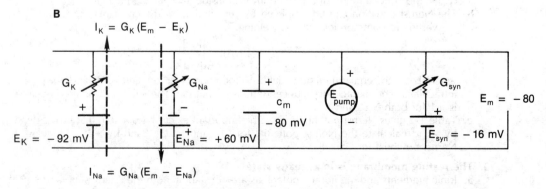

Figure 1-8. (A) Although Na^+ and K^+ continuously leak down their electrochemical gradients, their concentrations inside the cell are maintained at a constant value by the Na^+-K^+ pump. (B) The electrical analog of the cell membrane is used to analyze how changes in ionic concentrations and conductances affect the membrane potential (see II D).

b. Upstroke. The rapid depolarization of the membrane after threshold is reached in the **depolarization phase,** or upstroke, of the action potential. **The upstroke is produced by the flow of Na$^+$ into the cell.**

c. Overshoot. The portion of the action potential during which the membrane is positive is the overshoot. The peak of the action potential is the overshoot potential.

d. Downstroke. The rapid return of the membrane towards its resting potential is the **repolarization phase,** or downstroke, of the action potential. **The downstroke is produced by the flow of K$^+$ out of the cell.**

e. Undershoot. The membrane potential becomes more negative than its resting value at the end of the action potential. This is the **hyperpolarization phase,** or undershoot, of the action potential.

2. **All-or-none response.** The action potential is an all-or-none response to a stimulus. That is, if the stimulus is strong enough to reach threshold, the changes in membrane potential that characterize the action potential are always the same. This feature of the action potential is based on the regulatory gates that cover the Na$^+$ and K$^+$ channels (Figure 1-9).
   a. Regulation of the Na$^+$ channel
      (1) The **m gate** covers the extracellular side of the Na$^+$ channel and the **h gate** covers the intracellular side of the Na$^+$ channel.
      (2) Both the m and the h gates must be open for Na$^+$ to flow through the Na$^+$ channel.
         (a) When the m gate is open, the channel is said to be activated.
         (b) When the h gate is closed, the channel is said to be inactivated.
   b. Regulation of the K$^+$ channel
      (1) The K$^+$ channel is regulated by a single gate—the **n gate**—which is located on the extracellular side of the channel.
      (2) The n gate must be open for K$^+$ to flow through the K$^+$ channel.
         (a) When the n gate is open, the K$^+$ channel is activated.
         (b) The K$^+$ channel does not have an inactivation gate.

3. **Mechanism of gating action.** The m, n, and h gates have two qualities that permit an action potential to be generated when the cell membrane is depolarized by a stimulus: voltage dependence and time dependence.
   a. Voltage dependence
      (1) The m and n gates close when the membrane potential is polarized (–80 mV) and open when the membrane potential is depolarized (made more positive).
      (2) In contrast, the h gates open when the membrane potential is polarized and close when it is depolarized.
      (3) Thus, when the membrane is at its resting potential, almost all of the Na$^+$ and K$^+$ channels are closed by the m and n gates, respectively (see Figure 1-9).
         (a) Although only a few channels actually are open when the cell is at its resting potential, the number of open K$^+$ channels far exceeds the number of open Na$^+$ channels.
         (b) Although the Na$^+$ channels are closed when the membrane is at its resting potential, the channels are not inactivated because the h gates are open.

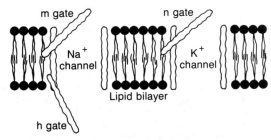

Figure 1-9. A diagrammatic representation of the gates covering the Na$^+$ and K$^+$ channels. The negative resting potential tends to keep the m and n gates closed and the h gate opened.

b. Time dependence

    (1) Time is required for the gates to respond to a change in membrane potential. For example, when the membrane potential is depolarized:

        (a) The **m gates open first,** activating the Na^+ channels

        (b) The **h gates then close,** inactivating the Na^+ channels

        (c) The **n gates then open,** activating the K^+ channels

    (2) Time dependence of the gates is essential for the production of the action potential.

        (a) If the h gates closed as rapidly as the m gates opened (i.e., if the Na^+ channel was inactivated and activated at the same time), Na^+ could not flow into the cell.

        (b) Similarly, if the n gates opened as fast as the m gates opened (i.e., if Na^+ and K^+ activation occurred at the same time), the upstroke could not occur.

c. The gating mechanism produces the phases of the action potential (Figures 1-10 and 1-11).

    (1) Upstroke. A **positive feedback,** or **regenerative, process** is responsible for the flow of Na^+ into the cell during the upstroke (see Figure 1-10).

        (a) When a stimulus depolarizes an excitable membrane, m gates on some of the Na^+ channels open, allowing Na^+ to enter the cell (see Figure 1-11A).

        (b) The flow of Na^+ into the cell causes the membrane to depolarize further, which causes more m gates to open (the Na^+ conductance increases) and allows more Na^+ to enter the cell.

        (c) Thus, the response of the membrane to a stimulus (opening of the m gates) causes an effect (membrane depolarization) that produces an even greater response.

    (2) Downstroke. Inactivation of the Na^+ channels and activation of the K^+ channels occur when the membrane potential is depolarized, causing the downstroke (see Figure 1-11B).

        (a) The Na^+ channels are inactivated by the closing of the h gates. Inactivation stops the flow of Na^+ into the cell. If inactivation did not occur, repolarization would be slowed because Na^+ would flow into the cell during the downstroke.

        (b) The K^+ channels are activated by the opening of the n gates. Activation allows K^+ to leave the cell, which produces the downstroke of the action potential.

        (c) The inactivation of the Na^+ channels must be removed before another action potential can occur. The h gates open during the downstroke, removing the inactivation from the Na^+ channels (see Figure 1-11C).

    (3) Undershoot is caused by the slow closing of the K^+ channels.

        (a) The n gates close slowly when the membrane is repolarized during the downstroke.

        (b) Consequently, the K^+ conductance is higher at the end of the action potential than during the resting state. The high K^+ conductance causes the membrane to **hyperpolarize** (become more negative than the resting membrane potential).

        (c) Eventually, the n gates close and the membrane potential returns to its resting level.

4. Refractory period refers to an interval during which it is more difficult to elicit an action potential. There are two refractory periods (see Figure 1-11).

    a. Absolute refractory period. During this interval, another action potential cannot be elicited, regardless of the strength of the stimulus.

    (1) The absolute refractory period begins at the start of the upstroke and extends into the downstroke.

        (a) During the upstroke, a second action potential cannot occur because the m gates are opening as fast as possible (see Figure 1-11A).

        (b) During the early portion of the downstroke, an action potential cannot occur because the Na^+ channels are inactivated by the h gates (see Figure 1-11B).

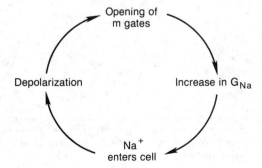

Figure 1-10. When the cell is depolarized to threshold, Na^+ channels open, causing an increase in the conductance for Na^+ (G_{Na}). This allows Na^+ to enter the cell, causing further depolarization. The positive feedback system represented by this cycle is responsible for the upstroke of the action potential.

(2) The absolute refractory period ends when the number of inactivated Na^+ channels is few enough to allow another action potential to occur. Inactivation of the Na^+ channels is removed when the h gates open during the downstroke.

b. Relative refractory period. During this interval, a second action potential can be elicited if the stimulus is sufficient. The stimulus must be greater than normal, because some Na^+ channels are still inactivated and more K^+ channels than normal are still open (see Figure 1-11C).

(1) The relative refractory period begins when the absolute refractory period ends.

(2) The action potential elicited during this interval has a lower upstroke velocity and a lower overshoot potential than the normal action potential.

    (a) These changes result from the increased number of inactivated Na^+ channels and activated K^+ channels that exist during the relative refractory period as compared to the resting state.

    (b) The changes do not violate the all-or-none principle of the action potential but demand that the principle be revised to state: If a stimulus is sufficient to bring the membrane to threshold, the strength of the stimulus will not affect the magnitude and time-course of the action potential.

5. Propagation of the action potential. Once generated, the action potential must be propagated (conducted) along the axon. Propagation occurs because the action potential generated at one location on the axon acts as a stimulus for the production of an action potential on the adjacent region of the axon.

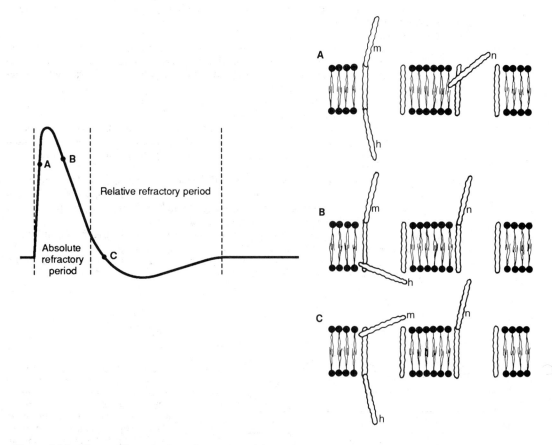

Figure 1-11. Diagram illustrating how the position of the Na^+ and K^+ gates change during the action potential. (*A*) Both the m and h gates are opened during the upstroke of the action potential, allowing Na^+ into the cell. (*B*) The closing of the h gates (inactivation), which stops the flow of Na^+ into the cell, and the opening of the n gates, which allows K^+ to flow out of the cell, are responsible for the downstroke. (*C*) During the undershoot, the n gates are open, the m gates are closed, and the h gates are open. The high K^+ conductance causes the cell to hyperpolarize.

a. Propagation in unmyelinated axons (Figure 1-12)

    (1) The process. During the overshoot of the action potential in an unmyelinated axon, the membrane potential becomes about +40 mV, creating an electrical potential difference between the area of membrane on which the action potential is generated and the adjacent, polarized area (see Figure 1-12A).

      (a) Because of this potential difference, current flows passively between the two areas, causing the adjacent region to become depolarized. If the adjacent area is depolarized to threshold, an action potential is generated (see Figure 1-12B).

      (b) The depolarization of the membrane produced by the new action potential spreads passively, and the entire process is repeated.

      (c) Thus, propagation of the action potential involves the generation of action potentials on contiguous patches of membrane along the axon.

    (2) The magnitude of the action potential does not change as it is conducted along the axon, because new action potentials are being generated constantly. This is different from the spread of an electrotonic potential (see II F), which diminishes in size along the axon.

    (3) The speed of propagation is proportional to the square root of the fiber diameter. To increase the speed of propagation, fiber diameter must be increased. However, there is a practical limit to how large axons can become.

      (a) For example, a typical unmyelinated fiber is 1 μ in diameter and propagates at about 1 m/sec.

      (b) For such a fiber to propagate at 50 m/sec (the typical velocity of a myelinated neuron), it would need to be 2.5 mm in diameter. Imagine how large a typical motor nerve (containing 1000 axons) would be if each axon were 2.5 mm in diameter.

b. Propagation in myelinated axons

    (1) The process

      (a) In principle, propagation in myelinated axons is the same as in unmyelinated axons. However, the cell membrane in myelinated axons is exposed to the ECF only at the nodes of Ranvier. Nodes occur every 100–500 μ. In general, as the diameter of the axon increases, the internodal distance increases.

      (b) The membrane area between the nodes is covered by an insulating sheath of myelin formed from Schwann cell membranes.

      (c) The action potentials are generated only at the nodes. That is, an action potential generated at one node becomes the stimulus for the generation of an action potential at the adjacent node. The propagation of the action potential from node to node is called **saltatory conduction**.

    (2) The speed of propagation is proportional to the diameter of the axon and the internodal distance. Because the action potential can spread from node to node, instead of having to produce an action potential at each contiguous patch of membrane, propagation of the action potential is faster than in unmyelinated nerves.

      (a) Myelinated fibers are used by the nervous system when high speeds of conduction are necessary. For example, the axons of the major ascending and descending tracts of the spinal cord, the sensory axons used for fine tactile discrimination, and the motor axons all are myelinated.

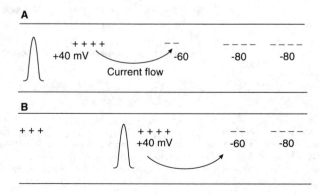

Figure 1-12. (A) The positive potential (+40 mV) produced during the overshoot of the action potential causes current to flow toward the negative, resting portion of the axon. The flow of current acts as a stimulus depolarizing the axon toward threshold. (B) When threshold is achieved, an action potential is elicited and the entire process is repeated, causing the action potential to be propagated along the axon.

(b) Myelinated fibers are 1–20 μ in diameter. The largest fibers conduct action potentials at speeds of up to 120 m/sec.

D. Electrical analog of membrane potentials. An electrical analog of the cell membrane, such as the one shown in Figure 1-8B, can be used to analyze and calculate the membrane potential during the resting state or the action potential.

1. The **electrochemical gradient** pushing an ion through the membrane is the difference between the membrane potential (E_m) and the equilibrium potential for the ion (E_{ion}), which is expressed as

$$\text{driving force} = E_m - E_{ion}$$

where the driving force is the electrical potential energy (mV).

2. The **ionic current** can be calculated for each ion using **Ohm's law** for solutions

$$I_{ion} = G_{ion}(E_m - E_{ion})$$

where I_{ion} = current (milliamperes), and G_{ion} = ionic conductance (mhos or Siemans).

3. The **net current** across the membrane in the steady state is zero. Thus, the current flowing into the cell must be equal and opposite to the current flowing out of the cell.
   a. The **inward current** is due to the flow of Na^+ into the cell and is given by

$$I_{ion} = - G_{ion}(E_m - E_{ion})$$

   b. The **outward current** is due to the flow of K^+ out of the cell and is given by

$$I_{ion} = + G_{ion}(E_m - E_{ion})$$

4. The **transference equation** can be derived by setting the inward and outward currents equal to each other and solving for E_m

$$E_m = \left(E_{Na} \times \frac{G_{Na}}{G_{Na} + G_K}\right) + \left(E_K \times \frac{G_K}{G_{Na} + G_K}\right)$$

where G_{Na} and G_K = the conductance for Na^+ and K^+, and the ratios $G_{Na}/(G_{Na} + G_K)$ and $G_K/(G_{Na} + G_K)$ = the transferences for Na^+ and K^+ (T_{Na} and T_K). Thus, the transference equation can be simplified to

$$E_m = (T_{Na} \times E_{Na}) + (T_K \times E_K)$$

5. The electrical analog of the cell can be used to illustrate how transference affects the membrane potential (see Figure 1-8B).
   a. The arrows drawn through the conductors, G_K and G_{Na}, represent the transferences of K^+ and Na^+.
      (1) If the only open channels in the membrane are K^+ channels (i.e., if T_K =1 and T_{Na} = 0), then E_m will equal the value of the K^+ battery (E_K, or –92 mV).
      (2) If the only open channels in the membrane are Na^+ channels (i.e., if T_K = 0 and T_{Na} = 1), then E_m will equal the value of the Na^+ battery (E_{Na}, or +60 mV).
      (3) If 90% of the open channels in the membrane are K^+ channels (i.e., if T_K = 0.9 and T_{Na} = 0.1), then E_m will be much closer to the K^+ battery (E_K) than to the Na^+ battery (E_{Na}).
   b. These examples illustrate that **E_m is determined primarily by the equilibrium (Nernst) potential of the ion with the highest transference (conductance).**

E. Calculation of resting membrane potential

1. **Transference equation**
   a. Using the transference equation, the resting membrane potential is calculated as follows:
      (1) Since G_K is 10 times G_{Na}

$$T_K = \frac{10}{10 + 1} = \frac{10}{11} = 0.91$$

      (2) Similarly

$$T_{Na} = \frac{1}{10 + 1} = \frac{1}{11} = 0.09$$

(3) Substituting these values and the values for E_K and E_{Na} (from Table 1-1), the transference equation yields a membrane potential of –79 mV

$$(0.91 \times -93) + (0.09 \times +60) = -79$$

b. The electrogenic nature of the Na^+-K^+ pump makes the actual membrane potential a few millivolts more negative than the value calculated by the transference equation.
   (1) Since three Na^+ ions are pumped out of the cell for every two K^+ ions that are pumped in, the activity of the pump produces a small negative potential. The Na^+-K^+ pump is labeled "pump" in the electrical analog of the membrane (see Figure 1-8B).
   (2) Thus, the transference equation should be written

$$E_m = (T_{Na} \times E_{Na}) + (T_K \times E_K) + E_{pump}$$

   (3) Because the pump potential usually is small, it generally is ignored.

2. Goldman-Hodgkin-Katz (GHK) equation. The GHK equation, like the transference equation, is derived for a steady-state condition in which the inward and outward currents are equal and opposite. However, the GHK equation expresses ionic currents according to the laws of **electrodiffusion,** rather than Ohm's law.
   a. The GHK equation is stated as

$$E_m = -\frac{RT}{F} \log \frac{P_K[K]_{in} + P_{Na}[Na]_{in}}{P_K[K]_{out} + P_{Na}[Na]_{out}}$$

   where P_K and P_{Na} represent the permeabilities of K^+ and Na^+.
   b. Note that when the membrane is permeable to only one ion (i.e., the permeability of the membrane to the other ion is zero), the GHK equation reduces to the Nernst equation (see II B 2 a).

3. Both the GHK and transference equations, thus, predict that E_m will be determined primarily by the equilibrium potential of the ion that is able to cross the membrane most easily (i.e., the ion with the highest membrane permeability, or conductance).

F. Electrotonic potentials are changes in membrane potential that do not propagate.

1. Types of electrotonic potentials
   a. A **local (subthreshold) response** is produced when a stimulus does not open enough m gates to elicit an action potential. Because of the small number of open m gates, the amount of Na^+ entering the cell is insufficient to initiate a positive feedback cycle.
   (1) The local response is **graded**; the magnitude and duration of the response vary with the size and strength of the stimulus.
   (2) The local response is **nonpropagated**; that is, its magnitude is insufficient to generate another local response.
   b. Other graded, nonpropagated responses are produced on nerve and muscle membranes.
   (1) For example, the generator and receptor potentials produced by sensory stimuli on receptors (see V B 1) and the excitatory and inhibitory potentials produced by neurotransmitters on synaptic membranes [see III A 3 b (2)] are graded and nonpropagated.
   (2) The receptor and synaptic potentials differ from the local response in that the membrane channels producing them are not voltage- or time-dependent.

2. Cable properties of the membrane. The membrane changes produced by graded, nonpropagated responses spread passively, or electrotonically, along the membrane. Cable properties of the membrane determine the time-course and voltage changes of an electrotonic potential. The electrical equivalent of the cable properties of an axon is illustrated in Figure 1-13A.
   a. Time constant. When a current is applied to a membrane by a stimulus, a charge is added to the capacitor, causing a potential difference to develop across the membrane.
   (1) The potential difference, V_t, develops at a rate that is determined using the equation

$$V_t = V_{max} \times (1 - e^{-t/\tau})$$

   where V_t = the voltage at time t; V_{max} = the voltage of the applied stimulus; τ = the time constant (seconds), or $r_m c_m$, where r_m = membrane resistance (ohms) and c_m = membrane capacitance (farads).

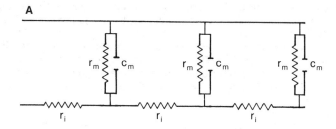

A

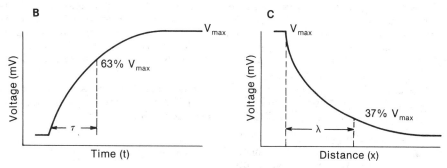

B

Voltage (mV)

V_{max}

63% V_{max}

τ

Time (t)

C

Voltage (mV)

V_{max}

37% V_{max}

λ

Distance (x)

Figure 1-13. (*A*) Equivalent circuit of an axon. Each patch of axon contains a resistor representing the conductive pathways through the membrane (r_m), a capacitor representing the lipid bilayer of the membrane (c_m), and a resistor representing the intracellular pathway for the flow of ions along the axon (r_i). (*B*) When a stimulus is applied to a nerve membrane, the membrane depolarizes exponentially according to the equation $V_t = V_{max} \times (1 - e^{-t/\tau})$. When $t = \tau$ (the time constant), $V_\tau = 0.63\ V_{max}$. (*C*) The magnitude of the depolarization decreases as the distance from the stimulus increases according to the equation $V_x = V_{max} \times e^{-x/\lambda}$. When the distance from the stimulus equals λ (the space constant), $V_\lambda = 0.37\ V_{max}$.

(2) The time constant, τ, of the membrane is equal to the time required for the membrane voltage to reach about 63% ($1 - 1/e$) of V_{max} (see Figure 1-13B).

b. Space constant. The potential produced by the stimulus is spread passively along the membrane.

(1) At any point along the membrane, the potential, V_x, is given by the equation

$$V_x = V_{max} \times e^{-x/\lambda}$$

where x = distance (mm), and λ = the space constant.

(2) The space constant, λ, is the distance from the stimulus to the point at which the applied voltage falls to about 37% ($1/e$) of V_{max} (see Figure 1-13C).

(3) In an axon, the space constant is equal to r_m/r_i and, thus, increases if either of the following occurs:

(a) The membrane resistance increases

(b) The axoplasmic resistance decreases

c. The cable properties of a neuron play an important role in determining the ability of a stimulus to elicit an action potential and the propagation velocity of action potentials. Graded, nonpropagating responses (e.g., synaptic and receptor potentials) must spread passively from the patch of membrane where they are produced to a patch of membrane that is able to produce an action potential.

(1) If the graded response is produced too far from the action potential–producing portion of the membrane (e.g., at the end of a long dendrite), it will decay too much to be able to depolarize the action potential–generating portion of the membrane to threshold.

(2) If the membrane resistance is reduced (e.g., by an inhibitory synaptic transmitter), the space constant will be reduced, and the ability of an excitatory response to spread passively to the action potential–generating region of the membrane will be reduced.

(3) If the time constant is increased, it will take longer for the action potential produced at one point along the axon to depolarize its adjacent region to threshold, and propagation velocity will slow.

    (a) In demyelinating diseases, such as multiple sclerosis, the loss of myelin increases the membrane capacitance, which increases the time constant.

    (b) The increase in the time constant causes action potential propagation to fail, producing the sensory and motor deficits characteristic of multiple sclerosis.

III. SYNAPTIC TRANSMISSION

III. **SYNAPTIC TRANSMISSION** is the process by which nerve cells communicate among themselves and with muscles and glands. The **synapse** is the anatomic site where this communication occurs. Most synaptic transmission is carried out by a chemical called a **neurotransmitter**. The neurotransmitter is released from a **neuron** (the **presynaptic cell**) and diffuses to its target (the **postsynaptic cell**), where it produces a **postsynaptic response**. This response is an action potential (if the target cell is another neuron), contraction (if the target cell is a muscle), or secretion (if the target cell is a gland). The space between the presynaptic and postsynaptic cells is called the **synaptic cleft**. The neurotransmitter may be **inhibitory,** in which case it decreases the activity of the postsynaptic cell. In some instances, synaptic transmission may be **electrical** and occur through **gap junctions**.

A. **Neuromuscular transmission** refers to synaptic communication between an alpha motoneuron and a skeletal muscle fiber. Neuromuscular transmission also occurs between autonomic efferent fibers and both smooth and cardiac muscle cells (see III B).

    1. Physiologic anatomy (Figure 1-14)

        a. Light microscopic appearance. Figure 1-14A is a drawing of the neuromuscular synapse as viewed with a light microscope.

        (1) The alpha motoneuron branches as it approaches the muscle, sending axon terminals to several skeletal muscle fibers. Each skeletal muscle fiber receives only one axon terminal. The number of skeletal muscle fibers innervated by an alpha motoneuron depends on the type of muscle fiber involved.

          (a) Alpha motoneurons innervating large muscles used primarily for strength or postural control innervate hundreds to thousands of skeletal muscle fibers.

          (b) Alpha motoneurons innervating muscles used for precision movements innervate just a few skeletal muscle fibers.

        (2) The axon terminal lies in a groove called the **synaptic trough,** which is formed by an invagination of the skeletal muscle fiber.

        (3) Synaptic transmission occurs at the **end-plate region** of the skeletal muscle fiber.

        b. Electron microscopic appearance. The details of the presynaptic and postsynaptic (end-plate) membranes are visible in the electron microscopic view of the synaptic junction in Figure 1-14B.

        (1) Synaptic vesicles (about 50 nm in diameter) containing the neurotransmitter **acetylcholine (ACh)** are found in the presynaptic nerve terminal concentrated around specialized presynaptic membrane structures called **dense bars**. The postsynaptic membrane contains receptor sites to which ACh binds.

        (2) The synaptic cleft (about 60 nm wide) is filled with an amorphous network of connective tissue called the **basal lamina,** in which the enzyme **acetylcholinesterase (AChEase)** is bound. AChEase is responsible for degrading ACh after it has produced its effect on the end-plate membrane of the skeletal muscle fiber.

        (3) The postsynaptic membrane contains numerous **junctional folds,** which are membrane invaginations located opposite the dense bars. The receptor sites for ACh are found on the membranes of the junctional folds.

    2. Synthesis, storage, and release of ACh

        a. ACh is synthesized in the nerve terminal from **choline** and **acetyl coenzyme A (acetyl-CoA)** by the enzyme **choline acetyltransferase (CAT)**.

        b. Newly synthesized ACh is stored within the synaptic vesicles. Approximately 5000–10,000 molecules of ACh are stored within each vesicle.

        c. Spontaneous release of ACh occurs by **exocytosis** (Figure 1-15) whenever a vesicle binds to an attachment site on one of the dense bars.

        (1) In exocytosis, the vesicle fuses with the presynaptic membrane, exposing its content to the ECF.

A

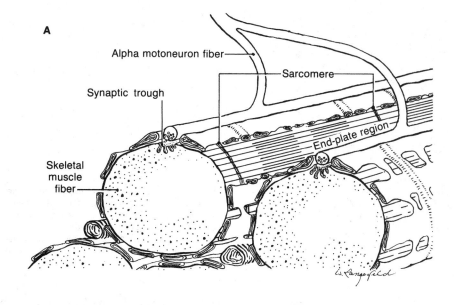

Alpha motoneuron fiber

Sarcomere

Synaptic trough

End-plate region

Skeletal
muscle
fiber

W. Langerfeld

B

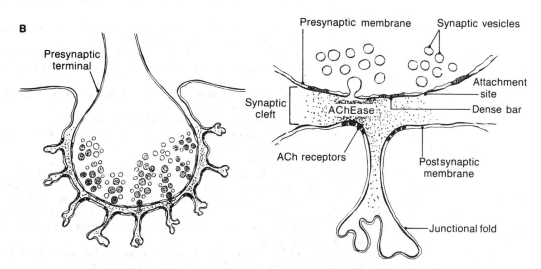

Presynaptic membrane

Synaptic vesicles

Presynaptic
terminal

Synaptic
cleft

AChEase

Attachment
site

Dense bar

ACh receptors

Postsynaptic
membrane

Junctional fold

Figure 1-14. (*A*) Light microscopic view of a neuromuscular junction. The alpha motoneuron branches as it reaches the muscle, and each branch forms a synapse with a single muscle fiber at the end-plate region of the muscle. The axon terminal lies in the synaptic trough. Junctional folds increase the surface area of the postsynaptic membrane. (*B*) Electron microscopic view, showing the synaptic vesicles within the presynaptic terminal, the acetylcholinesterase (*AChEase*) within the synaptic cleft, and the postsynaptic receptor sites. *ACh* = acetylcholine.

    (2) ACh diffuses out of the vesicle into the synaptic cleft and the vesicle merges with the presynaptic membrane.

    (3) Later, new vesicles are formed from the presynaptic membrane by endocytosis.

        (a) The presynaptic membrane forms invaginations that eventually bud off to form new vesicles.

        (b) The new vesicles are refilled with ACh and, once again, are available to release their contents into the ECF.

  3. Events in synaptic transmission

    a. Release of neurotransmitter. Synaptic transmission begins when an action potential is propagated into the nerve terminal.

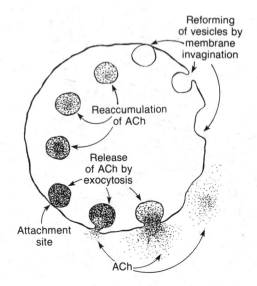

Figure 1-15. Diagram illustrating the life cycle of a synaptic vesicle. After binding to its attachment site, the vesicle releases its transmitter, acetylcholine (*ACh*), by exocytosis and then merges with the membrane. Later, a new vesicle is formed from invaginations of the synaptic membrane. These vesicles are filled with transmitter and can be used again.

(1) Depolarization of the nerve terminal by the action potential causes Ca^{2+} channels (which are located next to the dense bars) to open, allowing Ca^{2+} to enter the cell down its electrochemical gradient.

(2) The increase in intracellular Ca^{2+} concentration causes about 200–300 vesicles to bind to attachment sites and release their contents into the synaptic cleft.

b. **Postsynaptic response.** The ACh released into the synaptic cleft binds to ACh receptors on the end-plate membrane, where it causes a postsynaptic response called the **end-plate potential (EPP)**.

(1) The **ACh receptor** (Figure 1-16) is a transmembrane protein consisting of five subunits that form an aqueous channel within the lipid bilayer.

   (a) Two of the subunits, called α **subunits,** contain binding sites for ACh.

   (b) When the two α subunits are occupied by ACh, the proteins undergo a conformational change that opens a channel within the receptor.

   (c) The channel is equally permeable to Na^+ and K^+.

(2) The channel associated with the ACh receptor is a **chemically activated channel**.

   (a) Unlike the Na^+ and K^+ channels found on electrically excitable membranes (which are activated by changes in membrane voltage), the ACh-activated channel is opened by the binding of the neurotransmitter to the receptor and not by membrane depolarization. These channels are sometimes called **receptor-activated,** or **receptor-occupied, channels**.

   (b) The magnitude of the EPP is proportional to the number of channels opened by ACh (i.e., the EPP is a graded response and it is not propagated).

(3) Opening the channel causes the cell to depolarize.

   (a) When the channel is opened, Na^+ enters the cell down its electrochemical gradient, while K^+ leaves the cell down its electrochemical gradient.

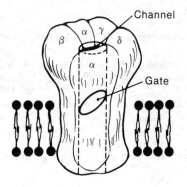

Figure 1-16. A diagrammatic view of the acetylcholine (ACh) receptor. The receptor contains five subunits, two of which (the α subunits) contain binding sites for ACh. When both subunits are occupied, the channel gate opens, allowing Na^+ and K^+ to pass through the membrane. (Adapted from Anholdt R, Lindstrom J, Montal M: In *Enzymes of Biological Membranes.* Edited by Martonosi A. New York, Plenum Press, pp 335–401, 1985.)

(b) The Na^+ current (I_{Na}) and the K^+ current (I_K) can be calculated using Ohm's law (see II D 2) as follows

$$I_{Na} = G_{Na} \times (E_m - E_{Na}) \text{ and } I_K = G_K \times (E_m - E_K)$$

(c) Since the electrochemical gradient for Na^+ is greater than that for K^+ (and the conductances for Na^+ and K^+ are equal), the amount of Na^+ entering the cell exceeds the amount of K^+ leaving the cell, and the cell depolarizes.

(4) The **reversal potential** is the potential at which no net current flows through the channel (i.e., when the Na^+ and K^+ currents are equal and opposite to each other).

    (a) These currents become equal and opposite to each other when the membrane potential becomes –16 mV. (This can be verified by substituting the appropriate values of $E_{Na} = +60$ and $E_K = -92$ into the above equations and solving for E_m when $I_{Na} = I_K$. Since $G_{Na} = G_K$, the conductance terms cancel out.)

    (b) If the only open channels in the membrane were the ACh synaptic channels, the membrane potential would be equal to the reversal potential.

        (i) This is analogous to what happens when the K^+ channel is the only open channel [see II D 5 a (1)]. Under these conditions, the membrane potential is equal to the equilibrium potential for K^+ (E_K).

        (ii) The analogy is illustrated by the equivalent circuit diagram (see Figure 1-8B), in which the synaptic battery (representing the reversal potential for ACh) is given the value –16 mV.

    (c) However, as the equivalent circuit diagram illustrates, the ACh channel is not the only open channel in the membrane.

        (i) If a **single ACh channel** is opened (as occurs when two ACh molecules bind to it—one to each α unit), the membrane will depolarize by only a few microvolts (μV).

        (ii) If a **single vesicle** releases its contents of 5000–10,000 ACh molecules, the membrane will depolarize by about 1 mV. The small (1 mV) depolarization caused by the spontaneous release of one vesicle is called a **miniature end-plate potential (MEPP)**. Spontaneous release of vesicles occurs at a rate of about 1/sec. Thus, MEPPs occur every second or so. The MEPPs may be important in maintaining the integrity of the muscle fiber, because denervation of skeletal muscle fibers leads to muscle atrophy.

        (iii) The **200–300 vesicles** that release their contents when an action potential invades the presynaptic nerve terminal produce a depolarization of about 50 mV. This depolarization is the **EPP**.

c. The EPP depolarizes the muscle fiber membrane to threshold.

    (1) The EPP is a graded, nonpropagated response that acts as a stimulus for the production of an action potential on the muscle membrane contiguous to the end-plate membrane.

    (2) The ACh molecules contained within the several hundred vesicles released during neuromuscular transmission depolarize the membrane from its resting potential of –90 mV to about –40 mV.

        (a) The depolarization of the end-plate membrane causes the contiguous membrane to be depolarized [see II C 5 a].

        (b) When the membrane next to the end-plate membrane reaches threshold (about –60 mV), an action potential is generated.

        (c) The action potential is propagated along the muscle membrane and is responsible for initiating a muscle contraction (see IV A 2).

    (3) An action potential cannot be generated on the end-plate because there are no electrically excitable K^+ and Na^+ channels present on the end-plate membrane. ACh receptors are so densely packed into the end-plate that there is no room for any other membrane proteins.

    (4) Each time an alpha motoneuron fires an action potential, it causes all of the muscle fibers that it innervates to contract.

d. ACh is degraded rapidly.

    (1) After binding to the ACh receptor, the ACh dissociates from the receptor and is hydrolyzed by the **AChEase** in the synaptic cleft.

    (2) Degradation of ACh is necessary to prevent it from causing multiple muscle contractions.

    (3) Enzymatic destruction is a unique method for inactivating the transmitter and occurs only at ACh synapses. Inactivation of the transmitter at all other synapses occurs when the transmitter diffuses out of the synaptic region or is actively transported back into the nerve terminal.

B. Autonomic synaptic transmission. The autonomic nervous system is divided into the **parasympathetic** and **sympathetic** systems, each of which has a **preganglionic neuron** in the **central nervous system (CNS)** and a **postganglionic neuron** in the **peripheral nervous system (PNS)**. In both divisions of the autonomic nervous system, ACh is the transmitter used to communicate between the pre- and postganglionic fibers. ACh also is used by the parasympathetic postganglionic fibers, whereas **norepinephrine** is the transmitter used by the sympathetic postganglionic fibers. Many postganglionic fibers, particularly those within the ganglia of the gastrointestinal tract (the **enteric nervous system**), do not use ACh or norepinephrine as the transmitter substance. Because the identity of the transmitter is not clear, it often is referred to as a **noncholinergic, nonadrenergic transmitter**.

1. **Ganglionic transmission** within the sympathetic and parasympathetic divisions of the autonomic nervous system is essentially the same as that at the neuromuscular junction. ACh is released from the preganglionic (presynaptic) fiber, diffuses across the synaptic cleft, and binds to receptors on the postganglionic fiber, causing it to depolarize.
   a. In almost all respects, this process is identical to that of skeletal muscle neurotransmission (see III A 3). However, a single preganglionic fiber does not release enough neurotransmitter to depolarize the postganglionic fiber to threshold. The postganglionic fiber can discharge an action potential only when there is a **summation** of the postsynaptic responses (see III C 2).
   b. The ACh receptors on the postganglionic fibers are called **nicotinic receptors** because they are activated by nicotine. The ACh receptors on the skeletal muscle end-plate also are nicotinic receptors.
      (1) The nicotinic receptors on the autonomic ganglion postganglionic membranes are not identical to those on the end-plates of skeletal muscle because the pharmacologic agents needed to block the two receptors are different.
         (a) **Hexamethonium** is used to block ganglionic transmission.
         (b) **Curare** is used to block neuromuscular transmission.
      (2) Nicotine is extremely toxic: it causes vomiting, diarrhea, sweating, and high blood pressure, primarily due to its action on the autonomic nervous system.
      (3) Similar to the receptors on skeletal muscle, the ACh receptors on postganglionic fibers cause the membrane to depolarize by opening channels that are equally permeable to K^+ and Na^+.

2. **Parasympathetic postganglionic fibers** can have either an excitatory or an inhibitory effect, depending on the postsynaptic receptor that is activated.
   a. **Excitatory effects** are produced on a variety of smooth muscles (e.g., those within the stomach, intestine, bladder, and bronchi) and on glands. ACh can produce its excitatory effect by a variety of mechanisms.
      (1) ACh can bind to receptors that cause the membrane to depolarize in a mechanism similar to that occurring on the skeletal muscle end-plate.
      (2) ACh can bind to receptors that increase Ca^{2+} conductance. The increase in Ca^{2+} conductance does not produce a major effect on membrane potential. However, the Ca^{2+} entering the cell through the channels opened by ACh can be used to initiate contraction in smooth muscles (see IV C 2).
      (3) ACh can bind to receptors that activate the membrane-bound protein **guanosine triphosphate (GTP) binding protein,** or **G protein** (Figure 1-17A). When the G protein is activated (see Figure 1-17B), it initiates a series of membrane and intracellular events leading to muscle contraction.
         (a) The activated G protein (see Figure 1-17C) activates a membrane-bound lipase called **phospholipase C.**
         (b) Phospholipase C causes **phosphatidylinositol diphosphate (PIP$_2$)** to break down into two other lipids, **diacylglycerol (DAG)** and **inositol triphosphate (IP$_3$).**
            (i) **DAG** activates **protein kinase C,** a cytoplasmic enzyme that activates a series of cytoplasmic proteins by phosphorylating them. These proteins then cause a physiologic effect such as opening an ionic channel.
            (ii) **IP$_3$** diffuses through the cytoplasm and binds to the **sarcoplasmic reticulum,** causing Ca^{2+} to be released into the cytoplasm. The increase in intracellular Ca^{2+} then leads to muscle contraction.

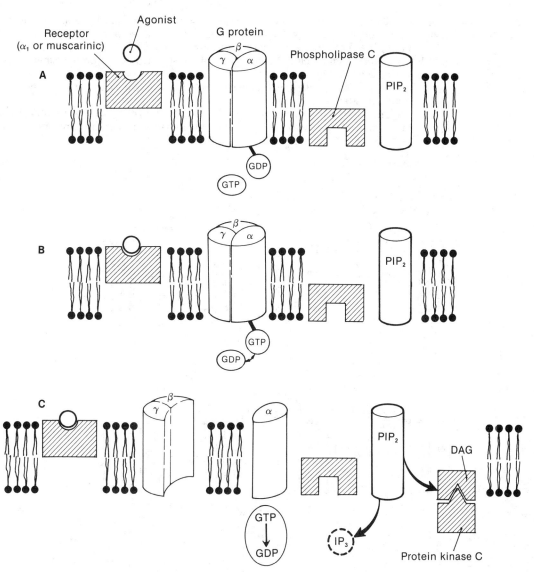

Figure 1-17. (*A*) When a neurotransmitter or other agonist binds to a muscarinic or α_1 receptor, (*B*) the agonist-receptor complex induces the G protein to exchange guanosine diphosphate (*GDP*) for guanosine triphosphate (*GTP*). (*C*) The presence of GTP causes the α subunit to separate from the G protein. The α subunit migrates within the membrane, eventually forming a complex with and activating phospholipase C. The phospholipase C, in turn, causes the breakdown of phosphatidylinositol diphosphate (*PIP$_2$*) to inositol triphosphate (*IP$_3$*) and diacylglycerol (*DAG*). IP$_3$ releases Ca^{2+} from the sarcoplasmic reticulum; DAG activates protein kinase C. The α subunit contains a GTPase that catalyzes the breakdown of GTP to GDP, inactivating the G protein.

    b. Inhibitory effects of parasympathetic fibers are produced on the heart, primarily on the pacemaker regions of the sinoatrial (SA) node (decreasing the heart rate) and atrioventricular (AV) node (slowing conduction of the action potential from the atria to the ventricles) [see Ch 2 II B 3 c (5) (b), C].

        (1) When ACh binds to a receptor on the heart, the receptor causes a K$^+$ channel to open.

        (2) K$^+$ flows out of the cell down its electrochemical gradient, causing the cell to hyperpolarize.

        (3) The hyperpolarization acts to decrease the rate of pacemaker activity and to slow the conduction of the action potential through the AV node.

    c. The postganglionic receptors are called **muscarinic** because they respond to the drug **muscarine** and not to nicotine. Muscarinic receptors are blocked by the drug **atropine**.

3. **Sympathetic postganglionic fibers** that release norepinephrine can have an excitatory or inhibitory effect, depending on the type and location of the receptor activated.
   a. **Norepinephrine receptors** are divided into **beta** (β) and **alpha** (α) **receptors**.
      (1) When norepinephrine binds to the β **receptors,** it activates a G protein similar to that activated by ACh (Figure 1-18).
         (a) However, in this case, the activated G protein activates a different membrane-bound protein: **adenylate cyclase**.
         (b) Adenylate cyclase stimulates the formation of **cyclic adenosine 3′,5′-mono-phosphate (cAMP)** from ATP. The G protein that stimulates the formation of cAMP is called the **G_s protein**.

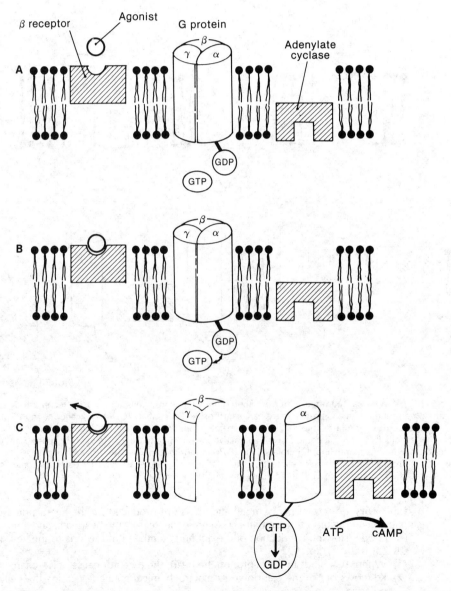

Figure 1-18. (*A*) When an agonist binds to the β receptor, (*B*) the agonist-receptor complex activates a G protein, which, in turn, (*C*) activates adenylate cyclase. The adenylate cyclase catalyzes the conversion of ATP to cyclic adenosine 3′,5′-monophosphate (*cAMP*). cAMP affects intracellular proteins by phosphorylating them. As described in Figure 1-17, the G protein is inactivated when guanosine triphosphate (*GTP*) is converted to guanosine diphosphate (*GDP*).

(c) cAMP then activates the cytoplasmic enzyme **protein kinase A** (or **cAMP-dependent protein kinase**), which, in turn, produces a variety of physiologic responses by phosphorylating an assortment of intracellular proteins.

 (i) In heart cells, protein kinase A phosphorylates Ca^{2+} channels, which increases the amount of Ca^{2+} entering the cell with each action potential and, thus, increases the force of contraction.

 (ii) In bronchiole smooth muscle cells, protein kinase A phosphorylates the Ca^{2+} pump on the sarcoplasmic reticulum, enhancing the pump's activity. The increased activity of the sarcoplasmic reticular Ca^{2+} pump removes Ca^{2+} from the cytoplasm and, thus, decreases the force of contraction.

(2) When norepinephrine binds to the α_2 **receptor** (one subtype of α-adrenergic receptors), it activates a G protein that inhibits adenylate cyclase, thus reducing the amount of cAMP in the cell. The G protein that inhibits the formation of cAMP is the G_i **protein.**

(3) When norepinephrine binds to the α_1 **receptor** (another subtype of the α-adrenergic receptors), it activates a G protein that, similar to its action at the muscarinic receptor, results in the formation of IP_3 and DAG.

(4) **Adrenergic receptors can be distinguished** by the types of drugs that activate and inhibit them.

 (a) α-**Adrenergic receptors** are activated preferentially by epinephrine and are blocked by phenoxybenzamine.

 (b) β-**Adrenergic receptors** are activated preferentially by isoproterenol and are blocked by propranolol.

b. Norepinephrine metabolism

(1) Biosynthesis of norepinephrine occurs in the sympathetic nerve terminals. Norepinephrine is released from synaptic vesicles by exocytosis in a manner identical to that by which ACh is released.

(2) Inactivation of norepinephrine occurs by active transport into the nerve terminal and diffusion out of the synaptic cleft. In contrast to ACh, enzymatic degradation is not an important mechanism for inactivating norepinephrine.

4. **The enteric nervous system** is an independent component of the autonomic nervous system. It is composed of the ganglia found within the wall of the gastrointestinal tract and is responsible for coordinating the activity of gastrointestinal smooth muscle and gastric secretions.

a. The enteric nervous system receives synaptic input from the sympathetic and parasympathetic postganglionic fibers.

b. Many of the **neurotransmitters** used by the enteric nervous system have not been identified. Among those known to be active are serotonin, the enkephalins and endorphins, somatostatin, vasoactive intestinal peptides (VIP), and the purines ATP and adenosine.

C. The central nervous system

1. **Synaptic mechanisms.** An enormous variety of synaptic connections occur within the CNS, and, with few exceptions, these synapses use chemical transmitter mechanisms. In a few locations (e.g., within the retina and olfactory bulb), synaptic transmission is accomplished by passive electrotonic spread of current between two cells.

a. **Electrical synaptic transmission** between two cells occurs at specialized junctions called **gap junctions** (Figure 1-19).

(1) Only 2 nm separate the pre- and postsynaptic membranes at the site of gap junctions.

(2) Gap junctions are formed by membrane bridges containing aqueous channels through which small molecules and ions can pass from one cell to another, establishing cytoplasmic continuity.

 (a) When an action potential propagating along the membrane in one cell reaches the gap junction, an electrical current flows passively through the gap from one cell to another.

 (b) Electrical current can pass through the gap in both directions, allowing either cell to serve as the pre- or postsynaptic cell.

(3) Although gap junctions are not common in CNS synaptic transmission, they play an important role in coordinating muscle contraction in the heart and viscera. Gap junctions rapidly transmit an action potential that is generated in one cell to all of the other cells within the organ. This permits the entire tissue to act as a syncytium and contract in a coordinated fashion.

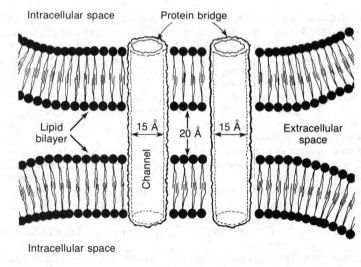

Figure 1-19. A gap junction forms a low-resistance electrical pathway between two cells. Although the two cells remain separate, they maintain cytoplasmic continuity via a channel within a protein bridge.

   b. Chemical synaptic transmission within the CNS occurs by a mechanism similar to that at the neuromuscular junction.

      (1) The synaptic transmitter is synthesized in the nerve terminal, stored in vesicles, and released by exocytosis when an action potential invades the nerve terminal.

      (2) After being released from the presynaptic terminal, the transmitter diffuses across the synaptic cleft, binds to a postsynaptic receptor, and causes the opening of channels through which ions can flow. Both **excitatory** and **inhibitory receptors** exist on the postsynaptic cell.

      (3) The transmitter is inactivated in one of three ways.

         (a) It diffuses out of the synaptic cleft.

         (b) It is actively transported into the presynaptic terminal.

         (c) It is enzymatically degraded (if the transmitter is ACh).

  2. Summation. The postsynaptic cell integrates the information it receives from thousands of presynaptic terminals. This is illustrated in Figure 1-20, using the alpha motoneuron found in the ventral horn of the spinal cord as an example. (Although synaptic junctions cover most of the cell body and proximal dendrites, only a few of these are shown in Figure 1-20.)

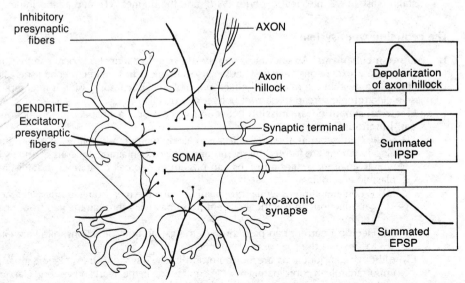

Figure 1-20. Inhibitory and excitatory synapses are formed on an alpha motoneuron. When the amplitude of the summated excitatory postsynaptic potentials (*EPSP*) exceeds the amplitude of the summated inhibitory postsynaptic potentials (*IPSP*), the axon hillock is depolarized to threshold, and an action potential is generated.

a. **Postsynaptic response.** Because the presynaptic membrane is only about 1–2 μ in diameter, only a few synaptic vesicles can attach to the dense bars within the presynaptic terminal and release their contents into the synaptic cleft when an action potential invades the nerve terminal. As a result, the postsynaptic response is only a few millivolts in amplitude. The postsynaptic response can be either inhibitory or excitatory.

  (1) **The excitatory postsynaptic potential (EPSP)** is the depolarization produced by an excitatory neurotransmitter. In order for an EPSP to depolarize an alpha motoneuron to threshold, the EPSPs must summate.

    (a) Although the synaptic channel is open for only a few milliseconds, the membrane capacitance causes the EPSP to decay more slowly. As a result, the duration of the EPSP is about 15 msec.

    (b) **Temporal summation** occurs if another action potential invades the nerve terminal before the first EPSP has disappeared. The second EPSP adds to the first, producing a larger response. If the presynaptic nerve terminal fires frequently enough, the EPSPs can summate sufficiently to depolarize the alpha motoneuron to threshold.

    (c) **Spatial summation** occurs if several nerve terminals fire synchronously.

  (2) **Inhibitory postsynaptic response (IPSP).** The IPSPs produced by an inhibitory neurotransmitter, like the EPSPs, can summate to produce a larger negative response.

b. **Axon hillock.** The potentials produced by the excitatory and inhibitory neurotransmitters spread passively to the axon hillock, where the action potential is generated.

  (1) The threshold for producing an action potential at the axon hillock is about –65 mV.

  (2) The channel opened by the excitatory neurotransmitter allows Na$^+$ and K$^+$ to flow through it and, like the end-plate potential (EPP), has a reversal potential of about –16 mV.

  (3) The channel opened by the inhibitory neurotransmitter allows Cl$^-$ to flow through it and has a reversal potential of about –82 mV.

  (4) If the summated EPSPs are large enough to depolarize the axon hillock to threshold, an action potential will be generated.

  (5) The summated IPSPs can prevent the axon hillock from being depolarized to threshold by hyperpolarizing the cell.

3. **Presynaptic inhibition** is produced by an axo-axonic synapse (Figure 1-21).

  a. The number of vesicles binding to the dense bar and releasing their contents into the synaptic cleft depends on the amount of Ca^{2+} entering the presynaptic terminal, which, in turn, depends on the magnitude of the action potential invading the nerve terminal.

    (1) The presynaptic nerve terminal of the axo-axonic synapse (*neuron 1*) releases a transmitter that opens Cl$^-$ channels on the postsynaptic membrane of the axo-axonic synapse (*neuron 2*).

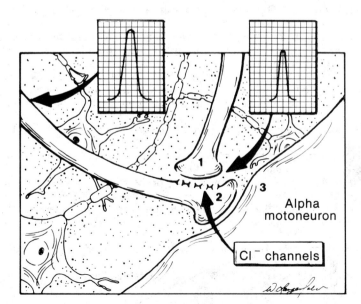

Alpha motoneuron

Cl$^-$ channels

Figure 1-21. Diagram illustrating the axo-axonic synapse responsible for presynaptic inhibition. *Neuron 1* releases GABA, which opens Cl$^-$ channels on *neuron 2*. The increased Cl$^-$ conductance reduces the amplitude of the action potential as it approaches the nerve terminal. Because the action potential is smaller, less Ca^{2+} enters the nerve terminal, less transmitter is released from the nerve terminal, and the magnitude of the EPSP produced on the postsynaptic membrane of *neuron 3* is reduced.

(2) When an action potential invades the postsynaptic neuron of the axo-axonic synapse (*neuron 2*), Cl$^-$ flows into the neuron through the opened Cl$^-$ channels. The flow of a negative ion into the neuron reduces the magnitude of the overshoot.

(3) Because the overshoot is smaller, less Ca^{2+} enters the nerve terminal, and the amount of neurotransmitter released by neuron 2 is diminished.

(4) The reduction in transmitter release reduces the size of the EPSP produced on the alpha motoneuron (*neuron 3*).

    b. In presynaptic inhibition, the excitability of the postsynaptic cell is not diminished, whereas an IPSP reduces the effectiveness of all excitatory input to a cell. By using presynaptic inhibition, a particular excitatory input can be inhibited without affecting the ability of other excitatory synapses to fire the cell.

IV. MUSCLE CONTRACTION.
Muscle fibers are divided into two types based on their appearance in light micrographs. **Striated muscle,** which includes **skeletal** and **cardiac** muscle, is characterized by alternating light and dark bands. **Smooth muscle** has no distinguishing surface features. Although all muscle types function in a similar way, several important differences exist. In the following discussion, the structural and contractile properties of skeletal muscle are noted first, with the major differences in cardiac and smooth muscle noted afterward.

A. Skeletal muscle

1. Structure (Figure 1-22). Skeletal muscle fibers vary in diameter from about 10 μ to 100 μ and can be several centimeters in length.

    a. Fascicles. Skeletal muscle fibers are grouped into **fascicles** of about 20 fibers by the **perimysium,** a connective tissue sheath that is continuous with the connective tissue surrounding the entire muscle.

        (1) The perimysium is continuous with the **endomysium** surrounding each muscle fiber.

        (2) The endomysium is continuous with the **sarcolemma,** a glycoprotein-containing sheath that closely envelops the true cell membrane of the muscle fiber.

        (3) The tight connection between the cell membranes and the surrounding connective tissue structures makes it possible for the force developed by the muscle fibers to be transmitted effectively to the tendons.

    b. Myofibrils. Each muscle fiber is divided into **myofibrils** by a tubular network called the **sarcoplasmic reticulum** (see Figure 1-22B).

        (1) The myofibrils are about 1 μ in diameter and extend from one end of the muscle fiber to the other.

        (2) The myofibrils are divided into functional units, or **sarcomeres,** by a transverse sheet of protein called the **Z disk.**

        (3) The Z disks of neighboring myofibrils are lined up with each other so that in histologic slides a **Z line** spans the entire width of the fiber.

    c. Filaments. The myofibrils contain thick and thin filaments composed of contractile proteins.

        (1) **Thin filaments** contain the proteins **actin, tropomyosin,** and **troponin** and are about 50 Å in diameter and 1 μ in length.

            (a) One end of the thin filament is attached to the Z disk, so that filaments from opposing Z disks extend longitudinally into the center of the sarcomere toward each other.

            (b) The thin filaments appear to be held in place by a thin protein that connects their free ends.

        (2) **Thick filaments** contain the protein **myosin** and are about 110 Å in diameter and 1.6 μ in length.

            (a) The thick filaments are interspersed between the thin filaments at the center of the sarcomere. They are not attached to the Z disk.

            (b) Projections from the thick filaments called **cross-bridges** extend toward the thin filaments. Cross-bridges play a fundamental role in muscle contraction.

        (3) **Banding pattern.** The interdigitating thick and thin filaments create the pattern of light and dark bands that characterizes light microscopic views of skeletal muscle (see Figure 1-22A).

            (a) The dark areas in the center of the sarcomere are called **A bands.** These contain the thick filaments.

            (b) The light areas on either side of the Z disk are called **I bands.** They contain the thin filaments.

A

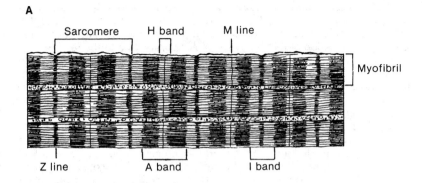

B

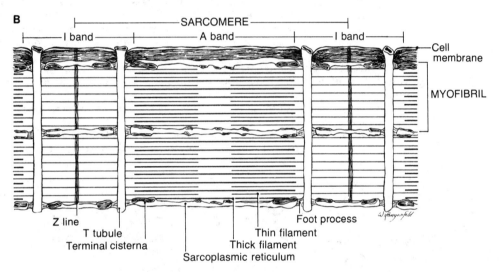

Figure 1-22. High- and low-power light microscopic view of human skeletal muscle. (*A*) Under low power, the alternating A and I bands can be seen. The functional unit of the muscle fiber, the sarcomere, is bounded by Z lines. (*B*) Under high power, the T tubules can be seen at the junction of the A band and I band. The terminal cisternae of the sarcoplasmic reticulum form specialized junctions with the T tubules called foot processes, or junctional feet.

    (c) The thick and thin filaments overlap to some extent in the A band. The area of the A band without any thin filaments is called the **H band**. At the center of the H band is the **M line**. This is the region of the thick filaments that does not contain any cross-bridges.

    (d) The amount of overlap between thick and thin filaments varies with sarcomere length and determines how much force skeletal muscle will develop when it is stimulated. When the muscle is stretched or shortened, the thick and thin filaments slide past each other, and the I band increases or decreases in size.

  d. Tubules. Two tubular networks are present in skeletal muscle fibers (see Figure 1-22B).

    (1) The **transverse (T) tubule** is formed as an invagination of the surface of the muscle membrane.

      (a) In mammalian skeletal muscle, the T tubules are located at the junction of the A and I bands. (In frog muscle, the T tubules are at the Z disk.)

      (b) An action potential spreading over the surface of the muscle membrane is propagated into the network of T tubules.

    (2) The T tubule forms specialized contacts with the **sarcoplasmic reticulum (SR),** the internal tubular structure that runs between the myofibrils.

      (a) The SR has a high concentration of Ca^{2+}, which is used to initiate muscle contraction when the muscle is stimulated.

      (b) The ends of the SR expand to form **terminal cisternae (TC),** which make contact with the T tubule. Small projections, or **foot processes,** span the 200 Å separating the two tubular membranes. These foot processes contain Ca^{2+} channels.

2. Excitation-contraction (EC) coupling is the process by which an action potential initiates the contractile process. EC coupling involves four steps: the propagation of the action potential into the T tubule and release of Ca^{2+} from the TC, the activation of the muscle proteins by Ca^{2+}, the generation of tension by the muscle proteins, and the relaxation of the muscle.

    a. Release of Ca^{2+}. Depolarization of the T tubule by the action potential causes the Ca^{2+} channels on the foot processes to open. Ca^{2+} flows out of the TC and into the cytoplasm.

    b. Activation of muscle proteins. For a muscle to contract, the thick and thin filaments must interact. When the cell is at rest, this interaction is inhibited. Ca^{2+} removes this inhibition by binding to troponin on the thin filament.

        (1) Thin filament proteins (Figure 1-23A)

            (a) The backbone of the thin filament is formed by two chains of **actin** molecules, which wind around each other. Each actin molecule has a myosin binding site.

            (b) When not stimulated, the myosin binding site is covered by another thin filament protein: **tropomyosin.** This long molecule coils around the actin chain.

            (c) The position of tropomyosin on the thin filament is controlled by another protein: **troponin.** The troponin molecule contains a binding site for Ca^{2+}. When the binding site is occupied, troponin undergoes a conformational change that causes tropomyosin to move away from its position covering the myosin binding site on actin.

            (d) Once uncovered, the binding site on actin combines with the cross-bridges from the thick filament, and contraction begins.

        (2) Thick filament protein. About 300 **myosin** molecules coalesce to form the thick filament.

            (a) Myosin is a complex protein consisting of a globular head, the cross-bridge, and a long tail.

            (b) Rotation can occur at two points in the myosin molecule. One, located in the tail region, is used to rotate the molecule outward when the spacing between the thick

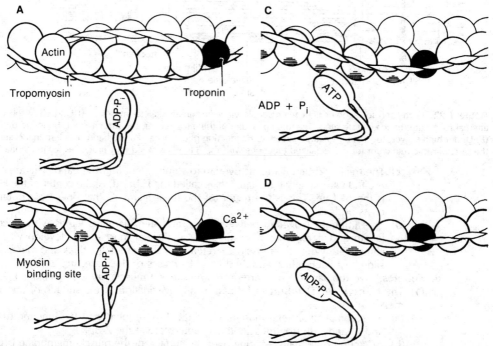

Figure 1-23. (A) The thin filament is composed of three proteins. Actin forms the backbone of the thin filament. At rest, the tropomyosin is covering the myosin binding site on actin. When Ca^{2+} binds to troponin, the troponin undergoes a conformational change, pulling tropomyosin away from the myosin binding site and allowing the cross-bridge cycle to begin. (B) In the first step of the cycle, myosin binds to actin. (C) In the second step, the cross-bridge bends and the thin filament slides over the thick filament. During this step, the products of hydrolysis, adenosine diphosphate (ADP) and inorganic phosphate (P_i), are released from the cross-bridge. (D) In the third step, a new molecule of ATP binds to the cross-bridge and the cross-bridge detaches from the thin filament. The cross-bridge then stands up, and a new cycle begins. Cycling continues for as long as intracellular Ca^{2+} concentration remains high.

and thin filaments changes. The other, located at the junction of the head and tail, is the site where the cross-bridge bends when generating tension.

    (c) Located on the cross-bridge is the enzyme **myosin-ATPase,** which is used to hydrolyze ATP and, thus, provide the energy for muscle contraction.

c. Generation of tension. Tension is generated by the cycling of the cross-bridges, which occurs after they bind to the thin filament.

    (1) The **first step** in the cross-bridge cycle is the binding of actin and myosin (see Figure 1-23B). This occurs spontaneously, after Ca^{2+} binds to troponin, and tropomyosin moves away from its position blocking the myosin binding site on actin.

        (a) At rest, the intracellular Ca^{2+} concentration is less than 10^{-7} mol/L.

        (b) When stimulated, enough Ca^{2+} is released from the TC to raise the intracellular Ca^{2+} concentration to 10^{-5} mol/L. At this concentration, all of the muscle protein is activated.

    (2) The **second step** is the bending of the cross-bridge and the sliding of the thin filament across the thick filament (see Figure 1-23C).

        (a) The energy used to bend the cross-bridge and generate tension is obtained from ATP.

        (b) Both the ATP molecule and its hydrolyzing enzyme ATPase are attached to the cross-bridge.

    (3) The **third step** is the detachment of the cross-bridge from the thin filament. This occurs after the cross-bridge has bent (see Figure 1-23D).

        (a) For detachment to occur, the products of ATP hydrolysis [i.e., adenosine diphosphate (ADP) and inorganic phosphate (P_i)] must be removed from the cross-bridge and a new molecule of ATP put in their place.

        (b) If no ATP is available, the thick and thin filaments cannot be separated. (This is the cause of rigor mortis.)

        (c) As soon as myosin separates (or, perhaps, while it is separating) from actin, the initial steps of ATP hydrolysis occur. The product of these metabolic reactions is a high-energy ATP intermediate: myosin · ADP · P_i.

        (d) Complete dissociation of the phosphate from the ADP does not occur until after myosin has bound to actin and completed its bending cycle.

    (4) In the **final step,** the cross-bridge returns to its original upright position. Once there, it can participate in another cycle. Cycling continues as long as Ca^{2+} is attached to troponin.

d. Relaxation occurs when the Ca^{2+} is removed from the cytoplasm by Ca^{2+} pumps (Ca^{2+}-ATPase) located on the SR membrane. When the intracellular Ca^{2+} concentration falls below 10^{-7} mol/L, troponin returns to its original conformational state, tropomyosin moves back to cover the myosin binding site on actin, and cross-bridge cycling stops.

3. Mechanical properties. The force developed by the bending of the cross-bridge is transmitted through the thin filament to the Z disk and then through the sarcolemma and tendinous insertions of the muscle to the bones. The contractile properties of the muscle can be studied in two types of mechanical conditions, isometric and isotonic contractions.

a. In **isometric contractions,** the muscle length remains constant during the contractile event and the force developed during the contraction is measured.

    (1) Muscle twitch. Figure 1-24 illustrates the intracellular Ca^{2+} concentration and the force developed by a muscle during a single **twitch.**

        (a) A single electrical stimulus is used to depolarize the muscle membrane to threshold and produce an action potential, which causes the release of Ca^{2+} from the TC.

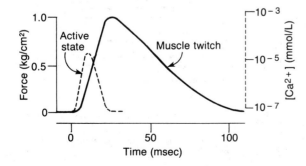

Figure 1-24. When intracellular Ca^{2+} concentration (*right ordinate*) rises above 10^{-7} mmol/L, cross-bridge cycling causes an increase in muscle force (*left ordinate*). The duration of cross-bridge cycling is the active state, and the resulting force development is the muscle twitch.

     (b) Ca^{2+} initiates the cross-bridge cycle, which lasts until the Ca^{2+} is resequestered by the Ca^{2+} pumps on the SR membrane.

     (c) The period during which the intracellular Ca^{2+} concentration is above resting values and the cross-bridges are cycling is called the **active state**.

  (2) **An increased strength of electrical stimulation** causes an increase in the isometric force developed by the muscle.

     (a) The increased force is the result of more muscle fibers being activated. Although this is a predictable result, it is important to note because it is one of the two major methods used by the motor/control system to increase the force of skeletal muscle contraction.

     (b) If the initial conditions are the same, then the force developed by each muscle fiber during a twitch remains constant. This is because the amount of Ca^{2+} released from the TC activates all of the muscle protein, and the time for the SR Ca^{2+} pump to resequester the Ca^{2+} remains the same. However, the force developed by the muscle can be changed by varying the length of the sarcomeres and the frequency of stimulation.

  (3) Figure 1-25 illustrates the relationship between muscle length and force development during a muscle twitch. This is called the **length-tension relationship**. The figure shows both the actual muscle length and the length of the individual sarcomeres. The sarcomere length plays an important role in the length-tension relationship.

     (a) The initial length of the muscle, called the **preload,** is set by a load (or force) applied to the muscle before it is stimulated.

        (i) Maximal force is obtained when the preload is set at a sarcomere length of 2.2 μ. At this length, the overlap between thick and thin filaments is optimal, since every cross-bridge from the thick filament is opposite an actin molecule on the thin filament. When contraction is initiated, each cross-bridge binds to a thin filament and contributes to the contractile force.

        (ii) Increasing the preload causes a decrease in the force developed by the muscle fiber. At preloads greater than 2.2 μ, the overlap between thick and thin filaments is decreased. Thus, some cross-bridges do not have actin filaments with which to combine, and the force developed by the muscle is diminished.

        (iii) It is not so obvious why decreasing the preload below 2.2 μ causes a decrease in force development, since every cross-bridge has an actin site with which to combine. However, at short sarcomere lengths the thin filaments bump into each other, making it more difficult for the muscle to develop force. Also, it is possible that the amount of Ca^{2+} released by the TC is reduced as a function of sarcomere length.

     (b) Varying the preload is not an important method of varying the contractile force of skeletal muscle. Often, the muscle length is determined by the particular motor task being performed. However, if muscle length is not constrained by the motor activity, more force can be obtained by holding the muscle at its optimal length (i.e., where the sarcomere length is 2.2 μ).

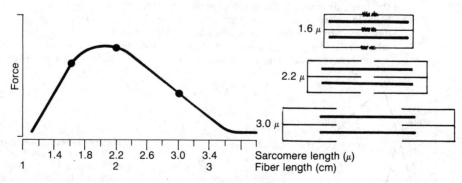

Figure 1-25. The length-tension relationship results from the overlap between thick and thin filaments. At a sarcomere length of 2.2 μ, overlap is optimal and force development is maximal. At lengths greater than 2.2 μ, force decreases because cross-bridge overlap is lessened. At lengths less than 2.2 μ, force is less because the thin filaments meet at the center of the sarcomere, causing an increased resistance to shortening.

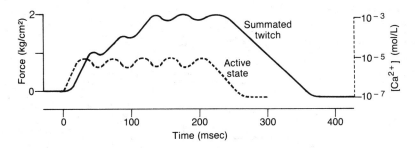

Figure 1-26. When the duration of the active state is increased by repetitive firing of the muscle, force development increases because there is sufficient time for the series elastic component to be stretched completely.

    (4) **An increased frequency of electrical stimulation** also increases the force of muscle contraction.
        (a) Varying the frequency of muscle stimulation is the other major method used by the motor control system to vary the force of muscle contraction. This is illustrated in Figure 1-26. Note that the Ca^{2+} concentration does not increase above 10^{-5} mol/L but that the duration of the active state is increased due to the repetitive release of Ca^{2+} from the TC.
        (b) The increase in muscle force due to repetitive stimulation is called **summation**. When the frequency of stimulation is rapid enough to allow the force to rise smoothly to a maximum, the response is called a **tetanus**. Figure 1-27 is a simple mechanical analog of the contractile properties of the muscle, which can be used to explain the phenomena of summation and tetanus. In this figure, the thick and thin filaments are represented by the **contractile component (CC),** and the tendons and other compliant structures of the muscle are represented by a **series elastic component (SEC).**
          (i) In order for the force developed by the cross-bridges to be transmitted to the bones, the SEC must be stretched. The more the SEC is stretched, the more force is transmitted.
          (ii) Since each cross-bridge is only 100 Å long, its rotation will move the thin filament about 75 Å. At least 30 cross-bridge cycles are required to stretch the SEC sufficiently to transmit all the force of the attached cross-bridges to the bone.
          (iii) Since each cycle lasts about 1 msec in fast-twitch fibers [see XII B 1 c (1)], the cross-bridges must cycle for at least 30 msec. This cannot happen in a single twitch because the Ca^{2+} is resequestered too rapidly. By repeatedly stimulating the muscle, however, the Ca^{2+} remains in the cytoplasm for a longer time and allows the SEC to be stretched sufficiently.
  b. In **isotonic contractions**, the skeletal muscle shortens.
    (1) Figure 1-28 illustrates the development of force and the change in muscle length that occurs during an isotonic contraction.
        (a) In order to shorten, the muscle must lift a weight, called the **afterload,** which is applied after the muscle begins to contract.
        (b) When the muscle is stimulated, the cross-bridges begin to cycle, the SEC is stretched, and force is applied to the afterload.
          (i) During the initial part of the contraction, the force is less than the afterload and shortening does not occur. This is the isometric portion of the isotonic contraction.
          (ii) After the SEC has been sufficiently stretched, the force equals the afterload and the muscle begins to shorten.
          (iii) While the muscle is shortening, the length of the SEC remains constant and the force remains equal to the afterload, thus the term, "isotonic" contraction.

Contractile
component

Series
elastic
component

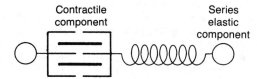

Figure 1-27. The contractile properties of a muscle can be explained using a mechanical analog consisting of a force-generating contractile component in series with an elastic element.

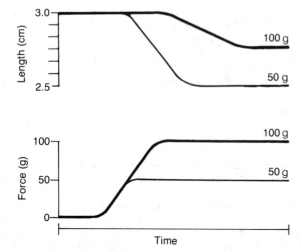

Figure 1-28. During an isotonic contraction, sufficient force must be generated in order for the muscle to shorten against the afterload on the muscle. Once sufficient force is developed, the muscle shortens at a constant velocity. Increasing the afterload increases the amount of force that must be developed before shortening can begin and decreases the velocity and extent of shortening. *Heavy lines* represent an afterload of 100 g; *light lines* represent an afterload of 50 g.

(iv) The velocity of shortening also remains constant during the contraction.

(2) As shown in Figure 1-28, the magnitude of the afterload affects the characteristics of the contraction.

(a) As the afterload increases, the duration of the isometric portion of the contraction increases. This occurs because the SEC must stretch more to transmit the force required to lift the greater afterload.

(b) The amount of shortening that can occur decreases as the afterload increases due to the length-tension relationship. As the sarcomere length is reduced below 2.2 μ, the force that can be developed decreases. At some length, the maximal force that can be developed by the muscle becomes less than the afterload, and further shortening is impossible. The greater the afterload, the sooner this length is reached.

(c) Finally, the velocity of shortening decreases with an increase in the afterload. This **load-velocity relationship** is an important characteristic of muscle because it indicates that the greatest velocity of shortening is generated when the afterload on the muscle is zero. The peak velocity of shortening is an indication of the cross-bridge cycling speed.

B. Cardiac muscle, like skeletal muscle, is striated. Although it is similar to skeletal muscle in structure and behavior, cardiac muscle displays some important differences.

1. Structure

a. Cardiac muscles are smaller than skeletal muscles. Each fiber is about 15–20 μ wide, about 100 μ long, and only about 5 μ thick, making its shape more ribbon-like than cylindrical.

b. The striation pattern of a cardiac muscle fiber is similar to that described for a skeletal muscle fiber.

c. Cardiac muscle also has both T tubular and SR systems. The T tubule is larger than in skeletal muscle and is located at the Z disk rather than at the junction of the A and I bands. The SR makes contact with the T tubule and with the cell membrane.

2. EC coupling of cardiac muscle also differs in significant ways from that of skeletal muscle. Most importantly, the amount of Ca^{2+} in the cytoplasm is never enough to activate all of the muscle protein. As a result, the force of contraction can vary with the amount of Ca^{2+} entering the cell. Several mechanisms may be involved in regulating Ca^{2+} entry during the action potential.

a. As in skeletal muscle, Ca^{2+} is released from the TC by the action potential. However, the Ca^{2+} is not released directly by the depolarization accompanying the action potential. Instead, the Ca^{2+} entering the cell during the cardiac action potential triggers the release of Ca^{2+} from the TC. This is called **Ca^{2+}-induced Ca^{2+} release**.

b. The Na^{+}-Ca^{2+} exchange mechanism is an important regulator of intracellular Ca^{2+} concentration. Under resting conditions, this exchange mechanism removes Ca^{2+} from the cell. However, when the cell depolarizes, the amount of Ca^{2+} removed is reduced, or the exchange mechanism may even reverse direction and actually add Ca^{2+} to the cytoplasm.

3. Mechanical properties of cardiac muscle differ from those of skeletal muscle due to the duration of the action potential and the ability of cardiac muscle to regulate the amount of Ca^{2+} entering the cell.

    a. The action potential and the period of time during which Ca^{2+} remains in the cytoplasm (the active state) are nearly equal in duration; thus, summation and tetanus are not possible. This is not a physiologic disadvantage for the heart, which must relax after each beat so blood can enter it.

    b. The sarcomere length in a cardiac muscle prior to contraction (the preload) depends on how much blood has entered the heart. Because this amount is under physiologic control, it is an important regulator of the force of cardiac muscle contraction.

    c. The force of contraction can vary at a given sarcomere length if the amount of Ca^{2+} entering the cell is changed. This also is under physiologic control and, thus, is an important regulator of cardiac muscle contractile force.

C. Smooth muscle differs greatly from striated muscle in appearance and in its mechanism of EC coupling.

    1. Structure

        a. Smooth muscles are thinner than cardiac muscles. They are about 5–10 μ wide and 10–500 μ long.

        b. Smooth muscles are not divided into sarcomeres with interdigitating thick and thin filaments, so they lack striations.

            (1) Smooth muscle proteins are contained in thick and thin filaments, but these filaments are not organized into sarcomeres. Instead they are dispersed throughout the cell.

            (2) The thin filaments are attached to **dense bodies**. Some of these bodies are anchored to the cell membrane, but most are floating within the cytoplasm.

            (3) Cross-bridges extend from the thick filaments, and the cross-bridge cycle appears similar to that of striated muscle.

        c. T tubules are absent and unnecessary in smooth muscle, because the cell is small enough for a stimulus present on the cell surface to be effective in activating the contractile machinery.

    2. EC coupling. Smooth muscle differs most from striated muscle in its mechanism of activation. In striated muscle, the cross-bridge is prevented from binding to actin by the troponin-tropomyosin regulatory complex. In smooth muscle, there is no troponin-tropomyosin complex inhibiting cross-bridge cycling. However, the myosin cross-bridges cannot bind to actin until they are activated, and this activation requires that one of the light-chain proteins on the myosin head be phosphorylated.

        a. Myosin phosphorylation is catalyzed by **myosin light-chain kinase (MLCK),** which is inactive at rest. MLCK is activated by combining with another enzyme, **calmodulin,** which cannot occur until calmodulin has combined with Ca^{2+}. Thus, the role of Ca^{2+} in smooth muscle is to initiate a sequence of reactions ending in the phosphorylation of myosin.

        b. Another enzyme, **myosin phosphatase,** removes the phosphate from the myosin. The amount of myosin active at any time depends on the relative activity of the MLCK and the phosphatase. Relaxation occurs when the intracellular Ca^{2+} concentration falls and the MLCK becomes inactive. At this point, the phosphatase removes phosphate from the myosin light-chain protein, and cross-bridge cycling ceases.

    3. Mechanical properties of smooth muscle differ from those of striated muscle.

        a. The speed of shortening, or the rate of cross-bridge cycling, is dependent on the phosphorylation of the myosin light chain. When the light chains are dephosphorylated by the phosphatase enzyme, the speed of shortening decreases.

        b. However, the dephosphorylated cross-bridges remain attached to actin and are able to sustain the tone developed by the muscle early in its contraction. These attached, but dephosphorylated, cross-bridges are called **latch-bridges**. Latch-bridges provide smooth muscle with the ability to maintain tone with little energy consumption. Since the latch-bridges do not cycle, or cycle very slowly, they do not use much ATP.

V. GENERAL SENSORY MECHANISMS

A. Classification of receptors. The function of receptors is to transform (transduce) stimulus energy into electrical energy and to transmit information about the nature of the stimulus to the CNS. Receptors can be classified on the basis of the following criteria.

1. **Source of stimulus**
   a. **Exteroceptors,** such as the eye, ear, and taste receptors, receive stimuli from outside the body.
   b. **Enteroceptors,** such as the chemoreceptors for measuring blood gases and the baroreceptors for measuring blood pressure, receive stimuli from within the body.

2. **Type of stimulus energy**
   a. **Mechanical receptors** detect skin deformation and sounds.
   b. **Thermal receptors** detect environmental temperatures.
   c. **Photoreceptors** detect light.
   d. **Chemoreceptors** detect substances that produce the sensations of smell and taste.

3. **Type of sensation.** Receptors are classified as touch, heat, cold, pain, light, sound, taste, and olfactory receptors.

4. **Rate of adaptation**
   a. **Slowly adapting** receptors (**tonic** or **static** receptors) fire action potentials continuously during stimulus application.
   b. **Rapidly adapting** receptors (**phasic** or **dynamic** receptors) fire action potentials at a decreasing rate during stimulus application.

5. **Neuron type**
   a. **Classification.** Neurons are classified on the basis of size and propagation velocity.
      (1) **Size.** Neurons fall into one of four groups (I–IV) depending on size.
         (a) Group I fibers are 12–20 μ in diameter, group II fibers are 6–12 μ, group III fibers are 1–6 μ, and group IV fibers are less than 1 μ in diameter. All but group IV fibers are myelinated.
         (b) Figure 1-29A is a histogram showing the size distribution of the neurons within a nerve trunk. The peaks in the distribution correspond to each of the three groups, indicating that they are not arbitrary divisions.
      (2) **Propagation velocity.** Figure 1-29B illustrates the recording of a **compound action potential** from a nerve trunk. The compound action potential is the summed activity

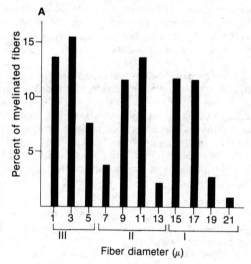

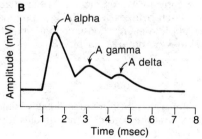

Figure 1-29. (*A*) Bar graph illustrating the size distribution of myelinated fibers within a nerve. The three peaks in the distribution correspond to the three functional groups of fibers, I, II, and III. (*B*) The compound action potential recorded from a nerve bundle has three peaks, which correspond to the functional grouping of myelinated nerve fibers according to conduction velocity (i.e., A alpha, A gamma, and A delta fibers).

of all the fibers within the nerve bundle. The action potentials are generated at one end of the nerve trunk and propagated into the other end, where the recording electrodes are placed.

    (a) The fibers with the fastest conduction velocities reach the recording electrodes first and are responsible for the first wave of the compound action potential. Subsequent peaks occur when the fibers with slower conduction velocities reach the recording electrodes.

    (b) The fibers responsible for generating the first peak are called **A alpha fibers,** those producing the second peak are called **A gamma fibers,** and those producing the final peak are called **A delta fibers.** All are myelinated fibers. The slowest conducting fibers, called **C fibers,** are unmyelinated (not shown in Figure 1-29).

  b. Functional distribution. Certain important sensory and motor fibers are identified with particular neuron types.

    (1) Muscle spindle afferent neurons belong to the group I and group II fibers.

    (2) Pain and temperature afferent neurons belong to the A delta fibers, if myelinated, and to the C fibers, if unmyelinated.

    (3) Alpha motoneurons are so-called because they are part of the A alpha group of fibers.

    (4) Similarly, gamma motoneurons are part of the A gamma fiber group.

B. Transduction of stimulus energy by receptors. Each receptor is specialized to receive a particular type of stimulus called an **adequate,** or **appropriate, stimulus.** The receptor is exquisitely sensitive to its adequate stimulus. For example, the retina can detect the presence of a single photon of light. However, the receptor can respond to other forms of stimulus energy if the stimulus is strong enough. For example, a mechanical stimulus such as rubbing the eyes can produce a sensation of light. The functional components of transduction in a simple mechanical receptor are illustrated in Figure 1-30 and described here.

  1. The **transducer region** of this receptor is at the end of the nerve terminal. In contrast, in the eye, ear, and tongue, the transducer region is on a receptor cell that is separate from the afferent neuron, and the two cells communicate by synaptic transmission.

    a. The transducer region of the receptor is responsible for converting the stimulus energy into an electrical signal called the **receptor,** or **generator, potential.**

    b. Each receptor has a particular mechanism for producing a receptor potential. In the mechanical receptor, the stimulus produces a receptor potential by deforming the nerve terminal.

      (1) The deformation opens channels that are permeable to Na^+ and K^+ and causes the membrane to depolarize.

      (2) As illustrated in Figure 1-30, the receptor potential follows the time-course of the mechanical stimulus.

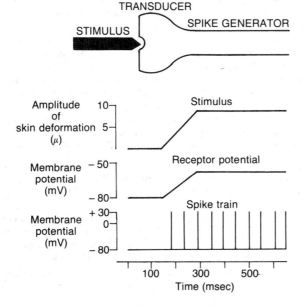

Figure 1-30. When a stimulus is applied to a sensory nerve ending, a generator potential results. The magnitude of the generator potential and the frequency of action potential discharge are proportional to the magnitude of the stimulus.

2. The **spike generator region** of the receptor is responsible for converting the receptor potential into a train of action potentials.

  a. Figure 1-30 illustrates the repetitive discharge of action potentials by the receptor axon in response to a receptor potential.

  b. The receptor potential spreads passively from the transducer region to the spike generator region. If the spike generator is depolarized to threshold, an action potential is generated.

  c. At the end of the action potential, the receptor potential again causes the spike generator membrane to depolarize toward threshold. If threshold is reached again, another action potential is generated.

C. Adaptation. The most obvious function of adaptation is to decrease the amount of sensory information reaching the brain. However, the brain stem mechanisms responsible for consciousness and attention are capable of keeping the information flow to the brain within tolerable limits so that sensory adaptation is not required for this purpose. The major role of sensory adaptation is to permit the receptor to encode information about the rate of stimulus application.

  1. Adaptation occurs by two major **mechanisms**.

   a. In one, the transducer mechanism fails to maintain a receptor potential despite continued stimulus application.

   b. In the other, the spike generator fails to sustain a train of action potentials despite the presence of a receptor potential. This occurs because the excitability of the spike generator membrane decreases. Although the mechanism of this decreased excitability is not known, it may be due to an increase in:

   (1) The membrane conductance to K^+

   (2) The activity of an electrogenic Na^+-K^+ pump

   (3) The inactivation of Na^+ channels

  2. The **discharge rate** in a phasic receptor increases as the rate of stimulus application increases.

   a. When a stimulus is applied too slowly, adaptation of the spike generator occurs before an action potential can be generated.

   b. When a stimulus is applied rapidly enough, excitation exceeds adaptation, and a steady discharge of action potentials occurs.

D. Sensory coding. The intensity, location, and quality of a stimulus are encoded by the sensory system.

  1. Intensity is encoded by the firing frequency of a sensory neuron.

   a. As the intensity of a stimulus is increased, the magnitude of the receptor potential is increased. This relationship is expressed by the **Steven's power law function**

$$V = k \times I^n$$

   where V = the magnitude of the receptor potential; k = a constant; I = the intensity of the stimulus; and n = a constant. Figure 1-31 is a plot of this relationship in which the maximal value of the receptor potential is about 50 mV.

   (1) Assuming that the sensory system requires a receptor potential of at least 1 mV to detect the presence of a stimulus, the maximal stimulus that can be detected is 50 times as great as the minimally detectable stimulus.

   (2) The power law relationship between stimulus intensity and receptor potential magnitude enables this relatively small range of receptor potential amplitudes (50 to 1) to encode a range of stimulus intensities that is greater than 1 million to 1.

   b. The graph in Figure 1-31 also illustrates that sensory perception and the firing frequency of a sensory neuron are related to stimulus intensity by the Steven's power law. This substantiates the concept that stimulus intensity is encoded by the rate of sensory nerve discharge.

  2. Stimulus location is encoded primarily by the location of the sensory projection to the cerebral cortex. This mechanism of encoding, called **topographic representation,** is used by the visual and somatosensory systems to localize the point of stimulus application. The auditory system uses a different mechanism (see VII D 2).

   a. Each sensory neuron receives information from a particular sensory area called its **receptive field**. The smaller the receptive field, the more precise the encoding of stimulus localization.

   (1) For example, the receptive fields of the fovea are the smallest within the eye.

   (2) Similarly, the receptive fields within the fingertips are much smaller than those on the hands and back, where it is difficult to localize precisely the point of a mechanical stimulus.

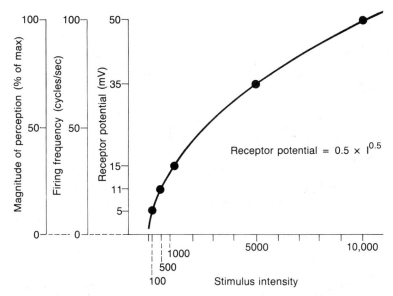

Figure 1-31. The proportionality between generator potential and stimulus can be expressed as a power function. This same relationship can be used to describe the proportionality between stimulus magnitude and firing frequency and between the magnitude of the stimulus and its perceived intensity. I = intensity.

  b. Stimulus localization can be made more precise by **lateral inhibition**. This is demonstrated in Figure 1-32, which shows three overlapping receptive fields in the skin.
  - **(1)** Two stimuli are applied to the skin, and the sensory task is to indicate whether one or two stimuli are present. The closest that two stimuli can be to each other and still be distinguished as separate stimuli is called the **two-point threshold**.
  - **(2)** Without lateral inhibition (see Figure 1-32A), the two stimuli are recognized as separate only if they are placed in receptive fields that are separated from each other by a nonstimulated receptive field.
  - **(3)** Lateral inhibition (see Figure 1-32B) can decrease the two-point threshold by reducing the discharge of the neuron innervating the receptive field in the center and, thus, making it apparent to the CNS that two stimuli are present.

3. Stimulus quality is encoded by a variety of mechanisms.
  a. The simplest mechanism uses a **labeled line,** in which the stimulus is encoded by the particular neural pathway that is stimulated.
  - **(1)** The basic sensory modalities are encoded in this way. Thus, the sensation of touch is elicited whether the receptors on the skin are excited by mechanical deformation or by electrical stimulation. Similarly, light always is evoked no matter how the retina is activated, and sound always results from stimulation of the cochlea.
  - **(2)** The same type of sensation results no matter where along the sensory pathway the stimulus is applied. For example, stimulating electrodes placed on the visual cortex evoke the sensation of light, while seizures within the olfactory cortex produce sensations of smell.
  b. A more complex mechanism of coding uses the **pattern of activity** within the neural pathway that is carrying information to the brain.
  - **(1)** In **temporal pattern coding,** the same neuron can carry two different types of sensory information depending on its pattern of activity. For example, cutaneous cold receptors indicate temperatures below 30° C by firing in bursts and temperatures above 30° C by firing without bursts.
  - **(2)** In **spatial pattern coding,** the activity of several neurons is required to elicit a sensation. For example, three neurons may be required to encode different taste sensations. A sour taste may result if all three neurons are activated, while a salty taste may result if only two neurons fire.
  c. The most sophisticated mechanism of sensory coding uses **feature detectors**. These are neurons within the brain that integrate information from a variety of sensory fibers and fire to indicate the presence of a complex stimulus.

A

Receptive fields

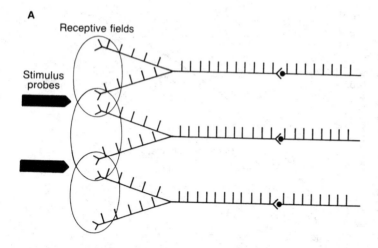

Stimulus
probes

B

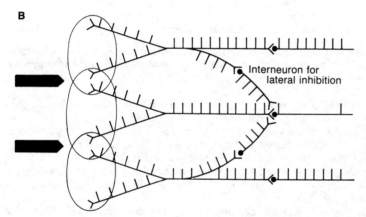

Interneuron for
lateral inhibition

Figure 1-32. In order for two distinct stimuli to be perceived, they must stimulate receptive fields separated by an unstimulated receptive field. (*A*) Without lateral inhibition, the stimuli produce equal amounts of discharge in all three neurons. (*B*) In the presence of lateral inhibition, however, the neuron with the receptive field in the center is presynaptically inhibited by collaterals from the neurons with receptive fields located laterally. As a result, the receptive field in the center does not fire, and two stimuli are perceived.

    (1) For example, the location of an object in space can be encoded by cortical cells receiving information from a single eye. However, special feature detectors receiving information from both eyes are required to specify the depth of an object in space.

    (2) Similarly, the location of a sound in space requires integration of information from both ears by feature detectors within the brain stem.

VI. CUTANEOUS SENSATION. About 1 million sensory nerve fibers innervate the skin. Most of these are unmyelinated nerve fibers that are responsible for crude somatosensory mechanical sensation. Although far fewer in number, the large myelinated (group II) sensory fibers encode the important sensory qualities of touch, vibration, and pressure. The sensations of temperature and pain are encoded by small myelinated (group III, A delta) fibers and unmyelinated (group IV, C) fibers.

    A. Mechanical sensation. The sensation produced by mechanical stimulation of the skin can be divided into three submodalities: vibration, touch, and pressure. Each of these sensations is encoded by a different set of mechanical receptors.

        1. Vibration is encoded by the **pacinian corpuscle** (Figure 1-33A).

           a. The pacinian corpuscle is a rapidly adapting **encapsulated receptor**.

               (1) Its capsule is an onion-like lamellar structure about 1 mm in diameter.

               (2) A single, large (group II) sensory axon enters the capsule.

           b. When a mechanical stimulus deforms the outer lamellae of the capsule, the deformation is transmitted through the capsule to the nerve terminal.

               (1) The deformation of the nerve terminal increases the membrane permeability to Na^+ and K^+, producing a depolarizing receptor potential.

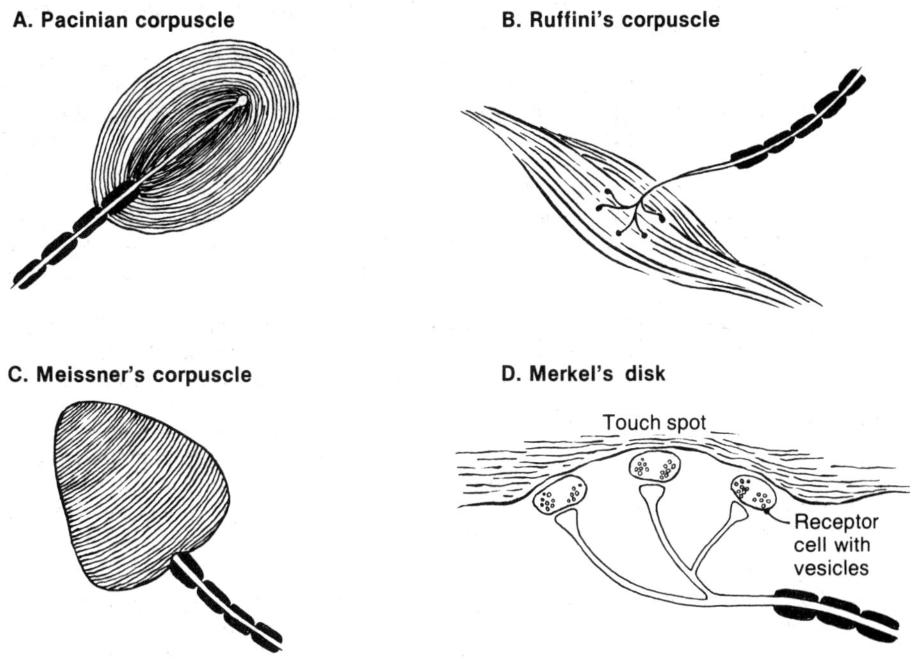

A. Pacinian corpuscle

B. Ruffini's corpuscle

C. Meissner's corpuscle

D. Merkel's disk

Touch spot

Receptor cell with vesicles

Figure 1-33. The encapsulated organs of the skin all are innervated by a single, large afferent fiber. (*A*) The pacinian corpuscle is a rapidly adapting receptor that encodes vibration. (*B*) The Ruffini's corpuscle is a slowly adapting receptor that encodes pressure. (*C*) The Meissner's corpuscle is a rapidly adapting receptor responsible for detecting the speed of stimulus application. (*D*) The Merkel's disk is a slowly adapting receptor capable of encoding the location of a stimulus. Its receptor cells communicate with the primary afferent neurons by synaptic transmission.

      (2) The size of the receptor potential increases in proportion to the magnitude of the deformation. However, because the pacinian corpuscle is a very rapidly adapting receptor, only a few action potentials will be generated, regardless of stimulus intensity.

   c. Adaptation occurs because the deformation of the nerve terminal is not maintained during continuous stimulus application.

      (1) Only rapid deformations are transmitted effectively to the core of the capsule.

      (2) If the stimulus is applied slowly or is left in place, the inner lamellae become rearranged so that they no longer deform the nerve terminal. This is an example of adaptation caused by a failure to maintain the receptor potential.

   d. A steady discharge of the pacinian corpuscle can be generated by a vibratory stimulus.

      (1) Each time the stimulus is removed and reapplied, the pacinian corpuscle discharges another action potential.

      (2) The frequency of discharge by the pacinian corpuscle equals the frequency of a vibratory stimulus in the range of 50–500 Hz (cycles/sec).

         (a) This is about the same range of frequencies identifiable by humans, indicating that the frequency of firing is encoding the frequency, not the intensity, of vibration.

         (b) Intensity is encoded by the number of action potentials generated by each deformation (limited to two or three) and the number of pacinian corpuscles responding to the vibration.

2. Pressure is encoded by the **Ruffini's corpuscle** (see Figure 1-33B).

   a. The Ruffini's corpuscle also is encapsulated. Its capsule is a liquid-filled collagen structure, in which a single group II sensory axon terminates. Collagen strands within the capsule make contact with the nerve fiber and the overlying skin. Thus, any deformation or stretch of the skin causes the nerve terminal to depolarize and generate action potentials.

   b. The Ruffini's corpuscle is a slowly adapting receptor. The train of action potentials continues for as long as the stimulus is present.

   c. The magnitude of the stimulus is encoded by the firing frequency of the Ruffini's corpuscle.

3. Touch. Both the Meissner's corpuscle and the Merkel's disk contribute to the encoding of touch.

a. The **Meissner's corpuscle** is a rapidly adapting encapsulated receptor that receives a single group II afferent nerve fiber (see Figure 1-33C).

(1) Its frequency of firing is proportional to the rate of stimulus application; thus, the Meissner's corpuscle encodes the velocity of the stimulus application.

(2) Rapid deformation occurs when the skin is jabbed quickly with a probe and when the fingers move over a rough object. This use of velocity information (i.e., identification of the contour of objects palpated by the fingers and hands) is especially important to blind individuals reading braille.

b. The **Merkel's disk** is a unique type of cutaneous receptor because the transducer is not on the nerve terminal but on the epithelial cells that make up the disk (see Figure 1-33D).

(1) The epithelial sensory cells form synaptic connections with branches of a single group II afferent fiber.

(2) The disk is about 0.25 mm in diameter and can be stimulated only if the stimulus is applied directly to the disk. The small receptive field of the Merkel's disk makes it an ideal receptor to encode information about the location of the stimulus.

B. Temperature sensation is encoded by thermoreceptors located on the free endings of small myelinated (A delta) and unmyelinated (C) fibers. Separate receptors with discrete receptive fields exist for encoding warm and cold sensations. Warm fibers respond only when a warm object is applied to them, and cold fibers respond only to cool objects. Warm fibers are not as numerous as cold fibers and, almost exclusively, are unmyelinated.

1. Warm fibers

a. Figure 1-34 shows the static firing rate of a typical warm fiber as a function of skin temperature. (The firing rates are measured after the skin has adapted to the ambient temperature.) Warm fibers begin to respond at a skin temperature of 30° C, reach a maximal response at about 45° C, and cease activity at about 47° C.

b. A warm fiber always increases its firing rate phasically when the skin is warmed and phasically decreases it when the skin is cooled. Soon after the initial response to the change in temperature, the fiber adapts its firing rate to the steady level indicated on the graph in Figure 1-34.

2. Cold fibers. The response of cold fibers is analogous to that of the warm fibers (see Figure 1-34).

a. Cold fibers begin to display a static discharge at skin temperatures of about 15° C, reach a maximal response at about 25° C, and cease firing at about 42° C.

b. If the skin temperature is raised above 45° C, the cold fibers begin to fire again. This is the threshold for activating pain fibers. Thus, if the skin is heated to a temperature above 45° C, the sensation of pain is accompanied by a feeling of coolness. This is referred to as **paradoxical cold.**

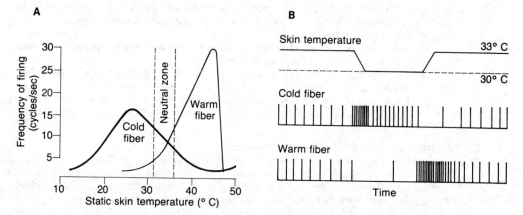

Figure 1-34. (*A*) Graph illustrating the tonic level of firing in both warm and cold fibers as a function of temperature. (*B*) Spike trains illustrating the dynamic responses of warm and cold fibers to a change in temperature. When the temperature decreases, cold fibers increase their rate of firing and then adapt to the firing rate indicated by the graph. Similarly, when the temperature increases, warm fibers phasically increase their firing rate before adapting to the rate indicated by the graph.

c. Cold fibers always increase their rate of firing transiently when the skin temperature is cooled and decrease it when the skin temperature is warmed. Also, like warm fibers, cold fibers quickly adapt their firing rate to the static levels indicated in Figure 1-34.

3. **Sensory coding** by thermoreceptors is complicated by the fact that the hypothalamic thermoreceptors that monitor the body's core temperature also contribute to the perception of the thermal environment.

    a. At skin temperatures within the **neutral** or **comfort zone** (31–36° C), complete perceptual adaptation occurs (i.e., awareness of temperature disappears). Persistent warmth and coolness are felt at temperatures above and below this zone. Temperatures below 25° C are described as unpleasantly cold, while those below 15° C and above 45° C are painful. This is due to stimulation of pain receptors and not to the stimulation of thermoreceptors.

    b. The skin is most sensitive to temperature changes when the skin temperature is 31–36° C (the neutral zone). At these temperatures, changes of less than 0.5° C are detected easily.

        (1) Transient temperature changes can be encoded by labeled-line mechanisms because cold fibers always increase their firing rate when the skin is cooled, and warm fibers always increase their firing rate when the skin is warmed.

        (2) Coding is more difficult under static conditions, however, because cold fibers have the same firing rate at temperatures above and below 30° C, and warm fibers have the same firing rate at temperatures above and below 37° C. Temperatures that produce the same firing rates can be discriminated by two coding mechanisms.

            (a) **Spatial coding** can be used since at temperatures between 30° and 42° C both cold and warm fibers are firing, at temperatures below 30° C only cold fibers are firing, and at temperatures above 42° C only warm fibers are active.

            (b) **Temporal coding** can be used by cold fibers because action potentials fire in bursts at temperatures below 30° C but not at temperatures above 30° C.

C. **Pain sensation** is different from other sensations because its purpose is not to inform the brain about the quality of a stimulus but rather to indicate that the stimulus is physically damaging. Although unpleasant, pain is a useful sensation if it leads to removal of the damaging stimulus. However, the sensation of pain can endure long after the stimulus is removed and the injury is healed. This sort of **chronic pain** is an extremely debilitating condition that is very difficult to treat. Much more information is required about the physiologic and psychological mechanisms of pain before its elimination becomes a routine part of medical practice.

    1. The **peripheral mechanisms** of pain sensation are reasonably well understood.

        a. The **receptors for pain,** which are called **nociceptors** to indicate that they respond to noxious stimuli, are on the free nerve endings of small myelinated (A delta) and unmyelinated (C) fibers.

            (1) Nociceptors are specific for painful stimuli, responding to damaging or potentially damaging mechanical, chemical, and thermal stimuli. Cutaneous receptors that respond to nonpainful levels of these stimuli do not elicit pain sensations no matter how intense the stimulus.

            (2) Although the adequate stimulus for nociceptors is not known, it is assumed that a chemical such as histamine or bradykinin is released from cells damaged by the pain stimulus and that the chemical substance activates the nociceptors.

        b. Two **types of pain** sensation result from the application of a strong, noxious stimulus to the skin.

            (1) **Fast,** or **initial, pain** is a discrete, well-localized, pin-prick sensation that results from activating the nociceptors on the A delta fibers.

            (2) **Slow,** or **delayed, pain** is a poorly localized, dull, burning sensation that results from activating the nociceptors on the C fibers.

        c. The two types of pain sensation use different **pathways** to reach the centers of consciousness in the brain.

            (1) Action potentials that are propagated by the fast pain fibers travel faster and, thus, reach the brain before those conducted by the slow pain fibers. The sensory fibers for fast pain have small receptive fields, travel to the cortex through the spinothalamic tract, and are topographically represented on the cortex—all factors that account for the ability of these fibers to encode the location of the stimulus producing the fast pain.

            (2) Slow pain, which has a more diffuse pathway, travels to the brain through the spinoreticulo-thalamic system. Collaterals of this system pass through the reticular formation to activate fiber tracts that produce the emotional perceptions accompanying

pain sensations. These pathways account for the intense unpleasantness associated with slow pain.
  d. The two types of pain sensation elicit different **reflexes.**
    (1) Fast pain evokes a withdrawal reflex (see XI A) and a sympathetic response including an increase in blood pressure and a mobilization of body energy supplies.
    (2) Slow pain produces nausea, profuse sweating, a lowering of blood pressure, and a generalized reduction in skeletal muscle tone. (Pain sensations originating in the muscles, blood vessels, and viscera produce similar reflexes.)

2. The **central mechanisms** of pain sensation are not well known.
  a. Generally it is assumed that pain sensation is conveyed to the CNS through the anterolateral quadrant; however, surgical section of the cord through this area is not very successful in relieving chronic pain.
    (1) Failure to relieve pain with this procedure may result from the existence of parallel pain pathways outside the anterolateral tracts.
    (2) Alternatively, chronic pain may result from the spontaneous activity of pain centers within the CNS.
      (a) Reverberating circuits that develop because of continuous pain input may fail to stop firing when the input is removed.
      (b) Denervation supersensitivity may develop in pain centers subsequent to the removal of pain fiber input. (Denervation supersensitivity is an increased sensitivity to circulating neurotransmitters, which results when the normal synaptic input is removed from a neuron.)
  b. **Treatment of chronic pain** is based on attempts to remove the pain area within the brain by surgery or to reduce the activity of the pain pathways by activating inhibitory pathways projecting to the pain areas.

3. **Referred pain**
  a. Pain originating in visceral organs is referred to sites on the skin. These sites are innervated by nerves arising from the same spinal segment as the nerves innervating the visceral organs. Because of this anatomic relationship, diagnosis of visceral disease can be made based on the location of the referred pain. For example, ischemic heart pain is referred to the chest and the inside of the arm.
  b. Most likely, referred pain occurs because the visceral and somatic pain fibers share a common pathway to the brain. Since the skin is topographically mapped and the viscera are not, the pain is identified as originating on the skin and not within the viscera.

4. **Projected pain.** Referred pain should not be confused with projected pain, which occurs as a result of directly stimulating fibers within a pain pathway.
  a. Because a labeled-line mechanism is used to encode the location of the pain, stimulation anywhere along the pathway will result in the same perception. For example, striking the elbow causes pain to be projected to the hand.
  b. Amputees often have sensations that appear to come from the severed limb. This is known as **phantom limb sensation** and presumably results from activation of the sensory pathway either at the site of amputation or within the CNS. Occasionally, the phantom sensation is one of pain, but with no obvious source of stimulation, it is difficult to eliminate the pain.

VII. AUDITION

A. Physics of sound

1. **Sound waves** are produced by vibrating objects that cause alternating phases of compression and rarefaction in the medium through which the sound is transmitted.
  a. In air, sound travels at about 330 m/sec (1100 ft/sec or 700 mi/hr).
  b. In water, sound travels much faster, at about 1500 m/sec.

2. **Characteristics of sound waves**
  a. The **frequency** of sound is measured in **hertz (Hz).**
    (1) The range for human hearing is about 20–20,000 Hz.
    (2) The human voice produces sounds of about 1000–3000 Hz when speaking.
  b. The amplitude or **intensity** of sound is measured using a logarithmic scale. The unit of intensity is the **decibel (db).**

(1) Decibels are measured using the formula

$$db = 20 \log \frac{P_{sound}}{P_{SPL}}$$

where db = the number of decibels; P_{sound} = the pressure of the sound stimulus; and P_{SPL} = the sound pressure level (SPL) at the threshold for human hearing.

(a) Thus, sound intensities are measured on a ratio scale using a subjective intensity (threshold) as a base.

(b) The actual SPL intensity is 0.0002 dynes/cm².

(2) According to the formula for calculating sound pressures in decibels, a stimulus that has a pressure 10 times as great as another will be 20 db greater in intensity than the other stimulus.

(a) The sound pressures used during normal conversation are about 1000 times as great as threshold, or 60 db. (This value can be confirmed using the above equation.)

(b) An airplane produces a sound pressure of about 100,000 times that of threshold, or 100 db.

(c) Sound pressures above 140 db (10^7 times threshold) are painful and damaging to the auditory receptors.

B. Structure of the ear. The ear consists of three divisions: the external ear, the middle ear, and the inner ear (Figure 1-35).

1. The **external ear** consists of two parts that function to direct sound waves into the auditory apparatus.

a. In lower animals, the cartilaginous **pinna** (auricle) can be moved by muscular action in the direction of a sound source to collect sound waves for the receptor organ. In humans, these muscles have little action. By transforming the sound field, however, the convoluted pinna does help identify sound sources. For example, the pinna aids in distinguishing a sound source placed in front of the head from one placed behind the head.

b. The **external auditory canal** extends from the pinna to the tympanic membrane of the middle ear. Its outer portion is cartilaginous and its inner portion is osseous.

2. The **middle ear** (tympanic cavity) is an air-filled cavity within the temporal bone. It contains the **tympanic membrane** (eardrum) and the **auditory ossicles** (ossicular chain). Sound stimuli pass through the pinna and strike the tympanic membrane, causing it to vibrate. The major function of the ossicular chain is to convey the sound stimulus from the tympanic membrane to the inner ear. Its presence is essential for normal hearing.

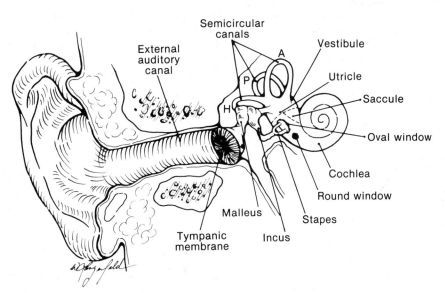

Figure 1-35. The major anatomic components of the human ear. The bony labyrinth consists of the sac-like otolith organs (i.e., the saccule and utricle) and three semicircular canals: the horizontal (*H*), anterior (*A*), and posterior (*P*) canals.

a. **Components of the ossicular chain**
   (1) **Malleus.** This small bone resembles a mallet. The **manubrium** (handle) of the malleus is connected to the inner surface of the tympanic membrane.
   (2) **Incus.** This bone, which resembles an anvil, articulates with the head of the malleus.
   (3) **Stapes.** This ossicle resembles a stirrup. The head of the stapes articulates with the incus, and the oval **footplate** contacts the membrane of the **oval window** of the cochlea.

b. **Impedance matching.** To reach the auditory receptors, the sound must be transferred from the air-filled middle ear to the fluid-filled inner ear.
   (1) Effective transfer of sound energy from air to water is difficult because most of the sound is reflected. This is due to the different mechanical (i.e., elastic, resistive, and inertial) properties of the two media and is described as an impedance mismatching.
   (2) The middle ear functions as an impedance matching device primarily by amplifying the sound pressure.
      (a) Most of this amplification occurs because the area of the tympanic membrane (55 mm^2) is about 17 times as great as the stapes–oval window surface area. This concentrates the same force over a smaller area and, thus, amplifies the pressure.
      (b) A small additional amount of amplification is obtained by the mechanical advantage that is due to the leverage of the middle ear bones.
      (c) Together these effects increase the sound pressure by 22 times (27 db).

c. **Minimum audibility curve** (Figure 1-36). The middle ear does not amplify all sounds equally well.
   (1) Amplification is greatest for sounds between 2000 and 5000 Hz. These are the frequencies used for speech.
   (2) Sounds below 20 Hz or above 20,000 Hz are not amplified at all.
   (3) The ability of the middle ear to amplify some sounds better than others accounts for the shape of the minimum audibility curve.

d. **Middle ear muscles**
   (1) **Structure and innervation**
      (a) **Tensor tympani.** This elongated muscle inserts on the manubrium of the malleus and is innervated by the trigeminal nerve (cranial nerve V).
      (b) **Stapedius.** This small muscle inserts on the neck of the stapes and is innervated by the facial nerve (cranial nerve VII).
   (2) **Action.** These muscles contract reflexly in response to intense sounds. They act on the mobility and transmission properties of the ossicular chain to reduce the pressure of sounds reaching the inner ear.
   (3) **Functional importance.** Although the middle ear muscles may act to prevent receptors from being damaged by high-intensity sounds, contraction occurs too long after the stimulus to provide much protection. Muscle contraction also occurs just prior to vocalization and chewing, which suggests that the middle ear muscles may act to reduce the intensity of the sounds produced by these activities.

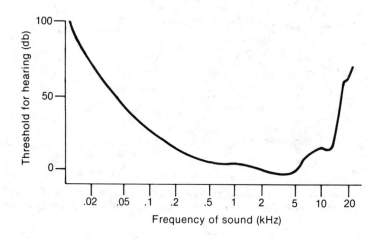

Figure 1-36. The minimum audibility curve traces the threshold for hearing at different frequencies in the human hearing range. Maximum sensitivity occurs at about 4 kHz. The actual threshold at this frequency is 0.0002 dynes/cm^2 or 0 decibels.

3. The **inner ear** consists of two components. The **bony labyrinth** is a network of cavities within the petrous part of the temporal bone. The bony labyrinth consists of three divisions: the semicircular canals, the cochlea, and the vestibule. Within the bony labyrinth is a series of interconnected membranous ducts called the **membranous labyrinth**. The membranous labyrinth is surrounded by a fluid called **perilymph,** which is rich in Na^+. The fluid within the membranous ducts is **endolymph,** which has a high concentration of K^+.

   a. The **cochlea** is a spirally arranged tube (two and one-half turns in humans) that functions as the auditory analyzer of the ear. Figure 1-37 depicts the cochlea as an uncoiled, linear structure. The cochlea contains the following components.

   (1) The **cochlear duct** is an elongated endolymph-filled inner tube-like structure that lies within the coiled length of the cochlea. The open space within this membranous duct is the **scala media**. Four membranes enclose the scala media.

   (a) The **spiral ligament** is a thickening of the endosteum of the cochlea, which forms the peripheral edge of the cochlear duct.

   (b) **Reissner's membrane** is an extremely delicate membrane that forms the roof of the cochlear duct. It separates the scala media from the **scala vestibuli**.

   (c) The **limbus** forms the central, narrow edge of the cochlear duct.

   (d) The **basilar membrane** represents the floor of the cochlear duct and separates the scala media from the **scala tympani**. It also is the seat of the **organ of Corti**. The basilar membrane is broader at the apex (near the **helicotrema**) than it is at the base (near the **oval window**) and is 100 times more stiff at the base than it is at the apex.

   (2) The **scala vestibuli** is the space above the Reissner's membrane.

   (3) The **scala tympani**, the space below the basilar membrane, is continuous with the scala vestibuli at the very apex of the cochlear duct. This junction is termed the **helicotrema**. The scala vestibuli and scala tympani contain perilymph.

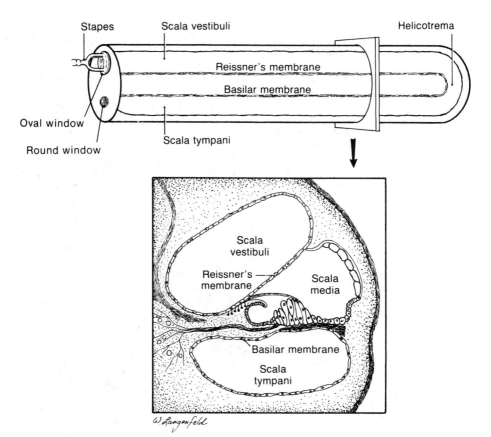

Figure 1-37. The components of the cochlea, shown as it would appear if uncoiled (*upper diagram*) and shown in cross section (*lower diagram*).

      (4) The **oval window** is the membrane-covered opening of the scala vestibuli, which joins the footplate of the stapes.

      (5) The **round window** is the membrane-covered opening of the scala tympani.

   b. The **organ of Corti** is the specialized region that contains the sensory receptor cells of the ear. It lies on the basilar membrane, adjacent to the limbus, and contains the following structures (Figure 1-38).

      (1) Hair cells that lie within the organ of Corti are the sensory receptors for the auditory system. The essential excitatory event in auditory transduction is the deflection of the cilia protruding from the surface of hair cells. When the cilia are deflected, transmitter substance is released from the cells.

      (2) Tectorial membrane. This flap of tissue rests on the tips of the cilia of the hair cells and is attached to the limbus at one end. Any movement of the basilar membrane causes the free end of the tectorial membrane to drag or push the cilia of the hair cells.

      (3) Nerve fibers. The peripheral endings of the afferent fibers form synapses with the hair cells, and the cell bodies form the spiral ganglion. Axons of these neurons join the vestibulocochlear nerve (cranial nerve VIII).

C. Auditory transduction is the process by which an auditory receptor potential is generated by sound waves entering the ear.

   1. Sound wave propagation. Sound waves are propagated through the **cochlea**.

     a. Sound waves entering the inner ear from the oval window spread along the scala vestibuli as a **traveling wave**. Most of the sound energy is transferred directly from the scala vestibuli to the scala tympani. Very little of the sound wave ever reaches the helicotrema at the apex of the cochlea.

     b. When the pressure of a sound wave causes the oval window to bulge inward, the pressure (after passing through the incompressible fluids of the scala vestibuli, scala media, and scala tympani) forces the round window to bulge outward.

   2. Movement of the organ of Corti

     a. The sound waves cause the organ of Corti to vibrate up and down, which, in turn, causes the stereocilia to bend back and forth.

      (1) When the sound wave pushes the organ of Corti downward, the stereocilia bend toward the limbus. When the sound wave pulls the organ of Corti upward, the stereocilia bend away from the limbus (Figure 1-39).

      (2) The attachment of the stereocilia to the tectorial membrane makes it possible for the up-and-down movement of the organ of Corti to be translated into the back-and-forth movement of the stereocilia.

        (a) The bottoms of the hair cells are anchored to the basiliar membrane, while the stereocilia are connected to the overlying tectorial membrane.

        (b) Because the tectorial and basilar membranes are attached at different points on the limbus (see Figure 1-39A), they slide past each other as they vibrate up and down, causing the stereocilia on the hair cells to bend back and forth.

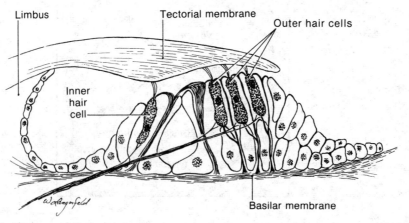

Figure 1-38. The organ of Corti, showing the connection between the tectorial membrane and the cilia of the hair cells.

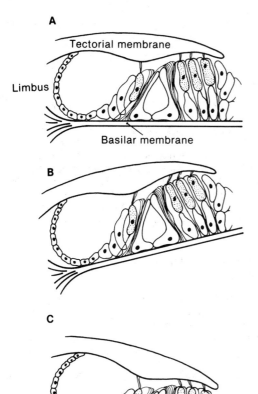

A

Tectorial membrane

Limbus

Basilar membrane

B

C

Figure 1-39. Diagram showing how the up-and-down movement of the basilar and tectorial membrane causes the stereocilia extending from the hair cells to bend back and forth. (*A*) The tectorial and basilar membranes are attached to the limbus at different points. (*B*) When the membranes rotate upward, the tectorial membrane slides forward relative to the basilar membrane bending the stereocilia away from the limbus; (*C*) when the membranes rotate downward, the stereocilia bend toward the limbus.

      (i) When the organ of Corti moves up, the stereocilia bend away from the limbus (see Figure 1-39B).

     (ii) When the organ of Corti moves down, the stereocilia bend toward the limbus (see Figure 1-39C).

   b. The stereocilia are **polarized**.

     (1) Bending the stereocilia away from the limbus causes them to depolarize.

     (2) Bending the stereocilia toward the limbus causes them to hyperpolarize.

     (3) Thus, when the organ of Corti is pushed downward, the hair cells hyperpolarize, and when it is pulled upward, the hair cells depolarize.

3. Mechanism of depolarization

  a. Movement of K^+ into the cell. The electrochemical gradient for K^+ favors the movement of K^+ into the cell.

    (1) The endolymph contains a high concentration of K^+ (135 mEq/L) and is electrically positive in comparison to the perilymph.

    (2) Hair cells, like all other cells, contain a high concentration of K^+ and are electrically negative when compared to the perilymph.

    (3) Because the endolymph is electrically positive and the hair cell is electrically negative, a very large potential difference (in excess of -100 mV) exists across the hair cell membrane.

    (4) The large negative potential and lack of a K^+ concentration difference between the inside and outside of the hair cell create a driving force, pushing K^+ into the cell.

  b. K^+ channels. Bending the hair cells causes K^+ channels to open or close.

    (1) When the stereocilia bend away from the limbus, they cause K^+ channels to open. K^+ then flows into the cell and the hair cell depolarizes.

(2) When the stereocilia bend toward the limbus, they cause K^+ channels to close and the hair cell hyperpolarizes.

D. Encoding of auditory information

1. **Release of synaptic transmitter**
   a. When the hair cell depolarizes, a Ca^{2+} channel opens, allowing Ca^{2+} to enter the cell.
   b. Ca^{2+} initiates the release of a synaptic transmitter.
   c. An auditory nerve fiber is stimulated by the synaptic transmitter.

2. **Place principle of frequency discrimination.** The frequency of sound that activates a particular hair cell depends on the location of the hair cell along the basilar membrane. The basilar membrane is narrowest and stiffest at the base of the cochlea (near the oval and round windows) and widest and most compliant at the apex of the cochlea (near the helicotrema).
   a. The energy contained in high-frequency sounds passes through the organ of Corti near the base of the cochlea, because the stiff portion of the basilar membrane resonates with high frequencies.
   b. In contrast, the energy in low-frequency sounds passes through the organ of Corti near the apex of the cochlea, because the compliant portion of the basilar membrane resonates with low frequencies.

3. **Encoding of frequency.** The auditory nerve fiber activated by a particular sound frequency is similarly dependent on the location of the hair cell it innervates.
   a. There are about 30,000 nerve fibers in each auditory nerve.
      (1) Over 90% of these fibers innervate inner hair cells. Each inner hair cell receives innervation from about 10 auditory nerve fibers; each auditory nerve fiber innervates only one hair cell.
      (2) The outer hair cells, which are far more numerous, are innervated by the remaining 10% of the sensory fibers.
   b. For low-frequency sounds, the auditory nerve fibers can fire at the same frequency as the sound wave. This mechanism of sensory encoding is called the **volley principle of frequency discrimination**.
   c. The sound frequency producing the greatest response in an auditory nerve fiber is called the **characteristic frequency** of that nerve fiber.

4. **Encoding of intensity.** The frequency of firing in an auditory nerve fiber increases as the intensity of the sound wave increases. In addition, a larger portion of the basilar membrane is vibrated as the sound intensity increases so that more auditory nerve fibers are activated. Thus, sound intensity is encoded by the frequency of auditory nerve discharge and by the number of auditory nerve fibers that are active.

5. **Inhibitory innervation.** The hair cells receive a very prominent efferent innervation from the superior olivary nucleus via the olivocochlear bundle. Although such a large pathway probably plays an important role in auditory transduction, its purpose has not been deciphered.

E. Physiology of central auditory neurons

1. **Tonotopic organization.** Generally, any neuron of the auditory pathway can be tested with tones of different frequencies to determine a best frequency for that cell. Neurons responding best to low-frequency tones will be located at one end of a nucleus, while neurons responding best to high-frequency tones will be represented at the opposite end of the nucleus. This orderly arrangement of frequency sensitivity, termed tonotopic organization, resembles the retinotopic organization of the visual system and the somatotopic organization of the somatosensory system. The tonotopic map reflects the methodical arrangement of frequency sensitivity along the length of the basilar membrane from base to apex. Tonotopic organization is prominent in the cochlear nuclei but becomes less precise in more rostral structures of the auditory pathway.

2. **Feature detection.** Higher auditory centers respond to particular features of sound stimuli. For example, cortical neurons may respond specifically to a shift from high- to low-frequency notes, which is why lesions of the auditory cortex may not impair the ability to discriminate frequency. Instead, lesions of the auditory cortex cause a loss of ability to recognize a patterned sequence of sounds. In addition, the ability to identify the position of a sound source is impaired.

3. **Localization of sound in space.** Detection of the position of a sound source depends on the ability of the CNS to compare intensity differences and phase differences. A sound source located behind the head, closer to one ear than the other, produces a slightly more intense sound in the near ear than in the remote ear. Sound absorption by the tissues of the head attenuates sound intensity in the remote ear. In addition, at low frequencies there may be phase differences in the sound waves striking the two ears. These minute phase and intensity differences between the two ears are detected and discriminated by the auditory system. The extensive decussation and complex circuitry of the auditory pathway provide the CNS with the information necessary to identify a sound source.

F. **Hearing impairments**

1. **Tinnitus** is a ringing sensation in the ears caused by irritative stimulation of either the inner ear or the vestibulocochlear nerve.

2. **Deafness.** Two classes of deafness are distinguished based on the location of the abnormality or lesion.
   a. **Conduction deafness** is caused by interference with the transmission of sound to the sensory mechanism of the inner ear. Conduction deafness represents a defect of the external or middle ear.
   b. **Nerve deafness** results from defects of either the inner ear or the vestibulocochlear nerve. Deafness due to a lesion in the CNS structures generally is not found clinically because of the redundancy and bilateral nature of the central auditory pathways.

VIII. VISION

A. **Physiologic optics.** In order for an object to be perceived by the visual system, a clear image of the object must be formed on the retina.

1. **Principles of image formation.** Several physical principles govern the process of image formation by a lens system.
   a. **Light rays** emanating from a luminous object **travel in all directions** (Figure 1-40A). However, only those rays entering the lens are used to form an image.
   b. **Refraction** is the process of bending rays of light.
      (1) A **converging (positive) lens** causes rays of light to be bent toward the center of the lens (see Figure 1-40B). A converging lens can form an image.
         (a) The **object distance** is the distance from the object to the lens.
         (b) The **image distance** is the distance from the lens to the image.
      (2) A **diverging (negative) lens** causes rays of light to be bent away from the center of the lens (see Figure 1-40B). A diverging lens cannot form an image.
   c. **Focal plane**
      (1) If the rays of light entering a converging lens are parallel, they will form an image **on the focal plane** (see Figure 1-40C).
         (a) The **focal distance** is the distance from the lens to the focal plane.
         (b) The **power of the lens** is defined as

$$P = \frac{1}{f}$$

where P = the power of the lens [diopters (D)] and f = the focal length (m).
      (2) If the rays of light entering a converging lens are diverging from each other when they enter the lens, they will form an image **behind the focal plane**.
      (3) The **lens formula** defines the relationship between the image, object, and focal distances (Figure 40D)

$$\frac{1}{o} + \frac{1}{i} = \frac{1}{f} = P$$

where o, i, and f = object, image, and focal distances (m), respectively, and P = lens power (D).
      (4) The **image and object size** have the following relationship

$$\frac{\text{image size}}{\text{object size}} = \frac{\text{object distance}}{\text{image distance}}$$

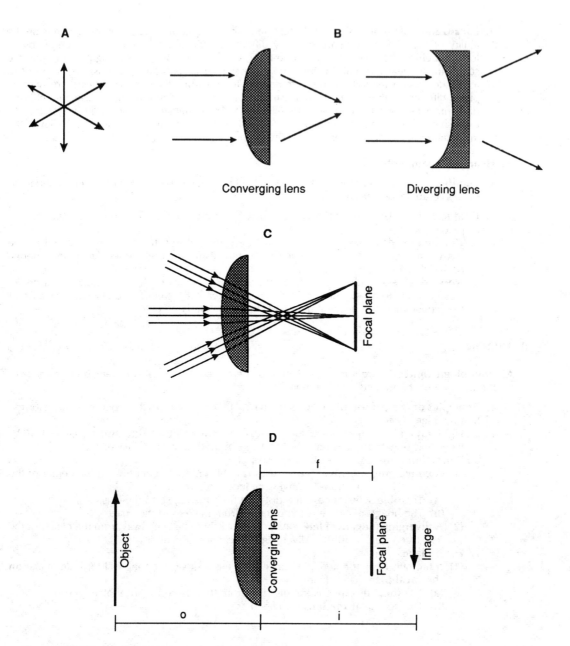

Figure 1-40. Diagram illustrating some fundamental optical principles. (*A*) Although light rays travel in all directions from a source of light, only those passing through the lens are used to form an image. (*B*) Converging lenses focus all rays of light coming from a point on an object onto a single image point. Diverging lenses cannot form an image. (*C*) If the rays of light entering a converging lens are parallel, their image will be formed on the focal plane. (*D*) If the rays of light are not parallel, their image will be formed behind the focal plane. The relationship between the object distance, focal distance, and image distance is given by the lens formula [see VIII A 1 c (3)]. *o* = object distance; *i* = image difference; *f* = focal distance.

2. **Refractive media of the eye.** A horizontal section through the human eyeball is shown in Figure 1-41. To reach the retina, light must pass through four refractive interfaces: the anterior surface of the cornea, the posterior surface of the cornea, the anterior surface of the lens, and the posterior surface of the lens.

  a. **Cornea**

    (1) **Transparency**

      (a) The absence of blood vessels in the cornea allows light to pass through unhindered.

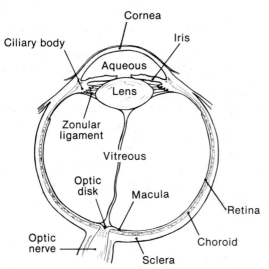

Figure 1-41. Horizontal section of the eyeball.

The cornea, consequently, must receive oxygen and nutrients via diffusion through the aqueous humor.

 (b) Elevated intraocular pressure, a condition termed **glaucoma,** can compromise the transparency of the cornea. Intraocular pressure is measured clinically by a tonometer, which records the resistance to a calibrated indentation of the cornea.

(2) Refractive power. The greatest amount of refraction of the eye occurs at the anterior surface of the cornea, because the difference between the indices of refraction of air and of the cornea is larger than the difference between the indices of refraction of the other ocular media. The cornea contributes about two-thirds, or + 40 D, of the eye's refractive power. The amount of refraction contributed by the cornea is fixed.

 b. Lens

  (1) Transparency

   (a) The lens, like the cornea, is avascular. It consists of a capsule, a layer of epithelial cells, and a system of transparent fibers. New fibers are continually laid down.

   (b) As an individual ages, the lens becomes harder and less malleable. **Cataracts** are opacities of the lens that reduce transparency. One cause of cataracts is an elevated glucose level in the aqueous humor, as occurs in uncontrolled diabetes.

  (2) Refractive power. The refractive power of the lens, unlike that of the cornea, is adjustable. The unaccommodated lens contributes about one-third of the eye's refractive power. The lens is suspended by the tough **zonular fibers,** which stretch the lens and flatten it. The tension on the zonular fibers is released by contraction of the **ciliary muscle**. Contraction results in the constriction of this ring of muscle fibers, which lies anterior to the attachment of the zonular fibers. The release of tension on the zonular fibers results in an increase in the curvature of the anterior surface of the lens. The increased convexity of the lens increases its refractive power. The lens can add, at most, +12 D of power to the ocular apparatus.

3. The "reduced eye" model. The optical properties of the eye are very complicated. To simplify the application of image formation and lens equations to the eye, a model of the eye, called the **reduced eye** (Figure 1-42), is used.

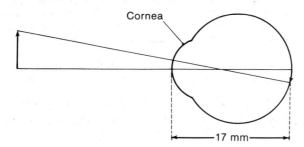

Figure 1-42. Image formation in the reduced eye.

a. **How the model works**

(1) In the reduced eye, the refractive surfaces of the cornea and lens are combined into a single surface.

(a) The **axial length** of the reduced eye (the distance from the lens surface to the retina) is considered to be 17 mm. In reality, the axial length of an adult human eye is typically about 25 mm.

(b) The **refractive power** of the reduced eye is approximately 58.8 D.

(2) According to the lens formula, the image of an object placed 6 m or more from the eye will be formed on the retina since

$$\frac{1}{i} = \frac{1}{f} = 58.8 \text{ D or}$$

$$i = 0.017 \text{ m} = 17 \text{ mm}$$

(3) If the object is closer than 6 m, the image will be formed behind the retina.

b. **The near point and far point** define the range of distances over which a clear image can be formed by the eye.

(1) The **near point** is the nearest point at which an object can be seen clearly. A typical young adult can increase his or her refractive power by 12 D. At a maximum lens power of 70.8 D, the near point is about 8.3 cm from the eye.

(2) The **far point** is the distance from the eye that an object must be placed so that it can be seen clearly without accommodation. For a normal person, this is 6 m.

4. **Accommodation.** A clear image of the object can be formed if the eye accommodates for near vision (objects closer than 6 m). Accommodation occurs by a reflex that has three components.

a. **Bulging of the lens.** Contraction of the ciliary muscle releases tension on the zonular fibers (see Figure 1-41). The elastic capsule surrounding the lens retracts, increasing the convexity (and thus the power) of the lens.

b. **Pupillary constriction.** By reducing the area through which light can enter the eye, spherical aberration is reduced. Reducing spherical aberration increases the depth of focus (the degree to which the image can form in front of or behind the retina and still appear to be in focus). Nearsighted persons typically squint to increase their depth of focus.

c. **Convergence.** The gaze of the two eyes shifts toward the center of the head in order to keep both eyes focused on the object.

5. **Refractive errors,** in which distant objects cannot be seen clearly without accommodation, are due to variations in the axial length or the refractive power of the eye.

a. **Myopia (nearsightedness)** results from an axial length that is too long for the refractive power of the eye. In this case, the focal point is in front of the retina; thus, distant objects cannot be focused on the retina.

(1) The object can be seen clearly if it is moved closer to the eye so that the image forms on the retina (but behind the focal plane). In severe myopia, the far point may be only 10–15 cm from the eye.

(2) As the object is moved closer to the eye, the lens accommodates for near vision. Myopes have near points very close to the eye. Thus, they are able to do fine work without magnifying glasses.

(3) A diverging lens can be placed in front of a myopic eye, reducing its refractive power. When wearing such a lens, the myope can see distant objects clearly.

b. **Hyperopia (farsightedness)** results from an axial length that is too short for the refractive power of the eye. In this case, distant objects cannot be focused clearly because the focal point is in back of the retina.

(1) A distant object can be seen clearly if the hyperope increases his or her refractive power by accommodating. If the degree of hyperopia is not very great, the person will usually not be aware of the refractive error. However, if a large amount of accommodation is required to focus distant objects, the constant contraction of the ciliary muscle will cause eye strain.

(2) Because some amount of accommodation is used to see distant objects, there is not so much available for near vision. Thus, the near point in hyperopes is further from the retina than normal.

(3) A converging lens can be placed in front of a hyperopic eye to increase its refractive power. When wearing a diverging lens, a hyperope can see distant objects clearly without accommodation.

c. Presbyopia is a loss of accommodative power associated with aging. As an individual ages, the elastic recoil of the lens decreases, and it is unable to bulge as much when the ciliary muscles contract.
  (1) Because the accommodative power is lost in presbyopia, the near point moves farther from the eye.
  (2) A converging lens is used to substitute for the lost accommodative power. The use of these lenses in reading glasses makes near vision possible.
d. Astigmatism is due to an uneven cornea. Normally, the cornea has a spherical surface; in astigmatism, it has more of an egg-shaped surface. As a result, the power of the lens is different in different axes.
  (1) Images formed with an astigmatic lens are distorted. For example, a point is seen as a line, and a line appears to have a halo on either side of it.
  (2) Cylindrical lenses can be worn by an astigmatic individual. These lenses add refractive power in only one axis and thus allow the lens system to behave as a spherical surface.

B. Retina. The **retina** is the receptor mechanism of the eye and the ultimate tool of vision. Light and color-sensitive elements of the retina communicate their messages, via **neural pathways,** to the **visual cortex** of the brain.

  1. Morphology and function of retinal components
    a. Layers, fovea, and optic disk. The retina is divided into 10 histologically defined **layers** (Figure 1-43). In the center of the retina is a yellowish region termed the **macula lutea.** The **fovea** is a central depression in the macula lutea, about 0.3 mm in diameter, where some of the retinal layers are displaced and the retina, consequently, is thinned. The fovea is the visual center of the eye and the area of highest resolution. About 3 mm to the nasal side of the fovea is the **optic disk,** where the ganglion cell axons turn and pierce the retina to form the optic nerve. Receptor elements are absent at the optic disk, which creates a **blind spot** on the retinal field.
    b. Cells. The retina is composed of the following cells.

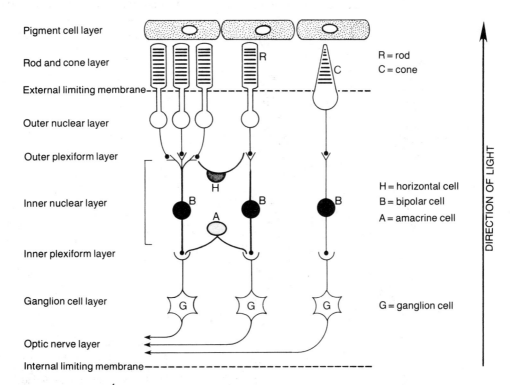

Figure 1-43. Cytoarchitecture of the retina. The three-neuron circuit of the rod system differs from that of the cone system. Convergence of receptors onto bipolar cells is seen only in the rod system. Horizontal cells and amacrine cells link adjacent retinal circuits.

(1) Pigment cells contain melanin and are arranged in a single lamina. Their major functions include:
   (a) Absorption of stray light
   (b) Phagocytosis of disks that have been sloughed from the outer segments of receptor cells
(2) Receptor cells divide into two distinct types: **rods** and **cones** (Figure 1-44).
   (a) The **outer segment** of rods and cones invaginates to form membrane disks.
      (i) In rods, but not in cones, the disks pinch off and become independent of the cell membrane. In rods, the disks undergo constant renewal. Disks are formed initially at the basal pole of the outer segment and migrate through the outer segment until they reach the apical pole. Here the disintegrating disks are phagocytosed by the pigment cells.
      (ii) Cones show a more diffuse pattern of disintegration.
      (iii) The photopigment molecules are embedded in the disk membranes.
   (b) The **inner segment** contains numerous mitochondria as well as the nucleus of the rod or cone.
   (c) The **receptor terminal** contains synaptic vesicles.
   (d) Functions of rods and cones
      (i) Color vision. Only cones are involved with color vision.
      (ii) Sensitivity to light. Cones function at high light intensities and, therefore, are responsible for day **(photopic)** vision. Cones are not as sensitive as rods to low levels of illumination. Rods function at low light intensities and are responsible for night **(scotopic)** vision.
      (iii) Visual acuity. Cones have a high level of visual acuity and are present in the fovea. Rods have a much lower level of acuity; they are lacking in the fovea and are confined to the peripheral retina.
      (iv) Dark adaptation. When an individual moves into a darkened environment, the eyes adapt slowly to lower levels of illumination. The threshold for sensation of low-intensity light stimuli falls progressively, as shown in Figure 1-45. The dark adaptation curve has two limbs, representing adaptation of cones followed by

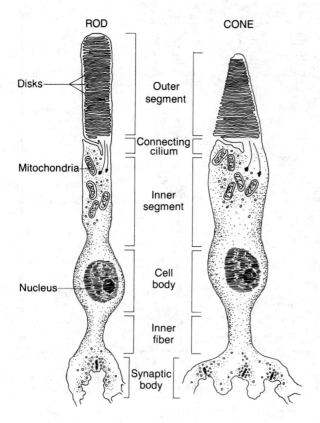

Figure 1-44. Morphology of rod and cone receptor cells. Cones, which are responsible for color perception and high visual acuity, are found in the fovea. Rods, which are responsible for night vision, are located in the peripheral retina.

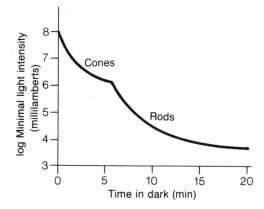

Figure 1-45. Dark adaptation. The eye becomes more sensitive to low-intensity light stimuli with increasing time in the dark. The ordinate indicates threshold light intensities.

that of rods. Rod function is suppressed during day vision and comes to the fore only when background illumination is low.

(3) Müller cells are glial elements that have cell bodies in the **inner nuclear layer**. Their processes extend to form the **internal limiting membrane** as well as the **external limiting membrane**. The processes also fill the spaces between receptor cells and serve as a connecting and supporting tissue.

(4) Bipolar cells and ganglion cells

    (a) Bipolar cells synapse with the receptor terminals of rods or cones in the **outer plexiform layer**. Bipolar cells in turn terminate on ganglion cells in synapses located in the **inner plexiform layer**.

    (b) Ganglion cells. The dendrites of these neurons receive their input from the bipolar cells. Their axons course along the retina to the optic disk and form the **optic nerve**.

    (c) Function in light sensitivity and visual acuity. The low sensitivity but high acuity of the cone system results from the connection of single cones to individual bipolar cells and subsequently to single ganglion cells projecting to the brain (see Figure 1-43). The rod system has a higher sensitivity because a large number of these receptors converge onto an individual bipolar cell, and a photon striking any one of these rods is effective in activating the bipolar cell. The bipolar cell of the cone is less sensitive, since a photon must strike a particular receptor location to be effective. Because stimuli at several adjacent rod locations all can result in an identical response in the second-order neuron, acuity is sacrificed in the rod system.

(5) Local circuit neurons

    (a) Horizontal cells. The processes of these cells are confined to the **outer plexiform layer**. They form a triad synapse at the junction of the receptor terminals and the bipolar cells.

    (b) Amacrine cells form complex synapses in the **inner plexiform layer** at the junction of the bipolar cells and the dendrites of the ganglion cells.

2. Photochemistry of receptor cells

    a. Bleaching of rhodopsin. When light strikes the eye, it is absorbed by the photopigments of the retina, causing a photochemical transformation to occur (Figure 1-46). During this

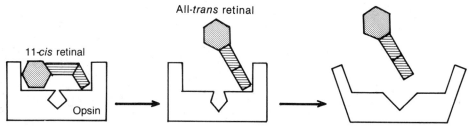

Figure 1-46. Action of light on the rhodopsin molecule. Initially, the chromophore moiety, 11-*cis* retinal, fits closely into the opsin moiety. Light causes the conformation of the 11-*cis* retinal to change to the elongated all-*trans* retinal, causing the opsin configuration to change.

process, the color of the photopigment changes from purple to pink; consequently, this process is called **bleaching**.

(1) Light causes a photoisomerization of 11-*cis* **retinal** into an intermediate form that, after going through a series of spontaneous changes, finally is converted into all-*trans* retinal.

(2) The all-*trans* retinal dissociates from the opsin and then is converted into all-*trans* **retinol** (vitamin A).

b. **Regeneration of rhodopsin.** All-*trans* retinol is enzymatically isomerized to 11-*cis* retinol and then enzymatically oxidized to 11-*cis* retinal. The 11-*cis* retinal then spontaneously combines with opsin, reforming rhodopsin.

(1) The conversion of the retinol to retinal is dependent on the pigment epithelium.

(2) Deficiency of vitamin A prevents the regeneration of rhodopsin. For this reason, lack of vitamin A leads to **night blindness**.

c. **Absorption spectrum of visual pigments**

(1) Although rhodopsin absorbs light over most of the visible spectrum, it displays a maximal absorption at about 500 nm (the wavelength of green light).

(2) There are three types of cones, each of which contains a photopigment that absorbs light best at a particular wavelength (Figure 1-47).

(a) **Red cones** have a maximal absorption at a wavelength of 570 nm.

(b) **Green cones** have a maximal absorption at a wavelength of 535 nm.

(c) **Blue cones** have a maximal absorption at a wavelength of 445 nm.

(3) **Color blindness** is the inability to produce one or more of the cone photopigments.

(a) **Monochromats** have no cone photopigments and must rely entirely on rods for their vision. They are unable to see any colors.

(b) **Protonopes** lack red pigment and, thus, make color confusions in the red-green portion of the spectrum.

(c) **Deuteranopes** lack green pigment and, like the protonopes, are unable to distinguish colors in the red-green portion of the spectrum.

(d) **Tritanopes** lack blue pigment and, thus, have difficulty distinguishing colors in the blue portion of the spectrum.

(e) Protonopes and deuteranopes are called red-green color-blind because they cannot see colors in this region of the spectrum normally. However, they are not truly blind in that they are able to see colors. Similarly, loss of the blue pigment does not prevent the seeing of blue light, because light in the blue end of the spectrum is absorbed to some extent by the red and green pigments.

(4) **Color coding.** Over most of the visible spectrum, light is absorbed by all three cone pigments (see Figure 1-47). Thus, the encoding of color depends on the relative amount of activity in each of the three cones. For example, if the red cones are stimulated most strongly, the CNS interprets this activity as red light. If the red and green cones are stimulated equally, then a sensation of yellow is generated by the CNS.

3. **Electrophysiology of the retina**

a. **Receptor cell potential**

(1) Receptor cells have a resting membrane potential of about −40 mV due to a high Na$^+$ conductance.

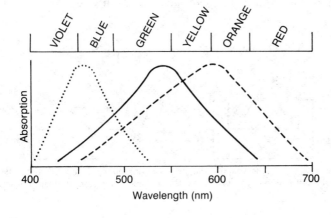

Figure 1-47. Spectral sensitivity of the three types of cones. The photopigment that is present in the outer segment of each cone type is responsible for the variance in the maximal light absorption and sensitivity in these cells. The *dotted line* indicates cones with blue-sensitive pigment; the *solid line* indicates cones with green-sensitive pigment; and the *dashed line* indicates cones with orange-sensitive pigment.

(2) The high Na$^+$ conductance is maintained by the presence of cyclic guanosine mono-phosphate (cGMP).

(3) When light strikes the eye, photoisomerization of 11-*cis* retinal is accompanied by the activation of a G protein [similar to the G protein involved in synaptic transmission; see III B 2 a (3)] called **transducin,** which causes cGMP levels to decrease.

    (a) The G protein activates a phosphodiesterase that breaks down cGMP to GMP.

    (b) The reduction in cGMP levels causes the Na$^+$ channels to close.

(4) The closing of the Na$^+$ channel causes the receptor to hyperpolarize. This is different from other receptors, which are depolarized when activated by a stimulus.

(5) Increasing the light intensity increases the receptor hyperpolarization (Figure 1-48). The amplitude of the hyperpolarization varies with the logarithm of the light intensity.

b. **Organization of ganglion cell receptive fields** (Figure 1-49)

    (1) **Central circuit.** In the peripheral retina, each ganglion cell receives synaptic input from many **bipolar cells,** which, in turn, receive input from a large number of receptor cells. All of the receptor cells that connect directly to a ganglion cell through bipolar cells form the receptive field **center** for that ganglion cell.

        (a) In the dark, the receptor cells are depolarized and, thus, release synaptic transmitter. Because the transmitter is inhibitory, it keeps the bipolar cell hyperpolarized. As a result, the ganglion cell is not stimulated.

        (b) When light strikes the receptor cells, it causes them to hyperpolarize. As a result, receptor cell transmitter release is decreased, the bipolar cell depolarizes, and the ganglion cell is stimulated.

        (c) Thus, when any of the cells in the receptive field center of a ganglion cell are stimulated by light, it causes the ganglion cell to fire.

    (2) **Surrounding circuit.** The ganglion cell also is influenced by receptors that connect to the bipolar cells through **horizontal cells**. Receptor cells lying within an annulus surrounding the receptive field of the ganglion cell form the receptive field **surround**. When light strikes any of the receptors in the receptive field surround, the ganglion cell is inhibited through the horizontal-to-bipolar-to-ganglion-cell pathway.

    (3) This organization of the ganglion cell receptive field is called an **on-center, off-surround** organization to indicate that light hitting the center of the field causes ganglion cell excitation, while light hitting the surround of the field causes ganglion cell inhibition. There is a complementary type of organization that has **off-center, on-surround** characteristics. Ganglion cells with this type of receptive field are inhibited by light hitting the center of the field and are excited when light shines on the cells within the area surrounding the center.

        (a) Because of the center-surround organization of the receptive fields, the retina transmits information only about changes in levels of illumination within the visual field.

        (b) For example, a bright spot surrounded by darkness will cause a large increase in the firing of an on-center ganglion cell, whereas a dark area surrounded by light will cause the ganglion cell to stop firing. If both the center and the surround are bathed by light or placed in darkness, then the ganglion cell firing rate will not change.

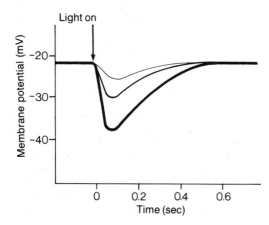

Figure 1-48. Response of a receptor cell to light. Responses to three brief light flashes of increasing intensity are shown. The receptor response is a graded hyperpolarization that increases with increasing light intensity. The *heaviest line* indicates the light flash of highest intensity.

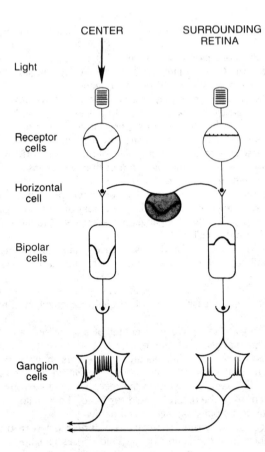

CENTER   SURROUNDING
                RETINA

Light

Receptor
cells

Horizontal
cell

Bipolar
cells

Ganglion
cells

Figure 1-49. Organization of retinal circuits. Light striking one spot on the retina (center circuit) causes hyperpolarizing responses in the receptor cells, depolarization
of the bipolar cells, and spikes in the ganglion cells of this
circuit. By the action of horizontal cells, bipolar cells in
the surrounding retinal circuits are depolarized, resulting
in the hyperpolarization of the ganglion cells of these
circuits. Note that of the cell types shown, true action
potentials are seen in only the ganglion cells.

C. Neural pathways of vision

   1. Optic tract. The axons of the ganglion cells are rearranged at the **optic chiasm**. Axons from
   the right half of each retina project to the **right lateral geniculate,** and axons from the left half
   of each retina pass to the **left lateral geniculate**. This means that fibers of the **temporal** retina
   on each side pass through the chiasm without crossing, while fibers of the **nasal** half of each
   retina decussate at the chiasm.

   2. Lateral geniculate nucleus. The inputs from each eye remain segregated in the lateral geniculate; uncrossed fibers go to layers 2, 3, and 5, and crossed fibers go to layers 1, 4, and 6.
   Neurons in the lateral geniculate also retain the on-center, off-surround (or the off-center,
   on-surround) organization of the ganglion cells.

D. Primary visual cortex. Neurons of the cortex are responsible for visual perception. This is accomplished by cells whose input configuration allows for **feature detection** (i.e., sensitivity to stimuli
of specific shape and orientation).

   1. Feature detectors are divided into three types: **simple cells, complex cells,** and **hypercomplex cells**.
      a. Simple cells
         (1) Response characteristics. Like ganglion and lateral geniculate cells, simple cortical
         cells have antagonistic center-surround organizations. However, rather than circular
         receptive fields, simple cells have elongated receptive fields. The best simple cell re-
         sponse is obtained with light or dark bars. The long axis of the bar must fall on a specific
         place on the retina and be oriented at a particular angle.
         (2) Synaptic input. The response characteristics of simple cells arise from the input they
         receive. Several adjacent ganglion cells, arrayed in a straight line at a particular orienta-
         tion on the retina, project (via the lateral geniculate) onto a single simple cell of the
         visual cortex. When the ganglion cell array is activated with a bar of light, the cortical
         neuron is activated simultaneously. The simple cortical cell receiving input from this

ganglion cell array can be activated only by (i.e., recognize) a bar of light that is at the proper retinal position and orientation.

  b. **Complex cells**
  (1) **Response characteristics.** Like simple cells, complex cortical cells respond best to light lines or dark bars of a particular orientation. The complex cell, however, responds even if the oriented stimulus does not fall on a particular retinal position. The stimulus can be delivered to a larger area of the visual field and still evoke a response.
  (2) **Synaptic input.** The response of a complex cell can be interpreted as a response to a **series of simple cells,** representing adjacent portions of the retina. The antagonistic center-surround organization is less prominent in complex cells.
  c. **Hypercomplex cells.** These units respond only to light lines of particular length or to edges. They receive and **integrate input from a variety of complex cells**. Hypercomplex cells are more prominent in the **prestriate cortex** (areas 18 and 19) than in the primary visual cortex (area 17).

2. **Cortical columns.** Cortical cells respond to lines, bars, and edges. The orientation of the linear stimulus is critical in determining whether a cell in the visual cortex will respond. Cells that respond to stimuli of a particular orientation are grouped together in an **orientation column,** which is perpendicular to the cortical surface.

IX. TASTE, or **gustation,** is the sensory modality mediated by the chemoreceptors of the tongue, mouth, and pharynx. Taste should be distinguished from flavor, which includes the olfactory, tactile, and thermal attributes of food in addition to taste. The taste chemoreceptors are sensitive to dissolved chemicals, both organic and inorganic.

  A. **Psychophysics of taste**

  1. **Taste qualities.** There are four primary taste qualities; all taste sensations are assumed to result from various combinations of these four primaries.
  a. **Sweet** sensation is produced by various classes of organic molecules including sugars, glycols, and aldehydes. The tip of the tongue is the area most sensitive to sweet stimuli.
  b. **Bitter** sensation is produced by alkaloids, such as quinine and caffeine. The back of the tongue is the area most sensitive to these stimuli.
  c. **Salt** sensation is produced by the anions of ionizable salts. The greatest sensitivity to salty tastes occurs in the front half of each side of the tongue.
  d. **Sour** sensation is produced by acids; this sensation relates, to some degree, to the pH of applied stimulus solutions. The greatest sensitivity to these stimuli occurs in the posterior half of each side of the tongue.

  2. **Threshold.** The lowest concentration of a sapid substance that can be discriminated is the threshold concentration of that substance. Increasing the area of the tongue exposed to the solution reduces the threshold concentration. Individual differences in sensitivity are genetically determined. For example, some individuals cannot detect phenylthiourea (a compound used in genetics to test ability to taste) unless it exists in very high concentrations.

  3. **Adaptation.** The intensity of a sensation resulting from continuous application of a taste stimulus declines with time until the solution becomes tasteless. The firing rate of afferent nerve fibers from the tongue similarly declines.

  B. **Peripheral taste receptors**

  1. **Lingual papillae.** These specialized protuberances on the surface of the tongue are of four morphologic types.
  a. **Filiform papillae** are mechanical, nongustatory structures.
  b. **Fungiform papillae** have 8–10 taste buds on each papilla.
  c. **Circumvallate papillae** are arranged in a V-shaped row of 7–12 on the posterior part of the tongue. Each papilla has approximately 200 taste buds. The taste buds are located on the sides of these large structures.
  d. **Foliate papillae,** located on the lateral border of the tongue anterior to the circumvallate papillae, have numerous taste buds.

2. Taste buds. Each taste bud represents a cluster of 40–60 taste cells (Figure 1-50) as well as supporting cells. Taste buds are located on the tongue papillae, hard palate, soft palate, epiglottis, and in the pharynx.

3. Taste cells
   a. Structure. These chemosensitive receptor cells are elongated cells with microvilli at the apical surface. Dissolved substances enter the taste pore of the taste bud to contact the microvilli. The taste cells are innervated by afferent neurons.
   b. Turnover. Each taste cell has a life cycle of only a few days. The degenerating taste cell is replaced by a cell that arises from the supporting epithelial cells. Contact with the afferent neuron converts an epithelial cell into a taste cell. Conversely, nerve transection causes the disappearance of taste cells.

C. Innervation of taste cells is by branches of the facial, glossopharyngeal, and vagus nerves (cranial nerves VII, IX, and X, respectively). Nerve fibers terminate on receptor cells to form synapses. [The tactile and temperature receptors of the mouth, tongue, and pharynx are innervated by the trigeminal nerve (cranial nerve V).]

   1. The taste buds in the **anterior two-thirds of the tongue** are innervated by lingual branches of the **facial nerve;** the lingual nerve, which branches from the chorda tympani, is part of the facial nerve. The cell bodies are located in the geniculate ganglion, and the nerve terminals end in the **nucleus solitarius** of the medulla.

   2. The taste buds in the **posterior third of the tongue** are innervated by the **glossopharyngeal nerve**. The cell bodies lie in the superior and inferior ganglia of this nerve. The fibers relating to taste sensation terminate in the nucleus solitarius.

   3. Taste receptors in the **pharyngeal aspect of the tongue** and on the hard palate, soft palate, and epiglottis are innervated by fibers of the **vagus nerve**. The cell bodies are located in the superior and inferior ganglia of the vagus nerve and terminate in the nucleus solitarius.

D. Taste coding. Each nerve fiber in the gustatory nerves responds to more than one taste stimulus. However, each fiber responds best to one of the four primary taste qualities. Thus, the coding of a gustatory sensation is not a simple, labeled-line, chemical sensory system. Taste sensation depends on the **pattern** of nerve fibers activated by a particular stimulus.

X. OLFACTION, like taste, is a chemical sense involving receptors that are sensitive to chemicals in solution. For olfaction, the chemicals initially are airborne; however, they must dissolve in the mucous layer lining the nose before they can come in contact with olfactory receptors.

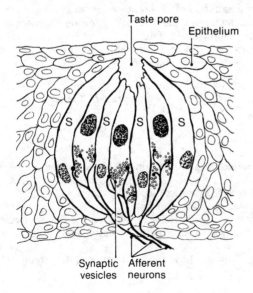

Taste pore

Epithelium

Synaptic vesicles   Afferent neurons

Figure 1-50. Structure of a taste bud. Taste receptor cells contain synaptic vesicles and receive afferent nerve terminals. Supporting cells (*S*) are not innervated.

A. Psychophysics of olfaction

1. **Characteristics of odorants.** To be an effective odorant, a substance must be:
   a. **Volatile**, since the olfactory receptors respond to chemicals transported by air into the nose
   b. **Water soluble** (to some degree) in order to penetrate the watery mucous layer lining the nasal epithelium to reach the receptor cell membrane
   c. **Lipid soluble** (to some degree) in order to penetrate the cell membranes of the olfactory receptor cells to stimulate those cells

2. **Threshold.** The olfactory receptors have varying sensitivity to substances. For many substances, the sensitivity is so high that a few molecules interacting with a receptor are sufficient to produce excitation.

3. **Odor discrimination.** The olfactory system has the capacity to discriminate various chemical compounds. This ability depends on the receptor sites on the membranes of receptor cells, which vary in their capability to accept odorant molecules of complementary conformation. In some instances, the olfactory system can discriminate dextro- and levorotatory forms as well as *cis*- and *trans*- conformations of a molecule. In contrast to gustatory sensation, no primary olfactory qualities can be identified.

4. **Dynamic range.** The olfactory system has limited ability to discriminate differences in odorant concentration in ambient air. Perceived odor intensity conforms to a power function with an exponent of less than 1 (see V D 1).

5. **Adaptation.** Olfactory sensation decreases very rapidly with continued exposure to an odorant.

B. Olfactory mucosa.
The olfactory mucosa refers to the specialized part of the nasal mucosa that contains the olfactory receptor cells. It is distinguished from the surrounding respiratory mucosa by the presence of tubular **Bowman's glands,** the absence of the rhythmic ciliary beating that characterizes the respiratory mucosa, and its distinctive yellow-brown pigment. A mucous layer covers the entire epithelium.

1. **Location.** The olfactory mucosa is located on the superior nasal concha, adjacent to the nasal septum. Because of its superior position in the nasal cavity, the olfactory mucosa is not directly exposed to the flow of inspired air entering the nose. Odorant molecules come in contact with the olfactory mucosa by sniffing (i.e., by short, forceful inspirations). Sniffing produces turbulence in the air flow and thereby transports molecules to the receptor cells.

2. **Innervation** of the olfactory mucosa is by the olfactory nerve (cranial nerve I) and some branches of the trigeminal nerve (cranial nerve V). The irritative character of some odorants is due to stimulation of the free nerve endings of the trigeminal nerve.

3. **Cells.** The olfactory mucosa is composed of three cellular elements (Figure 1-51).
   a. **Receptor cells** are bipolar neurons with a single dendrite that extends to the surface and terminates in a knob. Projecting from the knob of each sensory neuron are cilia that have

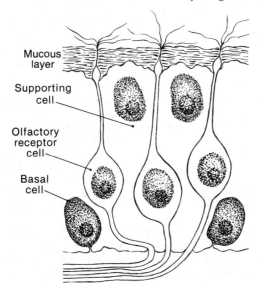

Mucous layer

Supporting cell

Olfactory receptor cell

Basal cell

Figure 1-51. The olfactory epithelium. Olfactory receptor cells are situated among supporting cells. The cilia lie within the mucous layer covering the epithelium. New receptors are generated from basal cells. Bowman's glands, which contribute to the mucus secretion, are not shown.

the familiar microtubule configuration of two central microtubules surrounded by a ring of nine microtubules. The cilia reach the outer surface of the mucous layer. The axon that leaves the basal end of the receptor cell is unmyelinated. Receptor cells degenerate and have a life span of about 60 days.

   b. Supporting cells have a columnar shape. Microvilli extend from the surface of these cells into the mucous layer covering the nasal mucosa.

   c. Basal cells are stem cells from which new receptor cells are formed. There is a continuous replacement of receptor cells by mitosis of basal cells.

C. Receptor cell electrophysiology

   1. **Receptor potential.** The adsorption of odorant molecules to the plasma membrane of the cilia of the receptor cells generates a receptor potential in the receptor cell.

   2. **Electro-olfactogram (EOG)** is a tracing of the slow monophasic negative potential detected by an electrode placed on the olfactory mucosa. It reflects the local, graded receptor potentials evoked in the population of olfactory receptors by an odorous stimulus. The amplitude of the EOG increases with increases in the intensity of the stimulus.

   3. **Impulse initiation.** Neural impulses are generated by the receptor potential. Spikes originate at the point in the olfactory receptor where the cell narrows to form the **olfactory fiber** or **axon**.

   4. **Spike activity.** The resting activity in olfactory fibers is extremely low. Odorous stimulation causes excitation of most fibers, but a reduction in firing rate, or inhibition, is noted in some cases. Conduction velocity is slow because the fibers are unmyelinated and small in diameter.

   5. **Sensory coding.** A specific olfactory receptor does not respond to a particular compound or category of compounds; instead, an individual receptor responds to many odors. Furthermore, no two receptor cells have identical responses to a series of stimuli. Sensory perception, therefore, is based on the pattern of receptors activated by the stimulus.

XI. SPINAL REFLEXES. A reflex is an automatic response to a stimulus carried out by a relatively simple neuronal network consisting of a receptor, an afferent pathway, and an effector organ. Spinal cord reflexes are classified as originating from cutaneous receptors or from muscle receptors.

   A. Cutaneous reflexes. The most important of the cutaneous reflexes is the **withdrawal, flexor,** or **pain reflex,** by which a body part is withdrawn from the site of a painful stimulus. Several reflex pathways within the spinal cord act together to coordinate the activity of all of the muscles necessary to produce a smooth movement.

   1. **Receptors for the withdrawal reflex** are the nociceptors located on the free nerve endings of A delta and C fibers. Upon entering the spinal cord, the pain fibers synapse on many interneurons. Some of these convey information to the CNS. Others form several reflex pathways that coordinate the withdrawal of the limb.

      a. The major afferent pathway travels through several interneurons before synapsing on the alpha motoneurons innervating the muscles used to withdraw the limb from the painful stimulus. Because several interneurons are interposed between the afferent and efferent neurons, the reflex is described as **multisynaptic** or **polysynaptic**. The interneurons form several pathways of different lengths, through which the pain stimulus can reach the alpha motoneuron.

         (1) Reverberating circuits are a consistent feature of the withdrawal reflex pathways. In these circuits, a branch from the axon of an interneuron in the reflex pathway feeds back onto previously excited interneurons, causing them to become reexcited. Reverberating circuits cause activity in the reflex pathway to continue even after the sensory receptor has stopped firing. This is called **afterdischarge** .

         (2) Under most circumstances, only the minimal number of muscles necessary to withdraw the limb are activated. This is called **local sign**. For example, if the hand accidentally touches a hot stove, just the hand may move away.

         (3) When the stimulus is very strong, however, the withdrawal reflex may require the involvement of more muscle groups. This is called **irradiation**. Thus, if the hand picks up a hot coal, not only will the fingers open to drop it, but the entire arm will withdraw and the individual may leap away from the site of painful stimulation.

b. Another important pathway formed by the interneurons within the spinal cord terminates on the antagonistic motoneurons. This is an **inhibitory** pathway.

(1) By inhibiting the motoneurons that innervate muscles antagonistic to those withdrawing the limb, this pathway ensures that the flexion movement is not impeded by contraction of the extensors.

(2) This type of neuronal organization, in which the reflex pathway activating one group of alpha motoneurons also inhibits its antagonistic motoneuron, is quite common within the spinal cord and is called **reciprocal innervation**.

c. Finally, the **interneurons form pathways that cross the spinal cord** to innervate the extensor motoneurons on the contralateral side. This is called a **crossed extensor reflex**.

2. Integration of the withdrawal reflex occurs on the alpha motoneuron, the **final common pathway** through which all the afferent fibers act. If the excitatory pathways dominate, the alpha motoneuron discharges a train of action potentials; if the inhibitory pathways dominate, the neuron does not fire.

3. Effector organs of the withdrawal reflex are the skeletal muscles that cause withdrawal of the limb. Although they are called flexors, these muscles are flexors in the physiologic, not the anatomic, sense. For example, the muscles that cause the fingers to open in order to drop a hot coal, although anatomically referred to as extensors, are considered to be flexors because they are involved in the withdrawal reflex.

4. Behavioral responses. The anatomic organization of the withdrawal reflex produces the following characteristic behavior.

a. The withdrawal reflex has a relatively long latency because the afferent pathway uses small, slowly conducting fibers and involves many synapses.

b. The afterdischarge that results from the parallel pathways and reverberating circuits causes the response to outlast the stimulus. This keeps the affected limb away from the painful stimulus while the brain determines where to put it.

c. The crossed extensor reflex produces a patterned response in which the affected limb flexes while the contralateral limb extends. In the lower limbs, this allows the contralateral limb to support the body while the other limb is raised off the ground.

B. Muscle reflexes. Two important reflexes originate in the muscles: the **stretch reflex** and the **lengthening reaction**. The stretch reflex causes the reflex contraction of a muscle that is stretched. For example, when the patellar tendon is tapped by a reflex hammer, it causes the quadriceps muscle to be stretched. The stretched muscle contracts reflexly, causing the leg to be elevated (called the knee jerk reflex). The lengthening reaction causes inhibition of alpha motoneurons innervating muscles that are under tension, allowing them to lengthen.

1. The stretch reflex

a. The receptor for the stretch reflex is the muscle spindle. This complex, spindle-shaped, encapsulated receptor contains muscle fibers that have both sensory and motor innervation (Figure 1-52).

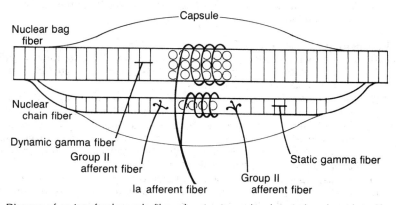

Figure 1-52. Diagram of an intrafusal muscle fiber, showing its nuclear bag and nuclear chain fibers. The afferent innervation (Ia and II fibers) and efferent innervation (gamma dynamic and gamma static fibers) of the intrafusal muscle fiber also are illustrated.

(1) The number of spindles in each muscle depends on the task performed by the muscle. Muscles involved in precision movements contain many more spindles than muscles used to maintain posture. For example, hand muscles have about 80 spindles, which is about 20% of the number of spindles contained in back muscles weighing 100 times as much.

(2) The muscle fibers within the spindle are called **intrafusal muscle fibers** in contrast to the **extrafusal muscle fibers** responsible for generating tension. There are two types of intrafusal muscle fibers.

   (a) The **nuclear bag fibers** are 30 μ in diameter and 7 mm in length and are so-called because their nuclei appear to be bunched up in the center of the cell as if in a bag. About 2–5 nuclear bag fibers exist in a typical spindle.

   (b) The **nuclear chain fibers** are 15 μ in diameter and 4 mm in length and are so-called because their nuclei are lined up in a single file in the center of the fiber. About 6–10 nuclear chain fibers exist in each spindle.

   (c) The connective tissue surrounding the intrafusal fibers is continuous with that of the extrafusal fibers. As a result, when the extrafusal fiber contracts the intrafusal fiber is shortened, and when the extrafusal fiber stretches the intrafusal fiber is lengthened. The muscle spindle, thus, is **in parallel** with the extrafusal muscle fibers.

(3) Two sensory neurons emerge from the muscle spindle.

   (a) A single large fiber, called a **group Ia fiber** or **primary ending,** sends branches to every intrafusal fiber within the muscle spindle.

   (b) Several smaller neurons, called **group II fibers** or **secondary endings,** innervate the nuclear chain fibers.

   (c) The primary endings surround the center of the intrafusal muscle fiber; the secondary endings terminate on either side of the primary endings.

(4) The efferent fibers to the muscle spindle are called **gamma fibers** because their axons belong to the A gamma group of fibers. There are two types of gamma fibers.

   (a) **Static gamma fibers** primarily innervate nuclear chain intrafusal muscle fibers and produce tonic activity in the Ia afferent fibers.

   (b) **Dynamic gamma fibers** primarily innervate nuclear bag intrafusal muscle fibers and generate phasic activity in the Ia afferent fibers.

b. The Ia fiber enters the spinal cord through the dorsal root and sends branches to every alpha motoneuron that goes to the muscle from which the Ia originated.

   (1) In contrast to the withdrawal reflex, the stretch reflex has a rapidly conducting afferent fiber, is monosynaptic, has a short latency, and does not exhibit afterdischarge and irradiation.

   (2) In the stretch reflex, as in the withdrawal reflex, the alpha motoneuron is the final common pathway, serving as both an integrating center and efferent pathway.

   (3) Also, like the withdrawal reflex, the stretch reflex is characterized by reciprocal innervation. Thus, when a stretch reflex is elicited, the muscle antagonistic to the stretched muscle is inhibited, allowing the agonistic muscle to contract without interference.

c. Both extensor and flexor muscles exhibit stretch reflexes. These are elicited routinely during a neurologic examination to test for damage to either the spinal cord or the sensory or motor neurons. The gamma efferent fibers control the sensitivity of the receptors to stretch. Their role is discussed in XI C 2.

2. The lengthening reaction

a. The **receptors for the lengthening reaction** are the **Golgi tendon organs**.

   (1) These small (0.5 mm long) encapsulated receptors are located in the tendons, between the muscles and tendon insertions.

   (2) The Golgi tendon organs are stretched whenever the muscle contracts and, thus, in contrast to the muscle spindle, are **in series** with the extrafusal muscle fibers.

   (3) The Golgi tendon organs have neither muscle fibers nor an efferent innervation.

b. The afferent fiber innervating the Golgi tendon organ is called a **group Ib fiber**. It enters the dorsal root and forms a disynaptic pathway, which ends on the alpha motoneurons that send axons to the muscle from which the Ib fiber originated.

   (1) The disynaptic pathway of the Golgi tendon organ is inhibitory to the alpha motoneuron.

   (2) The Ib fiber, like all sensory fibers, releases an excitatory transmitter. To produce inhibition, an inhibitory interneuron must be activated.

c. Like the stretch reflex, the lengthening reaction occurs in both flexor and extensor muscles, lacks afterdischarge and irradiation, and displays reciprocal innervation.

d. Historically, the lengthening reaction has been described as a protective reflex in which a strong and potentially damaging muscle force reflexly causes the muscle to be inhibited, and, as a result, the muscle lengthens instead of trying to maintain the force and risking damage. Although this is true, it is now clear that the reflex plays a more important role in regulating tension during normal muscle activity. The lengthening reaction is described as **autogenic inhibition,** which indicates that the force generated when the muscle contracts is the stimulus for its own relaxation.

C. Role of the stretch reflex in the control of movement. The stretch reflex is used by the motor control system to aid in the performance of a movement. During activity generated by the motor command system, the Ia fibers from the muscle spindle inform the motor control system about the changes in muscle length and provide the alpha motoneuron with a source of excitatory input in addition to that coming from higher centers.

1. Figure 1-53 illustrates the effect of muscle stretch and contraction on Ia discharge when the gamma motoneurons are not activated.
   a. Stretching the extrafusal muscle fiber causes the intrafusal muscle fiber to be stretched. Most of the stretch occurs in the central, most compliant region of the intrafusal fiber, which lacks sarcomeres.
      (1) When the intrafusal muscle fiber is stretched, the Ia fiber terminal is deformed. The deformation causes a receptor potential, which generates a train of action potentials (see Figure 1-53B).
      (2) The action potentials initially discharge at a frequency proportional to the velocity of stretch and then adapt to fire tonically at a frequency proportional to the amount of muscle stretch.
   b. Shortening the extrafusal fiber causes the central region of the intrafusal muscle fiber to be compressed.
      (1) This reduces the deformation of the Ia fiber terminal, causing the Ia fiber to reduce its firing rate (see Figure 1-53C).

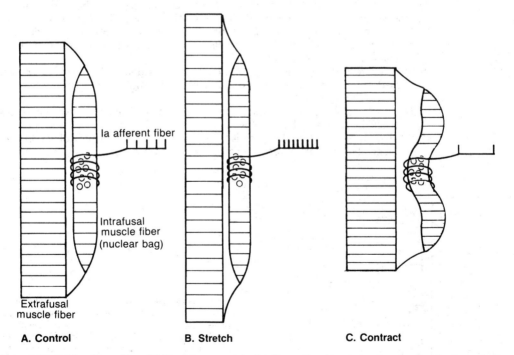

A. Control   **B. Stretch**   **C. Contract**

Figure 1-53. (*A*) Ia afferent fiber. (*B*) The firing rate of the Ia afferent fiber increases when the extrafusal muscle fiber is stretched because the intrafusal muscle fiber also is stretched. Most of the stretch occurs at the center region of the intrafusal muscle fiber. (*C*) When the extrafusal muscle fiber contracts, the Ia afferent discharge decreases because the intrafusal muscle fiber slackens (i.e., the intrafusal muscle fiber is unloaded).

(2) The reduction of firing that occurs during muscle contraction, called **unloading,** is functionally disadvantageous because the CNS stops receiving information about the rate and extent of muscle shortening.

2. Unloading can be prevented by the activity of the gamma efferent motoneurons, as shown in Figure 1-54. The gamma motoneurons cause the sarcomeres of intrafusal muscle fibers to shorten as the extrafusal muscle fiber shortens. As a result, the central region of the intrafusal muscle fiber remains stretched during muscle contraction, and unloading does not occur.
   a. When the dynamic gamma motoneurons are fired, only the nuclear bag intrafusal fibers shorten. Since the nuclear bag fibers are responsible for producing the phasic (i.e., velocity-sensitive) portion of the Ia response to stretch, stimulation of the dynamic gamma fibers causes an increase in phasic activity without affecting the tonic activity.
   b. When the static gamma motoneurons are fired, only the nuclear chain fibers shorten. Since nuclear chain fibers are responsible for the tonic component of the Ia response, stimulation of the static gamma fibers causes an increase in the tonic (length-sensitive) level of Ia firing without affecting the phasic level.

3. During a normal movement, the motor command system **coactivates** both the alpha and gamma motoneurons. This diminishes the amount of unloading that occurs during muscle contraction, allowing the CNS to determine if its motor commands are being carried out.
   a. For example, when the motor control system issues a command to lift a weight, the alpha and gamma motoneurons are coactivated.
      (1) As the extrafusal muscle fiber shortens, the intrafusal muscle fiber sarcomeres also shorten.
      (2) If the two muscle fibers shorten at the same rate, then the central region of the intrafusal fiber is neither compressed nor lengthened, thus keeping Ia activity at a constant level.
      (3) The constant level of Ia input to the CNS during a movement indicates that the motor command is being carried out.
   b. If the weight to be lifted is underestimated by the CNS, the motor command system does not activate a sufficient number of alpha motoneurons to lift the weight. Thus, the extrafusal muscle fibers do not shorten.
      (1) However, the intrafusal muscle fibers do shorten. But, because the tendon ends of the muscle cannot move, the central portion of the intrafusal fiber lengthens.
      (2) Stretching the central region of the fiber causes Ia activity to increase, indicating that the motor command is not being carried out. The CNS uses this information to readjust its command to the spinal cord.
      (3) Even before the CNS responds to the information provided by the Ia fibers, the Ia activity is used at the spinal cord level to adjust the alpha motoneuron activity to meet the unexpectedly high load.
         (a) Since the Ia fiber synapses on the alpha motoneuron, its activity increases the excitability of the alpha motoneuron.
         (b) The increase in excitability leads to an increase in the frequency of action potential generation and an increase in muscle force development.

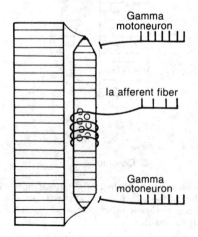

Figure 1-54. The firing rate of the Ia afferent fiber does not decrease if the gamma motoneuron discharge accompanies contraction of the extrafusal muscle fiber. Unloading is prevented because the intrafusal muscle fiber also shortens. Since the center region of the extrafusal muscle fiber does not contain sarcomeres, it does not shorten and, in fact, may lengthen if the gamma motoneuron causes sufficient shortening of the intrafusal muscle fiber sarcomeres.

4. **Gamma loop.** The CNS is theoretically capable of initiating movements directly by stimulating just the gamma motoneurons, using a pathway called the **gamma loop** (Figure 1-55).
   a. Increasing gamma motoneuron activity causes the intrafusal muscle fiber sarcomeres to shorten, which, in turn, leads to stretching of the central portion of the intrafusal fiber and activation of the Ia fiber. Firing the Ia fibers causes alpha motoneuron activity to increase, which results in an increased amount and force of skeletal muscle activity.
   b. Although the gamma loop can elicit movement on its own, it normally does not do so. However, because of coactivation, the gamma loop is activated during all movements and thus contributes to the excitability and firing rate of the alpha motoneurons.

XII. THE MOTOR CONTROL SYSTEM

A. **Movement.** The major components of the motor control system are shown in Figure 1-56. This system is a complex and highly integrated network in which all parts work together to produce a movement. The specific tasks assigned to each component can be analyzed as a basis for understanding how movements are coordinated.

1. The **idea for a movement** is generated within the cortical association areas of the parietal lobe.

2. Then, the idea is transferred to the motor areas of the frontal lobe, where it is organized into a **motor command**.

3. The command is sent to the spinal cord for **execution of the movement**. The basal ganglia, cerebellum, brain stem, and spinal cord all participate in the production of a coordinated movement by modifying the motor command.
   a. The basal ganglia provide the motor patterns necessary to maintain the postural support required for motor commands to be carried out properly.
   b. The cerebellum receives information from the motor cortex about the nature of the intended movement and from the spinal cord about how well it is being performed. This information is used to adjust the motor command so that the intended movement is executed smoothly.
   c. The brain stem is the major relay station for all motor commands except those requiring the greatest precision, which are transferred directly to the spinal cord. The brain stem also is responsible for maintaining normal body posture during motor activities.
   d. The spinal cord contains the final common pathways through which a movement is executed. By selecting the proper motoneurons for a particular task and by reflexly adjusting the amount of motoneuron activity, the spinal cord plays an important role in the coordination of motor activity.

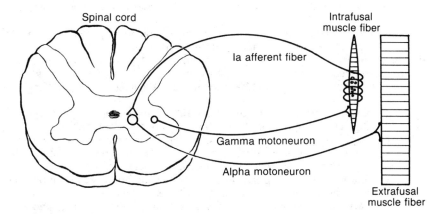

Figure 1-55. The gamma loop increases the firing of the alpha motoneuron during muscle contraction. The loop begins with the gamma motoneuron, which discharges to cause intrafusal muscle fiber contraction. This leads to an increase in Ia afferent fiber activity, which, in turn, causes increased alpha motoneuron discharge via a monosynaptic reflex.

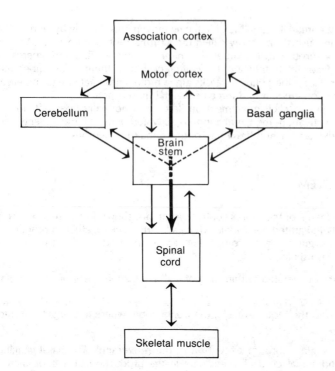

Figure 1-56. Diagram illustrating the extensive interconnections between the components of the motor control system. Note that all of the descending pathways except for the pyramidal tract (*thick arrow*) communicate with the spinal cord through the brain stem.

B. The spinal cord

1. **Physiologic anatomy**
   a. **Motoneuron pool.** The alpha and gamma motoneurons within the spinal cord that innervate a particular muscle are collected together into a motoneuron pool.
      (1) The alpha and gamma motoneurons are randomly distributed within the pool and overlap to some extent with cells from other motoneuron pools.
      (2) Usually, the motoneurons leave the spinal cord in several contiguous ventral roots and then combine into a single motor nerve containing the alpha and gamma motoneurons as well as the Ia, Ib, and II afferent fibers from the muscle.
   b. **Motor unit.** Each motoneuron and all the muscle fibers it innervates form a motor unit. The motor units within a muscle vary in size from a few muscle fibers to several thousand muscle fibers. Muscles that perform precise movements have smaller motor units than those responsible for large body movements and for maintaining posture.
      (1) Whenever an alpha motoneuron fires, all of the muscle fibers in its motor unit are activated.
      (2) The muscle fibers belonging to a single motor unit are dispersed throughout the muscle so that the force they produce is distributed evenly.
   c. **Muscle fibers.** All of the muscle fibers in a motor unit are of the same physiologic type and are categorized according to their histochemical and contractile characteristics.
      (1) **Fast-twitch fatigable (FF) fibers** contract quickly, fatigue easily, and have the following characteristics.
         (a) **Rapid contractile speeds.** This results from the high myosin-ATPase activity of their cross-bridges and the rapid sequestering of Ca^{2+} by their SR.
         (b) **Rapid fatigue.** This occurs because FF fibers have few mitochondria and, thus, cannot make use of oxidative metabolism. FF fibers rely on glycolysis for their ATP supply and fatigue when their glucose stores are depleted.
         (c) **Sparse capillary supply.** Because FF fibers do not make use of oxidative metabolism, the growth of surrounding capillaries is limited.
         (d) **Large size.** Although FF fibers cannot sustain activity for long periods of time, they can generate large contractile forces. Their large size is not a disadvantage from a diffusional point of view because they do not make use of oxidative metabolism.
      (2) **Slow-twitch (S) fibers** contract slowly, are virtually untiring, and have the following characteristics.

(a) **Slow contractile speeds.** This is due to the low myosin-ATPase activity of their cross-bridges and the slow sequestering of Ca^{2+} by their SR. These characteristics reduce the amount of ATP used by S fibers and, thus, contribute to their resistance to fatigue. Their long contraction times make summation and tetanus possible at low frequencies of stimulation.

(b) **Great resistance to fatigue.** This is a consequence of the ability of S fibers to use oxidative metabolism as a primary source of ATP.

(c) **A rich capillary supply** provides the oxygen needed by S fibers.

(d) **Small size.** Although S fibers cannot produce a large amount of force, they can sustain force for a long time. Their small size is necessary for oxygen to diffuse into the center of the fiber and for waste products to diffuse out of the fiber.

(3) **Fast fatigue-resistant (FR) fibers** have characteristics that are intermediate to those of the FF and S muscle fibers.

2. **Task performance.** During the performance of a motor task, the small motor units, because they are more excitable, are recruited before the large ones. This **size principle** of motor unit selection has significant physiologic advantages.

   a. **To perform a precision movement requiring small amounts of force,** it is advantageous to use small motor units. When more force is required, larger motor units must be activated.

   (1) The proper motor units are selected automatically because the motoneuron pool is organized according to the size principle.

   (a) When a small amount of force must be applied, the motor cortex provides a minimal amount of input to the motoneuron pool, activating only the smallest motor units.

   (b) If more force is required, the motor cortex increases its input to the motoneuron pool, and larger motor units are recruited.

   (2) Because of the size principle, the motor cortex does not need to specify the particular motoneuron to be recruited during a movement. This simplifies the organization of the motor command structure by reducing the number of cortical neurons that are involved in generating a movement.

   b. **To perform a task requiring endurance,** the fatigue-resistant motor units must be recruited. When power is required, the larger motor units must be recruited.

   (1) Again, motor unit selection is accomplished automatically because of the size principle.

   (a) When an endurance movement must be performed, the motor cortex provides a minimal input to the spinal cord, and the smallest, most fatigue-resistant motor units are recruited.

   (b) When the amount of force being generated by the muscle is not sufficient to execute the movement, the motor cortex increases its input and recruits more motor units. In all cases, the most fatigue-resistant fibers are recruited first without requiring that the motor cortex determine which motoneuron to activate.

   (2) The relation between fatigue resistance and motoneuron size also is a consequence of the size principle. A muscle fiber's fatigability can be altered by its activity.

   (a) When a muscle fiber is contracted repeatedly, its oxidative capacity and blood supply increase, making the fiber more resistant to fatigue. Conversely, when a muscle fiber is required to contract against large afterloads, the fiber increases in size.

   (b) Small motor units are recruited first. Thus, they are involved in all movements and, consequently, develop fatigue resistance.

C. **The brain stem** contains the medulla, pons, midbrain, and parts of the diencephalon. Neuronal circuits within these areas control many physiologic functions including blood pressure, respiration, body temperature, sleep, and wakefulness. In addition, the **reticular formation** and **vestibular nuclei** are important components of the motor control system.

   1. The **reticular formation** is the relay station for all descending motor commands except those traveling directly to the spinal cord through the medullary pyramids.

   a. **Normal function.** The reticular formation receives and modifies motor commands to the proximal and axial muscles of the body and is involved in the performance of all motor activities except those fine movements performed by the distal muscles of the fingers and hands. The reticular formation also is responsible for maintaining normal postural tone.

   (1) Neurons within the pontine reticular formation send axons to the spinal cord in the medial reticulospinal tract and are excitatory to the alpha and gamma motoneurons innervating the extensor **antigravity muscles.**

(2) These neurons are prevented from firing too rapidly by inhibitory input derived from the cerebral and cerebellar components of the motor control system. The amount of inhibition is increased to reduce postural tone and is decreased to enhance postural tone. The withdrawal of inhibition, called **release of inhibition,** frequently is used to increase activity within the CNS.

b. Lesions

(1) If the brain stem is severed from the spinal cord, the syndrome of **spinal shock** results.

(a) The initial result of removing the spinal cord from the control of the brain stem is the complete loss of reflex activity, which can last for days in cats and for months in humans.

(b) When reflexes return, they no longer are under the influence of the brain stem and, therefore, do not follow their normal patterns. For example, the local sign that characterizes the withdrawal reflex disappears. Instead, even a light touch on the foot can cause activation of all the flexor muscles in the body.

(c) The excitability of the motoneurons ultimately becomes greater than normal, causing particular groups of muscles to contract continuously.

(2) If the cerebral and cerebellar inhibitory inputs to the reticular formation are removed by severing the brain stem above the pontine reticular formation, the condition of **decerebrate rigidity** results.

(a) Without the inhibitory input from these higher centers, the pontine reticular formation fires uncontrollably, subjecting both the alpha and gamma antigravity motoneurons to intense excitation.

(i) The excessive firing of the alpha motoneurons causes the antigravity muscles (i.e., the leg extensors and the arm flexors) to contract continuously.

(ii) The firing of the gamma motoneurons activates the gamma loop, causing increased discharge of the Ia afferent fibers. These reflexly add to the alpha motoneuron excitation produced by the reticulospinal tract.

(b) Decerebrate rigidity is analogous to the condition of **spasticity**.

(i) Spasticity normally is caused by lesions of the descending pathways from the motor cortex to the reticular formation.

(ii) Spasticity is characterized by an increased amount of antigravity muscle activity. This activity is increased further when the affected muscle is stretched, due to the increased activity of the gamma motoneurons. Because these fibers are firing at higher than normal rates, the muscle spindles become more sensitive to stretch.

(iii) Cutting the dorsal roots reduces the amount of Ia input to the spinal cord and reduces the amount of alpha motoneuron activity. This reduces the spasticity.

2. Vestibular system. Vestibular system reflexes are responsible for maintaining tone in antigravity muscles and for coordinating the adjustments made by the limbs and eyes in response to changes in body position.

a. Vestibular receptors. The receptors that initiate the vestibular reflexes are located within the **labyrinth** (see Figure 1-35)—a system of fluid-filled, membrane-bound canals and sacs that are continuous with the cochlea. The labyrinth consists of the sac-like **otolith organs** (i.e., the **saccule** and **utricle**) and three **semicircular canals**. The saccule communicates directly with the cochlea, which is beneath it, and the utricle, which is above it. Both ends of all three semicircular canals emerge from the utricle. Each canal has an expanded end, called the **ampulla,** which contains the receptor cells.

(1) Orientation

(a) The semicircular canals are responsible for detecting angular accelerations of the head. The **horizontal canal** lies in a plane that is approximately parallel to the earth when the head is held in a normal upright position. The vertical canals lie in vertical planes. The plane of the **anterior vertical canal** is oriented along a line from the center of the head toward the eye, while the plane of the **posterior vertical canal** is oriented along a line from the center of the head toward the ear. The orientation of the posterior canal on one side of the head is roughly parallel to the orientation of the anterior canal on the other side of the head.

(b) The otolith organs are responsible for detecting linear acceleration and the static position of the head. The receptors within the utricle are oriented in a horizontal plane, while those in the saccule are oriented in a vertical plane.

(2) Receptor cells. The receptor cells of the vestibular system are similar to those of the cochlea. However, unlike the auditory receptor cells, the cilia of the vestibular hair cells are polarized (Figure 1-57). A large cilium, called the **kinocilium,** is located at one end of the cell. All of the others are called **stereocilia.** Whenever the stereocilia are bent toward the kinocilium, the cell is depolarized. When the stereocilia are bent away from the kinocilium, the cell is hyperpolarized.

(a) The hair cells of the semicircular canals are located on a mass of tissue (the **crista**) within the ampulla. The cilia are embedded in a gelatinous structure called the **cupula,** which completely fills the ampullar space.

  (i) When the head begins to move, the fluid within the semicircular canals lags behind and pushes the cupula backward. This causes the cilia of the hair cells to bend, producing either a depolarization or hyperpolarization depending on whether the stereocilia are pushed toward or away from the kinocilium.

  (ii) After 15–20 seconds of continuous movement at a constant velocity (e.g., the movement of a twirling dancer), the velocity of fluid movement catches up to that of the head, and the cupula returns to its resting position. This causes the cilia to return to their upright position and the hair cell to return to its resting membrane potential. Thus, the semicircular canals signal changes in motion (acceleration) and are insensitive to movements at a constant angular velocity.

  (iii) When the head stops moving, the fluid within the semicircular canals continues to move, pushing the cupula forward. This causes the cilia to bend in the opposite direction. Thus, if the original movement caused the hair cell to depolarize, the hair cell hyperpolarizes when the movement ceases.

  (iv) The hair cells of the horizontal canals are oriented with the kinocilium located closest to the utricle. Head movements that bend the stereocilia toward the utricle (utriculopedal movements) cause the hair cells to depolarize, while movements that bend the stereocilia away from the utricle (utriculofugal movements) cause hyperpolarization. For example, when the head is rotated toward the left, the endolymph in the left horizontal semicircular canal pushes the cupula and stereocilia toward the utricle (i.e., the fluid lags behind the head movement and causes the cupula to move toward the right), causing the hair cells to depolarize (Figure 1-58). At the same time, the hair cells within the right horizontal canal are pushed away from the utricle and hyperpolarize. When the head stops rotating, the fluid continues to move, pushing the cupula in the opposite direction. This causes the hair cells in the left horizontal canal to hyperpolarize and those in the right semicircular canal to depolarize.

  (v) The vertical canals also work in pairs: when the anterior vertical canal on one side is stimulated, the posterior vertical canal on the other side is inhibited. In the vertical canals, the kinocilium is located on the side of the hair cells away from the utricle.

  (vi) Because each of the three canals is oriented in a different plane, movement of the head in any direction causes a unique pattern of activity to be generated by the semicircular canals. This information is used by the CNS to interpret the

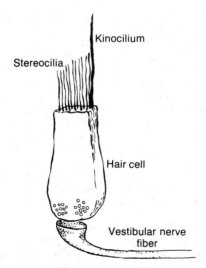

Figure 1-57. Diagram illustrating the polarized hair cells of the vestibular system. The large cilium is called the kinocilium. The smaller cilia are called stereocilia. When the stereocilia are bent toward the kinocilium, the cell depolarizes; when the stereocilia are bent away, the cell hyperpolarizes.

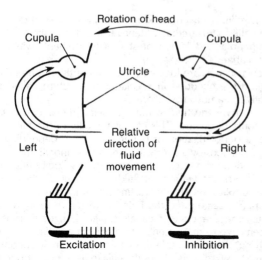

Figure 1-58. Diagram illustrating the movement of the endolymph within the semicircular canal when the head rotates to the left. The fluid lags behind the head and pushes the cupula toward the right. As indicated, the stereocilia are bent toward the kinocilium on the left and away from the kinocilium on the right. Thus, the hair cells within the left horizontal canal are stimulated and those on the right are inhibited.

speed and direction of head movement and to make the appropriate adjustments in posture and eye position.

    (b) The hair cells of the utricle and saccule are located on a mass of tissue called the **macula**.

        (i) The cilia are enmeshed in a gelatinous structure filled with small calcium carbonate crystals called **otoconia**.

        (ii) Because the otoconia are heavier than the fluid of the otolith organs, they tend to sink and, thus, can bend the cilia of the hair cells.

        (iii) When the head deviates from the horizontal position, the hair cells bend. The kinocilium of each hair cell is oriented in a different plane so that, regardless of the direction in which the head is tilted, some of the hair cells are maximally stimulated while others are maximally inhibited.

        (iv) Linear acceleration, which occurs when jumping down stairs or accelerating forward in a car, also causes the otoconia to be displaced and to stimulate the otolith organs.

  b. Neuronal pathways. The hair cells of the labyrinth synapse on the vestibulocochlear nerve (cranial nerve VIII), which projects to the vestibular nuclei within the brain stem. In addition, the vestibular nuclei receive inhibitory input from the cerebrum and cerebellum.

    (1) The vestibular nuclei send their axons into the spinal cord through a number of vestibulospinal tracts. This input is excitatory to antigravity alpha motoneurons.

    (2) If the inhibitory input from the cerebrum and cerebellum is removed, the vestibular nuclei greatly increase their firing rate and contribute to the syndrome of decerebrate rigidity.

        (a) The rigidity produced by the vestibular nuclei differs from that produced by the reticular formation in that the vestibular nuclei primarily affect the alpha motoneurons, rather than both alpha and gamma motoneurons.

        (b) Spasticity produced by the vestibular nuclei, therefore, is not reduced greatly by cutting the dorsal roots and eliminating the Ia input. This type of spasticity is called alpha rigidity to distinguish it from the type of spasticity caused by the reticular formation, which is called gamma rigidity.

        (c) When discussing the effects of the brain stem on antigravity muscles, the terms spasticity and rigidity are used interchangeably. Clinically, however, spasticity and rigidity are not alike. Spasticity refers to the condition in which the stretch reflexes of the antigravity muscles are increased due to increased activity of the alpha or gamma motoneurons. Rigidity refers to the condition seen in Parkinson's disease (see XII D 2 d), in which there is increased activity of all of the muscles at a joint.

  c. Vestibular reflexes. The vestibular system participates in a number of reflexes.

    (1) The most important of these reflexes maintains visual fixation during movement of the head.

        (a) For example, if the head is rotated to the left, the eyes move toward the right in order to prevent an image from moving off the fovea. When the eyes have rotated as far as they can, they are rapidly returned to the center of the socket. If rotation of the head continues, the eyes once again move in the direction opposite the head rotation. These movements of the eyes are called **nystagmus**.

   (b) The slow movement of the eyes to maintain visual fixation is initiated by the activity within the semicircular canals. When the head rotates to the left, the activity of receptors in the left horizontal canal causes the eyes to move toward the right.

   (c) After the body has been rotated and the movement ceases, the receptors within the right horizontal canal are stimulated, and these cause **postrotary** eye movements to occur in the opposite direction. That is, the eyes slowly move to the left until they reach the end of the socket, at which point they return quickly to the center. These movements continue until the cupula returns to its resting position.

   (d) Lesions within the vestibular pathways can cause nystagmus to occur spontaneously, while damage to the vestibular receptors can prevent nystagmus from occurring during rotation of the head.

(2) Another important reflex is initiated by receptors within the otolith organs and occurs when an individual walks down stairs or jumps from a platform.

   (a) When making such a descent, the muscles in the leg begin to contract before the feet reach the ground in order to cushion the force of impact. The otolith receptors responsible for this reflex are stimulated by the linear acceleration of the head that occurs during the descent.

   (b) Individuals lacking otolith reflexes are prone to leg injuries because of the large contact forces that occur when descending from a height (e.g., stepping off a bus).

D. The basal ganglia are interconnected nuclei within the cerebrum and do not make direct connections with the spinal cord. It appears that the major function of the basal ganglia is to aid the motor cortex in generating commands concerned with controlling proximal muscle groups during a movement. For example, when the hand is used to write on a blackboard, the large muscles of the arm and shoulders are used to hold the hand in its proper position for writing. The coordination of these muscle groups is believed to be under the control of the basal ganglia.

1. Physiologic anatomy. The basal ganglia consist of the **striatum** (which is composed of the **caudate nucleus** and **putamen**), the **globus pallidus** (also called the **pallidum**), the **subthalamic nucleus,** and the **substantia nigra**.

   a. These nuclei have **complex interconnections**.

   (1) The globus pallidus is in the center of three major feedback loops.

   (a) One pathway leaves the globus pallidus and travels through the thalamus, cortex, and striatum before returning to the globus pallidus.

   (b) Another pathway follows the same route from the globus pallidus to the thalamus, but travels to the striatum and back to the globus pallidus without going through the cortex.

   (c) The third pathway connects the globus pallidus with the subthalamic nucleus.

   (2) Another important feedback loop is established between the striatum and the substantia nigra.

   b. These pathways between the nuclei are described in a highly diagrammatic way because not enough is known about how the basal ganglia function to allow a more realistic and more detailed description.

2. Lesions in the basal ganglia produce characteristic deficits in motor behavior.

   a. Lesions in the globus pallidus result in an **inability to maintain postural support** of the trunk muscles. The head bends foward so that the chin touches the chest, and the body bends at the waist.

   (1) The motor deficits are not due to muscular weakness or failure of voluntary control because individuals with these lesions can stand upright when requested to do so.

   (2) Since the globus pallidus is the major outflow tract of the basal ganglia, it is possible that the motor deficits occur because the cortex is deprived of information it needs to automatically control the trunk muscles.

   b. Lesions in the subthalamic nucleus cause spontaneous, wild, flinging, ballistic movements of the limbs. This syndrome is called **hemiballismus**.

   (1) The movements, which are caused by a release of inhibition, appear on the side opposite the lesion.

   (2) Because the movements appear to be like those performed when an individual is thrown off balance, the subthalamic nucleus is believed to be involved with controlling the centers that issue the motor commands for balance.

          (a) Normally, the subthalamic nucleus responds to the need for initiating the balancing movement by momentarily withdrawing its inhibition from these centers.

          (b) When there is a lesion in the subthalamic nucleus, these centers no longer are under inhibitory control, and the movements are generated spontaneously.

    c. **Lesions within the striatum** produce a variety of motor syndromes related to a release of inhibition.

       (1) **Huntington's chorea** is characterized by continuous, uncontrollable, quick movements of the limbs.

       (2) In **athetosis** the limbs, fingers, and hands perform continuous, slow, irregular, twisting motions.

       (3) **Dystonia** is typified by twisting, tonic-type movements of the head and trunk.

    d. **Lesions within the substantia nigra produce Parkinson's disease**.

       (1) This disease is characterized by **rigidity, hypokinesia** (reduction in voluntary movement), and **tremor**.

          (a) The rigidity in Parkinson's disease involves all of the muscles at a joint and, thus, is different from spasticity that is associated with cortical lesions. It has been described as **lead-pipe rigidity** because the rigid limb, when moved, remains where it is placed. The rigidity seen in Parkinson's disease also has been described as **cogwheel rigidity**. When an examiner tries to move the limb, the limb periodically gives way and then reestablishes its resistance to movement like cogs on a wheel.

          (b) The hypokinesia is not due to a loss of muscle strength or power because normal movements occur under certain conditions. It also is not caused by rigidity because some cases of hypokinesia occur in the absence of any rigidity. The hypokinesia reduces the movement patterns normally associated with motor activity. For example, the arms do not swing during walking, and facial expressions do not change during conversation.

          (c) The tremor associated with Parkinson's disease occurs at a frequency of about 4–7 cycles/sec. This is slower than the normal physiologic tremor, which has a frequency of about 10 cycles/sec. The tremor in Parkinson's disease occurs at rest and usually disappears during voluntary activity.

       (2) The lesion of Parkinson's disease involves a pathway that uses **dopamine** as its neurotransmitter. Some success has been achieved in treating Parkinson's disease with L-dopa, a precursor of dopamine that can cross the blood-brain barrier. Dopamine inhibits the striatum sufficiently to reduce some of the clinical signs of the disease.

E. **The cerebellum** is intimately associated with all aspects of motor control. Its removal produces no deficits in emotional or intellectual function but causes profound disturbances in the ability to produce smooth, coordinated movements. The functions of the cerebellum appear to be related to the control of the timing, duration, and strength of a movement.

    1. **Anatomy**

      a. Two transverse fissures divide the cerebellum into three lobes: the **anterior, posterior,** and **flocculonodular lobes**. These lobes have developed at different times during evolution.

       (1) The oldest lobe, the flocculonodular lobe, is called the archicerebellum.

       (2) Next to evolve was the anterior lobe, which is called the paleocerebellum.

       (3) The last lobe to evolve was the posterior lobe, which is called the neocerebellum.

      b. Another description of the cerebellum is based on the connections it makes with other components of the motor control system.

       (1) The entire anterior lobe, and those parts of the posterior lobe that receive information from the spinal cord, are called the **spinocerebellum**.

       (2) The remainder of the posterior lobe receives its input from the cerebral cortex and, thus, is called the **cerebrocerebellum**.

       (3) The flocculonodular lobe is functionally related to the vestibular apparatus and so is also called the **vestibulocerebellum**.

    2. **Lesions** in the various divisions of the cerebellum produce characteristic motor deficits.

      a. **Lesions in the vestibulocerebellum** cause deficits related to the loss of vestibular function, principally, a loss of equilibrium and **ataxia.** Individuals with this type of lesion are unable to maintain their balance and, so, tend to fall over when standing. When walking, these individuals keep their legs far apart and stagger from place to place.

b. Lesions in the spinocerebellum have no obvious effects in humans, probably because spinocerebellar functions can be assumed by the neocerebellum. In cats, lesions in the anterior lobe cause an increase in the tone of the antigravity muscles.

c. Lesions in the cerebrocerebellum cause small motor deficits unless a wide area of the cerebellar cortex is affected. If the outflow pathways are damaged, however, the ability to produce smooth, coordinated movements is lost.

    (1) A major sign of neocerebellar disease is **decomposition of movement.** Instead of all the muscles acting in a coordinated way to produce a movement, the muscles are used one at a time. For example, when reaching for an object, the arm first extends at the shoulder, followed by extension at the elbow, and finally by the movement of the hand.

    (2) **Dysmetria** (the inability to stop a movement at the appropriate time or to direct it in the appropriate direction) is another sign of neocerebellar disease. These effects result from the loss of the neuronal circuitry required to control the duration and strength of a movement.

    (3) The **intention tremor** that results from neocerebellar lesions also is related to the inability to time and sequence movements properly. The intention tremor is different from the resting tremor (spontaneous tremor) of Parkinson's disease and appears to occur because an entire movement cannot be directed by a single motor command. Instead, the movement is directed partway to the target and then halted. Several other motor commands, each directing the movement closer to the target, are required before the movement is completed.

    (4) **Adiadochokinesia** is the inability to make rapidly alternating movements such as turning the hands back and forth. This, too, appears related to the inability to time the duration of a movement, which occurs as a consequence of lesions within the neocerebellum.

F. The cerebral cortex contains the neuronal circuits responsible for the conception and generation of motor commands. Two parallel systems of descending pathways originate in the cerebral cortex: the **pyramidal** and **extrapyramidal** systems. Clinically, these systems are considered together because lesions within the cortex almost always involve both of them. These systems are functionally different, however, and should be considered separately.

1. Physiologic anatomy

    a. Several areas within the cortex are responsible for generating motor commands.

        (1) The primary motor area occupies the frontal cortex, just in front of the central sulcus. It is **topographically organized,** with the cortical areas that control the muscles of the hands and face occupying more space than the cortical areas that control the muscles of the limbs and trunk.

        (2) The supplementary motor area is located on the medial surface of the cortex, and it, too, has a complete representation of the entire body.

        (3) The secondary motor area occupies part of the parietal lobe, just across the central sulcus from the region of the primary motor area that controls the facial muscles. The neurons in this area affect movements on both sides of the body.

    b. All of these cortical areas send projections to both of the descending motor systems.

        (1) The axons of the pyramidal system travel directly from their origin in the cortex to their destination in the spinal cord.

            (a) However, they also send collateral fibers to all areas of the motor control system and, thus, communicate their motor command to the basal ganglia, cerebellum, and brain stem.

            (b) The cortical cells that synapse directly on cranial nerves controlling facial muscles perform the same function as pyramidal tract neurons and, thus, are considered part of the pyramidal system.

            (c) The pyramidal tract system is responsible for controlling muscles that make precision movements. These include the muscles that move the fingers and hands and the muscles that produce speech.

        (2) The axons of the extrapyramidal system end on relay neurons within the brain stem. These neurons influence the spinal cord through the reticulospinal tracts.

2. Lesions

    a. Lesions in the pyramidal system cause relatively small motor deficits considering the importance of this system and the large number of neurons involved (over 1 million in each pyramid). The major deficit resulting from pyramidal tract lesions is weakness and a loss of precision in the muscles controlling fine movements of the fingers.

    b. Lesions within the extrapyramidal system produce more profound conditions. These lesions cause the release of cerebral inhibition of the reticular formation, which leads to the production of spasticity [see XII C 1 b (2)].

XIII. SLEEP AND CONSCIOUSNESS. Many of the body's regulatory mechanisms vary in their activity during the day. For example, body temperature is about 1° C higher during the early evening than it is at dawn, and adrenocortical hormones are secreted at levels that are higher in the morning than they are at night. The cycle of these **diurnal,** or **circadian, rhythms** is roughly 24 hours. However, if an individual is isolated from the environmental stimuli that indicate the normal day-night periods, the cycle time lengthens, demonstrating that the circadian rhythms are not rigidly linked to the rotation of the earth but can be driven by an individual's internal **biological clock**. It is necessary to understand these normal variations in physiologic activities when evaluating pathologic functions. For example, it would be misleading to compare fevers or blood cortisol levels obtained at different times of the day. The most obvious, and probably most important, diurnal rhythm is the sleep-wake cycle.

A. Assessment of sleep states. When awake, an individual is able to perform all activities that are required for individual and species survival (e.g., eating, drinking, learning, procreating). When asleep, an individual is not aware of the environment and is unable to perform activities that require consciousness. The presence of sleep can be assessed by behavioral analysis. An individual who does not move and does not respond when spoken to or touched probably is asleep. However, a more accurate assessment of sleep can be obtained from an **EEG**.

    1. Obtaining the EEG
        a. The EEG is obtained by placing electrodes on the scalp. The location of the electrodes and the amplification and paper speed of the polygraph used for recording the EEG are standardized. (The same is true for the EKG, which measures cardiac electrical activity; see Ch 2 III C.)
        b. The brain waves recorded by an EEG represent the summated activity of millions of cortical neurons. The IPSPs, EPSPs, and the passive spread of electrical activity into the dendrites of these neurons, rather than their action potentials, form the basis for the EEG.

    2. Variations in the EEG during sleep and wakefulness. The two states of sleep—slow-wave and fast-wave sleep—have characteristic EEG patterns.
        a. When an individual is awake, the electrical activity recorded from the brain is asynchronous and of low amplitude (Figure 1-59). This type of brain electrical recording is called a **beta wave**.
        b. If an individual sits quietly for a while, the brain waves gradually become larger and highly synchronized. The typical resting EEG pattern has a frequency of 8–13 cycles/sec and is called an **alpha wave**. When the eyes open or when conscious mental activity is initiated, the EEG shifts from an alpha to a beta pattern. This is called **alpha blocking**.
        c. As consciousness is reduced still further, an individual enters a state of sleep called **slow-wave sleep (SWS)**. SWS progresses in an orderly way from light to deep sleep (see Figure 1-59).

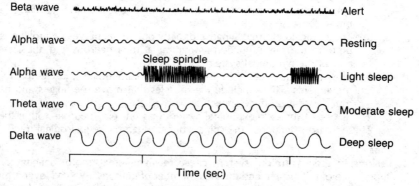

Figure 1-59. As an individual passes from wakefulness to deep sleep, the EEG wave increases in amplitude and decreases in frequency. Sleep spindles indicate the presence of light sleep.

(1) Light sleep is characterized by an EEG that shows high-amplitude waves of about 12–15 cycles/sec, called **sleep spindles,** which periodically interrupt the alpha rhythm.

(2) During moderate sleep, the EEG displays slower and larger waves called **theta waves.**

(3) The deepest stage of SWS produces an EEG pattern with very slow (4–7 cycles/sec) and large waves called **delta waves.**

(4) Behaviorally, SWS is characterized by a progressive reduction in consciousness and an increasing resistance to being awakened. Muscle tone is reduced, the heart and respiratory rates are decreased, and, in general, body metabolism is slowed.

  d. The other type of sleep, called **fast-wave** or **desynchronized sleep,** is characterized by the same high-frequency and low-amplitude EEG pattern that is seen in the waking state. In this case, however, the individual clearly is unresponsive to environmental stimuli and, thus, is asleep. For this reason, fast-wave sleep also is called **paradoxical sleep.**

    (1) Because this state of sleep is characterized by the presence of rapid eye movements, it also is called **rapid eye movement (REM) sleep.**

    (2) Because dreaming occurs during REM sleep, it is also called **dream sleep.**

    (3) Behaviorally, REM sleep is quite different from SWS.

      (a) It is as difficult to arouse an individual from REM sleep as it is from deep sleep. However, when awakened from REM sleep, the individual is immediately alert and aware of the environment.

      (b) The eyes are not the only organ that is active during REM sleep. The middle ear muscles are active, penile erection occurs, heart rate and respiration become irregular, and there are occasional twitches of the limb musculature. Fortunately, muscle tone also is reduced tremendously during REM sleep so the frequency and intensity of muscle twitching do not produce injuries or awaken the individual.

B. The sleep cycle. There is an orderly progression of sleep stages and states during a typical sleep period (Figure 1-60).

  1. When an individual falls asleep, the light stage of SWS is entered first. During the next hour or so, the individual passes into progressively deeper stages of sleep until deep sleep is reached. After about 15 minutes of deep sleep, the depth of sleep starts to decrease and continues to do so until the individual reenters the light stage of sleep (about 90 minutes after the start of the first sleep cycle). At this point, the individual passes from SWS to REM sleep.

  2. This cycle repeats itself about five times during the night. However, as can be observed (see Figure 1-60), after the second cycle, the intervals between periods of REM sleep become shorter and the duration of each period of REM sleep becomes longer. Also, as morning approaches, an individual spends less time in the deeper stages of SWS and periodically awakens.

  3. The sleep cycle indicated (see Figure 1-60) is typical of an adult. The cycle varies greatly with age.

    a. During infancy, about 16 hours of every day are spent asleep. This drops to 10 hours during childhood and to 7 hours during adulthood. Elderly individuals spend less than 6 hours of each day sleeping.

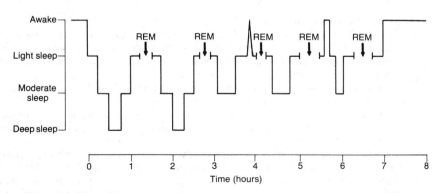

Figure 1-60. Diagram indicating the pattern of sleep during one sleep cycle. As the evening progresses, the depth of slow-wave sleep decreases and the duration and frequency of rapid eye movement (*REM*) sleep episodes increase. Note that occasional periods of wakefulness occur during the night.

b. It is interesting to note that prematurely born infants spend about 80% of their sleep time in REM sleep, whereas full-term infants spend about 50% (8 hours) in REM sleep. The total time spent in REM sleep is reduced to about 1.5–2 hours by puberty and remains unchanged thereafter.

c. During infancy and childhood, therefore, the reduction in sleep time from 16 to 10 hours occurs almost entirely by a reduction of the amount of time spent in REM sleep. In adulthood, the reduction in sleep time is caused by a reduction in the time spent in the deep stages of SWS.

C. Physiologic basis for sleep. Areas throughout the entire brain participate in the sleep-wake cycle.

1. **The waking state** is maintained by the ascending reticular activating system, which is a diffuse collection of neurons within the medulla, pons, midbrain, and diencephalon. Electrical stimulation anywhere within this area causes the EEG pattern to change abruptly from that of the sleep state to that of the waking state. This is called a **cortical alerting** or **arousal** response.

2. **The sleep state** does not result from the passive withdrawal of arousal. Two sleep centers exist in the brain stem; one is responsible for producing SWS, and the other produces REM sleep.
   a. The SWS center is located in a midline area of the medulla containing the **raphé nuclei.**
      (1) The neurons within these nuclei use serotonin (5-hydroxytryptamine) as a neurotransmitter.
      (2) Administration of serotonin directly into the cerebral ventricles of experimental animals induces a state of SWS, whereas lesions in this region induce a permanent state of insomnia.
   b. The REM sleep center is located in specific nuclei of the pontine reticular formation, including the **locus ceruleus,** which uses norepinephrine as a neurotransmitter. Lesions within this area eliminate the electrophysiologic and behavioral signs of REM sleep.

3. **Sleep disorders.** As noted previously, there is a cycling between SWS and REM sleep during a normal sleep period.
   a. In **narcolepsy,** REM sleep is entered directly from the waking state.
      (1) Individuals suffering from narcolepsy often report an intense feeling of sleepiness just prior to an attack, although sleep sometimes occurs without warning.
      (2) In some narcoleptics, the profound reduction in muscle tone characteristic of REM sleep can occur without loss of consciousness. During such an attack, called **cataplexy,** the individual suddenly becomes paralyzed, falls to the ground, and is unable to move.
      (3) Another symptom associated with narcolepsy is the presence of a dream-like state during wakefulness, which narcoleptics describe as a hallucination.
   b. Most of the other symptoms of sleep disorders are associated with SWS. These include sleepwalking (**somnambulism**), bed-wetting (**nocturnal enuresis**), and nightmares (**pavor nocturnus**), all of which occur during stages of SWS.
      (1) During a nightmare that occurs in SWS sleep, the individual wakes up screaming and appears terrified. However, no reason for the acute anxiety is recalled.
      (2) By contrast, terrifying dreams that occur during REM sleep are graphically remembered.

4. **Disturbances of consciousness**
   a. A lesion blocking the connection between the ascending reticular activating system and the thalamus produces a permanent state of sleep, or **coma.** In this situation, stimulation of sensory pathways can cause a momentary desynchronization of the EEG but does not produce any behavioral signs of arousal.
      (1) Coma is not simply a deep sleep state. It is characterized by a loss of consciousness from which arousal cannot be elicited.
      (2) Oxygen consumption by the brain is reduced during coma. This is in marked contrast to normal sleep, in which there is no change in brain oxygen consumption from the waking state.
   b. A transient pathologic loss of consciousness is called **syncope** (fainting). More persistent losses of consciousness are called **stupor,** from which arousal can be obtained.
   c. **Brain death** occurs when the brain no longer can achieve consciousness. Because of the desire to obtain organs for transplant operations and the desire to remove heroic life support systems, an objective standard for determining the presence of brain death has been developed.

(1) Brain death is said to occur when a loss of consciousness is accompanied by a flat EEG (i.e., an EEG with no brain waves) and a loss of all brain stem regulatory systems (e.g., those systems that control respiration and blood pressure).
(2) Moreover, these clinical signs must be due to traumatic or ischemic anoxia and not to hypothermia or metabolic poisons, from which later recovery is possible.
(3) Finally, the criteria for brain death must be present for 6–12 hours.

STUDY QUESTIONS

Directions: Each of the numbered items or incomplete statements in this section is followed by answers or by completions of the statement. Select the **one** lettered answer or completion that is **best** in each case.

1. If the flux of glucose across a cell membrane is directly proportional to the concentration gradient of glucose, the transport process is

(A) simple diffusion
(B) facilitated diffusion
(C) Na^+-coupled cotransport
(D) active transport

2. The diffusion of lipid-soluble particles across a membrane would increase with an increase in any of the following factors EXCEPT

(A) the temperature of the solution
(B) the area of the membrane
(C) the solubility of the particles in the membrane
(D) the size of the particles
(E) the concentration gradient of the particles

3. ATP is used directly in which of the following transport processes?

(A) Transport of glucose into a red blood cell
(B) Transport of Ca^{2+} out of a cell by Na^+-Ca^{2+} exchange
(C) Transport of K^+ out of a cell during an action potential
(D) Removal of norepinephrine from the synaptic cleft by the presynaptic nerve terminal
(E) Removal of Ca^{2+} from the cytoplasm by the sarcoplasmic reticulum

4. A red blood cell is placed in a saline (NaCl) solution, and the cell's volume increases to 1.5 times its original volume. This finding indicates that the Na^+ concentration in the saline solution is approximately

(A) 50 mEq/L
(B) 75 mEq/L
(C) 100 mEq/L
(D) 200 mEq/L
(E) 600 mEq/L

5. A decrease in the extracellular K^+ concentration will

(A) increase the transport of Na^+ out of a cell by the Na^+-K^+-ATPase active transport system
(B) increase the flow of Na^+ into a cell down its electrochemical gradient
(C) both
(D) neither

6. Plasma proteins, which have a concentration of 1.2 mmol/L and an oncotic pressure of 24 mm Hg, normally can prevent the net flow of fluid out of the capillaries. What concentration of glucose would be required to prevent the flow of water out of the capillaries, if the reflection coefficient of glucose is 0.8?

(A) 0.96 mmol/L
(B) 1.20 mmol/L
(C) 1.50 mmol/L
(D) 19.60 mmol/L
(E) 30.00 mmol/L

Questions 7–8

Use the following equilibrium potentials (E_K and E_{Na}) and conductances (G_K and G_{Na}) to answer the next two questions.

$$E_K = -90 \text{ mV} \quad G_K = 8 \text{ conductance units}$$
$$E_{Na} = +60 \text{ mV} \quad G_{Na} = 2 \text{ conductance units}$$

7. Based on these values, the resting membrane potential is approximately

(A) –60 mV
(B) –70 mV
(C) –75 mV
(D) –80 mV
(E) –90 mV

8. If the given G_K were decreased by half while the G_{Na} remained the same, what effect would this have on the membrane potential?

(A) It would depolarize by 20 mV
(B) It would depolarize by 10 mV
(C) It would hyperpolarize by 10 mV
(D) It would hyperpolarize by 20 mV
(E) None

9. A cell at the threshold for firing an action potential has Na$^+$ channels with

(A) most of the m gates open
(B) most of the h gates open
(C) both
(D) neither

10. Which phase of the action potential is caused by inactivation of Na$^+$ channels in a nerve axon?

(A) Upstroke
(B) Absolute refractory period
(C) Downstroke
(D) Undershoot phase
(E) Relative refractory period

11. The maximum firing rate of an axon is determined primarily by the

(A) number of Na$^+$ channels in the axon
(B) rate at which Na$^+$ channel m gates open during the upstroke
(C) rate at which Na$^+$ channel h gates open during the downstroke
(D) number of K$^+$ channels in the axon
(E) rate at which K$^+$ channel n gates open during the upstroke

12. What would be the result of removing the inactivation gate from the Na$^+$ channels of a nerve axon?

(A) Action potentials could not be elicited
(B) The rate of repolarization during the action potential would decrease
(C) Na$^+$ conductance would decrease
(D) The membrane would hyperpolarize

13. The propagation velocity of an action potential would be increased by

(A) an increase in the rate of depolarization during the upstroke of the action potential
(B) an increase in the diameter of the axon
(C) both
(D) neither

14. Which statement below correctly describes the channels located on the end-plate membranes of skeletal muscle?

(A) The channels are activated by ACh
(B) The channels are primarily permeable to Na$^+$
(C) When all of the postsynaptic channels are open, the membrane potential becomes positive
(D) The number of open channels increases when the membrane depolarizes

15. Which ionic channel would produce the greatest amount of membrane depolarization when it is opened by a neurotransmitter?

(A) A channel specifically permeable to Ca^{2+}
(B) A channel equally permeable to K$^+$ and Na$^+$
(C) A channel specifically permeable to K$^+$
(D) A channel equally permeable to K$^+$, Na$^+$, and Ca^{2+}
(E) A channel specifically permeable to Na$^+$

16. Which type of receptor is activated by ACh and blocked by atropine?

(A) Nicotinic
(B) Muscarinic
(C) Beta
(D) Alpha
(E) Noncholinergic, nonadrenergic

17. Ca^{2+} is released from the sarcoplasmic reticulum by

(A) diacylglycerol
(B) G protein
(C) phospholipase C
(D) inositol triphosphate
(E) adenylate cyclase

18. When activated by β-adrenergic receptors, the G protein

(A) activates phospholipase C
(B) activates adenylate cyclase
(C) activates protein kinase C
(D) converts GDP to GTP
(E) stimulates the release of Ca^{2+} from the sarcoplasmic reticulum

19. Action potentials are initiated on the axon hillock of CNS neurons because

(A) a large number of excitatory nerve terminals synapse on the axon hillock
(B) inhibitory nerve terminals are located far from the axon hillock
(C) both
(D) neither

20. Which of the following substances is important for skeletal muscle contraction but not for smooth muscle contraction?

(A) Actin
(B) Myosin-ATPase
(C) Troponin
(D) Myosin
(E) Ca^{2+}-ATPase

21. The twitch-to-tetanus ratio (i.e., the ratio of a force developed during a twitch to the force developed during a tetanus) is greater in a slow-twitch fiber than in a fast-twitch fiber because the

(A) activity of the sarcoplasmic reticular Ca^{2+} pump is less in slow-twitch fibers than in fast-twitch fibers
(B) amount of Ca^{2+} released from the sarcoplasmic reticulum is greater in slow-twitch fibers than in fast-twitch fibers
(C) amount of ATP used during a contraction is less in slow-twitch fibers than in fast-twitch fibers
(D) myosin-ATPase activity is greater in slow-twitch fibers than in fast-twitch fibers
(E) density of thick and thin filaments is greater in slow-twitch fibers than in fast-twitch fibers

22. The bones of the middle ear are primarily responsible for

(A) amplifying the sound waves reaching the ear
(B) detecting the presence of a sound stimulus
(C) locating the source of a sound
(D) discriminating among different frequencies of sound
(E) adapting to a prolonged monotonous sound

23. Voltage-sensitive channels are most important for the

(A) release of transmitter from nerve terminals during synaptic transmission
(B) postsynaptic depolarization of the end-plate on skeletal muscle
(C) release of Ca^{2+} from the sarcoplasmic reticulum of cardiac muscle cells
(D) transmission of information within the retina
(E) response of cutaneous mechanoreceptors to deformation of the skin

24. Which of the following is a feature of both the stretch reflex and the withdrawal reflex?

(A) Irradiation
(B) Afterdischarge
(C) Small afferent fibers
(D) Multiple spinal synapses
(E) Reciprocal innervation

25. In comparison to motor units that fire later during a movement, motor units that fire at the beginning of a movement

(A) can be tetanized at a lower frequency of stimulation
(B) generate a greater amount of force
(C) have a greater amount of glycogen stored within them
(D) fatigue more rapidly
(E) are innervated by larger alpha motoneurons

26. In which of the following sensory systems is the transducer and spike generator located on the same cell?

(A) Taste
(B) Vision
(C) Olfaction
(D) Hearing

27. Auditory hair cells and taste receptor cells are alike in which of the following ways?

(A) Both release neurotransmitter when stimulated
(B) Both can regenerate when damaged
(C) Both respond to a stimulus by opening K^+-selective channels
(D) Both have very high resting membrane potentials

28. Reciprocal innervation is most accurately described as

(A) inhibition of flexor muscles during an extension
(B) activation of contralateral extensors during a flexion
(C) reduction of Ia fiber activity during a contraction
(D) simultaneous stimulation of alpha and gamma motoneurons
(E) inhibition of alpha motoneurons during a contraction

29. True statements about the initial stages of a normal sleep period include which of the following?

(A) Sleep is initiated by a decrease in the activity of the ascending reticular activating system
(B) The first EEG sign of sleep is an increase in the amplitude of the EEG waves
(C) Both
(D) Neither

30. Which of the following conditions is most closely related to slow-wave sleep?

(A) Dreaming
(B) Atonia
(C) Bed wetting
(D) High-frequency EEG waves
(E) Irregular heart rates

31. A person with normal vision has a total refractive power (without accommodation) of 60 diopters (D). In order to focus an object placed 25 cm from the eye, the refractive power of the lens must increase by approximately

(A) 1 D
(B) 2 D
(C) 4 D
(D) 5 D
(E) 10 D

32. The variation in auditory threshold as a function of frequency (the minimum audibility curve) is related most to the properties of the

(A) outer ear
(B) auditory canal
(C) middle ear
(D) tympanic membrane
(E) basilar membrane

Directions: Each question below contains four suggested answers of which **one or more** is correct. Choose the answer.

A if **1, 2, and 3** are correct
B if **1 and 3** are correct
C if **2 and 4** are correct
D if **4** is correct
E if **1, 2, 3, and 4** are correct

33. The osmotic flow of water across a membrane will be increased by increasing

(1) the reflection coefficient of the membrane
(2) the concentration of particles on one side of the membrane
(3) the temperature of the solution
(4) the permeability of the membrane

34. A Donnan equilibrium between the plasma and interstitium is characterized by

(1) greater osmotic pressure in the plasma than in the interstitium
(2) more negative electrical potential in the plasma than in the interstitium
(3) greater concentration of diffusible cations in the plasma than in the interstitium
(4) greater concentration of diffusible anions in the plasma than in the interstitium

35. An action potential elicited during the relative refractory period will

(1) require a greater stimulus than normal to reach threshold
(2) have a smaller than normal overshoot potential
(3) depolarize more slowly than normal
(4) repolarize more slowly than normal

36. An increase in the afterload in an isotonically contracting muscle would produce which of the following effects?

(1) A decrease in the rate of cross-bridge cycling
(2) A decrease in the amount of stretch on the series elastic component
(3) An increase in the force of contraction
(4) An increase in the amount of shortening

37. Compared to cones, rods can detect light at lower intensities because

(1) rods are larger than cones
(2) rods can increase their sensitivity to light more quickly than cones
(3) rods have larger receptive fields than cones
(4) rods are more concentrated than cones in the fovea

38. When light strikes a visual receptor cell (a rod or cone), it causes which of the following reactions?

(1) An increase in the intracellular cGMP concentration
(2) A decrease in the conductance of the cell
(3) Release of an inhibitory neurotransmitter from the cell
(4) Hyperpolarization of the cell

39. Cerebellar lesions produce which of the following motor signs?

(1) Adiadochokinesia
(2) Dysmetria
(3) Ataxia
(4) Spontaneous tremor

40. Movement disorders related to the removal of inhibition are produced by which of the following components of the motor control system?

(1) Striatum
(2) Internal capsule
(3) Substantia nigra
(4) Medullary pyramids

Directions: Each group of items in this section consists of lettered options followed by a set of numbered items. For each item, select the **one** lettered option that is most closely associated with it. Each lettered option may be selected once, more than once, or not at all.

Questions 41–44

The diagram below illustrates the anatomic arrangement for the presynaptic inhibition of an excitatory neuron. Match each of the following descriptions with the appropriate lettered area of the diagram.

Question 45–49

Match each of the following descriptions with the appropriate lettered point on the action potential shown below.

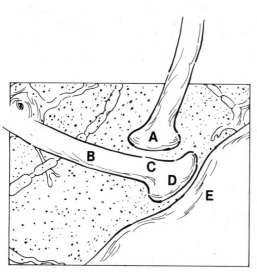

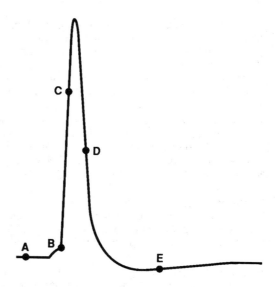

41. A decrease in the entry of Ca^{2+}
42. The release of GABA by exocytosis
43. A decrease in the size of the excitatory postsynaptic potential
44. An increase in the Cl^- conductance

45. The point at which K^+ conductance is highest
46. The point contained within the absolute refractory period
47. The point at which Na^+ conductance is highest
48. The point at which the net flow across the membrane is zero
49. The point at which the driving force for K^+ is greatest

ANSWERS AND EXPLANATIONS

1. The answer is A. [*I C 1 a*] In simple diffusion, the flux is directly proportional to the concentration gradient. Carrier-mediated transport differs from simple diffusion in that it displays saturation kinetics. Facilitated diffusion, active transport, and Na^+-coupled transport all are examples of carrier-mediated transport.

2. The answer is D. [*I C 1 a (3), (5)*] The diffusion of lipid-soluble materials obeys Fick's law of diffusion, which states that diffusion is directly proportional to the diffusion coefficient, the area of the membrane, and the concentration gradient. In general, the diffusion coefficient increases as the solubility of the material in the membrane increases and as the temperature increases, and it decreases as the size of the particle increases. Thus, diffusion decreases as the size of the particle increases.

3. The answer is E. [*IV A 2 d*] Ca^{2+} is transported into the sarcoplasmic reticulum by the Ca^{2+}-ATPase (calcium-adenosine triphosphatase) pump. The removal of norepinephrine from the synaptic cleft and the transport of Ca^{2+} are Na^+-dependent, indirect-energy-utilizing, active transport systems. K^+ flows down its electrochemical gradient during an action potential. Glucose is transported into red blood cells by facilitated diffusion.

4. The answer is C. [*I C 3 c, d*] The steady state volume of a cell placed in a solution containing an osmotic concentration different from the cell's can be calculated using the equation: $\pi_1 V_1 = \pi_2 V_2$. The osmolality of normal body fluids is approximately 285 mOsm. In this problem, a red blood cell is placed in a solution of unknown osmolality and swells to 1.5 times its original volume. Solving the equation for the osmolality of the extracellular fluid (π_2) yields 285/1.5, or 190 mOsm. A saline solution of this osmolality will have a Na^+ concentration of 95 mEq/L.

5. The answer is B. [*I C 2 a; II D 1–4*] A decrease in extracellular K^+ concentration will hyperpolarize a cell, increasing the driving force for Na^+. According to Ohm's law for solutions: $I_{Na} = G_{Na} (E_m - E_{Na})$, increasing the driving force ($E_m - E_{Na}$) will increase the flow of Na^+ current into the cell. The rate of pump activity will be lowered if K^+ drops below 1 mmol/L; thus, lowering the extracellular K^+ concentration will, if anything, lower the activity of the Na^+-K^+ pump.

6. The answer is C. [*I C 3 e, f*] According to the osmotic flow equation, the flow of water is proportional to the osmotic pressure difference across the membrane. Since glucose has a reflection coefficient of 0.8, a concentration difference of 1.5 mmol/L (1.2/0.8) would be required to equal the osmotic pressure of 1.2 mmol/L of plasma protein, which has a reflection coefficient of 1.

7–8. The answers are: 7-A, 8-A. [*II D 4*] The membrane potential can be calculated using the transference equation, if equilibrium potentials and transferences for the permeable ions are known. In this problem, the equilibrium potentials (E_K and E_{Na}) are known, and the transferences (T_K and T_{Na}) can be calculated from the conductances (G_K and G_{Na}), since $T_{ion} = G_{ion}/G_{total}$. Thus, the transference for K^+ is 8/10 (0.8), and the transference for Na^+ is 2/10 (0.2). Applying the transference equation, the resting membrane potential is calculated as

$$E_m = E_K T_K + E_{Na} T_{Na}$$
$$= (-90 \times 0.8) + (+60 \times 0.2)$$
$$= -60 \text{ mV}$$

If G_K were decreased by half while G_{Na} remained the same, then T_K would become 4/6 (0.67) and T_{Na} would become 2/6 (0.33). Using these values and applying the transference equation, the membrane potential is calculated as

$$E_m = E_K T_K + E_{Na} T_{Na}$$
$$= (-90 \times 0.67) + (+60 \times 0.33)$$
$$= -40 \text{ mV}$$

Thus, the membrane potential would depolarize by 20 mV (i.e., from -60 mV to -40 mV).

9. The answer is B. [*II C 2 a, 3 c*] When a cell is at rest, the m gates are closed and the h gates are open. At threshold, a few (not most) of the m gates begin to open, allowing some Na^+ to enter the cell. The flow of Na^+ into the cell is responsible for the upstroke of the action potential. The majority of m gates open during the upstroke. The h gates remain open at threshold and during most of the upstroke.

10. The answer is B. [*II C 4 a, b*] As long as the Na$^+$ channels are inactivated by the h gates, another action potential cannot be generated and the axon will remain in the absolute refractory period. The h gates begin to open during the downstroke, and most of them are open during the relative refractory period.

11. The answer is C. [*II C 4 a (2)*] The maximum firing rate is directly proportional to the duration of the absolute refractory period, which is related to the rate at which Na$^+$ channels recover from inactivation (i.e., the rate at which the h gates reopen) at the end of the action potential.

12. The answer is B. [*II C 3 c (2)*] The closing of the inactivation (h) gate is, in part, responsible for the repolarization phase of the action potential. If the h gate were removed, the rate of repolarization would be slowed. Although some of the Na$^+$ channels are inactivated at rest, removing them would have little or no effect on the Na$^+$ conductance. However, if the Na$^+$ conductance did increase slightly, the membrane would depolarize, and there might be a negligible increase in K$^+$ conductance. Action potentials could still be elicited, because the activation gates on the Na$^+$ channels would be intact.

13. The answer is C. [*II C 5*] The speed of action potential propagation is determined by the rate at which contiguous areas of the membrane are depolarized to threshold. An increase in the rate of depolarization during the upstroke is accompanied by an increase in inward current and a more rapid depolarization of contiguous areas. Similarly, increasing the diameter of the axon will increase the flow of current along the axon, which, in turn, will increase the rate of depolarization.

14. The answer is A. [*III A 3 b*] The end-plate channel on skeletal muscle is activated by acetylcholine (ACh). The channel is equally permeable to Na$^+$ and K$^+$ and is not opened when the membrane depolarizes. If all of the ACh channels were to open, the membrane would depolarize to –16 mV, which is the reversal potential for the channel.

15. The answer is A. [*III A 3 b (4); Table 1-2*] The reversal potential for a synaptic channel can be calculated using the transference equation. When a synaptic channel is opened by a neurotransmitter, the membrane potential moves toward the reversal potential for that channel. The reversal potentials for the channels given in the question are as follows: Ca^{2+} = +130 mV; Na$^+$ and K$^+$ = –16 mV; K$^+$ = –92 mV; K$^+$, Na$^+$, and Ca^{2+} = 33 mV; and Na$^+$ = 60 mV. Since the Ca^{2+} channel has the most positive reversal potential, it would produce the greatest amount of depolarization.

16. The answer is B. [*III B 2 c*] Although both muscarinic and nicotinic receptors are activated by ACh, only muscarinic receptors are blocked by atropine. Nicotinic receptors are blocked by curare. Alpha and beta receptors are adrenergic receptors; they are activated by norepinephrine and epinephrine, not by ACh.

17. The answer is D. [*III B 2 a (3)*] Inositol triphosphate (IP$_3$) causes Ca^{2+} to be released from the sarcoplasmic reticulum. IP$_3$ and diacylglycerol are formed from phosphoinositol diphosphate by phospholipase C, which is activated by the G proteins that are activated by muscarinic and α_1-adrenergic receptors.

18. The answer is B. [*III B 3 a*] When the G protein is activated by β-adrenergic receptors, the result is the activation of adenylate cyclase, which, in turn, converts adenosine triphosphate (ATP) to cyclic adenosine 3′,5′-monophosphate (cAMP). cAMP increases the activity of the sarcoplasmic reticular Ca^{2+} pump. The G protein has intrinsic guanosine triphosphatase (GTPase) activity and, thus, is able to hydrolyze guanosine triphosphate (GTP).

19. The answer is D. [*III C 2 b*] Action potentials are initiated on the axon hillock because it has a high concentration of the electrically excitable Na$^+$ and K$^+$ channels responsible for producing an action potential. There are very few, if any, excitatory synaptic channels on the axon hillock. Inhibitory synaptic channels tend to be located on the cell body, close to the axon hillock, where they can most easily inhibit the generation of an action potential.

20. The answer is C. [*IV A 2 b, d, C 2*] In skeletal muscle, contraction is initiated when Ca^{2+} binds to troponin. Smooth muscle contraction is initiated by the phosphorylation of the myosin light-chain proteins.

21. The answer is A. [*IV A 3 a (4); XII B 1 c (1), (2)*] The force developed during a twitch is related to the duration of the active state (i.e., the time during which Ca^{2+} concentrations are elevated). If Ca^{2+} remains in the cytoplasm longer, cross-bridge cycling continues for a longer time, the series elastic component is stretched further, and more force develops. More force is developed during a tetanic contraction because the active state is prolonged by repetitive stimulation. However, because the sarcoplasmic reticulum of slow-twitch fibers removes Ca^{2+} from the cytoplasm so slowly (i.e., its active state lasts for a long time), the force produced during a twitch is almost as great as the force produced during a tetanic contraction.

22. The answer is A. [*VII B 2 b*] When sound waves pass from air to water, most of the sound energy is lost. By amplifying the sound, the middle ear bones make it possible for the cochlea to detect sounds, even though the auditory sensory apparatus is within a fluid medium.

23. The answer is A. [*III A 3 a*] The release of synaptic transmitter occurs when voltage-sensitive Ca^{2+} channels open in response to the depolarization of the presynaptic nerve terminal by an action potential.

24. The answer is E. [*XI A 1 b, B 1 b*] The motor control system is organized so that neurons causing excitation of one muscle group will inhibit the alpha motoneurons innervating the antagonist muscle groups. This type of reciprocal innervation does not prevent agonist and antagonist muscle fibers from being coactivated. However, in such cases, the reciprocal inhibition must be overcome. Both the stretch reflex and the withdrawal reflex are characterized by reciprocal innervation.

25. The answer is A. [*XII B 2 b (2) (b)*] Smaller motoneurons are generally activated before larger motoneurons. These neurons have slower speeds of contraction and longer active states. Thus, they can be tetanized at lower frequencies of stimulation.

26. The answer is C. [*V B 1–2; X C 1*] The receptors for olfaction and the spike generator both are found on the end of the bipolar olfactory nerve (cranial nerve I) fibers. In the other sensory cells, the receptors are located on modified epithelial cells, and the spike generator is located on a cranial nerve fiber. The sensory receptor cell communicates with the cranial nerve fiber by synaptic transmission.

27. The answer is A. [*V B 1–2; VII C 3; IX B 3*] Both auditory hair cells and taste receptor cells are modified epithelial cells that communicate with their cranial nerve fibers by synaptic transmission. Hair cells (but not taste cells) have a very high resting potential because they are surrounded by an extracellular fluid (endolymph), which contains a high concentration of K^+. When hair cells are stimulated, K^+-selective channels are opened, allowing K^+ to enter the cell. Taste cells are able to regenerate when damaged but auditory hair cells cannot.

28. The answer is A. [*XI A 1 b (2)*] Reciprocal innervation is most accurately described as inhibition of the antagonist muscle when the agonist muscle is activated. For example, flexor muscles are inhibited during an extension. Reciprocal innervation allows extensor contraction to occur without interference from the flexor muscles that are being stretched during the movement. Under normal circumstances, stretching the flexors will elicit a stretch reflex leading to contraction of the flexor muscles. Inhibiting the alpha motoneurons that innervate the flexor muscles prevents the stretch reflex from interfering with the extension. Reciprocal innervation characterizes all movements, not just extension.

29. The answer is B. [*XIII A 2, C 1, 2*] Sleep is initiated by the activity of sleep centers within the brain stem. Although the ascending reticular activating system (ARAS) is inhibited during this time (and wakefulness cannot occur if the ARAS is damaged), inhibition of the ARAS is not the cause of sleep onset. Under normal circumstances, an individual falling asleep enters slow-wave sleep (SWS), which is characterized by low-frequency, high-amplitude EEG waves.

30. The answer is C. [*XIII C 3 b*] Bed wetting is a sleep disorder associated with SWS. Rapid eye movement (REM) sleep is characterized by high-frequency, low-amplitude EEG waves and is associated with dreaming, atonia, and irregular heart rate and breathing patterns.

31. The answer is C. [*VIII A 1 C (3)*] The total refractive power (P) required to focus an object can be calculated using the lens formula, which defines the relationship between object (o), image (i), and focal (f) distances as

$$\frac{1}{o} + \frac{1}{i} = \frac{1}{f} = P$$

In this problem, the object distance is 0.25 m, the image distance equals the focal length of the relaxed eye (i.e., 1/60, or 0.0167 m), and the total power of the lens required to focus the object clearly is equal to 1/f after accommodation for near vision has occurred. Solving the equation for P gives

$$\frac{1}{0.0167} + \frac{1}{0.25} = \frac{1}{f} = P$$

Thus, the lens must increase its refractive power by about 4 D (i.e., from 60 D to 63.8 D).

32. The answer is C. [*VII B 2 c*] The middle ear transfers sounds from the tympanic membrane to the cochlea. However, it does not transfer all sounds equally well. For example, sounds in the frequency range used for speech are transferred more efficiently than those at the low and high ends of the hearing range.

33. The answer is A (1, 2, 3). [*I C 3 e, f*] The osmotic flow of water equals $\sigma \times L \times A \times (\pi_1 - \pi_2)$. The osmotic pressure ($\pi$) is dependent on the number of particles in solution and the temperature of the solution, so that increasing either of these increases the osmotic pressure gradient and, thus, the osmotic flow of water. Increasing the reflection coefficient (σ) also increases the osmotic flow of water. However, increasing the permeability of the membrane to the particles in solution, decreases σ and, thus, decreases the flow of water.

34. The answer is A (1, 2, 3). [*I D*] A Donnan equilibrium is established between the plasma and interstitium because the plasma contains negatively charged, nondiffusible proteins. When equilibrium is established, the plasma has a higher osmolarity and is negatively charged compared to the interstitium. The negative charge increases the concentration of cations and decreases the concentration of anions in the plasma relative to the interstitium.

35. The answer is E (all). [*II C 4 b*] During the relative refractory period, K^+ conductance is higher than normal so a greater stimulus is required to depolarize the nerve membrane to threshold. Also, because of the higher than normal K^+ conductance, the overshoot potential and rate of depolarization during the upstroke are smaller. Because the overshoot potential is smaller, less n gates open, and, consequently, repolarization occurs more slowly.

36. The answer is B (1, 3). [*IV A 3 b*] In an isotonically contracting muscle, the force of contraction is equal to the afterload. Thus, increasing the afterload increases the force of contraction. Also, when the load on the muscle increases, the rate of cross-bridge cycling decreases. The stretch of the series elastic component is proportional to the force developed by the muscle, so increasing the afterload increases the stretch of the series elastic component.

37. The answer is B (1, 3). [*VIII B 1 b (4) (c)*] The higher sensitivity of rods in low light occurs because rods absorb more light than cones (because rods are larger than cones) and because their receptive fields are larger (more rods than cones converge on each ganglion cell). Although rods can increase their sensitivity to a much greater extent than cones, they adapt more slowly when placed in the dark. There are no rods in the fovea.

38. The answer is C (2, 4). [*VIII B 3 a*] When light strikes a photoreceptor cell, it initiates a series of reactions that cause the conductance of the cell to decrease and the membrane potential to hyperpolarize. Hyperpolarization reduces the amount of transmitter released by the cell. The chemical reactions initiated by the absorption of light results in the conversion of cyclic guanosine monophosphate (cGMP) to GMP. The decrease in intracellular cGMP concentration is responsible for the closing of Na^+ channels and the decrease in membrane conductance.

39. The answer is A (1, 2, 3). [*XII E 2*] Cerebellar lesions cause adiadochokinesia (the inability to make rapidly alternating movements), dysmetria (the inability to stop a movement at the appropriate time or to direct it in the appropriate direction), ataxia (failure of muscular coordination), and intention tremor (tremor that occurs when the limb is moved). Spontaneous tremor (tremor that occurs when the patient is at rest) is not observed.

40. The answer is A (1, 2, 3). [*XII D 2, F 2*] The striatum and substantia nigra are part of the basal ganglia. Lesions to these nuclei produce spontaneous movements due to the removal of their normal inhibitory effects. Lesions in the internal capsule damage fibers from the motor cortex that normally inhibit postural tone. Removal of this inhibitory effect results in spasticity. The medullary pyramids contain axons that originate from neurons in the motor cortex and terminate in spinal motoneuron pools. These neurons are normally excitatory, and, thus, when they are damaged, motor tone is reduced.

41–44. The answers are: 41-D, 42-A, 43-E, 44-C. [*III C 3; Figure 1-21*] In presynaptic inhibition, an inhibitory neuron forms an axo-axonic synapse with the neuron that is excitatory to the alpha motoneuron. The inhibitory neuron releases γ-aminobutyric acid [GABA] (A), which opens Cl^- channels on the postsynaptic membrane (C). The action potential invading the nerve terminal is reduced in size. As a result, less Ca^{2+} enters the nerve terminal (D), less transmitter is released, and the magnitude of the excitatory postsynaptic potential on the motoneuron is reduced (E). (B) is the portion of the axon that is not affected by presynaptic inhibition.

45–49. The answers are: 45-D, 46-C, 47-C, 48-A, 49-C. [*II C 1–4; Figure 1-7*] When the membrane potential is maintained at a constant value (e.g., the resting state, *point A*), the net current across the membrane is zero. The action potential begins when the membrane is depolarized from its resting potential to its threshold potential (*point B*) by a stimulus.

The upstroke of the action potential is caused by the inward flow of Na^+ following a large increase in Na^+ conductance. Na^+ is highest at *point C*. The driving force for K^+ (i.e., $E_m - E_K$) also is greatest at *point C* because E_m is furthest from E_K at this point.

Point C also is contained within the absolute refractory period, which begins when threshold is reached and lasts until enough Na^+ channels have recovered from their inactivation state to allow another action potential to be elicited.

Points D and E are in the relative refractory period. K^+ conductance is highest at *point D*. The downstroke of the action potential is caused by the outward flow of K^+ following an increase in K^+ conductance. *Point E* has a lower K^+ conductance than *point D* because the n gates close as the membrane repolarizes.

2
Cardiovascular Physiology

Joseph Boyle, III

I. VESSELS. The major functions of the cardiovascular system are to **distribute metabolites and oxygen (O_2)** to all body cells and to **collect waste products and carbon dioxide (CO_2)** for excretion. The cardiovascular system also is involved in **thermoregulation,** since heat is carried by the blood from active metabolic sites, where it is generated to the body surface, where it is dissipated. Blood flow to the skin and extremities is varied to enhance or retard heat loss to the environment. The circulation also functions to **distribute hormones** to distant sites. The **heart** provides the driving force for this system; the **arteries** serve as distribution channels to the organs; the **veins** serve as blood reservoirs and collect the blood to return it to the heart; and the **microcirculation** and **capillaries** serve as the exchange vessels.

A. Elastic arteries

1. **Structure.** The **aorta** and its large branches (**carotid, iliac,** and **axillary arteries**) are composed of a thick medial layer containing large amounts of **elastin** and some **smooth muscle** cells. The structure of these vessels makes them **distensible** so that they take up the volume of blood ejected from the heart with only moderate increases in pressure.

2. **Function**
   a. **Compliance (distensibility).** The elastic arteries serve as a reservoir; that is, they accommodate the stroke volume of the heart because they are distensible (see Ch 3 II D and E for further discussion of compliance and elastance). Figure 2-1 shows the pressure-volume relationships for the aorta in different age-groups.
      (1) Note that these curves are relatively linear in the middle, which is the normal working range. Note also that the distensibility is reduced with age (the vessels become stiffer).
      (2) This means that for any volume increase there is an increase in the pressure fluctuation as one ages. Clinically, this results in an increased pulse pressure in the older individual.
   b. **Elastic recoil.** In addition to the vessels being distensible, the elastic recoil during diastole provides potential energy (pressure) to maintain blood flow during the diastolic phase of the cardiac cycle.

B. Muscular arteries comprise most of the named arteries in the body.

1. **Structure.** As the distance from the heart increases, the amount of smooth muscle in the medial layer in these vessels increases.

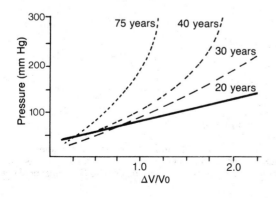

Figure 2-1. Average pressure-volume relationships of arterial systems from 20-, 30-, 40-, and 75-year-old individuals. Note the increasing elastance (decreased compliance) that occurs with advancing age. ΔV = volume change; V_0 = unstressed volume.

2. **Function.** These vessels serve as the distributing channels to the body organs and have a relatively large lumen-to-wall-thickness ratio. The large lumen minimizes the pressure drop due to resistance losses (Figure 2-2).

3. **Cross-sectional area and velocity of flow.** Figure 2-2 shows that the total cross-sectional area of the arterial tree increases markedly as it proceeds toward the periphery. Since the same volume of blood traverses each class of vessels per unit time, the velocity of flow is an inverse function of the cross-sectional area.

C. **Arterioles,** the stopcocks or valves of the circulation, control the volume of blood flowing to the various vessel beds. The site of the major pressure drop in the cardiovascular system occurs in the arterioles (see Figure 2-2).

1. **Structure.** The arterioles contain a thick layer of **smooth muscle** within the medial layer and have a relatively narrow lumen. The smooth muscle allows these vessels to alter the lumen diameter so that resistance and flow are readily controlled.

2. **Function**
a. **Control of blood flow.** The arterioles can markedly alter their diameter by contraction or relaxation of the smooth muscle in their walls. The variation in the lumen diameter allows a fine control over the distribution of the cardiac output to different organs and tissues. The smooth muscle tone varies depending on the activity of the sympathetic nerves, the arterial pressure, the local concentration of metabolites (caused by changes in metabolic rate), various hormones, and many other mediators such as prostaglandins, thromboxanes, and histamine. Increases in metabolic rate, such as occur in skeletal muscle during exercise, cause vasodilation, which increases blood flow. This phenomenon is termed **functional hyperemia**.
(1) **Autoregulation** is the mechanism whereby many organs and tissues—both of which have a constant metabolic rate—adjust their vascular resistance to maintain a constant blood flow in the presence of changes in arterial pressure. This phenomenon is depicted

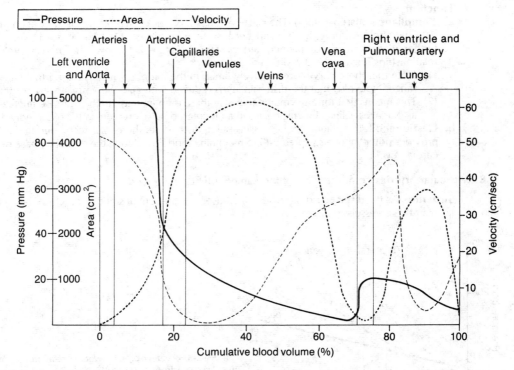

Figure 2-2. Relationships between velocity, area, volume, and pressure in various segments of the cardiovascular system. The same volume of blood must pass through each segment of the system per unit time; therefore, the velocity and area are inversely related. The resistance of the large vessels is minimal, so pressure loss also is minimal until the small arterioles are reached.

schematically in Figure 2-3, and several theories have been proposed to explain its mechanism.

   (a) Metabolic theory. An increase in blood pressure would initially increase blood flow to a tissue or organ. This increased blood flow would wash out any vasodilator substances that were present, vascular resistance would increase, and blood flow would return to the normal levels. Increased CO_2, H^+, adenosine, prostaglandins, K^+, or PO_4^{2-} or hypoxia can cause vasodilation. Autoregulation cannot be explained in all instances by changes in the concentration of any of these substances.

   (b) Myogenic theory. Vascular smooth muscle constricts when the pressure inside the vessel is increased. Vascular smooth muscle responds to wall tension, which is a function of pressure and wall radius according to the **law of Laplace**

$$T = P \times r$$

   where T = wall tension, P = pressure, and r = radius. Thus, an increase in arterial pressure raises wall tension; blood flow increases because of the higher driving pressure. Contraction of vascular smooth muscle returns wall tension to control levels by reducing the vascular radius. The narrowed lumen compensates for the increased arterial pressure so that blood flow returns to control levels.

   (2) Reactive hyperemia is a phenomenon that occurs after occlusion of the arteries to an organ or tissue. When the occlusion is released, the blood flow exceeds the control level with a magnitude and duration that depends on the duration of the occlusion. Reactive hyperemia seems to be caused by a metabolic mechanism that controls blood flow to these tissues.

   b. Dampening of pulsations. Arterioles convert the pulsatile flow in the arteries to a steady flow in the capillaries.

D. The microcirculation consists of those vessels smaller than 100 μ in diameter; it includes **metarterioles, arterioles, capillaries,** and **postcapillary venules.** Exchange of substances between extracellular fluid and the vascular system occurs primarily in the capillaries and postcapillary venules. In some tissues, especially the skin, there are short, low-resistance connections between the arterioles and the veins, which are termed **arteriovenous (A-V) shunts.** In the skin, the A-V shunts are involved in thermoregulation.

   1. Organization. The microcirculation is a meshwork of fine vessels that distribute blood to every cell in the body.

   a. Capillaries arise directly from arterioles or metarterioles. The **metarterioles** are relatively high-resistance conduits between arterioles and veins. A cuff of smooth muscle cells surrounds the origin of many capillaries and is termed the **precapillary sphincter.**

   b. Vasomotion. Not all of the capillaries in a tissue are functional at one time, so blood flow alternates among different capillaries from moment to moment. Under basal conditions, 1%–10% of the capillaries function at any moment. During high metabolic activities, many more capillaries carry blood simultaneously, which enhances the delivery of O_2 and metabolites to the tissues. The opening and closing of these vessels and the resultant changes in blood flow are termed **vasomotion.**

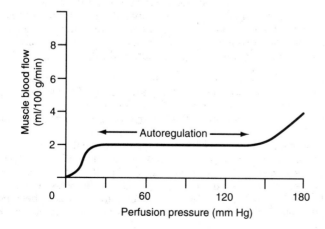

Figure 2-3. Autoregulation of blood flow. Note that blood flow remains relatively constant over a wide range of pressures. This can only be accomplished by changing the resistance to flow through the particular vascular bed in proportion to the change in pressure.

2. Function. The microcirculation is the area where exchange occurs between the blood and the interstitial fluids. The interstitium, in turn, equilibrates with the body cells. Many capillaries arise from each arteriole and metarteriole, which provides a total cross-sectional area of 0.4–0.5 m² for the microcirculation. This large area results in an average blood flow velocity of 0.3–0.4 mm/sec in the capillaries, but the velocity varies widely, from 0 to 1.0 mm/sec within short periods of time within the same capillary due to vasomotion.

3. Endothelial structure varies in different organs depending on function.
   a. Fenestrations
   (1) At one extreme, in the renal glomeruli, gastrointestinal tract, and some glands, large (200–1000 Å in diameter) transcellular fenestrations are present through the endothelial cells. These fenestrations are coupled with an incomplete basement membrane. A thin membrane may or may not cover the fenestrations.
   (2) At the other extreme, in the brain, the microcirculatory endothelium has no fenestrations, and a complete basement membrane is present. These structures are morphologic evidence of the **blood-brain barrier,** which retards or prevents the transfer of many substances between the blood and the brain.
   b. Gap junctions. Large gaps occur between the endothelial cells in the bone marrow, liver, and spleen. The postcapillary venules (20–60 μ in diameter) represent the most permeable portion of the microcirculation. This region represents leakage sites for large molecules after the administration of certain substances (e.g., histamine, bradykinin).
   c. Tight junctions. In most body tissues the clefts between endothelial cells appear fused, but anatomic and physiologic evidence indicates that pores (about 40 Å n diameter) exist between endothelial capillary cells. These pores are absent in cerebral capillaries, which is further evidence of a blood-brain barrier.

4. Mechanisms of exchange. The microcirculation provides a total surface area of about 700 m² for exchange between the circulatory system and the interstitial compartment.
   a. Diffusion is the principal mechanism of microvascular exchange. Passive diffusion results from the random movement of molecules from an area of high concentration to an area of low concentration and does not involve any carrier molecules or demonstrate saturation kinetics properties.
   (1) The **rate of diffusion** depends on the **solubility** of the substance in the tissues, the **temperature,** and the **surface area** available, and it is inversely related to **molecular size** and the **distance** over which diffusion must occur.
   (a) O_2, CO_2, water, and glucose rapidly equilibrate across the microvascular endothelium.
   (b) Larger molecules (e.g., albumin) cross the endothelial barrier very slowly or are impeded completely because the average pore size is smaller than albumin.
   (2) Diffusion path. The route of transport of most substances is still controversial. It is unclear whether transport occurs through gap junctions, fenestrae, intracellular vesicles (cytopempsis), or across the cells themselves.
   b. Bulk flow. A hydrostatic pressure difference across the endothelium results in the filtration of water and solutes from the capillaries into the tissues. The presence of large molecules (proteins, especially albumin) in the blood exerts a force (**oncotic pressure**) that counteracts filtration. The factors determining the balance between filtration and reabsorption were originally formulated by Ernest Starling, and the concept is termed the **Starling hypothesis** (Figure 2-4). The balance between the filtration and reabsorption of water depends on the difference in hydrostatic and oncotic pressures between the blood and the tissues and on how leaky the vessels are. The bulk flow of water (Q_w) is expressed as

$$Q_w = K [(P_c - P_i) - (p_c - p_i)]$$

where K = the filtration coefficient, which varies with the permeability of the vessels and the available exchange surface area; P_c = hydrostatic capillary pressure; P_i = hydrostatic pressure in the interstitium; p_c = oncotic pressure of the blood; and p_i = interstitial oncotic pressure.
   (1) Factors intrinsic to the blood vessel. Increased blood pressure within the microvessels (P_c) causes more fluid to be filtered from the capillaries; the accumulation of excess fluid in the tissues is termed **edema**. A loss of plasma proteins (decreased p_c), as in starvation, also can cause edema formation due to decreased oncotic pressure, which normally retards filtration. Injury to vessels or infection causes the vessels to become leakier (an increase in the filtration coefficient K), which leads to the accumulation of

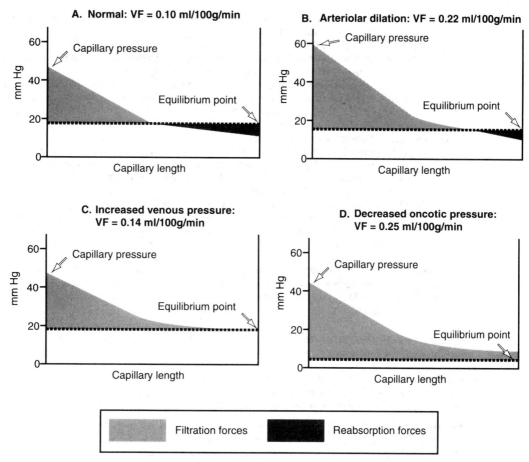

Figure 2-4. Schematic drawing of the intravascular pressures involved in determining the filtration of fluid through the microcirculation. Equilibrium point represents the pressure needed to counteract the oncotic pressures in the blood and interstitial fluid. Flow is in L/min. The effects of various interventions are shown in *panels A through D.* *VF* = volume filtered.

tissue fluids. Conversely, decreased pressure within the capillaries, as in shock, tends to draw fluid out of the tissues into the vascular system.

(a) **The filtration coefficient (K)** varies primarily due to changes in the number of microvessels having blood flow and also due to changes in vascular permeability. An increased permeability occurs when the vessels are injured by hypoxia and inflammatory products such as histamine. Dilation of arterioles, and especially precapillary sphincters, increases the number of functional capillaries (recruitment), since dilation transmits a greater fraction of the arterial pressure into the capillary. The greater capillary pressure dilates vessels so that perfusion increases. Hypercapnia, hypoxia, increased H^+ and K^+ concentrations, histamine, bradykinin, adenosine, and many other substances all can result in vasodilation and an increase in the number of perfused capillaries.

(b) **Capillary hydrostatic pressure (P_c)** represents a force tending to push fluid out into the interstitium. P_c depends especially on the venous pressure but also on the arterial blood pressure, the arteriolar resistance, and the state of the precapillary sphincter.

(i) Capillary pressure varies widely but ranges between 30 and 45 mm Hg in most tissues. Whenever blood flow is present, there must be a pressure gradient so that the capillary pressure must always exceed the venous pressure.

(ii) A low arteriolar resistance allows a greater proportion of the arterial pressure to be transmitted to the capillaries. For example, in the kidneys the glomerular capillary pressure is about 50–60 mm Hg, which provides a large force to produce glomerular filtration.

    (iii) Contraction of the precapillary sphincter can occlude the vascular lumen so that P_c equilibrates with the local venous pressure; under these conditions fluid may be reabsorbed along the entire length of the capillary.

  (c) The **oncotic pressure of blood (p_c)** is due to the osmotic pressure of blood proteins, which is only a small fraction (0.5%) of the total osmotic pressure of blood. The normal p_c is 25–27 mm Hg, which provides a force tending to reabsorb fluid into the capillaries.

  (2) Factors extrinsic to the blood vessel

    (a) The **interstitial pressure (P_i)** magnitude is controversial since different experimental techniques have shown it to be either slightly positive or slightly negative. P_i becomes greater if the interstitial fluid volume increases (**edema**), which hinders further fluid filtration from the capillary.

    (b) Oncotic pressure of the interstitium (p_i). Significant amounts of plasma proteins cross the endothelial barrier through either intercellular pores or fenestrae, depending on the tissue. Protein in the interstitial space produces a force directed toward outward filtration of fluid from the vascular system. The effective p_i in the interstitium is estimated as 5–10 mm Hg in most tissues.

  c. Cytopempsis. Some substances are transported across the endothelium by incorporation into **cytoplasmic vesicles,** which are then discharged into the interstitium. Transport of large, lipid-insoluble molecules has been proposed by this route, but the transport of water and small solutes would be negligible by this mechanism.

E. Lymphatic vessels start as closed endothelial tubes that are permeable to fluid and high-molecular-weight compounds.

  1. Function. The major circulatory function of the lymphatic system is to remove albumin and other macromolecules that have escaped from the microcirculation. In addition, fluid is removed from the interstitium to maintain a gel state.

  2. Mechanism. Lymphatic fluid is pumped out of the tissues by the contraction of the large lymph vessels and contiguous skeletal muscles. The lymph vessels have an extensive system of one-way valves to maintain the flow of lymph toward the heart.

  3. The **rate of lymph flow** varies in different organs and depends on tissue activity and capillary permeability. Normal lymph flow is about 2 L/day for the entire body, and this volume contains approximately 200 g of protein lost from the microcirculation. Most lymph comes from the gastrointestinal tract and the liver.

  4. Importance. The lymphatic system represents the only mechanism for returning interstitial proteins back to the circulatory system. Due to the concentration of albumin in the blood, back-diffusion of albumin into the capillaries is impossible. The lymph collected from the peripheral tissues is returned to the cardiovascular system via the thoracic duct, which empties into the left subclavian vein near its junction with the left internal jugular vein.

F. Veins

  1. Structure. The veins are relatively thin-walled structures whose walls contain **elastic tissue, collagen,** and **smooth muscle**. The venous system is a low-resistance, low-pressure, very distensible part of the vascular system. The veins provide a larger cross-sectional area than the arteries at equivalent distances from the heart; therefore, the resistance to flow and the velocity of flow are less in the veins than in the arteries.

  2. Function

    a. Conduits. The systemic veins carry deoxygenated blood from the tissues to the right atrium; the pulmonary veins collect the oxygenated blood from the lungs and return it to the left atrium.

    b. Reservoir. The veins serve as a fluid reservoir; 65%–75% of the circulating blood volume is in the veins at any one time. The veins normally are not fully distended, having an oval cross-sectional shape; thus, small changes in pressure can cause large changes in volume as they assume a rounded configuration. At normal pressures the veins are about 20 times more distensible (compliant) than the arteries.

    c. Venoconstriction is extremely important for increasing cardiac output. The smooth muscle of veins responds to **sympathetic stimulation** by constriction, which displaces blood toward the heart and raises the ventricular filling pressure. This results in an increase in cardiac output by increasing ventricular end-diastolic volume and ventricular stroke volume.

    d. Effects of gravity. Due to the high distensibility of the veins, small pressure changes can produce large changes in the volume of blood in the venous system.

        (1) Peripheral pooling of blood results whenever a person changes from the supine to the erect position. Standing causes a shift of about 500 ml of blood from the pulmonary circulation to the dependent veins of the legs, which is due to the hydrostatic pressure generated by the erect position causing a dilation of the leg veins and a peripheral pooling of blood. This peripheral pooling decreases venous return and consequently lowers cardiac output, which tends to decrease arterial blood pressure. These changes normally are compensated for by a reflex increase of sympathetic tone to the veins and the heart.

        (2) Hydrostatic effects. The arterial pressure in the head is reduced by about 30 mm Hg compared to the pressure at heart level due to the gravitational effects on the column of blood from the heart to the brain. These changes are compensated for by a reflex increase in the smooth muscle tone in the veins as well as increases in heart rate and peripheral vascular resistance, which maintain an adequate arterial pressure.

    e. Venous valves

        (1) The veins of the dependent parts of the body are equipped with a system of valves that prevent the backflow of venous blood. Venous valves also support the column of blood to minimize the increase in capillary pressure in the dependent parts of the body.

        (2) Insufficiency (backward leaking) of venous valves allows the full hydrostatic pressure to be exerted by the column of blood on the capillaries, leading to the formation of edema in dependent parts. This loss of fluid from the vascular compartment reduces venous return and tends to lower arterial pressure. This is the mechanism that causes soldiers who stand at attention for long periods of time to faint due to decreased cardiac output. Varicose veins represent a marked dilation and consequent valvular insufficiency of the dependent veins due to a familial or occupational predisposition.

    f. Skeletal muscle pump. Contraction of skeletal muscle aids in venous return by compressing the veins between the contracting skeletal muscle. This effect is aided by competent venous valves that prevent backflow during relaxation of the skeletal muscle.

    g. Respiration. The normal negative intrathoracic pressure tends to enhance venous return by increasing the pressure gradient between the heart and the peripheral vessels. Inspiration lowers the intrathoracic (interpleural) pressure, which increases venous return by enhancing the pressure gradient even further.

    h. Positive end expiratory pressure (PEEP), or a **Valsalva maneuver** (forced expiration against closed airways, commonly performed when lifting heavy objects or when moving one's bowels), impedes venous return and produces a complex set of direct and reflex changes in the cardiovascular system. PEEP is used to treat respiratory distress syndrome [see Ch 3 II E 2 c (3) (a)]; the positive alveolar pressure stabilizes alveoli and minimizes or eliminates atelectasis, which enhances gas exchange.

II. HEART. The heart has an intrinsic contraction rate because it contains its own pacemaker. The rate of the pacemaker is influenced by the **noradrenergic neurons,** which increase rate, and the **cholinergic neurons,** which decrease rate. The contractile phase of the heart is termed **systole,** and the relaxation or filling phase of the heart is called **diastole.**

    A. Myocardial syncytium. The heart muscle **(myocardium)** is composed of separate cardiac muscle cells that are electrically connected to one another by tight junctions. These connections are low-resistance pathways called **intercalated disks.**

        1. Intercalated disks occur at the **Z lines** of the **sarcomere** (Figure 2-5). Depolarization of one cardiac cell is transmitted to adjacent cells; thus, the myocardium is a functional **syncytium** (i.e., a mass of cytoplasm with numerous nuclei). Actually, the heart functions as two syncytia, one being the **atria** and the other the **ventricles.**

        2. The atrioventricular (AV) valve ring separates these two masses of muscle and insulates the electrical events of the atria from those of the ventricles. Normally, there is only one functional electrical connection between the atria and the ventricles; this is the **AV node** and, its extension, the **bundle of His.**

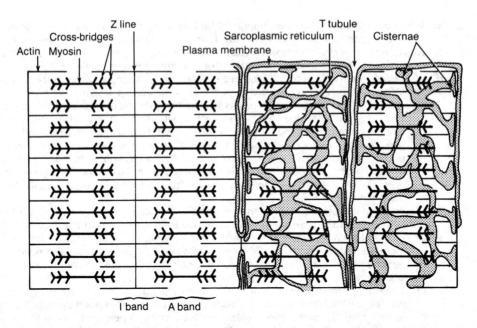

Figure 2-5. Schematic drawing of cardiac muscle showing sarcomeres with and without sarcoplasmic reticulum. The overlap of the actin and myosin varies as a function of the initial length and contraction of the muscle.

B. Electrical activity

1. **Resting membrane potential (RMP).** The myocardial cells maintain a voltage difference of 60–90 mV across their cell membranes; the inside of the cell is electrically negative compared to the outside. The RMP is generated because of differences in **membrane permeability** for different ions and a difference in ionic concentrations between the inside and the outside of the cell. The RMP can be predicted by means of the **transference equation** (see Ch 1 II D 4).
    a. The resting cell can be considered to be a K^+ battery, since the K^+ concentration is much higher on the inside than the outside, and the cell membrane is very permeable to K^+. At the same time, the resting membrane is rather impermeable to Na^+ [see Ch 1 II D 5 a (1)].
    b. The different ionic concentrations are maintained by membrane-bound ionic pumps. These ionic pumps obtain energy for their function by breaking down ATP to ADP. This energy is largely derived from the metabolism of foodstuffs within the mitochondria of the cells.

2. **Threshold potential.** Excitable cells, such as muscle or nerve cells, undergo rapid depolarization if the membrane potential is reduced to a critical level, which is called the **threshold potential**. Once the threshold potential is reached, the remainder of the depolarization is spontaneous. This property is termed the **all-or-none law**.

3. **Cardiac action potential (AP)** refers to the change in membrane potential that occurs after the cell receives an adequate stimulus. The myocardial cells are subdivided into slow and fast fibers depending on the shape and the velocity of conduction of the AP (Figure 2-6).
    a. **Slow fibers** are normally present only in the sinoatrial (SA) and AV nodes. The membrane potential of slow fibers usually is in the range of 50 to 70 mV. Fast fibers can be converted into slow fibers by the effects of ischemia or **hypoxia** (decrease in O_2 levels) or by certain drugs (e.g., tetrodotoxin, which blocks Na^+ channels). The AP in slow fibers is due to the inward movement of positive ions (Ca^{2+}), and this movement neutralizes the normally negative intracellular charge. During the AP the membrane potential approaches zero, and the cell is said to be **depolarized**.
        (1) **Slow channels.** Slow fibers apparently possess only "slow channels," through which ions enter the cells. These slow channels limit the rate at which ions enter the cells and produce a slow rate of cell depolarization. Thus, the upstroke of the AP requires about 100 msec in slow fibers compared to 1 msec in fast fibers.
        (2) The **conduction velocity** of the AP along the cells is directly related to the RMP and the rate of depolarization. Because the RMP is low, and the rate of depolarization is slow, the AP of slow fibers has a conduction velocity of 0.02–0.1 m/sec.

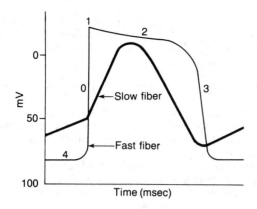

Figure 2-6. Schematic plot of the action potential from a slow and a fast fiber, with the different phases (*0 through 4*) of the action potential represented. Note the diastolic depolarization and the slow phase 0 in the slow fiber.

  (3) The **absolute refractory period (ARP)** lasts throughout the duration of the AP. The ARP is an interval in which no stimulus, no matter how strong, can cause another AP.
  (4) The **relative refractory period (RRP)** may last for several seconds in slow fibers. A stimulus applied during the RRP must be stronger than normal to elicit an AP. This AP has an even slower conduction velocity and lower amplitude than normal. This effect frequently blocks the conduction of impulses within the AV node so that all of the atrial depolarizations do not reach the ventricles; this is referred to as an **AV nodal block** (see IV C 2).
 b. Fast fibers. The normal atrial and ventricular myocardial cells as well as the specialized conducting tissues in the heart are fast fibers. The **RMP** in fast fibers is 80–90 mV; the inside of the cell is negatively charged with respect to the outside of the cell.
  (1) Na^+ channels. The AP of fast fibers exhibits an extremely rapid rate of rise, because once the threshold potential is reached, the cell membrane becomes extremely permeable to Na^+.
   (a) Since the Na^+ concentration is much higher outside the cell and because of the negative charge inside the cell, there is a very large electrochemical gradient for Na^+. Therefore, Na^+ rushes into the cell through channels that open once the threshold potential is reached. The cell essentially becomes a Na^+ battery during the AP.
   (b) The duration of the AP varies in different fast fibers and is longest in the **Purkinje** and **bundle of His** fibers. This property provides the heart some protection against certain arrhythmias (e.g., ventricular tachycardia).
  (2) Conduction velocities vary from 0.3–1.0 m/sec in myocardial cells to about 4 m/sec in Purkinje fibers. The fast conduction velocity insures that the entire myocardium is depolarized almost instantaneously, which improves the effectiveness of myocardial contraction.
  (3) The **ARP** in fast fibers lasts until the repolarization has reached a membrane potential of 50–60 mV.
  (4) The **RRP** ends when the RMP of 80–90 mV is reestablished.
 c. Phases of the AP. The AP can be divided arbitrarily into five phases (see Figure 2-6).
  (1) Phase 0. The resting membrane is relatively impermeable to Na^+, but once the threshold potential is reached the Na^+ channels open and allow Na^+ to rush into the cell along the Na^+ electrochemical gradient. This process is self-regenerative in that inward movement of Na^+ reduces the membrane potential, which opens more Na^+ channels, so that more Na^+ enters the cell. The movement of Na^+ is a current-carrying process and results in the depolarization of the cell membrane, which is termed phase 0. This process is extremely rapid in fast fibers. In slow fibers phase 0 coincides with an increase in the Ca^{2+} conductance so that Ca^{2+} is the current-carrying ion in these fibers.
  (2) Phase 1. During phase 1 the Na^+ conductance is reduced rapidly due to a closing of the Na^+ channels, while the membrane conductance for Ca^{2+} increases. The overall effect is a small change in the membrane potential toward repolarization, which prevents the membrane potential from reaching the Na^+ equilibrium potential of +40 to +60 mV.
  (3) Phase 2. The plateau of the AP in fast fibers coincides with an increased membrane conductance for Ca^{2+}. The inward movement of Ca^{2+} and the decreased efflux of K^+

maintain the membrane potential near zero during this phase of the AP. In cardiac muscle, the influx of Ca^{2+} is important in the **excitation-contraction (EC)** process.

   (4) **Phase 3** is a rapid repolarization due to a reduction in the inward Na^+ and Ca^{2+} currents and a large increase in the outward K^+ current.

   (5) **Phase 4**

   (a) Nonpacemaker cells (atrial and ventricular myocardium) exhibit a constant membrane potential during phase 4.

   (b) In pacemaker tissues, such as SA and AV nodes and Purkinje fibers, there is a slow diastolic depolarization during phase 4, which indicates the presence of **automaticity**. The **diastolic depolarization,** also termed a **pacemaker potential** or the slope of phase 4, brings the membrane potential toward threshold.

   (i) The cell whose membrane potential first reaches threshold is the pacemaker of the heart, and the depolarization then spreads to the remainder of the heart. The normal pacemaker of the heart is the SA node since its cells usually have the highest automaticity (steepest slope of phase 4).

   (ii) Elimination of SA node activity or an increased slope of phase 4 in other areas of the heart causes a pacemaker outside the SA node to initiate the depolarization process. This new pacemaker is termed an **ectopic pacemaker** since it is outside the normal place.

C. Conduction pathways

   1. The **SA node** is located near the junction of the superior vena cava and the right atrium (Figure 2-7). The SA node normally is the pacemaker of the heart since it has the highest automaticity.

   a. Intrinsic heart rate. The isolated SA node has a firing rate of 90–120 beats/min; this rate is higher in young individuals and declines with advancing age. The firing rate is increased by a rise in temperature, thyroid hormone, epinephrine, or norepinephrine.

   b. Effect of vagus nerve. The right vagus nerve densely innervates the SA node and liberates **acetylcholine (ACh)** from its nerve endings when stimulated. Normally, **vagal activity (vagal tone)** hyperpolarizes SA node fibers by increasing K^+ conductance. The hyperpolarization slows the firing rate of the SA node from its automatic rate of 90–120 beats/min to the actual heart rate of about 70 beats/min. Strong vagal stimulation can eliminate completely SA

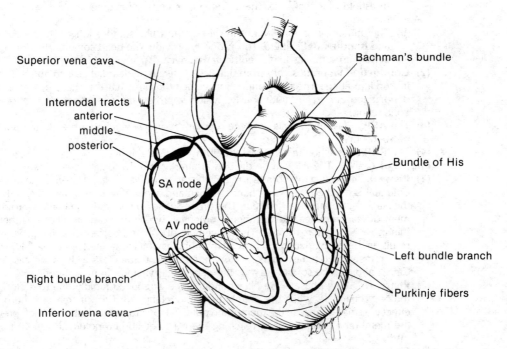

Figure 2-7. Schematic drawing of the specialized conducting tissues of the heart. *SA* = sinoatrial; *AV* = atrioventricular.

node impulses **(sinus arrest),** which causes **asystole (cardiac standstill)** until an ectopic pacemaker begins to function.

   c. Effect of sympathetic nerves. Stimulation of cardiac sympathetic nerves or the injection of sympathetic-like drugs (e.g., epinephrine) increases heart rate and the automaticity of ectopic sites. Stimulation of the cardiac sympathetic nerves or administration of sympathomimetic drugs markedly increases the force of contraction **(positive inotropic effect),** increases the rate of pressure developed by the ventricle, and shortens the duration of systole.

   d. Ions. The major effects of ions on automaticity are caused by changes in the concentrations of K^+ and Ca^{2+}.

   (1) K^+

   (a) Hyperkalemia reduces the RMP by decreasing the ratio of K^+_{in}/K^+_{out}. Hyperkalemia in excess of 8 mEq/L depolarizes the cell and inactivates the membrane Na^+ channels, which decreases the AP rate of rise and slows the conduction rate. These changes lead to conduction disturbances (e.g., atrial standstill, AV nodal block, reentry phenomena, increased ectopic activity), which can lead to ventricular fibrillation.

   (b) Hypokalemia causes a decreased membrane conductance for K^+, which reduces RMP; this alters automaticity in the same manner as in cases of hyperkalemia.

   (2) Ca^{2+}

   (a) Hypercalcemia slows heart rate and decreases the excitability of cardiac tissues while it increases the tension developed by cardiac muscle. Marked hypercalcemia results in cardiac arrest during systole.

   (b) Hypocalcemia increases the slope of phase 4 of the AP and reduces RMP; both effects increase heart rate.

2. The **interatrial tract (Bachman's bundle)** is a band of specialized muscle fibers that run from the SA node to the left atrium. The interatrial tract causes almost simultaneous depolarization of both atria since the conduction velocity through this tissue is faster than through the regular atrial muscle fibers.

3. Internodal tracts. Three bundles of specialized cells (the **anterior, middle,** and **posterior internodal tracts**) connect the SA and AV nodes. The internodal tracts are more resistant to the blockade of impulses than the atrial muscle, which makes it more likely that SA node impulses reach the AV node to initiate ventricular depolarization.

4. The **AV node** is located just beneath the endocardium on the right side of the interatrial septum near the tricuspid valve (see Figure 2-7). The AV node is normally the only path by which the ventricles are activated. It is richly supplied by fibers from both the sympathetic and vagus nerves, which can affect the conduction of impulses. Occasionally an alternative pathway develops between the atria and ventricles during the formation of the heart. Its presence results in abnormal excitation of the ventricles, which can be detected by an EKG tracing; the condition is called the **Wolff-Parkinson-White syndrome** (see IV C 3).

   a. AV nodal delay. The conduction velocity through the AV node is extremely slow (0.02–0.05 m/sec), which delays ventricular depolarization for 100–150 msec after atrial depolarization. This delay provides time for atrial contraction to occur, which enhances ventricular filling, especially at fast heart rates.

   b. AV nodal block. The AV node cells have long refractory periods and a very slow conduction velocity. This limits the number of impulses that can be conducted through the AV node to about 180 beats/min. The ventricles are protected from being driven at excessive rates, which can reduce the pumping effectiveness of the heart. Atrial impulses can be completely blocked from passing through the AV node by various diseases; an ectopic ventricular pacemaker usually functions to drive the ventricles but at a very slow rate. Under these conditions it may be necessary to implant an electronic device (pacemaker) in order to stimulate the ventricles to contract at an adequate rate.

   c. Autonomic effects

   (1) Sympathetic stimulation increases the conduction rate through the AV node and enhances the transmission of impulses.

   (2) Parasympathetic (vagal) stimulation to the AV node slows the conduction velocity and prolongs the refractory periods. These effects make it more likely that AV block will occur and are the result of the reduced slow inward Ca^{2+} current caused by ACh.

5. **Ventricular conduction.** Impulses conducted through the AV node are distributed to the ventricles over specialized fibers.
   a. The **bundle of His** is the continuation of the AV node and is located beneath the endocardium on the right side of the interventricular septum. The bundle of His divides into the right and left bundle branches.
   b. The **right and left bundle branches** proceed on each side of the interventricular septum to their respective ventricles. A block of the depolarization impulse can occur in either of the bundle branches, which delays the depolarization of the ventricle on that side. The affected ventricle is eventually depolarized by the spread of impulses through the regular ventricular muscle fibers. A right or left bundle branch block produces characteristic EKG changes.
   c. **Purkinje fibers** arise from both bundle branches and branch out extensively just beneath the endocardium of both ventricles (see Figure 2-7). These cells have the largest diameter in the heart (70–80 μ) and possess the highest conduction velocity of the AP (1–4 m/sec). This insures that both ventricles contract almost simultaneously, which increases the effectiveness of contraction.
   d. **Ventricular muscle depolarization** occurs from the endocardial surface to the epicardium. Depolarization initially reaches the surface of the heart at the apex and then spreads throughout the myocardium. The last area of the ventricles to be depolarized is the base of the left ventricle. The conduction velocity of the AP through the ventricular myocardium is 0.3–0.4 m/sec.

D. **Excitation-contraction (EC) coupling** is the term used to define the events that connect the depolarization of the cell membrane to the contraction of the muscle fibers.

1. **Structure of myocardial cells.** Each of the cardiac cells contains many bundles of protein strands called **myofibrils,** which, in cardiac muscle, are surrounded by an extensive network of tubules known as the **sarcoplasmic reticulum** (see Figure 2-5).
   a. The **sarcoplasmic reticulum** in cardiac muscle is much less developed than in skeletal muscle. It contains dilated terminals (**cysternae**), which are located next to the external cell membrane and T tubules. The sarcoplasmic reticulum and the cysternae contain high concentrations of ionized Ca^{2+}.
   b. **T tubules** are continuations of the cell membrane, and, therefore, the T tubules conduct the AP into the interior of the cell. T tubules are continuous with the sarcolemma and invaginate into the interior of the cell at the **Z line** of the sarcomere in mammalian cardiac cells. T tubule depolarization causes the release of Ca^{2+} from the sarcoplasmic reticulum and cysternae.
   c. **Myofibrils** are composed of thick and thin filaments called **myosin** and **actin,** respectively. The thin filaments also contain two other proteins, **troponin** and **tropomyosin.**
   d. **Sarcomeres** are the contractile unit of myofibrils (see Figure 2-5).

2. **Contraction mechanism**
   a. **Effects of Ca^{2+}** (Figure 2-8)
      (1) The concentration of Ca^{2+} within the cell is very low between contractions. Each muscle contraction is preceded by an AP, which depolarizes the cell membrane and T tubules.
      (2) The AP causes Ca^{2+} release from the sarcoplasmic reticulum, and additional Ca^{2+} diffuses into the cell across the T tubules (slow inward Ca^{2+} current). The AP raises the sarcoplasmic Ca^{2+} concentration about 10-fold compared to the resting conditions.
      (3) At high Ca^{2+} concentrations, troponin binds four Ca^{2+} ions per molecule. When troponin is saturated with Ca^{2+}, a shape change occurs in the troponin, which causes the tropomyosin molecule to uncover the cross-bridge sites. This allows cross-bridges to be formed between actin and myosin.
      (4) Under normal circumstances, all of the troponin is not saturated with Ca^{2+} so that factors raising intracellular Ca^{2+} concentration can provide more sites for cross-bridge formation.
      (5) An increase in the number of cross-bridges formed causes an increase in the force of the contraction.
   b. **Sympathetic effects.** Sympathetic nerve stimulation or the administration of drugs (e.g., epinephrine or norepinephrine) increases the concentration of Ca^{2+} within the cell and results in a more forceful contraction. Norepinephrine also causes a more rapid uptake of Ca^{2+} by the sarcoplasmic reticulum, which shortens the duration of both the AP and the contraction.

A. Resting muscle: low Ca²⁺

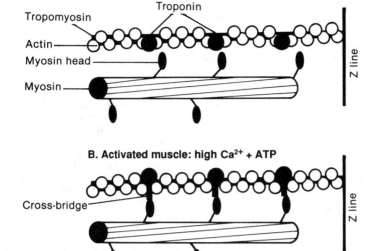

B. Activated muscle: high Ca²⁺ + ATP

Cross-bridge

C. Shortening: Ca²⁺ + ATP

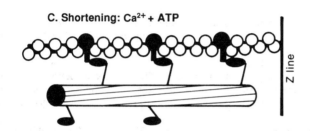

D. Dissociation: Ca²⁺ + ATP

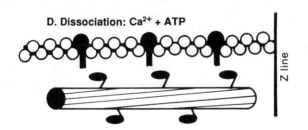

Figure 2-8. Proposed sequence of changes during the contractile process. (*A*) In resting muscle the Ca²⁺ concentration is low, and the troponin molecules shield the cross-bridge sites from the myosin. (*B*) During depolarization of the myocardial cells, Ca²⁺ concentration rises, which results in a configurational change in the troponin so that cross-bridge sites are uncovered. In the presence of ATP, cross-bridges are formed between the actin and myosin heads, and ATP is hydrolyzed. (*C*) Energy from ATP hydrolysis causes configurational change in the myosin heads so that actin is ratcheted about 10 nm per cycle in relationship to the myosin molecule. Note the movement of the Z line in the figure. (*D*) ATP is again required to allow dissociation of the cross-bridges to allow the muscle to relax. Each contraction involves numerous cycles through *steps C and D* as described in this figure.

   c. Digitalis is given to patients who are in heart failure, because it also increases the force of cardiac contraction by raising the intracellular Ca²⁺ concentration. Digitalis has been used for several hundred years to treat heart failure and was initially prescribed as a tea prepared from the leaves of a flowering plant (*Digitalis purpurea*, purple foxglove).

E. Frank-Starling effect (Starling's law of the heart). One of the major factors that controls the force of cardiac contraction is the initial length, or **preload,** of the muscle fibers (Figure 2-9). Changes in preload alter the number of available cross-bridges by changing the overlap of the thick and thin filaments. In the intact heart the preload depends on the volume of blood in the ventricles just before contraction begins (end-diastolic volume).

   1. An increase in end-diastolic volume lengthens the cardiac fibers, which, up to an optimum, exposes more of the myosin heads (cross-bridge sites) to the actin. The additional cross-bridges allow the heart to generate more force with each contraction. Thus, an increase in preload produces a stronger contraction and a greater volume of blood is pumped by the heart. These effects result in a higher pressure being generated by the myocardium.

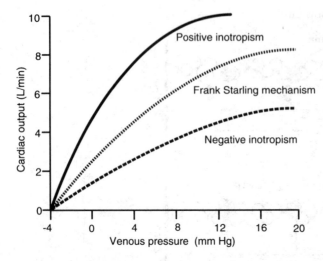

Figure 2-9. A series of Frank-Starling curves. Changes in cardiac output can be accomplished by alterations in the preload (venous pressure or ventricular end-diastolic volume) causing movement along one of the curves (Frank-Starling mechanism) or by shifting from one curve to another. The latter mechanism indicates changes in ventricular contractility (inotropism).

2. The Frank-Starling mechanism is important in balancing the output of the two sides of the heart over extended periods of time. In the intact heart, changes in stroke volume also depend on the contractility of the ventricles.

F. Contractility

1. **Increased contractility (positive inotropism)** is defined as an increased force of contraction at a constant preload. Positive inotropism can also be defined as an increase in the maximal velocity of shortening (V_{max}).

   a. This is in contrast to the Frank-Starling relationship, where an increase in the force of contraction is produced by an increase in preload. The difference between an increased contractility and an increase in preload is shown in Figures 2-9 and 2-10. Positive inotropic effects are produced by sympathetic stimulation and the effect of sympathomimetic drugs and digitalis glycosides.

   b. Figure 2-9 shows a family of ventricular function (Frank-Starling) curves and depicts the ventricular performance as a function of the ventricular end-diastolic volume or preload. Positive inotropic effects produce a shift toward the left; that is, the same cardiac work can be performed at a smaller end-diastolic volume.

2. **Decreased contractility (negative inotropism)** is caused by hypoxia, acidosis, and myocardial ischemia (lack of blood flow) or infarcts (death of tissue). Negative inotropism causes a shift toward the right in the ventricular function curves, which results in cardiac failure; less work is performed by the heart from any given end-diastolic volume or pressure. The intact heart shifts

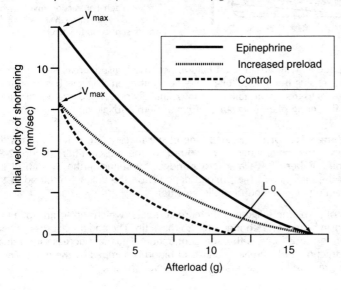

Figure 2-10. The velocity of shortening as a function of afterload. Changes in preload (venous pressure, ventricular end-diastolic pressure, or volume) do not cause an increase in the maximal velocity of shortening (V_{max}). Sympathomimetic substances, such as epinephrine, increase contractility and result in an increased V_{max}. L_0 = initial length of muscle.

from one curve to another (alters contractility) in response to various physiologic or pathologic events.

III. ELECTROCARDIOGRAPHY.

The science of electrocardiography is about 100 years old. The techniques of recording electrical activity from the heart originally were developed by Dutch physiologist, Willem Einthoven, who referred to the recording of this electrical activity as an *"Elektro Kardio Gramm"* (thus the abbreviation EKG). The EKG provides a method of evaluating excitation events, arrhythmias, tissue damage, the presence of ischemia or necrosis, and hypertrophy of the heart. The EKG does not provide any information on the mechanical performance of the heart, and at times even abnormal electrical activity may not be recordable.

A. Volume conduction

1. The body tissues function as an electrical conductor because they contain electrolytes. The heart is assumed to lie centrally within the thorax, and its electrical activity is conducted to the body surface through the body fluids.

   a. Attaching electrodes to the body surface allows voltage changes to be recorded within the body after adequate amplification of the signal. A **galvanometer** within the EKG machine is used as a recording device. Galvanometers record potential differences (voltages) between two electrodes.

   b. EKG recordings (**electrocardiographs**) are merely the differences in voltage between two electrodes located on the body surface.

2. **Zero potentials.** During diastole the cardiac cell membranes are polarized—positively charged on the outside and negatively charged on the inside. Electrodes on the skin do not detect any voltage differences since all parts of the heart are equally polarized. Thus, the recording shows no deflection from the zero potential line (Figure 2-11A).

3. **Action potentials**

   a. **Depolarization.** Excitation of a portion of the heart causes the myocardial cells to depolarize, which produces a reversal of membrane potential in this area. The outside of these cells is now negatively charged with respect to ground. Thus, a potential difference exists between the depolarized cells and the nonexcited cells (see Figure 2-11B). This potential difference can be recorded from surface electrodes, and the direction of its deflection will depend on the polarity of the electrodes. When the entire heart has been depolarized, all of the cells are now negatively charged outside, and both electrodes again "see" the same potential. The galvanometer again will read zero (see Figure 2-11C).

   b. **Repolarization.** Assuming that repolarization proceeds in the same direction as depolarization, the galvanometer will be deflected in the opposite direction (see Figure 2-11D and E) during the repolarization process. The resulting record is termed a **biphasic action potential** because there are two waves in opposite directions.

4. **Equivalent dipole.** The voltage differences between resting, depolarized, or repolarizing cells function as a battery. The various charges that are present can be represented as a point source of charge and are termed an equivalent dipole.

   a. The total charge that is present depends on the mass of tissue involved as well as the magnitude of the membrane potentials.

   b. The cardiac dipole is a vector quantity because it has both magnitude and direction. Vectors are represented as arrows with the arrowhead indicating the direction and the length of the arrow indicating the magnitude.

5. **Surface potential,** or the magnitude of the voltage recorded at the body surface, is a function of electrode position and the orientation and magnitude of the dipole. By convention, a wave of depolarization approaching the positive electrode results in an upward (positive) deflection of the EKG tracing. A wave of depolarization proceeding parallel to an electrode axis (the line connecting two electrodes) produces the maximal deflection for that dipole, whereas a depolarization wave perpendicular to the electrode axis produces no net deflection of the tracing (i.e., positive and negative waves are equal). Figure 2-12 depicts these relationships, which are essential to understand when analyzing EKG recordings.

B. EKG leads.

The standard EKG consists of recordings from 12 different leads. A **lead** consists of the recording from electrodes placed at specific sites on the body. Each lead shows the same cardiac events as other leads but from a different view. Additional leads are used in special circumstances.

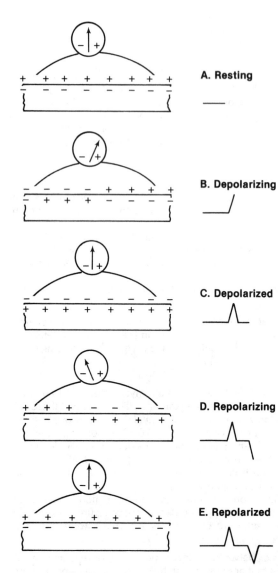

A. Resting

B. Depolarizing

C. Depolarized

D. Repolarizing

E. Repolarized

Figure 2-11. The effects of depolarization and repolarization of isolated, excitable tissue on a galvanometer's deflection, using standardized connections. Note that when repolarization and depolarization are in the same direction, the galvanometer deflections are oppositely directed. In the limb leads of an electrocardiogram, the QRS complex and the T wave normally are in the same direction.

1. **Einthoven's triangle.** The torso is considered to be an equilateral triangle with the right and left shoulders and left leg as the three apices. The right leg serves as a ground connector; all four electrodes must always be attached to the extremities, which merely serve as conductors from the torso.
   a. **Zero potential lines.** If lines are drawn perpendicularly from the center of each side of an equilateral triangle, they will meet at the center of the triangle. These lines represent the zero potential lines for the three sides of the triangle (Figure 2-13A).
   b. **Bipolar limb leads.** Three leads are formed by measuring the potential differences between any two of the limb electrodes. These leads are selected by a switch on all standard EKG machines.
      (1) **Lead I** measures the difference in voltage (potential) between the left-arm (LA) electrode and the right-arm (RA) electrode (LA − RA).
      (2) **Lead II** is the potential at the left leg (LL) minus the potential at RA (LL − RA).
      (3) **Lead III** is LL − LA.

2. **Unipolar (V) leads.** If the three limb leads are connected to a common terminal, the combined voltage from the three leads will be zero, theoretically. This common terminal can be attached to the negative pole of a galvanometer and a fourth, or exploring, electrode can be attached to the positive pole. The galvanometer can still only read the potential difference between two

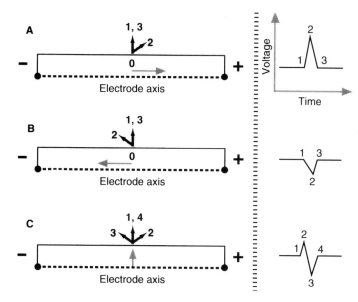

Figure 2-12. The deflection of a galvanometer needle using standard connections. (*A*) A wave of depolarization approaching the positive electrode causes an upward (positive) deflection. (*B*) A wave of depolarization approaching a negative electrode causes a negative deflection. (*C*) A wave of depolarization that is proceeding perpendicular to the electrode axis produces no net deflection.

points, but, if the common electrode is at zero volts, then the other electrode will provide the actual or absolute voltage at the body surface. This arrangement of connections is termed a **unipolar electrode** and is used to record the **precordial** or **chest leads** from standardized sites. There are six precordial leads, also termed **V leads,** in the standard EKG. These precordial leads measure electrical activity in the horizontal plane of the body.

a. **V₁** is in the fourth **intercostal space (ICS)** just to the right of the sternum.
b. **V₂** is in the fourth ICS just to the left of the sternum.
c. **V₄** is at the **midclavicular line (MCL)** in the fifth ICS.
d. **V₃** is halfway between V₂ and V₄.
e. **V₅** is in the anterior axillary line at the same level as V₄.
f. **V₆** is in the midaxillary line at the same level as V₄ and V₅.

3. **Augmented unipolar leads.** Any of the three limb electrodes can be used to record cardiac potentials in comparision to the common terminal. For example, the voltage recorded at RA can be determined by the equation: RA – (RA + LA + LL). The resulting voltage is small, because the potential difference is reduced by the RA potential in the common terminal.

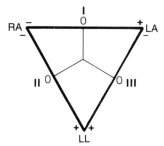

A. Einthoven's triangle

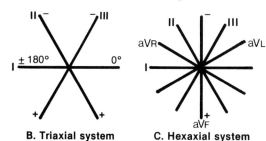

B. Triaxial system   **C. Hexaxial system**

Figure 2-13. (*A*) Einthoven's triangle showing connections, leads (*I, II,* and *III*), and lead polarity. Two other axis systems, the triaxial system (*B*) and the hexaxial system (*C*), also are shown. The electrical zero occurs at the center of Einthoven's triangle and at the center of the triaxial and hexaxial systems. *RA* = right arm; *LA* = left arm; *LL* = left leg; *aVR, aVL,* and *aVF* = augmented unipolar limb leads.

Disconnecting the RA lead from the common terminal increases the potential difference by 50% and results in the augmented unipolar limb lead, aVR.

a. **aVR** is the potential difference betwen RA and (LA + LL).

b. **aVL** is the potential difference between LA and (RA + LL).

c. **aVF** is the potential difference between LL and (RA + LL).

d. Augmented limb lead axes bisect the angles of Einthoven's triangle. The axis of aVR is from −150° to +30°; the axis of aVL is from −30° to +150°; and the axis of aVF is from +90° to −90°.

4. A **triaxial reference system** is obtained by moving the sides of Einthoven's triangle so they intersect at the center of the triangle. Lead I then divides the system into an upper (negative) and a lower (positive) hemisphere (see Figure 2-13B).

5. **Hexaxial reference system.** Superimposing the axes of the augmented unipolar limb leads on the triaxial system provides a hexaxial reference system with axes at every 30° (see Figure 2-13C). In this system lead I and aVF divide the system into quadrants with the zero isopotential point for all leads at the central intersection.

6. A **standard 12-lead EKG** consists of the three bipolar limb leads (I, II, and III), the three augmented limb leads (aVR, aVL, and aVF), and the six chest leads (V_1–V_6).

C. **Electrocardiography.** EKGs are the tracings of the surface cardiac potentials recorded against time. These tracings usually are made at a standard recording speed (25 mm/sec) and amplification (1 mV = 1 cm deflection). With standard EKG paper, each small horizontal division represents 0.04 second and each large division is 0.2 second, while each small vertical division represents 0.1 mV. Each lead indicates the same electrical events in the heart but from a different angle in any one subject unless there is a change in the sequence of excitation. This can be compared to looking at a house from the front, the side, and the back. It is all the same house, but it looks different from different angles (Figure 2-14).

1. **EKG waves and intervals** (Figure 2-15)

a. The **P wave** is due to atrial depolarization, and it is normally positive (upright) in the standard limb leads and inverted in aVR.

b. The **P-R interval** is measured from the onset of the P wave to the onset of the QRS complex, and it normally varies between 0.12 and 0.21 second depending on heart rate. The P-R interval is a measure of the **AV conduction time,** and it includes the delay through the AV node.

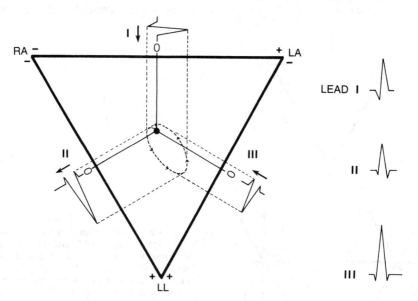

Figure 2-14. Reconstruction of the QRS complex from the depolarization loop for the three bipolar limb leads. *Arrows* indicate the direction of the recording. *RA* = right arm; *LL* = left leg; *LA* = left arm.

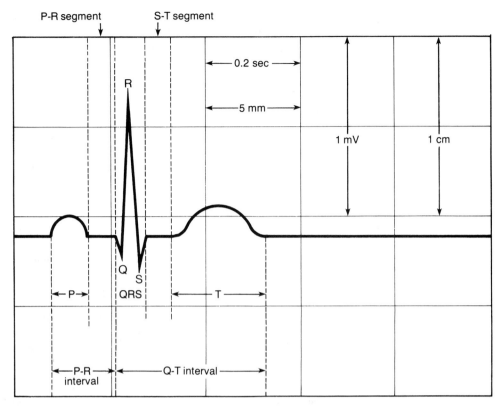

Figure 2-15. A typical electrocardiogram showing the various waves and segments as well as the standard voltages and times.

    c. The **QRS complex** is caused by ventricular depolarization, and its duration is normally less than 0.08 second. A prolongation of the QRS duration indicates an intraventricular conduction block caused by blockage of one of the bundle branches or the presence of an ectopic ventricular pacemaker.

    d. The **T wave** is caused by ventricular repolarization, and it is normally in the same direction as the QRS complex since ventricular repolarization follows a path that is opposite to depolarization.

    e. The **R-R interval** is the time between successive QRS complexes. The heart rate is equal to 60 divided by the R-R interval in seconds.

2. The **mean electrical axis (MEA)** is the average vector produced by any given wave in the EKG. The MEA can be derived for the P, QRS, and T waves by using any two standard limb leads or any two augmented limb leads. An augmented limb lead and a bipolar limb lead cannot be used together because of the difference in amplification of the two leads.

    a. Method of measurement. Clinically, the MEA of the QRS complex is determined by algebraically summing the heights of the Q, R, and S waves in each of two leads. Figure 2-16 shows this method.

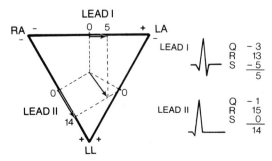

Figure 2-16. Vectorial analysis of mean electrical axis determination. The Q and S waves are summed and added algebraically to the R wave to give the magnitude of the vector in two leads. The magnitudes are plotted along the respective leads toward the appropriate polarity. Perpendiculars are drawn through the arrowheads, and the intersection marks the head of the mean electrical vector. The vector is drawn from the electrical zero (the center of the triangle) to the intersection of the perpendiculars. The mean electrical axis, in degrees, and the magnitude are given by the direction and length of the mean electrical vector (axis). *RA* = right arm; *LL* = left leg; *LA* = left arm.

    (1) Plotting of lead vectors. The vector from each lead is plotted on an appropriate scale, using either the Einthoven triangle, the triaxial, or the hexaxial reference system. The tail of the arrow is placed on the zero potential point, and the vector is drawn along the respective lead with the head of the arrow pointing toward the correct pole. Perpendiculars are then drawn to the heads of the two vectors.

    (2) Drawing the mean vector. The head of an arrow is drawn at the intersection of the two perpendiculars, and the tail is at the center of the triangle or the zero potential point. The length of this arrow represents the magnitude of the MEA, and its direction, in degrees, represents the electrical axis in the frontal plane.

  b. Axis deviation. The MEA of the QRS determines if axis deviation is present.

    (1) A **normal axis** is present if the MEA lies between −30° and +120°.

    (2) Right axis deviation (RAD) is present when the MEA lies between +120° and +180°. RAD is caused by right ventricular hypertrophy secondary to chronic lung disease or pulmonary valve stenosis or to delayed activation of the right ventricle as in right bundle branch block.

    (3) Left axis deviation (LAD) is present when the MEA lies between −30° and −90°. This condition is associated with obesity, left ventricular hypertrophy, or left bundle branch block.

    (4) An **indeterminate axis** is present if the MEA lies between −90° and −180°. Indeterminate axis may be due to either extreme right or extreme left axis deviation.

  3. Electrical orientation in the horizontal plane is determined using the precordial or chest leads.

    a. The **transition zone,** or **null point,** is defined as the electrode position in which the QRS is biphasic (i.e., when the algebraic sum of positive and negative waves is zero). The null point normally occurs between leads V_2 and V_3. A leftward shift in the null point is associated with left ventricular hypertrophy; a right shift is associated with right ventricular hypertrophy.

    b. The **electrical axis** in the horizontal plane is perpendicular to the null point.

D. Vector loops. The instantaneous cardiac vector represents the electrical vector generated by the cardiac dipole during the depolarization process. This vector begins at the **zero isopotential point** and inscribes a loop as the tissues are depolarized. Three loops can be recorded during one cardiac cycle. The **P loop** is caused by atrial depolarization, the **QRS loop** is caused by ventricular depolarization, and the **T loop** is due to ventricular repolarization (Figure 2-17A). Atrial repolarization cannot be recorded with standard techniques because of the prolonged time-course and the small voltages involved.

  1. The **P loop** is small and is directed leftward and inferiorly, resulting in a positive P wave in the three bipolar limb leads.

  2. The normal **QRS loop** is inscribed counterclockwise and is directed toward the left hip, posteriorly. The QRS loop generates the QRS complex on scalar EKGs and represents a specific sequence of activation. Ventricular depolarization is a continuous process, but it can be divided into stages for discussion (see Figure 2-17B).

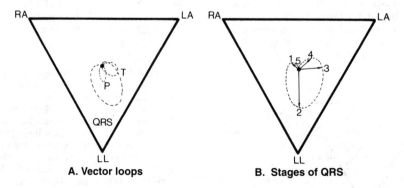

            A. Vector loops   **B. Stages of QRS**

Figure 2-17. (*A*) Schematic representation of the vectorcardiographic loops P, QRS, and T, which can be recorded with appropriate equipment and placement of electrodes. (*B*) A QRS loop that has been arbitrarily divided into five stages.

    a. Stage 1. The initial phase of ventricular depolarization involves the left endocardial surface of the interventricular septum. Depolarization spreads superiorly and to the right, resulting in a small vector directed toward the right shoulder. This initial vector produces small negative waves in leads I and II and small positive waves in the right precordial leads.

    b. Stage 2 represents depolarization of the remainder of the interventricular septum and the subendocardial areas of both ventricles. The resultant vector is directed inferiorly and produces large positive waves in leads II, III, and aVF.

    c. Stage 3 depolarization involves the remainder of the right ventricle and a large portion of the left ventricle. Since the left ventricle normally has a larger muscle mass than the right, the stage 3 vector is directed toward the left. This vector produces **R waves** in lead I and causes a progressive increase in the R waves in the precordial leads V_2 to V_6.

    d. Stage 4 represents the last portion of the ventricle to depolarize, which is the posterior base of the left ventricle. This vector is directed toward the left shoulder and produces small **S waves** in leads III and aVF.

    e. Stage 5 is the return of the cardiac potential to the zero potential point and the end of the depolarization process.

   3. T loop. The direction of the repolarization process is roughly opposite to that of depolarization. This occurs because the inner layers of the myocardium are slightly hypoxic due to vascular compression during systole, and hypoxia prolongs the relative refractory period and the onset of repolarization. The opposite direction of repolarization compared to depolarization results in T waves that normally are in the same direction as the QRS complex.

IV. CARDIAC RATE, RHYTHM, AND CONDUCTION DISTURBANCES

A. Cardiac rate

   1. Normocardia is a normal resting heart rate of 60–100 beats/min.

   2. Tachycardia is a heart rate in excess of 100 beats/min.

   3. Bradycardia is defined as a heart rate that is less than 60 beats/min, which commonly is seen in well-trained athletes.

B. Cardiac rhythms

   1. Sinus rhythm is present when the SA node is the pacemaker; sinus rhythm is assumed if each P wave is followed by a normal QRS complex, the P-R and Q-T intervals are normal, and the R-R interval is regular.

    a. Sinus arrhythmia is present if all of the above criteria apply except that the R-R interval (cardiac rate) varies with respiration. Heart rate normally increases during inspiration and slows during expiration due to variations in vagal tone that affect the SA node. Sinus arrhythmia is common in children and in endurance athletes with slow heart rates.

    b. Sinus tachycardia is a normal response to exercise and also occurs in the presence of fever, hyperthyroidism, and as a reflex response to low arterial pressures.

   2. Atrial rhythms

    a. Atrial tachycardia occurs when an ectopic atrial site (outside the SA node) becomes the dominant pacemaker.

     (1) Characteristics. Atrial rhythms are characterized by very regular rates ranging from 140 to 220 beats/min. Usually, atrial tachycardias have a very rapid onset and last only for a few seconds or minutes, which is why they are termed **paroxysmal atrial tachycardia (PAT)**. Occasionally, these tachycardias may last for hours or days. A prolonged bout of atrial tachycardia is termed a **supraventricular tachycardia** since it is usually not possible to determine the actual pacemaker site.

     (2) Causes. PATs may be precipitated by overindulgence in caffeine, nicotine, or alcohol, or they may occur during anxiety attacks. PAT may be due to discharge of a single ectopic site, or it may be due to a **reentry phenomenon.**

      (a) Reentry occurs when a cardiac impulse reexcites an area of the heart that previously had been depolarized. This phenomenon requires parallel paths of depolarization that have different conduction velocities and refractory periods.

      (b) Fast fibers regain excitability by the time the impulse arrives from the slow conducting fibers over connecting paths, and the fast fibers are reexcited.

     (3) Treatment. Attacks of PAT can usually be interrupted by increasing vagal tone (carotid sinus massage or a Valsalva maneuver); in persistent cases of PAT, administration of digoxin or verapamil may be required, or, infrequently, cardioversion may be necessary to break the attack.

  b. Premature atrial contraction (PAC)

     (1) Cause. PACs (also known as atrial extrasystoles) occur if an atrial ectopic site fires and becomes the pacemaker for one beat.

     (2) Significance. These beats are entirely normal, and the patient may note an occasional irregularity in the cardiac rhythm.

     (3) Diagnosis. The condition can be readily diagnosed if it occurs during an EKG tracing, since a premature P wave is present followed by a normal QRS complex and a T wave. The premature P wave may be of different configuration than normal, and the P-R interval may also be altered due to the different path of depolarization that occurs. The condition can usually be diagnosed at the bedside since the premature impulse discharges the SA node, which then must repolarize and fire after the normal interval. This results in a shift in cardiac rhythm (Figure 2-18A). This condition is benign and usually does not require treatment.

  c. Atrial flutter occurs when atrial rates are 220–350 beats/min. During atrial flutter the AV node is unable to transmit all of the atrial impulses so that a physiologic AV block develops, and the ventricular rate is half, one-third, or one-fourth of the atrial rate.

     (1) Cause. Atrial flutter may be due to a single ectopic focus or to a reentry phenomenon.

     (2) Characteristics. The ventricular rate may be normal and very regular, although rapid changes in ventricular rate may occur as the AV block changes from, for example, 4:1 to 3:1. The P waves produce a saw-toothed appearance on the EKG, which is virtually diagnostic of atrial flutter.

     (3) Significance. Atrial flutter differs from atrial tachycardia only in the atrial rate, and it has a similar prognosis and treatment.

  d. Atrial fibrillation is a totally irregular, rapid atrial rate in which there is contraction of only small portions of the atrial musculature at any one time, since large portions of the atria are still refractory to depolarization. The ventricular rate is completely (irregularly) irregular since only a fraction of the atrial impulses that reach the AV node are transmitted to the ventricles.

A. Premature atrial contraction

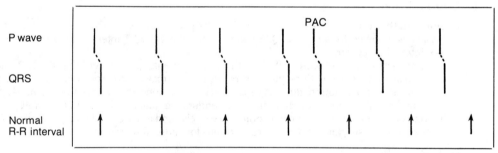

B. Premature ventricular contraction

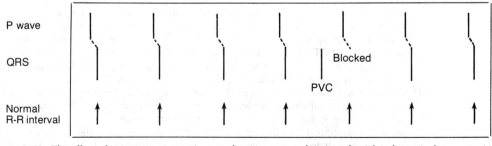

Figure 2-18. The effect of premature contractions on the sequence and timing of atrial and ventricular events. (*A*) A premature atrial contraction (*PAC*) causes a permanent shift in the atrial and ventricular events. (*B*) A premature ventricular contraction (*PVC*) causes no permanent shift in rhythm, since the PVC and the subsequent beat typically equal two normal cycle lengths.

(1) **Characteristics.** The baseline of the EKG shows small, irregular oscillations (**F waves**) due to the depolarization of small units of the atrial musculature. There are no recognizable P waves, and the R-R interval is irregularly irregular. The QRS and T waves are normal since the impulses that are transmitted through the AV node are normally conducted through the ventricles.

(2) **Significance.** Atrial fibrillation represents an even higher rate of atrial activity than atrial flutter. Atrial fibrillation frequently is associated with enlarged atria secondary to AV valve disease. The cardiac output is reduced due to the valve disease as well as to the loss of an effective atrial contraction. Prolonged periods of atrial fibrillation are associated with the presence of thrombi in the atrial appendages. **Atrial thrombi** may be the source of **pulmonary emboli** (right atrium) or **systemic emboli** (left atrium).

(3) **Treatment.** Atrial fibrillation should be converted to normal sinus rhythm by cardioversion (defibrillation) once the presence of atrial thrombi is ruled out. The prognosis of atrial fibrillation depends on the underlying cause.

3. **AV junctional (nodal) rhythms.** Pacemaker cells have been found only in the A-N and N-H zones of the AV node; thus, the term junctional rhythm, or junctional arrhythmia, is used for ectopic pacing from the AV node.

   a. **Junctional premature beats** are characterized by an inverted P wave and normal QRS complex. They have been subdivided into **high** or **low** junctional beats depending on whether the P wave precedes or follows the onset of the QRS complex. Junctional beats have the same significance as PAC. If the AV nodal tissues recover excitability, a retrograde impulse may return and cause a second ventricular depolarization, which is known as a **reciprocal beat**.

   b. **Transient or permanent junctional rhythms** may arise in normal subjects due to suppression of the SA node. Permanent junctional rhythms may arise secondary to a large number of organic heart diseases.

   c. **Junctional tachycardias** are similar to atrial tachycardias and may be indistinguishable by EKG; therefore, they should be included in the supraventricular tachycardias.

4. **Ventricular rhythms**

   a. A **premature ventricular contraction (PVC)** can arise from any portion of the ventricular myocardium, and, occasionally, it can occur in normal people. Frequent PVCs occur with any form of heart disease, especially coronary heart disease, since ischemia increases the irritability of the myocardium.

      (1) **Characteristics** of a PVC are a prolonged (greater than 0.1 second), bizarre QRS complex that does not have a preceding P wave. The T wave is usually oppositely directed from the QRS complex. The origin of the PVC can be determined by vectorial analysis using the initial portion of the QRS complex.

      (2) **Compensatory pause.** Retrograde transmission of depolarization to the atria usually does not occur with PVCs; thus, the atrial rate remains unaltered.

         (a) The atrial depolarization that follows a PVC usually arrives while the AV node is still refractory and, thus, is not conducted to the ventricles. This creates a pause in the ventricular rhythm.

         (b) This pause is usually fully compensatory so that the R-R interval of the beat preceding the PVC plus the PVC interval is equal to two normal cycle lengths (see Figure 2-18B). The beat following the PVC is stronger than normal, due to the added stroke volume, and is detectable by the patient as a **palpitation of the heart**.

      (3) **Interpolated beats.** If the sinus rhythm is slow, a PVC may occur without altering the normal R-R interval. The premature beat in this case is termed an interpolated beat.

   b. **Ventricular tachycardias** result from a rapid, repetitive discharge from a ventricular site, usually due to a reentry phenomenon. Alterations in vagal tone (carotid sinus massage, Valsalva maneuver) do not affect ventricular tachycardias since the ventricles do not receive any vagal innervation.

      (1) **Characteristics.** The EKG reveals wide, bizarre QRS complexes occurring at a rapid rate. The P waves are usually indistinguishable, although the SA node activity continues independently of the ventricles.

      (2) **Significance.** Ventricular tachycardias are almost always associated with serious heart disease or digitalis or quinidine toxicity. However, ventricular tachycardias can also be seen when intracardiac catheters stimulate the endocardial surface of the ventricles. A slight movement or withdrawal of the catheter may terminate this condition. The cardiac

output is reduced during ventricular tachycardias due to a reduced filling time and a decreased effectiveness of ventricular contraction due to asynchrony.

(3) **Treatment.** Cardioversion is frequently necessary to terminate an attack of ventricular tachycardia due to organic heart disease.

c. **Ventricular fibrillation** is due to rapid, irregular, ineffective contractions of small segments of the ventricular myocardium. The peripheral pulse is absent, and cardiac output is zero. The diagnosis of ventricular fibrillation must be made by EKG since this condition cannot be distinguished from cardiac standstill by physical signs.

(1) **Characteristics.** The EKG shows undulating waves of varying frequency and amplitude. Ventricular fibrillation is often precipitated by one or more PVCs, one of which falls on the vulnerable interval of the T wave.

(2) **Treatment.** Cardiopulmonary resuscitation must be started immediately until cardioversion can be carried out.

C. Conduction disturbances

1. **SA nodal block.** This condition, also termed the **sick sinus syndrome,** consists of the disappearance of the P wave for many seconds while an ectopic pacemaker, usually a junctional or ventricular site, drives the ventricles. SA block results in bradycardia and is seen especially in the elderly or after a coronary occlusion. Various drugs and implanted pacemakers have been used to increase the heart rate to normal levels.

2. **AV nodal block**
   a. **First-degree AV nodal block** is a prolongation of the P-R interval beyond 0.21 second due to a slowed conduction through the AV node. This condition has no effect on the pumping ability of the ventricles but indicates that some conduction disturbance is present. First-degree AV block can be caused by increased vagal tone, quinidine, acute infectious diseases, and many diseases affecting the heart.
   b. **Second-degree AV nodal block** is characterized by the failure of the AV node to transmit all of the atrial impulses.
      (1) **Wenckebach block (Mobitz type I)** is characterized by a progressive lengthening of the P-R interval in successive beats and finally a failure of one impulse to be transmitted.
      (2) **Periodic block (Mobitz type II)** is characterized by an occasional failure of conduction, which results in an atrial-to-ventricular rate of, for example, 6:5 or 8:7. The P-R interval is constant in this condition.
      (3) **Constant block.** This form of second-degree block represents a higher degree of block than the periodic type, and the atrial-to-ventricular rate is a constant small number ratio (e.g., 2:1, 3:1).
   c. **Third-degree (complete) AV block** causes the atria and ventricles to beat at independent rates due to a complete interruption of AV conduction.
   d. **Significance** of AV block depends upon the underlying cause. First-degree AV block due to increased vagal tone is of no consequence. Second- and third-degree AV blocks are usually due to organic heart disease. The onset of third-degree AV block may be associated with prolonged ventricular standstill until a ventricular focus begins firing. The cardiac arrest, if prolonged, can result in cerebral ischemia and syncope; this condition is **Stokes-Adams syndrome.**

3. **Wolff-Parkinson-White syndrome (ventricular pre-excitation or accelerated conduction)** is due to an aberrant conduction pathway between the atria and the ventricles. The aberrant pathway eliminates or reduces the normal delay between atrial and ventricular activation.
   a. **Characteristic pattern.** The EKG shows a shortened P-R interval, usually less than 0.1 second and, classically, a widened QRS complex with a slurred initial upstroke (**delta wave**). The P-J interval may be normal or shortened, depending on whether the accessory bundle inserts into the ventricular myocardium or the bundle of His, respectively.
   b. **Significance.** The major significance of the accessory bundle is that it predisposes to paroxysmal tachycardias due to a reentry phenomenon. The atrial impulse is normally conducted to the ventricles and returns to the atria via the accessory bundle, thus triggering the reentry mechanism.
   c. **Treatment.** Surgical ablation of the accessory bundle is possible if it can be adequately localized.

4. **Bundle branch block.** Blockage of the right or left common bundles or the left anterior or posterior fascicles results in an abnormal sequence of ventricular excitation.
    a. **Right bundle branch block** delays the activation of the right ventricle, and the QRS duration is prolonged beyond 0.12 second. The initial QRS vectors are shifted to the left due to unopposed activation of the left ventricle, and the final vectors are directed toward the right as excitation spreads to the right ventricle through the myocardium. This sequence of excitation produces the typical rSR′ pattern in the right precordial leads, a wide S wave in lead I, and a right axis deviation of the mean electrical axis. Repolarization of the right ventricle follows the same path as depolarization, which causes an inversion of the T wave in the right precordial leads.
        (1) **Incomplete right bundle branch block** produces similar QRS patterns as right bundle branch block, except that the QRS duration is 0.08–0.10 second.
        (2) **Significance.** Right bundle branch block can occur in a normal individual as a transient or a permanent manifestation. It occurs secondary to chronic pulmonary disease and may appear acutely as a consequence of pulmonary embolism. Right bundle branch block also occurs in conjunction with some forms of congenital heart disease such as interatrial septal defect.
    b. **Left bundle branch block** occurs in almost any form of heart disease, and its occurrence is rare in the absence of organic heart disease. It is common in association with coronary artery disease or conditions leading to left ventricular hypertrophy (hypertension, aortic stenosis). Left bundle branch block is best diagnosed using the left precordial leads, which show a QRS duration of greater than 0.12 second and also show initial and final QRS vectors directed toward the left. The initial vector is due to septal activation from right to left, and the final vector is due to delayed activation of the left ventricle.
    c. **Other forms of intraventricular block.** Intraventricular conduction block may involve all three fascicles of the bundle branches, a peri-infarction block, or a block in the peripheral Purkinje network.

D. **Other cardiac dysfunctions**

1. **Coronary insufficiency and myocardial infarction** are due to a narrowing or occlusion of a coronary artery, resulting in the reduction or absence of blood flow to an area of the myocardium. The most common cause of coronary artery narrowing is an **arteriosclerotic plaque,** and the most common cause of coronary occlusion is an **arteriosclerotic coronary thrombosis.** Reduction of coronary blood flow results in myocardial ischemia or infarction.
    a. **Myocardial ischemia**
        (1) **Symptoms.** The cardinal sign of coronary artery insufficiency is **anginal pain** related either to exertion or emotional stress. The pain may be precordial or may radiate to the shoulder, arm, or jaw. The pain is typically described as a tight band or weight upon the chest, which results in a strangling sensation; thus, the term **angina pectoris** is truly apt. The pain of coronary insufficiency typically lasts 1–10 minutes and is relieved by rest.
        (2) **Diagnosis** is usually made from patient history, since the resting EKG may be within normal limits. Twenty-four-hour EKG (Holter) monitoring may provide evidence of ischemic changes during anginal attacks. Stress (exercise) tests with EKG monitoring are used to precipitate ischemic episodes for diagnostic purposes.
        (3) **EKG pattern.** Myocardial ischemia results in S-T segment and T-wave changes. An ischemic area of the myocardium has a reduced membrane potential compared to normal regions of the heart, which then results in a current flow from the normal regions to the ischemic area. This current flow is termed a "current of injury" and results in an S-T segment elevation from leads overlying the ischemic area, whereas S-T segment depression occurs in leads on the opposite side of the heart. Ischemia also alters the path of repolarization and results in T-wave inversion.
        (4) **Treatment.** The major aim is to reduce the work of the heart either with **vasodilators (nitroglycerin)** or a **β-adrenergic blocker (propranolol),** which reduces heart rate.
    b. **Myocardial infarction** occurs when blood flow ceases or is reduced below a critical level. The left ventricle is almost always the site of a myocardial infarction.
        (1) **Symptoms** of myocardial infarction are similar to those of coronary insufficiency except that the pain is more intense, generally lasts for more than 15–20 minutes, may not be related to exertion or exercise, and is not relieved by nitrates.
        (2) **Diagnosis.** The history is critical in making the diagnosis of a myocardial infarction, while the EKG provides supportive evidence. Measurement of serum levels of cardiac

enzymes creatine phosphokinase (CPK), serum glutamic-oxaloacetic transaminase (SGOT), and lactic acid dehydrogenase (LDH) may aid in the diagnosis, but no enzyme is specific for the myocardium.

(3) EKG pattern. The EKG undergoes a series of changes following a myocardial infarction, and these changes should be recorded with daily EKG tracings for diagnostic purposes. A **transmural infarct** results in wide, deep Q waves from leads overlying the infarcted area since these leads essentially record an intracavitary potential. These same leads demonstrate an S-T segment elevation, and, early after the infarct, there is an inversion of the T wave.

(4) Treatment. The major aims are the relief of pain and the support of the circulation if cardiogenic shock develops. In addition, continuous EKG monitoring should be carried out in an intensive care unit to detect the presence of arrhythmias, especially PVCs, which may trigger bouts of ventricular tachycardia or ventricular fibrillation. Cardioversion may be necessary to interrupt arrhythmias, or drug therapy (membrane stabilizers) may be required to prevent the recurrence of arrhythmias.

2. **Ventricular hypertrophy** occurs if the work of one or both ventricles is increased sufficiently. Ventricular hypertrophy increases the diameter but not the number of individual myocardial cells. Thus, the diffusion distance increases for O_2 and other metabolites. The increased diffusion distance may lead to borderline ischemic conditions.

a. Diagnostic criteria for ventricular hypertrophy

(1) R wave. There is a direct correlation between the thickness of the ventricular wall and the height of the R wave in the overlying leads. The increased height of the R wave is a reflection of the increased magnitude of the depolarization vector.

(2) QRS duration is increased slightly because of the increased muscle mass present. The duration of the QRS complex is usually less than 0.12 second, but, occasionally, it may exceed this value.

(3) S-T segment and T-wave alterations. In the presence of ventricular hypertrophy there is S-T segment depression and inversion of the T wave in epicardial leads. The altered repolarization process may be related to endocardial fibrosis and mild ischemia, which are commonly present with ventricular hypertrophy. If changes in the S-T segment and T wave are the only manifestations, the condition is referred to as a **ventricular strain pattern**.

b. Diagnostic criteria for left ventricular hypertrophy

(1) Axis. The mean electrical axis is shifted toward the left and superiorly, and the transition zone of the precordial leads is shifted to the left.

(2) QRS configuration. Left ventricular hypertrophy is present if the R wave in V_5 or V_6 plus the S wave in V_1 is greater than 35 mm, the R wave in V_5 or V_6 is greater than 25 mm, the R in aVL is greater than 13 mm, or the R in aVF is greater than 25 mm. S-T segment and T-wave changes may or may not be present.

c. Diagnostic criteria for right ventricular hypertrophy

(1) Axis. The mean electrical axis is usually vertical or greater than $+110°$ in the frontal plane.

(2) QRS configuration. Right precordial leads generally show tall R waves rather than the normal S waves. The QRS is usually prolonged but less than 0.12 second, the S-T segment is depressed, and the T wave is inverted in the right precordial leads.

(3) Diagnostic criteria for right ventricular strain (acute cor pulmonale). The appearance of S-T segment and T-wave changes without abnormal R waves in the right precordial leads is consistent with acute cor pulmonale. This condition occurs with acute exacerbations of lung disease, and it is an important sign in the detection of pulmonary emboli.

V. THE HEART AS A PUMP

A. Pump components. The heart is two pumps in series (right and left sides), which are connected by the pulmonary and systemic circulations. Each side of the heart is equipped with two valves that normally maintain a one-way flow of blood.

1. The **AV valves** separate the atria from the ventricles and open during diastole to allow the blood to fill the ventricles. The AV valves close during systole to prevent the back flow (regurgitation) of blood from the ventricles into the atria. The right AV valve is called the **tricuspid valve;** the left AV valve is the **mitral or bicuspid valve.**

2. The **semilunar valves** (aortic and pulmonic) open to allow the ventricles to eject blood into the arteries during systole and close to prevent back flow of blood into the ventricles during diastole.

3. The **right ventricle** is well designed to pump relatively large volumes of blood at relatively low pressures through the pulmonary circulation.

4. The **left ventricle** is thick walled and can pump effectively against the higher pressures in the systemic circulation. The left ventricle works much harder than the right ventricle because of the high pressures and consequently is more commonly affected by disease processes than the right ventricle.

B. **Stroke volume** is the volume of blood that is ejected with each beat of the heart. Both ventricles must pump the same volume of blood during any significant time interval. Normally, the stroke volume of each ventricle averages 70 ml of blood, but the stroke volume varies with changes in ventricular contractility, arterial pressure (afterload), and end-diastolic volume (Figure 2-19).

1. **Effects of contractility**
   a. **Increased contractility (positive inotropism)** results in a stronger contraction so that the ventricles are able to eject more blood from the same diastolic volume, and stroke volume increases. An increased contractility occurs with sympathetic stimulation to the heart (e.g., during exercise).
   b. **Decreased contractility (negative inotropism)** occurs in heart failure and results in a decrease of stroke volume. In this condition, the ventricle increases in size because of the rise in residual ventricular volume. The increased ventricular volume tends to raise stroke volume toward normal through the **Frank-Starling mechanism** (see Figure 2-9). Cardiac output tends to be maintained by increasing the heart rate. In severe ventricular failure both cardiac output and stroke volume are decreased and the ventricular diastolic pressure is high. The increased ventricular diastolic pressure causes a high venous pressure, which is reflected back to the capillaries and produces leakage of fluid into the tissues (edema).
      (1) In **left heart failure,** fluid accumulates in the lungs (**pulmonary edema**), causing difficulty in breathing (**dyspnea**) and the inability to breathe when lying down (**orthopnea**).
      (2) During **right heart failure,** edema occurs in the feet and ankles (**dependent edema**) since the high venous pressures occur in the systemic veins.

2. **Arterial pressure** during systole affects the ejection of blood from the ventricle and is termed the **ventricular afterload.** The higher the arterial pressure, the greater the afterload that the ventricle must overcome. An increase in arterial pressure causes a reduction in the stroke volume (see Figure 2-19B) and a decrease in the velocity of ejection from the ventricle (see Figure 2-10).

3. **Ventricular end-diastolic volume** affects the stroke volume because of the Frank-Starling relationship. The greater the ventricular end-diastolic volume, the stronger the contraction and the larger the stroke volume. The ventricular end-diastolic volume increases if the pressure in the atria or the central veins increases, because the higher pressure can stretch the ventricular walls. The ventricular end-diastolic volume also increases if the distensibility of the ventricles increases or if the heart rate slows, which allows more time for the blood to enter the ventricles.

4. **Cardiac work** is proportional to the pressure generated by the ventricular contraction and to the stroke volume. Since both ventricles pump the same average stroke volume and since the right ventricular systolic pressure is about one-seventh that of the left ventricle, then the work of the right ventricle is about one-seventh that of the left ventricle.
   a. **Hypertension.** The work of the ventricles is markedly increased if the blood pressure is high (hypertension). Hypertension can occur in either the right or left side of the circulation. **Pulmonary,** or **right-sided, hypertension** almost invariably is caused by some chronic lung disease, whereas **systemic hypertension** most commonly is caused by an unknown mechanism and is termed **essential hypertension.** Hypertension can also be caused by endocrine, renal, or congenital conditions.
   b. **Hypertrophy.** The increase in the work of the heart causes the ventricular muscle to enlarge just like any other muscle if it is exercised sufficiently. The enlargement (hypertrophy) may be harmful if the blood supply that provides O_2 does not increase in proportion to the size of the muscle.

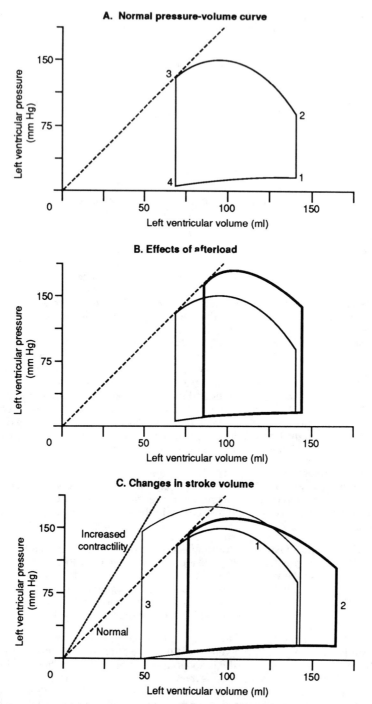

Figure 2-19. Ventricular pressure-volume (PV) relationships and the effects of changes in afterload and end-diastolic volume. (*A*) Normal PV loop. *1* = beginning of isovolumic contraction; *2* = onset of ejection; *3* = beginning of isovolumic relaxation; *4* = onset of ventricular filling. The *dashed line* represents the maximal PV relationship, which is determined by the level of contractility. (*B*) Increased afterload (*thick line*) results in a reduction of stroke volume since the myocardial shortening is abbreviated and the end-systolic volume is increased compared to normal. (*C*) *Curve 1* represents control conditions. An increase in end-diastolic volume (*curve 2*) results in an increased stroke volume but the maximal PV relationship is unchanged. The slight increase in end-systolic volume is caused by the increased aortic pressure resulting from the increased stroke volume. An increased contractility (*curve 3*) results in an increased stroke volume by reducing the end-systolic volume since the maximal PV line is shifted to the left and has a steeper slope.

    c. Myocardial O$_2$ consumption. The amount of O$_2$ utilized by the heart depends on the amount of work that is performed. The O$_2$ consumption can be calculated from the difference in O$_2$ content in the coronary arterial and venous blood and the coronary blood flow.

  5. Myocardial O$_2$ supply. The heart is able to function without any blood flow for only very short periods of time; thus, normal cardiac function requires a continuous delivery of O$_2$ to the heart. The O$_2$ delivery to any tissue is equal to the arterial O$_2$ content (the volume of O$_2$ carried per 100 ml of blood) multiplied by the blood flow. A reduction in O$_2$ delivery to the heart is almost always due to a partial or complete obstruction of a coronary artery by fatty deposits in the lining (**atherosclerosis**).

VI. CARDIODYNAMICS refers to the mechanical events that are associated with the contraction of the heart.

  A. Cardiac cycle events (Figure 2-20). The cardiac cycle includes both **electrical** (EKG) and **mechanical** (muscle contraction) events within the cardiovascular system. The pressure (potential energy) is generated by myocardial contraction and is converted into flow (kinetic energy) in order to supply all of the body cells with nutrients and to serve other circulatory functions. A full understanding of the events of the cardiac cycle is essential to understand the physical signs of cardiovascular disease that are elicited during the physical examination of a patient.

    1. EKG. The electrical events precede and initiate the corresponding mechanical events; thus, the P wave precedes atrial contraction, and the QRS complex precedes ventricular contraction.

    2. Atrial contraction. The two atria contract almost simultaneously due to the specialized interatrial tracts that carry the excitation between the atria. The increase in atrial pressure that results from atrial contraction is termed the **a wave** and is reflected back into the large veins, where it can be recorded from the jugular vein with appropriate transducers. The level of filling of the

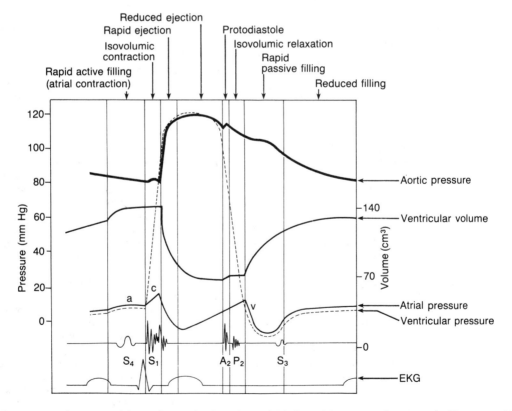

Figure 2-20. The events of the cardiac cycle. S_1, S_2 (A_2 and P_2), S_3, and S_4 represent heart sounds. The a wave (*a*) is caused by atrial contraction. The c wave (*c*) is caused by increased pressure due to the AV valves ballooning in the atria during isovolumic contraction. The v wave (*v*) indicates the opening of the AV valves.

jugular veins provides an estimate of the right ventricular filling pressure. The pressure of the blood in the ventricles at the termination of the atrial contraction is termed the **ventricular end-diastolic pressure;** this represents the maximal ventricular volume, which sets the preload for the next cardiac contraction. The ventricular end-diastolic pressure for the left ventricle is normally less than 12 mm Hg.

3. **Ventricular contraction (systole)**
    a. **AV valve closure**
       (1) Shortly after the QRS complex begins, the pressure in the ventricular cavities rises due to the onset of ventricular contraction. The rising ventricular pressure now exceeds the atrial pressure and causes the AV valves to close.
       (2) Closure of the AV valves is the major component in generating the **first heart sound (S_1).** The first heart sound is a low-frequency sound that occurs just after the onset of ventricular contraction and that signals the onset of ventricular systole.
    b. **Isovolumetric (isovolumic) contraction.** Once the AV valves close, the ventricular chambers are sealed from both the atria and the arteries. The ventricular volume remains constant (isovolumic) until the ventricular pressure exceeds the pressure in the respective artery (aorta or pulmonary artery); at this point ventricular ejection begins.
    c. **Rapid ejection.** The onset of ventricular ejection is indicated by the rise in aortic and pulmonary artery pressures and the decrease in ventricular volume. The initial part of ejection is rapid; about two-thirds of the stroke volume is ejected in the first third of systole. The end of the rapid ejection phase occurs at about the peak of ventricular and arterial systolic pressure. The right ventricular ejection begins before that of the left and continues after left ventricular ejection is complete. Since both ventricles eject (on the average) the same volume of blood, the velocity of right ventricular ejection is less than that of the left ventricle.
    d. **Reduced ejection.** During the latter two-thirds of systole the rate of ejection declines, and the ventricles begin to relax. Both ventricular and arterial pressures begin to decrease as the rate of blood flow through the peripheral vessels exceeds the rate of blood flow from the ventricles. The volume of blood remaining in the ventricles at the end of ejection is termed the **end-systolic volume.**
    e. **Second heart sound (S_2).** The S_2 is a relatively high-frequency sound that occurs as the semilunar valves snap closed. The valve closure stops the movement of the blood toward the ventricles and causes vibrations of the tissues that are heard as the S_2. The closure of the semilunar valves also causes the dicrotic notch in the downslope of the arterial pressure pulse, the **incisura.** The incisura or S_2 indicates the end of systole and the onset of the phase of diastole.

4. **Ventricular diastole** refers to the relaxation phase of the cardiac cycle.
    a. **Isovolumetric (isovolumic) relaxation.** Ventricular pressure falls rapidly as the ventricles continue to relax; isovolumic relaxation only lasts about 0.04 second. The ventricular volume remains constant from the closure of the semilunar valves (S_2) until the AV valves open, which occurs when ventricular pressure falls below atrial pressure.
    b. **Rapid passive filling.** During ventricular systole the venous return continues so that the atrial pressure is high when the AV valves first open. The high atrial pressure causes a rapid, initial flow of blood into the ventricles. Once the AV valves open, the atria and ventricles are a common chamber and the pressure in both cavities falls as ventricular relaxation continues.
    c. **Reduced filling and diastasis.** Toward the end of diastole, the pressure in the atria and ventricles rises slowly as blood continues to return to the heart. With very slow heart rates, ventricular filling virtually ceases since the ventricles reach their volume limit. This phase is termed **diastasis.** When there is an increase in heart rate the length of the cardiac cycle is obviously shorter. The shortening of the cycle occurs primarily by an abbreviation of the phase of diastasis or reduced filling. Thus, increases in heart rate to 140–150 beats/min do not cause a significant decrease in ventricular filling. Rapid heart rates produce an increased cardiac output until rates of 180–200 beats/min are achieved, because the increase in rate more than compensates for any slight decrease in ventricular filling. At heart rates in excess of 200 beats/min the cardiac output declines due to a decrease in the ventricular filling (ventricular end-diastolic volume is markedly reduced).
    d. **The third and fourth heart sounds (S_3 and S_4)** occur during diastole and are not normally heard in adults. S_3 occurs during the rapid passive filling, whereas S_4 occurs during atrial

contraction. S_3 can be heard in children, and S_4 is sometimes heard in people with congestive heart failure.

5. **Aortic and pulmonary arterial pressures.** The recordings of the pulsatile pressure changes in these vessels are termed **pressure pulses**.

   a. **Systole.** During the period of rapid ventricular ejection, the pressure in the aorta is slightly less than that in the ventricle. The peak of arterial pressure is the **arterial systolic pressure** and occurs at the end of the rapid ejection phase. The normal aortic systolic pressure is 120 mm Hg; pulmonary arterial systolic pressure averages 15–18 mm Hg. The incisura, during the downslope of the arterial pressure, indicates the end of ventricular systole.

   b. **Diastole.** During diastole the arterial pressure declines as blood continues to flow from the arteries out through the capillaries and into the veins. The lowest pressure in the artery during a cardiac cycle is termed the **arterial diastolic pressure,** and it occurs just prior to the onset of the next ventricular ejection. The average diastolic pressure in the aorta is about 80 mm Hg; pulmonary arterial diastolic pressure averages 8–10 mm Hg.

   c. **Pulse pressure** is the difference between the systolic and the diastolic pressures, which normally is 120 minus 80, or 40 mm Hg, for the aorta.

   d. **Mean arterial pressure (MAP)** can be obtained by integrating the area under the arterial pressure pulse and dividing by the cycle length. MAP can be approximated by adding one-third of the pulse pressure to the arterial diastolic pressure.

   e. **Recorded blood pressure** is normally written as arterial systolic pressure/diastolic pressure, or 120/80 mm Hg.

6. **Ventricular volume** is difficult to measure clinically, but **echocardiography** can be used to measure ventricular dimensions as well as ventricular wall and valve motion during the cardiac cycle.

   a. **Ventricular end-diastolic volume** is the volume of blood in the ventricle just prior to the onset of ventricular contraction. The normal left ventricular end-diastolic volume is 120–140 ml.

   b. **Ventricular end-systolic volume** is the volume of blood remaining in the ventricle at the end of ejection. The normal left ventricular end-systolic volume is 40–70 ml.

   c. **Stroke volume** is the volume of blood that is ejected with each beat and equals ventricular end-diastolic volume minus ventricular end-systolic volume, or 75–80 ml.

B. **Valve lesions and murmurs** (Figure 2-21). The cardiac valves may be abnormal due to either congenital or acquired heart disease. The circulatory effect of valve lesions depends on the reduction in cardiac output, the additional work imposed upon the heart by the extra pressure or volume load that is generated, and the back up of blood that is produced by a particular lesion. Abnormal cardiac valves produce signs of dysfunction that are recognizable during physical examination. The physical signs are due to the abnormal blood flow and pressure gradients.

1. **Stenosis of a valve** is a narrowing of the valve orifice that retards the forward flow of blood. Aortic or pulmonic valve stenosis causes the left or right ventricles, respectively, to generate higher pressures in order to eject blood through the narrowed orifice. In these conditions, the ventricular systolic pressure is much higher than the systolic pressure in the artery. This pressure difference can be detected during cardiac catheterization. Auscultation of the chest reveals a systolic, ejection-type murmur. These stenotic lesions can lead to ventricular hypertrophy and eventually produce ventricular failure.

2. **Insufficiency of a valve** allows the backward flow of blood; for example, with aortic insufficiency there is a flow of blood from the aorta to the left ventricle during diastole, which reduces the effective cardiac output. However, with aortic insufficiency, the left ventricle ejects a large volume of blood so that systolic pressure increases markedly. During diastole, much of the stroke volume returns to the ventricle, resulting in a very large pulse pressure, which is characteristic of aortic insufficiency. An insufficiency of an AV valve allows blood to flow from ventricle to atrium during systole. Dilation of one or more heart chambers occurs to compensate for the regurgitant blood and can eventually result in heart failure.

3. **Murmurs.** An important physical sign of a valve lesion is a murmur. Murmurs are caused by turbulent blood flow resulting from narrowing of a valve or changes in the direction of the blood flow. Normal flow is laminar or streamlined and is silent, whereas turbulent flow sets up vibrations in the tissues that can be heard as murmurs.

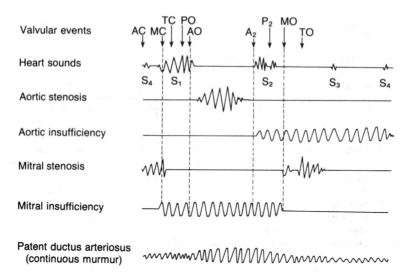

Figure 2-21. Schematic representation of the heart sounds (S_1 through S_4), valvular events, and murmurs generated by left-sided valve lesions. Right-sided valve lesions begin and end in association with right heart valve events (not shown). *AC* = atrial contraction; *MC* = mitral valve closure; *TC* = tricuspid valve closure; *PO* = pulmonic valve opening; *AO* = aortic valve opening; A_2 = aortic valve closure; P_2 = pulmonic valve closure; *MO* = mitral valve opening; *TO* = tricuspid valve opening.

    a. Timing. Murmurs may be either systolic, diastolic, or continuous throughout the cardiac cycle. Murmurs may also be classified as being presystolic, early diastolic, or middiastolic. The timing of the murmurs is associated with specific valve lesions.

    b. Location and radiation. Murmurs are best heard on the chest wall closest to their origin and in the downstream direction of blood flow. For example, the murmur of aortic stenosis is best heard at the right sternal border in the second interspace, and the murmur radiates to, and is also clearly heard over, the vessels of the neck.

VII. HEMODYNAMICS

A. Basic units

    1. Pressure gradient (dP). Fluid flow depends on a difference in total energy between two locations within the system. The fluid always flows from an area of high pressure to one of lower pressure; in other words, water (or blood) always flows downhill.

        a. The **total energy** of a system equals the sum of the potential energy (pressure) and the kinetic energy, which is due to the velocity of the blood. The energy can be converted between potential and kinetic energy. Figure 2-22 is a system of pitot tubes, which can be used to measure changes in flow velocity. A pitot tube facing upstream measures both potential and kinetic energy, whereas a pitot tube facing downstream measures the potential energy minus the kinetic energy. The difference in the fluid level in the two tubes provides a measure of the velocity of flow.

        b. Pressure is a force per unit area (dynes/cm²) and is usually expressed in terms of the height of a column of fluid that the pressure will support. The common units of pressure are **mm Hg** and **cm H$_2$O** (1 mm Hg = 1.36 cm H$_2$O).

    2. Flow (Q) is a volume movement per unit time, represented as the average velocity of movement (cm/sec) times the cross-sectional area of the tube (cm²). Flow equals the pressure gradient between two points in the system divided by the resistance to flow: Q = dP/R. This is similar to **Ohm's Law** in electricity, where Q would be amps, dP is voltage, and R is resistance. Thus, an increase in the pressure gradient causes an increase in flow, whereas an increase in resistance causes a decrease in flow.

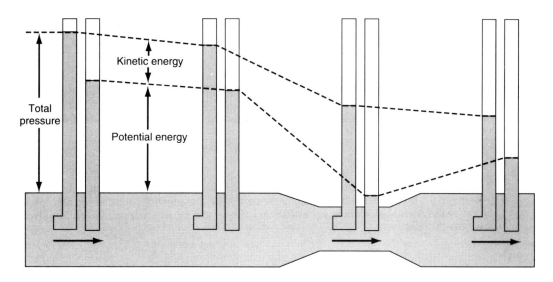

Figure 2-22. System of pitot tubes indicating the variation in total pressure, potential energy, and kinetic energy with change in flow velocity. The *arrows* represent the relative velocities in the segments of the system. The *narrow section*, a high-resistance segment, causes a large pressure drop. An increased flow velocity decreases lateral pressure (increases kinetic energy) and is termed a Bernoulli effect.

B. Poiseuille's law. From Poiseuille's law, for laminar flow

$$Q = \frac{dP}{R}$$

$$R = \frac{8 \times \mu \times L}{r^4}$$

where r = the radius of the tube, μ = the viscosity of the fluid, and L= the length of the tube.

1. **Resistance to blood flow.** The major factors that determine the resistance to flow are the radius of the vessels and the viscosity of the blood.
    a. **Effects of tube radius.** Since the tube radius (r) is raised to the fourth power, any change in tube radius produces a large change in resistance, which markedly alters the flow. For instance, for a constant pressure gradient (dP), if the radius is reduced to half, the flow will be one-sixteenth of its previous value. Blood flow to different organs or tissues is regulated primarily by changing the radius of the arterioles. Arteriolar radius is controlled by altering the sympathetic impulses to the blood vessels, by various drugs or hormones, or by the release of various substances from the tissues (e.g., histamine, adenosine, prostaglandins).
    b. **Viscosity** is the internal friction to flow in a fluid. Blood is a complex fluid because of the presence of cells and proteins, and its viscosity varies as a function of the flow rate. Blood with a normal hematocrit has a viscosity that is about three times that of water flowing through medium-sized tubes or vessels.
    c. **Hematocrit.** The amount of red cells present in blood is the major factor that alters the blood viscosity in the body. Blood with a hematocrit of 60 has approximately twice the viscosity of blood with a hematocrit of 40 (the relationship between hematocrit and viscosity is exponential, not linear). An increase in hematocrit requires a greater pressure to produce a given flow rate.

2. **Total peripheral resistance** is usually expressed as the ratio of the difference in pressure (dP) to the flow (Q) such that R = dP/Q, where R is resistance. Expressed in this manner, resistance has the units of mm Hg/L/min or peripheral resistance units (PRU). The normal systemic circulation exhibits a total peripheral resistance of approximately 20 PRU in contrast to the pulmonary circulation, where the resistance is about 2 PRU.
    a. Systemic circulation represents a system of parallel resistances, since blood is supplied to

the body organs by individual arteries. The total resistance in a system of parallel vessels is determined by adding the reciprocal of the resistance of each vessel

$$\frac{1}{R_{total}} = \frac{1}{R_1} + \frac{1}{R_2} \cdots \frac{1}{R_n}$$

Thus, the total resistance is less than the resistance of any of the individual resistances.

b. Each organ system represents a system of series resistances. A system of series resistances is additive. The resistance in any organ (R_o) is composed of the resistance in the artery (R_a), arteriole (R_{ar}), capillary (R_c), and veins (R_v), so that

$$R_o = R_a + R_{ar} + R_c + R_v$$

C. Arterial pressure is determined by the interaction of four parameters: the elasticity of the arterial system (E), the mean arterial blood volume (V), the pulse pressure (dP), and the systolic uptake of blood by the arteries (dV). These parameters are interrelated by an equation derived from **Hook's law,** which states

$$E = \frac{dP \times V}{dV}$$

1. Parameters

a. The **elastic constant** (E) is similar to Young's modulus of elasticity; E increases as an individual ages, and the vascular system becomes less compliant (see Figure 2-1). An increase in E, while stroke volume and arterial resistance remain normal, causes an increase in the pulse pressure. E is less in the pulmonary artery than it is in the aorta.

b. The **pulse pressure** (dP) is largely determined by the elastic constant and the systolic uptake of blood by the arteries. Increases in stroke volume, with other parameters constant, increase dP.

c. Systolic arterial uptake (dV) is less than the stroke volume, because some blood flows out of the arterial system during systole. Increases in stroke volume increase dV.

d. The **mean arterial blood volume** (V) normally is 10%–15% of total blood volume. Increases in V cause an increase in the arterial pressure by distending the vascular system, which then recoils on the contained blood with a greater force.

2. Physiologic determinants of arterial pressure. The body adjusts arterial pressure by altering heart rate, stroke volume, and peripheral resistance. It is important to understand the relationships between these parameters in order to analyze compensatory mechanisms during such stresses as hemorrhage, syncope, and arrhythmias.

a. Factors that increase arterial blood pressure

(1) Increased heart rate raises arterial pressure, since it reduces the time for blood to leave the arterial vessels (reduces diastole) and usually increases the cardiac output; both factors raise the arterial blood volume and arterial blood pressure.

(a) At the onset of an increase in cardiac output, the arterial pressure rises until the amount of blood leaving the arterial system per minute equals the amount entering.

(b) There now is a new equilibrium (i.e., inflow equals outflow), but there has been a transfer of blood from the venous to the arterial system, which has raised the arterial pressure so that the increased flow is forced through the systemic resistance.

(2) Increased peripheral resistance is caused by a contraction of the smooth muscle in the walls of the arterioles, which narrows their lumens. This raises the arterial volume by initially reducing the amount of blood that leaves the arterial system per minute. The arterial pressure rises until the new pressure is sufficient to overcome the added resistance to flow, and outflow again equals inflow.

(3) Increased stroke volume primarily raises the arterial systolic pressure so that the pulse pressure increases. Again, the arterial pressure increases until outflow equals inflow.

b. The **elastic constant** refers to the stiffness of the arterial system, which progressively increases from in utero until death. An increased stiffness or **elastance** of the arterial system causes an increase in the pressure for any increase in arterial volume. The increase in the elastic constant of the arteries is one of the factors that raises the arterial pulse pressure during the aging process. This increase in the arterial pressure raises the work of the ventricle during ejection and can reduce the ventricular stroke volume since it represents an increase in the ventricular afterload.

$F = \frac{dP}{R}$

VIII. CARDIAC OUTPUT AND VENOUS RETURN. Over any significant period of time, the venous return must equal the cardiac output. In the normal human, the cardiac output and the venous return are about 5 L/min. A complicated interaction of neural, humoral, and physical factors on the cardiovascular system maintains the flow rates at this level.

A. Vascular function curves refer to the relationship between venous return and venous pressure in the circulatory system. With each contraction of the heart, blood is removed from the venous side of the circulation and is ejected under higher pressure into the arteries. Increasing the cardiac output (and venous return) increases the transfer of blood from the venous to the arterial vessels; this causes the arterial pressure to rise and the venous pressure to fall. Thus, an increased venous return, associated with an increased cardiac output, results in a decreased venous pressure. This relationship is shown in Figure 2-23A.

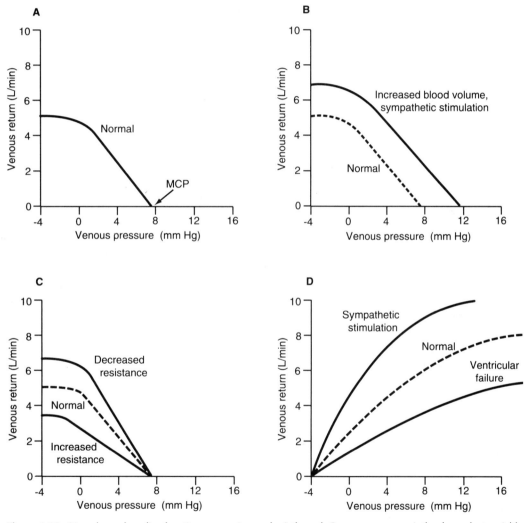

Figure 2-23. Vascular and cardiac function curves. In *graphs A through C,* venous pressure is the dependent variable but is plotted on the x-axis in contrast to the usual method of plotting the dependent variable along the y-axis. (*A*) Normal curve. Venous pressure decreases as venous return increases, since an increased flow results from blood being transferred from the venous to the arterial system. When venous return (vascular flow) is zero, then venous pressure is maximal and is referred to as mean circulatory pressure (*MCP*). (*B*) An increased blood volume or sympathetic stimulation, which decreases venous compliance, shifts the vascular function curve to the right. (*C*) Changes in arteriolar resistance alter the slope of the vascular function curve. An increased resistance results in a reduction of pressure transmitted through the microcirculation so that venous pressure is lower for any given flow rate. (*D*) Three cardiac function curves demonstrate the effect of changes in contractility on the relationship between venous pressure and cardiac output.

1. **Mean circulatory pressure (MCP).** If the heart is stopped, blood continues to flow from the arteries to the veins for the first few seconds until the pressures throughout the circulatory system become equal. Since the veins now contain more blood than they did when the heart was beating, the venous pressure is highest when the cardiac output (or venous return) is zero. This pressure is termed the mean circulatory pressure (see Figure 2-23A). The MCP varies with blood volume and the activity of the sympathetic nervous system.

   a. **Increased blood volume** increases the MCP and shifts the entire vascular function curve to a higher pressure, since the vascular system is more distended, just as the pressure in a tire increases if more air is added. These relationships are shown in Figure 2-23B.

   b. **Sympathetic stimulation** causes a contraction of the smooth muscle in the veins and arteries. The vessels now squeeze down on the blood with more force, raising the MCP and shifting the vascular function curve to the right.

2. **Increased resistance to flow** occurs when there is a constriction of the vessels, which creates more friction between the blood and the vessels. Because of the increased friction, there is a greater pressure drop between the arteries and the veins, resulting in a lower venous pressure at any venous return and shifting the curve left as shown in Figure 2-23C.

B. **Cardiac function curves** refer to the relationship between venous pressure and cardiac output. This is the **Frank-Starling relationship** or **Starling's law of the heart** (see II E). To summarize this relationship: An increase in venous pressure causes a greater filling of the ventricle during diastole resulting in a greater stretch of the cardiac muscle fibers (increased preload). The increased preload produces a stronger contraction and a greater ejection of blood during systole. Thus, an increased venous pressure produces an increased cardiac output. This relationship is shown in Figure 2-23D, which also depicts the effects on the heart of increased contractility (sympathetic stimulation) and decreased contractility (heart failure).

C. **Cardiovascular function curves.** Because both venous return and cardiac output depend on venous pressure, one can combine the normal curves from Figure 2-23A and D into one graph (Figure 2-24). The intersection of the venous return and the cardiac output curves is the only point where this system can function, since this is where the two flow rates are in equilibrium. Any change in cardiac contractility, blood volume, or vascular resistance will cause the equilibrium point to shift, either higher or lower, but again cardiac output and venous return must be equal.

   1. **Effects of exercise.** During exercise there is an increase in cardiac output due to a number of changes in the cardiovascular system. The effects of these changes are shown in Figure 2-25.

      a. **Activation of the sympathetic system** causes the cardiac output curve to shift to a higher level (see Figure 2-25, from *point A* to *point B*). Sympathetic stimulation also causes constriction of the venous smooth muscle, which increases the mean circulatory pressure, shifting the vascular function curve to the right (see Figure 2-25, *point C*).

      b. **Vasodilation** occurs in the skeletal muscles in order to increase the blood flow to provide more O_2 and metabolites. The vasodilation reduces the total peripheral resistance, which causes the vascular function curve to become steeper (see Figure 2-25).

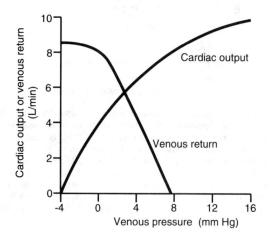

Figure 2-24. The vascular and cardiac function curves are superimposed. The only operating point on this graph is the intersection of the two curves, since this is the point where venous return equals cardiac output. In this graph the cardiac output (venous return) equals 5 L/min and venous pressure is 4 mm Hg.

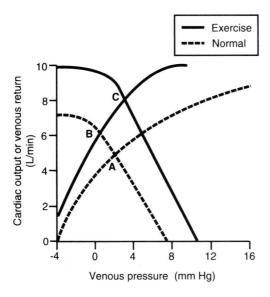

Figure 2-25. The effects of exercise on the vascular and cardiac function curves (see VIII C 1).

2. **Congestive heart failure** results in another series of changes that involves both the cardiac and vascular function curves (Figure 2-26).
    a. **Decrease in contractility of the heart** is the primary defect in congestive heart failure. The cardiac function curve is shifted downward, and cardiac output initially is reduced from *point A* to *point B* in Figure 2-26.
    b. **Increased blood volume.** Because of the reduced cardiac output, the kidneys retain fluids and electrolytes, so that, over a period of several weeks or months, the blood volume increases. This shifts the vascular function curve to the right, raising venous pressure, and cardiac output returns *toward* normal (see Figure 2-26, *point C*).
    c. **High venous pressure** invariably accompanies congestive heart failure. This high pressure causes an increased filtration of fluid from the vascular system into the tissues and results in edema formation. The edema may occur either in the lungs (pulmonary edema) or in the feet (swollen ankles) depending on whether it is the left or right ventricle that is failing.

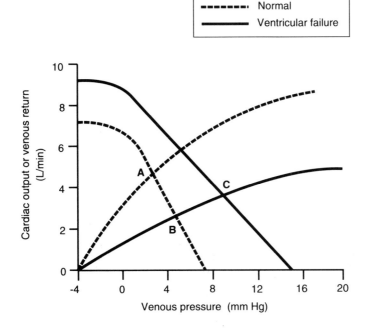

Figure 2-26. The effect of congestive heart failure on the vascular and cardiac function curves (see VIII C 2).

D. Measurement of cardiac output. The **cardiac output** (stroke volume × heart rate) averages about 5 L/min in a normal man and is about 20% less in a woman. Over any interval of time, the cardiac output must equal the venous return. The cardiac output is exactly controlled to provide an adequate mean blood pressure for delivery of nutrients and O_2. The cardiac output can be measured using either of the following techniques.

1. The **Fick principle** is an example of the law of conservation of mass, a widely applied principle in physiology. When applied to the cardiac output, the Fick principle states that the amount of O_2 delivered to the tissues must equal the O_2 uptake by the lungs plus the O_2 delivered to the lungs in the pulmonary artery. This is mathematically expressed as

$$Q = \frac{\dot{V}o_2}{Cao_2 - C\bar{v}o_2}$$

where Q = cardiac output, Cao_2 arterial O_2 content, $C\bar{v}o_2$ = mixed venous O_2 content, and $\dot{V}o_2$ = O_2 consumption (ml/min).

2. **Indicator dilution**
   a. **Theory**
      (1) If a suitable indicator is injected (A = amount injected) into an unknown volume of distribution (V), the V can be estimated from the resultant indicator concentration (c) with the equation: V = A/c.
      (2) Similarly, the flow of a fluid (Q) can be measured if the mean concentration of the indicator is determined for the time (t) required for that indicator to pass a given site. This flow is measured with the equation: Q = A/ct.
   b. **Technique.** In the body, the indicator is injected instantaneously but is dispersed in time due to different transit times between the injection and sampling sites. Usually, recirculation of some indicator occurs before passage of the indicator is complete (Figure 2-27). After the peak concentration is reached, the indicator concentration follows an exponential time-course, so that the curve can be extrapolated to zero concentration, and the duration of the indicator passage (t) can be estimated (see Figure 2-27).
   c. **Indicators** used for cardiac output measurement include Evans blue, Cardio-Green, hypertonic or hypotonic saline, ascorbate, and cold saline. Cold saline, used in thermodilution, is the most commonly used indicator in human cardiac output studies.
   d. **Thermodilution** uses a multilumened balloon catheter with a thermistor near the distal end. The catheter is inserted into a peripheral vein, and the tip is floated into a branch of the pulmonary artery; cold saline is injected through a proximal opening that is within the right ventricle. The saline mixes with the blood in the right ventricle; the temperature change is measured by the distal thermistor, which is downstream in the pulmonary artery. Thermodi-

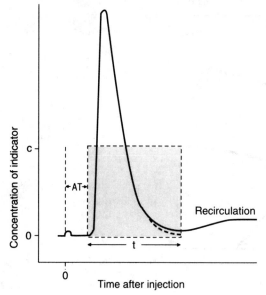

Figure 2-27. Indicator dilution curve, which can be used to measure cardiac output. *AT* = appearance time; *c* = mean dye concentration for one passage of the dye; *t* = passage time. The *dotted line* is an extrapolation of the indicator curve to zero concentration. The area inside the rectangle equals the area enclosed by the indicator dilution curve. If c = 1.0 mg/L, t = 0.5 min, and a = 2.5 mg, then Q = a/ct = 2.5/(1) (0.5) = 5 L/min.

lution is a preferred technique because little or no recirculation occurs, cold saline is relatively innocuous, and little temperature change occurs in the surrounding tissues.

IX. CONTROL OF THE CARDIOVASCULAR SYSTEM is designed primarily to maintain an adequate arterial pressure in order to provide sufficient blood flow to all of the body organs. To accomplish this, the system is designed to control both arterial pressure and blood volume.

A. Control of the blood volume represents a complex interaction of neural, endocrine, and renal mechanisms.

1. **Neural component of blood volume control.** The arterial baroreceptors, as well as stretch receptors in the atria (low-pressure baroreceptors), serve as sensors to detect increases in blood pressure or volume. These receptors are within the thoracic or cervical vascular system and therefore detect changes in central vascular volume and not changes in the volume of the extracellular space (e.g., standing causes fluid to shift from the thorax to the legs, which is interpreted by these receptors as a decrease in the vascular volume). Impulses from both high- and low-pressure receptors are transmitted to the hypothalamus where the neurohypophysial cells are inhibited from releasing **antidiuretic hormone (ADH)**.

2. **Endocrine component**
   a. **ADH** is released from the posterior pituitary by an increased osmolality of the blood perfusing the brain. The increased osmolality is detected by osmoreceptors located in the anterior hypothalamus. ADH acts on the kidney to increase the reabsorption of water from the distal convoluted tubule and especially the collecting duct; thus, water retention and dilution of the solutes occur.
   b. **The renin-angiotensin-aldosterone system** is the major control mechanism for electrolytes within the body. A decrease in the mean blood pressure or pulse pressure within the kidney causes the release of the enzyme **renin** from the juxtaglomerular cells of the nephron. Renin, within the plasma, reacts with a substrate called **angiotensinogen** to form the decapeptide, **angiotensin I (AI)**. AI is converted to **angiotensin II (AII),** an octapeptide, by **angiotensin-converting enzyme (ACE),** located primarily on the endothelial surface of the pulmonary vessels and also on the renal vasculature. AII is one of the most potent vasoconstrictive agents known. AII acts directly on arterioles to increase resistance and return blood pressure toward normal. In addition, AII causes the release of aldosterone from the zona glomerulosa of the adrenals. **Aldosterone** causes a reabsorption of electrolytes (primarily Na^+) and osmotically obligated water in the distal tubule of the kidney. The increased fluid reexpands circulating blood volume, which tends to increase blood pressure.
   c. **Atrial natriuretic factor (ANF)** has been found in the form of storage granules in certain cells of the left and right atria. This material, as well as natriuretic factors from other sources, is released into the circulation when the atria are distended by increases in blood volume. The ANF is transported to the kidneys where it dilates afferent arterioles, which increases the **glomerular filtration rate (GFR)**. An increase in GFR enhances the excretion of Na^+ and water, which then lowers blood volume and the pressure in the vascular system. As the blood volume decreases, the atrial distension is reduced, which lowers the rate of release of ANF.

B. Control of arterial pressure

1. **Baroreceptors** are highly branched nerve endings located in the carotid sinuses (at the bifurcation of the carotid arteries) and the aortic arch. Baroreceptors are stimulated by distension of the vessel walls; resultant nerve impulses are carried to the central nervous system (CNS) over the sinus and aortic nerves (branches of the glossopharyngeal and vagus nerves, respectively). The carotid sinus represents the most distensible area of the arterial system, which indicates that these receptors respond to stretch of the arterial walls rather than to pressure per se.

2. **Baroreceptor response.** The baroreceptors generate a receptor potential in response to distension of the vascular walls. The receptor potential results in the generation of action potentials in the afferent nerves, and the action potential frequency is linearly related to the receptor potential. The receptor potential has two distinct components, the **dynamic** and **static responses**.
   a. **The dynamic receptor response** is the initial receptor potential, which is proportional to the rate of change of the arterial pressure.

b. **The static receptor response** is proportional to the new steady pressure and continues without adaptation as long as the pressure is maintained. These responses provide information to the cardiovascular control centers in the CNS regarding the heart rate, the rate of change in blood pressure (related to the ventricular contractility), and the pulse pressure during each cardiac cycle.

c. **Baroreceptor (moderator) reflex.** Afferent neural activity over the sinus and the aortic nerves causes a reflex slowing of the heart mediated by the vagus nerve and a withdrawal of sympathetic tone to the arterioles, which results in vasodilation. Both of these effects lead to a reduction in aortic pressure. Massage of the carotid sinus is used clinically to induce a vagally mediated slowing of the heart to interrupt PAT. An increased sensitivity of the carotid sinus is seen in some older individuals who experience syncope secondary to the vagally mediated sinus arrest and a prolonged period of ventricular asystole. This condition is the **Stokes-Adams syndrome.**

3. **CNS centers**
   a. **Medulla.** Located within the reticular formation of the brain stem are diffuse collections of cells involved in the integration of cardiovascular function.
      (1) A **pressor area** is located in the lateral portion of the reticular formation, whereas stimulation of the medial portion of the reticular formation causes a depressor response.
      (2) The **depressor response** is due to the withdrawal of vasoconstrictor influence rather than an active vasodilation.
      (3) A **cardioaccelerator center** and a **cardioinhibitory center** are also located in the reticular formation. These latter two centers send impulses to the heart over the **cardiac sympathetic** and **vagal nerves**, respectively.
   b. **Hypothalamus.** The medullary cardiovascular centers receive input from higher levels of the brain. The effect of temperature changes on the hypothalamic centers is relayed to the medulla and results in vasoconstriction (heat conservation) or vasodilation (heat dissipation) in the vessels of the skin. Emotional stresses also influence heart rate and blood pressure, and these effects are relayed from the higher centers to stimulate or inhibit the medullary centers. In some animals, but apparently not in humans, the skeletal muscles receive a sympathetic cholinergic innervation as well as the α-adrenergic innervation that normally controls blood flow. Stimulation of these cholinergic fibers causes a vasodilation that occurs in anticipation of exercise. The vasodilation opens the thoroughfare channels in skeletal muscle, but, once exercise begins, the nutrient channels are dilated by the release of the local vasodilator substances.
   c. **Cortex.** Stimulation of various areas of the motor cortex leads to complex responses, including appropriate cardiovascular adjustments. These complex motor patterns involve autonomic responses. It is becoming more and more common to use biofeedback to attempt to control heart rate and blood pressure, which emphasizes the control that the higher centers can exert on the autonomic or vegetative nervous system. The sympathetic cholinergic outflow to the skeletal muscle arises from cerebral areas in lower animals and follows a separate descending tract compared to the adrenergic sympathetic outflow.

4. **Autonomic nervous system**
   a. **Sympathetic preganglionic fibers** arise from cell bodies in the intermediolateral gray columns of the spinal cord of the thoracodorsal regions. Axons of these cells exit the CNS through the ventral root and proceed to the paravertebral chain of ganglia where they synapse with the postganglionic cells of the sympathetic system. All of the preganglionic sympathetic fibers are cholinergic, while sympathetic postganglionic fibers may be either adrenergic or cholinergic. Sympathetic postganglionic cholinergic fibers innervate some exocrine glands and sweat glands and are the vasodilator fibers in skeletal muscles of certain species.
      (1) **Vasomotor tone.** Most of the sympathetic nerves to blood vessels exhibit some neural activity that induces about 50% of the maximal vasoconstrictor activity in most of the vascular beds. Thus, vasoconstriction or vasodilation is produced by increasing or decreasing the sympathetic neural activity.
      (2) **Denervation hypersensitivity** occurs following interruption of the sympathetic nerve supply to vascular smooth muscle and certain glands. Normally, receptors are located in the subjacent areas of the nerve terminals. However, following denervation, receptors develop over the entire membrane; thus, the end organs are extremely sensitive to any circulating catecholamines.

(3) **Receptors to sympathetic mediators** are of two major types.
  (a) α-**Receptors** are stimulated most strongly by epinephrine and norepinephrine and cause vasoconstriction.
  (b) β-**Receptors** are of two types, β_1 and β_2, both of which are most strongly stimulated by isoproterenol and least stimulated by norepinephrine. The β_1 receptors are located in the myocardium (not the coronary arteries), and activation of these receptors results in positive inotropic and chronotropic responses. The β_2 receptors are located in the other body tissues, and stimulation of these receptors leads to inhibition (e.g., bronchodilation or vasodilation). Pharmacologic agents are available that can stimulate or block the β_1 or β_2 receptors individually.

  b. The **parasympathetic nervous system** arises from cranial and sacral outflows of the CNS. The major effect of the parasympathetic system on cardiovascular function is due to its action, through the vagus nerve, on the SA and AV nodes and its direct reduction of atrial contractility. Marked efferent activity over the vagus nerve can result in extensive slowing of the heart, sinus arrest, and/or AV nodal block.

X. SPECIAL CIRCULATORY BEDS

A. Coronary circulation. The heart is supplied by the right and left coronary arteries; the left coronary artery divides after about 1 cm into the anterior descending and circumflex branches. Normally, the coronary arteries appear to function as end arteries; however, the presence of an arterial plaque or occlusion allows anastomoses to become functional. The heart can withstand a very limited O_2 debt so that adequate flow is essential to maintain an adequate supply of nutrients. The heart metabolizes all substrates in approximate proportion to their vascular concentration, with a preference for fatty acids.

1. **Coronary flow rate** represents approximately 5% of the resting cardiac output, or about 60–80 ml blood/100 g tissue/min.

2. **Myocardial O_2 consumption** averages 6–8 ml O_2/100 g tissue/min; normally hemoglobin releases approximately half of its arterial O_2 content to the myocardium.
   a. In contrast, for the remainder of the body **at rest,** hemoglobin releases only about 25% of the O_2 content in its transit through the systemic capillaries, which is equivalent to a P_{O_2} of 40 mm Hg. The coronary venous P_{O_2} is about 25–30 mm Hg under resting conditions and becomes even lower during exercise or stress.
   b. **During exercise,** the cardiac O_2 consumption may increase five- to six-fold in well-trained athletes partly due to increases in coronary flow and also by a widening of the arteriovenous O_2 difference.

3. **Coronary organization.** The major coronary vessels travel in the epicardium of the heart and subdivide, sending penetrating branches through the myocardium. It is these superficial vessels that are corrected by coronary bypass operations when they develop atherosclerotic narrowing. The penetrating branches further subdivide into arcades that distribute blood to the myocardium. It is important to understand that, during ventricular systole, myocardial wall tension increases and produces a compression of the vessels, thus increasing the resistance to flow. Consequently, the coronary flow pattern is complex due to this **extravascular compression** (Figure 2-28), and maximal flow in the left coronary vessels usually occurs during isovolumic relaxation, while arterial pressure is still relatively high and the myocardium is relaxed.
   a. **Left ventricular venous drainage** occurs primarily through the coronary sinus with some small contributions directly through the thebesian veins and coronary-luminal connections.
   b. **Right ventricular venous drainage** occurs through the multiple anterior coronary veins, and a larger proportion of flow occurs through the coronary-luminal connections than for the left ventricle.

4. **Control of coronary flow**
   a. **Adenosine.** The myocardium extracts large amounts of O_2 from the coronary blood even at rest; thus, during exercise, it is critical that coronary flow increases to maintain an adequate P_{O_2} in the myocardium. Recent evidence indicates that the major factor causing coronary vasodilation during hypoxic states is adenosine, which is derived from adenosine monophosphate by the reaction catalyzed by the enzyme 5'-nucleotidase. Release of adenosine into the myocardium results in an extremely strong vasodilator response.

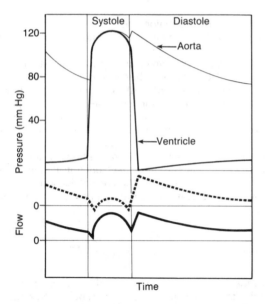

Figure 2-28. Diagram of the right coronary artery flow (*solid line*) and the left coronary artery flow (*dashed line*) in association with pressure pulses from the systemic circulation. Peak left coronary flow occurs at the end of isovolumic relaxation, when extravascular compression is minimal and perfusion pressure (aorta) is high.

    b. Sympathetic stimulation results in coronary vasodilation primarily because of an increase in the metabolic rate of the myocardium. An increase in the myocardial metabolism without a compensatory increase in coronary flow will cause the breakdown of ATP and the release of adenosine. The coronary vessels contain both α and β receptors, but the α vasoconstrictor activity is rather weak. There are no sympathetic cholinergic vasodilator fibers in the myocardium.

B. Cerebral circulation. The brain is surrounded by the rigid skull, but vascular diameter can be altered in response to changing blood flow needs since vasodilation on the arterial side can be compensated for by narrowing of the venous bed. Interruption of blood flow for only 5–10 seconds causes loss of consciousness, while circulatory arrest for only 3–4 minutes results in irreversible brain damage.

    1. Organization. The major vessels supplying the brain in humans are the **carotid arteries**. This is important since the presence of the carotid sinus provides a reflex input to maintain blood pressure at an adequate level to maintain flow to the brain.

    2. Cerebral autoregulation tends to be very effective in controlling blood flow over a range of arterial pressures of 80–180 mm Hg. **Hyper- or hypocapnia** produces a doubling or a halving, respectively, of cerebral blood flow compared to normal. The cerebral vasodilation produced by both hypoxia and hypercapnia appears to be due to the H^+ that arises from lactic or carbonic acids. The reduction in blood flow accompanying hyperventilation is sufficient to produce cerebral hypoxia, which results in dizziness and, occasionally, fainting (syncope).

    3. Neural control of cerebral blood flow seems to be limited to the larger vessels and those located within the pia. The cerebral vessels are supplied by both sympathetic and parasympathetic neurons, but the role of the neurons remains unknown.

    4. The **Cushing reflex** is a systemic vasoconstriction in response to an increase in the cerebrospinal fluid (CSF) pressure. This reflex produces a marked systemic hypertension in response to increases in the CSF pressure in order to maintain adequate blood flow to the brain. The reflex is initiated by the medullary centers, which become activated when they become hypoxic due to a limitation of blood flow.

C. Skeletal muscle comprises 40%–50% of the total body weight in the normal adult male. Because of this large mass, even resting skeletal muscle utilizes approximately 20% of the total resting O_2 consumption.

    1. Resting skeletal muscle receives a relatively large blood flow for its metabolic rate so that venous O_2 content is about 17–18 ml/dl. Blood flow in resting skeletal muscle varies from 1.5 to 6.0 ml/100 g/min, and the variability is largely due to the type of muscle. Muscles composed mainly of red fibers receive a larger blood flow than muscles composed mainly of white fibers.

2. **Exercising skeletal muscle** receives about 80 ml blood/100 g tissue/min and extracts about 80% of the O_2 from the arterial blood. Even with marked vasodilation, the blood flow is inadequate to maintain aerobic levels in the cells during moderate to heavy exercise. The **anaerobic threshold,** which is defined as the level of exercise that produces a significant rise in the level of lactic acid in the bloodstream, occurs at an O_2 consumption that is approximately 60% of the maximal.

D. **Fetal circulation** exhibits a number of differences compared to the postnatal condition. The **placenta** is a major organ of the fetus and receives more than 50% of the combined output of both fetal ventricles. At birth there are marked changes that occur in both the systemic and pulmonary circulations as the placental blood supply is lost.

1. **Pulmonary circulatory changes** are primarily due to a marked decrease in pulmonary vascular resistance as the lungs are inflated at birth. After inflation of the lungs, the pulmonary vascular resistance is only about 10% of its value during fetal life.
   a. **Foramen ovale.** The drop in pulmonary vascular resistance causes the pressure in the right atrium to become less than that in the left atrium, which is opposite to the situation present during fetal life. The pressure reversal causes the flap-like valve to close over the foramen ovale, which then normally fuses to the septum over the next few days.
   b. **The ductus arteriosus,** connecting the pulmonary artery and the descending aorta, conducts blood from the pulmonary artery to the aorta during fetal life. The flow through this vessel reverses after birth as the pulmonary resistance is reduced, and the pulmonary artery pressure becomes less than in the aorta. The ductus arteriosus begins to constrict shortly after birth perhaps due to the higher O_2 content of the blood. There is evidence that the patency of the ductus arteriosus is maintained by a dilator prostaglandin. Closure of this vessel can be facilitated by the administration of aspirin, which inhibits prostaglandin formation. The ductus arteriosus may remain patent for several days, during which time a continuous murmur may be heard over the upper chest due to turbulent blood flow through the narrowed structure.
   c. **Right ventricular wall thickness** is about equal to that of the left ventricle during fetal life. The thickness of the right ventricle diminishes over several months, and the electrical axis of the heart swings toward the left over a period of several years until it lies within the normal adult quadrant.

2. **Systemic circulatory changes** are caused by the loss of the placental circulation so that more blood flow is available for all of the other body organs. The umbilical arteries constrict in response to cooling, trauma, circulating catecholamines, and high Po_2. The loss of the placental circulatory bed after birth causes a large increase in the systemic vascular resistance. The increase in resistance raises the arterial pressure and consequently increases the afterload of the left ventricle. The higher afterload leads to increases in left-ventricular and left-atrial pressures, both of which contribute to the closure of the foramen ovale.

XI. RESPONSES TO STRESS

A. **Shock** is defined as a condition characterized by inadequate blood flow to critical organs such as the heart, brain, liver, kidneys, and gastrointestinal organs.

1. **Causes**
   a. **Primary shock or syncope (fainting)** is usually caused by a reflex that produces arteriolar dilation and/or cardiac slowing. Fainting is most commonly caused by pain, stress, fright, heat, or the sight of blood. The vasovagal reflex results in vasodilation, peripheral pooling of blood, and a decrease in the venous return. Without an adequate venous return, the cardiac output declines, and blood pressure decreases. If the person is sitting, or especially if standing, cerebral blood flow becomes inadequate because of the low blood pressure, and fainting occurs (consciousness is lost). It is imperative to keep the person lying down with feet raised to aid venous return to the heart, until the blood pressure has returned to normal.
   b. **Hypovolemic shock** is caused by a low blood volume and is caused by a number of disorders.
      (1) **External or internal hemorrhage** represents blood loss caused by injury, fractures, or ruptured vessels.

    (2) Fluid loss from the gastrointestinal tract (diarrhea or vomiting), kidneys (diabetes mellites, diabetes insipidus, or excessive use of diuretics), or skin (burns, sweating, or exudation) can dehydrate the body and reduce the circulating blood volume.

  c. Cardiac shock may be due to inadequate cardiac function following a heart attack, or it may be caused by too rapid or too slow a ventricular rate. The cardiac output is markedly decreased in either condition.

  d. Septic shock is due to infections in the bloodstream, especially gram-negative septicemia, or to bacterial toxins released from an infected site.

  e. Obstruction to blood flow. This type of shock is associated with such conditions as massive pulmonary emboli, tension pneumothorax, positive end-expiratory pressure respirations, and cardiac tamponade. All of these conditions block the circulation mechanically, so that cardiac output is reduced.

  f. Other causes of shock include endocrine failure such as Addison's disease or myxedema, severe hypoxia, severe allergic reaction, drug overdose, and trauma to the nervous system.

2. Symptoms of shock initially are agitation and restlessness, which later progress to lethargy, confusion, and coma.

3. Physical signs of shock are dependent on the stage of shock and the severity of the dysfunction.

  a. The **initial stage** can be characterized by a moderate reduction in cardiac output secondary to fluid loss or a negative inotropic effect on the heart. In general, compensatory reflexes may maintain the blood pressure at normal or low-normal levels. These reflex adjustments result in an increased heart rate, an increase in sympathetic discharge as evidenced by decreased blood flow to the skin (pallor), and a cold sweat.

  b. The **second stage** of compensation occurs with a loss in blood volume of 15%–25%. Intense arteriolar vasoconstriction results, which is usually not adequate to maintain a normal blood pressure since the cardiac output is moderately reduced.

  c. In the **third stage,** small additional losses of blood or cardiac function produce a rapid deterioration of the circulation with inadequate blood flow to critical body organs. If this state persists, a widespread tissue injury results due to the inadequate blood flow, which causes hypoxic damage to the tissues. This condition rapidly progresses to **irreversible shock,** which is fatal.

4. Treatment of shock requires attaining an adequate blood flow to all bodily tissues either by raising the circulating blood volume (if this is depleted), reestablishing normal rhythm to the heart, increasing the inotropic state of the ventricles with digitalis or β-sympathetic agonists, or administering vasodilators. The most useful measure for evaluating the extracellular fluid volume and function of the left ventricle is the **pulmonary wedge pressure.** Low pressures indicate a depletion of the extracellular fluid, while excessive pressures are associated with large fluid volumes and/or inadequate function of the left ventricle.

B. Hypertension is high blood pressure. Hypertension causes an increased work load on the heart and can cause damage to the blood vessels and other organs, especially the kidneys and the eyes. Hypertension is generally diagnosed when systolic pressure exceeds 140 and diastolic pressure exceeds 90 mm Hg. Recent evidence indicates that death rate and illness are increased in patients with diastolic pressures in excess of 85 mm Hg, which may prove to be a new borderline for the diagnosis of hypertension.

1. Etiology

  a. Essential hypertension accounts for more than 90% of all cases of hypertension. The cause of essential hypertension is unknown. Excessive salt intake, resetting of the baroreceptors, increased salt and fluid in the arteriolar walls, and long-standing exposure to stress are some of the theories that have been proposed to explain the increased systemic arterial pressures.

  b. Systolic hypertension with a wide pulse pressure can be caused by various malformations or conditions that decrease the compliance of the aorta (aging) or increase the cardiac output or stroke volume of the left ventricle. These latter conditions include thyrotoxicosis, patent ductus, A-V fistula, fever, and aortic insufficiency.

  c. Systolic and diastolic hypertension can be produced by a variety of causes.

    (1) Renal disease, if severe, results in systemic hypertension due to the release of renin and the formation of angiotensin II. Destruction of the kidney by a severe renal disease or narrowing of the renal arteries can result in hypertension.

    (2) Endocrine diseases such as Cushing's syndrome, primary hyperaldosteronism, pheochromocytoma, and acromegaly may result in hypertension.

(3) **Miscellaneous causes of hypertension** include psychogenic causes, coarctation of the aorta (important to diagnose because it can be cured surgically), polyarteritis nodosa, and excessive transfusion.

2. **Symptoms** of hypertension are generally absent until the disease is far progressed, and the hypertension has resulted in cardiac, renal, neural, or ocular damage. The disease is usually diagnosed as the result of a blood pressure taken during a routine physical examination.

   a. **Cardiac symptoms** are caused by the increased work required of the left ventricle. The ventricle initially hypertrophies and eventually, if the hypertension is severe, fails, resulting in pulmonary congestion and edema. Hypertension also increases the rate of formation of arteriosclerosis (hardening of the arteries), which reduces blood flow to critical vessels in the heart. The reduced blood flow can cause angina or a heart attack.

   b. **Neural symptoms** of hypertension include occipital headaches especially in the morning, dizziness, ringing in the ears, dimmed vision, and syncope. Permanent brain damage can result from complete blockage of a blood vessel by arteriosclerosis or by hemorrhage secondary to the stress imposed on the vessels by the hypertension.

   c. **Renal effects of hypertension** are due to atherosclerotic lesions that affect the afferent and efferent arterioles and the glomerular tufts. These lesions cause a decrease in GFR as well as renal tubular dysfunction. These pathologic alterations eventually result in renal failure.

   d. **Malignant hypertension** occurs when arterial pressure becomes markedly increased and vascular, CNS, renal, and cardiac deterioration occurs over weeks or months instead of years.

3. **Treatment** of hypertension is critically important to prevent or minimize damage to cerebral, ocular, renal, and cardiac tissues. Treatment of essential hypertension relies primarily on diuretics, β-adrenergic blocking drugs, or calcium channel blockers to lower the arterial blood pressure. Treatment of secondary hypertension also depends on treating the underlying disease.

C. Exercise

1. **Importance.** The cardiovascular and respiratory response to exercise has become of major interest due to the current concern with physical conditioning and prevention of cardiovascular disease. In addition, exercise is used as a stress test to evaluate the respiratory and cardiovascular systems.

   a. **Specificity.** Endurance (aerobic) training produces different effects than weight training, and only aerobic exercise produces cardiovascular conditioning. Training effects are specific for the particular muscle groups involved; for instance, rowing does not benefit running since different muscles are involved.

   b. **Characteristics.** Muscles conditioned by endurance training exhibit increases in capillary density, myoglobin concentration, glycogen, and mitochondrial enzymes of the citric acid cycle. These enzymes are involved in the breakdown of long-chain fatty acids, which serve as an energy source during exercise thus sparing muscle glycogen.

2. **Cardiac responses**

   a. **Heart rate** increases linearly with work rate up to a maximum, which is determined by a person's age. The maximal heart rate (HR_{max}) can be estimated from the following equation: $HR_{max} = 210 - [0.65 \times age (years)]$. Cardiovascular conditioning requires a heart rate of 60%–70% HR_{max} for 20–30 minutes, three to four times a week for at least 3 months to obtain optimum benefits.

   b. **Stroke volume.** Endurance training causes an increase in the stroke volume of the heart by increasing the volume of the ventricles. Thus, conditioned athletes can maintain any level of cardiac output at a lower heart rate than nonconditioned individuals.

   c. **Maximal cardiac output** in conditioned athletes is greater than in nonconditioned individuals because of the larger stroke volumes that are attained. Conditioning does not have any effect on the maximal heart rate for an individual. The maximal cardiac output may be on the order of 30–35 L/min in Olympic-class runners, while the maximal cardiac output in deconditioned adults is about 15–20 L/min.

   d. The **anaerobic threshold** is defined as the minimal level of exercise that produces a significant increase in the lactate concentration of the blood and occurs at approximately

60% of the maximal exercise level. The anaerobic threshold represents the point when muscle metabolism shifts from an aerobic to an anaerobic state. Training improves the absolute anaerobic threshold as well as the anaerobic threshold relative to the maximal O_2 consumption.

3. Respiratory responses

   a. Minute ventilation increases linearly with work rate until the anaerobic threshold is reached, when minute ventilation increases more steeply with work rate due to the additional respiratory drive imposed by the liberated lactic acid (Figure 2-29). Endurance training increases the maximal minute ventilation achieved during exercise, but the maximal voluntary ventilation does not improve with training.

   b. CO_2 output increases linearly with the work rate until anaerobic threshold, when the CO_2 output increases more steeply due to the increased respiration. Above the anaerobic threshold, the arterial PCO_2 declines as the body stores are depleted, since excretion exceeds production.

   c. O_2 consumption increases linearly with work rate and is exactly dependent on the work performed. The O_2 consumption does not change with training at any work load (i.e., training does not improve the body's efficiency unless muscle coordination is improved by practice). Conditioning produces a 5%–20% increase in the maximal O_2 consumption, which comes from an increase in the cardiac output and a widening of the arteriovenous O_2 difference at the maximal exercise level.

   d. The alveolar-to-arterial O_2 gradient normally remains at 5–10 mm Hg during moderate levels of exercise, but, beyond the anaerobic threshold the alveolar-to-arterial O_2 gradient widens slightly due to an increase in the alveolar PO_2.

   e. The respiratory exchange ratio ($R = \dot{V}CO_2/\dot{V}O_2$) rises to greater than 1 at work levels above the anaerobic threshold but normally never exceeds 1.25 even at maximal levels of exercise. The increase in R is due to the conversion of bicarbonate to carbonic acid in the buffering process of lactic acid and the liberation of the CO_2 in the lungs.

4. Pathologic responses (Figure 2-30)

   a. Chronic obstructive lung disease limits exercise mainly due to the onset of severe dyspnea. In these patients the heart rate does not reach the minimal level for training effects, and the anaerobic threshold occurs at very low levels of exercise due to the impairment of O_2 uptake. Thus, exercise training in these patients has only slight benefits, which seem to be due mainly to desensitization of symptoms. It appears to be more worthwhile to train the respiratory muscles in these patients.

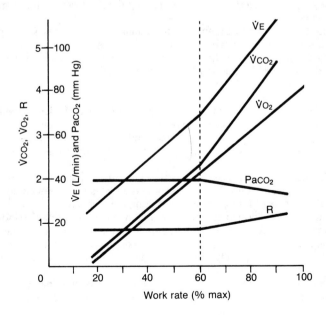

Figure 2-29. The effects of exercise on respiratory parameters of gas exchange. The *dashed line* indicates the onset of the anaerobic threshold. $\dot{V}E$ = minute ventilation; $\dot{V}CO_2$ = CO_2 excreted in expired gas (L/min); $\dot{V}O_2$ = O_2 utilization (L/min); $PaCO_2$ = arterial CO_2 tension; R = the respiratory exchange ratio.

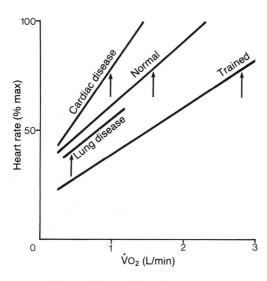

Figure 2-30. The effect of training, heart disease, and lung disease on the heart rate response to endurance (aerobic) exercise. *Vertical arrows* mark the onset of the anaerobic threshold, which occurs at about 60% of the maximal exercise level for every individual. $\dot{V}_{O_2} = O_2$ utilization.

b. **Cardiac patients** have an early onset of the anaerobic threshold due to the low cardiac output, and they reach maximal heart rates at much lower levels of exercise than normal individuals.

STUDY QUESTIONS

Directions: Each of the numbered items or incomplete statements in this section is followed by answers or by completions of the statement. Select the **one** lettered answer or completion that is **best** in each case.

1. In a recumbent person, the greatest difference in blood pressure would exist between the

(A) ascending aorta and brachial artery
(B) saphenous vein and right atrium
(C) femoral artery and femoral vein
(D) pulmonary artery and left atrium
(E) arteriolar and venous end of a capillary

2. Distribution of blood flow is regulated mainly by

(A) capillaries
(B) arterioles
(C) venules
(D) arteriovenous anastomoses
(E) postcapillary venules

3. Which of the following mechanisms is most important for maintaining an increased blood flow to skeletal muscle during exercise?

(A) An increase in aortic pressure
(B) An increase in α-adrenergic impulses
(C) An increase in β-adrenergic impulses
(D) A vasoconstriction in the splanchnic and renal areas
(E) A vasodilation secondary to the effect of local metabolites

4. Which of the following factors is most important to blood flow regulation at the local level?

(A) Vessel-tissue transmural pressure difference
(B) Metabolic activity of the organ or tissue
(C) Local neurotransmitters
(D) Circulating neurotransmitters
(E) Cardiac output

5. The most important mechanism for blood-tissue transport of small hydrophilic solutes is

(A) convection
(B) facilitated diffusion
(C) active transport
(D) passive diffusion
(E) cotransport

6. All of the following mechanisms will increase the transcapillary flux of liquids, solids, and gases EXCEPT

(A) an increase in capillary surface area
(B) an increase in the concentration gradient across the microvascular wall
(C) an increase in the diffusion distance
(D) an increase in the capillary transit time
(E) a decrease in arteriolar resistance

7. The rate of lymph flow in humans is approximately

(A) 10–20 L/day
(B) 100–200 ml/day
(C) 10–20 ml/hr
(D) 1.5–2 L/day
(E) 1–2 ml/hr

8. Venous return is enhanced during exercise by all of the following factors EXCEPT

(A) increased depth of respiration
(B) pumping action of skeletal muscles
(C) venoconstriction
(D) reduced arteriolar resistance
(E) an erect position

9. All of the following conditions would cause a decrease in cardiac output EXCEPT

(A) ventricular tachycardia
(B) electrical stimulation of the distal end of a cut vagus nerve
(C) electrical stimulation of the superior cervical ganglion
(D) external compression of the area over the carotid sinus
(E) the Valsalva maneuver

10. An EKG that reveals no P wave in any leads indicates a block of impulses coming from the

(A) SA node
(B) bundle of His
(C) Purkinje fibers
(D) left bundle branch
(E) ventricular muscle

11. Ventricular end-diastolic pressure can be used on the abscissa (x-axis) when plotting a Frank-Starling curve instead of the usual variable, which is

(A) cardiac output
(B) ventricular stroke volume
(C) cardiac work
(D) ventricular contractility
(E) ventricular end-diastolic volume

12. Increasing the preload of cardiac muscle will

(A) reduce the ventricular end-diastolic pressure
(B) reduce the peak tension of the muscle
(C) decrease the initial velocity of shortening
(D) decrease the time it takes the muscle to reach peak tension
(E) increase the ventricular wall tension

13. Which of the following events is represented on an EKG?

(A) SA node depolarization
(B) AV node depolarization
(C) Bundle of His depolarization
(D) Depolarization of Bachmann's bundle
(E) Atrial muscle depolarization

14. An increase in systemic blood pressure leads to which of the following effects?

(A) An increase in the velocity at which blood is ejected from the left ventricle
(B) An increase in cardiac output
(C) An increase in the residual volume of blood in the left ventricle
(D) A decrease in the time it takes for the left ventricular wall to develop peak tension
(E) A decrease in the maximal wall tension developed in the left ventricular muscle

15. The time from the upstroke of the carotid artery volume pulse to the dicrotic notch (incisura) is a measure of the period of

(A) atrial diastole
(B) ventricular ejection
(C) reduced ventricular filling
(D) rapid ventricular filling
(E) ventricular isovolumic relaxation

16. The pulmonic valve normally closes after the aortic valve because the

(A) diameter of the pulmonary artery is less than that of the aorta
(B) right ventricular contraction begins after left ventricular contraction
(C) velocity of ejection in the right ventricle is less than that in the left ventricle
(D) diastolic pressure in the pulmonary artery is less than that in the aorta
(E) leaflets of the pulmonic valve are stiffer and harder to close compared to those of the aortic valve

17. During diastole, blood flow into the ventricles sometimes produces

(A) a first heart sound
(B) a second heart sound
(C) a third heart sound
(D) an ejection click
(E) an ejection-type murmur

18. A patient with coronary artery disease undergoes coronary arteriography, which reveals a 50% decrease in the lumen of the left anterior descending coronary artery. For any arteriovenous pressure gradient, the flow through this artery (compared to normal) will decrease by a factor of

(A) 2
(B) 4
(C) 8
(D) 12
(E) 16

19. A decrease in heart rate (while stroke volume and peripheral resistance remain constant) will cause an increase in

(A) arterial diastolic pressure
(B) arterial systolic pressure
(C) cardiac output
(D) arterial pulse pressure
(E) mean arterial pressure

20. The hexaxial reference system consists of reference lines generated from

(A) leads V_1–V_6
(B) the coronal body plane
(C) the horizontal body plane
(D) standard and augmented limb leads
(E) all bipolar limb leads

21. The increased circulating fluid volume in chronic congestive heart failure results from all of the following factors EXCEPT

(A) increased sympathetic discharge to the kidney
(B) decreased rate of firing of atrial volume receptors
(C) decreased renal perfusion
(D) aldosterone activity
(E) stimulation of the arterial baroreceptors

22. The greatest resting arteriovenous difference in O_2 content is found in the

(A) liver
(B) skeletal muscle
(C) heart
(D) kidney
(E) lung

23. Stimulation of the high-pressure baroreceptors is associated with

(A) an increase in cardiac contractility
(B) an increase in heart rate
(C) an increase in the discharge rate of vagal efferent cardiac neurons
(D) a decrease in systemic blood pressure
(E) stimulation of the vasopressor center

Directions: Each group of items in this section consists of lettered options followed by a set of numbered items. For each item, select the **one** lettered option that is most closely associated with it. Each lettered option may be selected once, more than once, or not at all.

Questions 24–26

For each condition listed below, select the mechanism that is associated with it.

(A) Increase in contractility
(B) Increase in fiber length
(C) Both
(D) Neither

24. An increase in the maximal velocity of shortening (V_{max})
25. A shift to a new Frank-Starling curve
26. An increase in intracellular ionic calcium

Questions 27–31

For each of the conditions listed below, select the most appropriate response.

(A) Increased ventricular end-diastolic pressure
(B) Increased aortic diastolic pressure
(C) Both
(D) Neither

27. Myocardial tissue damage
28. Increased arteriolar resistance
29. Decreased ventricular contractility
30. Increased heart rate
31. Increased central venous pressure

Questions 32–35

Match the cardiac event with the interval of the cardiac cycle in which it occurs.

(A) Atrial contraction
(B) Isovolumic contraction
(C) Rapid ventricular ejection
(D) Reduced ventricular ejection
(E) Isovolumic relaxation

32. Second heart sound
33. Maximal ventricular volume
34. Closure of the AV valves
35. Aortic valve opening

Questions 36–40

Match each cardiovascular effect with the factor that is responsible for it.

(A) Functional hyperemia
(B) Histamine
(C) Hypertension
(D) P_{CO_2}
(E) Capillary pressure

36. Regulation of capillary filtration rate
37. Elevation of arterial diastolic pressure
38. Metabolic regulation of blood flow
39. Regulation of cerebral blood flow
40. Increase in microvascular permeability

ANSWERS AND EXPLANATIONS

1. The answer is C. [*I C; Figure 2-2*] The largest pressure drop in the vascular system (40–50 mm Hg) occurs as blood passes through the systemic arterioles; this segment of the vasculature would be represented by the region between the femoral artery and femoral vein. The large arteries provide little resistance to flow, so that there is practically no pressure loss from the aorta to any of the distributing arteries. The pressure loss across the venous system amounts to approximately 10 mm Hg, since the pressure in the large veins may be 8–10 mm Hg whereas the pressure in the right atrium may be near zero. The atrium is distended since the pressure outside the atrium (intrathoracic pressure) averages about –5 mm Hg, which means that the pressure across the wall of the atrium (transmural pressure) is 5 mm Hg. The pulmonary circulation is a low-resistance circuit, and the gradient across this vascular segment also is only about 8–10 mm Hg. The pressure at the arteriolar end of a capillary averages about 40 mm Hg, whereas the pressure at the venous end is 25–30 mm Hg. Thus, even the smallest vessels in the body, which should have the highest resistance, exhibit only a 10–15 mm Hg pressure gradient. This seeming paradox occurs because of the tremendous number of capillaries that exist, which are arranged in a parallel circuit. Remember that the total resistance to flow in a parallel circuit is the sum of the reciprocals of the resistance in each vessel, so that the greater the number of vessels, the lower the total resistance.

2. The answer is B. [*I C 2 a*] The arterioles often are referred to as the stopcocks of the circulation, since they act as valves to restrict or enhance blood flow to different tissues or organs. Small changes in the diameter of these vessels can markedly alter the resistance to flow since resistance varies as the fourth power of the radius in the arterioles, as in any other vessels. However, the arterioles have a thick muscular wall in comparison to the lumen, so that a relatively small amount of smooth muscle contraction can cause a marked change in lumen size.

3. The answer is E. [*I C 2 a*] Vasodilation in active skeletal muscles reduces the resistance to flow and results in an increased blood flow to the involved tissues. The increase in blood flow is dependent on maintaining the mean arterial blood pressure, which normally rises only slightly with exercise. Thus, the slight rise in arterial pressure coupled with a significant rise in blood flow indicates that the total vascular resistance must decline during exercise. The decrease in vascular resistance occurs because skeletal muscle represents about 50% of body weight in the normal adult, and a decrease in this tissue can more than compensate for an increased resistance in such tissues as the gastrointestinal tract, kidneys, and skin during periods of mild to moderate exercise.

4. The answer is B. [*I C 2 a (1) (a)*] Normally, the body adjusts blood flow by the release of vasodilator substances from the active tissue or organ. This eliminates the need for the body to supply receptors and afferent and efferent nerves in order to detect changes in metabolism in virtually every cell in the body. Thus, most tissues autoregulate their own blood flow dependent on their own metabolic activity. An increase in activity increases the release of vasodilators such as adenosine and potassium. The increased metabolism results in a reduced PO_2, which also produces vasodilation in the systemic circulation.

5. The answer is D. [*I D 4 a*] Small, hydrophilic molecules (e.g., O_2, CO_2, glucose, water) are thought to pass readily through the pores in the endothelial cells of capillaries and are not impeded significantly by capillary endothelium. The capillary pores are such that virtually all of the albumin (molecular weight = 64,000 daltons) is retained.

6. The answer is C. [*I D 4 a (1)*] Diffusion rate is inversely related to the distance over which it must occur, so that, while diffusion is a very effective means of moving substances over short distances, it is a very inefficient and slow process over long distances. A decreased arteriolar diameter enhances the flux of materials across capillaries, since it raises capillary pressure toward the arterial pressure, increasing the filtration forces and leading to increased bulk flow or convection of fluid.

7. The answer is D. [*I E 3*] The rate of lymph flow is about 1–2 L/day. Although this rate seems large, it is quite small compared with the flow through the entire vascular system. The entire blood flow per day is 5–6 L/min $\times$ 60 min/hr $\times$ 24 hr/day = 8000 L. Thus, approximately 99.98% of the blood flow returns to the heart via the veins, whereas 0.02% of the blood flow is returned to the heart via the lymphatics.

8. The answer is E. [*I F 2 d*] Standing erect exposes a column of blood to the effects of gravity, which raises the pressure in the dependent veins. Since the venous system is very compliant, the increase in pressure causes an increase in venous volume, which reduces venous return compared to the prone or horizontal position. Venoconstriction, which is caused by sympathetic outflow during exercise, will partially compensate for the peripheral pooling caused by the effects of gravity. An increased depth of respiration causes a more negative pressure in the throrax, which increases the pressure gradient along the venous system and improves venous return. Contraction of the skeletal muscles compresses the veins and displaces the blood toward the heart; this effect is enhanced if the venous valves are competent and prevent a backflow of blood between contractions. A reduced arteriolar resistance in the skeletal muscles raises capillary and venous pressures and increases the flow rate through the systemic circulation.

9. The answer is C. [*I F 2 c; II C 1 c; IV B 4 b*] Stimulation of the superior cervical ganglion increases the sympathetic outflow to the heart. This leads to increased heart rate and contractility, which raises cardiac output. Ventricular tachycardia reduces cardiac output, since it interferes with diastolic filling of the ventricles and markedly reduces stroke volume. Stimulation of the distal vagus nerve causes bradycardia or even cardiac arrest for a period of time, with an obvious reduction in cardiac output. Compression over the carotid sinus results in a stimulation of the baroreceptors and a reflex increase in vagal outflow to the heart, causing cardiac slowing and a decreased cardiac output. Cardiac output is reduced during a Valsalva maneuver since the intrathoracic pressure is increased, which impairs venous return. Since the heart can pump only the blood it receives, there is a decline in cardiac output during the Valsalva maneuver. Immediately following the Valsalva maneuver, however, there is a transient increase in cardiac output as the dammed up blood returns to the heart.

10. The answer is A. [*II C 1, 2; III C 1 a*] The sinoatrial (SA) node is the normal pacemaker of the heart, and its depolarization cannot be seen on the EKG because of the small mass of tissue whose depolarization cannot be seen on surface recordings. SA node depolarization normally spreads to the atrial muscle, causing depolarization, which is seen on the EKG as the P wave. The bundle of His, Purkinje fibers, left bundle branch, and ventricular muscle all lie below the atria and normally do not cause depolarization of the atria.

11. The answer is E. [*II E*] The x-axis usually represents the independent variable, so that, for the Frank-Starling mechanism, this would be the factor that defines the length of the sarcomere. In the intact heart, the sarcomere varies in length as a function of ventricular end-diastolic volume. Assuming that there is no change in compliance of the ventricular muscle, the sarcomere will also vary as a function of end-diastolic pressure. The factor plotted on the y-axis of such a graph represents the output of the heart, which can include such functions as cardiac output, stroke volume, and cardiac work.

12. The answer is E. [*II E 1*] An increase in both end-diastolic pressure and volume is synonymous with an increased preload. The increased preload causes a more forceful ventricular contraction, which results in an increase in the peak pressure generated by the ventricle as well as an increase in the stroke volume. This intrinsic property of the myocardium is termed the Frank-Starling mechanism. The increased preload results in an increased velocity of shortening, but, since the peak pressure increases, the time to peak pressure usually remains constant in these conditions.

13. The answer is E. [*III C 1 a*] Atrial muscle depolarization is seen as the P wave on a standard EKG. An EKG does not reveal depolarization of the SA node, atrioventricular (AV) node, bundle of His, or Bachmann's bundle, because the mass of the tissue involved is too small to cause a deflection on the surface electrograms that represent the EKG. Depolarization of the bundle of His can be recorded from electrode catheters located in the esophagus. Esophageal catheterization is frequently done to characterize supraventricular arrhythmias, which cannot be diagnosed using the standard EKG.

14. The answer is C. [*V B 2*] An increase in the systemic pressure, or afterload, that the ventricle must overcome reduces the velocity of shortening of the contractile elements in the muscle. An increase in afterload also causes a reduction in the extent of muscle shortening, which results in a decreased stroke volume and, consequently, an increased residual volume (i.e., the volume of blood left in the ventricles at the end of ejection). The time that it takes for the left ventricular wall to develop peak tension is not reduced, because the velocity of contraction would be reduced, and the peak tension would be increased. Wall tension is a function of the product of ventricular pressure and volume. Since pressure and volume are increased, wall tension also is increased in the presence of an increased afterload.

15. The answer is B. [*VI A 3 c*] The upstroke of the carotid artery pulse indicates the onset of ventricular ejection, while the dicrotic notch (caused by semilunar valve closure) indicates the end of ejection. Atrial diastole represents the entire cardiac cycle except for the period of atrial systole that is concomitant with the a wave in the atrial pressure pulse. The periods of ventricular filling represent diastolic events that can be determined from an atrial pressure pulse or the jugular volume pulse. The period of isovolumic relaxation occurs from the incisura, or S_2, to the peak of the atrial v wave, which represents the time when the AV valves open.

16. The answer is C. [*VI A 3 c, e*] Right ventricular ejection begins before that of the left ventricle and continues after left ventricular contraction ends. The normal sequence of semilunar valve closure, which represents the major components of the S_2, is closure of the aortic valve (A_2) followed by closure of the pulmonic valve (P_2). During inspiration, P_2 is delayed and A_2 occurs slightly earlier, which widens the interval between these events. This sequence can be appreciated during auscultation of the heart and is referred to as normal splitting of the S_2. If there is a delay in left ventricular activation, as in left bundle branch block, A_2 may follow P_2 and, during inspiration, there is a narrowing of the interval between these events. This is termed paradoxical splitting of the S_2.

17. The answer is C. [*VI A 4 d*] The first and second heart sounds (S_1 and S_2) indicate the beginning and end of systole, respectively. An ejection click and an ejection-type murmur are systolic sounds that are caused by blood leaving the ventricle. The third heart sound (S_3) occurs during mid-diastole when the ventricular wall becomes tense toward the end of the rapid filling phase. S_3 is not normally heard in adults but can be heard in children with thin chest walls.

18. The answer is E. [*VII B 1 a*] The effect of vessel radius on flow is so powerful because the flow rate is directly proportional to the fourth power of the radius (Poiseuille's law). Thus, decreasing the radius to half reduces the flow to one-sixteenth of the normal value.

19. The answer is D. [*VII C 2*] A constant stroke volume coupled with a decreased heart rate produces a decrease in cardiac output. The reduced flow in combination with a constant peripheral resistance means that mean arterial pressure must decline. The fact that pulse pressure increases is obtained from an analysis based on the vascular elastic modulus ($E = V \times dP/dV$: where E is the elastic modulus; V is arterial volume, which is determined by arterial pressure; dP is pulse pressure; and dV is arterial uptake). Arterial pressure and volume decrease, since cardiac output is reduced (heart rate decreased and stroke volume constant). The quantity dV remains constant, since it is primarily related to stroke volume, which is constant. Thus, dP must increase in order to cancel the decreased V in order to maintain E as a constant (E varies only as a function of age).

20. The answer is D. [*III A 5; Figure 2-13*] The hexaxial reference system indicates six lines, which would include the three bipolar limb leads and the three augmented limb leads. The three bipolar limb leads are separated by 60 degrees and, thus, lie at 0, 60, 120, 180, and 240 degrees. The augmented limb leads bisect the angles provided by the bipolar leads, which means that the hexaxial reference system provides reference lines at every 30 degrees. The hexaxial reference system is in the frontal plane, whereas the precordial (V) leads lie in the horizontal plane.

21. The answer is E. [*VIII C 2; IX A*] The arterial baroreceptors are stimulated by an increase in arterial pressure, not a decrease, as is likely in cardiac failure. The increased fluid volume in congestive heart failure occurs due to an increase in the release of renin from the juxtaglomerular cells of the kidney, which is due to a generalized increase in sympathetic discharge and to the reduced perfusion of the kidney. Renin, in turn, stimulates the release of aldosterone by the adrenal glands. There is evidence that accommodation of the atrial volume receptors occurs when these receptors are exposed to prolonged periods of vascular distension, causing decreased firing rate.

22. The answer is C. [*X A 2*] Under resting conditions, the heart removes approximately half of the O_2 from the blood, which results in a PO_2 of approximately 27 mm Hg. In contrast, blood flow through other tissues is much greater in proportion to the metabolic rate and results in an arteriovenous (A-V) O_2 difference of about 5 ml/dl blood. The mixed venous O_2 content, which represents the weighted average of blood flow to all parts of the body, is 15 ml O_2/dl blood, which represents a hemoglobin saturation of 75% and a PO_2 of 40 mm Hg.

23. The answer is C. [*IX B 2 c*] The high-pressure baroreceptors are stimulated by an increased arterial blood pressure. The afferent impulses are carried over the glossopharyngeal and vagus nerves, where they stimulate the cardioinhibitory and depressor center. The one portion of the efferent limb of the reflex causes an increased number of vagal impulses to the heart, which results in cardiac slowing leading to a decrease in cardiac output. This depressor response causes a decrease of sympathetic tone to the vascular system, resulting in decreased peripheral resistance and a pooling of blood in the venous system. Both of these peripheral effects also lead to a lowering of the arterial blood pressure.

24–26. The answers are: 24-A, 25-A, 26-A. [*II D 2 a, F 1*] An increased maximal velocity of shortening (V_{max}) is one definition of increased contractility, or positive inotropism. Another way of looking at increased contractility is an increase in stroke volume at a constant end-diastolic pressure or volume.

The presence of a new Frank-Starling curve implies a change in contractility. If the new curve is above and to the left of the control curve, this indicates an increased contractility. Changes along any one curve indicate a change in ventricular fiber length (changing ventricular volume or pressure), which is the mechanism for invoking the Frank-Starling mechanism of altering ventricular output, whether this is cardiac work or stroke volume.

Positive inotropism is associated with an increase in the intracellular concentration of calcium. The increased calcium allows a greater number of troponin molecules to become saturated, which then provides more cross-bridges to be formed during the contraction process. The increased number of cross-bridges increases the force of contraction and the velocity of contractile element shortening at any load compared to the control condition.

27–31. The answers are: 27-A, 28-C, 29-A, 30-B, 31-C. [*II E, F 2; V B 1 b; VII C 2 a (1), (2)*] Myocardial tissue damage results in a decline in ventricular force development, leading to a reduction in stroke volume. Transiently, the volume of blood pumped by the ventricle is less than the venous return, so that ventricular diastolic pressure and volume increase. The increase in ventricular volume represents an increase in the preload so that the remaining muscle generates a greater tension than before, as long as the ventricle remains on the ascending limb of the Frank-Starling curve. This mechanism is limited since the ventricle may exceed the peak of the Frank-Starling curve, at which point there is a decline in output as the preload increases further.

Increased arteriolar resistance results in a rise in arterial (aortic) pressure. If heart rate and stroke volume remain constant, then the pulse pressure must decline in accordance with the principles determining arterial pressure. The increased aortic diastolic pressure represents an increase in the ventricular afterload so that stroke volume will be reduced slightly, further decreasing the pulse pressure, as detailed in Figure 2-19.

Decreased ventricular contractility results in a reduction in cardiac output as the ventricle shifts to a lower Frank-Starling curve. As cardiac output decreases, there is a rise in the venous pressure so that ventricular diastolic pressure and volume increase due to an added inflow into the ventricle. As long as the ventricle remains on the ascending limb of the Starling curve, this will result in some restoration of the cardiac output. However, the ventricular radius is larger so that the myocardium must generate a higher wall tension in accordance with the Laplace equation. The one benefit of ventricular dilation is that the myocardium can produce a given stroke volume with a smaller amount of myocardial shortening because of the increased radius.

An increased heart rate results in an increased cardiac output and a decline in ventricular end-diastolic pressure, since both ventricular filling time and central venous pressure are reduced. An increase in cardiac output as the primary event decreases central venous pressure. This becomes more obvious if one thinks about the heart pumping blood from the venous to the arterial side of the circulation. Thus, with an increased cardiac output, an additional volume of blood is transferred from the venous to the arterial system, which raises the pressure in the arteries and lowers the pressure in the veins.

An increase in central venous pressure raises the gradient for ventricular filling, which increases the ventricular end-diastolic pressure and volume. This represents a Frank-Starling mechanism that increases stroke volume and consequently cardiac output. As a result of the increased cardiac output, there will be an increase in aortic diastolic and systolic pressures in accordance with the relationship between pressure, flow, and resistance: $[\uparrow]dP = \uparrow Q \times R$.

32–35. The answers are: 32-E, 33-A, 34-B, 35-C. [*VI A 2, 3 b, c, e*] S_2 is caused by closure of the semilunar valves and the consequent vibrations in the tissues caused by stopping the backward flow of blood in the aorta. Thus, the S_2 represents the beginning of isovolumic relaxation.

Ventricular filling begins as soon as the AV valves open in diastole and depends on the rate of venous return. Contraction of the atria, late in diastole, causes an additional increment in ventricular volume so that ventricular volume reaches a maximum. The increased filling caused by atrial contraction is particularly important at fast heart rates, since the time for ventricular filling is abbreviated due to the shortened diastolic interval.

Closure of the AV valves causes the left ventricular chamber to become isolated, since the aortic valves also are closed due to the diastolic aortic pressure. Since both inflow and outflow valves are closed, there is no change in ventricular volume for a time; hence the interval is called isovolumic (equal or constant volume).

Opening of the aortic valve occurs when the pressure in the left ventricle exceeds the pressure in the aorta, and blood begins to flow out of the ventricle. Since the rate of ventricular contraction is initially high, the blood rapidly leaves the ventricle during this interval, causing aortic pressure to increase rapidly in association with the pressure in the ventricle.

36–40. The answers are: 36-E, 37-C, 38-A, 39-D, 40-B. [*I C 2 a, D 4 b (1) (a); VI A 5 d; X B 2; XI B 1*] Of the factors listed, capillary filtration rate is most closely associated with capillary pressure. Capillary pressure and interstitial fluid oncotic pressure are direct determinants of capillary filtration rate, according to the Starling hypothesis. Capillary pressure is counteracted by the oncotic pressure of the blood and tissue pressure.

An elevated arterial diastolic pressure is a more important determinant of the mean arterial pressure than the systolic pressure. The arterial systolic pressure is largely determined by the stroke volume, whereas the arterial diastolic pressure reflects the resistance in the systemic circulation. The high arterial diastolic pressure leads to vascular changes in certain organs, especially the kidneys and eyes, which are characteristically affected by arterial hypertension.

During exercise, there is an increased blood flow to skeletal muscles, which can be defined as a functional hyperemia, or increased blood flow. This increased flow is due to a local decrease in vascular resistance brought about by an increase in tissue temperature, osmolality, PK and P_{CO_2}, and a decrease in the local P_{O_2}. None of these factors alone seems to be able to duplicate the effects of exercise.

P_{CO_2} is the most important factor that regulates the resistance in the cerebral vessels. Hyperventilation, which results in a reduction of P_{CO_2}, produces a marked vasoconstriction of cerebral vessels that, at times, can lead to hypoxia sufficient to cause dizziness and even fainting.

Histamine, released from mast cells in areas of inflammation, results in an increased permeability of microvessels, primarily of the postcapillary venules.

Respiratory Physiology

Joseph Boyle, III

I. INTRODUCTION. The primary function of the respiratory system is to **exchange oxygen (O_2)** and **carbon dioxide (CO_2)** between the body and the environment. Secondary roles of the respiratory system include aiding in **acid-base balance,** defending the body against **inhaled particles** such as bacteria and pollens, acting as a **circulatory filter** to prevent clots from entering the systemic circulation, and regulating several **hormonal** and **humoral concentrations**.

   A. External respiration is concerned with the **exchange** of gas between the environment and the lungs, the **transfer** of gas across the respiratory membrane, and the **transport** of the gas by the blood to and from the body cells. This chapter is concerned primarily with the process of external respiration.

   B. Internal respiration is concerned with the **oxygen consumption** by the body cells through metabolic transformations and is now considered to be the province of biochemistry.

II. RESPIRATORY MECHANICS refers to the study of factors involved in altering lung volume. These factors include the muscular forces needed to expand the respiratory system, the forces that impede expansion (resistance and elastance), and the determinants of lung volume.

   A. Inspiration refers to an **increase in lung volume requiring that a force be applied to the respiratory system**. The active force usually is generated by the inspiratory muscles, but this force also can be supplied by a mechanical respirator if the muscles are weakened by a disease such as polio or muscular dystrophy. The **lungs are passive structures** that follow any movements of the chest wall. The small amount of fluid normally found in the pleural space causes the visceral and parietal pleura to adhere to one another so that any movements of the chest wall cause similar changes in lung volume. Normally, no gas exists in the pleural space, but gas can enter the pleural space from a rupture of the lung or a penetrating wound of the chest wall. The presence of gas in the pleural space is termed a **pneumothorax.**

   1. **Inspiratory muscles** include the respiratory diaphragm, the external intercostals, and the accessory muscles of inspiration. The accessory muscles include the **sternomastoid** and other **strap muscles of the neck**.
      a. **The diaphragm,** the major muscle of inspiration, is a dome-shaped sheet of muscle separating the thoracic and abdominal cavities. The diaphragm acts similarly to a piston or a syringe, since contraction of the muscle causes the dome of the diaphragm to descend. Descent of the diaphragm enlarges the volume of the thoracic cavity from below (Figure 3-1). Contraction of the diaphragm also lifts the lower ribs since the diaphragm originates from these structures. Elevation of the ribs causes thoracic expansion laterally and anteriorly because the ribs are angled downward (see Figure 3-1). This is referred to as a **bucket-handle effect.**
      b. **The external intercostal muscle** fibers originate from an upper rib and insert on a lower rib more anteriorly. Contraction of these muscles causes an elevation of the ribs and enhances the bucket-handle effect.
      c. **The accessory muscles of respiration** originate from the neck and skull and lift the clavicles and the sternum. These muscles help to elevate the ribs and enlarge the thorax.

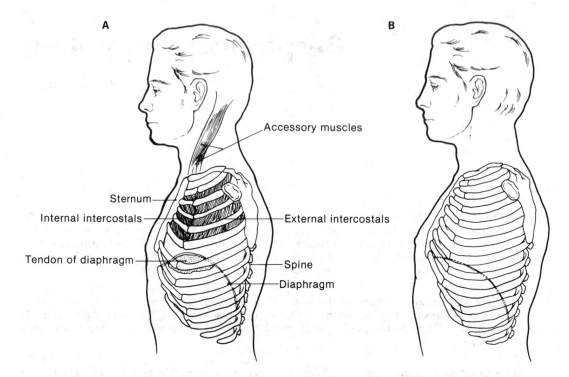

Figure 3-1. (*A*) The thorax and respiratory muscle relationships are shown during relaxation. (*B*) Maximal inspiration. Note the flattening of the diaphragm and the more horizontal position of the ribs denoting the increased volume of the thorax.

2. **Inspiratory force** is generated by contraction of the inspiratory muscles. This inspiratory force expands the volume of the respiratory system. **Boyle's law** states that the product of pressure times volume is a constant ($P \times V = C$). Thus, if the volume of a container is increased, the pressure will decrease so that the product of pressure and volume remains constant. Contraction of the inspiratory muscles expands the gas in the respiratory system, causing the pressure in the lungs to decrease below atmospheric pressure. If the airways are open, the gas can flow from an area of higher pressure (atmosphere) to lower pressure (alveoli) to renew the O_2 concentration in the alveoli.

   a. **Maximal inspiratory force** normally is –80 to –100 cm H_2O (80–100 cm H_2O pressure below atmospheric pressure). In a procedure called a **Müller maneuver,** one inspires as forcefully as possible, while holding the nose and closing the mouth. Note that the chest does not expand much but that negative pressure causes a suction effect felt in the ears.

   b. **Normal inspiration** requires pressures of only –3 to –5 cm H_2O to cause an adequate gas flow into the respiratory system. Thus, a tremendous reserve of force is available, which can be used to increase respiration during exercise or stress or if the forces that impede inspiration are increased by disease or injury.

   c. **Decreased inspiratory force** can be caused by many neurologic and muscular diseases such as muscular dystrophy and poliomyelitis, both of which impair the ability of the muscle to contract. Patients with these diseases may need respirators to maintain an adequate volume of inspiration.

B. **Expiration** refers to a **decrease in lung volume,** which is normally **a passive process**. During inspiration, the respiratory system is inflated above its equilibrium volume (see II C 4 a), just like blowing up a balloon. During relaxation (releasing the neck of the balloon), the elastic forces generated by the inflation compress the gas in the respiratory system, which raises its pressure (in accordance with Boyle's law), and the gas flows out until the system has reached its equilibrium volume. Expiratory muscle force may be needed when respiration must be increased during exercise or when disease has increased the resistance and elastance forces.

1. **Expiratory muscles** include the muscles of the **abdominal wall,** especially the rectus abdominus, and the **internal intercostal muscles.**
   a. **Contraction of the rectus abdominus muscle** decreases lung volume by pulling down on the lower ribs; contraction of the abdominal wall muscles compresses the abdominal contents and forces the diaphragm upward.
   b. **Contraction of the internal intercostal muscles** lowers the ribs, since these muscles originate from the lower ribs posteriorly and insert on upper ribs more anteriorly.

2. **Expiratory muscle force** compresses the gas in the respiratory system, and the gas pressure increases according to Boyle's law. Maximal contraction of the expiratory muscles can generate forces of 100–150 cm H_2O above atmospheric pressure. Maximal expiratory force can be generated from lung volumes near full inspiration when the nose and mouth are blocked. Forced expiration against closed airways is termed a **Valsalva maneuver** and is commonly performed when lifting heavy objects or when moving one's bowels.
   a. **Maximal expiratory pressure** can be measured with a mercury column or a blood pressure gauge (aneroid gauge). One takes in a maximum inspiration, holds the nose with one hand, and blows as forcefully as possible into the tubing to the gauge. The maximal pressure that can be sustained for 5 seconds is read. Since these gauges are calibrated in mm Hg, the gauge reading must be multiplied by 1.36 to convert it to cm H_2O pressure.
   b. **Normal expiration** is passive, since the elastic forces stored during inspiration are used to compress the alveolar gases. Thus, for expiration to occur, the inspiratory muscles gradually relax to provide a slow and controlled expiration.
   c. **Decreased expiratory force** also occurs in many neuromuscular diseases; however, since expiration normally is passive, a weakness of the expiratory muscles is not so critical as a weakness of the inspiratory muscles.

C. **Events of the respiratory cycle.** The two phases of respiration, **inspiration** and **expiration,** can be measured by a **spirometer,** which is a simple device that can measure and record changes in lung volume. Most spirometers use a water seal to collect the gas (Figure 3-2), which means that the gas is at room temperature and is saturated with water vapor. The gas in the spirometer occupies a different volume than it did in the body because of the change in the temperature and the amount of water vapor present. The volume that the gas occupied in the body can be calculated by using the **general gas law,** which is a combination of Boyle's and Charles' laws as shown in Table 3-1.

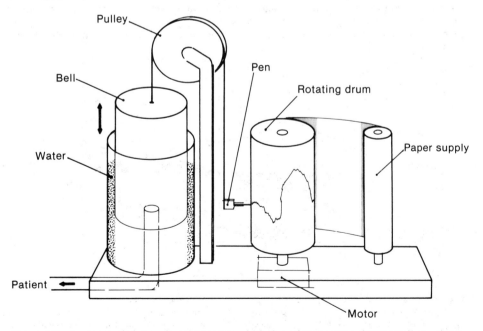

Figure 3-2. A water-sealed recording spirometer, which is used to measure changes in lung volume.

Table 3-1. The General Gas Law

The general gas law is stated as

$$\frac{P_1 \times V_1}{T_1} = \frac{P_2 \times V_2}{T_2}$$

To convert gas volumes from spirometer to body conditions, the general gas law is used in the following form

$$V_1 = \frac{V_2 (P_2 - PH_2O) T_1}{(P_1 - 47) T_2}$$

where V_1 = gas volume in the body; V_2 = gas volume in the spirometer; P_1 = barometric pressure; P_2 = standard pressure (760 mm Hg); T_1 = body temperature (310° K); T_2 = room temperature (° K); PH_2O = water vapor pressure at room temperature; and 47 = water vapor pressure at body temperature.

To identify the conditions for different lung volumes, the following definitions are used:

ATPS   =   atmospheric temperature, pressure, saturated (conditions in a spirometer)

BTPS   =   body temperature, pressure, saturated (gas volumes in the body)

STPD   =   standard temperature (0° C), pressure (760 mm Hg), dry (used to express O_2 and CO_2 volumes for metabolic equivalence)

1. **Alveolar pressure during inspiration** is negative compared to atmospheric pressure, as shown in Figure 3-3. This negative pressure in the alveoli creates the pressure difference between the atmosphere and the alveoli, which causes the gas to flow into the lungs. At the end of inspiration, the alveolar pressure becomes equal to atmospheric pressure, and the gas flow into the lungs ceases.

2. **Alveolar pressure during expiration** is greater than atmospheric pressure, because the elastic forces of the respiratory system and the contraction of the expiratory muscles compress the gas and raise the pressure. Thus, the higher pressure in the alveoli causes gas to flow out of the lungs, and lung volume returns to its initial level so that the next respiratory cycle can begin.

3. **Rate of gas flow** into or out of the lungs depends primarily on the pressure gradient between the alveoli and the atmosphere. The highest gas flow rates occur when the pressure difference between the alveoli and the atmosphere is greatest. Note, in Figure 3-3, that when the alveolar pressure becomes zero (i.e., equals atmospheric pressure), the gas flow also is zero. Since the

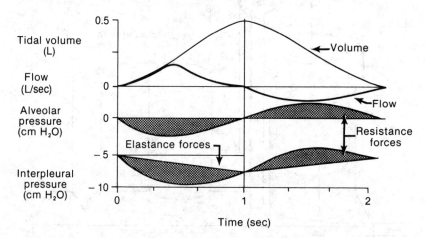

Figure 3-3. Events of a respiratory cycle. Resistance forces on the alveolar pressure curve indicate only airway resistance, while resistance forces on the pleural pressure curve include both airway and tissue resistance forces. Note that, at the peak of inspiration, flow is zero, resistance forces are zero, and the change in the interpleural pressure curve is due only to elastance forces.

pressure difference changes its direction between inspiration and expiration, the gas flow also reverses.

4. **Pressure in the interpleural space** normally is below atmospheric pressure (negative).
    a. **Negative interpleural pressure.** When the respiratory muscles are completely relaxed, the lungs contain 2–2.5 L of gas. The lungs are distended to this degree because the chest wall exerts a force tending to inflate the lungs. In turn, the elastic forces of the lungs try to compress the gas in the lungs and cause them to deflate. These opposing forces are equal but in opposite directions at this lung volume, which generates a negative pressure in the interpleural space. During complete relaxation of the respiratory muscles, the interpleural pressure is about –5 cm H_2O (Figure 3-4B). Because of the balanced forces under these conditions of relaxation, the volume of gas in the lungs is termed the **equilibrium volume**.
        (1) The **equilibrium volume** of the respiratory system represents the lung volume at the end of a normal, relaxed expiration. The equilibrium volume is more commonly called the **functional residual capacity** (see II H 2 c).
        (2) **Pleural fluid.** The lungs and the chest wall are coupled by the small amount of fluid in the interpleural space. This fluid acts as a lubricant so that the visceral pleura, which covers the lungs, slides easily over the parietal pleura, which lines the inside of the thoracic cavity. This system works similarly to two moistened pieces of glass. The glass pieces slide easily back and forth, but it is difficult to separate them because of the adhesion provided by the fluid.
    b. **During inspiration** the interpleural pressure becomes progressively more negative (see Figure 3-3), because the inspiratory muscles must stretch the lungs to take up the additional volume of gas inspired with each breath. Therefore, at the end of inspiration the lungs recoil away from the rib cage with more force. Also, the negative alveolar pressure is transmitted through the lungs into the interpleural space. The change in pleural pressure, from the beginning to the end of inspiration, measures the elastance forces for that breath. These forces can be used to measure the dynamic compliance of the lungs.
    c. **During expiration** the interpleural pressure returns to its resting level. During normal breathing, the interpleural pressure remains negative (i.e., below atmospheric pressure). However, with forced expirations, the interpleural pressure can become positive, which compresses the airways and slows the rate of expiration.

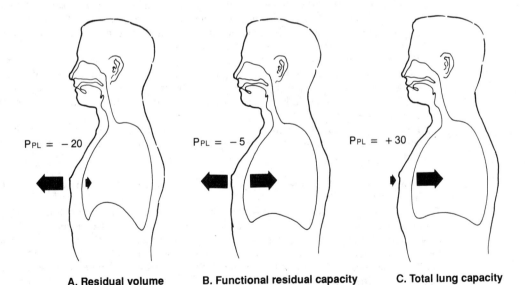

A. Residual volume   B. Functional residual capacity   C. Total lung capacity

Figure 3-4. Lung and chest wall forces at various lung volumes. The *arrow outside* the body indicates chest wall forces, and the *arrow inside* the body indicates the retractive force of the lungs. (*A*) At small lung volumes, the lung retractive forces are minimal, and the chest wall expansion forces predominate. The combination of the two forces results in a large, negative interpleural pressure. (*B*) At the equilibrium volume of the respiratory system, the lung retraction and chest wall forces are equal but in opposite directions. (*C*) At large lung volumes, both chest wall and lung forces are directed inward, resulting in a positive interpleural pressure. *PPL* = interpleural pressure.

D. Pressure-volume (P-V) relationships of the respiratory system depend on the passive (completely relaxed respiratory muscles) elastic properties of the chest wall and the lungs. These elastic properties are defined by the relationship between the gas volume in the lungs and the transmural pressure for the lungs and the chest wall under static conditions. Static conditions imply that volume is constant, and no gas is flowing; therefore, no resistance forces are present. Figure 3-5A indicates the anatomic locations that determine the transmural pressures for the lungs, chest wall (thorax), and the respiratory system.

1. **The transmural pressure (PTM)**, or the pressure across the wall, is the difference in pressure between the inside (PI) and the outside (PO) of any structure: $PTM = PI - PO$. PL refers specifically to the transmural pressure across the lung. The equilibrium volume of a structure is defined as the volume it contains when the PTM is zero, that is, when $PI = PO$. A positive PTM ($PI > PO$) is a distending force that enlarges a structure above its equilibrium volume, while a negative PTM is a force that tends to decrease the volume of a structure.

2. **Transpulmonary pressure (PL)** equals the pressure inside the lungs (PA) minus the pressure just outside the lungs, which is the pressure in the interpleural space (PPL): $PL = PA - PPL$. The pressure in the interpleural space is estimated in humans by measuring the esophageal pressure with a balloon-tipped catheter.
   a. The **lung** is a passive structure whose volume is determined by its transmural pressure (PL). Lung volume increases curvilinearly as PL increases until a **limiting volume** is approached at a PL of about **30 cm H$_2$O** when the compliance of the lung approaches zero. This limiting volume is thought to be determined by the complex arrangement of collagen fibers in the interstitium of the lung.
   b. The **equilibrium volume** of the lung normally is less than 10% of maximal lung volume. At negative PL, the airways close off so that a small volume of gas is trapped in the lungs—this is termed the **minimal volume**.
   c. **Compliance (distensibility)** is the change in volume of a structure for each unit change in pressure. The compliance is the slope (or steepness) of the P-V curve in Figure 3-5B, C, and D.
      (1) **Normal compliance** of the human lungs is about 0.2 L/cm H$_2$O. This means that, for each cm H$_2$O increase in transmural pressure, the volume increases 0.2 L of gas. The amount of gas inspired with each breath is about 500 ml or 0.5 L. In order to "stretch" the lungs to take up the normal tidal volume, an increase of about 2.5 cm H$_2$O in the transmural pressure across the lungs is required. The change in the transmural pressure normally is generated by the inspiratory muscles.
      (2) **Decreased compliance** of the lungs is caused by certain lung diseases (e.g., tuberculosis, abscess, respiratory distress syndrome) that produce scarring or **fibrosis** of the lungs or destroy functional lung tissue. This condition is termed **restrictive lung disease**. Patients with restrictive lung disease must generate forces much greater than normal to expand the lungs.
      (3) **Increased compliance** occurs in emphysema, which destroys some of the alveolar septa. This leads to a marked increase in airspace size and a reduction in the elastance forces of the lungs.

3. **Transthoracic pressure (PCW)** represents the transmural pressure across the chest wall; thus, $PCW = PPL - PBS$.
   a. The **chest wall** in the adult functions like a bellows that contains compression springs so that the equilibrium volume (see II D 2 b) is near maximal volume (see Figure 3-5C). If the lung volume is less than the chest wall equilibrium volume, then PCW is negative (PPL < PBS), which indicates that the chest wall has a tendency to expand. Thus, the relaxation P-V curve for the chest wall (PCW) is located largely on the negative side of the pressure axis in Figure 3-5C. The expansion force exerted by the relaxed chest wall helps to create the negative interpleural pressure and to keep the lungs expanded.
   b. At low volumes, the slope of the PCW line (compliance) approaches zero, which is caused by the thorax becoming essentially rigid.

4. **Transrespiratory pressure (PRS)** is the transmural pressure across the respiratory system. The inside pressure is PA and the outside pressure is PBS, so $PRS = PA - PBS$. Normally, the pressure

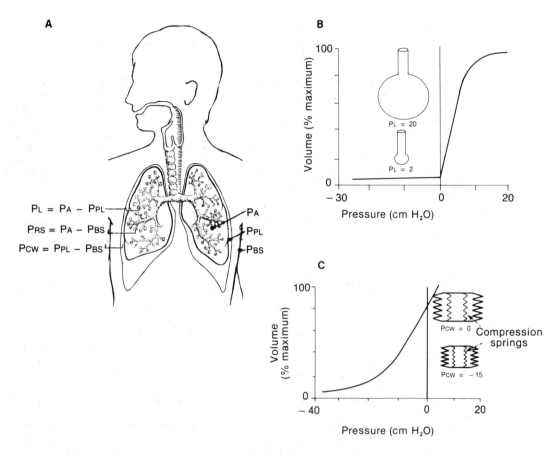

Figure 3-5. (*A*) Schematic relationships of the structures and pressures used to define the mechanical properties of the respiratory system. P_A = alveolar pressure; P_{PL} = interpleural pressure; P_{BS} = pressure at the body surface; P_{RS} = transrespiratory pressure; P_L = transpulmonary pressure; and P_{CW} = transthoracic pressure. (*B*) Pressure-volume curve of the lungs, which are represented as a simple balloon whose volume depends only on the transmural pressure. (*C*) Pressure-volume curve of the chest wall, which is represented as a bellows with internal compression springs that set its relaxation volume at about 80% of maximum. (*D*) Pressure-volume curve of the respiratory system, which is represented as the combination of the balloon and bellows. The system is shown at three different volumes (*1, 2, and 3*), with representative alveolar (P_A) and interpleural (P_{PL}) pressures.

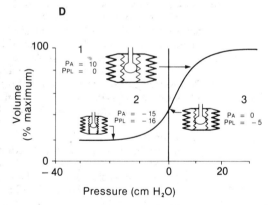

at the body surface (P_{BS}) is atmospheric, which is considered to be zero when dealing with differential pressures. It can be shown that $P_{RS} = P_L + P_{CW}$.*

a. The **respiratory system P-V curve** (see Figure 3-5D) is "S" shaped, steepest in the middle, and almost flat at both high and low lung volumes. Normally, the respiratory system functions near the middle of the curve, where the system is most distensible. This P-V curve is due to the summation of the elastic properties of the chest wall and the lungs, so that $P_{RS} = P_L + P_{CW}$.

---

*It can be proven that $P_{RS} = P_L + P_{CW}$ by substituting the expressions for P_L and P_{CW}; thus: $P_{RS} = P_A - P_{PL} + (P_{PL} - P_{BS})$. P_{PL} cancels out, leaving: $P_{RS} = P_A - P_{BS}$, which is the definition of the respiratory system pressure.

    b. The **equilibrium volume** of the respiratory system occurs at about 40% of maximal lung volume. This volume also is referred to as the **functional residual capacity** and represents the normal end-expiratory lung volume.

E. Elastance is the force generated by distension of any structure, whereas the **transmural pressure** represents the force that counteracts the elastance to keep a structure inflated. An example will clarify this relationship: If one stretches an elastic band between one's hands, the transmural pressure represents the force generated by the arms to lengthen the elastic band; in turn, the elastic band recoils against the hands, and this force is termed the **elastance**. The smaller the **compliance** (the thicker the elastic band), the greater the force (elastance) generated by any increase in volume. Thus, compliance (C) and elastance (E) are **inversely related** (C = 1/E). At a constant volume, the elastance force and the transmural pressure are equal but in opposite directions. **Elastic forces in the lungs** arise from two sources: tissue and surface forces.

    1. Tissue forces. The lung contains large amounts of collagen and elastin, which are connective tissue fibers whose roles in generating elastic (retractile) forces in the lungs are not fully understood. It is thought that elastic fibers are stretched at low and medium lung volumes, and that collagen prevents overdistension of the lungs at high lung volumes. Smooth muscle fibers in the lungs also may contribute to the tissue forces.

    2. Surface forces. Whenever an interface exists between two different phases (e.g., between a liquid and a gas), a surface force or surface tension is generated. The surface force tends to minimize the surface area of the interface; this is what causes water droplets to bead up on a waxed surface of a car. The surface force is caused by the unbalanced attraction of the surface molecules by the molecules below the surface (Figure 3-6A). The surface forces cause the lungs to retract or empty, which is counteracted by the transmural pressure. Filling the lungs with a liquid, such as saline, eliminates the surface forces and markedly reduces the recoil force compared to the air-filled lung (Figure 3-7). In the fluid-filled lung, the retractive forces are the result of only the tissue component. The conclusion from this experiment is that the surface forces, which are present when the lungs are air filled, represent a more powerful cause of the lungs' recoil pressure than the tissue forces.

    a. Surface tension. If the surface is spherical, as in an alveolus, then an opposing force is required to counteract the surface tension to keep the structure inflated. This opposing force is the **surface component** of the **transmural pressure** as shown in Figure 3-6C.

    b. Laplace's law gives the relationship between surface tension (ST), transmural pressure (P), and radius (r) of the spherical structure: $P = 2 \times ST/r$ (see Figure 3-6B). An increase in the surface tension or a decrease in the radius would require a higher pressure to keep the alveoli inflated. Since the lungs normally have a low transmural pressure, any decrease in lung volume or increase in alveolar surface tension leads to collapse of the alveoli. Collapse of the lung air spaces is termed **atelectasis**.

    c. Pulmonary surfactant lines the alveolar surface.

        (1) Composition. Pulmonary surfactant is a complex mixture made up of **phospholipids, cholesterol,** and a **specific protein;** all of these substances are formed by type II alveolar cells.

        (2) Functions

            (a) Reduction of surface tension. Adsorption of surface active molecules at the alveolar-air interface reduces the surface tension in the lungs. The reduction of the surface tension, which is the prime function of pulmonary surfactant, increases the compliance of the lungs and thereby decreases the work of respiration.

            (b) Increase in alveolar radius. Surfactant fills irregularities in the alveolar surface, which increases the mean alveolar radius. This increase reduces the transmural pressure required as shown by the formula for the law of Laplace. This effect also increases the compliance of the lungs and reduces the work of breathing.

            (c) Reduction of pulmonary capillary filtration. Normal surfactant, by lowering the alveolar surface tension, reduces the retraction forces in the lungs. This makes the interstitial pressure of the lungs less negative, which reduces the filtration forces across the pulmonary capillary (Frank-Starling mechanism; see Ch 2 V B 1 b). Thus, normal surfactant helps to keep the lungs from becoming edematous, which would interfere with gas exchange by increasing the thickness of the alveolar-capillary membrane or by causing alveolar flooding.

            (d) Stabilization of the alveoli. Alveolar stability would be enhanced if the alveolar surface tension changed in proportion to the size of the alveoli. The surface tension

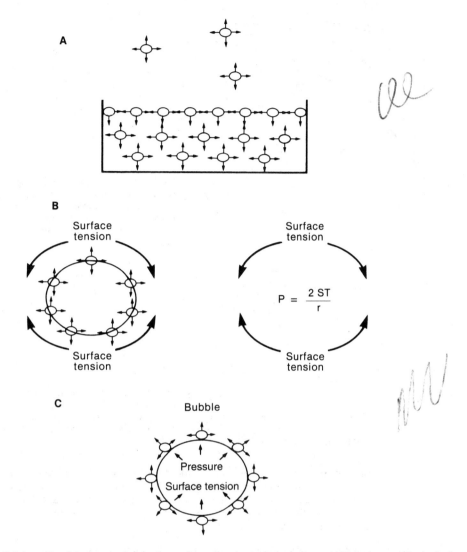

Figure 3-6. (*A*) Schematic of the intermolecular forces for molecules in the gas phase, at the surface and in the bulk phase of a liquid. Note that surface molecules have an unbalanced force tending to pull them into the liquid. (*B*) Surface forces in a hollow sphere tend to reduce the volume of the sphere. The relationship between surface tension and the surface component of transmural pressure is defined by Laplace's law. (*C*) To maintain a constant volume in a spherical structure, the transmural pressure must counterbalance the surface tension forces. P = pressure; ST = surface tension; and r = radius.

at an air-fluid interface changes when surface active agents are placed on a surface balance, and the surface area is cyclically changed. Lung extracts containing surfactant change their surface tension from 0 to 40–50 dynes/cm during such cycling (Figure 3-8). Whether such changes in surface tension occur in the alveoli is still controversial.

(3) Abnormalities

- **(a) Respiratory distress syndrome (RDS) of the newborn** is due to abnormal surfactant that is formed because of immaturity of the lung. These infants have inadequate gas exchange because of alveolar instability and consequent loss of surface available for exchange.
- **(b) Atelectasis** (aveolar collapse) occurs in several other conditions associated with abnormalities in surfactant.
- **(c) Shock, systemic infections,** and **trauma** can cause abnormal surfactant function in adults (adult respiratory distress syndrome).

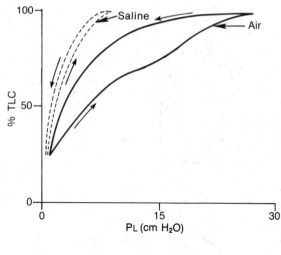

Figure 3-7. Pressure-volume loops for isolated lungs when filled with saline or air, showing the direction of volume change (*long arrows*) as a percent of the total lung capacity (*TLC*). Note the marked increase in pressures required during the air inflation. P_L = transpulmonary pressure.

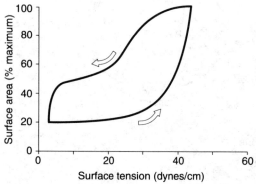

Figure 3-8. Surface tension as a function of surface area obtained from bronchoalveolar lavage fluid. Note the large hysteresis and that minimal surface tension approaches zero. The *arrows* indicate the direction of the recording.

    d. Tissue interdependence is another mechanism that stabilizes alveoli. Since the alveolar walls form a continuous network with one another and are under some tension, the collapse of one area is counteracted by the tension that is generated in adjacent alveolar walls (Figure 3-9). Thus, alveoli are not independent units that can change volume independently of the neighboring lung tissue.

F. Resistance forces are present whenever lung volume is changing. Resistance is caused by the friction of gas molecules between themselves and the walls of the airways (**airway resistance**) as well as by the friction of the tissues as they expand or contract (**tissue resistance**).

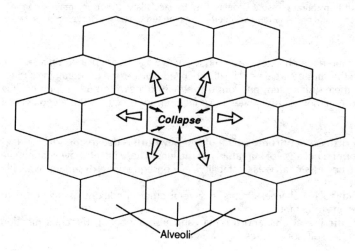

Figure 3-9. A schematic representation of the concept of tissue interdependence. Alveolar collapse occurs if surface tension increases or alveolar ventilation is reduced sufficiently. Tissue interdependence prevents alveolar collapse because of the tension generated in the surrounding alveolar septa (indicated by the *large open arrows*) and, therefore, stabilizes alveolar volume.

1. **Airway resistance** represents about 80%–90% of the total resistance forces during breathing. The airway resistance depends on the rate of gas flow and the diameter and length of the airways. Airway resistance can be calculated by dividing the alveolar pressure during respiration by the rate of gas flow from the lungs. The usual units of airway resistance are cm $H_2O/L/sec$.
   a. **Flow rate of gas** is determined by the velocity of the gas molecules as they pass through the airways. The airway system is organized so that the total area of the airways increases about 1000-fold as the gas proceeds from the trachea through about 23 generations of branchings to the alveoli. This is similar to a river that widens gradually into a lake. The velocity of flow is much higher in the river (trachea and bronchi) than it is in the lake (alveolar ducts), because the river is much narrower. The highest resistance to flow occurs in the intermediate-sized bronchi because of the velocity. The airways less than 2 mm in diameter represent only about 10% of the total airway resistance.
   b. **Airway diameter** is an extremely powerful factor that determines airway resistance. The smaller the airway the higher the resistance for any flow rate.* Actually, a decrease in airway diameter to half its original size causes the airway resistance to increase 16–32 times. People with **chronic obstructive pulmonary disease (COPD),** such as asthma, bronchitis, and emphysema, have a narrowing of their small airways and a very high airway resistance. The high airway resistance causes COPD patients to generate very high pressures with their respiratory muscles in order to breathe.
      (1) **Sympathetic nervous activity** to the nerves in the airways causes only minor changes, since these adrenergic nerves do not actually innervate the bronchial smooth muscle. However, circulating adrenergic substances (e.g., **epinephrine**) or **adrenergic-like drugs** cause a dilation of the airways, which reduces airway resistance. The dilation occurs since the bronchial smooth muscle contains largely β_2-receptors, which are activated by epinephrine or a β_2-adrenergic drug, which can be administered by inhalation.
      (2) **Vagus nerve (cholinergic) activity** to the lungs causes the smooth muscle in the walls of the bronchi and bronchioles to contract and increases the formation of mucus. This narrows the airways and increases the resistance to airflow. Cholinergic activity is a major component in the pathogenesis of asthma in some individuals.
      (3) **Nonadrenergic, noncholinergic activity.** The lungs are derived embryologically from the foregut. Recently, it has been recognized that many of the neuropeptides that control intestinal activity also control airway dimensions.
         (a) Noncholinergic excitatory compounds, such as **substance P** and **neurokinins A and B,** cause bronchoconstriction, mucus secretion, and increased vascular permeability. These compounds seem to be especially involved in the pathogenesis of asthma in certain individuals.
         (b) The nonadrenergic inhibitory system consists of postganglionic neurons that release vasoactive intestinal peptide (VIP) and other mediators that cause smooth muscle relaxation and inhibit mucus production.
   c. **Airway length** changes relatively little during respiration and disease processes and, thus, is not a significant factor that changes airway resistance.

2. **Tissue resistance** plays only a minor role in determining the forces required to breathe.

3. **Effort-independent flow**
   a. Effort-independent flow occurs during forced expiration at volumes less than 80% of the maximal lung volume. This relationship is shown in Figure 3-10, where expiratory flow is plotted as a function of the interpleural pressure at different lung volumes. Under these conditions, interpleural pressure can be used as a measure of the effort exerted during expiration. In Figure 3-10, note that the expiratory flow rate remains constant at each lung volume, while interpleural pressure varies over a large range (i.e., flow is independent of effort). The factor controlling flow under these conditions is the **transpulmonary pressure,** which is a function of lung volume.
   b. **Dynamic compression** of the airways (Figure 3-11) occurs during forced expirations and limits the flow rate during the terminal part of expiration. The amount of dynamic compression depends on several variables that determine the limitation of the expiratory gas flow.

---

*In the lungs, the smaller airways provide a larger total cross-sectional area, so that flow rates and airway resistance are reduced as gases proceed toward the alveoli.

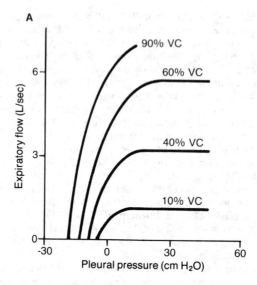

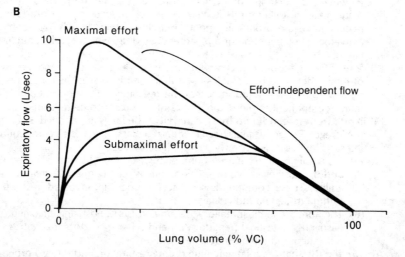

Figure 3-10. (*A*) Isovolume flow diagram. At volumes less than 80% of total lung capacity, expiratory flow rate is limited, and the flow rate is independent of effort as measured by the interpleural pressure. *VC* = vital capacity. (*B*) Flow-volume loop. Much information can be obtained about pulmonary mechanics by recording flow rate as a function of lung volume. The flow rate labeled *maximal effort* represents maximal flow rate and cannot be exceeded. The maximal flow rate at each volume depends on the state of the airways and the lung parenchyma and provides important diagnostic information. % *VC* = percent vital capacity.

(1) **Lung volume** is the most powerful modulator of expiratory gas flow. When lung volume is increased, airway diameter expands because of the enhanced radial traction and expiratory muscle force increases due to changes in the muscle length.

(2) **Elastic recoil** alters the diameter of the intrapulmonary airways also through radial traction on the airway walls. Elastic recoil is decreased by processes such as aging and emphysema and is increased in patients with restrictive lung disease.

(3) **Increased airway resistance,** caused by COPD, increases the pressure drop along the small peripheral airways. This loss of pressure produces a greater compressive effect on the airways, leading to a further narrowing and even greater reduction in the rate of gas flow at any given lung volume. Patients with obstructive lung disease can reduce the pressures needed to breathe by breathing slowly but with a larger tidal volume to maintain a normal alveolar ventilation. This breathing pattern reduces the velocity of gas flow, the pressures required to overcome resistance, and the work of breathing.

Eupneic expiration

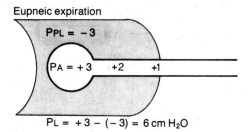

$PL = +3 - (-3) = 6\,cm\,H_2O$

Forced expiration

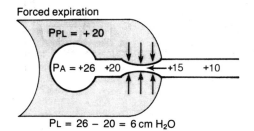

$PL = 26 - 20 = 6\,cm\,H_2O$

Figure 3-11. Schematic representation of the flow limiting segment in an airway during forced expiration. PPL = interpleural pressure; PA = alveolar pressure; and PL = transpulmonary pressure. During eupneic expiration, the transmural pressure of the airways is a distending force ($PA > PPL$), which dilates airways. During a forced expiration, $PPL >$ the pressure in the bronchi (indicated by the *arrows*), which narrows airways, increases airway resistance, and limits the expiratory flow rate.

G. **The work of respiration** equals the change in pressure needed to inflate the lungs times the volume change: work = $\Delta P \times \Delta V$. If either the resistance to breathing increases or the compliance decreases, then the respiratory muscles must generate more force and, thus, use more O_2 to overcome the added load.

 1. Respiratory failure. The respiratory muscles, just as any skeletal muscles, can become fatigued if exposed to a high work load for a long time. If the respiratory muscles cannot generate sufficient force, then pulmonary ventilation becomes inadequate, and respiratory failure occurs. In respiratory failure, there is an inadequate supply of O_2 to the body, and CO_2 excretion is impaired.

 2. The **minimal work for respiration** varies with the resistance and elastance of the respiratory system; these factors are largely determined by the size of the airways and the lungs. All animal species seem to use the combination of tidal volume and respiratory rate that creates the minimal work of respiration. The mechanism by which the brain selects this combination is unknown.

 a. Increased resistance work occurs when the resistance forces are increased. Resistance work can be minimized by breathing more slowly and deeply, since this reduces the flow rates of gas and thereby decreases the pressure gradients needed to generate gas flow.

 b. Increased elastance work occurs when the compliance is reduced (increased elastance). The elastance work of breathing is proportional to the tidal volume as well as to the elastance of the respiratory system. The elastance work of breathing can be minimized by decreasing the tidal volume but breathing more rapidly to maintain an adequate ventilation of the lungs.

H. **Lung volumes and capacities.** The amount of gas in the lungs can be divided into various compartments with specific functions. There are four lung **volumes** that are nonoverlapping fractions of the gas content and four **capacities** that are combinations of two or more of the lung volumes. Figure 3-12 shows how these volumes and capacities are organized.

 1. Lung volumes
 a. Tidal volume (V_T) is the volume of gas inspired or expired with each breath. The normal tidal volume is 500 ml (0.5 L).
 b. Inspiratory reserve volume (IRV) is the additional volume of gas that can be inspired above the tidal volume. The inspiratory reserve volume is in reserve and can be used if the tidal volume needs to be increased (e.g., during exercise).
 c. Expiratory reserve volume (ERV) is the volume of gas that can be expired after a normal expiration. The expiratory reserve volume also is partially used when tidal volume needs to be increased.
 d. Residual volume (RV) is the volume of gas that remains in the lungs after a maximal expiration. This gas prevents the alveoli from completely collapsing; complete collapse of

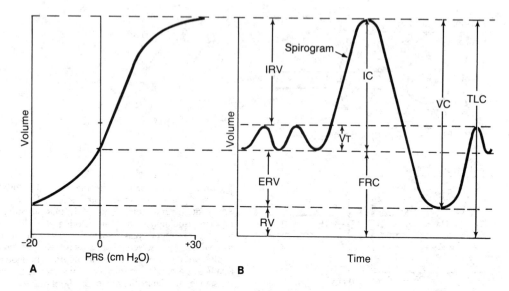

Figure 3-12. (*A*) A pressure-volume curve of the respiratory system. Residual volume (*RV*) and total lung capacity (*TLC*) are limited by the decreased compliance of the system, as indicated by the reduced slope of the curve. Functional residual capacity (*FRC*) is determined by the volume at which the transrespiratory pressure (*PRS*) is zero. (*B*) A spirogram showing several normal tidal volumes (*VT*); a maximal inspiratory effort followed by a maximal expiratory effort, termed a vital capacity (*VC*) maneuver; and the return to normal VT. This tracing can be used to define all of the subdivisions of lung volume. *ERV* = expiratory reserve volume; *IRV* = inspiratory reserve volume; and *IC* = inspiratory capacity.

an area of the lung (**atelectasis**) requires extremely large forces to reopen the alveoli. Since the residual volume cannot be expired into a spirometer, it must be measured by either a dilution method or a plethysmographic technique, as illustrated in the following examples.

(1) Dilution method

(a) Helium—an indicator gas that can be readily detected—is placed in a spirometer. [In this example, the fraction of helium is assumed to be 0.2 ($F_1 = 0.2$) of the total volume in the spirometer (V_{sp}), which is 3 L.]

(b) If a person begins to breathe from the spirometer at the end of a normal expiration, then the volume of gas in the lungs at the beginning of the test, or the **functional residual capacity (FRC)**, will be measured. The spirometer volume is maintained at 3 L by the addition of O_2.

(c) At the end of the test, when the helium is equilibrated between the lungs and the spirometer, the fraction of helium (F_2) is found to be 10%. The lung volume can be easily calculated using the following relationship

$$F_1 \times V_{sp} = F_2(V_{sp} + FRC)$$
$$FRC = F_1 \times V_{sp}/F_2 - V_{sp} = 0.2 \times 3/0.1 - 3 = 3 \text{ L}$$

(d) If the expiratory reserve volume is measured from the spirometer tracing and found to be 2 L, then the residual volume is 1 L.

(2) Plethysmographic method

(a) A person is placed in an airtight chamber (plethysmograph), and the pressures within the chamber and within the airways are measured while the person pants against a closed airway.

(b) With proper calibration of the system, it is possible to determine the volume of gas in the thorax by applying Boyle's law ($P_1V_1 = P_2V_2$), where the two subscripts indicate the pressure (P) and the volume (V) under the different conditions when the number of gas molecules remains constant.

2. Lung capacities

a. Total lung capacity (TLC) is the volume of gas in the lungs after a maximal inspiration. It represents the sum of the four lung volumes.

b. Vital capacity (VC) is the maximal volume of gas that can be expired after a maximal inspiration. Measurement of the vital capacity under different conditions is the most commonly performed pulmonary function test.

(1) Slow vital capacity is used to evaluate the size of the lungs. Many lung diseases can decrease the vital capacity by either reducing the total lung capacity or increasing the residual volume. A patient's vital capacity can be compared to predicted normal values, which are dependent on height, age, and sex.

(2) Forced vital capacity is used to evaluate the resistance properties of the airways and the strength of the expiratory muscles.

    (a) This test is performed by having a patient expire as rapidly as possible into a recording spirometer. Patients with high airway resistance (obstructive lung disease) are unable to achieve high expiratory flow rates and, therefore, require a longer time to expire their vital capacity (Figure 3-13).

    (b) A measurement that is commonly made is one that determines the percent of the vital capacity that can be expired in 1 second. This is called the %FEV$_1$ (forced expiratory volume in 1 sec/forced vital capacity $\times$ 100), which is normally greater than 75%–80%.

    (c) Figure 3-13 shows typical recordings of forced vital capacities and several measurements that can be obtained from normal, healthy patients and patients with obstructive and restrictive lung disease.

  c. FRC is the volume of gas that remains in the lungs after a normal expiration. The FRC represents a reservoir of gas that allows oxygenation of the blood between breaths. Since the FRC contains the residual volume, it also cannot be measured by spirometry alone.

  d. Inspiratory capacity (IC) represents the sum of the tidal volume and the inspiratory reserve volume and is the maximal volume that can be inspired after a normal expiration.

III. GAS EXCHANGE BETWEEN THE ATMOSPHERE AND THE BODY

  A. Composition of air. Air is a mixture of gases composed primarily of **N$_2$** and **O$_2$,** with a variable amount of **water vapor,** a negligible quantity of **CO$_2$** and a small amount of **inert gases.** These inert gases usually are lumped together with N$_2$. **Dry atmospheric air** can be considered to be composed of **79% N$_2$ and 21% O$_2$.**

    1. Partial pressure of gases. Gas molecules are in constant, rapid motion causing them to strike each other and the walls or surfaces of any container or structure. Because of this constant bombardment, these molecules exert a pressure, which is termed the **atmospheric** or **barometric pressure.** The **normal barometric pressure** at sea level is **760 mm Hg (760 Torr).** If there is a mixture of gases, then each gas exerts its own **partial pressure,** which is equal to its concentration (fraction of gas present) times the total pressure **(Dalton's law).**

    2. Partial pressure of N$_2$ (PN$_2$) in air is calculated by multiplying the fraction of N$_2$ in air by the barometric pressure. PN$_2$ = 760 $\times$ 0.79 = 600 mm Hg. The term **tension** is synonymous with partial pressure; thus, the N$_2$ tension = 600 mm Hg.

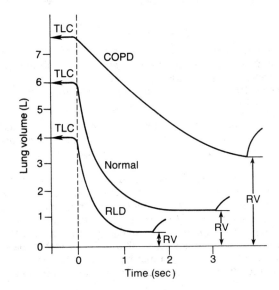

Figure 3-13. Forced vital capacity maneuvers from a normal individual and from patients with chronic obstructive pulmonary disease (*COPD*) and restrictive lung disease (*RLD*). The y axis represents the absolute lung volume. Note the differences in the rate of expiratory flow for the three subjects. *TLC* = total lung capacity; *RV* = residual volume.

3. **Partial pressure of O_2 (Po_2)** is determined the same way, so that the O_2 tension of air = $760 \times 0.21 = 160$ mm Hg. Note that the sum of the O_2 and the N_2 tensions equals the total barometric pressure.

B. Minute ventilation and alveolar ventilation

1. **Minute ventilation ($\dot{V}E$)** is the volume of air inspired or expired per minute, which equals the tidal volume (VT) multiplied by the respiratory rate or frequency (f): $\dot{V}E = VT \times f$. The tidal volume at rest averages 500 ml (0.5 L), and the normal respiratory rate is 12–15 breaths/min. Therefore, the normal minute ventilation is 6–7.5 L/min.

2. **Dead space ventilation ($\dot{V}D$).** The upper airways, from the nose and mouth to the terminal bronchioles, do not participate in gas exchange because of the thickness of their walls. This area of the respiratory system, which contains a gas volume of approximately 150 ml, is termed the **conducting zone,** or **dead space (VD),** because it does not participate in gas exchange. The **dead space ventilation** is the volume of gas per minute that enters the dead space of the respiratory system. For example, $\dot{V}D = VD \times f = 0.15 \times 15 = 2.25$ L/min.

   a. **Anatomic dead space** is the volume of gas that occupies the nose, mouth, pharynx, larynx, trachea, bronchi, and bronchioles. The anatomic dead space has several functions, including conditioning of air and removal of foreign materials.

      (1) **Conditioning of air.** During nasal breathing, the inspired gas is warmed to body temperature and is saturated with water vapor (has 100% relative humidity) by the time it reaches the trachea. The addition of water vapor to the inspired air dilutes the O_2 and N_2 concentrations slightly. Mouth breathing or the presence of a tracheostomy allows cool, dry air to reach the lower airways. This can result in the drying of airway secretions and the formation of mucous plugs.

      (2) **Removal of foreign material** occurs through the processes of filtration in the nose, impaction in the lower airways, or dissolving in the moist surface of the airways. Small particles (e.g., soot or pollen) impact on the surface of the airways and stick to the mucous lining. The foreign material is carried in the mucus toward the mouth, where it is either expectorated or swallowed. The mucus is propelled upward by the cilia of the respiratory epithelium that line the airways.

      (3) **Reaction to foreign material.** Foreign material in the inspired gas can stimulate irritant receptors in the airways, which causes coughing, secretion of mucus, and hypertrophy of the mucous glands. Prolonged breathing of air that contains foreign materials such as cigarette smoke and smog can cause **chronic bronchitis.**

   b. **Total dead space.** In normal persons, the anatomic dead space represents the entire dead space, but, in patients with lung disease, some alveoli do not receive any blood flow and therefore do not participate in gas exchange. These alveoli form the **alveolar dead space.**

      (1) **Bohr's method.** The total dead space equals the anatomic dead space plus the alveolar dead space. The total dead space can be determined by measuring CO_2 levels in the expired gases.

$$VD = VE \times (1 - PECO_2/PACO_2)$$

      where $PECO_2$ and $PACO_2$ represent PCO_2 in expired and alveolar gases, respectively. This method requires measuring the PCO_2 in mixed expired gases and in end-tidal samples of expired gases which, theoretically, represent pure alveolar gas.

      (2) **Fowler's method** of determining the dead space requires measuring expired N_2 tensions after inspiring 100% O_2 (see VI C 1 b).

3. **Alveolar ventilation ($\dot{V}A$)** is the volume of gas that participates in the exchange of O_2 and CO_2. The normal alveolar ventilation is 4.5–5 L/min (alveolar ventilation equals the minute ventilation minus the dead space ventilation). An adequate alveolar ventilation is critical since it determines the alveolar Po_2 and Pco_2.

   a. **Alveolar Po_2.** O_2 is continually removed from the alveoli by its diffusion into the pulmonary capillary blood. Inspiration brings fresh air into the alveoli, which maintains the alveolar Po_2 at about 100 mm Hg. The alveolar Po_2 (Table 3-2) is affected by changes in the barometric pressure (mountain climbing or diving), the fraction of O_2 inspired (supplemental O_2), and by the amount of CO_2 in the alveoli (hypoventilation or hyperventilation). The alveolar Po_2 at sea level can be raised above 600 mm Hg by administering 100% O_2 through a mask. The prolonged use of 100% O_2 results in lung damage so that high concentrations of O_2 must be used with caution, similar to any other powerful drug, when treating a patient.

Table 3-2. Alveolar Air Equation

The alveolar air equation can be used to calculate the average alveolar oxygen tension (PAO_2) and is stated as

$$PAO_2 = (PB - 47) \times FIO_2 - \frac{PACO_2}{R}$$

where PB = barometric pressure; 47 = vapor pressure of water at body temperature; FIO_2 = fraction of O_2 in inspired air; $PACO_2$ = alveolar CO_2 tension; and R = volume of CO_2 excreted per volume of O_2 absorbed per minute (R normally is 0.8).

Normal PAO_2 at sea level is

$$PAO_2 = (760 - 47) \times 0.21 - \frac{40}{0.8} = 100 \text{ mm Hg}$$

---

b. **Alveolar PCO_2**
  (1) Alveolar PCO_2 is determined by the ratio of the rate of CO_2 production ($\dot{V}CO_2$) to alveolar ventilation ($\dot{V}A$)

$$\text{alveolar } PCO_2 = K \times \dot{V}CO_2 / \dot{V}A$$

  where K is a constant.
  (2) Alveolar and arterial CO_2 tensions are essentially equal due to the high diffusibility of CO_2 and the shape of the CO_2 content curve. PCO_2 normally is very closely regulated to a value of 40 mm Hg, since changes in PCO_2 alter pH, which affects the rates of many enzymatic reactions.
      (a) An increase in alveolar ventilation causes the alveolar PCO_2 to decrease, since the CO_2 is washed out of the lungs by the increased volume.
      (b) A decrease in alveolar ventilation (hypoventilation) results in a rise in alveolar PCO_2 and is termed **hypercapnia** if it exceeds 45 mm Hg. The alveolar PCO_2 is determined by the ratio of the rate of CO_2 production to the rate of alveolar ventilation. Hypercapnia indicates that ventilation is inadequate for the metabolic rate and usually is accompanied by hypoxia.

C. **Diffusion.** O_2 and CO_2 move across the respiratory membrane between the alveoli and the pulmonary capillaries by diffusion. Diffusion is the movement of molecules from areas of high concentration to areas of low concentration by random motion. The rate of diffusion is determined by several factors as defined by **Fick's law of diffusion** (Table 3-3) [see Ch 1 I 1 C a (3)].

  1. **Diffusion capacity of the lungs (DL)** is a pulmonary function test that measures the diffusion properties of the respiratory system. The DL is measured with a dilute concentration of carbon monoxide (CO). The DL is proportional to the surface area of the lungs divided by the length of the diffusion path. Lung diseases can decrease the surface area for gas exchange or lengthen the diffusion path and reduce the diffusion properties of the lungs. These changes can be detected by the diffusion capacity test.

  2. **The respiratory membrane** consists of several components.
      a. Each layer of the respiratory membrane is extremely thin; the normal respiratory membrane averages about 0.5 μ (0.00002 inch) in thickness.

Table 3-3. Fick's Law of Diffusion

Fick's law of diffusion is stated as

$$V = \frac{K \times (P_1 - P_2) \times A}{d}$$

where V = volume of gas diffusing per minute; K = a constant that includes temperature and the diffusion coefficient of the gas; P_1 = partial pressure of the gas on one side of the membrane; P_2 = partial pressure of the gas on the other side of the membrane; A = surface area of the membrane; and d = diffusion distance between the two sides of the compartment.

---

    b. The **surface area** of the respiratory membrane is very large, averaging about 100 m² in the normal adult. The large surface area and the small distance separating the alveolar gas from the pulmonary capillary blood allows rapid diffusion of gases between the two compartments.

  3. Diffusion of O₂ occurs from the alveolar gas to the pulmonary capillary blood, because the alveolar P_{O_2} usually is 100 mm Hg, while the blood entering the pulmonary capillary normally has a P_{O_2} of 40 mm Hg.
    a. The O₂ molecules dissolve in the plasma, which raises the P_{O_2} and allows O₂ to diffuse into the red blood cell so that it can combine with hemoglobin. Normally, enough O₂ diffuses across the respiratory membrane so that the blood P_{O_2} becomes equal to the alveolar P_{O_2} in about 0.25 second.
    b. Pulmonary diseases that reduce the surface area or increase the thickness of the respiratory membrane can reduce the rate of O₂ transfer (Figure 3-14). These diffusion abnormalities can result in **hypoxemia** (i.e., a reduced O₂ level in the blood) if the P_{O_2} in the pulmonary capillary blood never becomes equal to the alveolar P_{O_2}. Under these conditions, O₂ exchange is termed **diffusion limited,** whereas O₂ exchange normally is limited by the rate of blood flow (**perfusion limited**).

  4. Diffusion of CO₂ occurs from the pulmonary capillary blood (P_{CO_2} = 46 mm Hg) to the alveoli (P_{CO_2} = 40 mm Hg). CO₂ diffuses much more rapidly than O₂, because it is about 25 times more soluble in body fluids than is O₂. This factor is included in the diffusion coefficient of the Fick equation. The increased solubility means there are more molecules dissolved in the liquid, which increases the rate of diffusion.

IV. GAS TRANSPORT. O₂ must dissolve in the plasma of the pulmonary capillaries before it can diffuse to the hemoglobin inside the red blood cell. The volume of gas that dissolves in any liquid is determined by **Henry's law.**

  A. Henry's law states that the volume of gas "x" that dissolves in a liquid equals the product of its **solubility coefficient** (Sx) and its **partial pressure** (Px): C = (Sx) (Px).

    1. Gas molecules dissolve in liquids the same way that sugar dissolves in coffee. In order to know how much sugar is dissolved in coffee, one needs to know how many scoops were put in as well as the size of the scoops. Similarly, the gas content of fluids depends on two factors. The solubility coefficient is similar to the size of the scoop, and the partial pressure is comparable to the number of scoops.

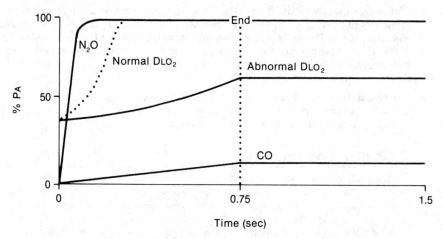

Figure 3-14. Changes in pulmonary capillary gas tensions, as a percent of the alveolar gas tension (% *PA*). The x axis represents the time the blood is in the pulmonary capillaries, which normally is 0.75 second. *D*$_{LO_2}$ = diffusion capacity of the lungs for O₂; *N₂O* = nitrous oxide; and *CO* = carbon monoxide. Normally, O₂ is a perfusion limited gas, as is N₂O, since there is equilibration between pulmonary capillary blood and alveolar gas tensions. CO represents a diffusion limited gas that is commonly used to measure the diffusion capacity of the lung.

Table 3-4. Solubility Coefficients of Some Important Gases in Respiratory Physiology

Temperature (°C)	Solubility Coefficient (ml gas/ml saline/mm Hg gas tension)			
	O_2	N_2	CO_2	CO
0	0.049	0.024	1.71	0.035
20	0.032	0.016	0.90	0.023
37	0.024	0.012	0.58	0.019

2. It is important to understand the difference between the partial pressure of a gas and the gas content of a liquid. The gas content represents the volume of the gas (in units of volume per unit volume of liquid) that is present, while the partial pressure of the gas represents the pressure it would exert in the gas phase. The solubility coefficient represents a proportionality constant that equates gas content and partial pressure.

3. Table 3-4 shows the solubility coefficients of some gases that are important in respiratory physiology. Note that the solubility coefficients for the various gases are different and that increasing the temperature reduces the solubility of gases.

B. Hemoglobin markedly increases the amount of O_2 that is carried in the blood (the size of the scoop is increased almost 50-fold). Each gram of hemoglobin can combine with 1.34 ml of O_2; normally, there are 12–15 g hemoglobin/dl blood. The arterial blood carries approximately 20 ml O_2/dl; about 98% of this O_2 is carried by hemoglobin, and 2% is dissolved in the plasma.

C. O_2 transport. O_2 dissolves in the plasma of the pulmonary capillaries after diffusing across the respiratory membrane of the lungs. O_2 then diffuses into the red blood cell, where it combines reversibly with the iron atom of hemoglobin and converts **deoxyhemoglobin (reduced hemoglobin)** into **oxyhemoglobin**. Normally, the Po_2 in the pulmonary capillary blood becomes equal to the alveolar Po_2 before the blood leaves the pulmonary capillary.

1. **Hemoglobin saturation** is the percent of hemoglobin that is combined with O_2. The amount of O_2 that combines with hemoglobin depends on the Po_2 as indicated by the **hemoglobin saturation curve** (Figure 3-15).
   a. **P_{50}** is the Po_2 that produces a 50% saturation of hemoglobin with O_2. The normal P_{50} for arterial blood is 27 mm Hg.
   b. **Hemoglobin affinity for O_2** is inversely related to the P_{50}. An increase in hemoglobin affinity means that O_2 combines more readily with hemoglobin at the lungs, while a decrease in the hemoglobin affinity causes hemoglobin to release O_2 more readily at the tissue level.
   c. **The loading zone** of the O_2-dissociation curve is the plateau above a Po_2 of about 60 mm Hg. The loading zone provides a margin of safety, since the Po_2 in the lungs can be reduced

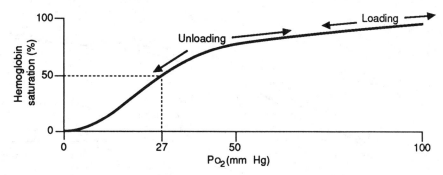

Figure 3-15. Hemoglobin saturation as a function of O_2 tension (Po_2) in the blood. Normally, the hemoglobin becomes 95%–98% saturated as it flows through the pulmonary capillaries (loading zone). When the blood enters the systemic capillaries, O_2 diffuses from the red cells into the tissues (unloading zone), and the hemoglobin saturation is reduced to about 75%, which is equivalent to a Po_2 of 40 mm Hg. At a hemoglobin saturation of 50%, the Po_2 is normally 27 mm Hg (P_{50} = 27 mm Hg).

substantially, and yet the O_2 content (hemoglobin saturation) remains relatively high. This allows one to climb mountains to moderate altitudes and suffer only a small decrease in the **arterial O_2 content** even though the **arterial Po_2** has decreased substantially.

d. **The unloading zone** is the steep portion of the dissociation curve below a Po_2 of about 60 mm Hg. The Po_2 in the tissues is very low since O_2 is consumed by metabolism. A gradient in Po_2 is established between the capillary blood and the tissues, which causes O_2 to diffuse from blood to tissue. As O_2 diffuses into the tissues, the capillary Po_2 is reduced. The capillary Po_2 is determined by the ratio of the blood flow to the rate of O_2 consumption; a low blood flow or a high O_2 consumption leads to a low tissue and capillary Po_2. The steep portion of the curve allows hemoglobin to release O_2 readily if the tissue O_2 consumption increases; this minimizes the decrease in Po_2 that would otherwise take place.

e. **Physiologic alterations in P_{50}** are produced by changes in temperature and by changes in the concentration of CO_2, H^+, other ligands such as diphosphoglycerate (DPG), ATP, ADP, and other organic phosphates (Figure 3-16).

f. **Bohr effect.** Hemoglobin is an allosteric enzyme that reacts with O_2, and the affinity between hemoglobin and O_2 is altered by these different ligands. The Bohr effect refers to the increased P_{50} of the O_2-dissociation curve caused by an increase in Pco_2.

2. **O_2 capacity** is determined by the hemoglobin concentration. The O_2 capacity equals hemoglobin concentration in g/dl of blood times 1.34 ml O_2/g. The normal O_2 capacity is 20.1 ml O_2/dl blood (15 g hemoglobin/dl times 1.34 ml/g). It is important to realize that the O_2 capacity varies only with the amount of hemoglobin in the blood. The O_2 capacity defines the maximal amount of O_2 that can be carried in the blood by hemoglobin.

3. **O_2 content** is the actual amount of O_2 being carried in the blood. The O_2 content depends on the hemoglobin concentration, the Po_2, and the P_{50} of the hemoglobin. Figure 3-17 shows the effect of varying these parameters on the O_2 content of blood.

a. The O_2 content usually is expressed as ml O_2/dl blood (also termed **volumes %, or vol%**). The normal arterial O_2 content is about 19.5 vol%, while the normal venous O_2 content is about 15 vol%. Thus, 4.5 vol% of O_2 are released from every deciliter of blood in the systemic capillaries to supply tissue metabolism.

b. The normal cardiac output is 5–5.5 L/min, so that 50–55 dl are pumped by the heart, and, if each deciliter supplies 4.5 ml of O_2, then the body consumes 225–250 ml O_2/min.

c. Hemoglobin saturation is calculated by dividing O_2 content by O_2 capacity, then multiplying by 100.

4. **Carboxyhemoglobin (HbCO)** is formed when CO combines with hemoglobin. This reaction occurs at the same site as O_2, which makes this hemoglobin molecule incapable of O_2 transport. CO has about 200 times the affinity for hemoglobin as does O_2; therefore, HbCO is a rather stable molecule.

a. The CO combines with the most reactive hemoglobin molecules, which are those molecules that have the highest P_{50}. Therefore, the hemoglobin available for O_2 transport has a low P_{50} (high O_2 affinity), which shifts the hemoglobin dissociation curve to the left and lowers the tissue Po_2 even further than expected.

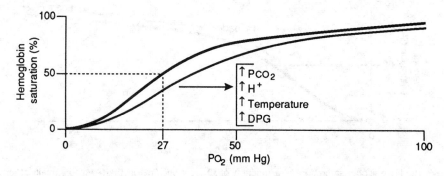

Figure 3-16. The effect of increased CO_2 tension (Pco_2), H^+ concentration, temperature, and diphosphoglycerate (*DPG*) concentration on the P_{50} of the hemoglobin saturation curve. The increase in the P_{50} indicates a reduced affinity of hemoglobin for O_2, which will raise the tissue O_2 tension (Po_2). Decreases in these parameters cause a shift of the curve to the left, which represents an increased affinity of the hemoglobin for O_2.

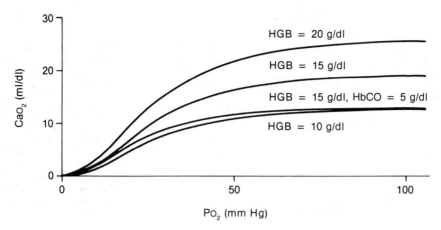

Figure 3-17. O_2 content of arterial blood (CaO_2) as a function of O_2 tension (PO_2) and hemoglobin (*HGB*) concentration. Note that carbon monoxide (*HbCO*) not only causes a reduction in O_2 content but also causes a decrease in the P_{50}.

    b. Because of the extreme affinity of CO for hemoglobin, a P_{CO} of approximately 0.5 mm Hg causes 50% of the hemoglobin to react with CO. The rapid loss of 50% of the functional hemoglobin can be fatal.

    c. Dissociation of CO from hemoglobin is enhanced by breathing 100% O_2. Administration of 100% O_2 provides an additional benefit, since significant amounts of O_2 (about 1.5 vol%) are dissolved in the plasma at this high PO_2. This dissolved O_2 helps to maintain an adequate O_2 delivery to the body tissues.

    d. HbCO gives a typical **cherry-red color** to the tissues.

  5. Methemoglobin is formed when hemoglobin iron is oxidized from the **ferrous** (Fe^{2+}) to the **ferric** (Fe^{3+}) state. Methemoglobin is incapable of carrying O_2 and has a bluish color that can impart a **cyanotic (blue) color** to tissues. Methemoglobin is formed at a slow rate within the red blood cell and is converted back to functional hemoglobin by the presence of reducing compounds produced in the red cell by metabolic reactions. Many drugs and foods (e.g., nitrites, phenacetin, fava beans) can produce methemoglobin, especially in people who have a genetic deficiency of glucose-6-phosphate dehydrogenase (G6PD).

  6. Abnormal hemoglobins occur as genetic variants, which may have abnormal affinities for O_2 or abnormal physical characteristics, as in **sickle cell anemia**. Sickle cell hemoglobin (hemoglobin S) becomes very insoluble in the deoxygenated state. These molecules cause the red cells to assume bizarre (sickle) shapes, causing them to lodge in capillaries and block the blood flow. The blocked blood flow results in **ischemic damage** to many different organ systems in affected individuals.

D. CO_2 transport. CO_2 is the usual end-product of oxidative metabolism and is formed in the tissues. CO_2 diffuses out of the cells and into the capillary blood, which raises the PCO_2 of the capillary and venous blood 5–6 mm Hg higher than the arterial PCO_2. CO_2 is transported in the blood in several different forms.

  1. Physically dissolved CO_2 accounts for about 5% of the total CO_2 in the blood.

  2. Carbamino-CO_2 refers to CO_2 that combines with terminal **NH_3** groups on protein side chains to form **NH_2-COOH**. Approximately 5% of the CO_2 is carried as carbamino compounds.

  3. Bicarbonate. Approximately 90% of the CO_2 enters the red cell, where, in the presence of **carbonic anhydrase,** it rapidly reacts with water to form **carbonic acid** (Figure 3-18). The carbonic acid dissociates into bicarbonate (HCO_3^-) and a hydrogen ion (H^+). The HCO_3^- diffuses into the plasma, whereas the H^+ is buffered by the deoxygenated hemoglobin, which is a weaker (less dissociated) acid than oxyhemoglobin. Although HCO_3^- is formed within the red cell, most of the CO_2 is carried in the plasma as HCO_3^-.

    a. Chloride shift. As the HCO_3^- is formed in red cells at the tissues, it diffuses out of the cells (see Figure 3-18). Since HCO_3^- is a negatively charged molecule, and the red cell membrane

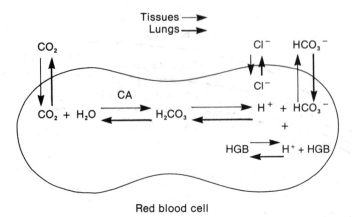

Figure 3-18. The chloride shift. CO_2 tension increases in the blood as it passes through the tissue capillaries, which causes CO_2 to diffuse into the red blood cell, where it is rapidly hydrated by carbonic anhydrase (*CA*). The hydrogen ion (*H+*), derived from the dissociation of carbonic acid (*H_2CO_3*), is buffered by the hemoglobin (*HGB*). Bicarbonate (*HCO_3^-*) diffuses out of the red cell and is replaced by chloride (*Cl^-*). Most of the CO_2 picked up from the tissues is transported in the plasma in the form of bicarbonate.

is relatively impermeable to cations, as HCO_3^- diffuses out of the cell it causes the inside of the cell to become less negatively charged. In order to neutralize this effect, the negatively charged chloride ion (Cl^-) diffuses from the plasma into the red cell to replace the HCO_3^-.
  b. Reversal of these reactions occurs in the pulmonary capillaries (see Figure 3-18).

4. **The Haldane effect** refers to the fact that deoxyhemoglobin combines with more CO_2 compared to oxyhemoglobin at any PCO_2. The Haldane effect minimizes the increase in PCO_2 that occurs in the venous blood, which, in turn, causes the venous blood to be less acid than it would be otherwise.

V. PULMONARY CIRCULATION.
The entire blood flow from the right ventricle normally is distributed to the pulmonary blood vessels. Since the output of the right heart must equal that of the left heart, the pulmonary blood flow equals the cardiac output (approximately 5 L/min).

A. Pulmonary artery pressure is much lower than the pressure in the aorta, since the resistance to blood flow in the pulmonary circulation is about one-tenth that of the systemic circulation. The normal systolic pressure in the pulmonary artery is about 20 mm Hg, whereas the diastolic pressure is about 10 mm Hg.

B. Flow distribution. Pulmonary blood flow varies in different parts of the lungs because of the low pressures that are present, the distensibility of the vasculature, and the hydrostatic effects of gravity.

1. **Hydrostatic pressure**
   a. **Definition.** Hydrostatic pressure is caused by the weight of the column of fluid and is equal to

$$h \times d \times g$$

where h = height of the column of fluid (cm), d = density of the fluid (g/cm³), and g = the gravitational constant. The hydrostatic pressure adds to or subtracts from the potential energy (pressure) at levels below or above the zero reference plane, respectively. The **zero reference plane** is taken at the level of the atria that is approximately at the middle of the lungs.
   b. **Example.** Assuming the lungs to be 40 cm from base to apex, the pulmonary artery pressure to be 20 cm H_2O, and the density of blood to be 1 g/cm³, then the blood pressure would be sufficient to lift the column of blood to a height of 20 cm. In this example, if the pulmonary artery pressure decreased 4 cm H_2O, then the top 4 cm of the lung would not be perfused, since the pressure would not be adequate. In the upright position, the apex of the lungs receives relatively little blood flow while the base of the lungs is well perfused.
   c. **Zones of the vertical lung.** Various zones for the vertical lung (Figure 3-19) have been defined as follows.
   (1) **Zone 1** is at the **lung apex** and is present if the regional pulmonary artery pressure (Pa) is less than alveolar pressure (PA). Under these conditions, the pulmonary capillaries are collapsed, and there is no blood flow (see Figure 3-19). Zone 1 is increased by factors that reduce Pa or increase PA [e.g., positive end-expiratory pressure (PEEP)]; it is de-

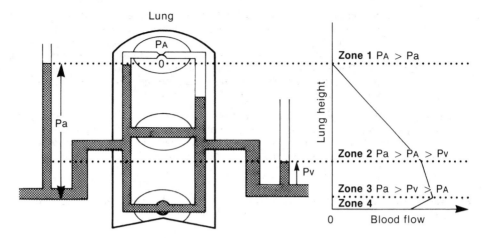

Figure 3-19. The effect of hydrostatic pressure and level on the distribution of pulmonary blood flow in a vertical lung. Pulmonary blood flow depends on the relationship between pulmonary artery pressure (*Pa*), alveolar pressure (*PA*), and pulmonary venous pressure (*Pv*). (Adapted from West JB, et al: Distribution of blood flow in the isolated lung: relation to vascular and alveolar pressure. *J Appl Physiol* 19:713, 1964.)

creased by factors that increase Pa or reduce the vertical height of the lung (e.g., lying down).

(2) Zone 2 is the region of the lung where Pa > PA > Pv (pulmonary venous pressure). In this region, the pressure gradient producing flow is Pa – PA, not the arteriovenous gradient, which is the usual mechanism. Since Pa increases 1 cm H_2O for each centimeter down the lung, there is an increase in the driving forces, and flow increases linearly in zone 2 (see Figure 3-19).

(3) Zone 3 occurs in the **lower part of the lung,** where Pa > Pv > PA. Blood flow increases down zone 3 as a result of distension and recruitment of vessels due to the increasing hydrostatic pressure, which reduces the resistance to flow.

(4) Zone 4 is present only when there is an abnormally high pulmonary venous pressure (e.g., due to left ventricular failure or mitral stenosis). Zone 4, when present, replaces part of zone 3 at the base of the lung and represents a zone of reduced blood flow (see Figure 3-19). Zone 4 results from edema occurring in the interstitium around blood vessels, which is an early event in pulmonary edema. This edema is termed **vascular cuffing** and results in an increased vascular resistance and a reduction in local blood flow.

2. Hypoxic pulmonary vasoconstriction (HPV) is caused by alveolar hypoxia (i.e., low alveolar PO_2). HPV is a localized response in that it occurs only in areas of hypoxia; it is potentiated by hypercapnia and acidosis.

    a. The **mechanism** of this vasoconstriction is unknown, but it is important to remember that hypoxia, hypercapnia, and acidosis produce **vasodilation** in the systemic circulation rather than vasoconstriction as in the pulmonary circulation.

    b. Function. HPV is an important mechanism of balancing ventilation and perfusion, since it results in an increased vascular resistance in hypoxic areas of the lung, which shifts blood flow to areas with a higher PO_2. If the hypoxia is generalized, as in hypoventilation or exposure to high altitude, then HPV increases the total pulmonary vascular resistance, which results in pulmonary hypertension.

    c. Pulmonary hypertension increases the work of the right ventricle, resulting in right ventricular hypertrophy, right axis deviation on the EKG, and, eventually, in right ventricular failure.

3. Changes in lung volume can alter markedly the pulmonary vascular resistance. Pulmonary vascular resistance is minimal at approximately functional residual capacity and increases at both higher and lower volumes. The effect of lung volume on pulmonary vascular resistance is complex, because two different categories of blood vessels are involved.

    a. Extra-alveolar vessels are the larger distributing arteries and veins in the lungs. These vessels are dilated at high lung volumes because of the **radial traction** exerted on their walls by connective tissue septa. The increased diameter reduces the flow resistance in these vessels.

b. **Alveolar vessels,** mainly **capillaries,** are located in the **alveolar septa** and are compressed by the alveoli at high lung volumes. During inspiration, the capillary vascular resistance increases while the larger vessels dilate. Thus, much of the increased right ventricular output during inspiration is stored in the pulmonary arteries until expiration, when it reaches the left ventricle. This alternate fluctuation in the stroke volume of the right and left ventricles prolongs the respective ventricular ejection and represents the major factor that alters the splitting of the second heart sound (see Ch 2 VI A 3 e).

4. **Local reduction in blood flow** results from the presence of diseases that can obstruct or destroy pulmonary vessels and alter the normal flow pattern. In the presence of pulmonary disease, the distribution of blood flow can be very uneven in adjacent areas of the lung rather than having the predictable pattern as shown in Figure 3-19.

VI. **THE DISTRIBUTION OF VENTILATION** is uneven in the lungs and is affected by the pleural pressure gradient, the time constant of the lung, and airway closure.

A. **Pleural pressure gradient** is caused by the effects of gravity on the lung within the intact thorax. While the cause of this variation in pleural pressure is incompletely understood, the lung behaves as if it were a low-density fluid (because of its gas content) when exposed to gravity. The density of the normal lung is about 0.25–0.3 g/cm³, which produces a hydrostatic pressure difference of 7.5–10 cm H_2O between the apex and the base of a vertical lung. The interpleural pressure in a standing adult increases from about –10 cm H_2O at the apex to about –2 cm H_2O at the base. Astronauts, exposed to zero-gravity conditions, have neither a pleural pressure gradient nor regional variations in lung volume.

1. **Regional lung volume** is affected by the local interpleural pressure, since this is one of the determinants of the **transpulmonary pressure** (PL = PA – PPL). Since PA is zero throughout the lung under static conditions, the PL varies from 2 cm H_2O at the base to 10 cm H_2O at the apex of the vertical lung. These two values of PL place the apex and the base of the lung on widely separated parts of the P-V curve, which causes the apex and the base to function at very different compliances as shown in Figure 3-20.

2. **Tidal volume** is unevenly distributed in the erect lung because of the variation in regional compliance due to the effects of gravity. In a vertical lung, most of the tidal volume goes to the base of the lung when inspiration starts at functional residual capacity. There is a linear reduction in regional tidal volume from base to apex under these conditions (see Figure 3-20). The distribution of the tidal volume is much more even in a supine individual because of the reduction in the interpleural pressure gradient.

B. **Time constant of the lung (RC)**—the product of resistance (R) and compliance (C)—sets the time course for acinar filling and emptying. The functional unit of the lung is the **acinus,** which consists of a respiratory bronchiole and all of the structures distal to it. The acinus can be modeled by a balloon and a tube; the balloon provides compliance, and the tube provides airway resistance. This type of system fills and empties exponentially; for each time constant the volume will change

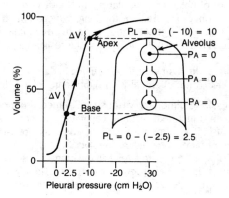

Figure 3-20. The transpulmonary pressure (*PL*) varies from the apex to the base of the upright lung due to the interpleural pressure gradient. The result is that basal areas of the lung are located on a more compliant portion of the pressure-volume curve than apical regions. During inspiration the transmural pressure of the lung increases and the lungs expand (volume at end-inspiration is indicated by *arrow heads*). Basal regions take up more of the tidal volume than apical regions, resulting in an uneven distribution of alveolar ventilation. ΔV = change in volume; and *PA* = alveolar pressure.

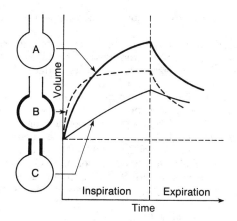

Figure 3-21. Models of acini showing the effect of normal compliance (*A*), decreased compliance (*B*), and high airway resistance (*C*) on the time course of filling. With a high airway resistance, the acini fill slowly, and the volume change varies with the duration of inspiration. A decreased compliance results in rapid filling but a reduction in the volume change.

63% toward the equilibrium value (Figure 3-21). The normal lung behaves as if all acini have the same time constant. Variations of the time constant due to disease can cause an even more unequal distribution of the tidal volume than the effect of the plural pressure gradient.

1. An **increased RC** is caused primarily by an **increased airway resistance** and results in a slower rate of acinar filling and emptying (see Figure 3-21C).
   a. The slower rates are caused by the high resistance (R), which reduces the flow rate ($\dot{V}$E) for any driving pressure (dP): $\dot{V}$E = dP/R. The presence of lung segments having increased RC results in the phenomenon of **frequency-dependent compliance,** since the extent of lung filling varies as a function of inspiratory time. When respiratory rate increases, lung filling decreases due to insufficient filling time, and the **dynamic lung compliance** decreases.
   b. The presence of frequency-dependent compliance indicates the presence of **obstructive airway disease** rather than parenchymal or restrictive lung disease as the name implies.

2. A **decreased RC** results primarily from a **decreased lung compliance** due to **lung fibrosis** or destruction of functional lung units. The decreased compliance allows these units to fill and empty rapidly but to a smaller extent than normal (see Figure 3-21B). The decreased compliance alters the distribution of the tidal volume and markedly increases the work of expanding the lungs.

C. **Airway closure** occurs when the regional transpulmonary pressure is reduced to a critical level. Transpulmonary pressure varies primarily as a function of lung volume. **Closing volume** is the volume of the vital capacity that is present when airway closure begins. **Closing capacity** is the closing volume plus the residual volume.

   1. The **single breath N$_2$ test** can be used to detect airway closure. In this test, an individual expires to residual volume, then inspires to total lung capacity using 100% O$_2$ and, finally, expires slowly and steadily to residual volume. The PN$_2$ in the expired gas is measured during the second expiration and is depicted in Figure 3-22, which shows four distinct phases.

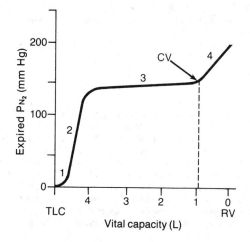

Figure 3-22. Record of a single-breath nitrogen (N$_2$) test. The tracing shows the expired N$_2$ tension (*P*N$_2$) following an inspiratory vital capacity of 100% O$_2$. *CV* = closing volume; *TLC* = total lung capacity; and *RV* = residual volume.

    a. Phase 1 of the expired P_{N_2} represents the first gas from the lungs, which is pure dead space gas (100% O_2).

    b. Phase 2 is a mixture of gas from the dead space and alveoli and represents a transition phase. The volume expired at the **midpoint** of this phase represents the **anatomic dead space**. Measuring the dead space in this fashion is referred to as **Fowler's method.**

    c. Phase 3 represents alveolar gas and is referred to as the **alveolar plateau**. The slope of phase 3 varies as a function of the unevenness of tidal volume distribution.

    d. The onset of **phase 4 (terminal rise)** indicates the closing volume (onset of airway closure). The higher P_{N_2} in phase 4 is due to the loss of the dilution effect of gases coming from the base of the lung. Recall that, at residual volume, the base of the lungs is almost completely deflated, while the apex is moderately inflated with gas containing 80% N_2. After inspiring 100% O_2 to total lung capacity, the base is maximally inflated by almost pure O_2 while the apex is maximally inflated but contains significant amounts of N_2. During the subsequent expiration, when basal airways begin to close, only gas from the apex continues to be expired, which has a higher P_{N_2}.

    e. Closing capacity is the closing volume plus residual volume.

  2. Early airway closure is considered to occur when airways close prematurely compared to the predicted value for a person's age and sex. Airway closure begins at the base of the lung where transpulmonary pressure and airway diameter are smallest. Early airway closure occurs in the presence of **airway disease, excess mucus, contraction of bronchial smooth muscle,** or **loss of radial traction** on the airways due to destruction of alveolar septa. In other words, anything that narrows the airways leads to early airway closure.

VII. VENTILATION:PERFUSION RATIO ($\dot{V}A/\dot{Q}$) is merely the ratio of the **alveolar ventilation** to the **pulmonary blood flow**. The $\dot{V}A/\dot{Q}$ at the acinar level determines the alveolar P_{O_2} and P_{CO_2} and is, therefore, critical in determining the gas exchange in each alveolus of the lung. The most efficient gas exchange occurs with $\dot{V}A/\dot{Q}$ ratios of about 1.0. Abnormal $\dot{V}A/\dot{Q}$ (either too high or too low a ratio) is by far the most common cause of hypoxia, which can result from many different pulmonary diseases.

A. Normal $\dot{V}A/\dot{Q}$

  1. Recall that the alveolar ventilation is normally 4.5–5.0 L/min and that the cardiac output is about 5 L/min. If the alveolar ventilation and pulmonary blood flow were both evenly distributed throughout the lungs, then the $\dot{V}A/\dot{Q}$ would be about 0.9.

  2. Even in normal lungs, both ventilation and blood flow vary in different regions (low at the apex and high at the base of a vertical lung). Due to the higher density of the blood compared to the lung tissue, the change in blood flow is much greater than the change in ventilation per unit distance (Figure 3-23). Thus, the lung base is **overperfused** compared to the ventilation, while the apex is **underperfused** compared to the ventilation; the base of the lung has a low $\dot{V}A/\dot{Q}$ while the apex has a higher $\dot{V}A/\dot{Q}$.

  3. The $\dot{V}A/\dot{Q}$ normally ranges from about **0.6 at the base** to about **3 at the apex** of the lung. This relatively narrow range impairs gas exchange only slightly. In the presence of pulmonary disease, the $\dot{V}A/\dot{Q}$ can vary from 0 to infinity.

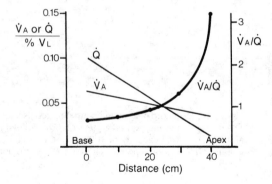

Figure 3-23. Diagram showing the distribution of alveolar ventilation ($\dot{V}A$), pulmonary blood flow ($\dot{Q}$), and the ventilation:perfusion ratio ($\dot{V}A/\dot{Q}$) in a normal lung. Because the blood flow gradient is steeper than the ventilation gradient, the base of the lung is overperfused, and the apex of the lung is overventilated. Thus, the $\dot{V}A/\dot{Q}$ is low at the base and high at the apex of the normal, vertical lung. % V_L = percent of the regional lung volume. (Adapted from West JB: *Ventilation/Blood Flow and Gas Exchange*. Oxford, Blackwell Scientific, 1970, p 33.)

B. Low $\dot{V}_A/\dot{Q}$ is caused either by a **low rate of ventilation** or by **excessive blood flow** to an area of the lung (Figure 3-24A). Either condition causes the alveolar P_{O_2} to decrease and P_{CO_2} to increase, since an inadequate amount of O_2 is brought into the lungs by ventilation, compared with the amount of O_2 that is carried away by the pulmonary capillary blood.

1. Due to the low alveolar P_{O_2}, the blood leaving this area of the lungs also contains a low P_{O_2} and consequently has a decreased O_2 content. When this blood mixes with blood from well-ventilated areas, the arterial O_2 content is reduced, which results in hypoxemia. The blood flow from low $\dot{V}_A/\dot{Q}$ areas is termed a **physiologic shunt**.

2. If there is no alveolar ventilation ($\dot{V}_A/\dot{Q} = 0$), then there is no gas exchange at all, and the blood flow is referred to as an **anatomic shunt**. An anatomic shunt may be intra- or extrapulmonary. Extrapulmonary anatomic shunts may be the result of congenital heart malformations resulting in a right-to-left shunt of blood. The presence of an anatomic shunt means that true venous blood is mixing with arterial blood, and even the administration of 100% O_2 will not correct this decrease in the arterial P_{O_2}.

3. The fraction of shunted blood (Q_s/Q_t) can be measured by means of the **shunt equation**
$$Q_s/Q_t = CiO_2 - CaO_2/CiO_2 - C\overline{v}O_2$$
where CiO_2 = ideal O_2 content, $C\overline{v}O_2$ = mixed venous O_2 content, and CaO_2 = arterial O_2 content.

C. High $\dot{V}_A/\dot{Q}$ is caused by excessive ventilation or inadequate blood flow (see Figure 3-24C).

1. The P_{O_2} in these areas of the lung is high, but O_2 exchange is inefficient for the level of ventilation because very little extra O_2 is added to the blood due to the plateau of the O_2-dissociation curve (see Figure 3-15). In addition, O_2 exchange is impaired since the high $\dot{V}_A/\dot{Q}$ is commonly due to a low blood flow so that there is little blood flow to carry the O_2 away. The extra ventilation needed to raise the P_{O_2} above normal can be considered ventilation of alveolar dead space.

2. If the blood flow to an area of the lung is zero ($\dot{V}_A/\dot{Q}$ = infinity), then there is no gas exchange in these alveoli, and all of the ventilation represents **alveolar dead space ventilation**. Any gas going to alveolar dead space is wasted and contributes to dead space ventilation.

D. The O_2-CO_2 diagram (Figure 3-25) is one method of visualizing the effects of ventilation-perfusion alterations on the blood gas tensions. This graph contains a $\dot{V}_A/\dot{Q}$ line that is defined by three points representing the venous blood composition ($\overline{v}$), ideal alveolar gas composition (i), and the inspired gas composition (I).

1. **R Lines** represent the P_{O_2} and P_{CO_2} that would be present under conditions where the rates of O_2 uptake and CO_2 release vary. R is the ratio of the volume of CO_2 released per minute ($\dot{V}_{CO_2}$) to the volume of O_2 absorbed per minute ($\dot{V}_{O_2}$). R lines can be determined for both the gas and blood phase, and these lines intersect the $\dot{V}_A/\dot{Q}$ line for each respective value of R.
   a. **Gas R lines** radiate from the inspired gas point and represent the change in gaseous O_2 and CO_2 tensions for a particular R value.
   b. **Blood R lines** radiate from the venous blood point and represent blood gas tensions for various R values.

2. **Alveolar dead space** can be represented in the gas R line as a movement of the alveolar gas point away from the ideal point. The alveolar gas point represents the **mean value** for all alveoli, so that addition of alveolar dead space, with a gas composition equal to the inspired gas, moves the mean alveolar value toward the inspired gas point. Unless the overall alveolar

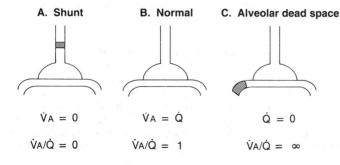

A. Shunt   **B. Normal**   **C. Alveolar dead space**

$\dot{V}_A = 0$ $\dot{V}_A = \dot{Q}$ $\dot{Q} = 0$

$\dot{V}_A/\dot{Q} = 0$ $\dot{V}_A/\dot{Q} = 1$ $\dot{V}_A/\dot{Q} = \infty$

Figure 3-24. Models depicting the effect of variations in alveolar ventilation ($\dot{V}_A$) and blood flow ($\dot{Q}$) on the ventilation:perfusion ratio. Areas of the lung with reduced alveolar ventilation result in a shunt effect or venous admixture (*A*), while other areas with reduced blood flow contribute to alveolar dead space (*C*).

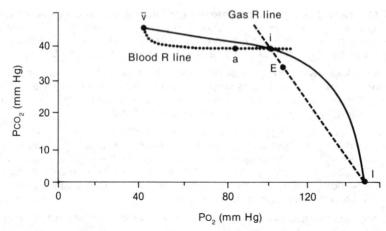

Figure 3-25. Ventilation:perfusion line, which is determined by the composition of mixed venous blood ($\bar{v}$) and inspired gas (*I*). The arterial blood gas composition (*point a*) is determined by the amount of shunt flow, while the mixed-expired gas (*point E*) is determined by the amount of dead space. *Point i* represents the average gas composition of the pulmonary capillary blood. *R lines* represent the ratio of carbon dioxide production to oxygen consumption, and the blood and gas R lines meet at the $\dot{V}A/\dot{Q}$ line.

ventilation is increased, an increase in alveolar dead space results in CO_2 retention as well as hypoxia. In the typical patient with increased alveolar dead space, the initial phase of hypercapnia stimulates the respiratory centers, so that minute ventilation increases, and the PCO_2 returns to normal, while the hypoxia remains due to the effects of a physiologic shunt.

   3. A **physiologic shunt** causes the arterial blood gas composition (see Figure 3-25, *point a*) to move along the blood R line toward the venous point. The more severe the $\dot{V}A/\dot{Q}$ abnormality, the further the arterial point will move to the left. Note that there is a marked reduction in PO_2 with only a minimal rise in the PCO_2 as one proceeds along the blood R line. This is typical of the blood gas values obtained from patients with $\dot{V}A/\dot{Q}$ abnormalities.

E. **Distribution of $\dot{V}A/\dot{Q}$ ratios in the lungs** normally can be represented by a log-normal distribution curve (Figure 3-26A) having a range that covers about 1 decade of $\dot{V}A/\dot{Q}$ values. Patients with $\dot{V}A/\dot{Q}$ abnormalities may have a single mode of distribution with a much wider range of values than normal (see Figure 3-26B), or there may be multiple modes of distribution (see Figure 3-26C). Either of these abnormalities causes a worsening of the blood gas values.

VIII. RESPIRATORY CONTROL

A. **Neural mechanisms.** Unlike the heart, which automatically generates its own rhythm, the respiratory muscles are typical skeletal muscles requiring electrical stimulation via somatic nerves to initiate contraction. The major muscle of inspiration, the **diaphragm,** is innervated by motor fibers in the **phrenic nerve.** Impulses reach this nerve from either **voluntary** or **involuntary** paths in the central nervous system (CNS). This dual pathway allows voluntary breath holding or control of breathing during talking, singing, or swimming. The involuntary pathway functions during most of the day and night, allowing humans to breathe automatically without conscious effort. (Hiccups represent involuntory spasms of the diaphragm frequently caused by irritation from surrounding structures such as the stomach or gallbladder.)

   1. **Medullary centers.** The automatic basic rhythm of respiration is generated within the **medulla,** but its exact source and mechanism of generation is unknown. Respiration continues in animals and humans as long as the medulla and the spinal cord are intact. Patients who are **"brain dead"** and require a respirator to breathe have destroyed the medullary centers as well as the higher brain functions. There are two groups of neurons on each side of the medulla that contain cells that discharge rhythmically during either inspiration or expiration. These groups of cells are called the **ventral respiratory group (VRG)** and the **dorsal respiratory group (DRG).** These centers are influenced by nervous activity in other areas of the brain such as the

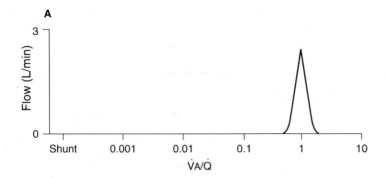

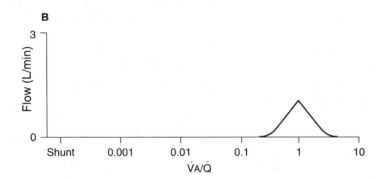

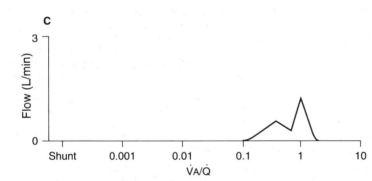

Figure 3-26. Distribution of $\dot{V}_A/\dot{Q}$ ratios. The y axis represents the amount of blood flow going to each $\dot{V}_A/\dot{Q}$ value. (*A*) Normal, showing a mean $\dot{V}_A/\dot{Q}$ of 1 and a standard deviation of the $\dot{V}_A/\dot{Q}$ population of 0.2, which results in a variation of the ratio from about 0.6 to 3. (*B*) An increased standard deviation of the population results in a wider and shorter pattern of flow distributions, while (*C*) shows a bimodal distribution of the flow. Both *B* and *C* produce a decrease in the arterial P_{O_2}.

pons, the **reticular activating system (RAS),** and the **cerebral cortex** as well as by afferent activity in the **vagus, glossopharyngeal,** and **somatic** nerves.

  a. The **DRG** is a bilateral group of cells that lies within the nuclei of the **tractus solitarius.** The DRG cells are primarily **inspiratory cells** (i.e., they discharge during inspiration).

    (1) Activity of the DRG. The DRG may be the **primary rhythm generator** for respiration, since the activity in these cells gradually increases during inspiration. The electrical activity of this center has been likened to a ramp, since the activity rises to a crescendo during inspiration and then rapidly disappears.

    (2) Afferent input to the DRG comes mainly from the **vagus** and **glossopharyngeal nerves,** which carry information from the peripheral chemoreceptors and mechanical receptors in the lungs. The DRG activity is influenced through these nerves by low P_{O_2}, high P_{CO_2}, and low pH as well as by changes in lung volume and the neural activity within the **RAS.** During sleep, the RAS activity is diminished, which results in a lessened inspiratory drive, a decrease in alveolar ventilation, and a slight rise in arterial P_{CO_2}.

    (3) Efferent connections from the DRG go to the **contralateral phrenic** and **intercostal motoneurons** and to the **VRG.**

  b. The **VRG** contains neurons that are active during both **inspiration** and **expiration.** The expiratory activity does not result in activation of the expiratory muscles during normal respiration (**eupnea**), since expiration is normally passive. The VRG is comprised of the **upper motor neurons** of the **vagus** and the nerves to the **accessory** muscles of respiration.

2. **Pontine centers** are areas of the brain stem that modify the activity of the medullary respiratory centers.

   a. The **pneumotaxic center** lies in the **upper part** of the **pons** and functions to inhibit the apneustic center. Stimulation of the pneumotaxic center shortens inspiration, leading to a shallower and more rapid respiratory pattern.

   b. The **apneustic center** lies in the **caudal area** of the **pons,** but it has not been identified with any specific collection of cells. The efferent outflow from the apneustic center stimulates **inspiration**. Stimulation of this area of the brain increases the duration of inspiration, which results in a deeper and more prolonged inspiratory effort. Consequently, the rate of respiration becomes slowed because of the greater depth of inspiration. Loss of inhibitory activity on the apneustic center, caused by vagotomy and destruction of the pneumotaxic center, results in prolonged periods of inspiration termed **apneusis**. The apneustic center is normally inhibited by impulses carried in the vagus nerves and also by the activity of the pneumotaxic center.

3. **Central chemoreceptors** are cells that lie just beneath the ventral surface of the medulla and that respond to the H^+ concentration in the **cerebrospinal fluid (CSF)** and the surrounding **interstitial fluid**.

   a. Charged ions do not readily cross the endothelium of the blood vessels in the brain, which constitutes the **blood-brain barrier** (Figure 3-27).

   b. However, CO_2 does cross this barrier, since it is a small uncharged molecule. An increase in CO_2 stimulates the central chemoreceptors after hydration to carbonic acid and dissociation into H^+ and HCO_3^-. A decrease in CO_2 inhibits respiration, since there is a reduction in the H^+ concentration within the CSF and the brain tissue.

   c. About 85% of the resting ventilatory drive is due to the stimulatory effect of CO_2 on the central chemoreceptors.

B. **Peripheral chemoreceptors** are located in the carotid and aortic bodies. These receptors respond to lowered PO_2, increased PCO_2, and increased H^+ concentration in the arterial blood. The peripheral chemoreceptors receive a tremendous blood flow for their size and thus can be considered to monitor the PO_2 of arterial blood. Decreases in the O_2 content of blood caused by anemia, methemoglobinemia, and carbon monoxide poisoning do not stimulate the peripheral chemoreceptors, since the PO_2, which is determined by the amount of dissolved O_2, remains normal.

1. **O_2.** The peripheral chemoreceptors are the only site in the body that detect changes in the PO_2. These receptors rapidly increase their firing rate as the arterial PO_2 falls below 100 mm Hg. Impulses from these receptors are carried to the brain via the **vagal** and **glossopharyngeal nerves** and result in an increased rate and depth of respiration.

2. **Decreased pH (increased H^+ concentration),** caused by the addition of fixed acids (e.g., lactate, acetoacetate, butyrate) into the blood, also stimulates the peripheral chemoreceptors and ventilation increases. The increased alveolar ventilation lowers the PCO_2 in the arterial blood and reduces the amount of acid in the blood, which tends to return the arterial pH

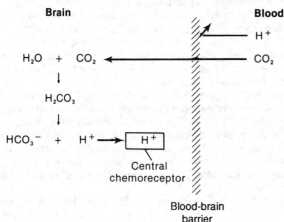

Figure 3-27. The effects of the blood-brain barrier on the transport of charged ions such as H^+ and the diffusibility of CO_2. CO_2 can readily cross the blood-brain barrier and serve as a source of H^+ to stimulate the central chemoreceptors.

toward normal. Acutely, the fixed acid cannot cross the blood-brain barrier, so that the resultant decrease in P_{CO_2} results in an alkalosis in the CSF. This central alkalosis inhibits respiratory drive for several days until CSF pH is returned to normal by ion transport by the meningeal tissues. The respiratory response to a low arterial pH is termed **respiratory compensation**.

3. **CO_2** also stimulates the peripheral chemoreceptors, but its major effect is on the central chemoreceptors as explained above.

C. Other respiratory reflexes

1. The **Hering-Breuer reflex** is initiated by inflation of the lungs and functions to **terminate inspiration**. Inflation of the lungs stimulates **stretch receptors** that are located in the small airways. These receptors send impulses to the respiratory centers via the **vagus nerves** that inhibit the **pontine** and **medullary** respiratory centers.

2. **Irritant receptors** are located in the large airways and are stimulated by **smoke, noxious gases,** and **particles** in the inspired air. These receptors initiate reflexes that cause coughing, bronchoconstriction, mucus secretion, and breath holding (**apnea**).

3. **J receptors** are stimulated by distension of the pulmonary vessels as in heart failure and pulmonary embolization. These receptors initiate reflexes causing rapid breathing (**tachypnea**).

4. **Chest wall receptors** can detect the force generated by the respiratory muscles during breathing. If the force required to distend the lungs becomes excessive (high airway resistance or low compliance) the information from these receptors gives rise to the sensation of **dyspnea** (difficulty in breathing).

D. Respiratory function tests. There are many tests that can be used to evaluate the function of the respiratory system. Some of the most commonly used tests are summarized below.

1. **Maximal voluntary ventilation** represents the maximum volume of gas that a patient can breathe. This test is very fatiguing and depends on a patient's complete cooperation (thus the term "voluntary" in the name). The patient breathes as rapidly and forcefully as possible while the volume changes are recorded with a spirometer. The test is carried out for 12 seconds, and the results are expressed in L/min. A normal value for a young adult male is about 200 L/min. The test evaluates all of the mechanical factors in breathing and results in the highest rate of ventilation for any condition.

2. **Respiratory responses to CO_2.** Gases containing different amounts of O_2 or CO_2 can be administered to evaluate the respiratory drive of the respiratory system. Increasing the P_{CO_2} in inspired gas causes a rise in the minute ventilation. A decrease in P_{O_2} is synergistic with hypercapnia, so that the respiratory response is much greater than with either stimulus alone (Figure 3-28). Patients vary in their response to this test, depending on, for example, the

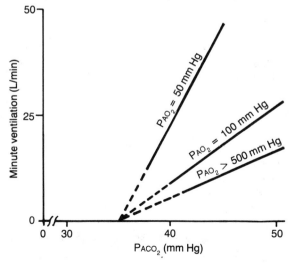

Figure 3-28. CO_2 response curves obtained at alveolar O_2 tensions ($P_{A O_2}$) of 50, 100, and > 500 mm Hg. $P_{A CO_2}$ = alveolar CO_2 tension.

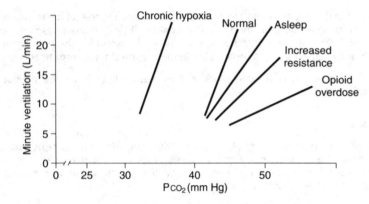

Figure 3-29. The effect of various pathologic conditions on CO_2 response curves.

sensitivity of their chemoreceptors, activity of the reticular activating system, airway resistance, and lung volume (Figure 3-29).

3. **Forced vital capacity** is a vital capacity maneuver performed with maximal effort. Many different measurements can be performed on the spirometer recording that is obtained. A decrease in the FEV_1 can be caused by either a high airway resistance or a lack of expiratory muscle strength. These two conditions can be differentiated by measuring the maximal expiratory force. The patient with a high airway resistance can generate a normal expiratory effort of about 100 cm H_2O, with the low flow rate due to the high airway resistance. In contrast, the patient with expiratory muscle weakness will be unable to generate a normal maximal expiratory pressure.

4. **Other tests.** Measurement of a patient's diffusion capacity, lung volumes (e.g., FRC or VC), and the concentration of O_2 and CO_2 in the arterial blood can provide physicians with valuable information about respiratory function.

E. **Dysfunction of respiratory control** can occur due to changes in the environment or to diseases affecting the respiratory system, cardiovascular system, or brain. Various respiratory patterns are seen clinically (Figure 3-30).

1. **Cheyne-Stokes respiration** is an abnormal respiratory pattern that occurs with depression of the brain due to disease, drug overdose, congestive heart failure, and hypoxia from other causes. Cheyne-Stokes respiration is characterized by periods of waxing and waning tidal volumes separated by periods of apnea.

2. **Biot's breathing** is another type of periodic breathing that consists of one or more large tidal volumes separated by periods of apnea. The condition occurs in many diseases producing brain damage.

Respiratory patterns

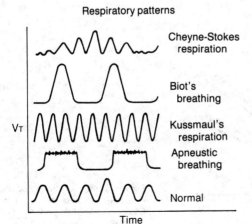

Figure 3-30. Illustration of various breathing patterns. V_T = tidal volume.

3. **Ondine's curse** gets its name from Greek mythology. Ondine was a water nymph who fell in love with a human. Ondine's father was king of the nymphs and placed a curse on the man that took away all of his automatic functions. Presumably, the man spent all of his time willing his heart to beat and his muscles to breathe and had little time for his lover. Automatic respiratory control is lost when there is a destruction of the involuntary neural pathways. Patients with this condition can breathe only by conscious effort and, therefore, cannot sleep without becoming apneic. This condition is treated with a **mechanical respirator** that maintains breathing while the patient sleeps.

4. **Sleep-apnea syndromes** have been recognized as disorders of respiratory control that affect large numbers of the population, especially elderly men. Many sleep centers have been established to study patients by recording physiologic parameters such as respiratory movements, air flow, EEG, and hemoglobin saturation during sleep.
   a. **Obstructed apnea**
      (1) Some individuals experience a marked loss of muscle tone in the pharyngeal muscles during REM (rapid eye movement) sleep. The loss of muscle tone results in partial or complete obstruction of the pharynx during inspiration. Partial obstruction is associated with snoring, as the inspired air causes these tissues to vibrate. During complete airway obstruction, there are contractions of the respiratory muscles, but no air can move because of the blocked airway. The person usually awakens because of the hypoxia that develops; muscle tone is reestablished, and sleep returns. These episodes may occur hundreds of times each night and are associated with a marked drop in the hemoglobin saturation of the arterial blood. The person wakes up in the morning still tired and unrested because of the poor sleep pattern. Frequently, these people fall asleep at work, in lectures, or even while eating because of the lack of restful nighttime sleep.
      (2) When this condition is associated with extreme obesity, it is referred to as the **Pickwickian syndrome** (read Dickens' *Pickwick Papers* to find out which character was affected).
      (3) Severe cases may require a tracheostomy in order to bypass the airway obstruction.
   b. **Nonobstructed (central) apnea** refers to a complete cessation of respiratory muscle activity due to a loss of rhythmic activity from the respiratory centers. These periods of apnea can last 30–60 seconds, with a marked drop in arterial hemoglobin saturation. Affected persons and some of their family members have been shown to have a decreased chemoreceptor sensitivity to O_2 and CO_2. Sleep-apnea syndrome has been proposed as one of many possible causes of **sudden infant death syndrome** (SIDS, or crib death).

IX. **HYPOXIA** is defined as an inadequate O_2 supply to the body tissues. The O_2 supply can be limited anywhere along the O_2 pathway from the atmosphere to the body tissues. Hypoxia will occur downstream (toward the tissues) of the limitation, whereas normal Po_2 may be present upstream (toward the environment).

A. **Symptoms of hypoxia** depend on how fast and how severely the Po_2 declines.

1. **Fulminant hypoxia** occurs within seconds upon exposure to Po_2 less than 20 mm Hg. This could happen if an aircraft lost cabin pressure at altitudes above 30,000 feet and there were no supplemental O_2 available. It also could happen if the O_2 were consumed by combustion in a closed space or was displaced by some other gas. Unconsciousness results in as little as 15–20 seconds, and brain death may follow in 4–5 minutes.

2. **Acute hypoxia** is produced by exposure to Po_2 equivalent to altitudes of 18,000–25,000 feet. Symptoms of acute hypoxia are very similar to the effects of ethyl alcohol; there is incoordination, slowed reflexes, slurred speech, overconfidence, and eventually unconsciousness. Coma and death can occur in minutes to hours if the compensatory mechanisms of the body are not adequate.

3. **Chronic hypoxia** occurs during exposure to arterial Po_2 of 40–60 mm Hg for long periods of time. Most clinical causes of hypoxia fall into this category. Chronic hypoxia produces symptoms similar to those of severe fatigue, and there is a sensation of difficulty in breathing (**dyspnea**) and shortness of breath. These patients may be bedridden or limited to sitting in a chair, because they are unable to increase their O_2 supply to the tissues due to either respiratory or cardiac disease. Respiratory arrhythmias (see VIII E) can occur in these people, especially during sleep, which can contribute to the hypoxic state.

B. Signs of hypoxia

1. **Cyanosis** is the bluish color of tissue caused by the presence of more than 5 g of deoxyhemoglobin/dl in the capillary blood. Patients with anemia may never develop cyanosis because of their inadequate hemoglobin concentration. **Methemoglobin,** because of its slate gray color, also can cause cyanosis. Cyanosis is most readily seen in the **nail beds, lips, mucous membranes,** and **ear lobes,** but it may not be recognized because of skin pigmentation or poor lighting. Cyanosis is **not a reliable sign of hypoxia.**

2. **Tachycardia** (rapid heart rate) occurs as a reflex response to the low P_{O_2} in the arterial blood. The hypoxia is detected by the aortic and carotid chemoreceptors, which then activate the sympathetic outflow to the heart. The increased heart rate is an attempt to increase cardiac output to enhance the O_2 delivery to the tissues.

3. **Tachypnea** (rapid breathing) and **hyperpnea** (deep breathing) also are reflex responses to hypoxia, which are activated by the arterial chemoreceptors. The resultant increase in minute ventilation is an attempt to raise the alveolar P_{O_2} toward its normal value of 100 mm Hg.

C. Causes of hypoxia. All types of hypoxia reduce the amount of O_2 that is available to the body tissues. The **O_2 delivery** is a measure of the amount of O_2 that is available to the body and is calculated as the product of cardiac output (L/min) and O_2 content of the blood/L. The normal O_2 delivery is about 1000 ml O_2/min (5 L/min × 200 ml O_2/L). Not all types of hypoxia cause a decrease in the O_2 delivery. Table 3-5 provides the characteristics of the different types of hypoxia and various tests that can be used to differentiate the types of hypoxia.

1. **Arterial hypoxia** (Figure 3-31A) results from inadequate oxygenation of the arterial blood, which is caused by breathing gas with a low P_{O_2} or by one of four pathophysiologic mechanisms. Notice in Figure 3-31 that **arterial hypoxia** is the only type of hypoxia in which there is a **decreased arterial P_{O_2}.** Arterial hypoxia is due to impaired gas exchange in the lungs. Pulmonary gas exchange is evaluated by using the **alveolar-to-arterial P_{O_2} difference.** A wide alveolar-to-arterial P_{O_2} indicates impaired pulmonary gas exchange and is caused by a decreased arterial P_{O_2}.

   a. **Hypoventilation** results from an inadequate rate and/or depth of respiration. There is a reduction in both alveolar and arterial P_{O_2} and an increase in alveolar and arterial P_{CO_2} (hypercapnia) because of the inadequate alveolar ventilation. **Hypercapnia** is synonymous with hypoventilation because of the relationship between arterial P_{CO_2} and alveolar ventilation.

   b. **Diffusion limitation** occurs if there is a severe loss of surface area of the alveolar membrane or a large increase in the diffusion distance in the lungs due to pulmonary disease. These changes reduce the amount of O_2 that diffuses into the pulmonary capillary blood. Consequently, the P_{O_2} in the pulmonary capillaries remains below the alveolar P_{O_2}, which results in the abnormally low arterial P_{O_2}.

   c. **Ventilation:perfusion ratio imbalance (physiologic shunt)** produces low P_{O_2} in areas of the lung with low $\dot{V}_A/\dot{Q}$ ratios. When the blood leaves these areas of the lungs and mixes with blood from other regions of the lung, the overall arterial O_2 content and P_{O_2} are

Table 3-5. Differentiating Types of Hypoxia

Type of Hypoxia	Pa_{O_2}	Pa_{CO_2}	$P\bar{v}_{O_2}$	Exercise Pa_{O_2}	100% O_2
Arterial hypoxia					
Hypoventilation	–	++*	–	+/–	CO_2++
Diffusion	–	n	–	–*	n
Ventilation:perfusion	–	n/–	–	+/–	n
Anatomic shunt	–	n/–	–	+/–	<500 mm Hg*
Anemic hypoxia	n	n	–*	n	n
Hypokinetic (ischemic) hypoxia	n	n	––*	+/–	n
Histotoxic hypoxia	n	n	+*	+/–	n

Pa_{O_2} = arterial O_2 tension; Pa_{CO_2} = arterial CO_2 tension; $P\bar{v}_{O_2}$ = mixed venous O_2 tension; n = normal; – = decreased; + = increased.
*Critical characteristic.

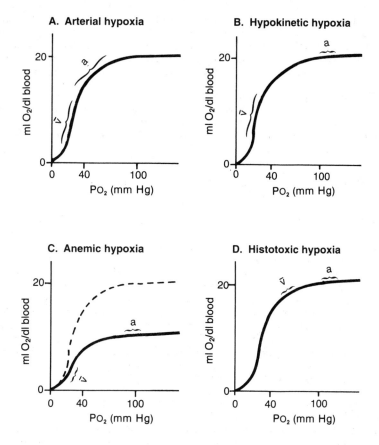

Figure 3-31. Oxyhemoglobin dissociation curves illustrating the four types of hypoxia; a = arterial point and $\bar{v}$ = mixed venous point. *Curve C* shows a normal curve (15 g/dl of hemoglobin) as the *dashed line*, whereas the *solid line* represents the effect of 7.5 g/dl of hemoglobin. Note that the arterial PO_2 is below normal only in arterial hypoxia, whereas the mixed venous point is below normal in all cases except in histotoxic hypoxia, where it is increased.

      reduced. Administration of 100% O_2 to these people can correct the hypoxia; this fact can be used as a diagnostic test to separate $\dot{V}A/\dot{Q}$ imbalance from anatomic shunts.

   d. **Anatomic shunts** are caused by blood flow that is never exposed to ventilated alveoli so that true venous blood enters the systemic (arterial) circulation. Anatomic shunts can occur through abnormal connections between the right and left sides of the heart caused by congenital heart disease, or it may represent blood flow through atelectatic (completely collapsed) areas of the lungs. The shunted blood represents venous blood that is dumped into the arterial circulation and results in a dilution of the normally oxygenated blood. Since the shunted blood is never exposed to ventilated alveoli, the administration of 100% O_2 does not raise the arterial PO_2 to its maximal levels.

 2. **Hypokinetic (ischemic) hypoxia** is due to an **inadequate blood flow**. The arterial PO_2 and content may be normal, but, because of inadequate blood flow, the tissues withdraw more O_2 from each unit of blood, so that the venous O_2 content is decreased (see Figure 3-31B). The reduced blood flow may involve the whole body, as in congestive heart failure, or it may involve only a localized area of the body due to abnormalities in a regional blood vessel. The most common cause of arterial obstruction is arteriosclerosis, where deposits of cholesterol and other lipids in the endothelium cause narrowing of a vessel's lumen, which reduces the blood flow. If this occurs in the coronary arteries, it can cause a **myocardial infarction** (heart attack), since a portion of the heart muscle dies from lack of O_2. This condition is fatal if a large portion of the heart muscle is affected, and cardiac output declines or arrhythmias occur.

3. **Anemic hypoxia** is caused by an **insufficient** amount of **functional hemoglobin**. The decrease in functional hemoglobin may be caused by deficiency of essential nutrients such as iron (iron deficiency anemia), or it may be due to the presence of abnormal amounts of methemoglobin or carboxyhemoglobin. Patients with anemic hypoxia have a **reduced O_2 capacity** and consequently a **decreased O_2 content,** whereas the arterial Po_2 remains normal (see Figure 3-31C).

4. **Histotoxic hypoxia** is caused by an inactivation of certain metabolic enzymes by chemicals such as cyanide. If these enzymes are not functioning, the tissues are unable to use O_2, even though there may be adequate O_2 in the tissues. In this condition, the O_2 delivery remains normal, and the venous Po_2 and O_2 content are high because the O_2 is not being consumed by the tissues (see Figure 3-31D).

D. **Physiologic responses to chronic hypoxia.** The body has many ways to compensate for decreases in the O_2 supply to the body.

1. **Accommodation** represents the immediate adjustments by the body to hypoxia. These mechanisms involve reflex adjustments of the cardiovascular and respiratory systems, whereas long-term adjustments can involve structural changes in the tissues.
   a. **Hyperventilation** occurs secondary to a stimulation of the peripheral chemoreceptors due to the low Po_2 in the arterial blood. The increased ventilation reduces alveolar Pco_2, which raises alveolar Po_2 proportionately. The reduced Pco_2 results in an alkalosis, which, in turn, lowers the respiratory drive. The alkalosis is slowly corrected by renal mechanisms over 1–2 weeks, so that respiration may continue to increase as the alkalosis is corrected.
   b. **Tachycardia** is a reflex response to carotid body stimulation by hypoxia. In humans, the cardiac output returns to normal after spending several weeks at high altitude.
   c. **Increased diphosphoglycerate (DPG)** concentration occurs during either hypoxia or alkalosis. The increased DPG concentration increases the P_{50} of the hemoglobin, which helps to maintain tissue Po_2 at slightly higher levels than it would be otherwise.

2. **Acclimatization** refers to changes in the body tissues in response to long-term exposure to hypoxia.
   a. **Polycythemia** is an abnormally high number of red cells/microliter of blood. This condition is also diagnosed if the hemoglobin level exceeds 18 g/dl in males or 16 g/dl in females or if the hematocrit exceeds 50%. The polycythemia usually is secondary to tissue hypoxia, which causes the release of renal erythropoietic factor, which acts on a plasma globulin to form erythropoietin. **Erythropoietin** stimulates the production of erythrocytes by the bone marrow, which eventually increases the number of circulating erythrocytes and the hematocrit. The increased number of red cells allows each unit of blood to carry additional O_2, which compensates for the decreased Po_2.
   b. **Pulmonary hypertension** is secondary to the generalized pulmonary hypoxia that results from **hypoxic pulmonary vasoconstriction**. The increased pulmonary artery pressure causes a more even distribution of the pulmonary blood flow, which can improve gas exchange by reducing the range of $\dot{V}A/\dot{Q}$ values. The increased pulmonary artery pressure can induce right ventricular hypertrophy, right-axis deviation on the EKG, right bundle branch block, and right ventricular failure.
   c. **Effects at the cellular and tissue level.** Chronic hypoxia has several effects including:
      (1) Increased enzyme activity
      (2) Increased mitochondrial density
      (3) Increased ability of cells to function at low Po_2
      (4) Increased capillary density, which occurs in skeletal and cardiac muscle
         (a) The increased number of capillaries reduces the diffusion distance.
         (b) This results in a decrease in the Po_2 diffusion distance from the blood into the cells.
   d. **Life-long exposure to hypoxia** causes additional alterations that seem to improve the body's function.
      (1) **Decreased respiratory drive** to hypoxia has been discovered in individuals exposed to hypoxia for prolonged periods. The reduced drive results in a higher Pco_2 and lower Po_2 but diminishes the work of respiration, which reserves more O_2 for use by other skeletal muscles.
      (2) **Total lung capacity** and **diffusing capacity** are increased in high-altitude natives compared to their sea-level controls. The increase in total lung capacity is apparent by the enlarged chest that high-altitude natives develop.

3. **Acute mountain sickness (AMS)** occurs in many individuals who stay at altitudes in excess of 9000–10,000 feet above sea level.
   a. Symptoms of AMS include fatigue, nausea, loss of appetite, headache, dyspnea, palpitations, and sleep disturbance. In some individuals, these effects progress and result in cerebral or pulmonary edema.
   b. Exercise during the first several days at high altitude can result in severe pulmonary edema due to the high blood flow coupled with the hypoxic pulmonary vasoconstriction. Pulmonary edema is very unevenly distributed throughout the lungs and is thought to occur in areas of the lung that are not fully vasoconstricted. In these areas, the pulmonary artery pressure is transmitted to the pulmonary capillaries, and the higher pressure increases the transudation of fluid into the interstitium and the alveoli.
   c. Treatment relies on increasing the P_{O_2} either by supplemental O_2 or, preferably, by evacuating the individual to lower altitude.

4. **Chronic mountain sickness (Monge's disease)** occurs in some long-term residents of high altitudes who develop extreme polycythemia, cyanosis, malaise, fatigue, and exercise intolerance. These individuals must be removed to lower altitude to prevent fatal pulmonary edema from rapidly developing.

STUDY QUESTIONS

Directions: Each of the numbered items or incomplete statements in this section is followed by answers or by completions of the statement. Select the **one** lettered answer or completion that is **best** in each case.

1. A 62-year-old male patient is known to have chronic lung disease and hypercapnia. He needs a major operation to remove an intestinal tumor. To insure that he has adequate alveolar ventilation while being anesthetized, which of the following should be available?

(A) Tank of 100% O_2
(B) Tank of 95% O_2, 5% CO_2
(C) Mechanical respirator
(D) Cardiac defibrillator
(E) EKG machine

2. All of the following can reduce vital capacity EXCEPT

(A) a decreased total lung capacity
(B) an increased residual volume
(C) a weakness of the inspiratory muscles
(D) a weakness of the expiratory muscles
(E) a decreased alveolar surface tension

3. Maximal inspiratory gas flow occurs when

(A) lung volume approaches total lung capacity
(B) lung volume approaches residual volume
(C) alveolar pressure is most negative
(D) interpleural pressure is approximately –5 cm H_2O
(E) the abdominal muscles are maximally contracted

4. Which of the following is true regarding the transmural pressure for the lungs?

(A) It always is negative
(B) It is equal to interpleural pressure minus atmospheric pressure
(C) It is equal to interpleural pressure minus alveolar pressure
(D) It is equal to alveolar pressure minus interpleural pressure
(E) It is independent of lung volume when the muscles are relaxed

5. During inspiration, as the diaphragm contracts, the pressure in the interpleural space becomes

(A) equal to zero
(B) more positive
(C) more negative
(D) equal to the pressure in the alveoli
(E) equal to the pressure in the atmosphere

6. The slope of a pressure-volume curve represents

(A) resistance
(B) compliance
(C) conductance
(D) reluctance
(E) inductance

7. Which of the following statements regarding the compliance of the respiratory system is true?

(A) It is greater than the compliance of the chest wall
(B) It is greater than the compliance of the lungs
(C) It is equal to the compliance of the chest wall
(D) It is equal to the compliance of the lungs
(E) It is less than the compliance of the chest wall

8. The respiratory system is at the equilibrium position in all of the following conditions EXCEPT

(A) at the end of a normal expiration
(B) when the transrespiratory pressure is zero
(C) when lung recoil is balanced by chest wall expansion
(D) when lung volume is at residual volume
(E) when the respiratory muscles are relaxed and the airway is open

9. A lack of normal surfactant, as occurs in infants with respiratory distress syndrome, results in

(A) an increased lung compliance
(B) stabilization of alveolar volume
(C) an increased retractive force of the lungs
(D) a reduced alveolar to arterial PO_2 difference
(E) a decrease in the filtration forces in the pulmonary capillaries

10. The primary defect responsible for respiratory distress syndrome in infants is

(A) a high airway resistance
(B) hypoventilation
(C) immature inspiratory muscles
(D) abnormal pulmonary surfactant
(E) left ventricular failure

11. Effort-independent flow occurs during

(A) normal tidal volume breathing
(B) a relaxed vital capacity maneuver
(C) forced inspiration
(D) the initial part of a maximal expiratory effort
(E) the terminal part of a maximal expiratory effort

12. The major area of airway resistance during breathing is located in the

(A) oropharynx
(B) trachea and large bronchi
(C) intermediate-sized bronchi
(D) bronchioles less than 2 mm in diameter
(E) alveoli

13. Airway resistance of the small peripheral airways (less than 2 mm in diameter) is

(A) high because of the small diameter of each tube
(B) high due to the high velocity of flow
(C) high due to the high viscosity of gas
(D) low because of the turbulent flow that is normally present
(E) low because of the low velocity of flow

14. The major factor that determines airway resistance is the

(A) length of the airways
(B) radius of the airways
(C) gas density
(D) gas viscosity
(E) alveolar pressure

15. Airway resistance can be reduced by

(A) increasing vagal impulses to the lungs
(B) administering a β-adrenergic blocking drug
(C) decreasing the radial traction exerted by lung tissue
(D) performing a maximal forced expiration
(E) increasing lung volume

16. During effort-independent flow, the flow rate of gas depends on the

(A) alveolar pressure
(B) interpleural pressure
(C) transmural pressure of the chest
(D) transmural pressure of the lungs
(E) transmural pressure for the respiratory system

17. A patient with restrictive lung disease typically has

(A) an increased forced expiratory volume in 1 second (FEV_1) and a normal lung compliance
(B) a decreased FEV_1 and an increased lung compliance
(C) a decreased FEV_1 and a decreased lung compliance
(D) an increased FEV_1 and an increased lung compliance
(E) an increased FEV_1 and a decreased lung compliance

18. During the effort-independent portion of a forced vital capacity, the expiratory flow rate

(A) varies as a function of the pleural pressure
(B) is limited by compression of the airways
(C) depends on the alveolar pressure
(D) is maximal for that individual
(E) is constant

19. An increased airway resistance is caused by all of the following factors EXCEPT

(A) increase in lung volume
(B) parasympathetic stimulation
(C) aging
(D) chronic bronchitis
(E) asthma

20. The timed vital capacity, FEV_1, is used to evaluate the

(A) flow resistance properties of the airways
(B) compliance properties of the lungs
(C) pulmonary blood flow resistance
(D) elastance properties of the lungs
(E) ventilation:perfusion ratio

21. The volume of gas in the lungs at the end of a normal expiration is referred to as the

(A) residual volume
(B) expiratory reserve volume
(C) functional residual capacity
(D) inspiratory reserve volume
(E) total lung capacity

22. The volume of N_2 dissolved in body fluids is greatest while breathing which of the following gas mixtures?

(A) Air at sea level
(B) Air at an altitude of 15,000 feet
(C) 20% O_2, 20% N_2, 60% He, while scuba diving at 2 atm of pressure
(D) 20% O_2, 30% N_2, 50% He, while scuba diving at 2 atm of pressure
(E) 20% O_2, 10% N_2, 70% He, while scuba diving at 5 atm of pressure

23. Increasing tidal volume, while keeping everything else constant, will result in an increase in

(A) dead space ventilation
(B) functional residual capacity
(C) inspiratory capacity
(D) alveolar ventilation
(E) alveolar P_{CO_2}

24. Which of the following statements is true regarding the fraction of O_2 in inspired (tracheal) gas?

(A) It equals 0.25 at sea level
(B) It decreases as a function of altitude
(C) It varies as a function of the weather
(D) It is less than the fraction of O_2 in the atmosphere
(E) It equals the fraction of O_2 in the alveoli

25. Which of the following statements regarding the P_{CO_2} in mixed expired gas is true?

(A) It is greater than the alveolar P_{CO_2}
(B) It is less than the alveolar P_{CO_2}
(C) It is equal to the alveolar P_{CO_2}
(D) It is equal to atmospheric P_{CO_2}
(E) It is greater than the P_{CO_2} in venous blood

26. Alveolar ventilation is equal to

(A) dead space ventilation
(B) tidal volume times respiratory rate
(C) minute ventilation
(D) minute ventilation minus dead space ventilation
(E) CO_2 production/min

27. A reduction in local alveolar ventilation is associated with

(A) an increase in regional pulmonary blood flow
(B) a decrease in regional alveolar P_{CO_2}
(C) a decrease in regional alveolar P_{O_2}
(D) an increase in regional tissue pH
(E) an increase in capillary hemoglobin saturation

28. Which of the following statements best characterizes the relationship between alveolar ventilation and alveolar P_{CO_2}?

(A) An increase in alveolar ventilation causes a decrease in alveolar P_{CO_2}
(B) An increase in alveolar ventilation causes an increase in alveolar P_{CO_2}
(C) A decrease in alveolar ventilation causes a decrease in alveolar P_{CO_2}
(D) Alveolar ventilation has no effect on alveolar P_{CO_2}

29. Arterial P_{CO_2} is increased in a normal individual while

(A) exercising
(B) breathing a gas mixture with a high P_{O_2}
(C) ascending a mountain
(D) hypoventilating
(E) scuba diving

30. The major sign of hypoventilation is

(A) cyanosis
(B) increased airway resistance
(C) hypercapnia
(D) dyspnea
(E) hypoxia

31. Which of the following statements regarding the normal alveolar P_{CO_2} is true?

(A) It is equal in all alveoli
(B) It is highest at the base of vertical lungs
(C) It is directly proportional to the inspired P_{O_2}
(D) It is directly proportional to the alveolar ventilation
(E) It is equal to 46 mm Hg

32. Which of the following gases normally is diffusion limited?

(A) O_2
(B) N_2O
(C) He
(D) CO_2
(E) CO

33. The diffusion coefficient of O_2, as compared to that of CO_2, is

(A) greater because O_2 combines with hemoglobin
(B) less because O_2 is less soluble
(C) greater because of a higher pressure gradient
(D) less because of the lower molecular weight of O_2
(E) essentially the same

34. The pH of venous blood, compared to the pH of arterial blood, is

(A) higher because of the additional CO_2
(B) lower because of the additional CO_2
(C) higher because of the removal of O_2
(D) lower because of the removal of O_2
(E) exactly the same

35. The Cl^- concentration in red cells of venous blood, compared to the Cl^- concentration in red cells of arterial blood, is

(A) lower due to the loss of Cl^- during capillary transit
(B) lower due to the increase in red cell volume
(C) unchanged
(D) higher due to an exchange with HCO_3^-
(E) higher due to an increase in Na^+ concentration

36. The distribution of pulmonary blood flow is

(A) equal throughout the lungs
(B) increased in the dependent portions of the lung
(C) not affected by changes in pulmonary vascular resistance
(D) not affected by changes in alveolar pressure
(E) not affected by changes in interpleural pressure

37. The variation in the ventilation:perfusion ratio in different areas of the normal lung is due primarily to the effects of

(A) neural control on the distribution of blood flow
(B) neural control on the distribution of tidal volume
(C) gravity
(D) the hemoglobin dissociation curve
(E) CO_2 on smooth muscle in the lungs

38. Areas of the lung near the base in a normal, standing human have which of the following ventilation:perfusion ratios ($\dot{V}_A/\dot{Q}$) and which volume?

(A) A $\dot{V}_A/\dot{Q}$ less than 1 and a volume less than that of the apex of the lung
(B) A $\dot{V}_A/\dot{Q}$ less than 1 and a volume greater than that of the apex of the lung
(C) A $\dot{V}_A/\dot{Q}$ greater than 1 and a volume less than that of the apex of the lung
(D) A $\dot{V}_A/\dot{Q}$ greater than 1 and a volume greater than that of the apex of the lung
(E) A $\dot{V}_A/\dot{Q}$ of 1 and a volume less than that of the apex of the lung

39. The alveolar P_{O_2} in the apex of a vertical lung is high because the apex of the lung

(A) has a low metabolic rate
(B) receives a high blood flow
(C) has a high ventilation:perfusion ratio
(D) receives the major portion of the tidal volume
(E) does not participate in gas exchange

40. If an area of the lung is not ventilated, but blood flow continues, then the blood leaving that area will have a composition equal to

(A) the inspired gas
(B) the normal systemic arterial blood
(C) the composition of blood in the pulmonary artery
(D) the composition of blood in the coronary arteries
(E) an indeterminate value because of an unknown ventilation:perfusion ratio

41. The most common cause of hypoxia is

(A) hypoventilation
(B) anemia
(C) ventilation:perfusion abnormalities
(D) anatomic shunt
(E) diffusion block between alveoli and pulmonary capillaries

42. The following figure shows two ventilatory patterns: one normal and the other abnormal. Which experimental maneuver listed below will create the abnormal pattern?

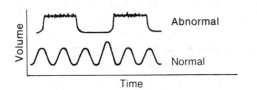

(A) Midpons transection with vagi intact
(B) Transection of the brain stem between the pons and medulla
(C) Midpons transection with vagi cut
(D) Transection rostral to the pons with vagi cut
(E) Transection rostral to the pons with vagi intact

43. P_{CO_2} affects respiration primarily by stimulating the

(A) carotid and aortic bodies
(B) J receptors
(C) medullary chemoreceptors
(D) baroreceptors
(E) hypoglossal nerve

44. Lactic acidemia can increase ventilation through its effect on receptors located in the

(A) small airways
(B) trachea and large bronchi
(C) medulla oblongata
(D) carotid bodies
(E) baroreceptors

45. Obstructed sleep apnea is characterized by airway occlusion at the level of the

(A) external nares
(B) naso- or oropharynx
(C) larynx
(D) trachea
(E) alveoli

46. Vital capacity is the sum of

(A) residual volume, tidal volume, and expiratory reserve volume
(B) residual volume, tidal volume, and inspiratory reserve volume
(C) residual volume, expiratory reserve volume, and inspiratory reserve volume
(D) expiratory reserve volume, inspiratory reserve volume, and tidal volume
(E) functional residual capacity and inspiratory capacity

47. Tissue hypoxia can be produced by all of the following factors EXCEPT

(A) decreased ventilatory drive
(B) decreased hemoglobin concentration in the blood
(C) decreased diffusion capacity of the lungs
(D) decreased CO_2 concentration in inspired air
(E) cyanide poisoning

48. The best measure of the gas exchange properties of the lungs is

(A) the alveolar-to-arterial difference in P_{O_2}
(B) arterial P_{CO_2}
(C) arterial P_{O_2}
(D) arterial blood pressure
(E) arterial hemoglobin saturation

49. A reduction of arterial P_{O_2} is typical of

(A) anemia
(B) CO poisoning
(C) moderate exercise
(D) cyanide poisoning
(E) hypoventilation

50. Venous P_{O_2} is higher than normal in which of the following conditions?

(A) Cyanide poisoning
(B) Exercise
(C) Decreased cardiac output
(D) Anemia
(E) CO poisoning

Directions: Each group of items in this section consists of lettered options followed by a set of numbered items. For each item, select the **one** lettered option that is most closely associated with it. Each lettered option may be selected once, more than once, or not at all.

Questions 51–55

The following graph shows a normal respiratory cycle followed by a maximal inspiration, a maximal forced expiration, and another normal respiratory cycle. Match each of the lung volumes listed below to the appropriate lettered arrow on the graph.

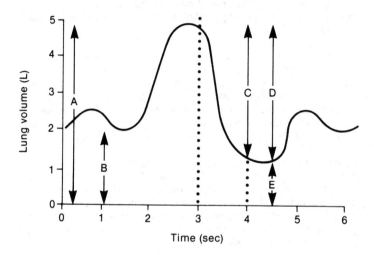

51. Vital capacity
52. Forced expiratory volume in 1 second
53. Functional residual capacity
54. Total lung capacity
55. Residual volume

Questions 56–59

Match each description of gas volume to the appropriate division of lung volume.

(A) Total lung capacity
(B) Residual volume
(C) Functional residual capacity
(D) Vital capacity
(E) Closing capacity

56. Lung volume that can be measured with just a spirometer
57. Lung volume at the end of a normal expiration
58. Maximal volume of gas that can be expired after a maximal inspiration
59. Lung volume at the end of a maximal expiration

ANSWERS AND EXPLANATIONS

1. The answer is C. [*III B 3 b*] Administering supplemental O_2 may correct the hypoxia, but, to maintain an adequate ventilation and correct hypercapnia, the lungs must be adequately ventilated. The only treatment option that provides ventilation is the respirator. The patient certainly does not need extra CO_2, because he is hypercapnic.

2. The answer is E. [*II A 2 a, B 2 a, H 1 d, 2 b*] A decreased alveolar surface tension will lead to an increased lung compliance, resulting in an increased total lung capacity and an increased vital capacity. Vital capacity equals total lung capacity minus residual volume, so that either a decrease in total lung capacity or an increase in residual volume can reduce the vital capacity. Expanding the lungs to the normal total lung capacity requires a strong inspiratory muscle force. Thus, weakness of the inspiratory muscles will produce a decrease in total lung capacity. Similarly, expiratory muscle force is required to decrease the lung volume to the normal level of the residual volume. A decrease in expiratory muscle force can result in an increase in the residual volume.

3. The answer is C. [*II C 3*] The driving force for gas flow is the pressure in the alveoli; negative pressures cause inspiratory flow, and positive pressures cause expiratory flow. It is the ratio of alveolar pressure to airway resistance that determines the actual flow of gas. As lung volume approaches total lung capacity or residual volume, much of the muscle force is expended in overcoming the low compliance of the respiratory system. When the interpleural pressure is approximately −5 cm H_2O, it is possible for alveolar pressure to be either positive or negative depending on muscle activity. The abdominal muscles are expiratory muscles and, when contracted, generate positive alveolar pressures and expiratory flow.

4. The answer is D. [*II C 4, D 1*] Transmural pressure represents the pressure across the wall of a hollow organ and is defined as the inside pressure minus the outside pressure. For the lungs, the internal pressure is the pressure in the alveoli or airways, while the pressure immediately outside the lungs is the pressure in the pleural space (interpleural pressure). In humans, the interpleural pressure is estimated by measuring the esophageal pressure, since the esophagus represents a flaccid tube that essentially traverses the pleural space. The interpleural pressure in animal experiments usually is measured by inserting a needle into an intercostal space and connecting it to a pressure gauge.

5. The answer is C. [*II C 4 b*] The increased negativity in the pleural space is caused by the chest wall being expanded by the inspiratory muscles. This negative pressure increases the transmural pressure of the lungs so that the lungs expand and fill with air, which enters through the airways.

6. The answer is B. [*II D 2 c*] The slope of a pressure-volume curve represents the change in volume for a unit change in pressure, which is the definition of compliance. If the graph were plotted as a volume-pressure relationship, then the slope would be elastance, which is the inverse of compliance.

7. The answer is E. [*II D 2 c; Figure 3-5*] The compliance of the respiratory system (C_{RS}) is determined by the compliance of the lungs (C_L) and the chest wall (C_{CW}) and can be calculated as $1/C_{RS} = 1/C_L + 1/C_{CW}$. Because of the need to add reciprocals, the compliance of the respiratory system always is less than the compliance of either of its parts. These relationships are apparent in Figure 3-5.

8. The answer is D. [*II D 3 b, 4 b, H 2 c*] At residual volume, the chest wall has a strong tendency to expand because it is far from its equilibrium position, which is about 80% of total lung capacity. At the same time, the recoil force of the lung is reduced since residual volume is close to the equilibrium position of the lung. Due to these unequal forces, either the expiratory muscles must be contracting in order to hold the respiratory system at that level, or the glottis must be closed to prevent gas from entering the airways. If the glottis is closed and the respiratory muscles are relaxed, then the strong expansion force of the chest wall will cause the gas in the airways to expand, the alveolar gas pressure will become less than atmospheric, and the transrespiratory pressure will be negative.

9. The answer is C. [*II E 2 c (3) (a)*] The lack of normal surfactant results in a high alveolar surface tension, which increases the retractile force of the lungs, resulting in a high transmural pressure. The high transmural pressure means that the lungs are less distensible, and the alveoli tend to collapse due to the increased surface forces. Due to the edema and atelectasis, there is an abnormal range of ventilation:perfusion ratios, which impairs gas exchange. The alveolar-to-arterial P_{O_2} difference is a good measure of the gas exchange capabilities of the lungs. This difference increases in the presence of ventilation:perfusion or diffusion abnormalities.

10. The answer is D. [*II E 2 c (3) (a)*] Alveolar lavage fluid from premature infants with respiratory distress syndrome shows abnormally high surface tension, which is thought to result in the alveolar collapse and edema characteristically found in the lungs of these infants. Several medical centers presently are running clinical tests to determine the effects of administering artificial or concentrated animal surfactants to these infants via the trachea. Preliminary results indicate that these substances are very effective in correcting the hypoxia and disability caused by lung immaturity.

11. The answer is E. [*II F 3*] As the name implies, effort-independent flow is not related to the expiratory force or effort that is expended. This phenomenon occurs when radial traction on the outside walls of the airways is overpowered by a negative transmural pressure. Thus, effort-independent flow does not occur at high lung volumes because of the large radial traction that occurs. Airway compression begins in the large airways at about 80% of the total lung capacity and proceeds peripherally as lung volume decreases. During normal tidal breathing, relaxed vital capacity maneuver, or forced inspiration, airway transmural pressure is always positive, so that compression of the airways and the formation of a flow-limiting segment is not feasible.

12. The answer is C. [*II F 1 a*] The airway tree is constructed such that each generation of airways is only slightly smaller in diameter than the parents. Thus, there is almost an exponential increase in cross-sectional area proceeding toward the periphery of the lung. Because of this relationship, the linear velocity of gas molecules decreases markedly as these molecules approach the terminal bronchioles. Therefore, very little pressure is required to achieve this velocity (i.e., there is a low resistance). Direct measurements indicate that bronchioles less than 2 mm in diameter represent less than 10% of the total airway resistance.

13. The answer is E. [*II F 1 a, b*] The small peripheral airways normally provide between 10%–20% of the total airway resistance. Although each airway is small in diameter, the large number of these airways provide a large total cross-sectional diameter. This results in a very low linear velocity of the gas molecules, which requires only a small pressure gradient. The viscosity of the gas in the small airways is no different than in the large airways.

14. The answer is B. [*II F 1 b*] While the length and radius of the airways, the gas density and viscosity, and the alveolar pressure all are involved in determining airway resistance, the most important factor is the radius of the airways. This is true because the flow rate varies as a function of the fourth or fifth power of the radius depending on whether flow is laminar or turbulent. Thus, small changes in the radius can produce large changes in the flow rate or the pressure needed to generate any flow.

15. The answer is E. [*II F 1 b, 3*] Increases in lung volume produce an increase in the radial traction forces that dilate the airways. This is one of the most powerful factors that can alter airway resistance. Increasing vagal impulses to the lungs or administering a β-adrenergic blocking drug increases airway resistance by narrowing the airways. The vagus nerve represents the motor nerve for bronchial smooth muscle that constricts the airways, while β-adrenergic blockers remove a dilating mechanism from the airways. A decrease in radial traction, as caused by the aging process, also narrows the airways by removing a mechanical dilating force on the outside of the airway walls. A forced expiration increases airway resistance, since the positive pleural pressure causes a compression of the large intrathoracic airways, which represent a flow-limiting segment under these conditions.

16. The answer is D. [*II F 3*] If the airways were rigid tubes, then gas flow would depend on the alveolar pressure, since this would determine the transairway pressure gradient. Since the airways are collapsible (and distensible), the presence of a positive pressure in the pleural space (caused by expiratory muscle contraction) represents a compressive force, since the transmural airway pressure is negative for the large airways. The driving pressure for airflow represents the difference between alveolar and pleural pressures, or transmural pressure of the lungs. The flow is limited by the narrowed segment, and the pressure downstream from this point (toward the mouth) has no influence on the flow rate.

17. The answer is C. [*II D 2 c, F 3, H 2 b (2); Figure 3-13*] A patient with restrictive lung disease has a decrease in the forced expiratory volume in 1 second (FEV_1) because of the reduced vital capacity. These patients typically can expire a larger fraction of their own vital capacity in 1 second because of the greater radial traction that results from the decreased compliance. The increased radial traction reduces airway resistance in the lungs. These patients have no difficulty breathing at high frequencies because of the low airway resistance and usually choose a high rate of respiration coupled with a reduced tidal volume to minimize the work of breathing.

18. The answer is B. [*II F 3 b*] The effort-independent portion of a forced expiration begins when lung volume reaches about 80% of total lung capacity. During the effort-independent period, the flow rate declines as a function of lung volume. Actually, the driving force for expiration during this period is the transpulmonary pressure, which is a function of lung volume. The term indicates that flow is independent of effort and that changes in expiratory force do not alter the flow rates. Therefore, the expiratory flow rate neither varies as a function of the pleural pressure, nor depends on the alveolar pressure, because these values are altered by expiratory effort. The expiratory flow rate is not maximal for the individual because maximal flow can be achieved only when lung volume is just below total lung capacity.

19. The answer is A. [*II F 3 b (1)*] Increases in lung volume reduce airway resistance because of the increased radial traction that is generated on the airway walls. Parasympathetic stimulation, chronic bronchitis, and asthma are associated with increased airway resistance because of bronchial smooth muscle contraction. Aging results in an increased airway resistance due to the destruction of alveolar septa, which reduces the radial traction and allows the airways to constrict.

20. The answer is A. [*II H 2 b (2)*] The FEV_1 represents the volume of gas expired in the first second of a maximally forced expiration. This volume is frequently expressed as a ratio to the forced vital capacity and is then termed the $\%FEV_1$. The $\%FEV_1$ compensates for different sized vital capacities and normally exceeds 80%. The $\%FEV_1$ is reduced either by a high airway resistance or by a reduced expiratory force due to muscle weakness or a less than maximal effort by the patient.

21. The answer is C. [*II H 2 c*] Since expiration is passive, the lung volume decreases during expiration until the equilibrium volume (functional residual capacity) is reached. The equilibrium volume represents the volume of a distensible structure when the transmural pressure (pressure inside minus pressure outside) is zero.

22. The answer is A. [*III A 2; IV A*] Henry's law states that the volume of gas dissolved in a liquid equals the partial pressure of the gas times the solubility coefficient. Since the partial pressure of N_2 (P_{N_2}) at an altitude of 15,000 feet would be less than the P_{N_2} at sea level, the amount of N_2 dissolved in the tissues also would be less. The P_{N_2} when breathing air at sea level is 0.79 atm. Gas equilibration requires approximately 12 hours after any change in pressure and occurs at different rates in different tissues. For N_2, equilibration takes the longest in the fatty tissues because of the high N_2 solubility and the low blood flow to this tissue. Scuba diving at 2 atm while breathing a gas with 20% N_2 provides 0.4 atm of P_{N_2}. The other two scuba conditions yield N_2 tensions less than 0.79 atm.

23. The answer is D. [*III B 1, 3*] If respiratory rate, dead space, and the ventilation:perfusion ratio remain constant, then an increase in tidal volume will increase minute ventilation and alveolar ventilation. Since dead space ventilation = dead space volume times respiratory rate, dead space ventilation is unchanged under these conditions. Functional residual capacity is not altered, while the inspiratory capacity is reduced by increases in tidal volume. Alveolar CO_2 tension (P_{CO_2}) is reduced by an increased alveolar ventilation.

24. The answer is D. [*III A 3, B 2 a (1)*] The fraction of O_2 in the air is constant from sea level to several hundred thousand feet altitude and equals 0.21. However, the P_{O_2} decreases with increased altitude, since the total pressure declines as one ascends. The partial pressure of any gas is given by the product of the mole fraction of the gas and the total or barometric pressure. Nasal breathing adds water vapor to inspired air, so that the O_2 content of the air is decreased somewhat by the time it reaches the trachea. Alveolar gas is diluted by the addition of CO_2 from the blood, so that alveolar O_2 is less than that in the trachea.

25. The answer is B. [*III B 2 b*] Mixed expired gas has a lower P_{CO_2} than alveolar gas because it is a mixture of alveolar and dead space gas. The composition of this gas varies depending on the ratio of dead space ventilation to alveolar ventilation. This fact can be used to determine the dead space tidal volume ratio (V_D/V_T) which is given by $1 - [\text{expired } P_{CO_2}/\text{alveolar } P_{CO_2}]$.

26. The answer is D. [*III B 3*] Alveolar ventilation is minute ventilation minus dead space ventilation. Minute ventilation is the volume of gas expired per minute, which is equal to the product of tidal volume times respiratory rate. Not all of the minute ventilation reaches the gas exchange region of the lungs, since some must occupy the conducting system of the lungs.

27. The answer is C. [*III B 3; V B 2*] A decrease in local or regional alveolar ventilation results in a decreased influx of gas to that region of the lung, so that, transiently, more O_2 is absorbed, and more CO_2 is released by the capillary blood. Consequently, alveolar Po_2 declines, while alveolar Pco_2 increases in this area of the lung. The hypoxia, hypercapnia, and resultant local acidosis cause pulmonary vasoconstriction, which results in a decrease in pulmonary blood flow. Due to the decrease in local Po_2, there is a reduction in the hemoglobin saturation of the blood that leaves this area of the lung. These changes represent the effects of a low ventilation:perfusion ratio on gas exchange in a small area of the lung.

28. The answer is A. [*III B 3 b*] Alveolar Pco_2 and alveolar ventilation are inversely related, so that a decrease in alveolar ventilation results in an increase in alveolar Pco_2. CO_2 can only be eliminated by adequate ventilation of the alveoli. It does not matter what type of gas is used to flush out CO_2, provided the flow is adequate. Of course, an adequate level of O_2 in the gas is required to maintain life.

29. The answer is D. [*III B 3 b*] Hypoventilation is synonymous with hypercapnia. The arterial Pco_2 is determined by the ratio of CO_2 production to alveolar ventilation. During exercise, the alveolar ventilation is adjusted so that arterial Pco_2 remains normal and may even decline under moderate and heavy exercise loads. During O_2 breathing there is little, if any, effect on alveolar ventilation in the normal individual. Mountain climbing exposes one to hypoxia with an increase in respiratory drive and a consequent decrease in Pco_2. Scuba diving, at least at moderate depths (i.e., less than 100 feet), should not alter the Pco_2, since alveolar ventilation should be maintained at a normal level.

30. The answer is C. [*III B 3 b*] Hypercapnia and hypoventilation are synonyms, since the alveolar Pco_2 is primarily determined by the alveolar ventilation. Cyanosis may be present with hypoventilation, but it is a sign (and not a very good one) of hypoxia, not hypoventilation. Both an increased airway resistance and dyspnea may be symptoms of airway disease but are not characteristic of hypoventilation.

31. The answer is B. [*III B 3 b; VII B*] Alveolar Pco_2 varies inversely as a function of the ventilation:perfusion ratio ($\dot{V}A/\dot{Q}$) or alveolar ventilation. With a high $\dot{V}A/\dot{Q}$, the ventilation exceeds blood flow so that CO_2 delivery to the lungs via the pulmonary blood flow is low and is diluted by the ventilation. The opposite is true at the base of the lungs. The Po_2 varies reciprocally with the Pco_2. At sea level, while breathing air, the sum of the two gas tensions cannot exceed 150 mm Hg, which is the inspired Po_2, because the remaining gas fraction is occupied by N_2, which is not utilized by the body. Alveolar Pco_2 is maintained at 40 mm Hg to maintain the pH of body fluids in the normal range.

32. The answer is E. [*III C 1, 4*] Normally only CO is diffusion limited, but, in the presence of severe lung disease, O_2 also may be diffusion limited if the diffusion capacity of the lungs is reduced sufficiently.

33. The answer is B. [*III C 4*] The diffusion coefficient concerns the movement of molecules in solution and is determined by the molecular weight and the solubility of any substance. O_2 has a lower molecular weight than CO_2 but is much less soluble—about 24 times less. The overall effect is that CO_2 is much more diffusible than is O_2, which is important because there is only about a 5 mm Hg gradient to cause CO_2 to diffuse from the pulmonary capillary blood to the alveoli.

34. The answer is B. [*IV D*] The pH of venous blood is lower than the pH of arterial blood because of the addition of CO_2 at the tissue level. CO_2 is an acid molecule because it becomes hydrated to carbonic acid (H_2CO_3), which can dissociate into H^+ and HCO_3^-.

35. The answer is C. [*IV D 3 a*] The Cl^- concentration in red blood cells is higher in venous blood than in arterial blood because of the chloride shift that occurs in the tissue capillaries. As blood picks up CO_2 at the tissues, some of it diffuses into the red cells, where it is hydrated to H_2CO_3, which then dissociates into H^+ and HCO_3^-. The H^+ is buffered by hemoglobin, and the HCO_3^- diffuses out of the red cells. In order to maintain electrical neutrality in the cells, the HCO_3^- is replaced by Cl^-, which diffuses in and raises the Cl^- concentration in the venous red cells.

36. The answer is B. [*V B*] The pulmonary vasculature is a low pressure–low resistance system that is very distensible. Gravity causes a linear loss of the driving force at locations above the heart, so blood flow decreases to apical areas. The dependent vessels are distended due to the higher hydrostatic forces and flow resistance decreases, so that flow increases toward the base of the lung. Changes in alveolar or interpleural pressure alter flow through the alveolar and extra-alveolar vessels in a complex fashion that alters the resistance in these vessels.

37. The answer is C. [*V B; VI A*] Gravity affects both the distribution of the tidal volume and pulmonary blood flow because of the hydrostatic effects on the driving pressure and the interpleural pressure. The autonomic nervous system has only a minor role in adjusting both pulmonary vascular and airway resistance under normal conditions. While an increased P_{CO_2} causes pulmonary vasoconstriction, and a decreased expired P_{CO_2} results in bronchoconstriction, neither of these responses represents a major factor under normal conditions. The hemoglobin dissociation curve is not involved in controlling blood or air flow.

38. The answer is A. [*VI A 1; VII A*] The ventilation:perfusion ratio ($\dot{V}_A/\dot{Q}$) is less than 1 at the base of the lungs because blood flow exceeds ventilation in this area. Blood flow increases more rapidly per vertical distance down the lung, since gravity has a greater effect on blood due to its higher density compared with lung tissue. The transmural pressure for the lung varies between base and apex because of the pleural pressure gradient. The pleural pressure is near zero at the base of the lung, whereas it is more negative at the apex. Thus, the transmural pressure is minimal at the base and greater at the apex, which results in the base of the lung being less distended than the apex.

39. The answer is C. [*VII A*] The P_{O_2} in the lungs depends on the regional ventilation:perfusion ratio. A high ratio results in a high alveolar P_{O_2} and a low alveolar P_{CO_2}. The high ventilation:perfusion ratio at the apex of a vertical lung occurs because the perfusion rate is reduced much more than the rate of ventilation. However, both ventilation and blood flow are reduced compared to what is present in the base of the lung.

40. The answer is C. [*VII B*] If an area of the lung is not ventilated, then there will be no gas exchange in that area. If there is no gas exchange, then the gas tensions will be equal to that in the venous blood. Since the blood in the pulmonary artery represents mixed venous blood from the systemic circulation, then blood leaving an unventilated area of the lungs will have a gas composition equivalent to that in the pulmonary artery.

41. The answer is C. [*VII*] Due to the complexity of factors involved in regulating ventilation and perfusion in the lungs, many diseases can lead to an imbalance in these factors. It has been estimated that 85% of patients with hypoxia suffer from ventilation:perfusion imbalance. Hypoventilation, anatomic shunt, and diffusion block represent three other abnormalities that result in arterial (hypoxic) hypoxia. Anemia is an additional mechanism of hypoxia that is less common than ventilation:perfusion imbalance.

42. The answer is C. [*VIII A 2 b; Figure 3-30*] This pattern of breathing is called apneustic breathing, or inspiratory breath holding. This respiratory pattern occurs when the inhibitory influences from the periphery via the vagus nerves and the pneumotaxic center are interrupted. The pneumotaxic center lies in the rostral pons, whereas the apneustic center (which enhances inspiratory drive) lies in the caudal pons. The vagus nerve carries impulses from stretch receptors in the airways that tend to inhibit inspiration after a certain threshold of tidal volume is exceeded.

43. The answer is C. [*VIII A 3*] Approximately 85% of the effect of CO_2 is mediated through the medullary chemoreceptors; only 15% of the effect comes from the carotid and aortic bodies. J receptors, baroreceptors, and the hypoglossal nerve are not affected by changes in CO_2.

44. The answer is D. [*VIII B 2*] H^+ affects ventilation by stimulating the peripheral chemoreceptors or carotid bodies. H^+ does not readily cross the blood-brain barrier and therefore cannot affect the central chemoreceptors in the medulla.

45. The answer is B. [*VIII E 4*] Obstructed sleep apnea is caused by a partial or complete obstruction of the upper airway. During sleep, the muscle tone of the pharyngeal dilator muscles is lost so that there is airway narrowing. This results in a higher velocity of gas flow through this narrowed area, producing a more negative pressure in the airway (Bernoulli effect). These very negative pressures result in fluttering of the tissues, causing the typical snoring indicative of intermittent closure of the airway in the oro- or nasopharynx.

46. The answer is D. [*Figure 3-12*] Vital capacity—which is made up of expiratory reserve volume, inspiratory reserve volume, and tidal volume—is the maximal amount of gas that can be expired after a maximal inspiration. Residual volume is not included in vital capacity.

47. The answer is D. [*IX C*] Air normally has only a negligible quantity of CO_2, so further reduction in the CO_2 will have no effect. Decreased ventilatory drive and decreased diffusion capacity of the lungs result in arterial hypoxia; decreased hemoglobin concentration in the blood represents hypemic hypoxia; and cyanide poisoning causes histotoxic hypoxia.

48. The answer is A. [*IX C 1*] Normally, the difference between the alveolar and arterial PO_2 is 5–10 mm Hg, which is due to a small anatomic shunt and to some ventilation:perfusion ($\dot{V}A/\dot{Q}$) imbalance. Any further increase in the difference between the alveolar and arterial PO_2 indicates that the gas exchange function is impaired by a $\dot{V}A/\dot{Q}$ imbalance (physiologic shunt), anatomic shunt, or diffusion limitation. Arterial PCO_2 depends primarily on the adequacy of alveolar ventilation. With hypoventilation, the alveolar PCO_2 and, consequently, the arterial PCO_2 rise while alveolar and arterial PO_2 fall; however, the difference between the alveolar and arterial PO_2 remains in the normal range. Arterial blood pressure is altered reflexly through the chemoreceptors by changes in the blood gases, while hemoglobin saturation is largely a function of the PO_2. The alveolar-to-arterial difference in PO_2 is the best measure of gas exchange.

49. The answer is E. [*IX C 1 a*] A normal arterial PO_2 is typical of anemia, CO or cyanide poisoning, and moderate exercise. It is either the tissue or the venous PO_2 that is reduced as a consequence of inadequate O_2 delivery. Hypoventilation results not only in hypoxia but also, even more characteristically, in hypercapnia.

50. The answer is A. [*IX C 4*] Cyanide poisoning causes venous PO_2 to be higher than normal, because the tissue oxidative enzymes are inactivated by cyanide and unable to use O_2. Therefore, less O_2 diffuses out of the blood in the systemic capillaries, which leaves a greater amount in the venous blood. During exercise, the tissues remove a greater amount of O_2 from the capillary blood, which results in a decreased venous PO_2. A decreased cardiac output, anemia, and CO poisoning all reduce O_2 delivery to the tissues so that more O_2 than normal is extracted from the capillary blood, and venous PO_2 declines.

51–55. The answers are: 51-D, 52-C, 53-B, 54-A, 55-E. [*II H; Figures 3-12, 3-13*] Vital capacity is the maximal volume of gas that can be expired after a maximal inspiration (indicated by the arrow labeled *D*). Vital capacity equals the sum of tidal volume, inspiratory reserve volume, and expiratory reserve volume.

The forced expiratory volume in 1 second (FEV_1) is the volume of gas that can be expired in 1 second during a maximal forced expiration. The forced expiration begins at the third second on the graph; the volume expired 1 second later is indicated by the arrow labeled *C*.

The functional residual capacity is the reservoir of gas that remains in the lungs after a normal expiration (indicated by the arrow labeled *B*). This gas buffers the changes in O_2 and CO_2 tensions in the blood that traverses the capillaries between inspirations.

The total lung capacity represents the amount of gas in the lungs after a maximal inspiration (indicated by the arrow labeled *A*).

The residual volume is the amount of gas in the lungs after a maximal expiration (indicated by the arrow labeled *E*). The residual volume cannot be removed from the lungs, because the chest wall becomes rigid at this volume and lung volume cannot be reduced further. In older individuals, the residual volume is set by closure of the airways, which prevents further emptying of the lungs.

56–59. The answers are: 56-D, 57-C, 58-D, 59-B. [*II H 1 d, 2 a–c; VI C 1 e*] A simple spirometer can only measure lung volumes that can be expired; thus, any division of lung volume that contains the residual volume (RV) requires the use of much more sophisticated equipment. The total lung capacity (TLC), functional residual capacity (FRC), and closing capacity all include the RV. (The closing capacity consists of the closing volume and the RV.)

The respiratory system normally reaches its equilibrium position at the end of a normal expiration. At this volume, the expansion force of the chest wall is counteracted by the retractile force of the lung, and the transrespiratory pressure is zero.

The maximal volume of gas that can be expired after a normal expiration is the vital capacity (VC). It is important that patients inspire maximally in order to achieve TLC and then expire maximally to reach RV. The VC equals TLC minus RV. Weakness of the inspiratory muscles will reduce TLC, while weakness of the expiratory muscles will increase the RV.

The RV represents the volume of gas that cannot be expired from the lungs with an intact chest. The RV is determined in young adults by the chest wall, which reaches a point where it cannot be compressed further. In older people and in patients with obstructive lung disease, the RV is set by airway closure. Once the airways close off, the gas distal to the closure cannot be expelled and represents the RV.

4
Renal Physiology
John Bullock

I. INTRODUCTION

A. Kidney function. The kidneys maintain the constant state of the body's internal environment by regulating the volume and composition of the extracellular fluids. To accomplish this, the kidneys balance precisely the intake, production, excretion, and consumption of many organic and inorganic compounds. This balancing requires that the kidneys maintain the volume and composition of body fluids by the conservation and excretion of water and solutes to meet bodily requirements.

1. **Water and electrolyte intake and metabolism**
   a. The **gastrointestinal system** provides the primary source for the normal oral intake of water and electrolytes.
   b. The **oxidation of food** provides a secondary but important source of water. The ordinary mixed diet will lead to the production of approximately 300 ml of metabolic water per day, mostly from the oxidation of fat.
   c. Metabolism also produces **urea,** a nontoxic product of protein metabolism; **uric acid,** the end product of purine metabolism; and **creatinine,** an endogenous anhydride of muscle creatine.

2. **Excretion of water and solutes**
   a. A nonperspiring young man in a basal metabolic state loses approximately 30 g of water/hr by insensible perspiration. This, together with water loss from the lungs, constitutes the **insensible water loss,** which can range between 800 and 1400 ml/day, depending on body surface area and metabolic rate. **Sensible sweat** is a hypotonic solution with a salt concentration ranging from 10 to 70 mEq/L of water.
   b. Less than 100 ml of fluid together with 1 mEq each of sodium and chloride are excreted in the feces per day.
   c. Normally, the kidneys excrete about 1500 ml of hypertonic urine in 24 hours.

3. **The formation and release of renin,** which is a major component of the renin-angiotensin-aldosterone mechanism, allows the kidneys to regulate **blood pressure**. In addition, the renal control of fluid volume is essential to the regulation of blood pressure.

4. **The formation and release of renal erythropoietic factor** (erythropoietin) increases the number of circulating erythrocytes. Erythropoietin is synthesized in the epithelial cells of the renal cortical glomerular tuft and the juxtaglomerular cells. The liver is the main source of erythropoietin in the prenatal period.

5. **Vitamin D activation.** Dietary vitamin D must undergo two hydroxylations in order to be useful to the body. The first step is performed by the liver. The final hydroxylation, performed by the kidneys, converts this vitamin to its most biologically active form (1,25-dihydroxy-cholecalciferol).

6. **Gluconeogenesis.** The kidney acquires the important ability to synthesize and secrete glucose produced from noncarbohydrate sources (glutamine) only in unusual circumstances such as prolonged starvation and chronic respiratory acidosis.

B. Kidney structure (Figure 4-1). The kidneys are paired organs that are located retroperitoneally in the upper dorsal region of the abdominal cavity. Each human kidney is composed of approximately 1 million nephrons, is about the size of a fist, and weighs about 150 g.

1. **Nephron.** This basic functional unit of the kidney is composed of a glomerulus, with its associated afferent and efferent arterioles, and a renal tubule.

   a. **Glomerulus.** The glomerulus consists of a tuft of 20–40 capillary loops protruding into **Bowman's capsule,** which is the beginning of the renal tubule. The capillary endothelium is fenestrated with an incomplete basement membrane, and together these structures provide a minimal resistance for filtration of plasma while providing a sieving function for retention of plasma proteins and blood cells.

   b. **Renal tubule** (Figure 4-2). The renal tubule begins as Bowman's capsule, which is an expanded, invaginated bulb surrounding the glomerulus. The epithelium of Bowman's capsule is an attenuated layer that is about 400 Å in thickness. The renal tubule consists of the **proximal convoluted tubule,** the **loop of Henle,** the **distal convoluted tubule,** and the **collecting duct** that carries the final urine to the renal pelvis and the ureter.

   c. **Types of nephrons**

      (1) **Cortical nephrons** comprise about 85% of the nephrons in the kidney and have glomeruli located in the renal **cortex.** These nephrons have short **loops of Henle,** which descend only as far as the outer layer of the renal **medulla.**

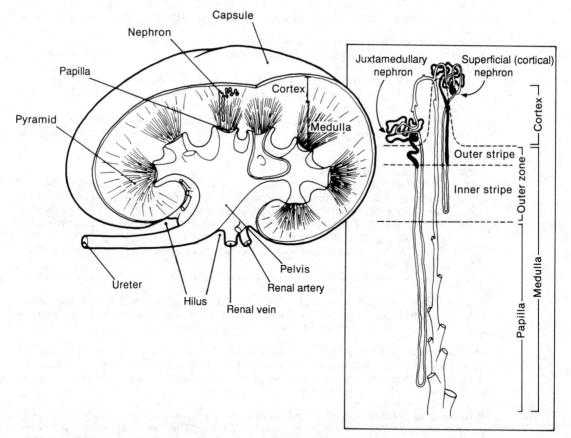

Figure 4-1. Structure of the human kidney, cut open to show the various zones. (Reprinted from Marsh DJ: *Renal Physiology.* New York, Raven, 1983, p 37.) **Inset:** Two principal nephron types and their collecting duct systems. Superficial (cortical) nephrons have their glomeruli near the surface of the kidney and do not possess long loops of Henle. Juxtamedullary nephrons have their glomeruli at the corticomedullary junction. Glomeruli and proximal tubules are shown as *solid black.* The remaining nephron is shown as *white.* (Reprinted from Brenner B, Coe FL, Rector FC: Transfer functions of renal tubules. In *Renal Physiology in Health and Disease.* Philadelphia, WB Saunders, 1987, p 29.)

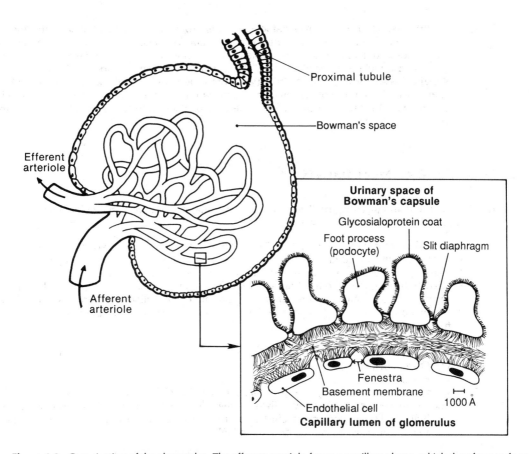

Figure 4-2. Organization of the glomerulus. The afferent arteriole forms a capillary plexus, which then fuses to form the efferent arteriole. The outer lining of the glomerulus, called Bowman's capsule, is continuous with the proximal tubule. (Reprinted from Marsh DJ: *Renal Physiology.* New York, Raven, 1983, p 41.) **Inset:** Glomerular capillary wall in cross-section. The luminal surface is covered by fenestrated endothelial cells. The basement membrane has a middle lamina densa surrounded by lamina rara interna and externa. Overlying this are the foot processes of epithelial cells, separated by small slit diaphragms. (Illustration by Nancy Lou Gahan Markris. Reprinted from Brenner BM, Beeuwkes R III: The renal circulations. *Hosp Pract* 13:35–46, 1978.)

      (2) Juxtamedullary nephrons are located at the junction of the cortex and the medulla of the kidney. Juxtamedullary nephrons have long loops of Henle, which penetrate deep into the medulla and sometimes reach the tip of the **renal papilla**. These nephrons are important in the **countercurrent system,** by which the kidneys concentrate urine.

2. Renal blood vessels

  a. Renal arteries. Each kidney receives a renal artery, which is a major branch from the aorta. The kidneys receive approximately 25% of the total resting cardiac output or about 1.25 L blood/min. The sympathetic tone to renal vessels is minimal at rest but increases during exercise to shunt renal blood flow to exercising skeletal muscles.

    (1) Afferent and efferent arterioles. Each renal artery subdivides into progressively smaller branches, and the smallest branches give off a series of afferent arterioles. Each afferent arteriole forms a tuft of capillaries, which protrudes into Bowman's capsule. These capillaries come together and form a second arteriole, the efferent arteriole, which divides shortly after to form the peritubular capillaries that surround the various portions of the renal tubule.

    (2) Peritubular capillaries differ in organization depending on their association with different nephrons.

      (a) The efferent arterioles of cortical nephrons divide into peritubular capillaries that connect with other nephrons, forming a rich meshwork of microvessels. This meshwork functions to remove water and solutes that have diffused from the renal tubules.

(b) Efferent arterioles of juxtamedullary nephrons also form peritubular capillaries, a special portion of which are the **vasa recta**. The vasa recta descend with the long loops of Henle into the renal medulla and return to the area of the glomerulus. The vasa recta form capillary beds at different levels along the loop of Henle.

b. Renal veins are formed from the confluence of the peritubular capillaries and exit the kidney at the **hilus**. The pattern of the renal venous system is similar to that found in the end arterial system except for the presence of multiple anastomoses between veins at all levels of the venous circulation.

II. BODY FLUID COMPARTMENTS

A. Water content and distribution (Figure 4-3). The two major fluid compartments are the intracellular fluid and extracellular fluid volumes.

1. **Total body water (TBW)** constitutes 55%–60% of the body weight in young men and 45%–50% of the body weight in young women. (The lower percentage in women largely is due to the relatively greater amount of adipose tissue in women than in men.)

   a. Distribution. About one-third of the total body water is in the extracellular fluid compartment, and the remaining two-thirds is in the intracellular fluid compartment. (The TBW distribution in a young, 70-kg man is summarized in Table 4-1.) Distribution of TBW is as follows:

   (1) Muscle (50%)
   (2) Skin (20%)
   (3) Other organs (20%)
   (4) Blood (10%)

   b. Lean body mass. Although the percentage of TBW declines with advancing age and with obesity, this percentage for any individual (regardless of sex) represents a constant 70% of that individual's lean body mass (LBM or fat-free mass).* Based on this constant relationship, the amount of body fat can be determined as

   $$\text{body fat (\%)} = 100 - \frac{\text{percentage of TBW}}{0.7}$$

   and the LBM can be estimated as

   $$\text{LBM (kg)} = \frac{\text{TBW (L)}}{0.7}$$

2. **Extracellular fluid (ECF).** The ECF compartment includes several subcompartments.

   a. Plasma volume, which represents the fluid portion of the blood, is about 3.5 L in a young, 70-kg man and represents about 25% of the ECF.

Table 4-1. Distribution of Body Water in a Young, 70-kg Man

| | | Percent | | |
Compartment	Volume (L)	Body Weight	Lean Body Mass	Body Water
Total body water	42*	60†	70	100
Extracellular fluid	14	20	24	33
Plasma	3.5	5	6	8
Interstitial fluid	10.5	15	18	25
Intracellular fluid	28	40	48	67

*Total body water is 35 L in a young, 70-kg woman.
†Body water accounts for 50% of the total body weight of a young, 70-kg woman.

---

*Lean body mass is defined as 15% bone, 10% fat, and 75% tissue.

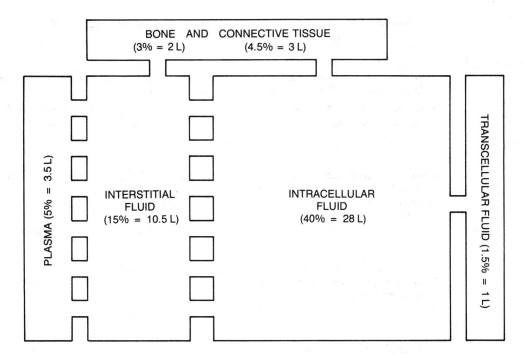

Figure 4-3. Distribution of the total body water in an average 70-kg man. The intracellular fluid compartment is not subdivided, but the extracellular fluid compartment consists of four subdivisions: plasma, interstitial fluid, transcellular fluid, and the water of bone and connective tissue. Relative volumes are expressed as percentages of the body weight, assuming a fat content of about 10%. Fat has relatively little water associated with it. (Reprinted from Bauman JW Jr, Chinard FP: Body fluids. In *Renal Function: Physiological and Medical Aspects*. St. Louis, CV Mosby, 1975, p 19.)

    (1) Blood volume constitutes about 8% of the total body weight and can be obtained from plasma volume and the hematocrit as

$$\text{blood volume (L)} = \text{plasma volume (L)} \times \frac{100}{(100 - \text{hematocrit})}$$

    (2) Blood volume occupies approximately 80 ml/kg of body weight.

  b. Interstitial fluid is the fluid between cells and includes lymph, which constitutes 2%–3% of the body weight. The volume of interstitial fluid in a young, 70-kg man is about 10.5 L and represents about 75% of the ECF. The interstitial fluid is also referred to as the internal environment, or **milieu interieur,** which surrounds all cells except the blood cells.

  c. Transcellular fluid volume is about 1 L in most humans and occupies about 15 ml/kg of body weight (1.5% of body weight). This ECF subcompartment represents fluid in the lumen of structures lined by epithelium and includes digestive secretions; sweat; cerebrospinal fluid (CSF); pleural, peritoneal, synovial, intraocular, and pericardial fluids; bile; and luminal fluids of the gut, thyroid, and cochlea.

    (1) Gastrointestinal luminal fluid constitutes about half of the transcellular fluid and occupies about 7.4 ml/kg of body weight.

    (2) CSF occupies about 2.8 ml/kg of body weight.

    (3) Biliary fluid volume is about 2.1 ml/kg of body weight.

  3. Intracellular fluid (ICF). The volume of the ICF compartment varies but usually constitutes 30%–40% of the body weight. This is the larger of the two major fluid compartments, containing about two-thirds of the TBW.

B. Volume measurement in the major fluid compartments

  1. Indicator dilution principle. The volume of water in each fluid compartment can be measured by the indicator dilution principle. This principle is based on the relationship among the amount of a substance injected intravenously (Q), the volume in which that substance is distributed (V), and the final concentration attained (C).

a. The equation for this relationship, based on the definition of concentration (C), is

$$C = \frac{Q}{V} \text{ or } V = \frac{Q}{C} \tag{1}$$

where V = the volume (in ml or L) in which the quantity, Q (in g, kg, or mEq), is distributed to yield the concentration, C (in g/ml or L or in mEq/ml or L).

Example. If 25 mg of glucose are added to an unknown volume of distilled water and the final concentration of glucose after mixing is 0.05 mg/ml, then the volume of solvent is

$$V = \frac{25 \text{ mg}}{0.05 \text{ mg/ml}} = 500 \text{ ml}$$

b. Volume measurement by the dilution principle requires that the introduced substance be distributed evenly in the body fluid compartment being measured.
  (1) Equation (1) must be altered if the solute leaves the compartment through one of the following mechanisms:
   (a) Excretion in the urine or transfer to another compartment where it exists in a different concentration
   (b) Metabolism of the solute
   (c) Vaporization of the solute from the skin and respiratory tract
  (2) The amount of substance lost from the fluid compartment, then, is subtracted from the quantity administered, as

$$V = \frac{Q \text{ administered} - Q \text{ removed}}{C}$$

Example. A 60-kg woman is infused with 1 millicurie (mCi) of tritium oxide (^3H_2O). After 2 hours, 0.4% of the administered dose is lost in the urine and by vaporization from the skin and respiratory tract. The radioactivity of a plasma sample is measured by liquid scintillation spectrometry and indicates a concentration of 0.03 mCi/L of plasma water. Since the concentration of ^3H_2O throughout the body fluids should be the same as in plasma after equilibration, the TBW can be calculated as

$$V = \frac{Q \text{ infused} - Q \text{ excreted}}{C}$$
$$= \frac{1 \text{ mCi} - (1 \text{ mCi} \times 0.004)}{0.03 \text{ mCi/L}}$$
$$= \frac{0.996}{0.03}$$
$$= 33.2 \text{ L}$$

c. ^3H_2O is an unstable isotope and the substance of choice for measuring TBW. It is a weak beta emitter with a **biologic half-life of 10 days** but a **physical half-life of 12 years**. Other substances used to measure TBW include:
  (1) Antipyrine and N-acetyl-4-amino antipyrine (NAAP), which rarely are used
  (2) Deuterium oxide (^2H_2O), which is a stable isotope

2. ECF volume
  a. Substances used to measure the ECF volume are of two types:
   (1) Saccharides such as inulin, sucrose, raffinose, and mannitol
   (2) Ions such as thiosulfate, thiocyanate, and the radionuclides of sulfate (SO_4^{2-}), chloride (Cl^-), bromide (Br^-), and sodium (Na^+)
  b. Interstitial fluid volume cannot be measured directly, because no substance is distributed exclusively within this compartment. To measure the interstitial fluid volume, the capillary membranes must be permeable to a substance injected intravenously, which becomes distributed throughout the ECF and not exclusively in the interstitium. Therefore, the interstitial fluid volume is determined as the difference between the ECF volume and plasma volume.
  c. Plasma volume is measured using either of two dilution methods.
   (1) The first method employs substances that neither leave the vascular system nor penetrate the erythrocytes. Such substances include:
    (a) Evans blue dye (T-1824)

(b) Radioiodinated human serum albumin (RISA)*

(c) Radioiodinated gamma globulin and fibrinogen, which generally do not leak out of the bloodstream

(2) The second method is based on the fact that the radioisotopes of phosphorus (^{32}P), iron ($^{55,\,59}Fe$), and chromium (^{51}Cr) penetrate and bind to erythrocytes. The tagged cells are injected intravenously, and their volume of distribution is measured. Plasma volume (PV), then, is calculated from the measured erythrocyte volume (BV) and hematocrit (Hct) as

$$PV \text{ (in L)} = BV \text{ (in L)} \times \frac{(100 - Hct)}{100}; \ BV \text{ (in L)} = \frac{PV \text{ (in L)}}{1 - Hct}$$

3. **ICF volume** cannot be measured directly by dilution, because no substance is confined exclusively to this compartment after intravenous injection of a marker substance. The ICF volume is obtained by subtracting the ECF volume from the TBW.[†]

C. **Disturbances of volume and concentration (osmolality) of body fluids** (Figure 4-4; Table 4-2)

1. **Terms and general concepts**
    a. The general clinical terms for volume abnormalities are **dehydration** and **overhydration**. Dehydration is an ambiguous term that does not distinguish between simple water loss and loss of Na^+. Both conditions are associated with a decrease in ECF volume. **The ECF volume is determined by the amount of Na^+ in the body (Na^+ content) and not by the Na^+ concentration in the plasma!**
        (1) A simple water deficit reduces the ECF and ICF proportionately.
        (2) A sodium chloride (NaCl) deficit will always cause a decrease in ECF volume.
    b. The adjectives **isosmotic, hyperosmotic,** and **hyposmotic** are used to describe the changes in volume (i.e., dehydration and overhydration) and refer to the osmolar concentration of ECF in its new steady state. Since the concentration of the body fluids is regulated by antidiuretic hormone, and the volume of the body fluid compartments is regulated by aldosterone, these two hormones attempt to reestablish the normal volumes and concentrations by an increase or decrease in their secretion.
    c. The distribution of body water in the ECF and ICF compartments is determined by **osmosis,** which is the diffusion of water across a membrane. Thus, water moves from a region of lower osmolality to a region of higher osmolality. Note that osmosis can also be defined as the movement of water from a region of lower solute concentration to a region of higher solute concentration.

2. **Dehydration (volume contraction) states**
    a. **Isosmotic dehydration** (see Figure 4-4A)
        (1) **Description**
            (a) Initially, fluid is lost from the plasma and then is repleted from the interstitial space. No major change occurs in the osmolality of the ECF; therefore, no fluid shifts into or out of the ICF compartment.
            (b) Finally, the volume of the ECF is reduced with no change in osmolality.
        (2) **Causes** of isosmotic dehydration include hemorrhage, plasma exudation through burned skin, and gastrointestinal fluid loss (e.g., vomiting, diarrhea).
    b. **Hyperosmotic dehydration** (see Figure 4-4A)
        (1) **Description**
            (a) Initially, fluid is lost from the plasma, which becomes hyperosmotic, causing a fluid shift from the interstitial fluid to the plasma.
            (b) The rise in interstitial fluid osmolality causes fluid to shift from the ICF back to the ECF compartment.
            (c) Finally, the ECF and ICF volumes both are decreased, and the osmolality of both major fluid compartments is increased.

---

*RISA slowly leaks out of the circulation into the interstitial fluid; therefore, the plasma volume is slightly overestimated when RISA is used.

[†]There is a good correlation between the ICF volume and total exchangeable K^+. ICF volume also is related to muscle mass, which decreases with age.

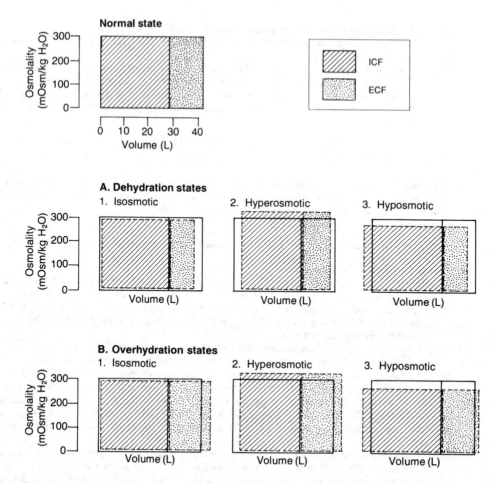

Figure 4-4. A Darrow-Yannet diagram representing the volume (*abscissa*) and osmolality (*ordinate*) of the intracellular and extracellular fluid compartments (*ICF* and *ECF*) in a 70-kg man. The combined plasma and interstitial fluid spaces are represented by the ECF compartment. This diagram is useful to simplify the clinical analysis of fluid balance. The area of each fluid compartment rectangle represents the total milliosmoles of solutes in that compartment. In all diagrams, the normal state is indicated by *solid lines* and the shifts from normality are indicated by *dashed lines*. (Reprinted from Valtin H: *Renal Function: Mechanism Preserving Fluid and Solute Balance in Health*, 2nd edition. Boston, Little, Brown, 1983, p 272.)

Table 4-2. Steady State Changes in Volume and Osmolal Concentration of Body Fluids

Type of Change	Volume (L)		Osmolality (mOsm/kg H$_2$O)	
	ICF	ECF	ICF	ECF
Contraction (dehydration)				
Isosmotic	0	↓	0	0
Hyperosmotic	↓	↓	↑	↑
Hyposmotic	↑	↓	↓	↓
Expansion (overhydration)				
Isosmotic	0	↑	0	0
Hyperosmotic	↓	↑	↑	↑
Hyposmotic	↑	↑	↓	↓

Note—The changes in volume and in osmolality refer to the **ECF compartment** in the new steady state.

(2) Causes of hyperosmotic dehydration include water deficits due to decreased intake, diabetes insipidus (neurogenic or nephrogenic), diabetes mellitus, alcoholism, administration of lithium salts, fever, and excessive evaporation from the skin and breath.

c. **Hyposmotic dehydration** (see Figure 4-4A)
   (1) **Description**
       (a) Initially, loss of NaCl causes a loss of water. This is followed by water retention but a continued loss of NaCl.
       (b) A net loss of NaCl in excess of water loss results in a decreased osmolality of the ECF and a subsequent shift of fluid from the ECF to the ICF compartment.
       (c) Finally, the ECF volume is decreased, the ICF volume is increased, and the osmolality of both major fluid compartments is decreased.
   (2) **Causes** of hyposmotic dehydration include loss of NaCl due to heavy loss of hypotonic sweat and renal loss of NaCl due to adrenal insufficiency [e.g., primary hypoadrenocorticalism (Addison's disease)].

d. **General signs of dehydration** include:
   (1) Decrease in plasma volume, decrease in blood volume and, in turn, a decrease in cardiac output with poor filling of veins below a venous tourniquet
   (2) Decrease in skin turgor ("tenting")
   (3) Sunken and soft eyeballs
   (4) Dry mucous membranes
   (5) Sunken fontanelles (in babies)
   (6) Gray, cool skin
   (7) Increased pulse rate

3. **Overhydration (volume expansion) states**
   a. **Isosmotic overhydration** (see Figure 4-4B)
      (1) **Description.** Isosmotic overhydration is characterized by an overall expansion of the ECF volume with no change in the osmolality of the ICF and ECF compartments.
      (2) **Causes** of isosmotic overhydration are edema and oral or parenteral administration of a large volume of isotonic NaCl.

   b. **Hyperosmotic overhydration** (see Figure 4-4B)
      (1) **Description**
          (a) Oral intake of large amounts of salt or intravenous infusion of a hypertonic saline solution leads to an increase in the plasma osmolality.
          (b) The rise in plasma osmolality causes a shift of water from the interstitium into the plasma, thereby initially increasing plasma volume.
          (c) Concomitantly, the increase in plasma salt concentration causes NaCl to diffuse into the interstitium. The net result is an increase in the osmolality of the ECF without a change in volume.
          (d) The increase in the osmolality of the ECF causes water to flow out of the ICF, which eventually decreases the volume of the ICF and increases the volume of the ECF. The osmolality of both major fluid compartments is increased.
      (2) **Cause.** Oral or parenteral intake of large amounts of hypertonic fluid causes hyperosmotic overhydration.

   c. **Hyposmotic overhydration** (see Figure 4-4B)
      (1) **Description**
          (a) Initially, water enters the plasma, causing a decline in the plasma osmolality, a shift of water into the interstitial space, and a decrease in the interstitial fluid osmolality.
          (b) The decrease in interstitial fluid osmolality causes water to shift from the ECF to the ICF compartment.
          (c) Finally, the ECF and ICF volumes increase and the osmolality of both major fluid compartments decreases.
      (2) **Causes** of hyposmotic overhydration include ingestion of a large volume of water and renal retention of water due to the syndrome of inappropriate **antidiuretic hormone (ADH)** secretion (SIADH).

D. Ionic composition of body fluids (Figure 4-5; Table 4-3)

1. **General considerations**
   a. **Ions** constitute about 95% of the solutes in the body fluids.
   b. The sum of the concentrations (in mEq/L) of the **cations** equals the sum of the concentrations (in mEq/L) of the **anions** in each compartment, making the fluid in each compartment **electrically neutral**.
   c. Physicians rely on the changes in electrolyte concentrations in the ECF compartment, particularly in the plasma, in the diagnosis and treatment of patients with fluid or electrolyte imbalances, or both.

2. **Cations**
   a. The **monovalent cations** Na^+ and K^+ are the predominant cations of the ECF and ICF compartments, respectively. Essentially all of the body K^+ is in the exchangeable pool, whereas only 65%–70% of the body Na^+ is exchangeable. Only the exchangeable solutes are osmotically active.
   b. The **divalent cations** Mg^{2+} and Ca^{2+} exist in body fluids in relatively low concentrations. Almost all of the body Ca^{2+} (in bone) and most of the body Mg^{2+} (in bone and cells) is nonexchangeable. After K^+, Mg^{2+} is the main cation of the ICF. After Na^+, Ca^{2+} is the main cation of the ECF.

3. **Anions.** The chief anions of the body fluids are Cl^-, HCO_3^-, phosphates, organic ions, and polyvalent proteins.
   a. Organic phosphates, proteins, and organic ions are the predominant anions in the ICF.
   b. Cl^- and HCO_3^- are the predominant anions in the ECF.

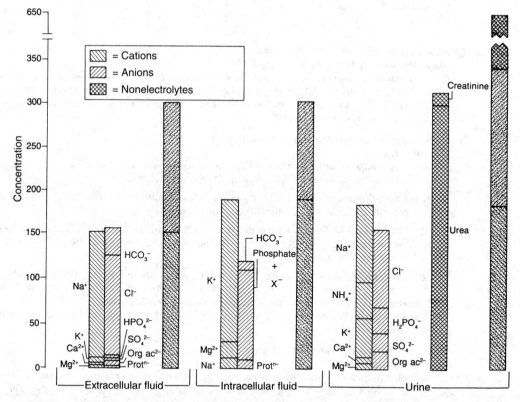

Figure 4-5. Millimolar and osmolal concentrations of body fluids. Total electrolyte concentrations are expressed in mmol/L to emphasize the qualitative heterogeneity of the body fluids. In the ECF, the cation and anion concentrations are almost equal and exert similar osmotic effects. In the ICF, the millimolar cation concentration is greater than the anion concentration, and cations exert a proportionately greater osmotic effect. However, the osmotic activity of the ECF and the ICF are equal. Although the ICF has a higher ion concentration than the ECF, there is no difference in osmolality because many of the cellular ions are bound to protein and are osmotically inactive. X^- = undetermined anions. The *shaded bars* represent concentrations measured in mOsm/kg water; the *unshaded bars* represent concentrations measured in mmol/L.

Table 4-3A. Electrolyte Concentration of Body Fluids

	Plasma			Interstitial Fluid	ICF	Urine
Ion	(mg/L)	(mEq/L)	(mmol/L)	(mEq/L)	(mEq/L)	(mEq/L)
Cations						
Na^+	3266	142	142	145	10	50–130
K^+	156	4	4	4	145	20–170
Ca^{2+}	100	5	2.5	3	. . .	5–12
Mg^{2+}	24	2	1	2	40	2–18
NH_4^+	. . .	. . .	. . .	. . .	. . .	30–50
Total cations	3546	153	149.5	154	195	107–380
Anions						
Cl^-	3692	104	104	117	5	50–130
HCO_3^-	1525	25	25	27	10	. . .
Phosphate*	106	2	1.1	2	100	20–40
Protein	65,000	15	1.1	1	60	. . .
$SO_4^{2-\dagger}$	16	1	0.5	1	20	30–45
Organic acids	175	6	3	6	. . .	20–50‡
Total anions	70,514	153	134.7	154	195	120–265
Total electrolytes	74,060	306	284.2	308	390	227–645

Note—At the pH of body fluids, the proteins have multiple charges per molecule (average valence of –15). Hence, the ICF has more total charges than does the ECF. The total concentration of cations in each compartment must equal the total concentration of anions in each compartment when expressed in mEq/L. (After Frisell WR: Intra- and extracellular electrolytes. In *Acid-Base Chemistry in Medicine.* New York, Macmillan, 1968, p 5.)

*HPO_4^{2-} (ECF, ICF); $H_2PO_4^-$ (urine).
†As free sulfur.
‡Assuming average valence of –2.

Table 4-3B. Nonelectrolyte Concentration of Body Fluids

	Plasma		Urine	
Solute	(mg/L)	(mmol/L)	(mg/L)	(mmol/L)
Urea	150	2.5	12,000–24,000	200–400
Creatinine	11	0.08	678–2260	6–20
Osmolarity	. . .	. . .	. . .	500–800

III. OVERVIEW OF RENAL TUBULAR FUNCTION. The constancy of the body's internal environment is maintained, in large part, by the continuous functioning of its roughly 2 million nephrons. As blood passes through the kidneys, the nephrons **clear the plasma of unwanted substances** (e.g., urea) while simultaneously retaining other, essential substances (i.e., water). Unwanted substances are removed by **glomerular filtration** and **renal tubular secretion** and are passed into the urine. Substances that the body needs are retained by **renal tubular reabsorption** (e.g., Na^+, HCO_3^-) and are returned to the blood by **reabsorptive processes**.

A. **Glomerular filtration** is the initial step in urine formation. The plasma that traverses the glomerular capillaries is filtered by the highly permeable **glomerular membrane,** and the resultant fluid, the **glomerular filtrate,** is passed into Bowman's capsule. **Glomerular filtration rate (GFR)** refers to the volume of glomerular filtrate formed each minute by all of the nephrons in both kidneys.

B. **Renal tubular secretion and reabsorption** (Table 4-4) refer to the direction of transport and not to the differences in the underlying mechanisms of transport (see III D).

1. **Secretion** refers to the transport of solutes from the peritubular capillaries into the tubular lumen.

2. **Reabsorption** denotes the active transport of solutes and the passive movement of water from the tubular lumen into the peritubular capillaries.

Table 4-4A. Tubular Function: Proximal Tubules

| | Reabsorption | | | Secretion* | |
	Active	Passive	Nonreabsorption	Active	Passive
Na^+	Urate	Cl^-	Inulin	$H^{+\dagger}$	NH_3
K^+	Glucose	$HCO_3^{-\ddagger}$	Creatinine	Urate	
Ca^{2+}	Amino acids	HPO_4^{2-}	Sucrose	PAH	
Mg^{2+}	Protein	Water	Mannitol	Penicillin	
HPO_4^{2-}	Acetoacetate	Urea		Sulfonamides	
SO_4^{2-}	β-Hydroxybutyrate			Creatinine	
NO_3^-	Vitamins				

*Most secretion involves active transport into the proximal tubule.
†Coupled with Na^+ secretion.
‡HCO_3^- reabsorption occurs by H^+ secretion, and, in this sense, HCO_3^- is active. However, CO_2, formed by the association of H^+ and HCO_3^- in the tubular lumen, diffuses into the tubular cell passively and forms HCO_3^- within the cell, which is passively reabsorbed. HCO_3^- is largely reabsorbed into the tubular cell as CO_2.

Table 4-4B. Tubular Function: Distal Tubules

| Reabsorption | | | Secretion | |
Active	Passive	Nonreabsorption	Active	Passive
Na^+	Cl^-	Urea*	$K^{+\dagger}$	$K^{+\ddagger}$
Ca^{2+}	HCO_3^-		$H^{+\dagger}$	NH_3
Mg^{2+}	Water			
Urate				

*Urea is passively reabsorbed from the thick segment of the ascending limb of the loop of Henle and from the lower part of the collecting tubule.
†Coupled with Na^+ reabsorption.
‡K^+ secretion in the distal tubule occurs primarily by passive electrical coupling of Na^+-K^+ on the luminal side and secondarily by active Na^+-K^+ exchange on the peritubular side of the cell.

C. Capillary membrane transport in the kidney

1. **Convection** is the bulk transport of a fluid with its dissolved small solutes across a membrane.
   a. Convection is caused by a **hydrostatic pressure difference** between the glomerular capillaries and Bowman's capsule.
   b. Since there are no concentration gradients for inorganic ions, glucose, amino acids, and insulin, the concentrations in plasma and the ultrafiltrate in Bowman's space are the same.
   c. In the absence of concentration gradients there can be no diffusion.
   d. The filtered load* of the solutes is proportional only to the GFR and to the forces that affect filtration, namely hydrostatic and oncotic pressures.

2. **Simple diffusion** is the movement of solutes from areas of higher concentration to areas of lower concentration. The movement of the individual solute particles is random, and the speed of the movement increases with increased temperature. While diffusion ceases when equilibrium is reached, the random motion of the particles continues.
   a. Diffusion is the principal mode of transport in the interstitial space and within cells.
   b. For an ion, the driving force for diffusion consists of two gradients: a concentration gradient and an electric (voltage) gradient.

D. Renal epithelial transport (Figure 4-6)

1. **Basolateral (abluminal) membrane transport systems**
   a. **Primary active transport mechanism of the Na^+-K^+-ATPase pump system.** Na^+ transport through the basolateral membrane is mediated principally by the Na^+-K^+-ATPase

*Filtered load is the amount of solute transported across the glomerular membranes per unit time. Mathematically, it is equal to the GFR times the plasma concentration of that solute.

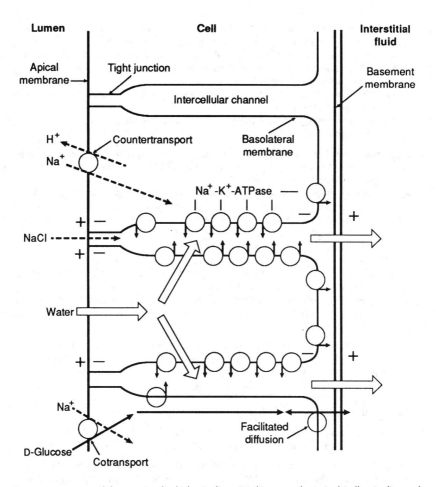

Figure 4-6. Transport systems of the proximal tubule. Sodium (*Na$^+$*) enters the apical (adluminal) membrane and is actively transported across the basolateral (abluminal) membrane by Na$^+$-K$^+$-ATPase. Although Na$^+$ influx in the tissue is favored by the electrochemical potential difference, it is mainly carrier-mediated. Water also follows this transcellular route taken by Na$^+$. Other solutes (*D-glucose*) enter the cell against a concentration gradient by coupling to a Na$^+$ carrier in the apical membrane. This transport of glucose is called **secondary active transport** because the uphill transport of glucose is driven by the energy provided by the transport of Na$^+$ down its electrochemical gradient. (Reprinted from Marsh DJ: *Renal Physiology*. New York, Raven, 1983, p 68.)

pump, which transports Na$^+$ against an electrochemical potential from the cell interior and maintains a low intracellular Na$^+$ concentration.*

    b. Facilitated diffusion. This system translocates glucose from the intracellular fluid across the membrane to the interstitial fluid. Facilitated diffusion involves the **passive transport** of a substance by a protein carrier from a region of higher concentration to a region of lower concentration.

2. Apical (adluminal) membrane transport systems. The apical surface of the renal epithelium of the proximal tubule differs from the other cell membranes in that simple diffusion plays almost no role in the transport of Na$^+$, K$^+$, Cl$^-$, H$^+$, glucose, or amino acids. The movement of Na$^+$ across the luminal membrane (influx) is favored by the electrochemical gradient, but the bulk of Na$^+$ transport is mediated by specific membrane transport proteins and not by simple diffusion. The apical surface does possess two transport mechanisms for Na$^+$.

    a. A **diffusion** mechanism is available for Na$^+$ to pass through the tight junction in association with Cl$^-$. Most of the Cl$^-$ that is reabsorbed does not enter the cell at all.

---

*Active transport of Na$^+$ (efflux) across all cells of the body consumes 30%–50% of the energy derived from metabolism in most cells.

    b. Carrier-mediated mechanism. The transport of Na^+ down its gradient provides energy for active (uphill) transport of other solutes (e.g., glucose, amino acids). Once inside the cell, glucose diffuses to the **basolateral membrane,** where it enters the interstitial fluid by facilitated diffusion. Na^+-dependent transport mechanisms are of two types and are also called **secondary active transport systems.**

      (1) Cotransport (symport) denotes the transport of two substances by a protein carrier in the **same direction.**

      (2) Countertransport (antiport) defines the transport of two substances by a protein carrier in **opposite directions.**

  3. Transepithelial transport. The differences between apical and basolateral membrane properties account for the transepithelial transport of all solutes, which can occur via the transcellular or the paracellular pathway.

    a. The transcellular pathway is used for active transepithelial transport. Approximately two-thirds of the Na^+ transport is active and transcellular. There are three principal mechanisms of Na^+ transport in the proximal tubule. All involve protein carriers that also combine with Na^+ and thus are **secondary active transport processes.** These are:

      (1) Electrogenic Na^+ cotransport (symport) with solutes such as sugars and amino acids

      (2) Electroneutral Na^+ countertransport (antiport) with H^+

      (3) Electroneutral Na^+ cotransport (symport) with phosphate and organic anions such as acetate, citrate, and lactate

    b. The **paracellular (intercellular) pathway** is used for passive transepithelial transport; this pathway consists of lateral intercellular spaces. About one-third of NaCl transport is passive and paracellular.

  4. Osmosis is the process of net diffusion of water from a region of higher water activity (or lower solute concentration) to a region of lower water activity (or higher solute concentration).

    a. Osmosis is the only important mechanism of water movement across cell membranes.

    b. Virtually all cells are permeable to water.

E. Energy requirements for membrane transport

  1. The energy required for the secondary active transport of glucose, amino acids, H^+, and organic ions is derived from the Na^+ concentration gradient.

  2. The energy for the work of primary active transport is provided by the ATP. The Na^+-K^+-ATPase pump maintains the Na^+ gradient across the membrane by maintaining the Na^+ concentration difference across the membrane.

IV. RENAL CLEARANCE (Figure 4-7)

A. Definitions

  1. Clearance is a measure of the volume of plasma completely freed of a given substance per minute by the kidneys. It is the **efficiency** with which the plasma is cleared of a given substance.

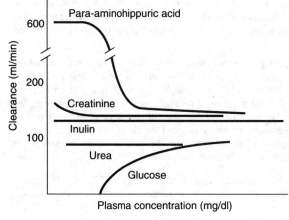

Figure 4-7. Clearances of several substances plotted against their plasma concentrations. Units of plasma concentrations vary over a wide range so that values on the abscissa are relative only. Note that the ordinate is interrupted, and para-aminohippuric acid (PAH) clearance at low plasma concentration is greater than 600 ml/min. (Reprinted from Bauman JW Jr, Chinard FP: Measurement of renal integrity. In *Renal Function: Physiological and Medical Aspects.* St. Louis, CV Mosby, 1975, p 43.)

2. **Renal clearance** of a given substance is the ratio of the renal excretion rate of the substance to its concentration in the blood plasma. This ratio is expressed as

$$C_x = \frac{U_x \times \dot{V}}{P_x} \qquad (2)$$

where C_x = clearance of the substance (ml/min); U_x and P_x = concentration (mg/ml) of the substance in urine and plasma, respectively; $\dot{V}$ = urine minute volume or volume of urine output per minute (ml/min); and $U \times \dot{V}$ = amount of substance excreted per unit time (mg/min). [Equation (2) is an expression of the indicator dilution principle, which is discussed in II B 1.]

  a. Equation (2) can be used to measure the clearance of any substance that is present in the plasma and excreted by the kidneys.

  b. Plasma concentrations rather than whole blood concentrations are used in calculations of renal clearance, because only the plasma is cleared by filtration. An analysis of this relationship demonstrates that the unit for clearance is ml/min, as

$$C_x = \frac{U_x \times \dot{V}}{P_x} = \frac{(mg/ml)\,(ml/min)}{(mg/ml)} = ml/min$$

  c. Because substances can be cleared by filtration, tubular secretion, or a combination of these, the magnitude of the tubular clearance depends on the plasma concentration and the tubular transport capacity (Tm) for reabsorption or secretion. (The phenomenon of Tm is discussed in more detail in VI A 1.)

3. **Amount filtered and excreted.** Clearance is a comparison of the amount of substance removed from plasma per unit time ($C_x \times P_x$) with the amount of substance excreted in the urine per unit time ($U_x \times \dot{V}$), which is demonstrated by a simple rearrangement of equation (2); that is

$$C_x \times P_x = U_x \times \dot{V} \qquad (3)$$

B. **Inulin clearance.** Inulin, a fructo-polysaccharide that does not occur naturally in the body, can be used in a test for determining renal function.

1. **As a measure of GFR**
  a. Inulin clearance (C_{in}) is a measure of GFR because the volume of plasma completely cleared of inulin per unit time equals the volume of plasma filtered per unit time (i.e., C_{in} = GFR). The following characteristics of inulin account for this quality.
    (1) Inulin is only freely filtered. Because no inulin is secreted or reabsorbed, all excreted inulin must come from the plasma.
    (2) Inulin is biologically inert and nontoxic.
    (3) Inulin is not bound to plasma proteins, and it is not metabolized to another substance.
  b. Although inulin is used most commonly, other substances can measure GFR. The most frequently used agents include mannitol, sorbitol, sucrose (intravenous), iothalamate, radioactive cobalt-labeled vitamin B_{12}, ^{51}Cr-labeled edetic acid (EDTA), and radioiodine-labeled Hypaque. (**Endogenous creatinine** clearance is used clinically as an **estimate of GFR,** as discussed in IV D 2 a, because a small amount of creatinine is secreted in humans.)

2. **As an indicator of plasma clearance mechanisms.** A comparison of the clearance of a given substance (C_x) with the clearance of inulin (C_{in}) provides information about the renal mechanisms used to remove the substance from plasma.
  a. When **C_x equals C_{in},** excretion is by **filtration alone.** In terms of equation (3), this means that the mass of the substance excreted in the urine per unit time equals the mass of the substance filtered during the same time, or

$$U_x \times \dot{V} = C_x \times P_x*$$

---

*The mass filtered per unit time (also called the **amount filtered** or **filtered load**) is equal to GFR $\times$ P_x (or $C_{in} \times P_x$) and is expressed in mg/min (see VI A 3).

    b. When C_x **is less than** C_{in}, excretion is by **filtration and reabsorption.** In this case, the mass of the substance excreted in the urine is less than the mass of the substance filtered during that time, or

$$U_x \times \dot{V} < C_x \times P_x$$

    c. When C_x **is greater than** C_{in}, excretion is by **filtration and secretion.** In this case, the mass of the substance excreted in the urine is greater than the mass of the substance filtered during the same time, or

$$U_x \times \dot{V} > C_x \times P_x$$

C. Application of the clearance concept to tubular function (Figure 4-8)

  1. Clearance ratios

    a. Definition. Clearance ratio is the ratio of the clearance of any substance (C_x) to the clearance of inulin (C_{in}).*

    (1) Clearance ratio has the formula

$$\frac{C_x}{C_{in}} = \frac{U_x \times \dot{V}}{P_x} \div \frac{U_{in} \times \dot{V}}{P_{in}}$$

    (2) Clearance ratio is a **double ratio** (i.e., a ratio of two ratios), and it can be determined without measuring urine flow ($\dot{V}$) [see equation (2)]. Because $\dot{V}$ cancels out, the clearance ratio is reduced to

$$\frac{C_x}{C_{in}} = \frac{U_x}{P_x} \div \frac{U_{in}}{P_{in}} \qquad (4)$$

  b. Tubular clearance ratio

    (1) The local tubular clearance of a solute compared to that of inulin at a specific site along the nephron is termed the tubular clearance ratio. This relationship is a reexpression of equation (4)

$$\frac{C_x}{C_{in}} = \frac{TF_x}{P_x} \div \frac{TF_{in}}{P_{in}}$$

where TF = the tubular fluid concentration at the local site of micropuncture. Micropuncture techniques have made it possible to analyze fluids collected with micropipettes (outer diameter = 6–10 μm) from segments of the nephron. This technique permits the quantitative analysis of the tubular transport process along the length of a given single nephron segment.

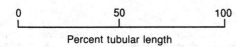

Percent tubular length

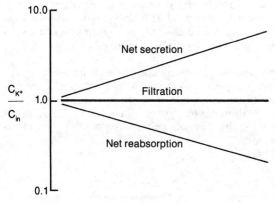

Figure 4-8. Idealized ratios of K^+ clearance (C_{K^+}) to the clearance of inulin (C_{in}), as a function of tubular length. Since inulin is neither secreted nor reabsorbed, an increase in the C_{K^+}/C_{in} ratio represents tubular reabsorption of K^+. In evaluating solute transport, the TF/P concentration ratio of a solute divided by the TF/P concentration ratio of inulin has the same significance as the clearance ratio. In fact, the TF/P clearance ratio is the clearance ratio for the site of sampling. When the TF/P concentration ratio of a solute is divided by the TF/P concentration ratio of inulin, it is called the **tubular clearance ratio.** (Reprinted from Rose BD: Evaluation of micropuncture data. In *Clinical Physiology of Acid-Base and Electrolyte Disorders.* New York, McGraw-Hill, 1977, p 67.)

---

*Creatinine clearance also can be used in the clearance ratio instead of inulin clearance (see IV D).

(2) The tubular clearance ratio is used for two purposes.

    (a) It is used to measure the fraction of a solute remaining in the lumen at a specific tubular segment along the nephron.

    (b) It is used to determine the nature of tubular solute transport (i.e., reabsorption or secretion). Changes in C_x/C_{in} ratios represent net solute reabsorption from, or solute secretion into, the lumen because inulin is neither reabsorbed nor secreted. Thus, to examine solute transport, the effect of water transport is eliminated by comparing the TF/P solute concentration to the TF/P concentration of a substance that remains in the tubule (i.e., inulin).

c. Interpretation of clearance ratios

    (1) Case 1: When $C_x/C_{in} = 1$. A ratio of 1 indicates that the amount of the substance excreted per unit time equals the amount of the substance filtered in the same time. This is simply a restatement of equation (3).

$$U_x \times \dot{V} = C_{in} \times P_x$$

    (a) A clearance ratio of 1 indicates that, **on a net basis,** the substance is neither reabsorbed nor secreted and, therefore, is only filtered.

    (b) Substances with clearance ratios close to 1 include mannitol, sorbitol, thiosulfate, ferricyanide, iothalamate, vitamin B_{12}, and sucrose (intravenous).

    (2) Case 2: When $C_x/C_{in} < 1$. A clearance ratio of less than 1 demonstrates that the amount of the substance excreted per unit time is less than the amount filtered per unit time, or

$$U_x \times \dot{V} < C_{in} \times P_x$$

    (a) A clearance ratio of less than 1 indicates that, **on a net basis,** the substance undergoes reabsorption.

    (b) Substances with clearance ratios of less than 1 include glucose, xylose, and fructose.

    (3) Case 3: When $C_x/C_{in} > 1$. A clearance ratio that is greater than 1 indicates the net excretion of the substance over time is greater than the quantity filtered over time, or

$$U_x \times \dot{V} > C_{in} \times P_x$$

    (a) A clearance ratio of more than 1 represents **net secretion** of the substance into the lumen; therefore, the substance is cleared by filtration and secretion.

    (b) Substances with clearance ratios greater than 1 include para-aminohippuric acid (PAH), phenol red, iodopyracet, certain penicillins, and creatinine (in humans).

2. Tubular fluid concentration ratio (Figure 4-9)

    a. Definition. A comparison of the tubular fluid (TF) solute concentration to the plasma (P) solute concentration is called the **TF/P concentration ratio**. The TF/P concentration ratio is not compared to that of inulin and is, therefore, a single ratio.

    b. Uses of the TF/P concentration ratio

        (1) This ratio **measures the tubular solute concentration along the nephron,** which is a function of solute transport (i.e., reabsorption or secretion) as well as water reabsorption. The following example illustrates this function.

        (a) The TF/P concentration ratio for K^+ (TF/P [K^+]) increases in the collecting duct, which indicates either water reabsorption or K^+ secretion.

        (b) However, the C_{K^+}/C_{in} (i.e., clearance ratio) remains constant, indicating that K^+ is neither secreted nor reabsorbed (see Figure 4-8).

        (c) Therefore, the elevated TF/P [K^+] represents water reabsorption in the collecting duct.

        (2) TF/P solute concentration ratios also can represent **total solute concentration** rather than the concentration of a single solute. When measuring the total solute concentration, the TF/P or urine-to-plasma (U/P) ratio is expressed in terms of osmolality (osmolarity) and can be interpreted as follows.

        (a) Case 1: When $TF_{osm}/P_{osm} = 1$ or $U_{osm}/P_{osm} = 1$, the tubular fluid or urine is **isosmotic** with respect to plasma.

        (b) Case 2: When TF_{osm}/P_{osm} or $U_{osm}/P_{osm} < 1$, the tubular fluid or urine is **hyposmotic** with respect to plasma.

        (c) Case 3: When TF_{osm}/P_{osm} or $U_{osm}/P_{osm} > 1$, the tubular fluid or urine is **hyperosmotic** with respect to plasma.

        (d) Example. States of hydropenia (dehydration) lead to ADH secretion. This, in turn, leads to increased water reabsorption and an elevation of U_{osm}, which has an upper

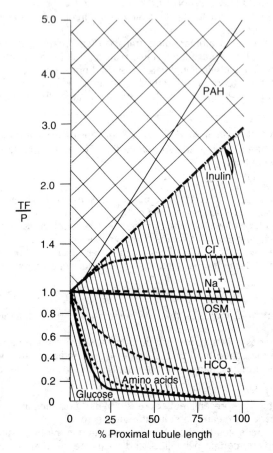

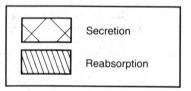

Figure 4-9. Transport of various solutes along the length of the proximal tubule. *TF/P* = tubular fluid-to-plasma concentration ratio for a particular solute. A TF/P for inulin approaches 3 at the end of the proximal tubule, indicating that approximately two-thirds of the filtered fluid is normally reabsorbed in the proximal tubule. Inulin is neither reabsorbed nor secreted. *PAH* = para-aminohippuric acid; *OSM* = osmolarity or osmolality, which denotes the total concentration of all solutes in the tubular fluid. (Reprinted from Brenner B, Coe FL, Rector FC: Transport functions of renal tubules. In *Renal Physiology in Health and Disease*. Philadelphia, WB Saunders, 1987, p 33.)

limit in humans of about 1400 mOsm/kg. With a P_{osm} of about 300 mOsm/kg, this is equivalent to a maximal U_{osm}/P_{osm} of approximately 4.7.

D. Creatinine clearance and creatinine concentration

    1. Normal levels and the effect of aging

        a. Creatinine clearance has a normal range of 80–110 ml/min per 1.73 m² body surface area (estimated average body surface area of 25-year-old humans) and declines with age in healthy individuals. Because creatinine clearance is an index of GFR (see IV D 2 a), normally GFR also declines with age.

        b. Plasma creatinine concentration remains remarkably constant through life, averaging 0.8–1.0 mg/dl in the absence of renal disease.

        c. Because creatinine clearance normally declines with age and plasma creatinine concentration does not, the decline in clearance is attributed to a parallel decrease in the excretion rate of creatinine. (The decline in creatinine clearance is not due to a decreased tubular secretion of creatinine, because the clearance ratio of creatinine to inulin is a quite constant 1.2, about 120% of inulin clearance.) Thus, urinary creatinine excretion exceeds the amount filtered, since approximately 20% of the urinary creatinine is derived from tubular secretion by the secretory pathway in the proximal tubule.

        d. The decline in creatinine excretion with age is due to a primary decline in renal function and a secondary decline in muscle mass.

            (1) The decline in GFR is due to declines in renal plasma flow, cardiac output, and renal tissue mass. (Decreased GFR is the most clinically significant renal functional deficit occurring with age.)

            (2) The decline in creatinine excretion with a decline in muscle mass over time accounts for the relatively constant plasma creatinine concentration among healthy individuals.

2. Estimating GFR

a. Creatinine clearance

(1) Although inulin clearance can be used to measure GFR, in clinical practice it is more common to determine the 24-hour endogenous creatinine clearance as an estimate of GFR.

(2) Creatinine clearance determinations do not require administration of exogenous creatinine, as creatinine is a product of muscle metabolism.

b. Creatinine concentration.
The tendency to rely on serum creatinine concentration rather than creatinine clearance as an estimate of renal function is a serious error, because **plasma creatinine concentration overestimates renal function** (i.e., GFR).

(1) Using serum creatinine concentration as a basis for drug dosages can lead to drug overdose, especially in elderly individuals. Overdose is a particularly serious risk with drugs that are cleared primarily by renal mechanisms (e.g., digoxin and aminoglycoside antibiotics). The dosages for these drugs frequently are based on serum creatinine concentration, although it is incorrect to do so.

(2) To avoid this error, patient age should be considered together with measurements of creatinine clearance and blood levels of the drug. Drug dosages should be adjusted to the 30%–40% decrease in GFR that normally occurs in individuals 30–80 years old.

V. GLOMERULAR FILTRATION AND GFR

A. Factors affecting glomerular filtration

1. **Role of hydrostatic and oncotic pressures.** Fluid movement is proportional to the membrane permeability and to the balance between hydrostatic and osmotic (oncotic) forces across the glomerular membrane. When hydrostatic exceeds oncotic pressure, filtration occurs. Conversely, when oncotic exceeds hydrostatic pressure, reabsorption occurs.

   a. **Effective filtration pressure (EFP)** refers to the **net driving forces** for water and solute transport across the glomerular membrane. EFP is a function of two variables: the **hydrostatic pressure gradient** driving fluid **out** of the glomerular capillary and into Bowman's capsule and the **colloid osmotic pressure gradient** bringing fluid **into** the glomerular capillary. Using these variables, EFP is expressed as

   $$EFP = (P_{cap} - P_{Bow}) - COP_{cap}*$$

   where $(P_{cap} - P_{Bow})$ = the outward forces promoting filtration, and COP_{cap} = the inward forces opposing filtration.

   b. The **filtration coefficient (K_f)** of the glomerular membrane is a function of the capillary surface area and the membrane permeability. K_f is expressed in ml/min/mm Hg.

   c. Using EFP and K_f, then, the GFR is expressed as

   $$GFR = K_f \times EFP$$

   (1) K_f normally equals about 12.5 ml/min/mm Hg.

   (2) EFP normally equals about 10 mm Hg, which can be determined from the normal values for P_{cap} (45 mm Hg), P_{Bow} (10 mm Hg), and COP_{cap} (25 mm Hg) using the equation for EFP.

   (3) The normal GFR for both kidneys is about 125 ml/min per 1.73 m^2 of body surface area.

2. **Role of capillaries.** Although the principle of opposing forces accounts for the magnitude and direction of flow, the main mechanism that alters intracapillary hydrostatic pressure probably is not the resistance along the length of the capillary but the contractile state of the **precapillary sphincters**.

B. Glomerular filtration and systemic filtration: a comparison.
The amount of filtration performed by the renal glomerular capillaries is better appreciated when this filtration is compared to that of the extrarenal (systemic) capillaries.

---

*The colloid osmotic pressure in Bowman's capsule is not expressed in this equation, although it is understood to be zero, because the glomerular filtrate normally contains little protein.

1. **Capillary exchange area and K_f**
   a. The **total glomerular capillary exchange area** is estimated to be 0.5–1.5 m²/100 g of renal tissue.
   b. In the adult human, if the entire systemic capillary bed were patent at a given time, it would afford a total capillary exchange area of 1000 m². In a human at rest, approximately 25%–35% of the systemic capillaries are open at any time, and the **effective pulmonary capillary surface area** is about 60 m².
   c. The K_f of the glomerulus is 50–100 times greater than that of a muscle capillary.

2. **GFR and systemic filtration rate**
   a. About 20 L of fluid are filtered daily from the systemic capillaries. Of this, 16–18 L/day are reabsorbed in the venular ends of the capillaries, and the remaining 2–4 L/day represent lymph flow. (**Diffusional exchange** across the entire systemic capillary bed, however, is about 80,000 L/day!)
   b. Approximately 180 L of fluid are filtered daily from the glomerular capillaries. Therefore, the transtubular flow of fluid out of the glomerular capillaries (GFR = 180 L/day) far exceeds the filtration from systemic capillaries (20 L/day).

C. **Determinants of GFR.** The major determinant of GFR is the hydrostatic pressure within the glomerulus. In addition, the **renal blood flow (RBF)** through the glomeruli has a great effect on GFR; when the rate of RBF increases, so does GFR. The following factors determine the RBF.

1. **Autoregulation of arterial pressure** (Figure 4-10)
   a. Over a wide range of renal arterial pressures (i.e., 90–190 mm Hg), the GFR and RBF remain quite constant. This intrinsic phenomenon observed in the renal capillaries also occurs in the capillaries of muscles and is termed **autoregulation**. Intrarenal autoregulation is virtually absent, however, at mean arterial blood pressures below 70 mm Hg.
      (1) Autoregulation has been observed to persist after renal denervation, in the isolated perfused kidney, in the transplanted kidney, after adrenal demedullation, and even in the absence of erythrocytes.
      (2) Autoregulation of RBF is necessary for the autoregulation of GFR.

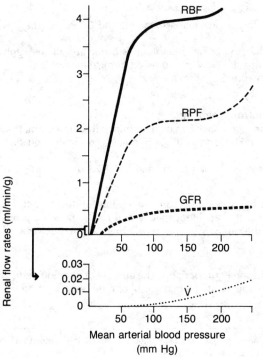

Figure 4-10. Autoregulation of renal blood flow (*RBF*) and glomerular filtration rate (*GFR*) and the effect of arterial blood pressure on RBF, renal plasma flow (*RPF*), GFR, and volume flow of urine (*V̇*). (Reprinted from Harth O: The function of the kidneys. In *Human Physiology*. Edited by Schmidt RF and Thews G. New York, Springer-Verlag, 1983, p 615.)

b. This phenomenon probably is mediated by changes in preglomerular (afferent) arteriolar resistance. The major factors that determine blood flow (F) are a pressure gradient (ΔP) and resistance (R). This relationship is expressed as

$$F = \frac{\Delta P}{R}$$

It becomes clear that to maintain a constant blood flow with a concomitant increase in renal perfusion pressure (ΔP), there must be a commensurate increase in renal vascular resistance.

2. **Sympathetic stimulation** (Table 4-5)
   a. In addition to the autoregulatory response to hypotension (i.e., vasodilation), the reflex increase in sympathetic tone results in both afferent and efferent arteriolar vasoconstriction and an increase in renal vascular resistance together with a decrease in GFR.
   b. Conversely, the autoregulatory response to increases in renal arterial blood pressure (i.e., vasoconstriction) causes the reflex decrease in sympathetic tone, which results in renal arteriolar vasodilation and a decrease in renal vascular resistance together with an increase in GFR.

3. **Arteriolar resistance**
   a. The **afferent arteriole** is the larger diameter arteriole and the major site of autoregulatory resistance. When resistance is altered in the afferent arterioles, GFR and RBF change in the same direction. Therefore, **changes in afferent arteriolar resistance do not affect the filtration fraction** [i.e., the ratio of GFR to renal plasma flow (RPF)].*
      (1) Constriction of the afferent arteriole decreases both RPF and GFR.
         (a) An increase in vascular resistance proximal to the glomeruli leads to a decrease in RPF.
         (b) A decrease in hydrostatic pressure within the glomerular capillaries leads to a decrease in GFR.
      (2) Dilation of the afferent arteriole increases both RPF and GFR.
         (a) A decrease in vascular resistance proximal to the glomeruli leads to an increase in RBF.
         (b) An increase in glomerular capillary perfusion pressure leads to an increase in GFR.
   b. When resistance is altered in the **efferent arterioles,** GFR and RPF change in opposite directions. Therefore, **changes in efferent arteriolar resistance do affect the filtration fraction**.
      (1) Constriction of the efferent arteriole decreases RPF and increases GFR.
         (a) An increase in vascular resistance distal to the glomeruli leads to a decrease in RPF.

Table 4-5. Effect of Changes in Renal Vascular Resistance with a Constant Renal Perfusion on GFR, RPF, and FF

Renal Arteriolar Vascular Resistance		GFR (ml/min)	RPF (ml/min)	FF
Afferent	**Efferent**			
↑	. . .	↓	↓	no change
↓	. . .	↑	↑	no change
. . .	↑	↑	↓	↑
. . .	↓	↓	↑	↑↓

Note—GRF = glomerular filtration rate, RPF = renal plasma flow, and FF = filtration fraction. Note that GFR and RPF exhibit parallel shifts with changes in afferent arteriolar resistance but exhibit divergent shifts with changes in efferent arteriolar resistance. Increases in vascular resistance always lead to a decline in RPF, and decreases in arteriolar resistance always lead to an increase in RPF. In the first two columns, upward arrows denote the effect of vasoconstriction, and downward arrows denote the effect of vasodilation.

*Filtration fraction is a measure of the fraction of the entering plasma volume (RPF) that is removed through the glomeruli as filtrate. Filtration fraction refers to the bulk volume of fluid (not solutes) and normally measures about 20% of RPF.

(b) An increase in the hydrostatic pressure within the glomerular capillaries leads to an increase in GFR.

(c) Following an increase in efferent arteriolar resistance, the filtration fraction increases.

(2) Dilation of the efferent arteriole increases RPF and decreases GFR.

(a) A decrease in vascular resistance distal to the glomeruli leads to an increase in RPF.

(b) A decrease in the hydrostatic pressure within the glomerular capillaries leads to a decrease in GFR.

(c) Following a decrease in efferent arteriolar resistance, the filtration fraction decreases.

VI. RENAL TUBULAR TRANSPORT: REABSORPTION AND SECRETION

A. Active transport processes

1. Renal tubular transport maximum (Tm) refers to the maximal amount of a given solute that can be transported (reabsorbed or secreted) per minute by the renal tubules.

a. The highest attainable rate of reabsorption is called the **maximum tubular reabsorptive capacity** and is designated **Tm** or **Tr**. Substances that are reabsorbed by an active carrier-mediated process and that have a Tm include phosphate ion (HPO_4^{2-}), SO_4^{2-}, glucose (and other monosaccharides), many amino acids, uric acid, and albumin, acetoacetate, β-hydroxybutyrate and α-ketoglutarate.

b. The highest attainable rate of secretion is called the **maximum tubular secretory capacity** and is designated **Tm** or **Ts**. Substances that are secreted by the kidneys and that have a Tm include penicillin, certain diuretics, salicylate, PAH, and thiamine (vitamin B_1).

c. Uric acid is the only organic substance to be both reabsorbed and secreted by the kidney. K^+ is the only inorganic cation that is both reabsorbed and secreted by the kidney.

d. Some solutes have no definite upper limit for unidirectional transport and, hence, have no Tm.

(1) The reabsorption of Na^+ along the nephron has no Tm.

(2) The secretion of K^+ by the distal tubules has no Tm.

2. Threshold concentration, which refers to the plasma concentration at which a solute begins to appear in the urine, is characteristic of that substance.

3. Filtered load and excretion rate

a. Definitions. Reabsorption and secretion are not directly measured variables but are derived from the measurements of the amount of solute filtered (filtered load) and the amount of solute excreted (excretion rate).

(1) The **filtered load** is the amount of a substance entering the tubule by filtration per unit time and is mathematically equal to the product of GFR (or C_{in}) and the plasma concentration of the substance (P_x), as

$$\text{filtered load} = \text{GFR} \times P_x; (\text{ml/min}) (\text{mg/ml}) = \text{mg/min}$$
$$= C_{in} \times P_x; (\text{ml/min}) (\text{mg/ml}) = \text{mg/min}$$

(2) The **excretion rate** is the amount of a substance that appears in the urine per unit time and is mathematically equal to the product of urine flow rate ($\dot{V}$) and the urine concentration of the substance (U_x), as

$$\text{excretion rate} = U_x \times \dot{V}; (\text{mg/ml}) (\text{ml/min}) = \text{mg/min}$$

b. Secretion. If the excretion rate exceeds the filtered load [or if the clearance of the substance (C_x) is greater than C_{in}], net tubular secretion of that substance has occurred. Thus, secretion is expressed as

$$U_x \times \dot{V} > C_{in} \times P_x \text{ or } C_x > C_{in}$$

c. Reabsorption. If the filtered load exceeds the excretion rate (or if C_x is less than C_{in}), net reabsorption of that substance has occurred. Thus, reabsorption is expressed as

$$C_{in} \times P_x > U_x \times \dot{V} \text{ or } C_x < C_{in}$$

d. Calculation of Tm (Tr or Ts). Tm is the difference between the filtered load ($C_{in} \times P_x$) and the excretion rate ($U_x \times \dot{V}$). Tr and Ts, then, are expressed as

$$\text{Tr} = C_{in} \times P_x - U_x \times \dot{V} \text{ (in mg/min)}$$
$$\text{Ts} = U_x \times \dot{V} - C_{in} \times P_x \text{ (in mg/min)}$$

B. **Graphic representation of renal transport processes: renal titration curves** (Figure 4-11)

    1. **Construction of renal titration curves**

        a. A titration curve is constructed by plotting the following pairs of variables:*

            (1) The filtered load ($C_{in} \times P_x$) against the plasma concentration (P_x)

            (2) The excretion rate ($U_x \times V$) against P_x

            (3) The difference between the filtered load and the excretion rate (i.e., Tr or Ts) against P_x

        b. The plotted renal titration curve, therefore, can be used to determine the plasma concentration (P_x) at which the renal tubular membrane carriers are fully saturated (Tm).

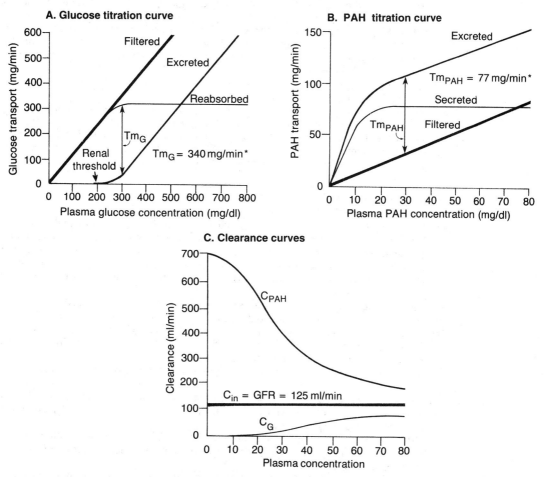

Figure 4-11. Relationships of renal titration curves and clearance curves to plasma concentration. The splay (curved portions) observed in the glucose reabsorption and excretion curves (*graph A*) and the splay in the para-aminohippuric acid (PAH) secretion and excretion curves (*graph B*) are due to the kinetics of tubular transport and the heterogeneity of the nephron population in terms of number, length, and functional variations in transport. *Graph C* demonstrates that the clearance of glucose (C_G) increases and approaches the clearance of inulin (C_{in}) asymptotically as the plasma glucose concentration increases beyond the tubular transport maximum for glucose (Tm_G). Conversely, *graph C* shows that the clearance of PAH (C_{PAH}) declines and approaches the C_{in} asymptotically as the plasma PAH concentration increases above the tubular transport maximum for PAH (Tm_{PAH}). In *graph C*, the plasma PAH concentration is in mg/dl and the plasma glucose concentration is plotted as one-tenth of the actual concentration (i.e., mg/dl $\times$ 10^{-1}). * = per 1.73 m² body surface area. (Adapted from Selkurt EE: Renal function. In *Physiology*, 5th edition. Edited by Selkurt EE. Boston, Little, Brown, 1984, pp 424 and 429.)

---

*The dependent variables (i.e., filtered load, excretion rate, and Tr or Ts) are plotted on the ordinate as a function of the independent variable (i.e., P_x).

2. Glucose transport: reabsorption
   a. Characteristics of glucose transport
      (1) Glucose is an essential nutrient that is actively reabsorbed into the proximal tubule by a Tm-limited process. The Tm for glucose (Tm_G) is about 340 mg/min (or about 2 mmol/min) per 1.73 m^2 body surface area.
      (2) Because the Tm_G is nearly constant and depends on the number of functional nephrons, it is used clinically to estimate the number of functional nephrons, or the tubular reabsorptive capacity (Tr).
   b. Glucose titration curve. Figure 4-11A illustrates that glucose transport and excretion processes are functions of the plasma glucose concentration (P_G).
      (1) Increasing the P_G results in a progressive linear increase in the filtered load ($C_{in} \times P_G$).
      (2) At a low P_G, the reabsorption of glucose is complete; hence, the clearance of glucose (C_G) is zero (see Figure 4-11C).
      (3) When P_G in humans reaches the renal threshold concentration of 180–200 mg/dl (10–11 mmol/L), glucose appears in the urine (glycosuria). (Note that the threshold concentration is not synonymous with the P_G that completely saturates the transport mechanism, which is about 300 mg/dl.)
      (4) As the Tm_G is approached, the urinary excretion rate increases linearly with increasing P_G (first-order reaction kinetics).
      (5) When the Tm_G is reached, the quantity of glucose reabsorbed per minute remains constant and is independent of P_G (zero-order reaction kinetics).
      (6) When the Tm_G is exceeded, the C_G becomes increasingly a function of glomerular filtration; therefore, the C_G approaches C_{in}, and the amount reabsorbed becomes a smaller fraction of the total amount excreted.

3. PAH transport: secretion
   a. Characteristics of PAH transport
      (1) PAH is a weak organic acid that is actively secreted into the proximal tubule by a Tm-limited process. PAH is a foreign substance that is not stored or metabolized and is excreted virtually unchanged in the urine. By filtration and secretion, PAH is almost entirely cleared from the plasma in a single pass through the kidney, if the Tm for PAH (Tm_{PAH}) has not been reached. (The Tm_{PAH} in humans is about 80 mg/min per 1.73 m^2 body surface area.)
      (2) Because about 10% of the plasma PAH is bound to plasma proteins, PAH is not entirely freely filterable, and the concentration of PAH in the plasma (P_{PAH}) is greater than that in the glomerular filtrate. However, PAH does not need to be dissolved in plasma to be transported (secreted) by the carrier into the tubular fluid, and protein binding does not diminish the effectiveness of tubular excretion.
      (3) Since the Tm_{PAH} is nearly constant, it is used clinically to estimate tubular secretory capacity (Ts).
   b. PAH titration curve. Figure 4-11B illustrates that the filtration and secretion of PAH are functions of the P_{PAH}. The amount of PAH excreted per minute ($U_{PAH} \times \dot{V}$) exceeds the amount filtered ($C_{in} \times P_{PAH}$) at any P_{PAH}.
      (1) When the P_{PAH} is low, virtually all of the PAH that is not filtered is secreted, and PAH is almost completely cleared from the plasma by the combined processes of glomerular filtration and tubular secretion.
      (2) When P_{PAH} is above 20 mg/dl, the transepithelial secretory mechanism becomes saturated, and the Tm_{PAH} is reached.
      (3) When the Tm_{PAH} is reached, the quantity of PAH secreted per minute remains constant and is independent of P_{PAH}.
      (4) When the Tm_{PAH} is exceeded, the C_{PAH} becomes progressively more a function of glomerular filtration; hence, the C_{PAH} approaches C_{in} (see Figure 4-11C), and the constant amount of PAH secreted becomes a smaller fraction of the total amount excreted.
   c. PAH clearance and RPF
      (1) Fick principle. According to the Fick principle, blood flow through an organ can be determined with the equation

$$BF = \frac{R}{A_x - V_x} \qquad (5)$$

where BF = the blood flow (in ml/min), R = the rate of removal (or addition) of a substance (in mg/min) from (or to) the blood as it flows through an organ, A_x = the concentration of that substance (in mg/ml) in the blood entering the organ, and V_x = the concentration of that substance (in mg/ml) in the blood leaving the organ.

(2) Since C_{PAH} **can be used to measure RPF,** equation (5) can be modified to

$$RPF = \frac{U_{PAH} \times \dot{V}}{A_{PAH} - V_{PAH}} \qquad (6)$$

where RPF = renal plasma flow (in ml/min), $U_{PAH} \dot{V}$ = the rate of PAH excretion (in mg/min), and A_{PAH} and V_{PAH} = concentrations of PAH (in mg/ml) in the renal artery and the renal vein, respectively.

   (a) At low concentrations of PAH in arterial plasma, the renal clearance is nearly complete; therefore, as a first approximation, the PAH concentration in renal venous plasma may be taken to be zero.

   (i) About 10%–15% of the total RPF perfuses nonexcretory (nontubular) portions of the kidney, such as the renal capsule, perirenal fat, the renal medulla, and the renal pelvis; therefore, this plasma cannot be completely cleared of PAH by filtration and secretion.

   (ii) Because about 10% of the PAH remains in the renal venous plasma, the RPF calculated from C_{PAH} underestimates the actual flow by about 10%. Accordingly, the C_{PAH} actually measures the **effective renal plasma flow (ERPF),** and

$$ERPF = \frac{U_{PAH} \times \dot{V}}{A_{PAH}} = C_{PAH} \qquad (7)$$

   (b) Since PAH is not metabolized or excreted by any organ other than the kidney, a sample from any peripheral vein can be used to measure arterial plasma PAH concentration, and the equation is written as

$$ERPF = \frac{U_{PAH} \dot{V}}{P_{PAH}} = C_{PAH}$$

(3) **Effective renal blood flow (ERBF)** is calculated from the relationship between plasma volume (PV), hematocrit (Hct), and blood volume (BV), as*

$$BV = \frac{PV \times 100}{(100 - Hct)}$$

and, therefore,

$$ERBF = \frac{ERPF \times 100}{(100 - Hct)}$$

(4) **True renal plasma flow (TRPF)** can be determined if the extraction ratio (E) is known, where

$$E = \frac{A_{PAH} - V_{PAH}}{A_{PAH}}, \text{ and}$$

$$TRPF = \frac{ERPF}{E} \; ; \; TRBF = \frac{TRPF \times 100}{(100 - Hct)}$$

C. Renal transport of common solutes and water (see Table 4-4)

   1. General considerations (see Figure 4-9)
   a. The proximal tubules reabsorb 70%–85% of the filtered Na^+, Cl^-, HCO_3^-, and water and virtually all of the filtered K^+, HPO_4^{2-}, and amino acids. In addition, glucose is reabsorbed almost completely by the proximal tubules and begins to appear in the urine when the renal threshold is exceeded.[†]
   b. Reabsorption of water is passive, and reabsorption of solutes can be passive or active; solute reabsorption generates an osmotic gradient, which causes the passive reabsorption of water (osmosis).

---

*In humans, the C_{PAH} is 600–700 ml/min and, when corrected to ERBF (assuming a hematocrit of 45%), is approximately 1100–1200 ml/min, which is about 20% of the resting cardiac output.

†i.e., approximately 200 mg glucose/dl arterial plasma, or 11 mmol/L.

2. Na⁺ reabsorption

a. **Proximal tubules.** Transtubular reabsorption of Na⁺ across the proximal tubules occurs in two steps (see Figure 4-6).

(1) **Step 1: Through the luminal membrane.** Although Na⁺ is transported through the luminal (adluminal) membrane passively down an electrochemical gradient, the transport is mediated by specific membrane proteins and is not by simple diffusion. This first step involves two processes.

(a) First, Na⁺ undergoes **passive cotransport** with the **active cotransport** of glucose or amino acids. Thus, the energy for the uphill transport of glucose from lumen to cell is derived from the simultaneous downhill movement (influx) of Na⁺ along a concentration gradient, which is maintained by the primary active Na⁺ pump.

(b) Next, **Na⁺ reabsorption** occurs, **coupled with H⁺ secretion** across the luminal membrane (**cation exchange**) to maintain electroneutrality. (Na⁺ transport in the proximal segment is not associated with K⁺ secretion.) Because tubular H⁺ is buffered by HCO_3^-, Na⁺ influx must be electrically balanced by an anion.* This is accomplished by passive inward diffusion of Cl^- down its concentration gradient or by regeneration of HCO_3^- from cellular CO_2 and water in the presence of carbonic anhydrase.

(2) **Step 2: Through the basolateral membrane**

(a) Na⁺ is **actively** transported through the basolateral membrane via the Na⁺-K⁺ pump. Na⁺ reabsorption is equivalent to Cl^- reabsorption + (H⁺ + K⁺) secretion. Approximately 67% of the proximal reabsorption of Na⁺ is based on the active transport process.

(b) The remaining percentage of Na⁺ reabsorption across the basolateral membrane of the proximal tubule occurs **passively** by solvent drag.

(i) The efflux of Na⁺ into the intercellular spaces creates an electrical gradient for the efflux of Cl^- and HCO_3^-, which, in turn, generates an osmotic gradient for water transport into the interspaces. About 20% of the Na⁺ movement is loosely coupled to the active secretion of H⁺, resulting in the reabsorption of HCO_3^-.

(ii) The osmotic movement of water generates a small increment in hydrostatic pressure, and some bulk flow of water containing Na⁺, Cl^-, and HCO_3^- proceeds into the peritubular capillaries.

(c) As a result of these active and passive transport processes, Na⁺ reabsorption together with anions along the proximal tubule allows for the isosmotic reabsorption of fluid.

b. **Distal and collecting tubules**

(1) **Na⁺ transport via cation exchange.** At the luminal membrane of the distal and collecting tubules, Na⁺ is reabsorbed by two electrically linked cation-exchange processes.†

(a) **Na⁺-H⁺ exchange** involves Na⁺ reabsorption and H⁺ secretion.

(b) **Na⁺-K⁺ exchange** involves Na⁺ reabsorption and K⁺ secretion. The quantity of Na⁺ reabsorbed distally is much greater than the amount of K⁺ secreted.

(c) These cation-exchange processes are competitive (i.e., K⁺ competes with H⁺ for Na⁺) and, therefore, are not one-to-one. Both cation-exchange processes, however, are enhanced by aldosterone (see IX B).

(2) **Na⁺ transport via Cl^- reabsorption.** Na⁺ also is transported by the distal and collecting tubules by reabsorption largely in association with Cl^- reabsorption.

(a) When the availability of either K⁺ or H⁺ is reduced, the exchange of the more available ion for Na⁺ is increased.

(i) **Example 1.** In hypokalemic alkalosis induced by chronic vomiting or hyperaldosteronemia, the exchange of H⁺ for Na⁺ is greater than the exchange of K⁺ for Na⁺, the urine becomes more **acidic,** and the plasma becomes more **alkaline**.

(ii) **Example 2.** In hypoaldosteronemia, the reduced secretion of K⁺ and H⁺ (due to the decreased rate of distal Na⁺ reabsorption) causes hyperkalemia and acidosis.

(b) The Na⁺-H⁺ exchange mechanism is not mandatory for H⁺ secretion.

c. **Nephron.** Na⁺ reabsorption along the nephron occurs in the following proportions:

(1) Proximal tubule (75%)

---

*The transcellular transport of Na⁺ is passive.

†Only a small amount of the filtered Na⁺ that is reabsorbed in the distal and collecting tubules is by cation exchange with H⁺ and K⁺.

(2) Ascending limb of the loop of Henle (22%)*
(3) Distal tubule (4%–5%)
(4) Collecting tubule (2%–3%)

3. K⁺ transport

a. General considerations

(1) K^+ is the only plasma electrolyte that is both reabsorbed and secreted into the renal tubules. Virtually all of the K^+ is actively reabsorbed by the proximal tubule, whereas K^+ secretion is a passive process and is largely a function of the distal tubule.

(2) The net secretion of K^+ by the nephron can cause the excretion of more K^+ than was filtered, as in hyperaldosteronemia and in metabolic alkalosis. (Elevated K^+ excretion also can occur, in part, as a result of decreased K^+ reabsorption, as observed in the osmotic diuresis seen with uncontrolled diabetes mellitus.)

(3) The distal tubule has the capacity for both net secretion and net reabsorption of K^+, depending on the plasma $[K^+]$ and the acid-base state of the body.

    (a) Metabolic acidosis usually is associated with cellular K^+ efflux, a decreased K^+ secretion, and hyperkalemia.

    (b) Metabolic alkalosis usually is associated with cellular K^+ influx, an increased K^+ secretion, and hypokalemia.

b. K⁺ reabsorption.
K^+ is completely filterable at the glomerulus, and tubular reabsorption occurs by active transport in the proximal tubule, ascending limb of the loop of Henle, distal tubule, and collecting duct.

c. K⁺ secretion
involves an active and a passive process and is influenced by aldosterone.

(1) The critical step is the active transport of K^+ from the interstitial fluid across the basolateral membrane into the distal tubular cell. This active transport step is associated with active Na^+ extrusion via the Na^+-K^+-ATPase system.

(2) The elevated intracellular $[K^+]$ achieved by the basolateral pump favors the net K^+ diffusion at the luminal membrane down a concentration gradient into the tubular lumen. The transcellular transport of K^+ is passive.

(3) Aldosterone increases the activity of the basolateral Na^+-K^+-ATPase pump, which increases K^+ entry into the distal cells and the $[K^+]$ gradient from cell to lumen. Aldosterone also increases the permeability of the luminal membrane to K^+, further increasing K^+ diffusion (secretion).

4. H⁺ secretion and HCO₃⁻ reabsorption
(Table 4-6; see Figures 5-3, 5-4). H^+ secretion is the process by which the filtered HCO_3^- is reabsorbed and the tubular fluid becomes acidified. The process of H^+ secretion and HCO_3^- reabsorption occurs throughout the nephron, except in the descending limb of the loop of Henle.

a. H⁺ secretion
into the tubular lumen occurs by active transport and is coupled to Na^+ reabsorption; therefore, for each H^+ secreted, one Na^+ and one HCO_3^- are reabsorbed. The renal epithelium secretes about 4300 mEq (mmol) of H^+ daily.

(1) Approximately 85% of the total H^+ secretion occurs in the proximal tubules.

(2) Approximately 10% of the total H^+ secretion occurs in the distal tubules, and about 5% occurs in the collecting ducts.

Table 4-6. Renal Regulation of H^+ and HCO_3^-

Ion	Filtered mEq/day	Reabsorbed mEq/day	Secreted mEq/day	Excreted mEq/day
HCO_3^-	4320	4318	None	2
HPO_4^{2-}*	144	124	None	20
NH_4^+†	None	...	...	40
H^+‡	<0.1	4315	4375	58

*Main constituent of titratable acid is $H_2PO_4^-$.
†Secreted mainly as NH_3 and NH_4^+.
‡Sum of titratable acid (HPO_4^{2-}) and NH_4^+ minus HCO_3^-.

---

*Na^+ is passively reabsorbed in the thin and thick segments of the ascending limb of the loop of Henle.

b. HCO$_3^-$ reabsorption
(1) Most of the H$^+$ secreted into the tubular fluid reacts with HCO$_3^-$ to form carbonic acid (H$_2$CO$_3$), which is dehydrated (via carbonic anhydrase) to form CO$_2$ and water. The CO$_2$ diffuses back into the proximal tubular cells, where it is rehydrated to H$_2$CO$_3$, which dissociates into HCO$_3^-$ and H$^+$.
(2) The buffering of secreted H$^+$ by filtered HCO$_3^-$ is not a mechanism for H$^+$ excretion. The CO$_2$ formed in the lumen from secreted H$^+$ returns to the tubular cell to form another H$^+$, and **no net H$^+$ secretion occurs.***
c. HCO$_3^-$ secretion by cortical collecting tubules has recently been shown.
d. Nonbicarbonate buffering of H$^+$. The H$^+$ secreted in excess of those required for HCO$_3^-$ reabsorption are buffered in the tubular fluid either by nonbicarbonate buffer anions (chiefly, HPO$_4^{2-}$) or by NH$_3$ to form NH$_4^+$.
(1) About 20 mEq (mmol) of H$^+$ per day are buffered by HPO$_4^{2-}$ (and organic ions) and excreted as **titratable acid**. The nonbicarbonate buffering of H$^+$ occurs mainly in the distal and collecting tubules.
(2) About 40 mEq (mmol) of H$^+$ per day are buffered by NH$_3$ and excreted as NH$_4^+$.
(3) The total amount of H$^+$ excreted daily by an individual with a normal diet equals the sum of titratable acid and NH$_4^+$ excreted, or about 60 mEq (mmol) of H$^+$ per day. Thus, only a minute concentration of free H$^+$ normally exists in the final urine despite the 4300 mEq (mmol) of H$^+$ secreted daily!

5. Cl$^-$ transport
a. General considerations
(1) Cl$^-$ reabsorption occurs at a rate of 20,000 mEq/day and is inversely related to HCO$_3^-$ reabsorption; that is, when HCO$_3^-$ reabsorption increases, Cl$^-$ reabsorption decreases and vice versa. Therefore, the plasma [Cl$^-$] varies inversely with the HCO$_3^-$ reabsorption rate.
(2) Cl$^-$ is the chief anion to accompany Na$^+$ through the renal tubular epithelium. Since Na$^+$ reabsorption is under control of aldosterone, the plasma [Cl$^-$] is secondarily influenced by aldosterone.
(3) Cl$^-$ is excreted with NH$_4^+$ to eliminate H$^+$ in exchange for Na$^+$ reabsorption.
b. Passive transport. When Na$^+$ is actively reabsorbed across the luminal membrane of the proximal tubule, Cl$^-$ is passively reabsorbed along an electrochemical gradient.
(1) Passive proximal diffusion (reabsorption) of Cl$^-$ is aided by the increased [Cl$^-$] within the luminal fluid of the late proximal tubule.
(2) This increased [Cl$^-$] is due to the reabsorption of NaHCO$_3$ and, therefore, water into the early proximal tubule.
c. Active transport. Cl$^-$ is actively reabsorbed in the thick segment of the ascending limb of the loop of Henle.
(1) Cl$^-$ reabsorption induces the passive reabsorption of Na$^+$ in the thick segment.
(2) The reabsorption of NaCl in the thick and thin segments together with the water impermeability of the ascending limb lead to the dilution of the tubular fluid.

6. Water reabsorption. Water movement across membranes is determined by hydrostatic and osmotic pressure gradients. Water is reabsorbed passively by diffusing along an osmotic gradient, which primarily is established by the reabsorption of Na$^+$ and Cl$^-$.
a. Proximal tubules
(1) In humans, about 75%–80% of the reabsorption of filtered water occurs in the proximal tubule.
(2) The proximal reabsorption of water is invariant and involves no change in osmolality.
b. Loop of Henle
(1) Unlike the proximal tubule, the loop of Henle reabsorbs considerably more solute than water. Only about 5% of the reabsorption of filtered water occurs here.
(2) The tubular fluid entering the descending limb of the loop of Henle always is **isosmotic,** regardless of the hydration state.
c. Distal and collecting tubules

---

*The reabsorption of HCO$_3^-$ across the luminal membrane does not occur in the conventional way (i.e., via an active HCO$_3^-$ pump); instead, HCO$_3^-$ is transported across the luminal membrane mainly as CO$_2$.

(1) The distal and collecting tubular reabsorption of water occurs only in the presence of ADH.*

(2) The tubular fluid entering the distal tubule always is **hyposmotic** to plasma, regardless of the hydration state. In the terminal distal tubule and the collecting ducts, the osmolality in the tubular fluid changes according to the water permeability of the tubule.

VII. RENAL CONCENTRATION AND DILUTION OF URINE

A. Functional considerations

1. Purpose. The kidney can alter the composition of the urine in response to the body's daily needs, thereby maintaining the osmolality of body fluids. When it is necessary to conserve body water, the kidney excretes urine with a high solute concentration. When it is necessary to rid the body of excess water, the kidney excretes urine with a dilute solute concentration.

2. Role of ADH. The principal regulator of urine composition is the hormone ADH. In the absence of ADH, the kidney excretes a large volume of dilute urine; when ADH is present in high concentration, the kidney excretes a small volume of concentrated urine.

3. Components of the concentrating and diluting system. The formation of urine that is dilute (hyposmotic to plasma) or concentrated (hyperosmotic to plasma) is achieved by the countercurrent system of the nephron (Figure 4-12.)† This system consists of:
  a. Descending limb of the loop of Henle (DLH)
  b. Thin and thick segments of the ascending limb of the loop of Henle (ALH)
  c. Medullary interstitium
  d. Distal convoluted tubule
  e. Collecting duct
  f. Vasa recta, which are the **vascular elements** of the juxtamedullary nephrons

4. Mechanisms of dilution and concentration. The kidney forms a dilute urine in the absence of ADH. In the presence of ADH, the kidney forms a concentrated urine via the functioning of the **countercurrent multipliers** (i.e., the loop of Henle and collecting duct) and the **countercurrent exchangers** (i.e., the vasa recta). Regardless of ADH, however, the fluid osmolality in the loop of Henle, vasa recta, and medullary interstitium always increases progressively from the corticomedullary junction to the papillary tip (see Figure 4-12D).
  a. The fundamental processes involved in the excretion of a dilute or concentrated urine include:
    (1) Variable permeability of the nephron to the passive back-diffusion (reabsorption) of water along an osmotic gradient and of urea along its concentration gradient
    (2) Passive reabsorption of NaCl by the thin segment of the ALH
    (3) Active reabsorption of Cl$^-$ and passive reabsorption of Na$^+$ by the thick segment of the ALH
  b. The formation of a hyperosmotic urine involves the following steps.
    (1) The medullary interstitium becomes hyperosmotic by the reabsorption of NaCl and urea.
    (2) The urine entering the medullary collecting ducts equilibrates osmotically with the hyperosmotic interstitium, resulting in the excretion of a small volume of concentrated urine in the presence of ADH.
  c. The final osmolality of the urine is determined by the permeability of the collecting ducts to water.
  d. The vascular loop (vasa recta) prevents the dissipation of the osmotic (hypertonic) "layering" in the interstitium (see VII C).

B. Countercurrent multipliers (see Figure 4-12)

1. General considerations
  a. The countercurrent multiplier system is analogous to a three-limb model consisting of the two limbs of the loop of Henle, which are connected by a hairpin turn, and the collecting

---

*In the presence of a maximal ADH effect, 99% of the water is reabsorbed; in the absence of ADH, 88% of the water is reabsorbed.

†Only the juxtamedullary nephrons, with their long loops of Henle, contribute to the medullary hyperosmolality.

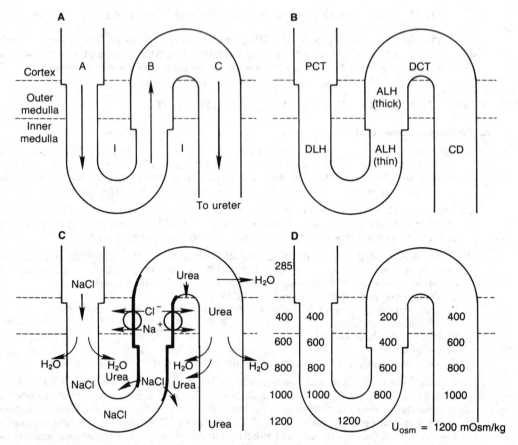

Figure 4-12. The three-limb model of the countercurrent multiplier system of the juxtamedullary nephron. *Model A* represents the directions of tubular flow (countercurrent); *A* = concentrating segment; *B* = diluting segment; *C* = collecting duct; and *I* = interstitium. *Model B* represents the major components of the nephron; *PCT* = proximal convoluted tubule; *DLH* = descending limb of the loop of Henle; *ALH* = ascending limb of the loop of Henle (thick and thin segments); *DCT* = distal convoluted tubule; and *CD* = collecting duct. *Model C* represents solute and solvent transfer; the *heavy line* indicates water impermeability. *Model D* represents the vertical (longitudinal) and horizontal (transverse) osmotic gradients in the presence of antidiuretic hormone. (After Jamison R, Maffly RH: The urinary concentrating mechanism. *N Engl J Med* 295:1059–1067, 1976.)

duct. The fluid flow through the three limbs is **countercurrent** (i.e., in alternating opposite directions; see Figure 4-12A).

  b. The countercurrent multiplier process has two important consequences (Figure 4-13; see Figure 4-12D).

   (1) At equilibrium in the presence of ADH, a maximal vertical gradient of 900 mOsm/kg is established between the corticomedullary junction (300 mOsm/kg) and the renal papillae (1200 mOsm/kg). This explains the origin of the term "countercurrent multiplier."

   (2) At any given level in the renal medulla, the osmolality is almost the same in all fluids except that in the ALH. The fluid in the ALH is less concentrated than that in either the DLH or the interstitium and becomes hyposmotic to plasma (see Figure 4-12D).

  2. Loop of Henle

   a. The **DLH** is the concentrating segment of the nephron (see Figures 4-12C, 4-12D, 4-13). The following characteristics of the DLH account for this.

   (1) The DLH is highly permeable to water. Solute-free water leaves the DLH, causing the fluid in the DLH to become concentrated to a degree that is consistently higher than that of the fluid in the ALH.

   (2) The DLH has a low permeability to NaCl and urea. The predominant solute is NaCl, with relatively small amounts of urea existing in the DLH.

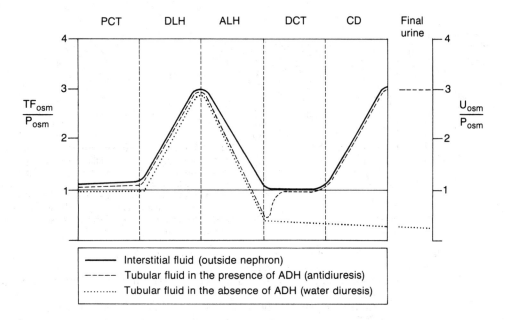

Figure 4-13. TF/P osmolality ratios as a function of length along the nephron. A ratio of 1 indicates isosmolality, a ratio of more than 1 indicates hyperosmolality, and a ratio of less than 1 indicates hyposmolality. Solute-free water reabsorption occurs in the early distal tubule, in the descending limb of the loop of Henle, and in the collecting duct. *PCT* = proximal convoluted tubule; *DLH* = descending limb of the loop of Henle; *ALH* = ascending limb of the loop of Henle; *DCT* = distal convoluted tubule; *CD* = collecting duct; and *ADH* = antidiuretic hormone. (Adapted from Koch A: The kidney. In *Physiology and Biophysics: Circulation, Respiration and Fluid Balance,* 20th edition. Edited by Ruch TC and Patton HD. Philadelphia, WB Saunders, 1974, p 445.)

    (3) The fluid in the DLH nearly attains the osmolality of the adjacent medullary interstitium. [The interstitial osmolality is maintained by solvent-free solute (NaCl) that is transported out of the ALH.]

  b. The **ALH.** The thick and thin segments of the ALH constitute the diluting segments of the nephron. Both segments are impermeable to water and both are permeable to NaCl; in addition, the thin segment of the ALH is permeable to urea.

    (1) As fluid passes up the thin segment of the ALH, NaCl diffuses down its concentration gradient into the interstitial fluid. This contributes to the increased interstitial osmolality and renders the tubular fluid hyposmotic to the peritubular interstitium.

    (2) The thick segment of the ALH actively transports Cl^- out of the lumen, with Na^+ following the electrical gradient established by Cl^- transport. The **active Cl^- transport from the ALH to the interstitium** is the most important single effect in the countercurrent system.

    (3) The active and passive transport of NaCl from the ALH to the interstitium forms a horizontal osmotic gradient of up to 200 mOsm/kg between the tubular fluid of the ALH and the combined fluid of the interstitium and the DLH (Figure 4-14; see Figures 4-12C, 4-12D). This leaves behind a smaller volume of hypotonic fluid rich in urea, which flows into the distal convoluted tubule and the cortical and medullary collecting ducts.

    (4) The fluid in the ALH becomes diluted by the net loss of NaCl in excess of the net gain of urea by diffusion. This efflux of NaCl causes the tubular fluid within the ALH to become hyposmotic and that in the interstitial fluid to become hyperosmotic.

3. Simultaneous events in the interstitium

  a. Fluid in the interstitium contains hyperosmotic concentrations of NaCl and urea.

    (1) As the collecting ducts join in the inner medulla, they become increasingly permeable to urea (especially in the presence of ADH), allowing urea to flow passively along its concentration gradient into the interstitium (see Figures 4-12C, 4-13).

    (2) The increase in medullary osmolality causes water to move out of the adjacent tubules, the terminal collecting ducts, and the DLH.

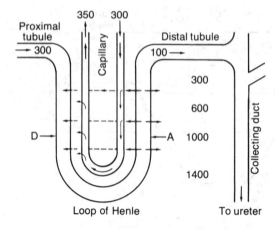

Figure 4-14. Countercurrent system for production of urine. *Numerals* denote osmolality; *D* = descending limb of the loop of Henle; *A* = ascending limb of the loop of Henle; and the *capillary* denotes the vasa recta countercurrent exchange system. (Reprinted from Wilatts SM: Normal water balance and body fluid compartments. In *Lecture Notes on Fluid and Electrolyte Balance.* Boston, Blackwell Scientific, 1982, p 19.)

   b. Water is reabsorbed in the last segment of the distal convoluted tubule and in the collecting duct in the cortex and outer medulla.
   c. In the inner medulla both water and urea are reabsorbed from the collecting duct (see Figure 4-12C).
      (1) Some urea reenters the ALH but at a slower rate than the efflux of NaCl.
      (2) This **medullary recycling of urea,** in addition to solute trapping by countercurrent exchange, causes urea to accumulate in large amounts in the medullary interstitium, where it osmotically abstracts water from the DLH and thereby concentrates NaCl in the DLH fluid.

   4. Distal convoluted tubule (see Figures 4-12, 4-14)
   a. General considerations. Before reaching the collecting duct, which is the next major component of countercurrent multiplication, the tubular fluid must pass through the distal convoluted tubule, where the role of ADH again is apparent.
   b. Role of ADH. Whether in the presence or the absence of ADH, fluid entering the distal convoluted tubule always is hypotonic to plasma in cortical as well as juxtamedullary nephrons (see Figure 4-13).
      (1) During diuresis (i.e., in the absence of ADH), this fluid remains hypotonic to plasma as it traverses the distal convoluted tubule and collecting duct; in fact, the solute concentration of the final urine may be lower than that of the fluid entering the distal convoluted tubule (see Figure 4-14). Thus, in the absence of ADH, Na$^+$ reabsorption is not retarded.
      (2) During antidiuresis (i.e., in the presence of ADH), the hypotonic fluid entering the distal convoluted tubule becomes isosmotic and the fluid entering the collecting duct is isosmotic (see Figure 4-13). This is because ADH increases the water permeability of the cells of the late distal convoluted tubule and collecting duct.
      (3) The osmolality of the glomerular filtrate, the cortical interstitial fluid, and the fluid leaving the distal convoluted tubule (in the presence of ADH) is isosmotic (300 mOsm/kg).

   5. Collecting duct. The cortical and upper medullary collecting ducts are relatively impermeable to water, urea, and NaCl.
   a. The relative impermeability to water occurs in the absence of ADH.
   b. The relative impermeability to NaCl permits the high interstitial concentration of NaCl to act as an effective osmotic gradient between the tubular fluid and the interstitium. (Some Na$^+$ and Cl$^-$ are reabsorbed from the distal convoluted tubule and collecting duct in the presence or absence of ADH.)
   c. When the kidney forms a concentrated urine, the collecting duct receives an isosmotic fluid from the distal convoluted tubule, the collecting duct is made permeable to water, and the urine equilibrates with the hyperosmotic medullary interstitium, resulting in the excretion of a low volume, hypertonic urine.
   d. If the collecting duct is impermeable to water (ADH absent), the dilute tubular fluid entering the collecting duct from the distal convoluted tubule remains hypotonic and is excreted as a higher volume, hypotonic urine.

C. Countercurrent exchangers (Figure 4-15; see Figure 4-14). The vasa recta function as countercurrent diffusion exchangers, making the concentrating mechanisms far more efficient.

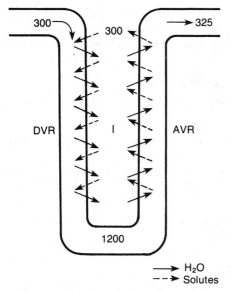

Figure 4-15. Vasa recta as countercurrent exchangers. *AVR* = ascending limb of the vasa recta; *DVR* = descending limb of the vasa recta; and *I* = interstitium. (Adapted from Pitts RF: *Physiology of the Kidney and Body Fluids*, 3rd ed. Chicago, Year Book, 1974, p 131.)

1. **Functional considerations of the vasa recta** (see Figure 4-15)
   a. The vasa recta are derived from the efferent arterioles of the juxtamedullary glomeruli and function to maintain the hyperosmolality of the medullary interstitium. The vasa recta are in juxtaposition with the loops of Henle.
   b. These vessels are permeable to solute and water and reach osmotic equilibrium with the medullary interstitium.
   c. The vasa recta return the NaCl and water reabsorbed in the loops of Henle and collecting ducts to the systemic circulation.
   d. The concentrations of Na^+ and urea in the medullary interstitium are kept high by the slow blood flow in the vasa recta.

2. **Countercurrent exchange effects of the vasa recta** (see Figure 4-15)
   a. Na^+, Cl^-, and urea are passively reabsorbed from the ALH, diffuse across the interstitial fluid into the descending limb of the vasa recta, and are returned to the interstitium by the ascending limb of the vasa recta.
      (1) These solutes recirculate in the vasa recta capillary loops and increase the medullary osmolality.
      (2) Na^+, Cl^-, and urea also recirculate in the loop of Henle.
   b. Water diffuses from the descending limb of the vasa recta across the interstitial fluid and into the ascending limb of the vasa recta.* As a result, the cortex receives blood that is only slightly hypertonic to plasma (see Figure 4-15).

D. **Role of urea.** The main function of urea in the countercurrent system is to exert an osmotic effect on the DLH, promoting the abstraction of water and raising the intraluminal concentration of NaCl.

1. In the medulla, ADH enhances both water and urea permeabilities of the collecting duct.
   a. Urea diffuses into the thin segment of the ALH via secretion but is more concentrated in the fluid entering the distal convoluted tubule (concentrated in a smaller volume) than in the filtrate entering the proximal convoluted tubule.
   b. In the presence of ADH, additional water is reabsorbed from the collecting duct, which further increases the urea concentration in the tubular fluid (urine).
   c. The **thin segment** of the ALH is less permeable to urea than to NaCl, and the **thick segment** of the ALH is impermeable to urea.

2. Urea recirculates in the loops of Henle and the vasa recta. The vasa recta are permeable to urea so that urea diffusing out of the papillary collecting duct is trapped in the medullary interstitium.
   a. Of the 1200 mOsm/kg of solute present in the renal papillary loop during **antidiuresis,** half is composed of NaCl and half of urea.

---

*Water is continually removed by the ascending limb of the vasa recta.

    b. During water **diuresis,** about 10% of the total medullary solute concentration is urea.

    c. The maximal urine osmolality (U_{osm}) cannot exceed that in the interstitium, and the ability to conserve water by excreting a concentrated urine is reduced when the papillary concentration is reduced.

E. Measurement of renal water excretion and conservation (Figure 4-16)

  1. Renal water excretion. The quantitative measure of the kidney's ability to excrete water is termed **free-water clearance.**

    a. General considerations

      (1) Free-water clearance is not a true clearance, because no osmotically free water exists in plasma.

      (2) Free-water clearance denotes the volume of **pure** (i.e., solute-free) **water** that must be removed from, or added to, the flow of urine (in ml/min) to make it isosmotic with plasma.

    b. Measurement

      (1) Free-water clearance (C_{H_2O}) is calculated using the equation

$$\dot{V} = C_{osm} + C_{H_2O}$$
$$C_{H_2O} = \dot{V} - C_{osm}$$

where

$$C_{osm} = \frac{U_{osm}\ \dot{V}}{P_{osm}}$$

Substituing for C_{osm} the equation becomes

$$C_{H_2O} = \dot{V} - \frac{U_{osm}\ \dot{V}}{P_{osm}}$$

and by factoring out $\dot{V}$, the final equation is

$$C_{H_2O} = \dot{V}\left(1 - \frac{U_{osm}}{P_{osm}}\right) \qquad (8)$$

where $\dot{V}$ = urine volume per unit time (in ml/min); U_{osm} = urine osmolality; P_{osm} = plasma osmolality; and C_{osm} = osmolal clearance, which is the volume of plasma (in ml) completely cleared of osmotically active solutes that appear in the urine each minute.

      (2) From equation (8), **during water diuresis** (i.e., in a hydrated state and in the absence of ADH), the U_{osm} is less than the P_{osm} (hyposmotic urine), and the U_{osm}/P_{osm} ratio is less than 1. Therefore, **C_{H_2O} is positive,** indicating that water is being eliminated by the

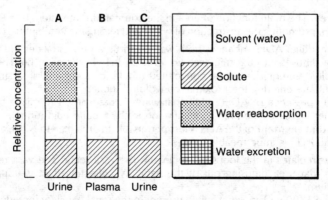

Figure 4-16. Renal free-water reabsorption (*A*) and free-water clearance (*C*). In column *A*, the *stippled area* represents the amount of free water that was removed from the urine by tubular reabsorption to **raise** the U_{osm} to the observed hypertonic value. In column *C*, the *cross-hatched area* denotes the amount of free water that was added to the urine by tubular excretion to **lower** the U_{osm} to the observed hypotonic value. Plasma osmolality (*B*) serves as the reference point for describing the osmolality of urine.

excretion of a large volume of dilute urine. The maximum C_{H_2O} in humans is 15–20 L/day (10–15 ml/min).

(3) From equation (8), **during antidiuresis** (i.e., in a dehydrated state and in the presence of ADH), the U_{osm} is greater than the P_{osm} (hyperosmotic urine), and the U_{osm}/P_{osm} ratio is greater than 1. Therefore, C_{H_2O} **is negative,** indicating that water is being conserved by the excretion of a small volume of concentrated urine. (To avoid the use of the term "negative free-water clearance," the symbol, $T^c_{H_2O}$, is used to denote **free-water reabsorption.**)

2. **Renal water conservation.** A quantitative measure of the ability of the kidney to reabsorb water is termed **free-water reabsorption**.
   a. **General considerations.** Free-water reabsorption denotes the volume of free water reabsorbed per unit time, or, the amount of free water that must be removed from the urine by tubular reabsorption to make it hyperosmotic with plasma.
   b. **Measurement**
      (1) **Free-water reabsorption ($T^c_{H_2O}$)*** is calculated using the equation

$$T^c_{H_2O} = C_{osm} - \dot{V}$$

where

$$C_{osm} = \frac{U_{osm}\ \dot{V}}{P_{osm}}$$

Substituting for C_{osm}, the equation becomes

$$T^c_{H_2O} = \frac{U_{osm}}{P_{osm}} - \dot{V}$$

and by factoring out $\dot{V}$, the final equation is

$$T^c_{H_2O} = \left(\frac{U_{osm}}{P_{osm}} - 1\right) \dot{V} \tag{9}$$

      (2) From equation (9), **during antidiuresis** (i.e., in a dehydrated state and in the presence of ADH), the U_{osm} exceeds P_{osm}, and the U_{osm}/P_{osm} ratio is greater than 1. Therefore, $T^c_{H_2O}$ **is positive,** indicating that water is being conserved by the elimination of a small volume of concentrated urine.

3. **Antidiuresis and diuresis** (Table 4-7) can be examined in terms of the volume of urine excreted per unit time ($\dot{V}$). Recall that $\dot{V}$ is the algebraic sum of osmolar clearance (C_{osm}) and either free-water clearance (C_{H_2O}) or free-water reabsorption ($T^c_{H_2O}$).
   a. **Antidiuresis** (i.e., a decrease in $\dot{V}$) can result from:
      (1) Water deprivation, which leads to an increase in plasma osmolality
      (2) Reduced circulating blood volume
   b. **Diuresis** (i.e., an increase in $\dot{V}$) can result from:
      (1) Reduced osmotic reabsorption of water, leading to increased solute-free water clearance and **water diuresis**
      (2) Reduced solute reabsorption (primarily Na^+ with associated anions), leading to increased osmolar clearance and **osmotic diuresis**

Table 4-7. Renal Water Excretion and Reabsorption

Renal State	$\dot{V}$ (ml/min)	C_{osm} (ml/min)	C_{H_2O} (ml/min)	$T^c_{H_2O}$ (ml/min)	U_{osm} (mOsm/kg)
Antidiuresis	↓	↑	. . .	↑	↑
Water diuresis	↑	↓	↑	. . .	↓
Osmotic diuresis	↑	↑	↑	. . .	↓

Note—$\dot{V}$ = urine excretion; C_{osm} = osmolar clearance; C_{H_2O} = free-water clearance; $T^c_{H_2O}$ = free-water reabsorption; and U_{osm} = urine osmolality.

*The superscript "c" in $T^c_{H_2O}$ signifies that the net reabsorption of water occurs in the collecting tubule.

VIII. ADH

A. Synthesis and chemical characteristics

1. ADH, also known as vasopressin,* is a **hypothalamic hormone** synthesized mainly by the cell bodies of the **neurosecretory neurons** of the **supraopticohypophysial tract** that terminate in the **pars nervosa** of the posterior lobe of the **pituitary gland**.

   a. **Neurosecretory neurons** are peptidergic neurons that conduct action potentials like all neurons and, unlike ordinary neurons, synthesize and secrete peptide hormones. The hypothalamic nuclei that represent the main source of ADH are the supraoptic nuclei, which are called **magnocellular neurosecretory neurons**.

   b. ADH is stored (but not synthesized) in the pars nervosa. In the absence of a pars nervosa, the newly synthesized hormone still can be released into the circulation from the ventral diencephalon (hypothalamus).

2. ADH is classified as a **polypeptide**. Specifically, it is an **octapeptide** or a **nonapeptide** with a molecular weight of approximately 1000.

   a. If the two cysteine residues are considered as a single cystine residue, ADH is an octapeptide.

   b. ADH is a nonapeptide if the two cysteine residues are numbered individually.

3. The neurophysins are the physiologic **carrier proteins** for the intraneuronal transport of ADH and are released into the circulation with the neurosecretory products (ADH and oxytocin) without being bound to the hormone.

4. The biologic half-life of ADH is 16–20 minutes.

B. Control of ADH secretion: stimuli and inhibitors

1. **Osmotic stimuli**

   a. **Hyperosmolality.** Under usual conditions, a 1%–2% increase in plasma osmolality is the prime determinant of ADH secretion, and the most common physiologic factor altering the osmolality of the blood is water depletion or water excess. (In clinical medicine, volume deficits are much more prevalent than volume excesses.)

      (1) The stimulation of the osmoreceptors located in the anterior hypothalamus causes the reflex secretion of ADH.

      (2) However, not all solutes (e.g., urea) stimulate the osmoreceptors, despite increasing plasma osmolality. Only those solutes to which cells are relatively impermeable make up the effective osmotic pressure, in response to which ADH is secreted.

      (a) Na^+ and mannitol, which cross the blood-brain barrier relatively slowly, are potent stimulators of ADH release. Since the plasma $[Na^+]$ accounts for 95% of the effective osmotic pressure, the osmoreceptors normally function as plasma Na^+ receptors.

      (b) Hyperglycemia is a less potent stimulus for ADH production and secretion than hypernatremia for the same level of osmolality.

   b. **Volume disturbances.** The control of ADH secretion by osmolality can be overridden by volume disturbances. For example, marked hyponatremia will be tolerated in order to maintain circulating blood volume.

2. **Osmotic inhibitors.** Expansion of the intracellular volume of the osmoreceptors secondary to hyposmolality of the ECF (water ingestion) inhibits ADH secretion.

3. **Nonosmotic stimuli**

   a. **Hypovolemia** (i.e., decreased effective circulating blood volume) is a more potent stimulus to ADH release than is hyperosmolality. A 10%–25% decrease in blood volume will evoke ADH release. A 10% decrease in blood volume is sufficient to cause the release of enough ADH to participate in the immediate regulation of blood pressure. Contraction of blood volume without an alteration in the tonicity of body fluids may cause ADH release.

---

*The term vasopressin denotes an excitatory action on the blood vessels, causing vasoconstriction of the arterioles and an increase in systemic blood pressure. This effect is observed only when relatively large quantities of vasopressin are released from the posterior lobe (e.g., during hemorrhage) or when pharmacologic amounts are injected. Therefore, the vasopressor effect usually is not considered to be a physiologic effect. The biologically active form of ADH in humans is **arginine vasopressin**.

(1) A lower osmotic threshold is required to cause ADH secretion in a volume-depleted state.

(2) A decrease in the stretch (tension) of the volume receptors (low-pressure baroreceptors) located in the left atrium, vena cavae, great pulmonary veins, carotid sinus, and aortic arch causes an increase in ADH secretion. Increased firing rates from these baroreceptors elicited by increased blood pressure exert an inhibitory effect on ADH secretion.

(3) Any reduction in intrathoracic blood volume (e.g., due to blood loss, quiet standing, upright body position, and positive-pressure breathing) leads to ADH secretion. Upon standing, the left atrial pressure falls markedly, leading to an increase in ADH release and antidiuresis.

b. **Other nonosmotic stimuli for ADH release** include:

(1) Pain

(2) Certain drugs (e.g., nicotine, morphine, barbiturates, acetylcholine, chlorpropamide, β-adrenergic agonists)

(3) Various pulmonary or CNS disorders (e.g., pneumonia, tuberculosis, stroke, meningitis, subdural hematoma)

(4) Angiotensin II

4. **Nonosmotic inhibitors**

a. Increased arterial pressure secondary to vascular or ECF volume expansion inhibits ADH release.

(1) Thus, ADH release is inhibited by increased tension in the left atrial wall, great veins, or great pulmonary veins secondary to increased intrathoracic blood volume due to hypervolemia, a reclining position, negative-pressure breathing, and water immersion up to the neck.

(2) In the recumbent position, the increase in central blood volume leads to an increase in left atrial pressure and inhibition of ADH release. During sleep, the production of a concentrated urine is by and large due to a reduction of blood pressure, which offsets the effect of reduced ADH secretion due to a change in body position.

b. **Other nonosmotic inhibitors of ADH release** include:

(1) Certain drugs (e.g., anticholinergic agents, ethanol, phenytoin, lithium, caffeine)

(2) CO_2 inhalation

C. **Role in regulation of renal water excretion**

1. **Renal effects.** The major site of action of ADH is the luminal membrane of the cortical and medullary collecting tubules, where ADH increases the permeability to water. ADH, through its effect on increased water reabsorption, leads to the production of a urine that has a decreased volume and an increased osmolality. ADH also decreases renal medullary blood flow.

2. **Extrarenal effects.** In addition, ADH stimulates the release of **adrenocorticotropic hormone (ACTH)** from the anterior lobe of the pituitary gland. ACTH plays a relatively small role in controlling aldosterone secretion (see IX D 1).

IX. **ALDOSTERONE** has the primary function of promoting Na^+ retention. In this way, aldosterone plays a major role in regulating water and electrolyte balance. Together with ADH, aldosterone influences urine composition and volume.

A. **Synthesis, secretion, and inactivation**

1. **Synthesis**

a. Aldosterone is a C-21 (21 carbon atoms) corticosteroid that is synthesized in the outermost area of the adrenal cortex, the **zona glomerulosa**.

b. Aldosterone represents less than 0.5% of the corticosteroids, and as all of the corticosteroids, it is stored in very small quantities. However, aldosterone is the major **mineralocorticoid** in humans.

c. The circulatory half-life of aldosterone is about 30 minutes in humans during normal activity.

d. **Cholesterol (esterified)** is the precursor for steroidogenesis and is stored in the cytoplasmic lipid droplets in the adrenocortical cells. (Free plasma cholesterol appears to be the preferred source of cholesterol for corticosteroid synthesis.)

2. Secretion. Most of the secreted aldosterone is bound to **albumin,** * with a lesser amount bound to **corticosteroid-binding globulin** (CBG; transcortin). CBG preferentially binds **cortisol,** which is a **glucocorticoid.**

3. Inactivation. Aldosterone is metabolized mainly in the liver, where greater than 90% of this corticosteroid is inactivated during a single passage.

  a. Most aldosterone inactivation is by **saturation (reduction)** of the double bond in the A-ring.

  b. The major metabolite is **tetrahydroaldosterone,** most of which is conjugated with glucuronic acid at the carbon-3 position of the A-ring. The resultant **glucuronides** are more polar and, therefore, more water-soluble, making them readily excreted by the kidney.

B. Physiologic effects

  1. Conservation of Na$^+$

  a. Aldosterone stimulates Na$^+$ reabsorption in the connecting segment of the distal tubule and in the cortical and medullary collecting tubules.

  b. Of the amount of filtered Na$^+$ (25,000 mEq/day), only 1%–2% (250–500 mEq or 6000–12,000 mg/day) are actively reabsorbed via an aldosterone-dependent mechanism in the distal nephron.†

  c. Aldosterone also promotes Na$^+$ reabsorption in the epithelial cells of the sweat glands, salivary glands, and the gastrointestinal mucosa.

  d. By restricting the renal excretion of Na$^+$, which is the main determinant of plasma osmolality, aldosterone regulates the ECF volume. Therefore, **aldosterone regulates the total body Na$^+$ content,** while ADH regulates the **plasma Na$^+$ concentration.**

  2. Secretion and excretion of K$^+$

  a. Aldosterone promotes K$^+$ secretion as a secondary effect of its action on Na$^+$ reabsorption.

   (1) In the distal tubule, the linked Na$^+$ reabsorption–K$^+$ secretion is referred to as the distal **Na$^+$-K$^+$ exchange process.**

   (2) Although K$^+$ appears to be secreted in exchange for Na$^+$, the distal secretion of K$^+$ is only indirectly related to Na$^+$ reabsorption. Distal Na$^+$ reabsorption is linked to the secretion of both K$^+$ and H$^+$.‡

   (3) Aldosterone increases the [K$^+$] in sweat glands and saliva.

  b. Stimulation of K$^+$ excretion greatly depends on dietary Na$^+$, as indicated by a lack of K$^+$ excretion after aldosterone administration in animals with Na$^+$-deficient diets.

  c. More than 75% of the K$^+$ excreted in the urine is attributed to distal K$^+$ secretion.

   (1) Excess aldosterone secretion causes a decline in the urinary Na$^+$/K$^+$ concentration ratio (from a normal value of about 2) because it decreases Na$^+$ excretion and increases K$^+$ excretion.

   (2) Elevated aldosterone release increases the plasma Na$^+$/K$^+$ concentration ratio (from a normal value of 30) due to the increased excretion of K$^+$. Aldosterone can cause an isotonic expansion of the ECF volume with no change in the plasma [Na$^+$] or a hypertonic expansion of the ECF volume with a rise in the plasma [Na$^+$]. In any case, the total amount of body Na$^+$ is increased.

  3. Water excretion and ECF volume regulation

  a. The effect of aldosterone on water excretion is not important physiologically, as aldosterone treatment fails to correct the impaired water excretion by patients with hypofunctional adrenal glands (due to Addison's disease or adrenal insufficiency) and by patients who have undergone adrenalectomy.

  b. Aldosterone has no direct effect on GFR, RPF, or renin production; however, by stimulating Na$^+$ reabsorption aldosterone causes water retention, and the resultant expansion of the ECF volume then leads to an increase in GFR and RPF and a decrease in renin production.

  c. A high circulating aldosterone level is a common finding in edema and is due primarily to the increased aldosterone secretion induced by the depletion of the effective circulating blood volume.

   (1) It is unlikely that the plasma [Na$^+$] is a major regulator of aldosterone secretion because the plasma [Na$^+$] during Na$^+$ depletion is normal in humans.

---

*Protein-bound hormones are biologically inactive.

†Filtered load refers to the amount of Na$^+$ in the glomerular filtrate and not in the distal nephron.

‡Na$^+$ transport in the proximal tubule is not associated with K$^+$ or H$^+$ exchange processes.

(a) **Hyponatremia** often is a consequence of ADH secretion and is accompanied by an increase in the ECF volume, which tends to suppress rather than stimulate aldosterone secretion. When ADH and water are given to Na^+-depleted individuals, aldosterone secretion falls despite a fall in plasma $[Na^+]$.

(b) By and large, hyponatremia in the clinical setting is due to **excess body Na^+** caused by a decreased effective blood volume. The low blood volume leads to an increase in ADH secretion and aldosterone secretion which, in turn, lead to edema. **Over 90% of the cases of hyponatremia should be treated by restriction of NaCl and water.**

(2) Therefore, it is the volume of the ECF rather than the plasma $[Na^+]$ that influences aldosterone secretion in most circumstances.

C. **Aldosterone and acid-base balance.** Aldosterone affects acid-base balance through its control of K^+ secretion. As stated above, aldosterone promotes increased distal tubular **secretion** and, therefore, **excretion** of K^+. Aldosterone also promotes the excretion of H^+ and NH_4^+.

1. **Hyperaldosteronemia** (i.e., increased aldosterone secretion) is one cause of K^+ depletion, a condition termed **hypokalemia.**
   a. Hypokalemia is characterized by an increase in intracellular $[H^+]$, which favors distal tubular secretion of H^+ over K^+ and results in a state of **metabolic alkalosis.**
   b. Conversely, if metabolic alkalosis is the primary event, there is a decrease in intracellular $[H^+]$. This results in an increase in distal tubular intracellular $[K^+]$, which favors increased urinary loss of K^+ over H^+ and results in a state of **hypokalemic metabolic alkalosis** (as occurs in hyperaldosteronemia).

2. **Hypoaldosteronemia** (i.e., decreased aldosterone secretion) is one cause of excess K^+, a condition termed **hyperkalemia.**
   a. Hyperkalemia is characterized by an increase in intracellular $[K^+]$, which favors distal tubular secretion of K^+ over H^+ and results in a state of **metabolic acidosis.**
   b. On the other hand, when metabolic acidosis is the primary event there is an increase in intracellular $[H^+]$. This results in an increase in distal tubular intracellular $[H^+]$, which favors increased secretion of H^+ over K^+ and results in a state of **hyperkalemic metabolic acidosis** (as occurs in hypoaldosteronemia).

3. **Hypokalemia** causes the following conditions.
   a. **Impaired renal concentrating ability** leads to the formation of hyposmotic urine and polyuria that is resistant to ADH.
   b. **Reduced carbohydrate tolerance** leads to a decline in insulin secretion.* As a rule, fasting hyperglycemia is not present with reduced insulin secretion.

D. **Control of aldosterone secretion.** At least three well-defined mechanisms control aldosterone secretion: ACTH, plasma $[K^+]$, and the renin-angiotensin system.

1. **Extrarenal control mechanisms.** The following mechanisms cause the release of aldosterone by direct action on the adrenal cortex.
   a. **Hypothalamic-hypophysial-adrenocortical axis.** Under normal conditions, ACTH is not a major factor in the control of aldosterone synthesis or secretion. However, the pituitary gland plays an important role in the maintenance of the growth and biosynthetic capacity of the zona glomerulosa.
      (1) **ACTH,** also known as **corticotropin,** is a 39-amino-acid polypeptide that supports steroidogenesis in the zona glomerulosa. ACTH enhances aldosterone production by stimulating the early biosynthetic pathway (i.e., the 20,22-desmolase enzyme complex that catalyzes the conversion of cholesterol to pregnenolone). ACTH also plays a minor role in mediating the diurnal rhythmic secretion of all of the corticosteroids.
      (2) **Corticotropin releasing hormone (CRH)** is a hypothalamic **hypophysiotropic** polypeptide made up of 41 amino acid residues. CRH is secreted into the hypophysial portal system and causes release of ACTH from the pituitary gland. Since CRH secretion is regulated by higher brain centers (e.g., the limbic system), these centers play a role in Na^+ balance.

---

*The reduction in carbohydrate tolerance due to hypokalemia occurs in only about half of patients with elevated plasma aldosterone.

    (a) Hypophysectomized patients and those with pituitary insufficiency exhibit normal aldosterone secretion on a moderate salt intake; however, these individuals demonstrate a suboptimal aldosterone response to Na$^+$ restriction.

    (b) Normal individuals injected chronically with ACTH show an acute rise in aldosterone secretion followed by a return to control level or below in 3–4 days, despite ACTH administration over a period of 7–8 days.

    (c) When **aldosterone** is administered for several days to normal individuals, the kidney "escapes" from the Na$^+$-retaining effect but not from the K$^+$-excreting effect. The escape phenomenon prevents the appearance of edema in individuals treated with aldosterone for prolonged periods and in patients with primary aldosteronism.

  b. Hyperkalemia. A 10% increase in plasma [K$^+$] can stimulate the synthesis and release of aldosterone by a direct action on the zona glomerulosa. In the anephric human, K$^+$ appears to be the major regulator of aldosterone even though aldosterone levels are low.

    (1) This release probably occurs by the depolarization of the glomerulosa cell membrane by the elevated plasma [K$^+$].

    (2) Stimulation of aldosterone secretion by K$^+$ loading will be limited by the simultaneous reduction in renin release.

    (3) K$^+$ stimulates an early step in the biosynthetic pathway for aldosterone synthesis.

    (4) K$^+$ loading increases the width of the zona glomerulosa layer in experimental animals; prolonged aldosterone administration results in atrophy of this layer due to depressed renin secretion.

  c. Hyponatremia. A 10% decrease in plasma [Na$^+$] also appears to stimulate the synthesis and release of aldosterone directly at the level of the zona glomerulosa. However, this effect usually is overridden by changes in the effective circulating volume. Thus, aldosterone secretion is increased in the hyponatremic patient who is volume-depleted but is reduced in the hyponatremic patient who is volume-repleted.

2. Intrarenal control mechanism. Aldosterone secretion also is regulated by the renin-angiotensin system, the major component of which is the **juxtaglomerular apparatus (JGA)**. The renin-angiotensin-aldosterone system is regulated by the **sympathetic nervous system**.

  a. Anatomy of the JGA (Figure 4-17). The JGA is a combination of specialized tubular and vascular cells located at the vascular pole where the afferent and efferent arterioles enter and leave the glomerulus. The JGA is composed of three cell types.

    (1) Juxtaglomerular (JG) cells are specialized **myoepithelial** (modified vascular smooth muscle) cells located in the **media** of the afferent arteriole, which synthesize, store, and release a proteolytic enzyme called **renin**. Renin is stored in the granules of the JG cells.

      (a) The JG cells are **baroreceptors** (tension receptors) and respond to changes in the transmural pressure gradient between the afferent arteriole and the interstitium. They are innervated by sympathetic nerve fibers.

      (b) These vascular "volume" receptors monitor renal perfusion pressure and are stimulated by hypovolemia, or decreased renal perfusion pressure.

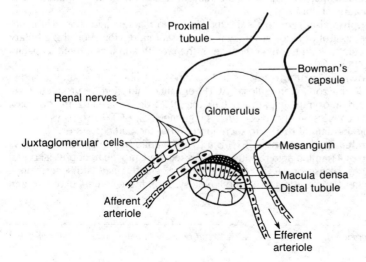

Figure 4-17. The anatomic components of the juxtaglomerular apparatus. (Reprinted from Yates FE, et al: The adrenal cortex. In *Medical Physiology*, 14th edition. Edited by Mountcastle VB. St. Louis, CV Mosby, 1980, p 1590.)

(2) **Macula densa cells** are specialized renal tubular epithelial cells located at the transition between the thick segment of the ALH and the distal convoluted tubule (see Figure 4-17).

    (a) These cells are in direct contact with the mesangial cells, in close contact with the JG cells, and contiguous with both the afferent and efferent arterioles as the tubule passes between the arterioles supplying its glomerulus of origin.

    (b) The macula densa cells are characterized by prominent nuclei in those cells on the side of the tubule that is in contact with the mesangial and vascular elements of the JGA.

    (c) The macula densa cells function as **chemoreceptors** and are stimulated by a decreased Na^+ (NaCl) load. This inverse relationship between Na^+ load and renin release provides a reasonable explanation for the clinical problems involving a decreased filtered load of Na^+ and Cl^- in association with increased renin release.

    (d) Macula densa cells are not innervated.

(3) **Mesangial cells** also are referred to as the **polkissen** (asymmetrical cap) and are the interstitial cells of the JGA. Mesangial cells are in contact with both the JG cells and the macula densa cells. A decreased intraluminal Na^+ load, Cl^- load, or both in the region of the macula densa stimulates the JG cells.

b. Role of the sympathetic nervous system. The sympathetic nervous system plays an important role in the control of renin release via the **renal nerves**.

(1) The JG cells of the afferent arterioles are innervated directly by the sympathetic postganglionic fibers (unmyelinated). In the absence of renal nerves, the renal response to Na^+ depletion is attenuated.

(2) Circulating catecholamines (i.e., epinephrine and norepinephrine) and stimulation of the renal nerves produce vasoconstriction of the afferent arterioles, which causes renin release by a decrease in perfusion pressure.

    (a) This renin response caused by catecholamines and renal nerve stimulation is mediated via the β-adrenergic receptor and can be elicited by the synthetic sympathomimetic amine, **isoproterenol,** which is a β-agonist.

    (b) Renal denervation and β-adrenergic receptor blockade by **propranolol** inhibit the release of renin.

(3) Renal innervation is not a requisite for renin release because the denervated kidney can adapt to a variable salt intake.

(4) In humans, exercise for assuming an upright posture increases renal sympathetic activity, which produces renal arteriolar vasoconstriction and an increase in renin release. Thus, the sympathetic nervous system, by modulating the secretion of renin, has an **indirect effect on aldosterone secretion**.

c. Role of renin: stimuli for release

(1) Renin has a circulatory half-life of 40–120 minutes in humans. The common denominator for renin release by the intrarenal mechanism is a decrease in the effective circulating blood volume, which is induced by:

    (a) Acute hypovolemia associated with hemorrhage, diuretic administration, or salt depletion

    (b) Acute hypotension associated with ganglionic blockade or a change in posture (postural hypotension)

    (c) Chronic disorders associated with edema (e.g., cirrhosis with ascites, congestive heart failure, nephrotic syndrome)

(2) Renin release is increased by K^+ depletion, epinephrine, norepinephrine, isoproterenol, and standing.

d. Role of renin: inhibition of renin secretion

(1) Renin release is inhibited by angiotensin II, angiotensin III, ADH, hypernatremia, hyperkalemia, and atrial natriuretic factor.

(2) **K^+ loading** leads to the inhibition of renin release and to the direct stimulation of the glomerulosa cells to secrete aldosterone. In contrast to its effect on normal individuals, aldosterone administration to patients with heart failure, cirrhosis with ascites, or nephrosis causes Na^+ retention without K^+ excretion, because of greater proximal reabsorption of Na^+ with less Na^+ available for the distal exchange with K^+.

e. Angiotensin synthesis

(1) Renin is secreted into the bloodstream, where it combines with the renin substrate, angiotensinogen, which is an α_2-globulin synthesized in the liver.

(a) Renin is not saturated with its substrate in normal plasma; the same amount of renin generates more angiotensin I if the substrate concentration is increased above normal.

(b) Oral contraceptives increase plasma angiotensinogen concentration and decrease plasma renin concentration.

(2) The only physiologic effect of renin is to convert angiotensinogen to the biologically inactive decapeptide, angiotensin I.

(3) Angiotensin I is converted primarily in the lung by pulmonary endothelial cells to the physiologically active octapeptide, angiotensin II (Figure 4-18).

(a) The enzyme that forms angiotensin II is a peptidase (dipeptidyl carboxypeptidase) called angiotensin-converting enzyme. It is found chiefly in pulmonary tissue and to a lesser degree in renal tissue and blood plasma.

(b) The converting enzyme is identical to kininase II, which converts the nonapeptide vasodilator, bradykinin (kallidin-9), to inactive peptides, thereby diminishing the circulating levels of a vasodepressor substance and enhancing the vasoconstrictive action of angiotensin II.

(4) **Angiotensin II,** with a circulatory half-life of 1–3 minutes, has several important physiologic actions.

(a) It functions as the tropic hormone for the zona glomerulosa and stimulates the secretion (and synthesis) of aldosterone. Angiotensin II is the **aldosterone-stimulating hormone**.

(b) It is a potent **local vasoconstrictor (vasopressor)** of the renal arterioles at low plasma concentrations; at higher concentrations, angiotensin II exerts a general vasopressor effect on the smooth muscle cells of arterioles throughout the cardiovascular system, leading to an elevation of systemic mean arterial blood pressure.

(c) Angiotensin II stimulates the secretion of ADH and ACTH.

(d) Angiotensin II stimulates thirst, which leads to increased fluid consumption. It also stimulates the release of epinephrine and norepinephrine from the adrenal medulla.

(5) Angiotensin II is inactivated by two angiotensinases (peptidases), and it can be converted to angiotensin III, which is a heptapeptide and a very potent stimulator of aldosterone secretion. Angiotensin III is not an effective vasoconstrictor in contrast to angiotensin II.

(6) Renin, converting enzyme, angiotensinogen, and angiotensin II have been found in brain tissue.

f. **Plasma renin activity and plasma renin concentration. Plasma renin activity** is defined as the rate of angiotensin I formation when plasma renin acts on **endogenous substrate**. **Plasma renin concentration** is measured when **exogenous substrate** is added to plasma to saturate the enzyme and increase the velocity of angiotensin I formation to a maximal rate.

(1) Oral contraceptive administration is known to increase plasma renin activity and aldosterone secretion via a marked increase in renin substrate concentration, whether the woman is normotensive or hypertensive prior to the therapy.

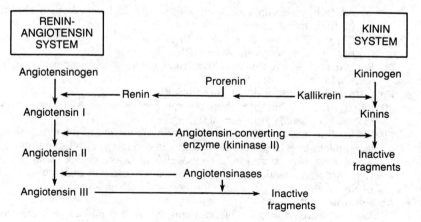

Figure 4-18. The renin-angiotensin system and its relationship to the kinin system.

(2) As a result of the increased angiotensin II formation, plasma renin concentration is suppressed. Thus, the increase in plasma renin activity in this situation occurs without an increase in renin concentration.

(3) Oral contraceptives are a cause of hypertension in women through this mechanism.

E. Aldosteronism refers to a condition of excessive aldosterone secretion.

1. Primary hyperaldosteronism (Conn's syndrome)
   a. Etiology. Primary hyperaldosteronism results from a tumor of the **zona glomerulosa.**
   b. Characteristics of primary hyperaldosteronism include:
   (1) Elevated plasma (and urinary) aldosterone
   (2) Hypertension due to Na^+ and water retention
   (3) Hypokalemic alkalosis with a K^+ excretion rate of greater than 40 mEq/day*
   (4) Decreased levels of angiotensin and renin
   (5) Decreased hematocrit due to the expansion of the plasma volume
   (6) Polyuria and dilute urine due to secondary nephrogenic diabetes insipidus
   (7) Absence of peripheral edema
   (8) Decreased plasma colloidal osmotic (oncotic) pressure due to ECF expansion

2. Secondary hyperaldosteronism
   a. Etiology. Secondary hyperaldosteronism is caused by the following extra-adrenal factors:
   (1) Diuretic therapy, which is the most common cause
   (2) Extravascular loss of Na^+ and water, which is associated with edema (due to such underlying factors as nephrosis, cirrhosis, and congestive heart failure), an increase in the total ECF volume, and a loss of effective blood volume
   (3) Hyperreninism caused by a tumor of the JG cells
   (4) Renovascular disease (e.g., renal artery stenosis)
   b. Characteristics. Secondary hyperaldosteronism is characterized by edema and Na^+ retention. Urinary K^+ excretion is not increased because there is a reduced flow of fluid into and through the distal segments of the nephron. This low fluid flow reduces K^+ secretion and offsets the stimulating effect of aldosterone. Also, with decreased Na^+ and water delivery to the distal tubule, the quantity of K^+ (and H^+) secreted in the urine is limited. Additional characteristics of secondary hyperaldosteronism include:
   (1) Increased plasma (and urinary) aldosterone
   (2) Hypertension with edema (due to Na^+ retention and water accumulation in the interstitial fluid compartment) and a decrease in plasma volume
   (3) Hypokalemic alkalosis
   (4) Increased angiotensin and plasma renin activity[†]
   (5) Peripheral edema

3. Chronic licorice ingestion in excessive amounts can mimic primary aldosteronism, because licorice contains the salt-retaining substance, glycyrrhizinic acid. Patients with this condition present with:
   a. Hypertension and hypokalemic alkalosis
   b. Suppressed plasma renin levels
   c. Reduced aldosterone secretion due to chronic volume expansion

F. Aldosterone antagonism: spironolactone

1. Renal effects. Spironolactone is a steroidal aldosterone antagonist, which competitively blocks the Na^+-retaining and K^+-excreting effects that aldosterone exerts on the distal renal tubule. As a result, spironolactone leads to an increase in urinary Na^+ excretion and a decrease in K^+ excretion. This antagonist is efficacious only in the presence of aldosterone or another mineralocorticoid; spironolactone is without effect in adrenalectomized individuals.

2. Blood pressure effect. Because the drug enhances Na^+ diuresis, spironolactone is effective in potentiating the action of many antihypertensive drugs whose dosage should be reduced in its presence.

---

*Hypokalemia is not always concomitant with hypermineralocorticoidism (e.g., in 11β-hydroxylase deficiency).
[†]The elevated renin level is the characteristic that differentiates secondary aldosteronism from the primary form.

3. **Clinical application.** Spironolactone is useful in the differential diagnosis of primary and secondary hyperaldosteronism.
    a. If both plasma [K^+] and blood pressure are returned to normal with spironolactone, primary hyperaldosteronism is suspected.
    b. If spironolactone causes the plasma [K^+] to return to normal without the antihypertensive effect, secondary hyperaldosteronism is suspected.

X. **ATRIAL NATRIURETIC FACTOR (ANF)** refers to a group of polypeptides (also abbreviated ANP) produced by the atrial muscle cells. ANF exerts a hormonal influence on the kidney, which results in changes in intrarenal hemodynamics through its effects on fluid volume, electrolyte (Na^+) balance, and blood pressure homeostasis.

A. **Synthesis**

1. In mammals, ANF is synthesized, stored, and released from **atrial cardiocytes**. The secretory activity of these cells is evidenced by the presence of membrane-bound storage granules with electron-dense cores.

2. The peptides that comprise the ANF group have molecular weights ranging from about 2500 to 13,000 daltons and lengths ranging from 21 to 73 amino acid residues. All are derived from a common 126-amino-acid precursor called **pro-ANP (atriopeptigen),** which is the predominant form of ANF in the atrium.

3. In humans, ANF is synthesized in a **prepro** (i.e., **pre-atriopeptigen**) form containing 151 amino acid residues.

4. The predominant circulating form of ANF is the 28-amino-acid peptide.

B. **Secretion.** Stimuli for ANF release include atrial distension (hypervolemia), epinephrine, arginine vasopressin (ADH), and acetylcholine.

C. **Physiologic effects of ANF**

1. **Renal and adrenal responses**
    a. **An increased GFR** is associated with constriction of the efferent arteriole, increasing the glomerular hydrostatic pressure. It is important to note that ANF induces relaxation (dilation) of precontracted renal arteries.
    b. **Dilation of the afferent arteriole** increases the hydraulic pressure in the glomerular capillary.
    c. **Natriuresis** (i.e., increased urinary Na^+ excretion) occurs primarily as a result of the increase in GFR, the increase in renal medullary or papillary blood flow, or both.
    d. **Inhibition of aldosterone secretion.** ANF blocks aldosterone secretion that was prestimulated by Na^+ depletion, angiotensin II, ACTH, K^+, and cyclic adenosine 3',5'-monophosphate (cAMP).
    e. **Inhibition of renin secretion** occurs via the increase in NaCl delivery to the macula densa or the increase in hydrostatic pressure at the JGA due to afferent arteriolar vasodilation.

2. **Cardiovascular effects**
    a. **Decreases in mean systemic arterial blood pressure** occur due to vasorelaxation or suppression of renin secretion.
    b. **Reduction in cardiac output** occurs primarily due to bradycardia.

STUDY QUESTIONS

Directions: Each of the numbered items or incomplete statements in this section is followed by answers or by completions of the statement. Select the **one** lettered answer or completion that is **best** in each case.

1. The renal threshold for glucose excretion normally corresponds to a plasma glucose concentration of

(A) 80 mg/dl
(B) 100 mg/dl
(C) 120 mg/dl
(D) 160 mg/dl
(E) 180 mg/dl

2. The following renal function data were collected for substance x: urine flow rate = 90 ml/hr, concentration of substance x in urine = 480 mg/ml, and concentration of substance x in plasma = 6 mg/ml. What is the renal clearance of substance x?

(A) 12 ml/min
(B) 120 ml/min
(C) 120 mg/min
(D) 240 ml/min
(E) 480 ml/min

3. The following data were derived from a renal function test in a normal person: GFR = 180 L/day, plasma uric acid concentration = 5 mg/dl, and uric acid excretion rate = 700 mg/day. What conclusion can be reached regarding renal processing of uric acid?

(A) Uric acid is secreted and reabsorbed but at different rates
(B) Uric acid is only secreted
(C) Uric acid is only filtered
(D) Uric acid is not metabolized by the kidney
(E) Uric acid is filtered and reabsorbed but not secreted

4. The primary renal site for the secretion of organic ions (e.g., PAH, urate, creatinine) is the

(A) proximal tubule
(B) thin segment of the loop of Henle
(C) thick segment of the loop of Henle
(D) distal tubule
(E) collecting duct

5. Transcellular fluid is found in the lumen of structures lined by epithelial cells. An example of transcellular fluid is

(A) plasma
(B) CSF
(C) interstitial fluid
(D) edematous fluid
(E) intravascular fluid

6. The renal transport maximum (Tm) for a substance is defined as the maximal

(A) GFR
(B) urinary excretion rate
(C) tubular reabsorption or secretion rate
(D) renal clearance rate
(E) amount of a substance filtered by the glomeruli per minute

7. Of the renal systems available for the excretion of H^+, the one with the greatest activity is

(A) H^+ secretion
(B) NaH_2PO_4 excretion
(C) SO_4^{2-} excretion
(D) titratable acid excretion
(E) NH_4^+ excretion

8. Which of the following substances undergoes both reabsorption and secretion in the kidney?

(A) PAH
(B) Lactate
(C) Urea
(D) Uric acid
(E) Creatinine

9. When the secretion of PAH reaches the Tm, a further increase in plasma PAH concentration causes its clearance to

(A) increase in proportion to its plasma concentration
(B) approach glucose clearance asymptotically
(C) remain constant
(D) approach inulin clearance asymptotically
(E) increase in proportion to GFR

10. All of the following statements concerning renin are true EXCEPT

(A) renin release from the JG cells is inversely related to the degree of stretch in the wall of the afferent arteriole
(B) renin is a secretory product of the JG cells
(C) renin substrate is a hepatic globulin
(D) the renin response to Na^+ depletion is attenuated in the absence of renal nerves
(E) renin secretion is a prerequisite for aldosterone secretion

11. All of the following conditions will evoke an increase in plasma ADH level EXCEPT

(A) dehydration
(B) trauma
(C) hemorrhage
(D) hyponatremia
(E) pain

12. Ions constitute approximately 95% of the solutes in body water. The body fluid compartment with the highest protein concentration is the

(A) ICF
(B) interstitial fluid
(C) plasma
(D) CSF
(E) transcellular fluid

13. Which of the following statements regarding Na^+ transport is correct?

(A) Active transport of Na^+ across all cells consumes most of the energy derived from cellular metabolism
(B) The Na^+ concentration is highest in the intracellular fluid
(C) Na^+ reabsorption across proximal tubular cells is mainly active and transcellular
(D) The Na^+ concentration gradient provides energy for the cotransport of H^+
(E) The transport of Na^+ across the luminal membrane of the nephron is an active transport process

14. All of the following statements regarding K^+ transport are true EXCEPT

(A) K^+ is the only inorganic electrolyte that is both reabsorbed and secreted by the kidney
(B) the filtered load of K^+ is more than that of Na^+
(C) insulin stimulates cellular uptake of K^+
(D) urinary K^+ excretion is primarily dependent on K^+ secretion from the distal tubular cell into the lumen

15. Small solutes are transported across the glomerular capillaries by the process of

(A) simple diffusion
(B) convection
(C) facilitated diffusion
(D) primary active transport
(E) secondary active transport

16. The renal threshold for a solute denotes the

(A) maximum filtration rate
(B) maximum reabsorption rate
(C) plasma concentration at which a solute begins to appear in the urine
(D) maximum secretion rate
(E) maximum tubular secretory capacity

17. Which of the following conditions causes a decrease in the ECF volume, an increase in the ICF volume, and a decrease in the osmolar concentration of both compartments?

(A) Hyperosmotic dehydration
(B) Hyposmotic dehydration
(C) Isosmotic dehydration
(D) Hyperosmotic overhydration
(E) Hyposmotic overhydration

18. The volume of the ECF is regulated primarily by the reabsorption and excretion of

(A) sodium lactate
(B) sodium bicarbonate
(C) sodium chloride
(D) sodium phosphate
(E) sodium citrate

19. If the GFR remains unchanged, a decrease in the urine-to-plasma inulin concentration ratio (U_{in}/P_{in}) indicates that

(A) free-water clearance has decreased
(B) urine flow has increased
(C) inulin clearance has decreased
(D) plasma inulin concentration has decreased
(E) creatinine clearance has increased

20. The major contributor to the osmotically active solutes in the ECF is

(A) sodium
(B) albumin
(C) urea
(D) creatinine
(E) potassium

21. Which of the following substances has the lowest renal clearance?

(A) Glucose
(B) Urea
(C) Inulin
(D) Creatinine
(E) PAH

22. Stimulation of renin secretion will cause an increase in the

(A) K^+ concentration in the blood
(B) volume of the ECF
(C) hematocrit
(D) plasma colloid osmotic pressure
(E) H^+ concentration in the blood

23. A drug that completely blocks the secretion of PAH is given to a laboratory animal. The clearance of which of the following substances will be the same as the PAH clearance under this condition?

(A) Water
(B) Sodium
(C) Urea
(D) Inulin
(E) Bicarbonate

24. Given a GFR of 125 ml/min, a plasma glucose concentration of 400 mg/100 ml, a urine glucose concentration of 75 mg/ml, and a urine flow of 2 ml/min, what is the Tm for glucose?

(A) 300 mg/min
(B) 350 mg/min
(C) 400 mg/min
(D) 500 mg/min
(E) 550 mg/min

25. The following graph shows the tubular loss of glucose (excretion rate) plotted against the rate at which glucose is filtered at the glomerulus (filtered load). Which lettered point on the curve corresponds to the Tm for glucose?

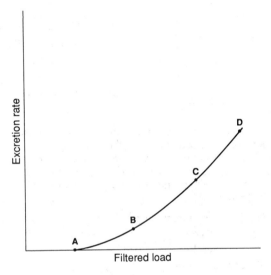

26. Which of the following factors can best explain an increase in the GFR?

(A) Increased arterial plasma oncotic pressure
(B) Increased hydrostatic pressure in Bowman's capsule
(C) Increased glomerular capillary hydrostatic pressure
(D) Decreased net filtration pressure
(E) Vasoconstriction of the afferent arteriole

27. Which of the following forces opposes ultrafiltration at the glomerulus?

(A) Plasma colloid osmotic pressure
(B) Glomerular capillary hydrostatic pressure
(C) Colloid osmotic pressure in Bowman's capsule
(D) Concentration of NaCl in Bowman's space
(E) Crystalloid osmotic pressure of the final urine

28. Measurements taken after an intravenous injection of inulin indicate that the substance appears to be distributed throughout 30%–35% of the total body water. This finding suggests that inulin most likely is

(A) excluded from the cells

(B) distributed uniformly throughout the total body water volume

(C) restricted to the plasma volume

(D) neither excreted nor metabolized by the body

(E) not freely diffusible through capillary membranes

Directions: Each question below contains four suggested answers of which **one or more** is correct. Choose the answer.

A if **1, 2, and 3** are correct
B if **1 and 3** are correct
C if **2 and 4** are correct
D if **4** is correct
E if **1, 2, 3, and 4** are correct

29. There is an inverse relationship between the rate of ADH secretion and the rate of discharge in afferent neurons from stretch receptors in the vascular system. These pressure receptors are found in which of the following regions of the vascular system?

(1) Great pulmonary veins

(2) Left atrium

(3) Aortic arch

(4) Carotid sinuses

30. A 38-year-old male patient presents with serum Na^+ concentration of 155 mEq/L together with a urine osmolality of 50 mOsm/kg H_2O. These findings could be explained by

(1) a lack of ADH

(2) volume expansion with isotonic saline

(3) an increase in free-water clearance

(4) excessive ingestion of water

31. The proximal tubular reabsorption of Na^+ would be markedly reduced by

(1) removal of the pituitary gland

(2) a decrease in plasma glucose concentration

(3) an increase in H^+ secretion

(4) a deficiency of luminal membrane transport proteins

32. Vasoconstriction of the renal artery can lead to an elevation in mean arterial pressure by

(1) increased renin secretion

(2) increased angiotensin II formation

(3) increased aldosterone secretion

(4) decreased plasma bradykinin levels

Directions: Each group of items in this section consists of lettered options followed by a set of numbered items. For each item, select the **one** lettered option that is most closely associated with it. Each lettered option may be selected once, more than once, or not at all.

Questions 33–37

Match each of the following descriptions with the most appropriate lettered region of the nephron pictured below.

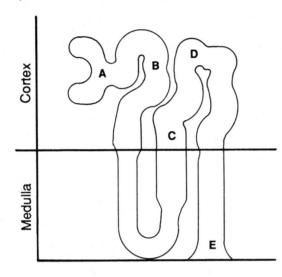

33. Tubular fluid always is hyposmotic at this site
34. The TF/P ratio for glucose is 1.0 at this site
35. The urea concentration is highest at this site
36. The U_{osm} can reach 1200–1400 mOsm/L at this site
37. The macula densa is closest to this site

Questions 38–42

Match each of the following filtered substances with the appropriate renal transport process.

(A) Reabsorption
(B) Secretion
(C) Both
(D) Neither

38. K^+
39. Mannitol
40. HCO_3^-
41. Creatinine
42. Urea

Questions 43–45

Match each description with the most appropriate morphologic segment of the nephron.

(A) Glomerulus
(B) Proximal convoluted tubule
(C) Thin descending limb of the loop of Henle
(D) Thick ascending limb of the loop of Henle
(E) Collecting tubule

43. Major site of ADH-induced water permeability
44. Site where largest fraction of water is reabsorbed
45. Concentrating segment of the nephron

Questions 46–51

Match each of the following substances with the appropriate lettered TF/P curve on the graph below.

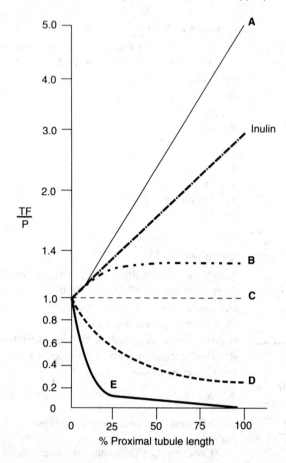

46. Na^+
47. Glucose
48. Glycine

49. Cl^-
50. HCO_3^-
51. PAH

ANSWERS AND EXPLANATIONS

1. The answer is E. [*VI A 2; Figure 4-11A*] The renal threshold for glucose is the plasma glucose concentration at which glucose begins to appear in the urine and is equal to 180 mg/dl. It is important to note that this concentration is not the plasma glucose concentration that completely saturates the renal glucose transport mechanism (Tm = 340 mg/min); glycosuria occurs long before the transport mechanism is fully saturated.

2. The answer is B. [*IV A 1, 2*] The renal clearance of a substance (C_x) is defined as the plasma volume from which a given substance is removed per minute. Thus, it is an empiric measure of the volume of plasma that contains the same amount of the substance as is excreted in urine in 1 minute. The clearance equation is

$$C_x = \frac{U_x \times \dot{V}}{P_x}$$

where U_x and P_x = the urinary and plasma concentrations of substance x in mg/ml, respectively; $\dot{V}$ = the urine flow rate in ml/min; and C_x = the clearance of substance x in ml/min. Using the data given

$$C_x = \frac{480 \text{ mg/ml} \times 1.5 \text{ ml/min}}{6 \text{ mg/ml}} = 120 \text{ ml/min}$$

Notice that the units of concentration cancel out, leaving the units for clearance as ml/min.

3. The answer is A. [*VI A 1 a, 3 a*] Clearance of a substance provides no information about the mechanism by which the kidney removes a substance from the plasma. In relation to this question, the data can be analyzed as follows:

$$\text{input} - \text{output} = \text{amount transported}$$
$$(\text{GFR} \times P_{\text{uric acid}}) - (U_{\text{uric acid}} \times \dot{V}) = \text{amount transported}$$

Substituting,

$$(180 \text{ L/day} \times 50 \text{ mg/L}) - 700 \text{ mg/day} = \text{amount transported}$$
$$9000 \text{ mg/day} - 700 \text{ mg/day} = 8300 \text{ mg/day}$$

Notice that approximately 8% of the filtered urate is excreted in the urine. Since the amount of uric acid filtered is higher than the amount excreted, this substance on a net basis is filtered and reabsorbed. However, in mammals (including humans), the tubular transport of uric acid is bidirectional in the proximal tubule, with reabsorption and secretion occurring within the same tubular cell. Uric acid is the end product of purine metabolism, with approximately 70% of the uric acid eliminated by the kidney. In humans, the bulk of the excreted urinary nitrogen is in the form of urea and ammonia, with excretion of uric acid accounting for only 5% of the total. Normally, more than 90% of the filtered uric acid is reabsorbed in the proximal tubule. Hence, tubular secretion must account largely for the bulk of uric acid excreted in the urine.

4. The answer is A. [*VI B 3 a, b*] The proximal tubule is the most important site of organic ion secretion because it contains an abundance of protein carriers. Secretion refers to the active or passive transport of a substance from the peritubular blood, interstitium, or tubular cell into the tubular lumen. The term "secretion" excludes the entry of substances into the tubular lumen by glomerular filtration. Para-aminohippuric acid (PAH) has a higher clearance than almost any other known substance. It is a weak organic acid, actively secreted primarily into the proximal tubule by a Tm-limited process. Creatinine, an endogenous product of muscle metabolism, enters the tubular lumen by the organic secretory pathway in the proximal tubule. Like PAH, creatinine is freely filtered and is not reabsorbed, synthesized, or metabolized by the kidney. Uric acid, derived from the metabolism of purine nucleotides, exhibits bidirectional tubular transport mainly in the proximal tubule, with reabsorption exceeding secretion.

5. The answer is B. [*II A 2 c*] Transcellular fluid is the extracellular fluid in the lumen of structures lined by epithelium. The largest component of the transcellular fluid is the intraluminal gastrointestinal water; other transcellular fluids include the cerebrospinal fluid, bile, synovial fluid, and ocular fluid. Transcellular fluid accounts for about 1 L of body fluid, which is equivalent to 2.5% and 1.5% of the total body water and body weight, respectively. The plasma, interstitial, edematous, and intravascular fluids all are extracellular fluids but do not contribute to the transcellular fluid compartment.

6. The answer is C. [*VI A 1 a, b, 3 d*] Tubular maximum or transport maximum, both abbreviated as Tm, is defined as the upper limit for the unidirectional rate of active transport, either reabsorptive or secretory, depending on the direction of solute transport. The units for Tm are mg/min. Thus, there is a maximal rate of reabsorption called maximum tubular reabsorptive capacity and a maximal rate of secretion called maximum rate of tubular secretory capacity. Tubular maxima vary with the substance involved. Renal reabsorption of glucose and secretion of PAH are examples of actively transported solutes exhibiting tubular transport maxima. An example of a substance that is not Tm-limited is the reabsorption of Na^+ along the nephron. Tubular maximum is the difference between the filtered load and the rate of excretion of a solute. Glomerular filtration rate (GFR) is defined as the volume of plasma filtered per minute (ml/min), while clearance is the virtual volume of plasma from which a substance is removed per minute (ml/min).

7. The answer is E. [*VI C 4*] The concentration of free H^+ at urinary pH above 4.4 is negligible; therefore, acid must be excreted in a buffered form. The two main urinary buffers are phosphate (HPO_4^-) and ammonia (NH_3). The amounts of secreted H^+ excreted bound to ammonia and phosphate are measured as ammonium (NH_4^+) and titratable acidity, respectively. The sum of ammonium and titratable acidity minus the amount of excreted bicarbonate (HCO_3^-) equals net acid excretion, which normally is approximately 1 mEq/kg body weight/24 hr. Ammonia production occurs primarily in the proximal tubule, and it represents the major adaptive mechanism available to the kidney for the increased excretion of H^+ during states of acidosis. In contrast, the phosphate buffer enters the tubular lumen by glomerular filtration. The normal kidney excretes almost twice as much acid combined with ammonia (40 mEq H^+/day) than it excretes titratable acid (20 mEq H^+/day). Approximately 4300 mEq of H^+ must be secreted per day to accomplish the reabsorption of 4300 mEq of bicarbonate. In this process, most of this secreted H^+ is reabsorbed in the form of water. Sulfate (SO_4^{2-}) in combination with ammonium forms a neutral salt [$(NH_4)_2SO_4$], which makes the tubular urine less acidic.

8. The answer is D. [*VI A 1 c*] Uric acid is the only organic substance that is both reabsorbed and secreted—two carrier-mediated processes that occur mainly in the proximal tubule. The reabsorption of uric acid exceeds its secretion. Lactate normally is completely reabsorbed by active transport. Urea, the nitrogenous end product of protein catabolism, is partially reabsorbed (35%–60%) from the tubular fluid by simple diffusion in many regions of the nephron and, therefore, does not exhibit a tubular maximum. Para-aminohippuric acid (PAH) and creatinine are secreted mainly by the proximal tubules. The secretion of PAH involves an active, carrier-mediated transport system and, therefore, exhibits a tubular (transport) maximum. The clearance of PAH is higher than that of almost any other known substance. When the plasma PAH concentration is low, virtually all the PAH that escapes filtration is secreted.

9. The answer is D. [*VI B 3; Figures 4-7, 4-11C*] At low plasma concentrations of PAH, PAH is almost completely cleared from the plasma by a combination of glomerular filtration and tubular secretion. When plasma concentrations of PAH are elevated beyond 30 mg/dl, the secretory mechanism becomes saturated, and the tubular transport maximum is reached. As the tubular secretory mechanism becomes saturated and is exceeded by progressive increments in plasma PAH concentration, the clearance of PAH declines and becomes more a function of glomerular filtration. The PAH clearance asymptotically approaches the inulin clearance. Thus, the amount of PAH secreted becomes a smaller fraction of the total amount of PAH excreted. The clearance of PAH always is greater than the clearance of inulin, since some PAH is always secreted.

10. The answer is E. [*IX D 2 c*] Renin secretion is not always required for aldosterone secretion. Renin is a proteolytic enzyme produced by the juxtaglomerular (JG) cells, which are modified smooth muscle cells in the renal afferent arterioles. The major stimulus for the release of renin is a decrease in the perfusion pressure of blood traversing these afferent arterioles. Thus, the JG cells function as low-pressure baroreceptors (volume receptors) where the secretion of renin varies inversely with the degree of stretch in the wall of the afferent arteriole. The major extrinsic factor influencing renin secretion is the renal postganglionic sympathetic vasomotor nerves, which are not necessary for renin secretion but exert a stimulatory effect (via β-adrenergic receptors) on the magnitude of renin release in response to a given stimulus. Increased K^+ concentration elicits aldosterone secretion by depolarizing the zona glomerulosa membranes; hyperkalemia inhibits renin secretion. Renin substrate, called angiotensinogen, is an α_2-globulin of hepatic origin containing a tetradecapeptide moiety that serves as the prohormone for the angiotensins.

11. The answer is D. [*VIII B 1 a, 3 a, b*] Normally, a 1%–2% increase in plasma osmolality causes an increase in plasma antidiuretic hormone (ADH) level. The central hypothalamic osmoreceptors are the primary regulators of ADH synthesis and release. The osmoreceptors communicate with the ADH-producing cells, mainly in the supraoptic nucleus. The renal regulation of blood volume and osmolality involves factors that modify the synthesis and release of ADH and aldosterone. Dehydration (hydropenia), trauma, hemorrhage (hypovolemia), pain, and anxiety are stimuli of ADH release. With massive blood loss, osmolality is sacrificed in order to maintain the volume of the circulation. Hyponatremia represents a decline in the osmolar concentration of plasma and, therefore, an inhibitory signal for ADH secretion. Hypertonic solutions of saline, sucrose, and mannitol stimulate ADH release, while increases in plasma osmolality due to glucose or urea have little or no effect.

12. The answer is A. [*II D 3; Figure 4-5; Table 4-3A*] The body fluid compartment with the highest protein concentration is the ICF, which has a protein concentration of about 60 mEq/L (4 mmol/ L). In decreasing order, the concentration of protein in the body fluid compartments is as follows:

Plasma	15 mEq/L	1 mmol/L
Interstitial fluid	1 mEq/L	0.06 mmol/L
Cerebrospinal fluid (CSF)	0.05 mEq/L	0.003 mmol/L

CSF is part of the transcellular fluid compartment. The milliequivalent concentrations of proteins in the body fluids are based on an average valence of -15 and an average molecular weight of 65,000 daltons.

13. The answer is C. [*III D 3; VI C 1, 2 a; Table 4-4*] The Na^+-K^+-ATPase enzyme in all cell membranes is the carrier molecule responsible for the active Na^+ transport (efflux) and the active K^+ transport (influx). This enzyme maintains the low intracellular Na^+ concentration and high extracellular Na^+ concentration. Active Na^+ transport consumes 30%–50% of the energy derived from metabolism in most cells. Although the influx of Na^+ from the tubular lumen to the proximal tubular cell is in the direction favored by the electrochemical potential, this transport is mediated by specific membrane carrier proteins and not by simple diffusion. These membrane proteins couple the active movement of other solutes to the passive movement of Na^+. Examples include Na^+-glucose and Na^+-amino acid cotransporters and a Na^+-H^+ countertransporter. In each case, the potential energy released by the downhill transport of Na^+ is used to power the uphill transport of the other substance. These transport systems are referred to as Na^+-coupled, secondary active transport processes. Most of the proximal reabsorption of Na^+ occurs by active transport and is transcellular. The transcellular pathway consists of luminal (apical) and basolateral membranes. It is essential to understand that reabsorption includes not only Na^+ influx from the tubular lumen but also active transport of Na^+ out of the cell into the bloodstream.

14. The answer is B. [*VI C 3 a–c*] Since the plasma K^+ concentration is approximately one-thirtieth that of Na^+, the filtered load of K^+ is much less than that of Na^+. K^+ and uric acid transport are bidirectional in the proximal tubule, with reabsorption and secretion occurring within the same tubular cell. Thus, K^+ is the only inorganic cation that is both reabsorbed and secreted by the kidney. Most of the filtered K^+ is actively reabsorbed in the proximal tubule, whereas the excreted K^+ is derived mainly from K^+ secretion in the distal tubule. Secretion involves the active pumping of K^+ into the distal tubular cells across the peritubular membrane followed by the passive diffusion across the luminal membrane into the tubular lumen. Insulin stimulates K^+ uptake by cells.

15. The answer is B. [*III C 1*] Small solutes are transported across glomerular capillaries by the passive process of convection (bulk flow). Convection is caused by a hydrostatic pressure difference causing the fluid with its dissolved material to flow from a region of high pressure to one of lower pressure. As there are no concentration gradients for substances across the glomerular membrane, there can be no simple diffusion. Primary active transport denotes a carrier mechanism for the movement of a solute against a concentration difference or a combined concentration and electrical potential difference, which is directly coupled to metabolic energy. Secondary active transport denotes a process that mediates the uphill movement of solutes, which is not directly coupled to metabolic energy expenditure; the energy required is derived from the concentration gradient of another solute (Na^+).

16. The answer is C. [*VI A 2; Figure 4-11A*] The renal threshold for a substance denotes the plasma concentration at which the solute begins to appear in the urine. It is not the plasma concentration that completely saturates the transport mechanism either for reabsorption or secretion. When the plasma concentration of a solute exceeds the renal threshold, the amount of that solute transported through the nephron exceeds the tubular transport maximum, and the solute appears in the urine in increasing amounts. The amount of a substance entering the tubule by filtration per unit time defines the filtered load.

17. The answer is B. [*II C 1–3; Figures 4-3, 4-4; Table 4-2*] Hydration states are named in terms of the ECF compartment. Overhydration, or fluid and salt retention, results from excessive influx of water and NaCl. Dehydration, or fluid and salt depletion, usually involves both the ECF and the ICF. The state of overhydration or dehydration is named only in terms of the volume of the ECF compartment that increases or decreases. The volume of the ECF compartment is determined by the Na^+ content and not by the Na^+ concentration. The situation described in the question represents a state of dehydration because of the contraction of the ECF volume. Since the solute concentration of the ECF also is decreased, there is a hyposmotic dehydration (e.g., as occurs with excessive salt loss in adrenal insufficiency). A net loss of salt in excess of water loss leads to hyposmolality of the ECF and a shift in water from the ECF to ICF. Thus, the volume of the ECF is decreased, the volume of the ICF is increased, and the osmolality of both is decreased.

18. The answer is C. [*II C 1 a; Figure 4-5*] The major factor regulating ECF volume is the reabsorption and excretion of sodium chloride. More than 90% of the osmotically active solutes in the ECF volume are salts of sodium. Most of these salts (e.g., sodium lactate, sodium citrate) exist in concentrations too small to affect the volume of the ECF. The concentrations of other sodium salts (e.g., sodium bicarbonate, sodium phosphate) are maintained within very narrow limits in order to maintain homeostatic functions other than ECF volume regulation.

19. The answer is B. [*IV B 1, C 2 b (1), (2); Figures 4-8, 4-9, 4-13*] The urine-to-plasma concentration ratio (U/P) reflects the concentration gradient between the urine and the plasma. Since inulin is neither reabsorbed from nor secreted into the nephron, the inulin concentration increases as water is reabsorbed. In fact, the inulin concentration in the tubular fluid or urine will be solely a function of the amount of filtered water that is reabsorbed. Normally, 70% of the filtered water is reabsorbed in the proximal tubules, and more than 99% is reabsorbed by the entire tubular system. A decrease in the urine-to-plasma inulin concentration ratio (U_{in}/P_{in}) with a constant GFR indicates that less water was reabsorbed, and the urinary inulin concentration was decreased. Thus, the urine flow was elevated.

20. The answer is A. [*II C 1 a, D 2 a*] Osmotic forces are the primary determinant of water distribution in the body. Consequently, the distribution of the total body water within the ICF and ECF compartments is determined by the number of osmotically active particles in each compartment. Sodium salts are the principal extracellular osmotically active solutes and act to hold water in the extracellular space. Urea is an ineffective osmole, while potassium salts account for almost all of the intracellular osmoles. The plasma concentration of protein is low (1.1 mmol/L or 15 mEq/L) as is that of creatinine (0.1 mmol/L).

21. The answer is A. [*IV C 1 c (1)–(3); Figure 4-7*] Renal clearance can be measured for any solute present in the plasma and excreted by the kidney; clearance best represents the rate of elimination of a substance by the kidney. If the kidney removes a substance (e.g., inulin) from the plasma by filtration only, the tubular clearance will be zero and the excretion rate will be proportional to the plasma concentration. If the clearance involves filtration and either tubular reabsorption or secretion, then clearance depends on the plasma concentration and the tubular transport capacity for reabsorption or secretion. Thus, clearance is higher for substances that are filtered and secreted than for substances that are filtered and reabsorbed. The lowest clearance is observed with substances that are completely reabsorbed, such as glucose. Approximately 50% of the filtered urea is reabsorbed. Creatinine and PAH are filtered and secreted, which increases their clearance compared to inulin.

22. The answer is B. [*IX B 3, D 2 c, d*] The only known physiologic effect of renin is to cause the formation of angiotensin I from its plasma substrate, angiotensinogen. Angiotensin I, in turn, appears to serve only as a specific substrate for angiotensin-converting enzyme (ACE), a peptidase that converts angiotensin I to angiotensin II, which does have significant biologic activity. Angiotensin II exerts a potent vasoconstrictive action on the vascular smooth muscle of peripheral arterioles, which causes an increase in the mean arterial blood pressure. It also stimulates the release of aldosterone from the zona glomerulosa of the adrenal cortex, causing an increase in the ECF volume through the increased active reabsorption of Na^+ by the distal tubules and collecting ducts. The increased ECF volume accounts for the decreased concentration of K^+, the decreased hematocrit, and the decreased plasma oncotic pressure. Aldosterone also acts on the distal tubule and collecting duct to increase the net reabsorption of Na^+ in exchange for the secretion of K^+ and H^+. Thus, aldosterone tends to produce hypokalemia and metabolic alkalosis.

23. The answer is D. [*IV B 1 a; VI B 3 a, b; Figures 4-7, 4-9*] Secretion refers to the transport of a substance from the peritubular blood, to the interstitium, to the tubular cell, and then into the tubular lumen regardless of whether the movement is active or passive. Secretion does not refer to the movement of substances across the lumen by glomerular filtration. Secretion is an active transport process that occurs mainly in the proximal tubule. PAH is actively secreted by the proximal tubular cells and is excreted without being metabolized. If a substance blocks the secretion of PAH, which is only filtered and secreted, then PAH is only filtered and neither reabsorbed nor secreted. Such a substance would resemble inulin, which is only filtered.

24. The answer is B. [*VI A 3; Figure 4-11A*] Using the data given, the Tm for glucose is found to be 350 mg/min. Tm refers to the maximal rate of transport of a substance. All substances that are reabsorbed or secreted have a Tm, which is expressed as an amount per minute. To determine whether the excretion of a substance involves either reabsorption or secretion (in addition to glomerular filtration), it is necessary to compare the excretion rate of the substance with the rate at which the substance is filtered at the glomerulus (filtered load). The algebraic difference between filtered load and the amount excreted determines which of the two processes is used by the kidney to excrete a substance. In the case of glucose, the amount excreted is less than the amount filtered, which is consistent with reabsorption, as

$$Tm = \text{amount filtered} - \text{amount excreted}$$
$$= (125 \text{ ml/min} \times 4 \text{ mg/ml}) - (75 \text{ mg/ml} \times 2 \text{ ml/min})$$
$$= 500 \text{ mg/min} - 150 \text{ mg/min}$$
$$= 350 \text{ mg/min}$$

25. The answer is B. [*VI B 1–3; Figure 4-11A*] When the filtered load of a substance exceeds the Tm for that substance, the excess is not reabsorbed, and, thus, proportionately more of the substance is excreted. In this example, the Tm for glucose (*point B* on the curve) is determined by extrapolation of the linear portion of the curve (from *point C* to *point D*) to the abscissa (filtered load). The splay of the curve (from *point A* to *point C*) indicates that all of the tubules are not uniform in length, number of glucose transporters, or renal threshold.

26. The answer is C. [*V A 1 a, C 1–3*] The major determinant of GFR is the hydrostatic pressure within the glomerulus. An increase in glomerular capillary hydrostatic pressure is a major outward directed force for fluid to leave the glomerular capillary and enter Bowman's capsule. In glomerular capillaries, the net movement of fluid is primarily or solely out of the capillaries, whereas in systemic capillaries the change in the balance of Starling forces is such that net movement out of the capillaries is nearly balanced by net return of fluid into the vessels. Increases in plasma oncotic pressure, hydrostatic pressure in Bowman's space, and afferent arteriolar resistance lead to a decrease in GFR, as does a decrease in net filtration pressure.

27. The answer is A. [*V A 1 a*] Ultrafiltration separates water and nonprotein constituents (the "crystalloids") of plasma—which enter Bowman's space—from the blood cells and protein macromolecules (the "colloids")—which remain in the blood. The glomerular ultrafiltration mechanism is governed by the same passive Starling's forces that determine the translocation of fluid across other capillaries in the body, namely, the imbalance between transcapillary hydrostatic and colloid osmotic (oncotic) pressures. The hydrostatic pressure of the afferent arteriolar blood exceeds the sum of the antifiltration forces—tubular hydrostatic pressure plus glomerular capillary oncotic pressure—thereby causing net filtration. Since albumin is the main component of the plasma oncotic pressure, which serves to hold fluid within the vascular space, the glomerular impermeability to albumin contributes to the maintenance of plasma volume by preventing renal loss of albumin.

28. The answer is A. [*II A 2; Figures 4-3, 4-4; Table 4-1*] The ECF volume constitutes approximately one-third of the total body water, or about 12–19 L. This compartment includes two subcompartments separated by the capillary membrane: intravascular fluid or blood plasma, and interstitial fluid. The ECF volume is measured with a test substance that does not penetrate the cells. Therefore, it has become useful to measure the volume distribution of a specific substance and refer to it as, for example, the inulin space, if the test substance is inulin. All substances used to measure ECF volume must cross capillaries and distribute at the same concentration in plasma and interstitial fluid.

29. The answer is E (all). [*VIII B 3 a (2)*] The major stimuli for ADH (vasopressin) secretion are an increase in the plasma osmolality and a decrease in the effective circulating blood volume. There is an inverse relationship between ADH secretion and the nerve impulse frequency in the autonomic (visceral) afferent neurons from the baroreceptors (also called volume receptors, pressoreceptors, or stretch receptors) in both low- and high-pressure regions of the vascular system. Thus, a decrease in vagal impulses (left atrium and aortic arch) and glossopharyngeal nerve impulses (carotid sinus) results in an increase in ADH secretion. The afferent neurons from the left atrium, carotid sinus, and aortic arch form a primary synapse in the nucleus tractus solitarius. The left atrial receptors are more sensitive to small changes in blood volume than are the carotid sinus and aortic arch baroreceptors.

30. The answer is B (1, 3). [*VII E; Figure 4-16*] The signs of hypernatremia and the excretion of a hyposmotic urine are consistent with a lack of ADH. The polyuria accounts for the water diuresis observed in this patient. The lack of ADH increases free-water clearance and results in the tendency to increase the concentration of the serum electrolytes, including Na^+.

31. The answer is C (2, 4). [*VI C 2 a (1)–(2), 4 a, b; Figure 4-6*] Although the influx of Na^+ across the luminal membrane is in the direction favored by the electrochemical gradient, Na^+ transport is mediated by specific membrane carrier proteins and is not by simple diffusion. In the proximal tubule, Na^+ combines with a glucose carrier and glucose is transported in the same direction against a concentration gradient. The Na^+ and glucose transport system is an example of cotransport (symport). The energy for the active transport of glucose comes from the Na^+ concentration gradient. Thus, a reduction in the intraluminal glucose concentration or a reduction in the number of carrier proteins in the proximal tubular cells will reduce Na^+ reabsorption. Na^+ reabsorption also is dependent on active H^+ secretion. Removal of the pituitary gland does not have a major effect on Na^+ reabsorption, because adrenocorticotropic hormone (ACTH) is not necessary for aldosterone secretion.

32. The answer is E (all). [*IX D 2*] The release of renin from the JG cells is inversely related to the degree of stretch in the wall of the afferent arteriole containing these cells. Thus, any factor that decreases renal perfusion pressure (e.g., hemorrhage, renal arterial vasoconstriction) will tend to reduce the renal perfusion pressure. The resultant increase in renin secretion will lead to the formation of angiotensin I which, in turn, is converted to angiotensin II by ACE, an enzyme that is identical to kininase II. Kininase II inactivates bradykinin, which is a vasodilator. The angiotensin II stimulates aldosterone secretion from the zona glomerulosa of the adrenal cortex.

33–37. The answers are: 33-C, 34-A, 35-E, 36-E, 37-C. [*III C 1 b; VII B 1 b, 2, 3 a, c, 4 b, D 2 a; IX D 2 a (2); Figures 4-12D, 4-13, 4-17*] The solute concentration in the ascending limb of the loop of Henle (*site C*) is less than that in any segment of the descending limb. The tubular fluid leaves the ascending limb at a lower concentration than it had when it entered the descending limb. Thus, the fluid presented to the distal tubule always is hyposmotic, regardless of the body's state of hydration.

Ultrafiltration separates water and nonprotein constituents (the "crystalloids") of plasma from the blood cells and protein macromolecules (the "colloids"). Except for proteins and lipids, the concentrations of crystalloids (e.g., Na^+, glucose) in the plasma and in Bowman's space (*site A*) are nearly the same.

The wall of the ascending limb of the loop of Henle is relatively impermeable to water. Therefore, NaCl in this segment is reabsorbed to the virtual exclusion of water, a process that renders the medullary and papillary interstitium hyperosmotic to plasma. The medullary interstitial osmolality is higher in antidiuresis than diuresis, due largely to urea. Thus, the highest osmolality exists in the papillary interstitium. With continued reabsorption of water, urea becomes even more concentrated at the terminals of the collecting ducts (*site E*).

During dehydration with maximal ADH secretion, the U_{osm}/P_{osm} approaches 4 to 1 (at *site E*) because of the increased free-water reabsorption.

The macula densa is located at the junction of the thick segment of the ascending limb of the loop of Henle and the distal convoluted tubule (*site C*).

38–42. The answers are: 38-C, 39-D, 40-C, 41-B, 42-A. [*IV B 1 b, C 1 c (1) (b), D 1 c, 2 a; VI C 3 a– c, 4; VII D*] K^+ is the only plasma inorganic electrolyte that is both reabsorbed from and secreted into the renal tubule. K^+ is largely actively reabsorbed by the proximal nephron, whereas the excreted K^+ is derived mainly from K^+ secretion in the distal nephron. Secretion involves active pumping of K^+ into the distal tubular cells across the peritubular membrane followed by passive diffusion across the luminal membrane into the tubular lumen.

Mannitol is a 6-carbon sugar alcohol that is filtered but not reabsorbed or secreted. Therefore, the intravenous administration of mannitol raises the osmolality of tubular fluid and decreases the reabsorption of water, so that an osmotic diuresis ensues.

HCO_3^- is completely filterable at the glomerulus. In a normal person, virtually all of HCO_3^- is actively reabsorbed, mainly by the proximal tubule. The remainder of the HCO_3^- reabsorption occurs in the loop of Henle and distal tubule. HCO_3^- also can be secreted by the cortical collecting tubules.

For practical purposes, endogenous creatinine clearance is equal to inulin clearance. However, exogenous creatinine is not only freely filtered but also is secreted by the proximal tubule. Thus, creatinine clearance exceeds inulin clearance by 20%–30%.

Urea is filtered and passively reabsorbed by the nephron. The reabsorption of urea is not limited to the proximal tubule. The net reabsorption of urea ranges between 40% and 60%.

43–45. The answers are: 43-E, 44-B, 45C. [*VI C 6 a; VII A 4; VIII C 1; Figures 4-12, 4-13*] ADH, a hormone primarily synthesized by the supraoptic nucleus of the diencephalon, is the major determinant of water permeability in the entire cortical and medullary collecting duct system. ADH also increases the passive permeability of the inner medullary collecting duct to urea.

Two-thirds of the glomerular filtrate is reabsorbed isosmotically by the proximal convoluted tubule. In this process, the total osmolality remains essentially unchanged.

The thin descending limb of the loop of Henle is relatively impermeable to urea but is highly permeable to water. Thus, water is reabsorbed from the descending limb in response to an osmotic gradient established by urea. The thin descending limb also is relatively impermeable to NaCl. Thus, as water (but not NaCl) is reabsorbed from this limb, the NaCl concentration of the tubular fluid increases until it exceeds the NaCl concentration in the interstitium. This latter process provides for the passive reabsorption of NaCl from the ascending limb of the loop of Henle.

46–51. The answers are: 46-C, 47-E, 48-E, 49-B, 50-D, 51-A. [*III D 3 a; VI B 2, 3 a, b, C 2 a, 4 a, b, 5 a, b; Figures 4-6, 4-7, 4-9, 4-11; Table 4-6*] Two-thirds of the filtered Na^+ is reabsorbed by the proximal tubule. Since two-thirds of the water also is reabsorbed proximally, the concentration of Na^+ remains unchanged.

The rate of urinary glucose excretion always is less than the rate of glucose filtration at the glomerulus. Thus, there is a net reabsorption of glucose that occurs solely in the proximal tubule. The tubular fluid-to-plasma concentration ratio (TF/P) for glucose falls to a value of 0.1, indicating that 90% of the filtered glucose is reabsorbed in the early portion (first quarter) of the proximal tubule. The filtered glucose is reabsorbed by an active, carrier-mediated process with Tm-limited characteristics.

Like glucose, glycine is transported from the tubular fluid into the proximal tubular cell by specific carrier molecules that also combine with Na^+. Thus, glycine transport is a Na^+-coupled, secondary active transport process. The TF/P for amino acids also falls to a value of 0.1, indicating that about 90% of the filtered glycine is reabsorbed in the initial 25% of the proximal tubule. The active reabsorption of amino acids also involves a Tm-limited process.

Early in the proximal tubule, the reabsorbed anion is chiefly HCO_3^-, leaving behind a fluid enriched in Cl^-. The rise in tubular fluid Cl^- concentration creates a gradient favoring the diffusion of Cl^- from tubular fluid to the interstitial space. Most of the Cl^- that is absorbed does not enter the cell at all but moves primarily via the paracellular pathway (i.e., through the tight junctions into the intercellular channel). About 60% of the filtered Cl^- is reabsorbed proximally.

Early in the proximal tubule, the reabsorbed anion is chiefly HCO_3^- and constitutes about 90% of the filtered HCO_3^-. The remaining 10% is reabsorbed in the distal tubule and the collecting ducts. Reabsorption of HCO_3^- is accomplished by the active transport of H^+ from the cell to the lumen. HCO_3^- added to the peritubular capillary is derived from the intracellular dissociation of H_2CO_3, whereas the filtered HCO_3^- is reabsorbed from the tubular fluid as CO_2 and H_2O. There is no absolute Tm for HCO_3^- because the reabsorption of HCO_3^- depends on Na^+ reabsorption.

PAH is an organic anion secreted by the proximal tubule, where there is an abundance of protein carriers. At low plasma concentrations, this exogenous substance is completely cleared by the kidney via filtration and secretion.

5
Acid-Base Physiology
John Bullock

I. ACID PRODUCTION AND ELIMINATION. Although the body produces large amounts of acid in two forms—carbonic (volatile) and noncarbonic (nonvolatile, or fixed) acids—the pH of the body fluids is maintained in an alkaline state (7.4). Most of the hydrogen ion (H^+) is formed as an end product of metabolism. The pathways for acid removal include the kidneys, lungs, and gastrointestinal tract.

A. Carbonic acid (H_2CO_3) formation. Because carbon dioxide (CO_2) can be formed from H_2CO_3 and, in turn, CO_2 can be eliminated by the lungs, H_2CO_3 is called a **volatile acid**.

    1. Unfortunately, the proton donor/acceptor terminology of Brønsted prevents the classification of CO_2 as an acid, but CO_2 functions as the single most important weak acid in the body fluids.

    2. Most of the CO_2 is derived from oxidative metabolism.

B. Noncarbonic acid formation. Much smaller amounts of fixed acids are produced, and, because they cannot be converted to CO_2, they are called **nonvolatile acids**. Noncarbonic acids are derived from three sources: diet, intermediary metabolism, and stool bicarbonate (HCO_3^-) loss.

    1. Diet. A high protein diet, which is common in Western countries, accounts for the formation of more acids than bases.

        a. Foodstuffs such as glucose and triglyceride are not acids in the body fluids but are converted to CO_2 during the course of their metabolism, and much of this CO_2 is hydrated to form H_2CO_3, which then dissociates into H^+ and HCO_3^-.

        b. A vegetarian diet will produce an excess of alkali, which must be excreted by the kidneys as HCO_3^-.

    2. Intermediary metabolism. The metabolism of foodstuffs does not always tend to acidify the body fluids; some foodstuffs have an alkalizing action. For example, the ingestion of large amounts of salts of organic acids found in fruit (e.g., lactate, isocitrate, citrate) alkalinize the body fluids because these organic ions are metabolized to CO_2 and H_2O, a process that involves the consumption of H^+. Between 40 and 60 mmol of inorganic and organic acids that are not derived from CO_2 are produced daily. About half of the metabolically produced acids are neutralized by bases in the diet, but the remainder must be neutralized by the buffer systems of the body.

        a. The metabolism of foodstuffs is a significant **source of noncarbonic acids**.

            (1) Lactate is produced from the anaerobic metabolism of glucose or glycogen. In cases of exercise or hypoxia, the excessive production of lactic acid can result in a transient increase in noncarbonic acid production.

            (2) Acetoacetic and β-hydroxybutyric acids are produced from the metabolism of triglycerides. These are the ketone bodies that are noncarbonic acids produced by normal subjects during fasting. Upon eating, acetoacetic and β-hydroxybutyric acids are further catabolized to CO_2 and H_2O. Excessive production of ketone bodies contributes to the noncarbonic acid "pool."

            (3) Phosphoric acid, which is produced from the metabolism of phospholipids, provides a major source of H^+.

            (4) Sulfuric acid is produced from the catabolism of proteins containing sulfhydryl groups in their cysteine, cystine, and methionine residues.

            (5) Uric acid is produced from the metabolism of nucleoproteins.

   b. Acetic acid (vinegar) functions transiently as a noncarbonic acid because humans can convert it rapidly to CO_2 and H_2O.

   3. Stool HCO_3^- loss
   a. The diet also contains organic cationic and anionic salts that may be metabolized to yield noncarbonic acids and bases (HCO_3^-). Organic anions that are metabolized in the body to HCO_3^- include acetate, citrate, and—in the presence of insulin—the anions of the ketoacids.
   b. Digestive processes result in the loss of 20–40 mmol of alkali in the stool, and this loss is equivalent to the addition of nonvolatile acid in the body.

C. Acid elimination. There are three processes available for maintaining H^+ concentration within normal limits:

   1. Combination of H^+ with a chemical buffer such as HCO_3^-, protein, phosphate, or hemoglobin

   2. Elimination of CO_2 by pulmonary ventilation

   3. Excretion of H^+ by the kidneys

II. THE HYDROGEN ION AND pH

A. Fundamental chemistry (Table 5-1). H^+ is a proton (i.e., a hydrogen atom without its orbital electron); H^+ in aqueous solution exists as a hydrated proton called the hydronium ion, or H_3O^+. pH refers to the negative Briggsian logarithm of the H^+ concentration. The gain and loss of protons constitutes acid-base chemistry. The currently accepted model of acid-base relationships is that proposed by Brønsted.

   1. An **acid** is a substance that acts as a proton donor.

   2. A **base** is a substance that accepts protons (i.e., H^+) in solution. Thus, bicarbonate ion (HCO_3^-), phosphate ion (HPO_4^{2-}), ammonia (NH_3), and acetate ion (CH_3COO^-) all are bases.

   3. Some substances are nearly equally divided between the acidic and basic forms at the normal H^+ concentration of the body. For example, the imidazole side groups of hemoglobin undergo the following reaction:

$$HHb \rightleftharpoons H^+ + Hb^-$$

   The acid, deoxyhemoglobin (HHb), dissociates to form H^+ and the conjugate base, Hb^-. HHb and Hb^- occur in about equal concentrations in blood cells.

B. Concept of pH and H^+ concentration (Table 5-2)

   1. **H^+ concentration is expressed in two different ways,** either directly as $[H^+]$ or indirectly as **pH**. (The symbol, $[H^+]$, refers to H^+ concentration in mol/L or Eq/L.) The relationship between $[H^+]$ and pH can be expressed as:

   a. $pH = \log_{10} \dfrac{1}{[H^+]}$

   b. $pH = -\log_{10} [H^+]$

   c. $[H^+] = 10^{-pH}$

Table 5-1. Important Buffer Acids at Physiologic $[H^+]$

Proton Donor (Conjugate Acid)*		Proton (H^+)		Proton Acceptor (Conjugate Base)*
$(CO_2)H_2CO_3$	$\rightleftharpoons$	H^+	+	HCO_3^-
$H_2PO_4^-$	$\rightleftharpoons$	H^+	+	HPO_4^{2-}
H · Protein	$\rightleftharpoons$	H^+	+	Proteinate$^-$
$HHbO_2$	$\rightleftharpoons$	H^+	+	HbO_2^-
HHb	$\rightleftharpoons$	H^+	+	Hb^-

*A conjugate acid can be an anion or a cation; a conjugate base usually is an anion; HCO_3^-, a conjugate base, also can be an acid.

Table 5-2. Relationship between pH and [H$^+$]

pH	[H$^+$] (nEq/L)
7.70	20
7.40 (plasma)	40
7.30 (CSF)	50
7.10 (ICF)	80
7.00	100
6.90	126

CSF = cerebrospinal fluid; ICF = intracellular fluid.

2. **pH is a dimensionless number** and should be treated as such; it should not be referred to in concentration units or as "pH concentration." In fact, the quantity whose logarithm determines the pH is a volume per equivalent, which is the inverse of concentration. Thus, pH could be correctly conceptualized as a logarithmic expression of the volume required to contain 1 equivalent of H$^+$. In human plasma at pH 7.4 [i.e., [H$^+$] of 40×10^{-9} mol (Eq)/L or 40 nmol (nEq)/L], that volume is 25 million L!

3. Because pH is the logarithmic expression of [H$^+$], it permits a graphic representation of a wide range of [H$^+$] values. (It is important to note that **pH and [H$^+$] are inversely related**.) Another advantage of the pH concept is that when the pK' of a buffer system is known, it is immediately possible to determine the effective pH range of the buffer.*

4. **A disadvantage of the pH system** is that it both inverts and uses the logarithmic scale to express [H$^+$]. For example, it is not immediately apparent that a decrease in pH from 7.4 to 7.1 represents a doubling of the [H$^+$] from 40 nmol/L to 80 nmol/L.

C. **H$^+$ concentration of body fluids** (Figure 5-1; Table 5-3)

1. **Blood and plasma.** With regard to [H$^+$], the body fluid compartment most studied is arterial blood plasma. The term **Blood pH** always refers to **plasma pH** (7.4), which is higher than the intracellular pH of the erythrocyte (7.2).

   a. In normal individuals, the [H$^+$] is approximately 40 nmol (nEq)/L, which is equivalent to pH 7.4. The range of [H$^+$] that is compatible with life is 20–126 nEq/L, which is equivalent to a pH range of 7.7–6.9.

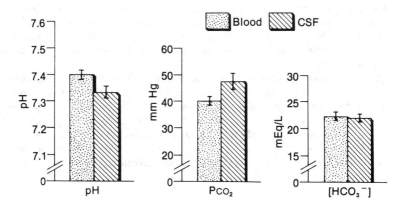

Figure 5-1. Comparison of the relationship between pH, Pco$_2$, and [HCO$_3^-$] in arterial blood and cerebrospinal fluid (*CSF*) in normal adult humans. Each vertical bar represents ± 1 SD. (Reprinted from Seldin DW, Giebisch G (eds): Acid-base balance in specialized tissues: central nervous system. In *The Regulation of Acid-Base Balance*. New York, Raven, 1989, p 109.)

---

*K = the ionization or dissociation constant; pK = the negative logarithm of K (– log K) and is equal to the pH at which half of the acid molecules are dissociated and half are undissociated; pK' = the apparent pK (see II D 2 a).

Table 5-3. H⁺ Concentration ([H⁺]) and pH of Biologic Fluids

Fluid	pH	[H⁺] (nEq/ or nmol/L)*	[H⁺] (Eq/L or mol/L)
Pure water	7.0	100	1×10^{-7}
Blood			
Normal mean	7.40	40	3.98×10^{-8}
Normal range	7.36–7.44	44–36	$4.36 \times 10^{-8} – 3.63 \times 10^{-8}$
Acidosis (severe)	6.9	126	1.26×10^{-7}
Alkalosis (severe)	7.7	20	2.00×10^{-8}
CSF (normal range)	7.36–7.44	44–36	$4.36 \times 10^{-8} – 3.63 \times 10^{-8}$
Pure gastric juice (normal)	1.0	100,000,000	1×10^{-1}
Urine			
Normal average	6.0	1000	1×10^{-6}
Maximum acidity	4.5	31,600	3.16×10^{-5}
Maximum alkalinity	8.0	10	1×10^{-8}
ICF (muscle)	6.8	158	1.58×10^{-7}

CSF = cerebrospinal fluid; ICF = intracellular fluid. (Adapted from Brobeck JR (ed): Regulation of hydrogen ion concentration in body fluids. In *Best and Taylor's Physiological Basis of Medical Practice*, 10th edition. Baltimore, Williams & Wilkins, 1979, p 5–13.)
*n = nano- = 10^{-9}. Thus, 100 nEq/L = 100×10^{-9} Eq/L, where nEq = nmol of a monovalent ion.

    b. In normal individuals at rest, the pH of mixed venous blood is 7.38 compared to 7.41 for arterial blood because of the uptake of CO_2 by blood as it perfuses the tissues.

    c. The [H⁺] of plasma is very small compared to that of other ions in plasma (e.g., the plasma Na⁺ concentration is about 142 million nmol/L, and the K⁺ concentration is about 4 million nmol/L).

  2. Cerebrospinal fluid (CSF) is essentially a bicarbonate buffer with a negligible concentration of protein (between 2×10^{-2} g/dl and 4×10^{-2} g/dl) or other nonbicarbonate buffers.

    a. The arterial pH (7.4) is higher than that of CSF (7.32), because the CO_2 tension (P_{CO_2}) is about 48 mm Hg in the CSF and about 40 mm Hg in the arterial blood (see Figure 5-1).

    b. Generally, it is possible to characterize acid-base disturbances precisely in terms of the blood data; however, the extracellular changes do not always reflect the intracellular changes. Also, acid-base alterations in arterial blood may produce similar or opposite changes in the CSF depending on whether the blood [H⁺] changes are due to respiratory or metabolic abnormalities.

      (1) An **increase in arterial P_{CO_2} (respiratory acidosis)** leads to a parallel rise in the CSF P_{CO_2}, with a resulting increase in the [H⁺] of both arterial blood and CSF.

      (2) A **decrease in arterial P_{CO_2} (respiratory alkalosis)** leads to a parallel decline in CSF P_{CO_2}, with a resulting decrease in the [H⁺] of both arterial blood and CSF.

      (3) An **increase in arterial [H⁺] (metabolic acidosis)** stimulates ventilation, lowering the P_{CO_2} of the arterial blood and CSF, resulting in CSF alkalosis and blood acidosis.

      (4) A **decrease in arterial [H⁺] (metabolic alkalosis)** decreases ventilation, raising the P_{CO_2} of the arterial blood and CSF, resulting in CSF acidosis and blood alkalosis.

D. Henderson-Hasselbalch equation

  1. Significance. Acid-base balance is maintained primarily through the control of two organ systems. The lungs control the P_{CO_2} through the regulation of alveolar ventilation, and the kidneys control the HCO_3^- concentration ([HCO_3^-]). The classic description of the acid-base state is based on the **Henderson-Hasselbalch equation,** which is an expression of three variables (pH, P_{CO_2}, and [HCO_3^-]) and two constants (pK' and S).

2. Definition of parameters. The Henderson-Hasselbalch equation could functionally be written as

$$pH = pK' + \log \frac{kidneys}{lungs} \qquad (1)*$$

However, the equation is expressed more usefully as

$$pH = pK' + \log \frac{[HCO_3^-]}{S \times Pco_2} \qquad (2)$$

or, using specific values for pK' and S (defined in II D 2 a and b), as

$$pH = 6.1 + \log \frac{[HCO_3^-]}{0.03 \times Pco_2} \qquad (3)$$

From equation (3) it is clear that the value of arterial pH depends on the ratio of $[HCO_3^-]$ to $S \times Pco_2$, not on the individual value of each variable. In clinical medicine, pH, Pco_2, and $[HCO_3^-]$ can be measured directly. With equation (3), however, any one of the variables can be calculated if the other two are known.

a. pK is defined as the negative logarithm of the $[H^+]$ at which half the acid molecules are undissociated and half are dissociated. When equimolar concentrations of weak acid and conjugate base exist, the pH value equals the pK (i.e., the log of 1 is 0).

   (1) The **actual dissociation constant (K)** for carbonic acid (H_2CO_3) in dilute aqueous solution at 38° C is 1.6×10^{-4} mol/L (pK = 3.8). Thus, H_2CO_3 is almost completely dissociated in the body where $[H^+] = 4 \times 10^{-8}$ mol/L, and it exists in quantities that are too small to be analyzed (i.e., 2.4×10^{-4} mEq/L). The formation and dissociation of H_2CO_3 is expressed as

$$\underset{\substack{\text{alveolar} \\ \text{gas}}}{CO_2} \rightleftharpoons \underset{\text{plasma}}{CO_2} + H_2O \underset{500:1}{\overset{\substack{\text{carbonic} \\ \text{anhydrase}}}{\rightleftharpoons}} H_2CO_3 \underset{4000:1}{\rightleftharpoons} H^+ + HCO_3^- \qquad (4)$$

At equilibrium there are approximately 500 mmol of CO_2 for every 1 mmol of H_2CO_3 and approximately 4000 mmol of H_2CO_3 for every 1 mmol of H^+. Because of the presence of carbonic anhydrase, equilibrium between CO_2 and H_2CO_3 is rapid and constant.†

   (2) Because the denominator of equation (3) is increased by a factor of 500, the **apparent dissociation constant (K')** for the CO_2/HCO_3^- buffer system in plasma at 38° C is correspondingly smaller (8×10^{-7} mol/L or 800 nmol/L), and the **apparent pK (pK')** for this same buffer pair is correspondingly larger (6.1). As a rule, the optimal buffer region of a buffer pair system is within a range of ± 1 pH units of its pK value.

   (3) It is important to note that, because CO_2 increases $[H^+]$, as shown in equation (4), it is considered an acid even though it is an acid **anhydride**. Thus, dissolved CO_2 is present as a potential H^+ donor, and its concentration is proportionate to the true donor, H_2CO_3. For these reasons, it is more meaningful to characterize acid-base disturbances in terms of the CO_2/HCO_3^- buffer system instead of the H_2CO_3/HCO_3^- system. For all practical purposes, the H_2CO_3/HCO_3^- buffer pair can be considered to be composed of HCO_3^- (conjugate base) and dissolved CO_2 (conjugate "acid").

b. S is defined as the solubility constant for CO_2 in plasma at 38° C and is equal to 0.03 mmol/L/mm Hg. S represents the proportionality constant between CO_2 and Pco_2. For blood plasma at 38° C, the amount of dissolved CO_2 is expressed as

$$\text{dissolved } CO_2 = 0.03 \times Pco_2$$

where CO_2 = the millimoles of dissolved CO_2 per liter of plasma. At an arterial Pco_2 of 40 mm Hg, the CO_2 concentration $([CO_2])$ is expressed more usefully as

$$0.03 \text{ mmol/L/mm Hg} \times 40 \text{ mm Hg} = 1.2 \text{ mmol/L}$$

Multiplying the proportionality constant by milliliters of CO_2 per millimole (22.3 ml/mmol)

---

*The kidneys primarily regulate $[HCO_3^-]$, the numerator; the lungs mainly regulate Pco_2, the denominator.

†Carbonic anhydrase is found in erythrocytes, gastric parietal cells, renal tubular cells, pancreatic and pulmonary tissue, bone, and the eye; it is not found in muscle, peripheral nerves, or skin.

yields a solubility constant of 0.67 ml/L/mm Hg for CO_2 in plasma. At this solubility constant and at a P_{CO_2} of 40 mm Hg, the concentration of dissolved CO_2 is expressed as

$$0.67 \text{ ml/L/mm Hg} \times 40 \text{ mm Hg} = 26.8 \text{ ml/L*}$$

c. **[HCO₃⁻]** denotes the bicarbonate ion concentration, which is expressed in millimolarity rather than molarity. The normal value of $[HCO_3^-]$ in plasma is 24 mmol (mEq)/L.

3. **Calculations with the Henderson-Hasselbalch equation**
   a. **Traditional calculation of pH.** pH in normal arterial plasma = 7.4; normal P_{CO_2} = 40 mm Hg, and normal $[HCO_3^-]$ = 24 mmol/L (or 24 mEq/L). Applying the Henderson-Hasselbalch equation, pH is calculated as

$$pH = 6.1 + \log \frac{[HCO_3^-]}{0.03 \times P_{CO_2}}$$

$$= 6.1 + \log \frac{24 \text{ mmol/L}}{0.03 \times 40 \text{ mm Hg}}$$

$$= 6.1 + \log \frac{24 \text{ mmol/L}}{1.2 \text{ mmol/L}}$$

$$= 6.1 + \log 20$$

$$= 7.4$$

   b. **Simple calculation of [H⁺] and pH**
      (1) **Conversion of pH to [H⁺]: a close approximation**
         (a) **Equation.** The $[H^+]$ in plasma is expressed in nmol/L. The equation for $[H^+]$ now becomes

$$\log [H^+] - 9 = - pH \text{ or } [H^+] = \text{antilog } (9 - pH)$$

         (b) **Examples** (see Table 5-2)
            (i) At pH 7.4

$$[H^+] = \text{antilog } (9 - 7.4)$$
$$= \text{antilog } 1.6$$
$$= 39.6 \text{ nmol/L}$$

            (ii) At pH 6.9

$$[H^+] = \text{antilog } (9 - 6.9)$$
$$= \text{antilog } 2.1$$
$$= 126 \text{ nmol/L}$$

      (2) **Conversion of [H⁺] to pH**
         (a) **Equation.** The $[H^+]$ in plasma is expressed in nmol/L. The equation for pH is

$$pH = 9 - \log [H^+]$$

         (b) **Examples** (see Table 5-2)
            (i) At $[H^+]$ = 40 nmol/L

$$pH = 9 - \log 40$$
$$= 9 - 1.6$$
$$= 7.4$$

            (ii) At $[H^+]$ = 126 nmol/L

$$pH = 9 - \log 126$$
$$= 9 - 2.1$$
$$= 6.9$$

E. **Henderson equation**

1. **Use of the Henderson equation.** This equation provides a simple means for converting pH to $[H^+]$, since P_{CO_2} and $[HCO_3^-]$ will have been provided. Additionally, this equation provides a simple **arithmetic estimate** of $[HCO_3^-]$, because the value of $[H^+]$ often will have been approximated from the pH value reported by the laboratory.

---

*This amounts to about 5% of the total amount of CO_2 carried in the arterial blood.

2. **Calculations with the Henderson equation.** The Henderson equation represents the non-logarithmic (arithmetic) method of determining $[H^+]$ or $[HCO_3^-]$ as

$$[H^+] = K' \frac{P_{CO_2}}{[HCO_3^-]} \text{ or } [HCO_3^-] = K' \frac{P_{CO_2}}{[H^+]} \tag{5}$$

   a. A value for K', is derived by converting the apparent pK for carbonic acid (6.1) into a dissociation constant (expressed in nmol/L) and multiplying that value by the solubility constant for CO_2 in plasma at 38° C (i.e., 0.03). K' has a value of 800 nmol/L [see II D 2 a (2)], which, when multiplied by 0.03, equals 24.

   b. Using normal values for P_{CO_2}, $[H^+]$, and $[HCO_3^-]$, the Henderson equation is applied as

$$[H^+] = 24 \frac{P_{CO_2} \text{ (mm Hg)}}{[HCO_3^-] \text{ (mmol/L)}}$$

$$= 24 \frac{40}{24} = 40 \text{ nmol/L}$$

Similarly,

$$[HCO_3^-] = 24 \frac{P_{CO_2} \text{ (mm Hg)}}{[H^+] \text{ (nmol/L)}}$$

$$= 24 \frac{40}{40} = 24 \text{ mmol/L}$$

Additionally,

$$P_{CO_2} = \frac{[H^+] \times [HCO_3^-]}{24}$$

$$= \frac{40 \times 24}{24} = 40 \text{ mm Hg}$$

III. **BODY BUFFER SYSTEMS.** A buffer is a solution consisting of a weak acid and its conjugate base. Buffering is the primary means by which large changes in $[H^+]$ are minimized.

   A. **Types of body buffer systems**

   1. **Blood buffers** (Table 5-4)
      a. It is important to note that the blood buffers are not solely plasma buffers, but that hemoglobin, HCO_3^-, and phosphate are found in erythrocytes and act as important blood buffers.
         (1) In blood, the chief H^+ acceptor is HCO_3^-, which exists in a concentration of 20–30 mEq/L.

Table 5-4. Whole Blood Buffers

Buffer Type	Buffering Capacity of Whole Blood (%)
Bicarbonate	
Plasma	35
Erythrocyte	18
Total bicarbonate	53
Nonbicarbonate	
Hemoglobin and oxyhemoglobin	35
Plasma proteins	7
Organic phosphate	3
Inorganic phosphate	2
Total nonbicarbonate	47

Adapted from Brobeck JR (ed): Regulation of hydrogen ion concentration in body fluids. In *Best and Taylor's Physiological Basis of Medical Practice*, 10th edition. Baltimore, Williams & Wilkins, 1979, p 5–14.

(2) H^+ acceptor is available in the hemoglobin system, where reduced hemoglobin (deoxyhemoglobin) is a stronger base than oxyhemoglobin (i.e., deoxyhemoglobin has a stronger capacity to combine with H^+).

(3) The hemoglobin buffer system is quantitatively as important as the bicarbonate buffer system.

  b. The major buffer anions of whole blood—HCO_3^-, protein, and hemoglobin—have a total concentration of approximately 48 mEq/L [see VII B 1 a (2)].

  c. With the bicarbonate and hemoglobin systems taken together, 5 L of blood of a normal adult have sufficient buffer capacity to combine with almost 150 mmol of protons (150 ml of 1 *N* HCl) before the pH of body fluids becomes dangerously acidic.

  2. Tissue buffers. The major buffer capacity of the body is not in the blood but in the H^+ acceptors found in other tissues, principally in the muscle and in bone. These tissues can neutralize about five times as much acid as the blood buffers.

  a. Muscle

    (1) Since skeletal muscle represents about half of the cellular mass, most intracellular buffering presumably occurs in muscle.

    (2) For any individual, the body HCO_3^- concentration averages 13 mEq/kg body weight. Muscle cells contain HCO_3^- at a concentration of about 12 mEq/L, and most other cells contain it at higher concentrations. The intracellular fluid (ICF) and extracellular fluid (ECF) compartments each contain about 50% of the total body HCO_3^-.

  b. Bone plays an important role in buffering H^+.

    (1) The total body store of bone carbonate is about 50 times the amount of HCO_3^- in the ICF and ECF compartments together. Bone carbonate appears to be the predominant source of base for neutralizing excess noncarbonic acid in the ECF. Indeed, it has long been recognized that chronic noncarbonic acidosis causes bone dissolution (resorption) through the loss of calcium carbonate ($CaCO_3$). The early carbonate release is in the form of sodium carbonate (Na_2CO_3).

    (2) Bone contains about 80% of the total CO_2 (including CO_3^{2-}, HCO_3^-, and CO_2) in the body. About two-thirds of this CO_2 is in the form of CO_3^{2-} complexed with Ca^{2+}, Na^+, and other cations located in the lattice of the bone crystals. The other third consists of HCO_3^- and is located in the hydration shell of the hydroxyapatite crystal, an inorganic compound found in the matrix of bone and teeth.

B. Distribution of body buffer systems in major fluid compartments

  1. Blood, in regard to its buffering activity, usually is considered as a whole rather than in terms of its components. Blood buffers are described here in terms of separate compartments (i.e., plasma and erythrocytes) for didactic purposes only.

  a. Important concepts

    (1) Whole blood is an excellent buffering system for noncarbonic acids because of its nonbicarbonate buffers as well as its bicarbonate buffer system.

    (2) More than 90% of the blood's capacity to buffer carbonic acid is attributed to the hemoglobin buffer system. Thus, the nonbicarbonate buffer in the erythrocyte is quantitatively more important than the bicarbonate buffer in that compartment. The bicarbonate buffer system remains quantitatively important, however, in the erythrocyte (see Table 5-4).

    (a) The bicarbonate buffer system does not function as a buffer for carbonic acid.

    (b) The nonbicarbonate buffer systems can buffer both noncarbonic and carbonic acids.

  b. Buffer capacity of blood components

    (1) Plasma, which contains three buffer systems, has a considerable capacity for buffering noncarbonic acids but a much smaller capacity for buffering carbonic acid.

      (a) Bicarbonate buffer system (HCO_3^-/H_2CO_3 or HCO_3^-/CO_2). This buffer exists in a concentration of 24 mmol/L of plasma. When **noncarbonic acid** is added to normal plasma, more than 75% of the buffering capacity of plasma is due to the HCO_3^-/CO_2 system. Most of the remaining buffering of noncarbonic acid involves the plasma protein buffers and, to a small degree, the phosphate buffer system. Again, the HCO_3^-/CO_2 system plays no role in the buffering of carbonic acid.

      (b) Nonbicarbonate buffer systems

        (i) Plasma protein (Protein$^-$/H · Protein). Plasma is a salt solution containing 7% protein, which exists as polyanions at the pH of plasma. Plasma protein H^+

acceptors exist in a concentration of 1.5 mmol/L of plasma and account for less than one-sixth of the total buffering capacity of whole blood.

    (ii) **Inorganic orthophosphate** ($HPO_4^{2-}/H_2PO_4^-$). Because this buffer system exists in a concentration of only 0.66 mmol/L of plasma, it contributes little to the total buffering activity of plasma.* At a plasma pH of 7.4, the concentration ratio of $HPO_4^{2-}/H_2PO_4^-$ is 4:1. Therefore, 80% of the inorganic phosphate exists as disodium phosphate, and 20% is in the form of monosodium phosphate. The $HPO_4^{2-}/H_2PO_4^-$ system is a major elimination route for H^+ via the urine, which has a relatively high phosphate content.

  (2) Erythrocytes. Although the blood contains other cells, the erythrocyte is the only important cellular component of the blood buffer system. The erythrocyte contains four buffer systems.

    (a) Bicarbonate buffer system. The HCO_3^-/CO_2 system exists in a concentration of 15 mmol/L of erythrocytes, compared to a concentration of 21 mmol/L and 25 mmol/L of whole blood and plasma, respectively.

    (b) Nonbicarbonate buffer systems

      (i) Hemoglobin buffers (Hb^-/HHb and $HbO_2^-/HHbO_2$) [see III C]. One L of **erythrocytes** contains 334 g (5.1 mmol) of hemoglobin. One L of **whole blood** contains 150 g (2.3 mmol) of hemoglobin.

      (ii) Organic phosphate. Although the erythrocyte contains a significant amount of organic phosphate buffer, this amount is quantitatively small compared to the bicarbonate and hemoglobin buffer concentrations in the erythrocyte.

      (iii) Inorganic orthophosphate. The $HPO_4^{2-}/H_2PO_4^-$ system exists in a concentration of 2 mmol/L of erythrocytes.

2. Interstitial fluid (including lymph)

  a. Bicarbonate buffer system. The HCO_3^-/CO_2 system is quantitatively the most important buffer of noncarbonic acid. The $[HCO_3^-]$ of the interstitial fluid is 27 mmol/L, which is similar to, or about 5% higher than, the $[HCO_3^-]$ of plasma. It is important to consider that, in humans, the interstitial fluid volume is about three times that of plasma; therefore, the total capacity of the interstitial fluid to buffer noncarbonic acid is considerably greater than that of the total blood volume to buffer these acids. **On a per unit of volume basis,** however, **the interstitial fluid has nearly the capacity of plasma to buffer noncarbonic acids**.

  b. Nonbicarbonate buffer system. The $HPO_4^{2-}/H_2PO_4^-$ system exists in a concentration of 0.7 mmol/L of interstitial fluid; therefore, this compartment has little capacity to buffer carbonic acid. The interstitial fluid is essentially free of protein.

3. ICF (excluding erythrocytes)

  a. Bicarbonate buffer system. The ICF contains only about 12 mmol of HCO_3^-/L in skeletal and cardiac muscle.

  b. Nonbicarbonate buffer systems. Protein and organic phosphate compounds exist in quantitatively significant amounts in the ICF, giving this compartment the capacity to effectively buffer both noncarbonic and carbonic acids as well as alkali. The ICF concentrations of these major buffer anions are:

    (1) $HPO_4^{2-}/H_2PO_4^-$ (skeletal muscle) = 6 mmol/L

    (2) Protein$^-$/H · Protein (skeletal muscle) = 6 mmol/L

    (3) Organic anions (skeletal muscle) = 84 mmol/L

C. Hemoglobin: an "extracellular" buffer

1. Important concepts. Although hemoglobin is found intracellularly, it is more conventionally regarded as extracellular and, therefore, part of the extracellular buffer system because:

  a. Hemoglobin is confined to the erythrocyte, which is a cellular component of the ECF.

  b. Hemoglobin is readily available for the buffering of extracellular acids.

  c. Hemoglobin is the primary nonbicarbonate buffer of the blood.

---

*The pK' of the acid form (i.e., $H_2PO_4^-$) is 6.8. With this pK', the $HPO_4^{2-}/H_2PO_4^-$ system would be a more effective buffer than the HCO_3^-/CO_2 system (pK' = 6.1) if it were present in an appreciable concentration. Many of the organic phosphate compounds found in the body have pK' values within half of a pH unit from 7.0.

2. Hemoglobin as a buffer. Like all proteins, hemoglobin is a buffer. At pH 7.2 (i.e., the pH of normal arterial erythrocytes), the buffering action of hemoglobin is due mainly to the imidazole groups of the histidine residues.

a. The titration curves of deoxyhemoglobin and oxyhemoglobin in Figure 5-2 illustrate the basis for the ability of hemoglobin to neutralize H^+ formed subsequent to the diffusion of CO_2 into the erythrocyte. Most of the H^+ is buffered by hemoglobin and most of the HCO_3^- diffuses into the plasma.*

b. Oxyhemoglobin dissociates more completely than does deoxyhemoglobin, and, as a result, deoxyhemoglobin produces less H^+ at a given pH than does oxyhemoglobin, which is a stronger acid. Thus, hemoglobin becomes a more effective buffer when CO_2 and, hence, H^+ are added from the tissues. This is important because the diffusion of CO_2 from the tissues to the capillary blood is accompanied by the simultaneous reduction of oxyhemoglobin.

c. As the uptake of CO_2 depends on H^+ acceptors, this increase in H^+ acceptors in the form of deoxyhemoglobin facilitates the uptake and buffering of the H^+ generated by the hydration of CO_2 and the dissociation of H_2CO_3. As a result of these reactions, CO_2 is converted into HCO_3^- within the erythrocyte.

(1) For each mmol of oxyhemoglobin that is reduced, about 0.7 mmol of H^+ can be taken up and, consequently, 0.7 mmol of CO_2 can enter the blood without a change in pH (see Figure 5-2, *point A* to *point C*).

(2) A reaction that causes no change in $[H^+]$ (or pH) is called **isohydric buffering**.

d. Respiratory quotient (R.Q.). When the body is at rest, the rate of O_2 consumption ($\dot{V}O_2$) under standard conditions is 250–350 ml/min, and the rate of CO_2 production ($\dot{V}CO_2$) is 200–250 ml/min. (The dot over the symbol V denotes **volume per unit time**.) R.Q. represents the ratio of $\dot{V}CO_2$ to $\dot{V}O_2$. Normally, on a mixed diet, the $\dot{V}O_2$ exceeds the $\dot{V}CO_2$, and the R.Q. is less than 1.

(1) The metabolic R.Q. equals the **molar ratio** of CO_2 production rate to the corresponding O_2 consumption rate by metabolizing tissues.

(2) If the R.Q. is 0.7, then, for 1 mmol of O_2 consumed, 0.7 mmol of CO_2 is produced, which, when converted to H_2CO_3, yields 0.7 mmol of H^+ upon dissociation.

(3) The complete deoxygenation of 1 mmol of oxyhemoglobin to liberate 1 mmol of O_2 results in the neutralization of 0.7 mmol of H^+ without a change in pH. Thus, all of the H^+ produced when the R.Q. is 0.7 can be buffered by deoxyhemoglobin with no change in pH (see Figure 5-2).

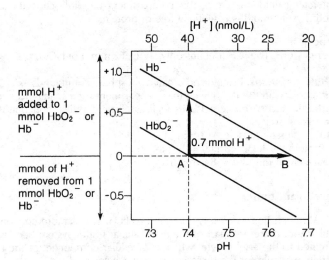

Figure 5-2. Titration curves of oxyhemoglobin (HbO_2^-) and deoxyhemoglobin (Hb^-), illustrating the importance of hemoglobin as a buffer. The complete deoxygenation of 1 mmol of HbO_2^- to liberate 1 mmol of O_2 results in the neutralization of 0.7 mmol of H^+ without a change in pH. *Arrow AC* indicates the amount of H^+ that can be added during the reduction of hemoglobin without causing a pH change. *Arrow AB* represents the pH change that would occur if the oxyhemoglobin at pH 7.4 was completely reduced. The reduction of HbO_2^- to Hb^- would cause a large increase in pH if CO_2 and, hence, H^+ were not added simultaneously to the system. Reduction of hemoglobin denotes the O_2-free state without a change in the valence of iron. (Adapted from White A, et al: Hemoglobin and the chemistry of respiration. In *Principles of Biochemistry*, 5th edition. New York, McGraw-Hill, 1973, p 843.)

*The buffer capacity of nonbicarbonate buffers is dependent primarily on the hemoglobin concentration.

e. At pH 7.2, about 84% of the deoxyhemoglobin is in the form of HHb, whereas only about 23% of the oxyhemoglobin is in the form of $HHbO_2$. Of the total oxyhemoglobin that causes O_2 to form deoxyhemoglobin:

   (1) 23% was already combined with H^+,

   (2) 16% will not combine with H^+, and

   (3) 61% will take up H^+ before a pH decrease occurs.

f. CO_2 entering the erythrocytes rapidly undergoes two reactions.

   (1) CO_2 is hydrated to form H_2CO_3, a reaction that is catalyzed by carbonic anhydrase in the erythrocyte. The H_2CO_3 dissociates to form HCO_3^- and H^+, which is buffered primarily by the hemoglobin buffers. Much of the HCO_3^- formed within the erythrocyte diffuses into the plasma in exchange for Cl^-.

   (2) CO_2 combines with the amino groups of deoxyhemoglobin to form carbaminohemoglobin. The carbaminohemoglobin dissociates to a carboxylate anion and H^+, which is buffered primarily by the hemoglobin buffers. In summary:

$$Hb \cdot NH_2 + CO_2 \rightleftharpoons Hb \cdot NHCOO^- + H^+$$

g. The unloading of O_2 from oxyhemoglobin to the tissues causes the formation of deoxyhemoglobin that is better able to tie up the H^+ produced by the simultaneous uptake of CO_2. The loss of O_2 from hemoglobin facilitates the uptake of CO_2 in the form of carbaminohemoglobin by the erythrocytes. Oxyhemoglobin contains about 0.1 mmol and deoxyhemoglobin about 0.3 mmol of carbaminohemoglobin per mmol.

IV. GENERATION AND ELIMINATION OF H^+.

The greatest source of H^+ is the CO_2 produced as one of the end products of the oxidation of glucose and fatty acids during aerobic metabolism.

A. Carbonic acid (H_2CO_3) is an acid that forms a volatile end product upon dehydration and can, therefore, be excreted by the lungs.

   1. Hydration-dehydration: dissociation-association reaction

   a. CO_2 can form H^+ according to the following series of reactions:

$$CO_2 + H_2O \underset{\text{dehydration}}{\overset{\text{hydration}}{\rightleftharpoons}} H_2CO_3 \underset{\text{association}}{\overset{\text{dissociation}}{\rightleftharpoons}} H^+ + HCO_3^-$$

   HCO_3^- is formed through the hydration of CO_2 and the subsequent dissociation of H_2CO_3. A significant concentration of HCO_3^- cannot be formed by this reaction because the dissociation of H_2CO_3 forms equal numbers of H^+ ions and HCO_3^- ions. Therefore, the significant concentration of HCO_3^- can only be attained after the H^+ formed from H_2CO_3 is buffered by blood buffer anions (i.e., reduced hemoglobin, protein, and phosphate) as:

$$H^+ + HCO_3^- + Na^+ + \text{buffer} \rightleftharpoons H \cdot \text{buffer} + Na^+ + HCO_3^-$$

   b. Thus, for every H_2CO_3 molecule that has its H^+ taken up by a buffer, one HCO_3^- appears in the blood. These HCO_3^- ions are distributed between the erythrocytes and the plasma.

   2. CO_2 production (Table 5-5). CO_2 is the chief product of metabolism and represents the greatest portion of the acid that is continuously eliminated from the body by the lungs.

   a. Most CO_2 in the body is produced from the decarboxylation reactions of the tricarboxylic (citric) acid cycle.

   b. HCO_3^- is an excellent buffer with respect to noncarbonic acid added to the blood by diet, metabolism, and disease, but the HCO_3^- buffer system plays no role in the buffering of carbonic acid.

Table 5-5. CO_2 Production under Basal Conditions*

Time Elapsed	Volume (L)	Molarity (mmol)	(mol)
1 minute	0.2	9	9×10^{-3}
1 hour	12.0	540	0.54
1 day	300	13,500	13.5

*Data for a respiratory quotient of 1.0.

    c. Comparing CO_2 and HCO_3^-, there is more CO_2 than HCO_3^- produced metabolically in the body. However, the ECF contains a preponderance of the HCO_3^- form.

    d. Under basal conditions, with a R.Q. of 0.82, the average adult produces about 300 L (13 mol)* of CO_2 per day.

    e. The lungs and kidneys are the principal routes for the elimination of protons and for maintaining normal $[HCO_3^-]/S \times P_{CO_2}$ ratios.

        (1) In the course of a day, the equivalent of 20–40 L of 1 N acid (20,000–40,000 mEq H^+)† are eliminated via the lungs, or, more correctly, the amount of CO_2 produced during a day in a normal individual is potentially capable of forming 20–40 Eq of H^+.

        (2) During a 24-hour period, the equivalent of 50–150 ml of 1 N acid (50–150 mEq H^+) is excreted via the kidneys.

    f. CO_2 must also be removed from the blood by the lungs at a rate of 200 ml/min during resting conditions.‡ With a resting cardiac output of 5 L/min, 40 ml of CO_2 must be added to each liter of blood per minute (i.e., 5 L/min $\times$ 40 ml/L = 200 ml/min).

3. Buffering of H_2CO_3 is primarily by the intracellular buffers (i.e., proteins, organic and inorganic phosphates, and hemoglobin). Most of the buffering of H_2CO_3 occurs in the erythrocyte, where the hemoglobin buffer system is quantitatively the most important buffer.

B. Noncarbonic acids are acids that do not form a volatile end product and, therefore, must be excreted by the kidneys (Table 5-6).

    1. Source

    a. The primary source of noncarbonic acid in humans is the **metabolism of exogenous protein.** The body produces 30 mEq of H^+ daily from the catabolism of proteins such as phosphoprotein and methionine, which are metabolically equivalent to phosphoric and sulfuric acids, respectively.

    b. Other noncarbonic acids are the **products of intermediary metabolism.**

        (1) Lactic acid, a product of anaerobic intermediary metabolism, accumulates in the ECF in large quantities during heavy exercise. This excess acid must be buffered until excreted or metabolized to CO_2 and water.

            (a) Although most organs generate lactic acid, the largest amount of lactic acid is produced by skeletal muscle, erythrocytes, and the skin. (Liver, kidney, and muscle can convert lactic acid to HCO_3^-.)

            (b) Tissue hypoxia leads to **hyperlacticemia,** which reduces the plasma $[HCO_3^-]$ and increases the anion gap (see VII B 2 for discussion of anion gap). The increment in the anion gap in this situation is a measure of serum lactate. Clinical conditions associated with tissue hypoxia include cardiac arrest, shock, severe cardiac failure, and severe hypoxemia.

            (c) Alkalosis stimulates glycolysis and generates lactic acid mainly by activation of phosphofructokinase. Respiratory alkalosis is a more effective stimulus for glycolysis, because CO_2 crosses cellular membranes more readily.

        (2) Acetoacetic acid and β**-hydroxybutyric acid** also are products of intermediary metabolism. These acids are formed almost exclusively by the liver during fasting.

Table 5-6. Sources of Noncarbonic (Nonvolatile) Acids in Humans

Source	Amount (mEq/day)	Noncarbonic Acid
Diet	30	Phosphoproteins, sulfur-containing amino acids, chloride salts
Intermediary metabolism	30	Ketoacids, lactic acid
Stool loss	30	Loss of HCO_3^-

*1 mol of CO_2 = 22.26 L at STPD (standard conditions of temperature and pressure, dry).

†*Depending on the level of physical activity; during resting metabolism, a normal individual is potentially capable of forming 13,000 mEq of H^+.

‡Equal to CO_2 production at rest.

    c. **Excess fluid intake** (hydration) reduces the $[HCO_3^-]$, which causes a corresponding increase in $[H^+]$ if P_{CO_2} is constant. This type of **noncarbonic acidosis is corrected more rapidly by the renal excretion of excess water** than by renal secretion of the apparent excess of H^+. Dehydration has the opposite effect.

2. Excretion

    a. Lactic acid production usually is transient; that is, after exercise is completed or the hypoxemic condition is removed, the lactic acid is metabolized to CO_2 and water. Lactic acid, therefore, can be eliminated by metabolic destruction of the acid rather than by renal excretion under normal conditions.

    b. Phosphoric and sulfuric acids cannot be further metabolized to CO_2, and the elimination of these acids and the H^+ that they yield can only be accomplished by the kidneys.

    c. Even noncarbonic acids that can be catabolized to CO_2 and water are excreted by the kidneys when present in large excess, as are **β-hydroxybutyric and acetoacetic acids** in diabetes mellitus. This condition is called **ketonuria.** Clinically more important is the acidosis of **uncontrolled diabetes mellitus,** where the accumulation of acetoacetic and β-hydroxybutyric acids in the ECF may produce coma and death.

3. Buffering of noncarbonic acids. In contrast to the bicarbonate buffer system, which can buffer only noncarbonic acids, the nonbicarbonate buffer systems can buffer both noncarbonic and carbonic acids.

    a. The plasma bicarbonate system is quantitatively the most important buffer in the ECF for noncarbonic acids.

    b. When noncarbonic acids are being buffered in the erythrocytes, more than 60% of the buffering occurs by the hemoglobin and more than 30% by the bicarbonate system in the erythrocytes. Approximately 10% of the buffer capacity in erythrocytes is attributed to the organic phosphate esters.

    c. The bodily response to mineral acids (e.g., HCl) is different in that lactic acid is distributed throughout the body water and is metabolized, whereas mineral acids are confined to the ECF and are buffered and excreted without being metabolized.

V. RESPIRATORY REGULATION OF ACID-BASE BALANCE. It is not immediately apparent that the actual quantities of CO_2 and O_2 transported in the blood are far greater than the amounts of these gases in physical solution, because these gases are transported mainly in the form of chemical derivatives. Further, it is even less apparent that there is far more total CO_2 than O_2 in every liter of blood. At the level of the lung, the HCO_3^- combines with the potential protons (i.e., the protons associated with $HHbO_2$) to produce HbO_2^- and H_2CO_3 resulting in the elimination of the protons as CO_2 and water.* Thus, free H^+ never appear to any appreciable extent but are in the form of HHb or H_2CO_3. **The entire respiratory cycle can be regarded as an exchange of a HCO_3^- for a HbO_2^-,** with the HCO_3^- transported from the tissues to the lungs and the HbO_2^- transported to the tissues from the lungs.

A. Blood forms of CO_2

    1. CO_2 is transported in three forms. However, H_2CO_3 ultimately must be converted to CO_2 in order to be eliminated via the lungs.

    a. HCO_3^- accounts for 90% of the total CO_2 in plasma.

    b. Carbamino compounds (i.e., carbamates of hemoglobin and protein) represent the combination of CO_2 with free NH_2 groups of blood proteins according to the equation

$$R \cdot NH_2 + CO_2 \rightleftharpoons R \cdot NHCOO^- + H^+$$

About 5% of the CO_2 normally carried in arterial blood is in the form of carbamino compounds (carbamates).

    c. A small amount of CO_2 gas (5%) is physically dissolved in plasma and also hydrated as H_2CO_3. Also, approximately 500 mol of CO_2 exist for every 1 mol of H_2CO_3 [see II D 2 a (1)].

---

*$HHbO_2$ and HbO_2^- are forms of oxygenated hemoglobin and constitute a buffer pair where the protonated form, $HHbO_2$, is the conjugate acid and the unprotonated form, HbO_2^-, is the conjugate base.

2. It is important to recognize that the term "bicarbonate" is used interchangeably with the term "total CO_2 combining power" or "total CO_2." This must never be confused with the term "partial pressure of CO_2," which is the PCO_2 of the arterial blood gas measurement.

3. In blood with an arterial PO_2 of 100 mm Hg and a mixed venous PO_2 of 40 mm Hg, there are 200 ml O_2/L and 150 ml O_2/L of arterial and mixed venous blood, respectively. However, the arterial PCO_2 and mixed venous PCO_2 are associated with 480 ml CO_2/L and 520 ml CO_2/L of arterial and mixed venous blood, respectively.

B. H_2CO_3 and other CO_2-forming acids

1. Source

a. The oxidation of glucose and triglyceride leads to the formation of CO_2, much of which is hydrated to H_2CO_3. H_2CO_3 in turn dissociates into H^+ and HCO_3^-.

b. CO_2 and water are the most abundant end products of metabolism.

(1) Lactic acid, a noncarbonic acid, normally is metabolized to CO_2 and water.

(2) The ketone bodies, β-hydroxybutyric acid and acetoacetic acid, also are noncarbonic acids, which are produced primarily by the liver during fasting. Upon reingestion, these acids that have accumulated are further catabolized to CO_2 and water by the extrahepatic tissues.

2. Elimination

a. Respiration accounts for the greatest portion of acid eliminated continuously from the body.

b. The lungs eliminate H_2CO_3 in the dehydrated (anhydrous) form of CO_2. Indeed, the unique property of the CO_2/HCO_3^- buffer system lies in the ability of the lungs to eliminate undissociated H_2CO_3 as nonionizable CO_2.

3. Buffering of H_2CO_3. Because CO_2 readily penetrates cellular membranes, H_2CO_3 is buffered by the entire body.

a. Most of the buffering of H_2CO_3 occurs via the nonbicarbonate buffer systems within erythrocytes. The H^+ derived from the dissociation of H_2CO_3 is buffered primarily by the hemoglobin buffer system.

b. The erythrocyte bicarbonate buffer and the plasma bicarbonate buffer play no role in the buffering of H_2CO_3.

VI. RENAL REGULATION OF ACID-BASE BALANCE

A. Overview

1. Normal acid-base conditions met by the kidneys

a. The kidneys are responsible for ridding the body of metabolically produced noncarbonic acids.

(1) Among these are sulfuric and phosphoric acids and smaller amounts of hydrochloric, lactic, uric, β-hydroxybutyric, and acetoacetic acids. The most abundant of the weak acid waste products of metabolism is acid phosphate ($H_2PO_4^-$).

(2) About half of these metabolically produced acids are neutralized by base in the diet. The other half must be neutralized by buffer anion systems of the body. Of those noncarbonic acids buffered in the ECF, 97%–98% are buffered by reacting with HCO_3^-.

b. Type of diet is a major determinant of the daily acid-base conditions that must be regulated by the kidneys.

(1) High-protein diets contain large amounts of sulfur in the form of sulfhydryl groups. The sulfur is oxidized to sulfate ion (SO_4^{2-}), a process that tends to lead to metabolic acidosis.

(2) Vegetarian diets are associated with large intakes of lactate and acetate, and the metabolites of these anions tend to lead to metabolic alkalosis.

2. Kidney function

a. The primary role of the kidneys in acid-base regulation is to conserve major cations and anions in the body fluids. To maintain the total quantities and concentrations of the major electrolytes within normal limits, the kidneys perform two major functions.

(1) The kidneys stabilize the standard HCO_3^- pool by obligatory reabsorption (mainly by the proximal tubule) and by controlled reabsorption of filtered HCO_3^- (by the distal and collecting tubules).

(2) The kidneys excrete a daily load of 50–100 mEq of metabolically produced noncarbonic acid.* This represents a H^+ excretion of 1 mEq/kg of body weight/day.

    (a) In most cases, 25% of this noncarbonic acid (10–30 mEq/day) is excreted in the form of **titratable acid** (see VI D).

    (b) About 75% (30–50 mEq/day) is excreted in the form of acid combined with **ammonium** (i.e., NH_4^+; see VI E).

    (c) The normal urinary ratio of NH_4^+ to titratable acid is between 1 and 2.5.

  b. The major sites of urine acidification are the distal and collecting tubules.

    (1) Essentially all of the H^+ within the tubular lumen is from the tubular secretion of H^+ generated by metabolism. There is no significant contribution of H^+ from the glomerular filtrate, which accounts for less than 0.1 mmol of H^+ per day.

    (2) Since the lowest pH attainable in urine is 4.4 (i.e., a $[H^+]$ of 40×10^{-6} Eq/L) and the plasma $[H^+]$ is 40×10^{-9} Eq/L, the kidney can cause a 1000-fold $[H^+]$ gradient between plasma and urine.

  c. An important difference between the acidification of urine and free H^+ excretion is that the ability to reduce urinary pH (acidification) does not reveal a great deal about the amount of free H^+ excreted. This is because most of the H^+ that is excreted occurs in association with an anion (mostly as $H_2PO_4^-$) or in combination with ammonia (as NH_4^+).

  d. The majority of secreted H^+ is used to bring about HCO_3^- reabsorption and, therefore, is not excreted.

B. Basic ion exchange mechanisms (Figures 5-3; 5-4)

  1. Passive transport of Na^+. Entry of Na^+ into the luminal (adluminal) membrane of the tubular cell is operationally linked to the secretion of H^+ into the tubular lumen. (It is a cation-exchange process that maintains intracellular electroneutrality.)

    a. Although the movement of Na^+ across the luminal membrane is favored by the electrochemical gradient, the transport is mediated by specific membrane transport proteins (i.e., it is carrier-mediated) and is *not* by simple diffusion.

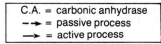

C.A. = carbonic anhydrase
--→ = passive process
—→ = active process

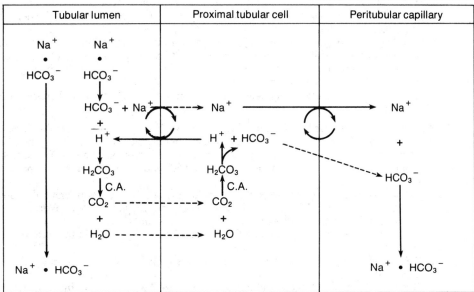

Figure 5-3. Mechanism of obligatory HCO_3^- reabsorption in the proximal tubule. The Na^+ is reabsorbed from the tubular lumen by a passive process and from the cell into the blood by an active process. Similarly, K^+ is reabsorbed as $KHCO_3$ or KCl. In contrast to Na^+, the transluminal reabsorption of K^+ is active, while transcellular K^+ transport is passive.

*In normal individuals in Western countries, a net of 40–60 mEq of noncarbonic acid is excreted daily.

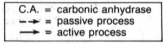

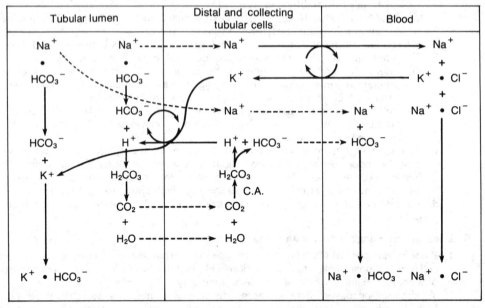

Figure 5-4. Mechanism of controlled HCO_3^- reabsorption in the distal and collecting tubules. Unlike the proximal tubule, the distal and collecting tubules lack luminal carbonic anhydrase. The exchange of K^+ for Na^+ leads to the reabsorption of 1 mol of NaCl for every 1 mol of $KHCO_3$ excreted. Note that intracellular carbonic anhydrase is found throughout the nephron, including the distal and collecting tubules.

    b. Secretion of H^+ is active in that it occurs against an electrochemical gradient.

    c. The process of H^+ secretion and HCO_3^- reabsorption occurs throughout the nephron, with the exception of the descending limb of the loop of Henle.

    d. The Na^+-H^+ antiporter is the major mechanism for H^+ secretion in the proximal tubule and thick ascending limb of the loop of Henle. This antiporter system is termed "secondary active" transport because the energy is derived from the Na^+ concentration gradient and not from the hydrolysis of adenosine triphosphate (ATP).

    e. The Na^+-H^+ antiporter exchanges one Na^+ ion for one H^+ ion and, therefore, is electroneutral.

  2. Active transport of Na^+. Subsequent to the passive Na^+ transport across the luminal membrane, Na^+ is actively reabsorbed across the basolateral (abluminal) membrane into the peritubular capillary in association with HCO_3^- that was formed within the cell. The secretion of 1 mol of H^+ leads to the reabsorption of 1 mol of Na^+ and 1 mol of HCO_3^- in the peritubular blood.

    a. H^+ for secretion originates from the hydration-dissociation reaction within the tubular cell [see II D 2, equation (4)] and from the splitting of water, which liberates hydroxyl ion (OH^-) within the cell; the OH^- then reacts with CO_2 to form HCO_3^-.

    b. H^+ secretion exceeds H^+ excretion; the amount of H^+ excreted in the form of free H^+ is negligible.

    c. The net effect of H^+ secretion is to facilitate the transfer of HCO_3^- from tubular lumen to tubular cell.

    d. It is probable that H^+ secretion along the entire tubule is mediated by an active transport process.

C. Reabsorption of HCO_3^- (see Figures 5-3; 5-4). The filtered HCO_3^- is not directly reabsorbed by the renal tubules; instead, it must first be converted to H_2CO_3, with the H^+ secreted into the tubular lumen in exchange for the Na^+ that is transported from the lumen. The HCO_3^- is reabsorbed indirectly by conversion to CO_2 within the tubular lumen, and most of the H^+ is reabsorbed following its conversion to water within the lumen.

1. **Proximal reabsorption.** Approximately 90% of the filtered HCO_3^- is reabsorbed into the proximal tubules as a result of the Na^+-H^+ exchange and represents more than 4000 mEq/day.
   a. The presence of carbonic anhydrase on the microvilli of the luminal border (brush border) of the proximal tubules catalyzes the rapid dehydration of H_2CO_3 to form CO_2 and water.
   b. CO_2 diffuses back into the proximal tubular cell, where it is rehydrated by intracellular carbonic anhydrase into H_2CO_3. The HCO_3^- formed by the dissociation of H_2CO_3 is passively reabsorbed into the peritubular blood along with equimolar amounts of Na^+, which is actively transported into the peritubular blood. The H^+ formed by the dissociation of H_2CO_3 serves as a source for another H^+ to be secreted.
   c. The entry of Na^+ into the tubular cell is balanced electrochemically by two mechanisms.
      (1) Cl^-, which is the only quantitatively important reabsorbable anion in the filtrate, is transported from lumen to cell by passive diffusion.
      (2) In the presence of carbonic anhydrase and with OH^- as its substrate, HCO_3^- is regenerated from the cellular CO_2 and water.
      (3) The net result of either the Cl^- diffusion into the cell or the regeneration of HCO_3^- is a net reabsorption of sodium chloride (NaCl) or sodium bicarbonate ($NaHCO_3$), respectively, into the peritubular capillaries. It is important to note that the Na^+ delivered to the peritubular capillaries comes from the filtrate in the lumen, but the HCO_3^- transported into the capillaries is synthesized within the proximal tubular cell.
   d. **Any secreted H^+ that combines with HCO_3^- in the lumen to bring about HCO_3^- reabsorption DOES NOT contribute to the urinary excretion of acid.** Thus, the majority of the secreted H^+ is used to accomplish HCO_3^- reabsorption.
   e. Referring to this process as HCO_3^- reabsorption, then, seems inaccurate, since the HCO_3^- that appears in the peritubular capillaries is not the same HCO_3^- that was filtered. Moreover, most of the H^+ that was secreted into the lumen is not excreted in the urine but is incorporated into water and reabsorbed.
   f. There is no absolute Tm for HCO_3^- since the reabsorptive capacity for HCO_3^- varies directly with the fractional reabsorption of Na^+.

2. **Reabsorption from the distal and collecting tubules**
   a. The remaining 10%–15% of the filtered HCO_3^- is reabsorbed by the distal and collecting tubules via a mechanism that involves the exchange of Na^+ for K^+ or H^+. As in the proximal tubules, every H^+ secreted into the distal tubular lumen leaves a HCO_3^- within the tubular cell.
   b. Except for the Na^+-H^+ antiporter or Na^+-K^+ exchange pump at the luminal border of the distal cellular membranes and the absence of carbonic anhydrase on the distal luminal border, the reabsorption pathways for HCO_3^- are analogous to those described for the proximal tubule.*

3. **Addition of new HCO_3^-.** In addition to the renal conservation of HCO_3^-, the kidneys add newly synthesized HCO_3^- to the plasma so that the quantity of HCO_3^- in the renal vein exceeds the amount that entered the kidneys.
   a. The addition of new HCO_3^- to the plasma does not involve the HCO_3^- reabsorbed into the tubule but the HCO_3^- generated within the tubular cell via the hydration of CO_2 and dissociation of H_2CO_3. This process is similar to the scheme for the reabsorption of filtered HCO_3^-; however, **the HCO_3^- that is generated within the tubular cell does not represent filtered HCO_3^-.**
   b. The renal contribution of new HCO_3^- is accompanied by the excretion of an equivalent amount of acid in the urine in the form of titratable acid, NH_4^+, or both.
   c. The amount of new HCO_3^- formed per day (approximately 70–100 mEq) is much less than the quantity of filtered HCO_3^- reabsorbed per day (more than 4300 mEq).

D. Excretion of titratable acid (Figure 5-5; Table 5-7)

1. **Filtration of HPO_4^{2-}.** Titratable acid denotes that portion of H^+ bound to filtered buffers and equals the amount of alkali, in the form of sodium hydroxide (NaOH), required to titrate urine back to the normal pH of blood (i.e., the number of milliequivalents of H^+ added to the tubular fluid that combined with phosphate or organic buffers).

---

*Intracellular carbonic anhydrase is found throughout the renal tubule, including the distal convoluted tubule and collecting duct. In the proximal convoluted tubule, carbonic anhydrase also is located in the luminal cell membranes.

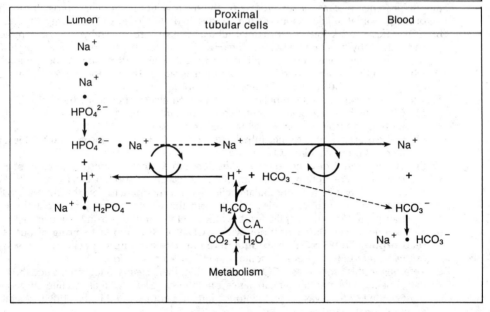

Figure 5-5. Mechanism of controlled excretion of titratable acid in the proximal tubules.

   a. Titratable acid is largely attributed to the conversion of HPO_4^{2-} to $H_2PO_4^{-}$. Additional buffer species that may contribute to titratable acid are creatinine, uric acid, β-hydroxybutyrate, and SO_4^{2-}.

   b. Titratable acid is a poor measure of the total amount of H^+ secreted by the tubules, since most of the acid produced by this secretion is H_2CO_3, which disappears from the urine as CO_2; however, the H^+ that is trapped by anions of noncarbonic acids remains in the final urine in the form of titratable acid.

   c. Titratable acid is a measure of the content of weak acids. It is not a measure of the H^+ that combines with ammonia (NH_3) to yield NH_4^+, because the pK' of the NH_3/NH_4^+ buffer system is high (9.2).

   d. The proximal tubule is the major nephron site where titratable acid is formed. Additional titratable acid is generated along the collecting duct by a H^+-ATPase pump.

  2. Excretion of $H_2PO_4^{-}$. The exchange of H^+ for Na^+ converts dibasic sodium phosphate (Na_2HPO_4) in the glomerular filtrate into dihydrogen phosphate (NaH_2PO_4), which is excreted in the urine as titratable acid. The H^+ secreted into the tubules, therefore, can react with filtered HPO_4^{2-} rather than the filtered HCO_3^{-}.

   a. Besides HCO_3^{-}, HPO_4^{2-} represents a major filtered conjugate base.* Furthermore, the $HPO_4^{2-}/H_2PO_4^{-}$ buffer pair provides an excellent buffer system because it has a pK' of 6.8.

Table 5-7. Urinary Acid Excretion in Health and Disease (in mEq H^+/day)

Urinary Acid	Normal Excretion	Excretion in Diabetic Ketoacidosis
Titratable acid	10–30	75–250
Ammonium	30–50	300–500
Total	40–80	375–750

*Since about 75% of the filtered HPO_4^{2-} is reabsorbed, only about 25% of the filtered HPO_4^{2-} is available for buffering.

b. The blood $HPO_4^{2-}/H_2PO_4^-$ molar ratio is 4:1. The H^+ secretory system converts much of the HPO_4^{2-} to $H_2PO_4^-$ in the tubular fluid, resulting in a urinary $HPO_4^{2-}/H_2PO_4^-$ molar ratio of 1:4. The acidification of the phosphate buffer system occurs significantly in the proximal tubule, and much of the H^+ generated by the formation of H_2SO_4 and H_3PO_4 during protein and phospholipid metabolism is excreted in this way.

c. H^+ secreted into the lumen that reacts with HCO_3^- is not excreted, whereas H^+ secreted into the lumen that reacts with a nonbicarbonate buffer remains in the tubular fluid and is excreted.

d. Urinary phosphate is the major anion contributing to urine acid excretion.

e. Virtually all of the filtered phosphate is reabsorbed by the Na^+-phosphate symporter of the proximal tubular brush border.

3. Role of aldosterone in H^+ excretion. Aldosterone exerts its major renal effect on the collecting duct.

   a. Aldosterone plays an important role in regulating collecting duct HCO_3^- reabsorption via its stimulatory effect on H^+ secretion.

   b. Aldosterone stimulates H^+ secretion directly on the tubular cell.

   c. Aldosterone stimulates H^+ secretion indirectly by increasing Na^+ reabsorption.

E. Excretion of ammonium (NH_4^+) [Figures 5-6; 5-7; see Table 5-7]. Unlike phosphate, NH_3 enters the tubular lumen not by filtration but by tubular synthesis and secretion, which normally are confined to the distal and collecting tubules.

1. Secretion of NH_3. NH_3 is synthesized mainly by the deamidization and deamination of glutamine in the presence of glutaminase. Most of the NH_4^+ excreted in the urine is produced in the proximal tubular cells from amino acids, primarily glutamine.

   a. The NH_4^+ that are formed are actively secreted into the lumen and are excreted in the final urine at a rate equal to their secretion rate.

   b. With regard to proximal NH_4^+ secretion, it must be emphasized that the traditional view that states that NH_3 combines with the H^+ derived from the dissociation of H_2CO_3 is incorrect. Instead, the H^+ produced in the metabolism of glutamine results in the formation of NH_4^+ within the proximal tubular cell (see Figure 5-6).

   c. The mechanism of distal nephron NH_4^+ excretion is different from that in the proximal tubule (see Figure 5-7).

2. Formation of NH_4^+. The NH_3/NH_4^+ system has a very high pK' (about 9.2), which means that, at the usual urine pH, virtually all the nonpolar NH_3 that enters the distal tubular lumen immediately combines with H^+ to form NH_4^+, which is nondiffusible because it is lipid insoluble. The renal excretion of NH_4^+ causes the net addition of HCO_3^- to the plasma.

   a. The important physiologic characteristic of the NH_3/NH_4^+ buffer pair is that, as H^+ is combined with intraluminal buffer (NH_3), H^+ is excreted in the urine as NH_4^+, a substance that does not cause the pH of urine to fall.

   b. Under most conditions, the excretion of NH_4^+ is quantitatively more important to the acid-base balance of the body than is the excretion of titratable acid. The normal kidney excretes almost twice as much acid combined with NH_3 (30–50 mEq/day) as titratable acid (10–30 mEq/day) [see Table 5-7].

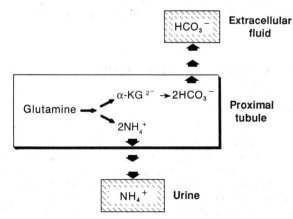

Figure 5-6. Formation and active secretion of NH_4^+ by the proximal tubules. $\alpha\text{-}KG^{2-}$ = α-ketoglutarate. (Reprinted from Seldin DW, Giebisch G (eds): New concepts in renal ammonium excretion. In *The Regulation of Acid-Base Balance*. New York, Raven, 1989, p 170.)

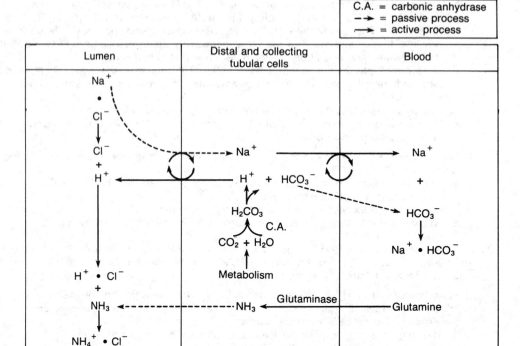

Figure 5-7. Mechanism of controlled excretion of NH_4^+ in the distal and collecting tubules.

   (1) In chronic severe metabolic acidosis, NH_3 serves as the major urinary buffer, and NH_4^+ excretion can increase from a normal value of 30 mEq/day to 500 mEq/day. However, in chronic acidosis, the excretion of $H_2PO_4^-$ may increase by only 20–40 mEq/day.

   (2) In diabetic ketoacidosis, the excretion of titratable acid may reach 75–250 mEq/day, while the excretion of acid combined with NH_3 may reach amounts of 300–500 mEq/day. However, the ratio of NH_4^+ to titratable acid remains within the normal range of 1 to 2.5.

  3. Excretion of H^+. The sum of titratable acid and urinary NH_4^+ excretion represents the net gain of HCO_3^- for the body fluids.

    a. The net amount of fixed acid excreted (in mEq/day) equals the sum of titratable acid and NH_4^+ minus the urinary $[HCO_3^-]$.

    b. Normally, the urine is free of HCO_3^- because all the HCO_3^- remaining in the distal and collecting tubules has combined with secreted H^+ and been reabsorbed.

    c. Aldosterone enhances NH_3 production via an effect on cellular metabolism.

VII. PRIMARY AND SECONDARY ACID-BASE ABNORMALITIES

  A. Definitions and basic concepts (Tables 5-8; 5-9)

   1. Acidosis and alkalosis

    a. Acidosis or **acidemia** is an abnormal clinical condition or process caused by the bodily accumulation of acid (or the loss of base) sufficient to decrease pH below 7.36 or to increase the $[H^+]$ above 43.6 nEq/L of blood in the absence of compensatory (secondary) changes.

    b. Alkalosis or **alkalemia** is an abnormal condition or process caused by the accumulation of base (or the loss of acid) sufficient to raise pH above 7.44 or to decrease the $[H^+]$ below 36.3 nEq/L of blood in the absence of compensatory changes.

Table 5-8. Characteristics of the Uncompensated Acid-Base Abnormalities

Acid-Base Abnormality	Primary Disturbance	Effect on:			Compensatory Response
		$[HCO_3^-]/S \times Pco_2$	$[H^+]$	pH	
Acidosis					
Respiratory	$\uparrow Pco_2$	< 20	$\uparrow$	$\downarrow$	$\uparrow [HCO_3^-]$
Metabolic	$\downarrow [HCO_3^-]$	< 20	$\uparrow$	$\downarrow$	$\downarrow Pco_2$
Alkalosis					
Respiratory	$\downarrow Pco_2$	> 20	$\downarrow$	$\uparrow$	$\downarrow [HCO_3^-]$
Metabolic	$\uparrow [HCO_3^-]$	> 20	$\downarrow$	$\uparrow$	$\uparrow Pco_2$

Note—These conditions are simple disturbances (i.e., they represent the effects of one primary etiologic factor) and, thus, are termed uncompensated or pure acid-base abnormalities. The compensatory response always occurs in the same direction as the primary disturbance.

> **2. Respiratory and metabolic**
> **a.** The adjective **respiratory** denotes that the primary abnormality involves impairment in alveolar ventilation, which results in an abnormally high or low [total CO_2] of the ECF.*
> > **(1) Respiratory acidosis** refers to a condition of abnormally high arterial Pco_2, which is termed **hypercapnia** (hypercarbia).
> > **(2) Respiratory alkalosis** refers to a condition of abnormally low arterial Pco_2, which is called **hypocapnia** (hypocarbia).
> **b.** The adjective **metabolic** denotes that the primary abnormality involves an abnormal gain or loss of noncarbonic acid by the ECF, which affects $[HCO_3^-]$.
> > **(1) Metabolic acidosis** refers to a disturbance that leads to the accumulation of noncarbonic acid in the ECF or to the loss of HCO_3^- from the ECF.
> > **(2) Metabolic alkalosis** refers to an imbalance characterized by a loss of noncarbonic acid or a gain of HCO_3^- by the ECF.
>
> **3. Primary and secondary factors**
> **a.** The factor in the $[HCO_3^-]/S \times Pco_2$ ratio (i.e., $[HCO_3^-]$ or Pco_2) that undergoes the greater degree of displacement (i.e., the larger proportional change) indicates the **primary abnormality** (see Tables 5-8; 5-9).
> > **(1)** If the $[HCO_3^-]/S \times Pco_2$ ratio becomes less than 20:1, by either decreasing the numerator or increasing the denominator, pH falls ($\uparrow[H^+]$), and acidosis occurs.

Table 5-9. Primary and Secondary* Changes in the CO_2/HCO_3^- System

	Acidosis		Alkalosis	
Respiratory	$\dfrac{[HCO_3^-]\ \uparrow}{S \times Pco_2 \ \blacktriangle}$	< 20:1	$\dfrac{[HCO_3^-]\ \downarrow}{S \times Pco_2 \ \blacktriangledown}$	> 20:1
	$[total\ CO_2]^{\dagger}\ \uparrow$		$[total\ CO_2]\ \downarrow$	
Metabolic	$\dfrac{[HCO_3^-]\ \blacktriangledown}{S \times Pco_2 \ \downarrow}$	< 20:1	$\dfrac{[HCO_3^-]\ \blacktriangle}{S \times Pco_2 \ \uparrow}$	> 20:1
	$[total\ CO_2]\ \blacktriangledown$		$[total\ CO_2]\ \blacktriangle$	

Note—*Dashed arrows* denote direction of compensatory responses. *Thin solid arrows* show changes in [total CO_2] during respiratory imbalances. *Wide arrows* depict direction of change of primary acid-base imbalance and direction of change in [total CO_2]. (Adapted from Christensen HN: *Diagnostic Biochemistry.* New York, Oxford University Press, 1959, p 106.)
*A compensatory (secondary) response in the alternate variable occurs in the same direction and counteracts the effect of the primary disturbance by returning the $[HCO_3^-]/S \times Pco_2$ ratio toward normal (20:1).
†[Total CO_2] does not always increase in alkalosis or decrease in acidosis. Dissolved CO_2 contributes little to the [total CO_2]. [Total CO_2] provides little information about pulmonary function.

---

*[Total CO_2] refers to the dissolved CO_2 ($S \times Pco_2$) plus the $[HCO_3^-]$.

(2) If the $[HCO_3^-]/S \times P_{CO_2}$ ratio becomes more than 20:1, by either increasing the numerator or decreasing the denominator, pH rises ($\downarrow[H^+]$), and alkalosis occurs.

    b. A **secondary response** in the alternate variable, which occurs in the same direction as the primary abnormality, counteracts the effect of the primary abnormality on the $[HCO_3^-]/$ $S \times P_{CO_2}$ ratio. The secondary change represents **compensation** and acts to minimize the pH alteration produced by the primary disorder.

   4. Simple and mixed acid-base disturbances
    a. Simple acid-base imbalances are caused by one primary factor.
    b. Mixed acid-base imbalances are caused by more than the primary factor.
      (1) Mixed-type disturbances are not uncommon. Sometimes a primary abnormality of one type is superimposed on a primary abnormality of another type.
        (a) A patient with respiratory acidosis from pulmonary emphysema may develop metabolic acidosis from uncontrolled diabetes or metabolic alkalosis from large doses of corticosteroids used in the treatment of an attack of status asthmaticus.
        (b) A patient with a metabolic acid-base disturbance, in turn, may develop a respiratory acid-base abnormality.
      (2) The changes in the variables may be difficult to interpret in mixed acid-base disturbances, because manifestations of one primary abnormality may be either cancelled out or augmented by those of the other primary abnormality. Therefore, **a normal pH may not necessarily mean a normal compensation in mixed acid-base imbalances.**

   5. Plasma $[HCO_3^-]$, P_{CO_2}, and [total CO_2]. For an adequate analysis of any acid-base disorder, it is necessary to measure pH, P_{CO_2}, and the total CO_2 content of blood.
    a. An acid-base disorder cannot be diagnosed with certainty from the plasma $[HCO_3^-]$ alone. Although a reduction in the plasma $[HCO_3^-]$ may be due to metabolic acidosis, it also can indicate a renal compensation for respiratory alkalosis. Similarly, an elevated $[HCO_3^-]$ can result from a metabolic alkalosis or the secondary response to respiratory acidosis. Since the aim of therapy is to bring about acid-base balance and not to normalize $[HCO_3^-]$, it is necessary to measure the pH (i.e., $[H^+]$) when acid-base imbalance is suspected.
    b. Low P_{CO_2} may be due to respiratory alkalosis or a respiratory compensation to metabolic acidosis. Similarly, a **high P_{CO_2}** may be caused by respiratory acidosis or a respiratory compensatory response to metabolic alkalosis.
    c. [Total CO_2] does not always increase in alkalosis or decrease in acidosis (Figure 5-8; see Table 5-9). About 90% of the [total CO_2] of plasma is contributed by $[HCO_3^-]$, and 5% is contributed by dissolved CO_2 and H_2CO_3; therefore, dissolved CO_2 contributes little to the [total CO_2]. The [total CO_2] gives little information about the functional status of the lungs.
    d. The physician accepts [total CO_2] and $[HCO_3^-]$ as essentially interchangeable terms. The [total CO_2] normally exceeds the $[HCO_3^-]$ by 1.2 mmol. Most laboratories measure the [total CO_2] and not the $[HCO_3^-]$.

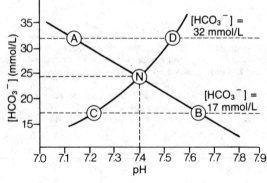

Figure 5-8. A pH-$[HCO_3^-]$ diagram illustrating that acidosis (*points A and C*) and alkalosis (*points B and D*) cannot be differentiated solely on the basis of $[HCO_3^-]$ or [total CO_2]. The *$[HCO_3^-]$* pertains to plasma. $A =$ respiratory acidosis; $B =$ respiratory alkalosis; $C =$ metabolic acidosis; $D =$ metabolic alkalosis; and $N =$ normal point. The slope of *line ANB* is a measure of the nonbicarbonate buffer capacity of whole blood. (After Davenport HW: *The ABC of Acid-Base Chemistry*, 5th edition. Chicago, University of Chicago Press, 1969, p 65.)

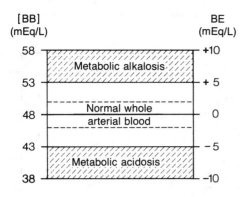

Figure 5-9. Diagram illustrating the relationship between base excess *[BE]* and the acid-base status of normal whole arterial blood. BE is the base concentration (in mEq/L), as measured by the titration with strong acid to pH 7.4, at P_{CO_2} of 40 mm Hg, and at 37° C. For negative values of BE, the titration must be carried out with strong base. BE measures the change in the concentration of the buffer base (*[BB]*) from its normal value, which is 48 mEq/L in normal whole arterial blood. [BB] is the sum of the buffer anions of blood or plasma. When observed [BB] is less than normal [BB], BE is a negative value; when observed [BB] is greater than normal [BB], BE is a positive value; and when observed [BB] equals normal [BB], BE equals zero. (After Winters RW, et al: *Acid-Base Physiology in Medicine.* Westlake, Ohio, London Co, 1967, p 45.)

B. Clinical expressions for evaluation of acid-base status

1. Base excess (BE) [Figure 5-9; Table 5-10]

    a. Definition. BE refers to the change in the concentration of buffer base [BB] from its normal value, or, the **observed [BB] minus the normal [BB]** in mEq/L of whole blood. The range of normality for BE is −2 to +2 mEq/L in arterial whole blood.*

        (1) The [BB] in normal whole blood is 48 mEq/L, which is arbitrarily set to 0, and deviations (i.e., ±BE) are measured from this reference point.

        (2) The normal [BB] of 48 mEq/L equals the sum of all the conjugate bases in 1 L of arterial whole blood. These bases and their concentrations are:

            (a) $[HCO_3^-]$ = 24 mEq/L

            (b) $[Protein^-]$ = 15 mEq/L

            (c) $[Hb^-/HbO_2^-]$ = 9 mEq/L

    b. Clinical application. BE refers principally to the $[HCO_3^-]$ but also to other bases in the blood (mainly plasma protein and hemoglobin). However, the $[HCO_3^-]$ or [total CO_2] is used in acid-base problems as the indicator of BE.

        (1) Because $[HCO_3^-]$ and BE are influenced only by metabolic processes, there are only two acid-base conditions associated with abnormalities in $[HCO_3^-]$ and BE.

            (a) Metabolic acidosis. When a metabolic process leads to the accumulation of non-carbonic acid in the body or the loss of HCO_3^-, the $[HCO_3^-]$ falls, and the BE value becomes negative. Metabolic acidosis is associated with a BE below −5 mEq/L.

            (b) Metabolic alkalosis. When a metabolic process causes a loss of acid or an accumulation of HCO_3^-, the $[HCO_3^-]$ rises above normal, and the BE value becomes positive. Metabolic alkalosis is associated with a BE above +5 mEq/L.

        (2) The BE value serves only as a rough guide for alkalosis and acidosis therapy.

2. Anion gap

    a. Definition

        (1) The ionic profile of normal serum is depicted in Figure 5-10. The **law of electroneutrality** states that the number of positive charges in any solution must equal the number of negative charges. If every ion present in serum were measured, the concentration of

Table 5-10. Base Excess and Metabolic Acid-Base Abnormalities

$[HCO_3^-]$	Base Excess	Metabolic Abnormality	Characteristics
↑	+	Metabolic alkalosis	Noncarbonic acid is lost; HCO_3^- is gained
↓	−	Metabolic acidosis	Noncarbonic acid is gained; HCO_3^- is lost

Adapted from Broughton JO: *Understanding Blood Gases.* Madison, Wisconsin, Ohio Medical Products, form no. 456, 1979, p 8.

*Because BE is a negative or positive value, the term "base deficit" is avoided.

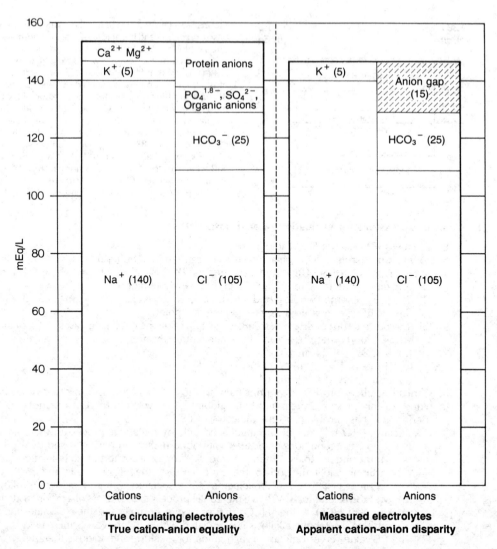

Figure 5-10. Normal plasma concentrations of anions and cations in the mEq/L show that circulating cations always counterbalance anions, thereby maintaining electroneutrality (*left*). Routine electrolyte assays only measure a portion of circulating anions, and an apparent cation-anion disparity exists (*right*). This difference, in mEq/L, is termed the anion gap. *Parentheses* indicate concentrations in mEq/L. (Reprinted from Narins RG, et al: Lactic acidosis and elevated anion gap (I). *Hosp Pract* 15:125–136, 1980. Original drawing by Albert Miller.)

cations would equal the concentration of anions if these concentrations were expressed in mEq/L. Routine serum electrolyte determinations measure essentially all cations but only a fraction of the anions. **This apparent disparity between the total cation concentration and the total anion concentration is termed the anion gap.** It is a virtual measurement and does not represent any specific ionic constituent.

(2) The anion gap, which has a normal value of 12 ± 4 mEq/L, reflects the concentrations of those anions actually present but routinely undetermined (i.e., other than $[Cl^-]$ and $[HCO_3^-]$), such as:
  (a) Polyanionic plasma proteins (primarily albumin)
  (b) Inorganic phosphates
  (c) Sulfate
  (d) Ions of organic acids (e.g., lactic, β-hydroxybutyric, and acetoacetic acids)
(3) When the anion gap is increased, unmeasured anions fill this gap. (Most of these anions usually are the products of metabolic processes that generate H^+.)

b. Determination of anion gap
   (1) The anion gap (AG) equals 16 mEq/L of anions other than HCO_3^- and Cl^- when determined by the following equation

$$AG = ([Na^+] + [K^+]) - ([HCO_3^-] + [Cl^-])$$
$$= (142 \text{ mEq/L} + 4 \text{ mEq/L}) - (25 \text{ mEq/L} + 105 \text{ mEq/L})$$
$$= 146 \text{ mEq/L} - 130 \text{ mEq/L}$$
$$= 16 \text{ mEq/L}$$

   (2) Often the anion gap is calculated with Na^+ as the major cation, because K^+ has a relatively minor quantitative contribution. The equation, then, becomes:

$$AG = [Na^+] - ([HCO_3^-] + [Cl^-])$$
$$= 142 \text{ mEq/L} - (25 \text{ mEq/L} + 105 \text{ mEq/L})$$
$$= 142 \text{ mEq/L} - 130 \text{ mEq/L}$$
$$= 12 \text{ mEq/L}$$

   (3) The plasma proteins contribute a significant amount of negative charges and, thus, anionic equivalence (16 mEq/L). The plasma proteins account, almost stoichiometrically, for the difference between the $[Na^+]$ and the sum of $[HCO_3^-]$ and $[Cl^-]$.

c. Clinical application
   (1) Acidification with acids other than HCl
      (a) Organic acids increase the anion gap. In this case, the lost (neutralized) HCO_3^- is *not* replaced by the routinely unmeasured anions of noncarbonic acids such as lactic acid and the ketoacids. Thus, the anion gap is increased by all metabolic acidoses except the **hyperchloremic acidoses**. With the accumulation of acid, there is rapid extracellular buffering by HCO_3^-. If the acid is HCl, then

$$HCl + NaHCO_3 \rightleftharpoons NaCl + H_2CO_3$$

   The net effect is the milliequivalent-for-milliequivalent replacement of extracellular HCO_3^- by Cl^-. Since the sum of $[Cl^-]$ and $[HCO_3^-]$ remains constant, the anion gap is unchanged in hyperchloremic acidosis.
      (b) It is apparent that the decrease in plasma $[HCO_3^-]$ equals the increase in the anion gap (see Figure 5-10). The presence of an increased anion gap usually indicates an excess of H^+ derived from noncarbonic acid. Whatever the size of the anion gap, **it is the retention of H^+—not the particular anion—that is responsible for the acidosis**.
         (i) An increased anion gap due to excess H^+ derived from noncarbonic acid occurs in **metabolic acidosis**; this also may result from a compensatory increase in lactic acid production in response to respiratory alkalosis via activation of the phosphofructokinase step in glycolysis.
         (ii) In **respiratory acidosis** the anion gap is not increased, because excess H^+ is derived from the H_2CO_3 pool, not the noncarbonic acid pool.
   (2) Conditions that increase the anion gap include diabetic and alcoholic ketoacidosis, intoxicant and lactic acidoses, and renal failure.
   (3) Conditions that cause metabolic acidosis without an increase in the anion gap are associated with a high serum $[Cl^-]$ and include diarrhea; pancreatic drainage; ureterosigmoidostomy; ileal loop conduit; treatment with acetazolamide, ammonium chloride, or arginine-HCl; renal tubular acidosis; and, rarely, intravenous hyperalimentation.

3. Osmolar gap
   a. Definition. Osmolar gap refers to the disparity between the measured serum (plasma) osmolality and the calculated serum osmolality.
   b. Clinical application. Osmolar gap measurement provides a reasonably good screening procedure for toxins. Other causes of metabolic acidosis do not affect the osmolar gap, since the metabolic acid simply replaces the HCO_3^- with another anion and HCO_3^- is lost as CO_2.
      (1) Serum osmolality is estimated by dividing blood urea nitrogen (BUN) and glucose concentrations (in mg/L) by their molecular weights,* thereby converting them to milliosmoles (mOsm)/L. Doubling the serum $[Na^+]$ (in mEq/L) provides an estimate of

---

*The molecular weight of urea (60) is not used because it is BUN that is measured, and urea contains 2 nitrogen atoms ($2 \times 14 = 28$).

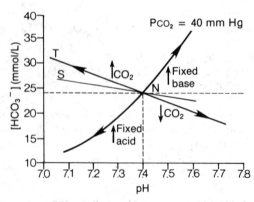

Figure 5-11. The pH-[HCO$_3^-$] diagram showing the buffer curves of separated plasma (*S*) and true oxygenated plasma (*T*). *Point N* denotes the intercept for a normal plasma [HCO$_3^-$] of 24 mmol/L and a normal pH of 7.4. The slope of the normal in vitro buffer line of true plasma is a function of the hemoglobin content of blood. *Lines S* and *T* are called nonbicarbonate buffer curves, and the steepness of the slopes of these lines is a quantitative assessment of the amount of nonbicarbonate buffer present in the system. These same two lines represent CO$_2$-titration curves. The steeper the buffer line, the smaller the pH change resulting from a given increase or decrease in Pco$_2$. *Vertical arrows* indicate increases and decreases in Pco$_2$, fixed acid, or fixed base. (Adapted from Davenport HW: *The ABC of Acid-Base Chemistry*, 5th edition. Chicago, University of Chicago Press, 1969, p 48.)

the serum ion concentration. With a BUN of 14 mg/dl, plasma glucose of 90 mg/dl, and a plasma [Na$^+$] of 142 mEq/L, serum (plasma) osmolality (P$_{osm}$) is estimated as

$$P_{osm} = 2 [Na^+] + [BUN (mg/L)/28] + [glucose (mg/L)/180]$$
$$= 2 (142) + 140/28 + 900/180$$
$$= 286 + 5 + 5 = 296 \text{ mOsm/kg}$$

(2) Circulating intoxicants increase measured serum osmolality without altering serum [Na$^+$]. Therefore, the measured serum osmolality exceeds the calculated serum osmolality, and the difference closely reflects the osmolar concentration of the circulating toxins.

C. Graphic evaluation of acid-base status: the pH-[HCO$_3^-$] diagram (Figures 5-11; 5-12)

1. Value of diagram. When acid-base data obtained from blood are plotted on the pH-[HCO$_3^-$] diagram, the physician has a tool that can be used to:
a. Diagnose the primary cause of an acid-base abnormality
b. Determine the degree of compensatory response of the kidneys and of the lungs by monitoring the [HCO$_3^-$] and Pco$_2$, respectively
c. Estimate the concentration of noncarbonic and carbonic acids in the ECF
d. Aid in the choice of therapy to correct an acid-base imbalance
e. Measure the nonbicarbonate buffer power of whole blood by the slope of the buffer line
f. Indicate not only the total buffer activity but also the distribution of the buffering activity between the bicarbonate and nonbicarbonate buffer systems

2. Construction of diagram: effects of H$^+$ and CO$_2$ on pH. The basis for plotting pH and [HCO$_3^-$] on cartesian coordinates is best understood from a consideration of the CO$_2$/HCO$_3^-$ buffer system

$$CO_2 + H_2O \rightleftharpoons H_2CO_3 \rightleftharpoons H^+ + HCO_3^- \qquad (6)$$

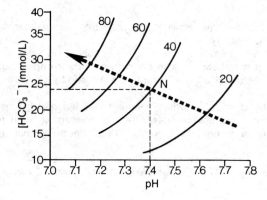

Figure 5-12. The pH-[HCO$_3^-$] diagram with arterial Pco$_2$ isobars for 20, 40, 60, and 80 mm Hg. Each arterial Pco$_2$ isobar is the titration curve of a HCO$_3^-$-H$_2$CO$_3$ solution with arterial Pco$_2$ held constant. *Point N* denotes the intercept for a normal plasma [HCO$_3^-$] of 24 mmol/L and a normal pH of 7.4. The *dashed arrow* represents the buffer line. (After Davenport HW: *The ABC of Acid-Base Chemistry*, 5th edition. Chicago, University of Chicago Press, 1969, p 45.)

a. Effect of fixed acid or base on pH

(1) Addition of H⁺ at a constant PCO_2

(a) When H⁺ ions, in the form of fixed (noncarbonic) acid, are added to the system depicted in equation (6), most of them combine with HCO_3^- to form H_2CO_3, which dehydrates to form CO_2 and water. Thus, the series of reactions in equation (6) is driven to the left.

(b) The decrease in $[HCO_3^-]$ closely approximates the amount of H⁺ added only if the solution has no other nonbicarbonate buffer substances. Otherwise, the amount of acid added would be greater than the decrease in $[HCO_3^-]$.

(2) Addition of base at a constant PCO_2

(a) When a base is added to the system depicted in equation (6), some of the base combines with H⁺, causing more H_2CO_3 to dissociate into H⁺ and HCO_3^-. Here the series of reactions in equation (6) proceeds to the right.

(b) The increase in $[HCO_3^-]$ determines the amount of base added.

(3) Principles (see Figure 5-11)

(a) When fixed acid or base is added under the conditions described, there is an inverse relationship between [H⁺] and $[HCO_3^-]$, and the primary acid-base disturbance is metabolic. Thus, when fixed acid or base is added, the [H⁺] and $[HCO_3^-]$ change in opposite directions.

(b) Or, if the direction of change in $[HCO_3^-]$ is the same as that for pH, the primary acid-base abnormality is metabolic.

b. Effect of CO_2 on pH

(1) Addition of CO_2 (see Figure 5-12)

(a) An increase in arterial PCO_2 titrates the blood (and ECF) in the acid direction. This is clear from equation (6), which shows that an increase in PCO_2 causes more CO_2 to hydrate to form H_2CO_3; the H_2CO_3 in turn dissociates into H⁺ and HCO_3^-, moving the reactions to the right.

(b) In the system shown in equation (6), every molecule of CO_2 that hydrates and dissociates forms one H⁺ and one HCO_3^-; therefore, the changes in [H⁺] and $[HCO_3^-]$ are exactly equal.

(c) The final [H⁺] depends not only on the change in PCO_2 but also on the buffers in the system. For every H_2CO_3 molecule that has its H⁺ taken up by a buffer ion, one HCO_3^- appears in the solution. For example, the reaction with hemoglobin is

$$H_2CO_3 + Hb^- \rightleftharpoons HHb + HCO_3^-$$

(d) If CO_2 is added to whole blood (which contains many buffer systems), the H⁺, rather than remaining in solution, are mostly bound to protein buffers (Pr⁻) as

$$CO_2 + H_2O \rightleftharpoons H_2CO_3 \rightleftharpoons HCO_3^- + H^+$$
$$+$$
$$Pr^-$$
$$\updownarrow$$
$$H \cdot Pr$$

Therefore, **the CO_2/HCO_3^- system alone is not effective in buffering the changes in pH that are induced by the addition of CO_2.**

(2) Removal of CO_2

(a) When CO_2 is removed from the system shown in equation (6), fewer CO_2 molecules are available for combination with water to form HCO_3^-.

(b) Thus, not only is CO_2 decreased but [H⁺] and $[HCO_3^-]$ also are decreased and the reactions shift to the left.

(3) Principles. The concentration of acid added to a buffer solution by a change in PCO_2 is equal to the change in $[HCO_3^-]$.*

(a) If CO_2 is added or removed, the [H⁺] and $[HCO_3^-]$ change in the same direction, and the primary acid-base disturbance is respiratory.

(b) Or, if the direction of change in $[HCO_3^-]$ is opposite to that for pH, the primary acid-base disturbance is respiratory.

---

*In humans, the serum $[HCO_3^-]$ is 24×10^{-3} Eq/L (24×10^{-3} mol/L), whereas the serum [H⁺] is 40×10^{-9} Eq/L (40×10^{-9} mol/L). Therefore, the $[HCO_3^-]$ is 6×10^5 greater than [H⁺], and any shift (left or right) in equilibrium has a much greater proportionate effect on [H⁺] than on $[HCO_3^-]$.

(c) As P_{CO_2} is increased or decreased, the values for pH and $[HCO_3^-]$ form coordinates that define a nearly straight line termed the buffer line or the CO_2 titration curve (see Figure 5-12).

(4) Summary

(a) When H^+ is added or removed, the $[H^+]$ and $[HCO_3^-]$ change in opposite directions and the acid-base imbalance is metabolic in origin.

(b) When CO_2 is added or removed, the $[H^+]$ and $[HCO_3^-]$ change in the same direction and the acid-base imbalance is respiratory in origin.

3. Properties of P_{CO_2} isobars

a. At constant $[HCO_3^-]$, P_{CO_2} is proportional to $[H^+]$. This property can be illustrated using the following form of the Henderson equation to solve for P_{CO_2}

$$[H^+] = 24\,\frac{P_{CO_2}}{[HCO_3^-]}$$

$$P_{CO_2} = \frac{[H^+][HCO_3^-]}{24}$$

b. At constant pH (i.e., along any vertical line in Figures 5-11 and 5-12), P_{CO_2} is proportional to $[HCO_3^-]$.

D. Interpretation of acid-base abnormalities using the pH-$[HCO_3^-]$ diagram (Figure 5-13). Acid-base imbalances are determined graphically with reference to the intercept of the P_{CO_2} isobar of 40 mm Hg and the normal buffer line. The intercept of these two curves marks the point of normality (*N*), which is associated with a pH of 7.4 (the *abscissa*) and a $[HCO_3^-]$ of 24 mmol/L (the *ordinate*). **Point N is the triple intercept that defines the pH, $[HCO_3^-]$, and P_{CO_2} of true arterial plasma of a normal individual.**

1. Points to the left of *point N* (*points A, C, E,* and *F*) indicate acidosis. **Points to the right** of *point N* (*points B, D, G,* and *H*) indicate alkalosis.

a. *Line NA* represents the direction of an individual's response to an increase in P_{CO_2}. Points to the left of the normal P_{CO_2} isobar (*points A, E,* and *F*) indicate **respiratory acidosis**.

(1) *Point A* denotes a condition of uncompensated respiratory acidosis, which is characterized by a low pH, high $[HCO_3^-]$, and high P_{CO_2}.

(2) *Point E* denotes respiratory acidosis ($\uparrow P_{CO_2}$) with metabolic acidosis ($\downarrow[HCO_3^-]$).

(3) *Point F* denotes respiratory acidosis ($\uparrow P_{CO_2}$) with metabolic alkalosis ($\uparrow[HCO_3^-]$).

b. *Line NB* represents the direction of an individual's response to a decrease in P_{CO_2}. Points to the right of the normal P_{CO_2} isobar (*points B, G,* and *H*) indicate **respiratory alkalosis**.

(1) *Point B* denotes a condition of uncompensated respiratory alkalosis, which is characterized by high pH, low $[HCO_3^-]$, and low P_{CO_2}.

(2) *Point G* denotes respiratory alkalosis ($\downarrow P_{CO_2}$) with metabolic acidosis ($\downarrow[HCO_3^-]$).

(3) *Point H* denotes respiratory alkalosis ($\downarrow P_{CO_2}$) with metabolic alkalosis ($\uparrow[HCO_3^-]$).

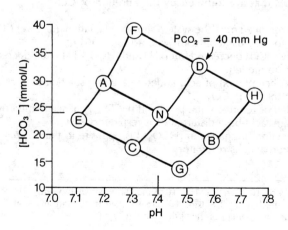

Figure 5-13. Pathways of acid-base imbalance. The *[HCO_3^-]* pertains to plasma. *Line ANB* represents the normal blood buffer line, and *line CND* represents the normal P_{CO_2} isobar (40 mm Hg). *N* = normal point; *A* = respiratory acidosis (uncompensated); *B* = respiratory alkalosis (uncompensated); *C* = metabolic acidosis (uncompensated); *D* = metabolic alkalosis (uncompensated); *E* = respiratory acidosis + metabolic acidosis; *F* = respiratory acidosis + metabolic alkalosis; *G* = respiratory alkalosis + metabolic acidosis; and *H* = respiratory alkalosis + metabolic alkalosis. (After Davenport HW: *The ABC of Acid-Base Chemistry*, 5th edition. Chicago, University of Chicago Press, 1969, p 65.)

2. **Points below** the normal buffer line (*points C, E,* and *G*) indicate conditions with a component of metabolic acidosis ($\downarrow$ [HCO$_3^-$]). **Points above** the normal buffer lines (*points D, F,* and *H*) indicate conditions with a component of metabolic alkalosis ($\uparrow$[HCO$_3^-$]).

    a. *Line NC* represents the direction of the development of metabolic acidosis in an individual; the [HCO$_3^-$] of this individual moves down the normal PCO$_2$ isobar.

        (1) *Point C* denotes a condition of uncompensated metabolic acidosis, which is characterized by low pH, low [HCO$_3^-$], and normal PCO$_2$.

        (2) *Point E* denotes metabolic acidosis ($\downarrow$[HCO$_3^-$]) with respiratory acidosis ($\uparrow$PCO$_2$).

        (3) *Point G* denotes metabolic acidosis ($\downarrow$[HCO$_3^-$]) with respiratory alkalosis ($\downarrow$PCO$_2$).

    b. *Line ND* represents the direction of the development of metabolic alkalosis in an individual; the [HCO$_3^-$] of this individual moves up the normal PCO$_2$ isobar.

        (1) *Point D* denotes a condition of uncompensated metabolic alkalosis, which is characterized by high pH, high [HCO$_3^-$], and normal PCO$_2$.

        (2) *Point F* denotes metabolic alkalosis ($\uparrow$[HCO$_3^-$]) with respiratory acidosis ($\uparrow$PCO$_2$).

        (3) *Point H* denotes metabolic alkalosis ($\uparrow$[HCO$_3^-$]) with respiratory alkalosis ($\downarrow$PCO$_2$).

3. **Mixed acid-base disturbances,** in which the primary state of acidosis or alkalosis has an additional abnormality with respect to PCO$_2$ and [HCO$_3^-$], are denoted by *points E, F, G,* and *H.*

VIII. COMPENSATORY MECHANISMS FOR PRIMARY ACID-BASE ABNORMALITIES (see Tables 5-8; 5-9)

A. Terminology

1. **Compensation** is the secondary physiologic process occurring in response to a primary acid-base disturbance by which the deviation of blood pH is ameliorated. Thus, abnormal pH is returned toward normal by **altering the component that is not primarily affected**.

    a. The terms secondary and compensatory may be used to describe a change in the composition of the blood or to describe a process.

    b. The terms secondary and compensatory may be used to describe a process, but they should not be used to describe acidosis or alkalosis.

2. **Correction** denotes the therapeutic course of action that is directed toward **counteracting or altering the factor that is primarily affected**. Correction involves the amelioration of the primary underlying abnormality in [H$^+$], [HCO$_3^-$], or PCO$_2$. In correction, as in compensation, the pH is returned toward normal.

B. Basic mechanisms of compensatory responses

1. **Buffers of the ECF** represent the body's first defense mechanism for neutralizing noncarbonic acid. H$^+$ is transferred across tissue cell membranes in a direction that normalizes blood pH.

2. A deviation in plasma pH acts on the respiratory center to change alveolar ventilation, which, in turn, alters the alveolar PCO$_2$ and arterial PCO$_2$ so as to counteract the change in plasma pH. This is called **respiratory compensation** of a primary noncarbonic acid excess or deficit; respiratory compensation usually is not complete (i.e., does not restore pH to normal).

3. **Carbonate ion is released from bone,** which is the predominant source of alkali for neutralizing excess noncarbonic acid added to the ECF.

4. Renal excretion of noncarbonic acid or base increases in order to eliminate an excess of noncarbonic acid or base or to compensate for the pH changes due to an abnormal arterial PCO$_2$. This is called **renal compensation** of a primary hyper- or hypocapnia.

5. The responses to respiratory acidosis and alkalosis differ from responses to metabolic acid-base disorders in that there is virtually no extracellular buffering of H$_2$CO$_3$ in respiratory acidosis because HCO$_3^-$ is not an effective buffer for H$_2$CO$_3$.

C. Principles of respiratory and renal compensatory responses

1. **General considerations**

    a. Physiologic compensation for major acid-base abnormalities rarely is complete. Therefore, the singular concern of clinical diagnosis is to differentiate the primary cause of the imbalance from the secondary (compensatory) response.

b. Most data indicate that the pH ($[H^+]$) is the most important factor in determining the bodily response to acid-base imbalances. (In both acid and base disturbances, the abnormal pH is returned toward normal.) However, in spite of its critical biologic significance, **a change in pH does not provide all the information needed for quantitative assessment and, thus, for planning therapy**.

   (1) The key determinants of the cause and compensation of acid-base abnormalities are Pco_2 and $[HCO_3^-]$, not pH.

   (2) The most important factor in determining the efficacy of the bodily responses to deviations from the normal acid-base status is pH or $[H^+]$.

2. Respiratory and renal processes (see Tables 5-8; 5-9)

   a. Both respiratory and renal compensatory responses tend to restore the abnormal $[HCO_3^-]/S \times Pco_2$ ratio toward its normal value of 20:1.

   (1) Primary acid-base disturbances of **metabolic origin** lead to secondary adjustment of the Pco_2 by changes in the rate of alveolar ventilation.

   (2) Primary acid-base disturbances of **respiratory origin** lead to secondary changes in blood $[HCO_3^-]$ by appropriate adjustment of the rate of H^+ secretion/excretion from the renal tubular cell into the tubular lumen.

   b. These two processes interact to control the $[H^+]$ of the ECF, and this integration is clearly understood using the mathematical relationship of the Henderson equation

$$[H^+] = 24 \frac{Pco_2}{[HCO_3^-]}$$

   From this equation it is apparent that:

   (1) An increase in the $[H^+]$, regardless of cause, can be reduced toward normal by a decrease in Pco_2, an increase in plasma $[HCO_3^-]$, or by both changes.

   (2) A decrease in the $[H^+]$, regardless of cause, can be increased toward normal by an increase in Pco_2, a decrease in plasma $[HCO_3^-]$, or by both changes.

D. Etiology of acid-base disorders and their compensatory responses

1. Metabolic acidosis is a disorder characterized by a low arterial pH ($\uparrow[H^+]$) or a reduced plasma $[HCO_3^-]$.

   a. Causes of metabolic acidosis include:

   (1) Gain of noncarbonic (fixed) acid by the ECF, such as excess quantities of the ketoacids (β-hydroxybutyric acid and acetoacetic acid) or lactic acid or the ingestion of alcohol, salicylates, or NH_4Cl

   (2) Loss of HCO_3^- and other conjugate bases, such as occurs with severe diarrhea and fistulas (which cause loss of bowel fluids containing HCO_3^-)

   b. Compensation for metabolic acidosis includes:

   (1) Extracellular buffering primarily by HCO_3^-

   (2) Intracellular buffering primarily by proteins and phosphates

   (3) Respiratory compensation by an increase in alveolar ventilation resulting in a decline in the Pco_2

   (4) Renal compensation by an increase in H^+ excretion and an increase in HCO_3^- reabsorption

2. Metabolic alkalosis is a disorder characterized by an elevated arterial pH ($\downarrow[H^+]$) or an increased plasma $[HCO_3^-]$.

   a. Causes of metabolic alkalosis include:

   (1) Decreased production of noncarbonic acid or a loss of noncarbonic acid via the kidneys or the gastrointestinal system (vomiting)

   (2) Excess HCO_3^- or other conjugate base by ingestion or infusion

   (3) Excessive renal reabsorption of HCO_3^-

   (4) Overtreatment with HCO_3^- or lactate

   b. Compensation for metabolic alkalosis includes:

   (1) Respiratory compensation by hypoventilation resulting in an increase in plasma Pco_2

   (2) Renal compensation by an increase in HCO_3^- excretion

3. Respiratory acidosis is a disorder characterized by a reduced arterial pH ($\uparrow[H^+]$), an elevated Pco_2 (hypercapnia), and a variable increase in the plasma $[HCO_3^-]$.

   a. Causes

   (1) The common denominator in respiratory acidosis is a reduction in alveolar ventilation.

      (2) The most common causes are chronic obstructive pulmonary disease and the overuse of respiratory depressant drugs.

  b. Compensation for respiratory alkalosis includes:
      (1) Increased renal reabsorption of HCO_3^-
      (2) Increased renal H^+ secretion and excretion

4. Respiratory alkalosis is a disorder characterized by an elevated arterial pH ($\downarrow[H^+]$), a low P_{CO_2} (hypocapnia), and a variable decrease in the plasma $[HCO_3^-]$.

  a. Causes of respiratory alkalosis include:
      (1) Hyperventilation (hyperpnea)
      (2) Anxiety and hysteria
      (3) Salicylate overdosage

  b. Compensation for respiratory alkalosis includes:
      (1) Decreased urinary H^+ excretion
      (2) Increased urinary HCO_3^- excretion

E. Pathways for compensation of primary acid-base disturbances: the pH-$[HCO_3^-]$ relationship (Figure 5-14)

1. Respiratory acidosis (*point A*). The system at fault in this condition is the respiratory system, and compensation occurs through metabolic processes.

  a. The kidneys excrete more acid and reabsorb more HCO_3^-, returning the $[HCO_3^-]/S \times P_{CO_2}$ ratio toward 20:1 and, therefore, returning pH toward normal (*point A_1*).

  b. If the P_{CO_2} is elevated but the pH is normal, the kidneys had time to retain HCO_3^- to compensate for the elevated P_{CO_2} and the process is not acute; that is, the condition has existed at least a few days to give the kidneys time to compensate (*point A_2*).

  c. Usually, the body does not fully compensate for respiratory acidosis.

2. Respiratory alkalosis (*point B*). In this condition, the acid-base abnormality again is respiratory, and compensation again occurs through metabolic means.

  a. The kidneys compensate by excreting HCO_3^-, thus returning the $[HCO_3^-]/S \times P_{CO_2}$ ratio toward 20:1; this compensation takes 2–3 days (*point B_1*).

  b. Of the four acid-base abnormalities, only in respiratory alkalosis is the body able to compensate fully; the $[HCO_3^-]/S \times P_{CO_2}$ ratio and pH, therefore, return entirely to normal (*point B_2*).

3. Metabolic acidosis (*point C*). In this condition, the major abnormality is low $[HCO_3^-]$ or negative BE, and the compensation is a respiratory process.

  a. By hyperventilation, the P_{CO_2} is lowered so that the $[HCO_3^-]/S \times P_{CO_2}$ ratio is returned toward 20:1 (*point C_1*).

  b. Since the compensatory system is the lungs, compensation can occur rapidly. If the metabolic acidosis is severe, however, the lungs may not be able to expel sufficient CO_2 to compensate fully.

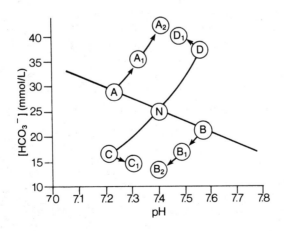

Figure 5-14. Effects of renal and respiratory compensation on the pH and plasma $[HCO_3^-]$. N = normal point; A = respiratory acidosis; A_1 = respiratory acidosis with partial renal compensation; A_2 = respiratory acidosis with complete renal compensation; B = respiratory alkalosis; B_1 = respiratory alkalosis with partial renal compensation; B_2 = respiratory alkalosis with complete renal compensation; C = metabolic acidosis; C_1 = metabolic acidosis with partial respiratory compensation; D = metabolic alkalosis; and D_1 = metabolic alkalosis with partial renal compensation. (After Davenport HW: *The ABC of Acid-Base Chemistry*, 5th edition. Chicago, University of Chicago Press, 1969, p 65.)

4. Metabolic alkalosis (*point D*). In this condition, the major abnormality is a high [HCO$_3^-$], and the compensation is a respiratory mechanism.
   a. By hypoventilaton, the Pco$_2$ is elevated so that the [HCO$_3^-$]/S $\times$ Pco$_2$ ratio is increased toward normal (*point D$_1$*).
   b. The body usually cannot compensate fully for metabolic alkalosis.

F. Summary: characteristics of uncompensated acid-base disturbances and their compensatory responses (see Table 5-8)

1. Metabolic acidosis
   a. An uncompensated metabolic acidosis is characterized by a low pH, a low CO$_2$ content, and a normal Pco$_2$.
   b. When a metabolic acidosis is compensated by a respiratory alkalosis, the pH tends to return to normal but the CO$_2$ content and Pco$_2$ decrease.

2. Metabolic alkalosis
   a. An uncompensated metabolic alkalosis is characterized by high pH, a high CO$_2$ content, and a normal Pco$_2$.
   b. When a metabolic alkalosis is compensated by a respiratory acidosis, the pH tends to return to normal but the CO$_2$ content and Pco$_2$ increase.

3. Respiratory acidosis
   a. An uncompensated respiratory acidosis is characterized by a low pH, a high CO$_2$ content, and a high Pco$_2$.
   b. When a respiratory acidosis is compensated by a metabolic alkalosis, the pH tends to return to normal but the CO$_2$ content increases; the Pco$_2$ remains unchanged.

4. Respiratory alkalosis
   a. An uncompensated respiratory alkalosis is characterized by a high pH, a low CO$_2$ content, and a low Pco$_2$.
   b. When a respiratory alkalosis is compensated by a metabolic acidosis, the pH tends to become normal but the CO$_2$ content decreases; the Pco$_2$ remains unchanged.

STUDY QUESTIONS

Directions: Each of the numbered items or incomplete statements in this section is followed by answers or by completions of the statement. Select the **one** lettered answer or completion that is **best** in each case.

1. Most of the body's total daily acid production is derived from

(A) protein catabolism
(B) triglyceride catabolism
(C) phospholipid catabolism
(D) oxidative metabolism

2. All of the following are intracellular buffers EXCEPT

(A) hemoglobin
(B) protein
(C) organic phosphate
(D) HCO_3^-
(E) carbonate

3. CO_2, which is not an acid, can lead to an increase in the $[H^+]$ of body fluids through the formation of

(A) HCO_3^-
(B) H_2CO_3
(C) lactic acid
(D) acetic acid
(E) phosphoric acid

4. Extracellular HCO_3^- is an effective buffer for all the following acids EXCEPT

(A) phosphoric acid
(B) lactic acid
(C) sulfuric acid
(D) carbonic acid
(E) β-hydroxybutyric acid

5. Which of the following statements about the pH of CSF is FALSE?

(A) It is the same as plasma pH during normal acid-base conditions
(B) It decreases during metabolic alkalosis
(C) It increases during metabolic acidosis
(D) It increases during hyperventilation
(E) It changes in the opposite direction as plasma pH during metabolic acidosis

6. In metabolic acidosis, the fall in arterial pH is associated with which of the following arterial blood conditions?

	$[HCO_3^-]$	P_{CO_2}
(A)	↑	normal
(B)	↑	↑
(C)	↓	↓
(D)	↑	↓
(E)	↓	normal

7. Respiratory alkalosis is characterized by which of the following arterial blood conditions?

	pH	P_{CO_2}
(A)	↑	↓
(B)	↓	↑
(C)	↓	↓
(D)	↑	↑
(E)	↑	normal

8. Which of the following sets of values is indicative of compensated metabolic alkalosis?

	$[HCO_3^-]$ (mEq/L)	P_{CO_2} (mm Hg)	pH
(A)	20	25	7.50
(B)	40	46	7.56
(C)	17	30	7.3
(D)	34	10	7.7
(E)	17	19	7.9

9. Compared to normal, the urinary excretion of acid in metabolic acidosis is best characterized by which of the following patterns?

	Urinary excretion of		
	HCO_3^-	NH_4^+	**Titratable acid**
(A)	↑	↑	↑
(B)	↓	↑	↓
(C)	↓	↑	↑
(D)	↓	↓	↑
(E)	↓	↓	↓

10. The following arterial blood data are obtained from a hospitalized 55-year-old man:

$[Na^+]$ = 140 mEq/L
$[HCO_3^-]$ = 15 mEq/L
$[Cl^-]$ = 113 mEq/L
pH = 7.2

From these data, which of the following statements about this patient is true?

(A) The patient has an elevated anion gap

(B) The patient has respiratory acidosis

(C) The $[H^+]$ of this patient is higher than 100 nmol/L

(D) The arterial P_{CO_2} of this patient is approximately normal

(E) The pH of this patient's CSF is decreased markedly

11. An hysterical, 35-year-old woman is admitted to the hospital and the following blood data are collected: $[HCO_3^-]$ = 22.2 mmol/L, P_{CO_2} = 30 mm Hg, and P_{O_2} = 98 mm Hg. From these data, the blood $[H^+]$ of this patient would be expected to be

(A) 17.4 nmol/L

(B) 22.5 nmol/L

(C) 28.1 nmol/L

(D) 32.4 nmol/L

(E) 36.7 nmol/L

12. A semi-comatose, 63-year-old male cigarette smoker enters the hospital complaining of labored breathing, chronic coughing, and drowsiness. He has had a "smoker's cough" for 15 years. Physical examination reveals wheezing and slowing of forced expiration as well as cyanosis and bilateral leg edema. Initial laboratory data include the following:

hemoglobin concentration = 18.5 mg/dl
serum pH = 7.32
P_{CO_2} = 68 mm Hg
P_{O_2} = 33 mm Hg

From the above findings on history, physical examination, and laboratory testing, the most likely diagnosis is

(A) metabolic acidosis

(B) metabolic alkalosis

(C) respiratory acidosis

(D) respiratory alkalosis

13. The following blood data are collected from a 58-year-old man: $[H^+]$ = 49 nEq/L, P_{CO_2} = 30 mm Hg, and P_{O_2} = 95 mm Hg. From these data, this patient's blood pH would be expected to be

(A) 4.23

(B) 7.31

(C) 8.24

(D) 3.56

(E) 6.17

Questions 14–15

The blood pressure falls during a surgical procedure on an anesthetized animal. The arterial blood becomes cyanotic despite the maintenance of normal ventilation. The arterial pH is 7.25, and the arterial P_{CO_2} is 40 mm Hg.

14. The acid-base status of the animal is most likely to be

(A) metabolic acidosis with respiratory compensation

(B) respiratory acidosis with metabolic compensation

(C) metabolic acidosis

(D) respiratory acidosis

(E) respiratory alkalosis with renal compensation

15. To increase pH toward normal, in which direction would the ventilation rate be changed and what would be the corresponding change in arterial P_{CO_2}?

	Ventilation Rate	Arterial P_{CO_2}
(A)	↑	↑
(B)	↑	↓
(C)	↓	↑
(D)	↓	↓
(E)	↑	no change

Questions 16–19

The following data were obtained from an arterial blood sample drawn from a hospitalized patient: pH = 7.55, Pco_2 = 25 mm Hg, and $[HCO_3^-]$ = 22.5 mEq/L.

16. This patient's arterial blood findings are consistent with a diagnosis of

(A) metabolic alkalosis
(B) respiratory alkalosis
(C) metabolic acidosis
(D) respiratory acidosis

17. These findings indicate that the ratio of $[HCO_3^-]$ to dissolved CO_2 is

(A) 5:1
(B) 10:1
(C) 20:1
(D) 30:1

18. The data indicate that the CO_2 content is approximately

(A) 22 mmol/L
(B) 23 mmol/L
(C) 24 mmol/L
(D) 25 mmol/L
(E) 26 mmol/L

19. The major compensatory response for this patient's acid-base disorder is

(A) hyperventilation
(B) hypoventilation
(C) increased renal HCO_3^- excretion
(D) increased H^+ excretion

Questions 20–23

A 25-year-old male patient suffers from severe diarrhea for about 1.5 hours and reports to the emergency room, where several tests are performed.

20. Based on his history, this patient's arterial pH would be expected to be approximately

(A) 7.32
(B) 7.40
(C) 7.42
(D) 7.45
(E) 7.50

21. The pH of this patient's CSF would be expected to be approximately

(A) 7.12
(B) 7.22
(C) 7.30
(D) 7.32
(E) 7.40

22. During the compensatory response to his acid-base disorder, this patient's arterial Pco_2 would most likely be

(A) 32 mm Hg
(B) 40 mm Hg
(C) 46 mm Hg
(D) 50 mm Hg
(E) 60 mm Hg

23. Following the compensatory response, a urinalysis of this patient would most likely indicate

(A) high pH
(B) decreased HCO_3^- excretion
(C) decreased titratable acid excretion
(D) increased Na^+ excretion

Directions: Each question below contains four suggested answers of which **one or more** is correct. Choose the answer.

> A if **1, 2, and 3** are correct
> B if **1 and 3** are correct
> C if **2 and 4** are correct
> D if **4** is correct
> E if **1, 2, 3, and 4** are correct

24. Statements that correctly describe noncarbonic acids include which of the following?

(1) They are called fixed acids
(2) They are classified as nonvolatile acids
(3) They are excreted by the kidneys
(4) They are buffered by both HCO_3^- and non-HCO_3^- buffers

25. Which of the following volatile acids can be transiently classified as a nonvolatile acid when present in high amounts?

(1) Lactic acid
(2) Acetic acid
(3) Citric acid
(4) Acetoacetic acid

26. The HCO_3^- buffer system is quantitatively the most significant buffer in the

(1) interstitial fluid
(2) tissue cells
(3) CSF
(4) erythrocyte

27. Acid-base disturbances associated with an elevated arterial CO_2 content include

(1) diabetic ketoacidosis
(2) respiratory alkalosis
(3) metabolic acidosis
(4) metabolic alkalosis

28. Chronic acid-base disorders associated with decreased arterial P_{CO_2} and $[HCO_3^-]$ include

(1) respiratory acidosis
(2) metabolic acidosis
(3) metabolic alkalosis
(4) respiratory alkalosis

Directions: Each group of items in this section consists of lettered options followed by a set of numbered items. For each item, select the **one** lettered option that is most closely associated with it. Each lettered option may be selected once, more than once, or not at all.

Questions 29–32

For each acid-base disorder listed below, select the appropriate pair of [HCO_3^-] and Pco_2 values.

	[HCO_3^-] (mmol/L)	Pco_2 (mm Hg)
(A)	30	65
(B)	10	25
(C)	20	20
(D)	40	45

29. Metabolic acidosis
30. Metabolic alkalosis
31. Respiratory acidosis
32. Respiratory alkalosis

Questions 33–37

For each patient described below, select the set of arterial blood values that coincides with that patient's acid-base condition.

Patient	pH	Pco_2 (mm Hg)	[HCO_3^-] (mmol/L)
(A)	7.00	70	16
(B)	7.10	27	8
(C)	7.34	70	39
(D)	7.50	46	36
(E)	7.56	26	23

33. Patient with metabolic alkalosis and partial respiratory compensation
34. Patient with metabolic acidosis partially compensated by hyperventilation
35. Patient with the highest CO_2 content
36. Patient with the highest [H^+]
37. Patient with the highest [HCO_3^-]/S × Pco_2 ratio

Questions 38–41

Points A–D on the pH-[HCO$_3$$^-$] diagram below indicate states of acid-base imbalance; point N indicates a normal acid-base state. Match each of the following conditions with the appropriate lettered point on the diagram.

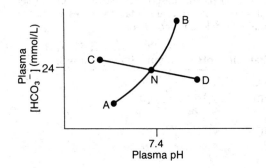

38. Hypocapnia
39. Hypercapnia
40. Ketoacidosis
41. NaHCO$_3$ ingestion

Questions 42–47

For each patient described below, select the set of arterial blood values that coincides with that patient's acid-base condition.

Patient	pH	[HCO$_3$$^-$] (mEq/L)	[Pco$_2$] (mm/Hg)	Po$_2$ (mm/Hg)
(A)	7.2	20	50	62
(B)	7.3	33	60	45
(C)	7.3	16	30	105
(D)	7.5	30	38	95
(E)	7.6	20	20	120

42. Patient with chronic respiratory acidosis and partial renal compensation
43. Patient with metabolic acidosis and partial respiratory compensation
44. Patient with uncompensated metabolic alkalosis
45. Patient with respiratory alkalosis and partial renal compensation
46. Patient with highest dissolved CO$_2$
47. Patient with highest CO$_2$ content

ANSWERS AND EXPLANATIONS

1. The answer is D. [*I A 2; IV A 2 a*] Most of the body's daily CO_2 production occurs from chemical reactions of the tricarboxylic acid cycle. From this metabolic activity, humans produce about 13,000 mmol of CO_2 daily, or, in acid-base terms, about 13,000 mEq of H^+ per day. The mammalian body produces large amounts of acids from two major sources. The volatile acid H_2CO_3 is produced from CO_2, the end-product of oxidative metabolism. A variety of nonvolatile acids (e.g., H_2SO_4, H_3PO_4) are produced from dietary substances.

2. The answer is E. [*III A 2 b, B 3, C 1*] Carbonate is a major buffer system in bone and, therefore, is not an intracellular buffer; bone contains approximately 35,000 mEq of carbonate that contributes to the buffering of acid and base loads. The intracellular fluid (ICF) does not have a high concentration of the bicarbonate (HCO_3^-) buffer system but does contain significant amounts of the non-HCO_3^- buffer systems. Intracellular protein, with its histidine residues, and organic phosphate are the quantitatively significant non-HCO_3^- buffer systems. Thus, the ICF functions to buffer both noncarbonic and carbonic acids well. The red blood cell compartment, by virtue of its hemoglobin content, is regarded in the physiologic context of body buffers as a part of the extracellular buffer system, although it is clearly intracellular. Hemoglobin is quantitatively more important than the erythrocyte HCO_3^- buffer.

3. The answer is B. [*I A*] Carbon dioxide (CO_2) is a hydrogen ion (H^+) generator. It is evident that the total reservoir of carbonic acid (H_2CO_3) [i.e., both dissolved CO_2 and H_2CO_3 termed the total "carbonic acid pool"] should be considered the acid component of the HCO_3^- buffer system, and HCO_3^- the conjugate base. The HCO_3^-/CO_2 buffer system is the most important ECF buffer pair. CO_2, which is not an acid, increases the acidity of a solution through the formation and dissociation of H_2CO_3. Students often are confused by the following series of reactions

$$CO_2 + H_2O \rightarrow H_2CO_3 \rightarrow H^+ + HCO_3^-$$

because both H^+ and HCO_3^- are formed simultaneously, and there should be no change in pH. However, the HCO_3^- concentration is several orders of magnitude (6×10^5) greater than the H^+ concentration and is correspondingly less affected by any change in the concentration of CO_2 or of H_2CO_3.

4. The answer is D. [*IV B 3; V B 3*] Extracellular HCO_3^- is not an effective buffer for H_2CO_3. The HCO_3^- buffer system plays no role in the buffering of H_2CO_3, but it is an effective buffer for noncarbonic acids. HCO_3^- cannot buffer H_2CO_3, because the combination of H^+ with HCO_3^- results in regeneration of H_2CO_3 as

$$H_2CO_3 + HCO_3^- \leftrightarrow HCO_3^- + H_2CO_3$$

Most buffering of H_2CO_3 occurs within the erythrocytes by hemoglobin. In contrast, the non-HCO_3^- buffer systems (hemoglobin, protein, phosphate) can buffer both noncarbonic acids and H_2CO_3. The HCO_3^-/CO_2 buffer system accounts for 97%–98% of the buffering in the extracellular fluid (ECF), which includes the interstitial fluid, lymph, and cerebrospinal fluid (CSF). A stipulation here is that HCO_3^- is a major buffer for noncarbonic acids.

5. The answer is A. [*II C 2*] The [HCO_3^-] of CSF and arterial blood are similar (about 24 mmol/L); however, the arterial pH (7.4) is higher than the pH of CSF (about 7.32), because the PCO_2 of CSF is about 48 mm Hg compared to 40 mm Hg in arterial blood. Acute metabolic alkalosis depresses ventilation, increasing the PCO_2 of the blood and CSF, resulting in CSF acidosis and blood alkalosis. Thus, in metabolic acid-base disturbances the [H^+] of CSF and blood change in opposite directions, and in respiratory acid-base disturbances the [H^+] of CSF and blood change in the same direction.

6. The answer is C. [*VIII D 1, E 3; Table 5-8*] Metabolic acidosis is an acid-base disturbance characterized by a decreased arterial pH (or increased [H^+]), a decreased plasma [HCO_3^-], and a compensatory hyperventilation resulting in a decreased arterial PCO_2.

7. The answer is A. [*VIII D 4, E 2; Table 5-8*] Respiratory alkalosis is an acid-base disturbance characterized by an increased arterial pH (or decreased [H^+]), a decreased PCO_2 (hypocapnia), and a variable reduction in arterial [HCO_3^-] due to renal compensation.

8. The answer is B. [*VII D 2 b; VIII D 2 b; Figure 5-13*] The hallmarks of metabolic alkalosis are elevated [HCO_3^-], which is the primary event, and elevated PCO_2, which is the compensatory event. Also, the

$[H^+]$ decreases and the $[HCO_3^-]$ increases. The only other patient set of values showing an increased $[HCO_3^-]$—*set D*—does not exhibit an increase in P_{CO_2}.

9. The answer is C. [*VI D 2, E 2 b, 3; VIII D 1 b; Table 5-7*] The kidney responds to an increased acid load by augmenting renal ammonia (NH_3) production and, consequently, ammonium (NH_4^+) excretion. There is a smaller increase in the excretion of titratable acid (primarily in the form of $H_2PO_4^-$). Thus, the decreased pH and decreased HCO_3^- excretion and increased H^+ excretion (in the form of NH_4^+ and $H_2PO_4^-$) are consistent with metabolic acidosis.

10. The answer is D. [*II D 3, E 2; Figure 5-13*] From the data given, this patient's arterial P_{CO_2} is determined to be 38 mmol/L, which is close to normal (40 mmol/L). The Henderson equation is used to calculate P_{CO_2}. The equation typically is expressed as

$$[H^+] = 24 \frac{P_{CO_2}}{[HCO_3^-]}$$

Note that when $[HCO_3^-]$ is low, as in this patient, a change in P_{CO_2} will have a greater effect on $[H^+]$ than when $[HCO_3^-]$ is normal or high.

To determine P_{CO_2}, the Henderson equation is rearranged as

$$P_{CO_2} = \frac{[H^+] [HCO_3^-]}{24}$$

First, the $[H^+]$ is calculated from the pH value as

$$[H^+] = \text{antilog } (9 - pH)$$
$$= \text{antilog } (9 - 7.22)$$
$$= \text{antilog } (1.78)$$
$$= 60 \text{ nmol/L}$$

Then, P_{CO_2} can be determined as

$$P_{CO_2} = \frac{[H^+] [HCO_3^-]}{24}$$

$$= \frac{(60)(15)}{24} = 38 \text{ mm Hg}$$

Thus, the P_{CO_2} of this patient is within normal limits and, therefore, respiratory compensation has not occurred. The acid-base disturbance of this patient is attributable to a hyperchloremic metabolic acidosis.

Notice that, in this type of acidosis, the anion gap (AG) is normal, as

$$AG = [Na^+] - [HCO_3^- + Cl^-]$$
$$= 140 \text{ mEq/L} - (15 \text{ mEq/L} + 113 \text{ mEq/L})$$
$$= 140 \text{ mEq/L} - 128 \text{ mEq/L} = 12 \text{ mEq/L}$$

The pH of the CSF in metabolic acidosis usually is elevated because of the compensatory hyperventilatory response to the metabolic acidosis. In this patient, there is no evidence of significant hyperventilation and, therefore, the pH of the CSF would not increase significantly. The axiom to remember is that metabolic acid-base alterations evoke opposite changes in the pH of the CSF and blood, while respiratory acid-base disturbances bring about parallel changes in the pH of CSF and blood.

11. The answer is D. [*II E 2 b*] From the data given, this patient's arterial $[H^+]$ is determined to be 32.4 nmol/L. The Henderson equation is used in this case, which is

$$[H^+] = 24 \frac{P_{CO_2}}{[HCO_3^-]}$$

It is necessary to keep in mind that the units for these factors are: $[H^+]$ (nmol/L), $[HCO_3^-]$ (mmol/L), and P_{CO_2} (mm Hg).
Substituting,

$$[H^+] = 24 \frac{30}{22.2}$$

$$= 32.4 \text{ nmol/L}$$

This patient has a partially compensated respiratory alkalosis. Note that her arterial $[H^+]$ and $[HCO_3^-]$ exhibit a parallel decrease.

12. The answer is C. [*II D 3, E 2; VIII D 3*] This patient has an acidosis caused by CO_2 retention (i.e., respiratory acidosis). The condition is partially compensated by renal retention of HCO_3^-. The $[HCO_3^-]$ can be determined using either the Henderson-Hasselbalch equation or the Henderson equation.

Using the Henderson-Hasselbalch equation, the $[HCO_3^-]$ is determined as

$$pH = pK + \log \frac{[HCO_3^-]}{S \times P_{CO_2}}$$

$$7.32 = 6.1 + \log \frac{[HCO_3^-]}{0.03 \times 68 \text{ mm Hg}}$$

$$7.32 = 6.1 + \log [HCO_3^-] - \log 2.04$$

$$\log [HCO_3^-] = 7.32 - 6.1 + 0.31$$

$$= 1.53$$

$$\text{antilog } 1.53 = 33.9 \text{ mmol/L}$$

Using the Henderson equation, the $[HCO_3^-]$ is determined as

$$[HCO_3^-] = 24 \frac{P_{CO_2} \text{ (mm Hg)}}{[H^+] \text{ (nmol/L)}}$$

$$= 24 \frac{68 \text{ mm Hg}}{47.9 \text{ nmol/L}}$$

$$= 34.1 \text{ mmol/L}$$

The patient's hypoxemia could be due to a pulmonary diffusion barrier. The cyanosis is caused by a greater than normal concentration of reduced hemoglobin (greater than 5 g/dl of arterial blood in the deoxy state). The leg edema could be due to the increased retention of HCO_3^- and Na^+ resulting in an increase in plasma volume. The hypoxemia causes polycythemia and constriction of vascular smooth muscle of the pulmonary circulation, leading to pulmonary hypertension, elevated right atrial pressure, and increased capillary pressure. The increase in capillary pressure causes fluid movement into the interstitial space and, ultimately, edema.

13. The answer is B. [*II D 3 a, E 2 b*] From the data given, this patient's blood pH is determined to be 7.31. To determine the arterial pH, it is first necessary to calculate the arterial $[HCO_3^-]$ using the Henderson equation, as

$$[HCO_3^-] = 24 \frac{P_{CO_2}}{[H^+]}$$

$$= 24 \frac{30}{49}$$

$$= 14.7 \text{ mmol/L}$$

Then, applying the Henderson-Hasselbalch equation, the arterial pH is determined as

$$pH = pK + \log \frac{[HCO_3^-]}{S \times P_{CO_2}}$$

$$= 6.1 + \log \frac{14.7 \text{ mmol/L}}{0.9 \text{ mmol/L}}$$

$$= 6.1 + \log 16.3$$

$$= 6.1 + 1.21$$

$$= 7.31$$

This patient has a partially compensated metabolic acidosis. It is important to note that the $[HCO_3^-]/S \times P_{CO_2}$ ratio is equal to 16.3, indicating a significant respiratory response (hyperventilation) to the metabolic problem. Thus, this metabolic acidosis is alleviated by a response that involves the respiratory system.

14–15. The answers are: 14-C, 15-B. [*VII D 2 a; VIII D 1 b*] The acid-base abnormality is metabolic acidosis without respiratory compensation. Since the respiratory response to metabolic acidosis is hyperpnea, hypocapnia would be expected, which is not observed in this case. Moreover, the imbalance cannot be attributed to respiratory acidosis, which would be characterized by an elevated arterial Pco_2. It is apparent that a mechanical respirator maintained the animal's Pco_2 at a constant level.

The pH of arterial blood is a function of two variables: $[HCO_3^-]$ and Pco_2. A compensatory response occurs in the alternate variable and in the same direction as the primary abnormality. In this case, the decline in $[HCO_3^-]$ (i.e., metabolic acidosis) would be associated with a parallel decline in Pco_2, the alternate variable. To bring this about, there would be a hyperventilatory response and a resultant decline in Pco_2.

16–19. The answers are: 16-B, 17-D, 18-B, 19-C. [*VII A 5 c, D 1 a; VIII D 4, E 2; Figure 5-8; Table 5-9*] The blood findings indicate that this patient has a respiratory alkalosis, an acid-base disturbance characterized by increased arterial pH (or decreased $[H^+]$), decreased arterial Pco_2 (hypocapnia), and decreased plasma $[HCO_3^-]$. It should be noted that both $[H^+]$ and $[HCO_3^-]$ are decreased in this patient, which is consistent with the axiom that $[H^+]$ and $[HCO_3^-]$ change in the same direction in respiratory acid-base imbalances. The decline in $[HCO_3^-]$ indicates that renal compensation has begun.

In alkalotic states, the $[HCO_3^-]/S \times Pco_2$ ratio exceeds the normal 20:1, due to either an increase in $[HCO_3^-]$ (metabolic alkalosis) or a decrease in Pco_2 (respiratory alkalosis). The normal ratio of 20:1 is derived as

$$\frac{[HCO_3^-]}{S \times Pco_2} = \frac{24 \text{ mmol/L}}{0.03 \times 40 \text{ mm Hg}} = \frac{24 \text{ mmol/L}}{1.2 \text{ mmol/L}} = \frac{20}{1}$$

In this alkalotic patient, the $[HCO_3^-]/S \times Pco_2$ ratio is 30:1. This ratio can be determined by substituting the patient's blood data into the above equation, as

$$\frac{[HCO_3^-]}{S \times Pco_2} = \frac{22.5 \text{ mmol/L}}{0.03 \times 25 \text{ mm Hg}} = \frac{22.5 \text{ mmol/L}}{0.75 \text{ mmol/L}} = \frac{30}{1}$$

The total CO_2 content for this patient is approximately 23 mmol/L. Total CO_2 equals the sum of all forms of CO_2 in the blood (i.e., HCO_3^-, H_2CO_3, dissolved CO_2, and CO_2 bound to proteins). Dissolved CO_2 content ($[CO_2]$) is calculated as

$$[CO_2] = 0.03 \times Pco_2$$
$$= 0.03 \times 40 \text{ mm Hg}$$
$$= 1.2 \text{ mmol/L}$$

Since $[H_2CO_3]$ is negligible, the total CO_2 content normally exceeds the $[HCO_3^-]$ by 1.2 mmol/L and, therefore, equals 25.2 mmol/L when normal plasma $[HCO_3^-]$ and Pco_2 values exist, as

$$\text{total } CO_2 \text{ content} = [HCO_3^-] + (S \times Pco_2)$$
$$= 24 \text{ mmol/L} + 1.2 \text{ mmol/L}$$
$$= 25.2 \text{ mmol/L}$$

Total CO_2 content is decreased in respiratory alkalosis. From this patient's blood data, the CO_2 content is calculated as

$$\text{total } CO_2 \text{ content} = [HCO_3^-] + S \times Pco_2$$
$$= 22.5 \text{ mmol/L} + 0.75 \text{ mmol/L}$$
$$= 23.25 \text{ mmol/L}$$

Respiratory alkalosis decreases the renal reabsorption of HCO_3^-, causing a transient HCO_3^- diuresis and a decline in net acid secretion. Since the major change in respiratory alkalosis is a decrease in arterial Pco_2, the compensation will be in the alternate variable (kidney) and in the same direction as the primary event. Thus, there is a decline in plasma $[HCO_3^-]$, which is indicative of a partial renal response (i.e., increased HCO_3^- excretion).

20–23. The answers are: 20-A, 21-E, 22-A, 23-B. [*II C 2; VIII D 1, E 3, F 1; Figures 5-1, 5-12, 5-13*] The fluid in the small and large intestine is relatively high in HCO_3^-. Therefore, severe diarrhea is a cause of metabolic acidosis, which is the likely diagnosis in this patient. Only one of the given arterial pH values—7.32—is consistent with acidosis.

Increased arterial $[H^+]$ stimulates ventilation, lowering the Pco_2 of the blood and CSF, resulting in blood acidosis and CSF alkalosis. Thus, in metabolic acid-base alterations the pH of the blood and CSF change in opposite directions, whereas in respiratory acid-base abnormalities the pH of the blood and CSF change in the same direction. Since the normal pH of CSF is 7.32, the pH would increase in metabolic acidosis. The only given pH value indicating a CSF alkalosis is 7.40.

Metabolic acidosis is compensated by hyperventilation, which reduces the P_{CO_2} and thereby attenuates the reduction in pH. This hyperventilatory response to metabolic acidosis is due to direct stimulation of the medullary respiratory center and of the peripheral chemoreceptors in the carotid and aortic bodies. Only one of the given P_{CO_2} values—32 mm Hg—is associated with hypocarbia.

In the absence of therapy with $NaHCO_3$, the renal compensation for a metabolic acidosis requires an increased HCO_3^- reabsorption together with an increased H^+ secretion and excretion. This adaptive response is accomplished by an increased NH_4^+ excretion with a limited ability to enhance titratable acidity via $H_2PO_4^-$ excretion. Thus, the urine will exhibit a low pH.

24. The answer is E (all). [*I B; III B 1–3; IV B*] Noncarbonic acids include all acids other than H_2CO_3, such as lactic, acetoacetic, β-hydroxybutyric, hydrochloric, and sulfuric acids. Some noncarbonic acids (lactic, acetoacetic, and β-hydroxybutyric acids) can be converted to CO_2 and eliminated by ventilation (the fate of "volatile" acids); however, when present in large amounts, these acids are eliminated by the kidney (the fate of "nonvolatile" acids). Sulfuric, hydrochloric, and phosphoric acids cannot be catabolized to CO_2 and H_2O and, thus, must be eliminated by renal excretion. Noncarbonic acids can be buffered by both HCO_3^- and non-HCO_3^- buffer systems. Fixed acid is another name for nonvolatile acid.

25. The answer is E (all). [*I B 2; IV B 1–2*] Acids that can be converted to carbonic acid are classified as volatile acids. In high amounts, many of these volatile acids are excreted by the kidneys without being converted to CO_2 and H_2O and, thus, are classified as nonvolatile acids. Examples of such acids are citric, isocitric, acetic, and lactic acids and the ketoacids. Acetic acid can increase following the ingestion of even small amounts of vinegar. All of these acids can be either eliminated by pulmonary ventilation or excreted by the kidneys. Thus, it is important to remember that nonvolatile (fixed) acids are dissociated at body pH, and the H^+ can be excreted via the urine.

26. The answer is B (1, 3). [*II C 2; III B 2 a*] In plasma, interstitial fluid, lymph, and CSF the HCO_3^- buffer system is the most important means of buffering noncarbonic acid. The ICF does not have a high $[HCO_3^-]$, but the intracellular protein and the large amount of organic phosphate compounds are quantitatively significant non-HCO_3^- buffer systems. Thus, the ICF can buffer both carbonic and noncarbonic acids. Additionally, noncarbonic acids are buffered in the erythrocytes by hemoglobin. It should be noted that the $[HCO_3^-]$ in plasma and CSF are similar.

27. The answer is D (4). [*VII A 5 b; Figure 5-7; Table 5-9*] The total CO_2 content is mainly a function of HCO_3^- and dissolved CO_2. Therefore, CO_2 content is elevated in metabolic alkalosis and respiratory acidosis. Conversely, CO_2 content declines with decreases in $[HCO_3^-]$ and $[CO_2]$, conditions that are associated with metabolic acidosis and respiratory alkalosis, respectively. Note that $[CO_2]$ denotes the concentration of dissolved CO_2 and is calculated as

$$[CO_2] = S \times P_{CO_2}$$
$$= 0.03 \times 40 \text{ mm Hg}$$
$$= 1.2 \text{ mmol/L (normal)}$$

28. The answer is C (2, 4). [*VII A 5 a–b; Table 5-9*] The changes in arterial $[HCO_3^-]$ and P_{CO_2} during various acid-base disturbances are illustrated, here, in terms of the $[HCO_3^-]/S \times P_{CO_2}$ ratio:

$$\frac{[HCO_3^-] \blacktriangledown}{S \times P_{CO_2} \downarrow}$$
metabolic acidosis

$$\frac{[HCO_3^-] \blacktriangle}{S \times P_{CO_2} \uparrow}$$
metabolic alkalosis

$$\frac{[HCO_3^-] \uparrow}{S \times P_{CO_2} \blacktriangle}$$
respiratory acidosis

$$\frac{[HCO_3^-] \downarrow}{S \times P_{CO_2} \blacktriangledown}$$
respiratory alkalosis

The large arrows depict the major changes, and the small arrows depict the compensatory responses. Changes in the numerator indicate metabolic acid-base imbalances, whereas changes in the denominator show respiratory acid-base disturbances. Clearly, both P_{CO_2} and $[HCO_3^-]$ decrease in respiratory alkalosis and metabolic acidosis, and both increase in respiratory acidosis and metabolic alkalosis. Note that the sum of $[HCO_3^-]$ and $S \times P_{CO_2}$ equals the total CO_2 content.

29–32. The answers are: 29-B, 30-D, 31-A, 32-C. [*VII A 1–3, D 1–2; VIII F; Figure 5-13; Table 5-8*] In analyzing these four acid-base disorders, it is necessary to consider both the primary abnormality (i.e., the variable that undergoes the greater degree of change) and the compensatory response (i.e., the alternate variable, which undergoes a lesser degree of change in the same direction).

In metabolic acidosis, there is a primary decrease in $[HCO_3^-]$ and a compensatory decrease in PCO_2. Only one pair of values (*B*) shows this pattern. In metabolic alkalosis, there is a primary increase in $[HCO_3^-]$ and a compensatory increase in PCO_2. Only one pair of values (*D*) shows this pattern. In respiratory acidosis, there is a primary increase in PCO_2 and a compensatory increase in $[HCO_3^-]$. Only one pair of values (*A*) shows this pattern. In respiratory alkalosis, there is a primary decrease in PCO_2 and a compensatory decrease in $[HCO_3^-]$. Only one pair of values (*C*) shows this pattern.

33–37. The answers are: 33-D, 34-B, 35-C, 36-A, 37-E. [*II D 3 b, E 2 b; VII A 3 a (2), 5 c, D 2; VIII D 1, 2, F*] *Patient D* has a metabolic alkalosis with partial respiratory compensation. Metabolic alkalosis can result from ingestion of alkaline substances (e.g., antacids) or from loss of acid (as occurs in vomiting). The total CO_2 content also is increased in metabolic alkalosis. The compensation for this condition is hypoventilation or excretion of an alkaline urine. Note that *patient D* shows partial compensation, as evidenced by the increased PCO_2. The $[HCO_3^-]/S \times PCO_2$ ratio and pH are increased.

Patient B has a metabolic acidosis with partial respiratory compensation. Metabolic acidosis can result from increased fixed (noncarbonic) acid production (as occurs in diabetic ketoacidosis), increased acid retention (as occurs in renal failure with subsequent accumulation of $H_2PO_4^-$), or loss of base (as occurs in diarrhea). The total CO_2 content also is decreased in metabolic acidosis. The compensation for this condition is hyperventilation or excretion of an acid urine. Note that *patient B* shows partial compensation, as evidenced by the decrease in PCO_2. The $[HCO_3^-]/S \times PCO_2$ ratio and pH are decreased.

Patient C has the highest plasma CO_2 content (i.e., $[HCO_3^-] + S \times PCO_2$). This patient has respiratory acidosis, a condition caused by CO_2 retention resulting from hypoventilation, which may be due to pulmonary disease. The compensation for this condition is an increase in both HCO_3^- reabsorption and in H^+ excretion in the form of $H_2PO_4^-$ and NH_4^+. Note that *patient C* shows nearly complete compensation, as evidenced by the increased $[HCO_3^-]$ and return of pH toward normal. The $[HCO_3^-]/S \times PCO_2$ ratio is only slightly decreased (i.e., 18:1 versus the normal 20:1).

Patient A has the lowest pH and, thus, the highest $[H^+]$. This patient has metabolic acidosis without compensation. $[H^+]$ can be calculated as

$$[H^+] = \text{antilog } (9 - pH)$$
$$= \text{antilog } (9 - 7)$$
$$= \text{antilog } 2$$
$$= 100 \text{ nmol/L}$$

or estimated as

$$[H^+] = 24 \frac{PCO_2}{[HCO_3^-]}$$
$$= 24 \frac{70}{16}$$
$$= 24 \ (4.38)$$
$$= 105 \text{ nmol/L}$$

Patient E has the highest $[HCO_3^-]/S \times PCO_2$ ratio (i.e., 29.5:1). This patient has respiratory alkalosis, a condition caused by loss of CO_2 due to anxiety or hysteria leading to hyperventilation. The compensation for respiratory alkalosis is excretion of an alkaline urine. This patient shows partial renal compensation, as evidenced by the slight decline $[HCO_3^-]$. *Patient E* has a slightly lower than normal CO_2 content.

38–41. The answers are: 38-D, 39-C, 40-A, 41-B. [*VII D 1–2; VIII D; Figure 5-13; Table 5-8*] In analyzing these four uncompensated acid-base disturbances, it is important to consider the PCO_2 and $[HCO_3^-]$ rather than the pH, as PCO_2 and $[HCO_3^-]$ are the key determinants of the cause of, and compensation for, these disturbances.

Point D represents a patient with increased pH ($\downarrow[H^+]$) brought about by respiratory alkalosis. This condition is characterized by a primary decrease in PCO_2 (hypocapnia) and a variable secondary decrease in plasma $[HCO_3^-]$. Metabolic acidosis also is characterized by declines in these two variables, but the pH is decreased ($\uparrow[H^+]$) as well. Respiratory alkalosis is defined as alveolar ventilation in excess of the existing need of the body to eliminate CO_2. This excess in alveolar ventilation, called hyperventilation, results in a reduced arterial PCO_2. Hyperpnea is the general term used to describe any increase in ventilatory effort. With respiratory alkalosis there is a decline in the $[total\ CO_2]$.

Point C represents a patient with decreased pH ($\uparrow$[H$^+$]) caused by respiratory acidosis. This clinical disorder is characterized by a primary increase in P_{CO_2} (hypercapnia) and a variable secondary increase in plasma [HCO$_3^-$]. The common denominator in respiratory acidosis is hypoventilation, which is defined as alveolar ventilation insufficient to excrete CO_2 rapidly enough to meet the existing needs of the body. With respiratory acidosis there is a relatively small increment in the [total CO_2], because the major fraction of the CO_2 content is comprised of HCO$_3^-$.

Point A represents a patient with diabetes mellitus, which is the most common cause of ketoacidosis. This overproduction of ketoacids is caused by a deficiency of insulin, which leads to: (1) increased lipolysis and an increased delivery of free fatty acids to the liver and (2) the preferential conversion of free fatty acids to ketoacids rather than to triglycerides. Thus, metabolic acidosis is characterized by a low arterial pH ($\uparrow$[H$^+$]), a reduced [HCO$_3^-$], and a compensatory hyperventilation resulting in hypocapnia. The renal compensatory response for respiratory alkalosis also diminishes the plasma [HCO$_3^-$], but the pH in that disorder is elevated ($\downarrow$[H$^+$]). Overproduction of ketoacids causes acidosis by two mechanisms: (1) a decrease in plasma [HCO$_3^-$] with an increase in the anion gap and (2) overloading of the renal capacity to excrete H$^+$ resulting in a loss of Na$^+$ and a failure to recover NaHCO$_3$. In metabolic acidosis, there is a decline in the [total CO_2]. Furthermore, ketoacidosis, like lactic acidosis, differs from other forms of metabolic acidosis in that the anion associated with H$^+$ can be metabolized back to HCO$_3^-$, as

$$\beta\text{-hydroxybutyrate}^- + O_2 \rightarrow CO_2 + H_2O + HCO_3^-$$

β-hydroxybutyrate represents about 75% of the circulating ketoacids in diabetic ketoacidosis. It can be seen from the above chemical equation that the metabolism of the β-hydroxybutyrate anion results in the regeneration of the HCO$_3^-$ that was neutralized in buffering the H$^+$. Since the HCO$_3^-$ is replaced by an anion that is metabolized back to HCO$_3^-$, there is no actual loss of HCO$_3^-$ from the body in ketoacidosis (or lactic acidosis). Insulin administration decreases the accumulation of β-hydroxybutyric acid and allows the metabolism of the acid ions back to HCO$_3^-$.

Point B represents a patient with metabolic alkalosis. Excessive ingestion of NaHCO$_3$ can result in metabolic alkalosis and an increase of pH ($\downarrow$[H$^+$]). Metabolic alkalosis is characterized by an increase in the plasma [HCO$_3^-$] and a compensatory increase in the P_{CO_2} produced by a decline in alveolar ventilation. Since elevation of plasma [HCO$_3^-$] can be due to the renal compensation for chronic respiratory acidosis, the diagnosis of metabolic alkalosis cannot be made without measuring the pH. Metabolic alkalosis is associated with a large increase in the [total CO_2].

42–47. The answers are: 42-B, 43-C, 44-D, 45-E, 46-B, 47-B. [*II D 2 b, c; VII A 5 c, D 1–2; VIII D–F; Figures 5-12, 5-13*] *Patient B* has respiratory acidosis with partial renal compensation. Respiratory acidosis is caused by increased P_{CO_2}, which reduces the [HCO$_3^-$]/S $\times$ P_{CO_2} ratio and, thus, the pH. Whenever P_{CO_2} rises, [HCO$_3^-$] also must increase somewhat because of the dissociation of H$_2$CO$_3$ formed by the hydration of CO_2. The kidney responds by conserving [HCO$_3^-$] via increased H$^+$ secretion, with acid excreted as H$_2$PO$_4^-$ and NH$_4^+$.

Patient C has metabolic acidosis with partial respiratory compensation. Metabolic acidosis means a primary decrease in [HCO$_3^-$], which decreases the [HCO$_3^-$]/S $\times$ P_{CO_2} ratio and, thus, the pH. The [HCO$_3^-$] may be lowered by the addition of H$^+$ or by the loss of HCO$_3^-$. Respiratory compensation occurs by increased ventilation via the action of H$^+$ in the peripheral and central chemoreceptors.

Patient D has uncompensated metabolic alkalosis. In this condition, the increase in [HCO$_3^-$] causes an increase in the [HCO$_3^-$]/S $\times$ P_{CO_2} ratio and, thus, the pH. Respiratory compensation occurs by a reduction in alveolar ventilation, which tends to raise P_{CO_2} (not evident in *patient D*).

Patient E has respiratory alkalosis with partial renal compensation. Respiratory alkalosis is caused by a decrease in P_{CO_2}, which increases the [HCO$_3^-$]/S $\times$ P_{CO_2} ratio and, thus, the pH. Renal compensation occurs by increased HCO$_3^-$ excretion.

Patient B has both the highest amount of dissolved CO_2 and the highest total CO_2 content. Dissolved CO_2 is determined by multiplying P_{CO_2} (in mm Hg) by the P_{CO_2} solubility constant, S, as

$$\text{dissolved } CO_2 = S \times P_{CO_2}$$
$$= 0.03 \times 60$$
$$= 1.8 \text{ mmol/L}$$

Total CO_2 content is the sum of [HCO$_3^-$] and dissolved CO_2. It is calculated in *patient B* as

$$\text{total } CO_2 \text{ content} = [HCO_3^-] + S \times P_{CO_2}$$
$$= 33 + 0.03 \times 60$$
$$= 33 + 1.8$$
$$= 34.8 \text{ mmol/L}$$

6
Gastrointestinal Physiology
Michael B. Wang

I. DIGESTIVE SYSTEM: OVERVIEW

A. Structure. The digestive system is composed of a long muscular tube—the gastrointestinal (GI) tract, or alimentary canal—and a set of accessory organs (Figure 6-1).

1. The **GI tract** consists of the oral cavity, pharynx, esophagus, stomach, small intestine, large intestine, rectum, and anal canal.

2. The **accessory organs** include the tongue, teeth, salivary glands, pancreas, liver, and gall-bladder.

B. Function. The digestive system is responsible for breaking down food and supplying the body with the water, nutrients, and electrolytes needed to sustain life. Before food can be used by the body, it must be ingested, digested, and absorbed—processes that involve coordinated movement of muscle and secretion of various substances.

1. **Ingestion** involves:
   a. Placing food into the mouth
   b. Chewing the food into smaller pieces (**mastication**)
   c. Moistening the food with salivary secretions
   d. Swallowing the food (**deglutition**)

2. **Digestion.** During **digestion,** food is broken down into small particles by the grinding action of the GI tract and then degraded by digestive enzymes into usable nutrients.
   a. Starches are degraded by amylases into monosaccharides.
   b. Proteins are degraded by a variety of enzymes (e.g., pepsin, trypsin) into dipeptides and amino acids.
   c. Fats are degraded by lipases and esterases into monoglycerides and free fatty acids.

3. **Absorption.** During **absorption,** nutrients, water, and electrolytes are transported from the GI tract (principally from the small intestine) to the circulation.

II. THE ORAL CAVITY, PHARYNX, AND ESOPHAGUS are involved in the chewing and swallowing, or **ingestion,** of food. Saliva helps ready the food to be moved from the mouth to the esophagus, and peristalsis is the muscle pattern primarily responsible for moving the food through the esophagus.

A. Mastication. After food is placed into the mouth, it is cut and ground into smaller pieces by chewing (mastication).

1. **The chewing reflex.** Although chewing is a voluntary act, it is coordinated by reflex centers in the brain stem that facilitate the opening and closing of the jaw.
   a. When the mouth opens, stretch receptors in the jaw muscles initiate a reflex contraction of the masseter, medial pterygoid, and temporalis muscles, causing the mouth to close.
   b. When the mouth closes, food comes into contact with buccal receptors eliciting a reflex contraction of the digastric and lateral pterygoid muscles, causing the mouth to open.
   c. When the jaw drops, the stretch reflex causes the entire cycle to be repeated.

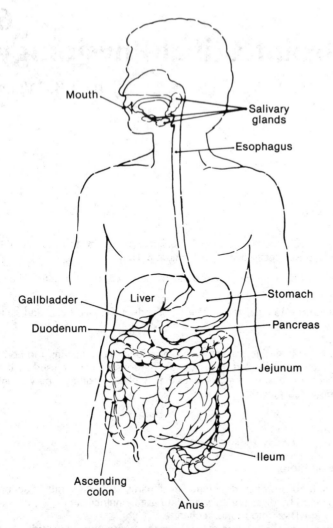

Mouth

Salivary glands

Esophagus

Gallbladder

Liver

Stomach

Duodenum

Pancreas

Jejunum

Ileum

Ascending colon

Anus

Figure 6-1. The gastrointestinal (GI) tract.

  d. The tongue contributes to the grinding process by positioning the food between the upper and lower teeth.

 2. Function of mastication
  a. Chewing **breaks food into smaller pieces,** which:
   (1) Makes it easier for the food to be swallowed
   (2) Breaks off the undigestible cellulose coatings of fruits and vegetables
   (3) Increases the surface area of the food particles, making it easier for them to be digested by the digestive enzymes
  b. Chewing **mixes the food with salivary gland secretions** (see II B), which:
   (1) Initiates the process of starch digestion by salivary amylase
   (2) Initiates the process of lipid digestion by lingual lipase.
   (3) Lubricates and softens the bolus of food, making it easier to swallow
  c. Chewing **brings food into contact with taste receptors** and **releases odors** that stimulate the olfactory receptors. The sensations generated by these receptors increase the pleasure of eating and initiate gastric secretions.

B. Saliva

 1. Salivary glands. Saliva is secreted primarily by three pairs of glands.
  a. The **parotid glands,** located near the angle of the jaw, are the largest glands. They secrete a watery fluid.
  b. The **submandibular** and **sublingual glands** secrete a fluid that contains a higher concentration of proteins and so is more viscous.

    c. Smaller glands are located throughout the oral cavity. Those in the tongue secrete lingual lipase.

  2. Composition of saliva. The salivary glands secrete a relatively high volume of fluid (0.5–1 L/day) containing electrolytes and proteins.

    a. Electrolytes. Electrolyte concentration and osmolality both vary with secretory flow rates, but generally, in comparison to plasma, saliva is hypotonic and contains higher concentrations of potassium (K^+) and bicarbonate (HCO_3^-) and lower concentrations of sodium (Na^+) and chloride (Cl^-).

    b. Proteins. Two types of proteins are found in saliva.

      (1) The **enzymes,** α-amylase (ptyalin) and lingual lipase, begin the process of starch and fat digestion.

      (2) Mucin is a glycoprotein that lubricates the food.

  3. Control of salivary secretion

    a. Salivary secretion is controlled entirely by **autonomic nervous system reflexes**.

      (1) Parasympathetic nerve stimulation causes the salivary gland cells to secrete a large volume of watery fluid that is high in electrolytes but low in proteins.

      (2) Sympathetic nerve stimulation causes the salivary glands to secrete a small volume of fluid containing a high concentration of mucus.

    b. Salivary reflexes are elicited by the thought, aroma, or taste of food or by the presence of food within the alimentary canal.

    c. Salivary gland metabolism and growth are both stimulated by increased autonomic nervous system activity.

  4. Functions of saliva. Salivary secretions perform a number of important functions.

    a. Protection. Salivary secretions protect the mouth by:

      (1) Cooling hot foods

      (2) Diluting any hydrochloric acid (HCl) or bile regurgitated into the mouth

      (3) Washing food away from the teeth and destroying harmful bacteria within the mouth

    b. Digestion. Salivary secretions begin the process of starch and fat digestion.

      (1) α-**Amylase** can digest most of the ingested starches into disaccharides before they reach the small intestine. α-Amylase is ultimately inactivated by the low pH of the stomach.

      (2) Lingual lipase begins to break down ingested fats while they are in the mouth, stomach, and upper portions of the small intestine.

    c. Lubrication. Salivary secretions lubricate the food, making swallowing easier, and moisten the mouth, facilitating speech.

C. Peristalsis

  1. Definition. Peristalsis is a coordinated pattern of smooth muscle contraction and relaxation.

  2. Function. Peristalsis helps move food through the pharynx and esophagus and within the stomach. Peristalsis plays a minor role in propelling food through the intestine.

  3. Mechanics. During peristalsis, contraction of a small section of proximal muscle is followed immediately by relaxation of the muscle just distal to it. The resulting wavelike motion moves food along the GI tract in a proximal (orad) to distal (caudad) direction (see also II D 1 c; III B 3, 6 c; IV B 2 b; V A 2 b).

D. Deglutition (swallowing)

  1. Phases of swallowing

    a. Oral (voluntary) phase. During the voluntary phase, the tongue forms a bolus of food and forces it into the oropharynx by pushing up and back against the hard palate.

    b. Pharyngeal phase

      (1) The pharyngeal phase is coordinated by a swallowing center in the medulla and lower pons. It begins when the food reaches the oropharynx and progresses as follows.

        (a) The nasopharynx is closed by the soft palate, preventing regurgitation of food into the nasal cavities.

        (b) The palatopharyngeal folds are pulled medially, forming a passageway for the food to move into the pharynx.

        (c) The glottis and vocal cords are closed and the epiglottis swings down over the larynx, guiding the food toward the esophagus and away from the airways.

(d) The bolus of food is pushed into the esophagus by the peristaltic contractions of the pharynx and the opening of the upper esophageal sphincter.

(2) Respiration is inhibited for the duration of the pharyngeal phase of swallowing (1–2 seconds).

c. Esophageal phase. After reaching the esophagus, food is propelled into the stomach by peristalsis. The strength of peristaltic contractions is proportional to the size of the bolus entering the esophagus.

(1) Sphincters involved in esophageal peristalsis. The esophagus is isolated from the oral cavity by the upper esophageal sphincter and from the stomach by the lower esophageal sphincter.

(a) The **upper esophageal sphincter** is composed of **striated muscle** and is completely under the control of the vagal fibers innervating the esophagus. Tone is maintained by the continuous firing of vagal postganglionic neurons. The transmitter released by these fibers is acetylcholine (ACh).

(b) The **lower esophageal sphincter** is composed of **smooth muscle**. Its tone is maintained by a myogenic process. Intrinsic nervous system neurons (under vagal control) cause the lower esophageal sphincter to relax during peristalsis. The synaptic transmitter released by these neurons is neither ACh nor norepinephrine and is thus called a nonadrenergic noncholinergic neural transmitter. Its identity is not known but may be either adenosine triphosphate (ATP) or vasoactive intestinal peptide (VIP).

(2) Types of esophageal peristalsis. There are two types of esophageal peristalsis, primary and secondary.

(a) Primary esophageal peristalsis is initiated by swallowing. It begins when food passes into the esophagus from the pharyngeal cavity.

(i) As soon as the food enters the esophagus, the upper esophageal sphincter contracts to prevent regurgitation of food into the mouth.

(ii) The peristaltic wave travels rather slowly (3–4 cm/sec), taking about 9 seconds to push the food from mouth to stomach (a distance of about 30 cm). The force of gravity causes liquids to fall through the esophagus at a much faster rate.

(iii) After food enters the stomach, the lower esophageal sphincter contracts to prevent regurgitation of food into the esophagus.

(iv) If swallowing is not accompanied by the passing of food into the esophagus the ensuing peristaltic wave will be very weak or may not occur at all.

(b) Secondary peristalsis is initiated by the presence of food within the esophagus.

(i) After primary peristalsis is completed, any food remaining in the esophagus stretches mechanical receptors, initiating another peristaltic wave.

(ii) Secondary peristaltic waves continue until all of the swallowed food is removed from the esophagus.

(3) Coordination of esophageal peristalsis

(a) Primary esophageal peristalsis is coordinated by vagal fibers emerging from the swallowing center within the medulla that are activated as part of the swallowing reflex. **Vagotomy,** which eliminates the efferent fibers emerging from the swallowing center, would prevent the initiation of primary esophageal peristalsis.

(b) Secondary esophageal peristalsis is coordinated by the intrinsic nervous system of the esophagus. Afferent fibers innervate stretch receptors within the wall of the esophagus and thereby activate the appropriate fibers in the intrinsic nervous system. Since intrinsic rather than vagal nerves are involved, **vagotomy** would reduce the intensity of secondary esophageal peristalsis but would not prevent it from occurring.

2. Disorders of swallowing

a. Esophageal reflux may occur if the intragastric pressure rises high enough to force the lower esophageal sphincter open, if the lower esophageal sphincter is unable to maintain its normal tone, or if the lower esophageal sphincter is forced through the diaphragm and into the thoracic cavity.

(1) During pregnancy, the growing fetus may push the top of the stomach into the thorax. The low intrathoracic pressure (compared to the higher intra-abdominal pressure) causes the lower esophageal sphincter to expand, allowing reflux to occur.

(2) Reflux of stomach acid causes esophageal pain (heartburn) and may lead to esophagitis.

b. Belching (eructation). Following a heavy meal or the ingestion of large amounts of gas (e.g., from carbonated beverages), the gas bubble that is usually in the fundus of the stomach is displaced to the cardia (see Figure 6-2). When the lower esophageal sphincter relaxes during the swallowing process, gas enters the esophagus and is regurgitated.

c. Achalasia is a neuromuscular disorder of the lower two-thirds of the esophagus that leads to absence of peristalsis and failure of the lower esophageal sphincter to relax. Food accumulates above this sphincter, taking hours to enter the stomach and dilating the esophagus.

III. THE STOMACH. Complex patterns of motility move food through the stomach where it can be broken down further by gastric secretions and then propelled into the small intestine. Only small amounts of food are digested or absorbed in the stomach.

A. Anatomy (Figure 6-2)

1. Functional components

a. The three functional parts of the stomach are the **fundus, corpus** (body), and **antrum**. Gastric contents are isolated from other parts of the GI tract by the lower esophageal sphincter proximally and by the pylorus (pyloric sphincter) distally.

b. The antrum and pylorus are anatomically continuous and respond to nervous control as a unit.

2. Musculature

a. As elsewhere in the gut, each muscle layer in the stomach forms a functional syncytium and therefore acts as a unit. In the fundus, where the layers are relatively thin, strength of contraction is weak; in the antrum, where the muscle layers are thick, strength of contraction is greater.

b. The stomach and duodenum (the uppermost part of the small intestine) are divided by a thickened muscle layer called the pyloric sphincter.

3. Innervation

a. Intrinsic. The interconnected **myenteric (Auerbach's) plexus** and **submucosal (Meissner's) plexus** within the stomach wall comprise the intrinsic innervation of the stomach, as they do elsewhere in the gut. They are **directly responsible for peristalsis** and other contractions. Because this system is continuous between the stomach and duodenum (see Figure 6-4), peristalsis in the antrum influences the duodenal bulb.

(1) The **myenteric plexus** is located between the layers of the circular and longitudinal muscles of the stomach.

(2) The **submucosal plexus** is located between the layers of the circular muscle and mucosa on the luminal surface of the stomach.

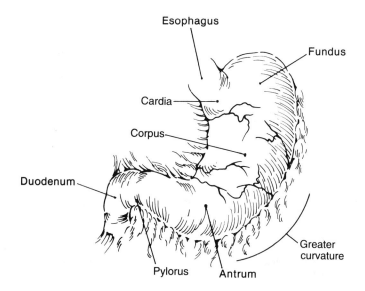

Figure 6-2. The stomach.

b. Extrinsic. Autonomic innervation is dual—both **sympathetic,** via the celiac plexus, and **parasympathetic,** via the vagus nerve. Sympathetic innervation inhibits motility, and parasympathetic innervation stimulates motility. Together, the two systems modify the coordinated motor activity that arises independently in the intrinsic system.

B. Motility

1. **Function.** Gastric motility serves three basic functions.
   a. **Storage.** When food enters the stomach, the orad region—primarily the fundus—enlarges to accommodate the food by a process referred to as **receptive relaxation**.
   b. **Mixing.** The presence of food in the caudad stomach—primarily the corpus and antrum—increases the contractile activity of the stomach.
      (1) The enhanced contractile activity (a combination of **peristalsis** and **retropulsion**) mixes the food with stomach acid and enzymes, breaking it into smaller and smaller pieces.
      (2) When the food is mixed into a pasty consistency, it is called **chyme**.
   c. **Emptying.** When the chyme is broken down into small enough particles, it is propelled through the pyloric sphincter into the intestine.

2. **Receptive relaxation.** When food is passed from the esophagus to the stomach, the contractile activity of the fundus is inhibited, enabling it to easily accommodate 1–2 L of food.
   a. **Innervation.** Vagal nerve fibers innervate the **intrinsic (enteric) nervous system** of the orad stomach to cause receptive relaxation. During this process, postganglionic fibers within the enteric nervous system release a noncholinergic nonadrenergic transmitter whose identity is unknown but may be ATP or VIP.
   b. **Initiation.** Receptive relaxation may be initiated as part of the peristaltic process causing swallowing and esophageal motility or in response to a bolus of food entering the stomach. Stretch receptors in the orad stomach detect the presence of food and initiate a vago-vagal reflex producing receptive relaxation.
   c. **Effects of vagotomy.** Sectioning the vagus nerve will prevent or greatly diminish receptive relaxation since the process is controlled by the enteric nervous system.

3. **Peristalsis.** Peristaltic contractions are initiated near the fundal-corpus border and proceed caudally, producing a peristaltic wave that propels the food towards the pylorus.
   a. **Mechanics of peristalsis.** Peristaltic contractions are produced by periodic changes in membrane potential, called **slow waves,** or the **basic electrical rhythm (BER)** [Figure 6-3]. These waves are responsible for the rhythm and force of gastric contractions.
      (1) Gastric slow waves originate in a pacemaker within the longitudinal muscle high on the greater curvature of the orad corpus.
      (2) Slow waves consist of an **upstroke** and **plateau** phase and occur at a rate of approximately 3–4/min.
      (3) The **velocity** of the waves is 1 cm/sec when they sweep over the corpus and increases to 3–4 cm/sec in the antrum.

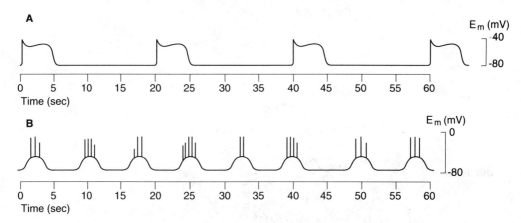

Figure 6-3. Basic electrical rhythm (BER), or slow waves, as recorded from smooth muscle cells of (*A*) the stomach and (*B*) the middle of the intestine. The slow waves of the stomach are 5–7 seconds in duration and occur at a rate of 3–5 waves/min, while the slow waves of the intestine are more frequent (12 waves/min in the duodenum and 8 waves/min in the ileum) and have spikes superimposed on their plateaus.

    (4) Although the electrophysiological basis for the slow waves is not entirely known, it is assumed that the upstroke is due to the flow of Na^+ and calcium (Ca^{2+}) into the cell and that the plateau is dependent primarily on the flow of Ca^{2+} into the cell.

  b. Force of peristalsis. The force of peristaltic contractions is regulated by gastrin and ACh. These hormones:

    (1) Increase the size of the slow wave plateau potential, which increases the amount of Ca^{2+} entering the cell from the extracellular fluid

    (2) Activate second messengers that release Ca^{2+} from the sarcoplasmic reticulum

4. Retropulsion. Retropulsion is the back and forth movement of the chyme caused by the forceful propulsion of food against the closed pyloric sphincter (Figure 6-4).

  a. The wave of peristaltic contraction reaches the pyloric sphincter before the chyme does. Thus, when the chyme reaches the sphincter, it is pushed back into the body of the stomach.

  b. The forward and backward movement of the chyme (caused by peristalsis and retropulsion) breaks the chyme into smaller and smaller pieces and mixes it with the gastric secretions present within the stomach.

5. Gastric emptying occurs when the chyme is decomposed into small enough pieces to fit through the pyloric sphincter.

  a. Each time the chyme is pushed against the pyloric sphincter, a small amount (2–7 mls) may escape into the duodenum.

  b. The amount of chyme passing through the pylorus depends on the size of the particles. If the particles are too large, none of the chyme will enter the duodenum.

  c. Thus, the rate of gastric emptying of solids depends on the rate at which the chyme is broken down into small particles.

  d. Liquids empty much faster than solids. The rate of liquids emptying is proportional to pressure within the orad stomach, which increases slowly during the digestive period.

6. Regulation of gastric emptying

  a. Local reflexes

    (1) Excitatory reflexes, initiated by expansion of the antrum and by the digestive products of food, are responsible for increasing gastric motility. Although these reflexes do not require the vagus nerve, vagotomy decreases the magnitude and coordination of stomach contractions.

    (2) Inhibitory reflexes. A variety of stimuli act on the duodenum to initiate **enterogastric reflexes** that slow the rate of gastric emptying.

      (a) Purpose. Enterogastric reflexes prevent the flow of chyme from exceeding the ability of the intestine to handle it.

      (b) Causes. High osmolarity, low pH, fat and protein digestion products, and distension of the duodenal wall all elicit an enterogastric reflex.

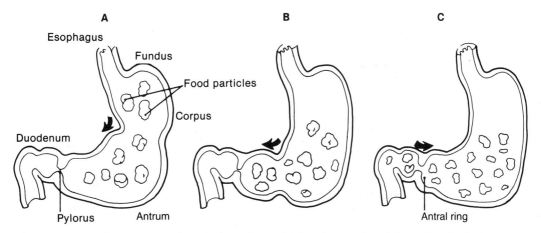

Figure 6-4. Peristaltic contractions begin in the midstomach (*A*) and proceed caudally, pushing the food toward the pylorus (*B*). When the food reaches the pylorus (where small enough pieces of food flow into the duodenum), a mass contraction of the terminal antrum pushes the food back toward the corpus through a narrow antral ring (*C*). The backward movement of the food is called retropulsion.

b. Hormones released from the stomach and intestine also influence gastric motility.
   (1) Excitatory effects. Gastrin, released into the circulation in response to antral distension or food breakdown products, enhances gastric contractions.
   (2) Inhibitory effects. A variety of intestinal hormones, collectively called **enterogastrones,** inhibit gastric contractions. **Cholecystokinin (CCK)** and **secretin** are two known enterogastrones. The identity and mode of action of other enterogastrones remain to be discovered.
      (a) CCK is released from the duodenum in response to fat or protein digestion products. CCK probably acts by blocking the excitatory effects of gastrin on gastric smooth muscle (see also IV C 3 c).
      (b) Secretin is released from the duodenum in response to the presence of acid. Secretin most likely has a direct inhibitory effect on smooth muscle (see also IV C 3 c).
c. Migrating motor complex (MMC). During the interdigestive period, any food left in the stomach is removed by the MMC.
   (1) The MMC is a peristaltic wave that begins within the esophagus and travels through the entire GI tract (see IV B 2 c).
   (2) The peristaltic wave occurs every 60–90 minutes during the interdigestive period.
   (3) The hormone **motilin,** which is released from endocrine cells within the epithelium of the small intestine, increases the strength of the MMC.

7. Vomiting (or **emesis**) is the forceful expulsion of the food from the stomach and intestine.
   a. Initiation. Vomiting may be initiated by direct activation of the **vomiting center** in the medulla or by activation of the **chemoreceptor trigger zone** within the area postrema of the brain stem.
      (1) The **vomiting center** may be directly activated by afferent fibers or by irritation due to injury or increases in intracranial pressure. When the vomiting center is directly activated, it causes **projectile vomiting**—a rapid, forceful emesis not accompanied by nausea.
      (2) The **chemoreceptor trigger zone** may be activated by afferent nerves originating within the GI tract or by circulating emetic agents such as apomorphine or copper sulfate. Vomiting caused by activation of the chemoreceptor trigger zone is accompanied by nausea.
   b. Mechanical sequence of vomiting
      (1) Vomiting begins with a deep inspiration followed by the closing of the glottis.
      (2) Next, a pressure wave originating in the intestine propels chyme into the orad stomach.
      (3) Finally, an increase in abdominal pressure forces the chyme into the esophagus and out of the mouth.
      (4) Retching may proceed vomiting. Retching involves all of the involuntary motions of vomiting but without the production of vomitus. The chyme is not ejected because the abdominal and thoracic pressures are not sufficient to overcome the resistance of the upper esophageal sphincter.

C. Gastric secretion

1. General considerations
   a. Function. Gastric secretions aid in the breakdown of food into small particles and continue the process of digestion begun by salivary enzymes. About 2 L/day of gastric secretions are produced.
   b. Phases of gastric secretion
      (1) The cephalic phase of gastric secretion is initiated by the thought, sight, taste, or smell of food. It is dependent on the integrity of the vagal fibers innervating the stomach.
         (a) Secretion of HCl from parietal cells, gastrin from G cells, and pepsinogen from chief cells is stimulated by vagal efferent fibers.
         (b) Almost half of the gastric secretions released during a meal occur as a result of cephalically induced vagal stimulation.
      (2) The gastric phase of secretion is initiated by the entry of food into the stomach.
         (a) Distension of the corpus, acting through local and vago-vagal reflexes, results in an increase in HCl secretion.
         (b) Distension of the antrum initiates vagally mediated and local reflexes that result in gastrin release from antral G cells. Gastrin release is inhibited at low pH (<3).
         (c) Low pH activates local reflexes, which enhance pepsinogen secretion.

    (d) Although the rate of gastric secretion during the gastric phase is less than during the cephalic phase, it continues for a longer time. Thus, the two phases contribute about the same amount of secretion.

  (3) The intestinal phase of secretion begins as the chyme begins to empty from the stomach into the duodenum. Overall, little gastric secretion occurs during the intestinal phase.

c. Gastric secretory cells are located on the surface of the stomach and in glands that are buried within the mucosa.

  (1) Oxyntic glands are located in the fundus and corpus of the stomach. They contain three types of secretory cells.

    (a) The **parietal (oxyntic) cells** secrete **HCl**. These cells are also responsible for the secretion of **intrinsic factor,** which is necessary for the absorption of vitamin B_{12} by the ileum of the small intestine (see III C 5; IV F 5 c).

    (b) Peptic (chief) cells secrete **pepsinogen,** the precursor for the proteolytic enzyme **pepsin**.

    (c) Mucous cells secrete mucus.

  (2) Pyloric glands are located in the antrum and pyloric regions of the stomach. They contain **G cells** and some mucous cells. **G cells** are responsible for the release of the hormone **gastrin.**

    (a) There are two forms of gastrin, G-17 (little gastrin, a 17-amino-acid peptide) and G-34 (big gastrin, a 34-amino-acid peptide). Although G-17 is more potent than G-34, the larger gastrin is found in higher concentrations within the circulation.

    (b) Gastrin is released from the basolateral surface of the G cells, enters the circulation, and travels to the orad stomach where it stimulates parietal-cell HCl secretion.

2. HCl secretion

  a. Functions of HCl

    (1) HCl participates in the breakdown of protein.

    (2) It provides an optimal pH for the action of pepsin.

    (3) It hinders the growth of pathogenic bacteria.

  b. Mechanism of HCl secretion (Figure 6-5)

    (1) HCl is secreted into the parietal cell **canaliculi** by a three-step process.

      (a) The active transport process is begun by the transport of K^+ and Cl^- into the canaliculi. Cl^- is transported either by a pump or through a channel. The flow of Cl^- creates a negative potential inside the canaliculi, causing K^+ to flow passively into the canaliculi.

      (b) Hydrogen ion (H^+) is then exchanged for K^+ by a H^+–K^+ ATPase.

      (c) Water enters the canaliculi down the osmotic gradient created by the movement of HCl.

    (2) The H^+ entering the canaliculi is supplied by the dissociation of carbonic acid (H_2CO_3) into H^+ and HCO_3^-.

      (a) H_2CO_3 is formed within the parietal cell from the reaction:

$$CO_2 + H_2O \rightarrow H_2CO_3 \rightarrow H^+ + HCO_3^-$$

      (b) The formation of H_2CO_3 from carbon dioxide (CO_2) is catalyzed by the enzyme **carbonic anhydrase (CA)**. Acetazolamide, a CA inhibitor, blocks the formation of HCl by the parietal cell.

      (c) The HCO_3^- diffuses back into the plasma (creating the alkaline tide associated with gastric secretion) in exchange for Cl^-, thus providing Cl^- for the initial step in the secretory process [see III C 2 b (1) (a)].

    (3) Most of the HCl that is secreted into the stomach is neutralized and reabsorbed within the small intestine. However, if the gastric contents are lost before they enter the small intestine (e.g., by vomiting), a severe alkalosis may ensue.

    (4) The active transport processes involved in the generation of HCl require a large amount of ATP. The ATP is generated by mitochondria found in very high concentration (40% of cell volume) within the parietal cell.

    (5) The pH of the parietal cell secretion can be as low as 0.8 (i.e., a H^+ concentration of approximately 150 mmol, or almost 4 million times as great as the H^+ concentration of plasma).

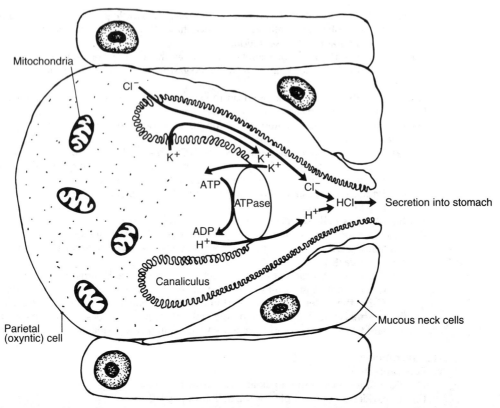

Figure 6-5. HCl being formed by the parietal cell. Cl$^-$ and K$^+$ are secreted into canaliculi by separate transporters, which may be channels or carriers. H$^+$ is exchanged for K$^+$ by an ATPase active transport system, allowing HCl to be secreted into the stomach. Numerous mitochondria provide energy for the active transport process. (Adapted from Guyton AC: *Textbook of Medical Physiology*, 8th edition. Philadelphia, WB Saunders, 1991, p 64.)

   c. Substances affecting HCl secretion
      (1) Stimulation of HCl secretion. ACh, histamine, and gastrin act directly on the parietal cell to stimulate HCl secretion (Figure 6-6). In addition, ACh and gastrin may directly stimulate the mast cell to secrete histamine.
         (a) ACh, a neurotransmitter, is released from nerve cells innervating the parietal cell.
         (b) Histamine is released from mast cells located within the corpus.
            (i) Histamine can stimulate HCl secretion directly or can potentiate the secretion produced by ACh or gastrin.
            (ii) Histamine is classified as a **paracrine agent** because it diffuses from its release site to the parietal cells (rather than traveling within the circulation as does a hormone).
            (iii) The most commonly used anti-ulcer drugs (i.e., cimetidine and ranitidine) are histamine antagonists that block the H$_2$ receptor on the parietal cell.
         (c) Gastrin is released from G cells in the distal stomach [see III C 1 c (2)]. Gastrin is classified as a hormone because it travels to its target cell through the circulation. A variety of substances affect gastrin secretion (see III C 3).
      (2) Inhibition of HCl secretion. Somatostatin inhibits HCl secretion by parietal cells and gastrin secretion by G cells. Somatostatin is released from interneurons within the enteric nervous system.
   d. Regulation of gastric acid (HCl) secretion
      (1) Stimulation during the cephalic phase. The vagus nerve stimulates the release of ACh and inhibits the release of somatostatin from interneurons within the enteric nervous system, thus enhancing the secretion of HCl (see Figure 6-6).
      (2) Stimulation during the gastric phase. The **amount of ingested protein** is the most important determinant of acid secretion during the gastric phase.
         (a) Protein is a good buffer and thus keeps the pH of the stomach at an optimal level for acid secretion.

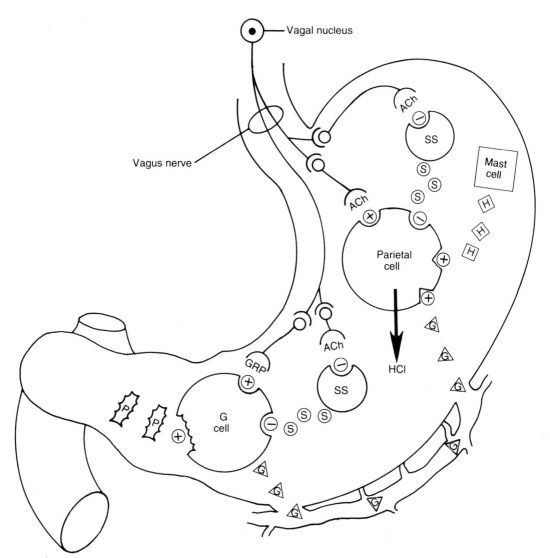

Figure 6-6. Many substances affect HCl secretion by the parietal cell. ACh from the vagus nerve, histamine (*H*) from mast cells, and gastrin (*G*) from G cells all stimulate the parietal cell directly to secrete HCl. The release of gastrin into the circulation, in turn, is stimulated by gastrin-releasing peptide (*GRP*) and protein digestion products (*P*). Somatostatin (*S*), released from somatostatin cells (*SS*), inhibits the release of both gastrin and HCl. Thus, the stimulation of vagal fibers, which causes the release of ACh and GRP but inhibits the release of somatostatin, has an amplified positive effect on parietal-cell HCl secretion. *Plus signs* = stimulation; *Minus signs* = inhibition. (Adapted from Johnson LR [ed]: *Gastrointestinal Physiology*, 3rd edition. St. Louis, CV Mosby, 1985, p 72.)

     (b) Amino acids and peptides directly stimulate parietal cells to secrete acid.

  (3) Inhibition during the gastric phase. The most potent inhibitor of HCl secretion during the gastric phase is the presence of acid in the stomach. If the pH of the stomach falls below 2, acid secretion stops. Acid secretion is inhibited by two mechanisms.

     (a) A low pH directly inhibits HCl and gastrin secretion.

     (b) Lowering the pH also releases somatostatin, which inhibits the secretion of gastrin by the G cells and HCl by the parietal cells (see Figure 6-6).

  (4) Stimulation during the intestinal phase. The presence of protein digestion products within the duodenum causes an increase in HCl secretion.

     (a) Although G cells have been identified within the duodenum, gastrin is not thought to cause the increase in acid secretion.

     (b) An as yet unidentified hormone, called **entero-oxyntin,** is postulated to be responsible for the increase in acid secretion.

(c) Amino acids circulating in the blood after being absorbed from the intestine may also stimulate HCl secretion.

(5) **Inhibition during the intestinal phase.** The inhibition of HCl secretion is accomplished by the same mechanisms responsible for inhibiting gastric motility.

(a) H^+, fatty acids, and increased osmolarity stimulate the release of enterogastrones from the duodenum.

(b) The most important of the enterogastrones may be **gastrin inhibitory peptide (GIP),** which inhibits both gastrin release and parietal cell secretion of HCl. GIP is thought to act by stimulating the release of somatostatin, which, in turn, inhibits the parietal and G cells.

3. Gastrin secretion

a. Functions of gastrin

(1) Gastrin stimulates HCl secretion.

(2) It increases gastric and intestinal motility.

(3) It increases pancreatic secretions.

(4) It is necessary for the proper growth of GI mucosa.

b. Substances affecting gastrin secretion

(1) **Stimulation of gastrin secretion. Bombesin [gastrin-releasing peptide (GRP)]** is most likely the neurotransmitter responsible for stimulating G cells to secrete gastrin. The vagus nerve increases the release of GRP during the cephalic phase (see Figure 6-6).

(2) **Inhibition of gastrin secretion.** Somatostatin inhibits gastrin secretion. The vagus nerve inhibits the release of somatostatin during the cephalic phase.

c. Regulation of gastrin secretion

(1) In general, gastrin secretion is regulated by the same mechanisms that regulate HCl secretion (i.e., vagal stimulation, pH, enterogastrones).

(2) In addition, several foods and food breakdown products (**secretagogues**) directly stimulate the release of gastrin. These include protein digestion products, alcohol, and coffee (both caffeinated and decaffeinated).

4. Pepsinogen secretion

a. Function of pepsinogen. Pepsin, the active form of pepsinogen, is a proteolytic enzyme that begins the process of protein digestion [see IV F 2 b (1)].

b. Regulation of pepsinogen secretion. Pepsinogen is released from the chief cells of the oxyntic glands during all three phases of digestion.

(1) **Cephalic stage.** During the cephalic stage of digestion, vagally stimulated cholinergic neurons within the enteric nervous system directly stimulate chief cells to release pepsinogen.

(2) **Gastric phase.** During the gastric phase of digestion, low pH activates local reflexes that enhance pepsinogen secretion. The low pH of the stomach is also responsible for converting pepsinogen into pepsin. Again, ACh is the transmitter that stimulates the chief cells.

(3) **Intestinal phase.** Secretin enhances pepsinogen release. Thus, the presence of H^+ within the duodenum during the intestinal phase of digestion may contribute to pepsinogen secretion.

5. Intrinsic factor

a. Definition. Intrinsic factor is a glycoprotein secreted by the parietal cells of the gastric mucosa, chiefly by those in the fundus.

b. Function. Intrinsic factor is required for the **absorption of vitamin B_{12}**.

(1) Intrinsic factor forms a complex with vitamin B_{12}.

(2) The intrinsic factor–B_{12} complex is carried to the terminal ileum, where the vitamin is absorbed (see IV F 5 c).

D. The **gastric mucosal barrier** protects the gastric lining cells from damage by intraluminal HCl, or **autodigestion**. Its chief component is a thick viscous alkaline mucous layer that measures over 1 mm thick and is secreted by the mucous cells. The mucous cells cover the surface between the various gastric glands and outnumber all other cell types found in the gastric mucosa.

1. The turnover rate of the gastric mucosa is extremely high; 5×10^5 mucosal cells are shed each minute, and the entire mucosa is replaced in 1–3 days.

2. Mild injury results in increased mucus secretion and surface desquamation followed by regeneration.

3. More serious injury denudes the mucosal surface, forming an ulcer, and produces bleeding. Ulceration results when damage to the mucosa (e.g., due to a highly concentrated HCl, 10% ethanol, salicylic acid, or acetylsalicylic acid) allows acid to penetrate the mucosal barrier and destroy mucosal cells. This liberates histamine, which increases acid secretion and produces increased capillary permeability and vasodilation. The latter two effects lead to edema. It is the exposure of mucosal capillaries to the digestive process that leads to bleeding.

4. The rate of repair of mucosal injury depends on the extent of injury and varies from as little as 48 hours for restricted desquamation to up to 3–5 months if damage has left only the deepest portions of the gastric pits intact.

E. Gastric digestion and absorption

  1. Digestion
    a. Carbohydrate digestion in the stomach depends on the action of salivary amylase, which remains active until halted by the low pH in the stomach.
    b. Protein digestion. About 10% of ingested protein is broken down completely in the stomach. Gastric pepsin facilitates later digestion of protein by breaking apart meat particles.
    c. Fat digestion is minimal due to the restriction of gastric lipase activity to triglycerides containing short-chain (< 10 carbons) fatty acids. Acid and pepsin break emulsions so that fats coalesce into droplets, which float and empty last.

  2. Absorption
    a. Nutrients. Very little absorption of nutrients takes place in the stomach. The only substances absorbed to any appreciable extent are highly lipid-soluble substances (e.g., the un-ionized triglycerides of acetic, propionic, and butyric acids). Aspirin at gastric pH is un-ionized and fat soluble; after absorption, it ionizes intracellularly, damaging mucosal cells and ultimately producing bleeding. Ethanol is rapidly absorbed in proportion to its concentration.
    b. Water moves in both directions across the mucosa. It does not, however, follow osmotic gradients. Water-soluble substances, including Na^+, K^+, glucose, and amino acids, are absorbed in insignificant amounts.

IV. THE SMALL INTESTINE is the major site of **digestion** and **absorption** of carbohydrates, proteins, and fats in the GI tract. The action and secretions of several **accessory organs** (see IV A 2) are essential to the digestive and absorptive functions of the small intestine. Nutrients and fluids that are not absorbed in the small intestine are passed on to the colon.

A. Anatomy

  1. Small intestine
    a. The small intestine has three parts: the **duodenum,** the **jejunum,** and the **ileum** (see Figure 6-1).
    b. Although the small intestine is approximately 5 m long, it has an **absorptive area** of over 250 m².
      (1) Its large surface area is created by numerous folds of the intestinal mucosa (**valvulae conniventes**); by densely packed **villi,** which line the entire mucosal surface; and by **microvilli,** which protrude from the surface of the intestinal cells.
        (a) The epithelial cells from which the microvilli protrude are called **enterocytes**.
        (b) The microvilli (about $1\mu m$ long and $0.1\mu m$ in diameter) give the intestinal mucosa its characteristic **brush border** appearance.
      (2) The **blood supply** of the villus is ideally organized to collect the nutrients after they are absorbed across the brush border membrane.
        (a) Each villus is supplied by an arteriole, which gives rise to a capillary tuft at the tip of the villus. The capillaries coalesce into venules, which drain into the portal vein. The portal vein carries the absorbed nutrients to the liver.
        (b) Branches of the lymphatics, called **lacteals,** also extend to the tip of the villus. These vessels carry absorbed fats to the thoracic duct from which they enter the general circulation.

2. The **accessory organs** involved in intestinal digestion and absorption are: the **pancreas,** the **liver,** and the **gallbladder.**

   a. The **pancreas** secretes various substances that aid in intestinal digestion, including HCO_3^-, which neutralizes the acidic content of chyme entering the small intestine.

   b. The **liver** secretes bile, which is necessary for fat digestion and nutrient absorption.

      (1) Bile travels from the liver through **bile ducts** to reach the duodenum of the small intestine.

      (2) The **sphincter of Oddi,** located at the distal end of the duodenum, forms the opening that connects the small intestine to the **common bile duct**.

         (a) This sphincter controls the flow of bile into the small intestine [see IV D 1 d, 5 b (1)].

         (b) When the sphincter is closed, bile cannot enter the small intestine and must be stored in the gallbladder.

      (3) The **portal circulation** carries bile that has been absorbed from the terminal ileum back to the liver (see IV D 3 and Figure 6-7).

   c. The **gallbladder** stores bile during the interdigestive period.

B. Motility

1. **Contractile activity**

   a. **Function.** Contractile activity of the smooth muscles lining the small intestine serves two major functions:

      (1) **Mixing the chyme** with the digestive juices and bile to facilitate digestion and absorption

      (2) **Propelling the chyme** from the duodenum to the colon

   b. **Transit time.** It usually takes about 2–4 hours for the chyme to move from one end of the small intestine to the other.

2. **Types of movements**

   a. **Segmentation** is the most common type of intestinal contraction.

      (1) During segmentation, about 2 cm of the intestinal wall contracts, forcing the chyme back toward the stomach (orally) and toward the colon (aborally).

      (2) When the muscle relaxes, the chyme returns to the area from which it was displaced.

      (3) This back-and-forth movement enables the chyme to become thoroughly mixed with the digestive juices and to make contact with the absorptive surface of the intestinal mucosa.

      (4) Segmentation contractions occur about 12 times/min in the duodenum and 8 times/min in the ileum. The contractions last for 5–6 seconds.

      (5) Segmentation occurs throughout the digestive period.

   b. **Peristaltic contractions** also occur in the small intestine.

      (1) Although peristaltic contractions occasionally propel food along the entire length of the intestine, they rarely involve more than a short segment of the intestine.

      (2) Peristalsis is not considered to be an important component of intestinal transit.

   c. The **MMC** (see III B 6 c) spreads over the intestine during the interdigestive period.

      (1) The MMCs sweep out the chyme remaining in the small intestine during the interdigestive period.

      (2) MMCs occur every 60–90 minutes and last for about 10 minutes.

3. **Propulsion of chyme.** During the digestive period, the higher frequency of segmentation in the proximal intestine (duodenum) than in the distal intestine (ileum) propels the chyme slowly toward the colon.

   a. Thus, when the chyme is pushed aborally, it is less likely to be forced back by a segmentation contraction.

   b. In contrast, when the chyme is pushed orally, it will be quickly pushed toward the colon again by a segmentation contraction in the more proximal region of the small intestine.

4. **Control of intestinal motility.** The frequency and strength of segmentation contractions in the intestine are controlled by the **slow waves** (see Figure 6-3).

   a. **Generation.** Segmentation contractions can occur only if the slow waves produce **spikes,** or **action potentials**. Spikes appear on the slow waves when the membrane potential is sufficiently depolarized.

   b. **Frequency**

      (1) The frequency of segmentation contractions is directly related to the frequency of the slow waves.

(2) Slow wave frequency is controlled by **pacemaker cells** within the wall of the intestine and is not influenced by neural activity or circulating hormones.

  c. Strength

  (1) The strength of a segmentation contraction is proportional to the frequency of the spikes generated by the slow wave. This frequency is controlled by the **amplitude of the slow wave**. Thus, the greater the slow wave amplitude, the greater the frequency of spikes generated and the greater the strength of the contraction.

  (2) Slow wave amplitude is controlled by the hormones released during digestion.

  (a) Gastrin, CCK, motilin, and insulin increase the slow wave amplitude.

  (b) Secretin and glucagon reduce the slow wave amplitude.

C. Pancreatic secretions

  1. Pancreatic cell types and their functions. The pancreas contains endocrine, exocrine, and ductal cells.

  a. The **endocrine cells,** arranged in small islets within the pancreas, secrete **insulin, glucagon, somatostatin,** and **pancreatic polypeptide** directly into the circulation.

  b. The **exocrine cells** are organized into acini that produce four types of digestive enzymes: **peptidases, lipases, amylases,** and **nucleases,** which are responsible for digesting proteins, fats, carbohydrates, and nucleic acids, respectively. In their absence, malabsorption syndromes develop.

  c. Each day, the **ductal cells** secrete about 1200–1500 ml of pancreatic juice containing a high concentration of HCO_3^-. The HCO_3^- neutralizes gastric acid and regulates the pH of the upper intestine. Failure to neutralize the chyme as it enters the intestine will result in duodenal ulcers.

  2. Composition of pancreatic secretions

  a. Electrolytes

  (1) Na^+ and K^+ concentrations in pancreatic juice are the same as those in plasma water (i.e., 142 mEq/L and 4.8 mEq/L, respectively).

  (2) HCO_3^- concentration in pancreatic juice is much higher than it is in plasma water (100 mEq/L as opposed to about 24 mEq/L).

  (3) Pancreatic juice also contains small amounts of other ions such as Ca^{2+}, magnesium (Mg^{2+}), zinc (Zn^{2+}), monohydrogen phosphate (HPO_4^{2-}), and sulfate (SO_4^{2-}).

  (4) Secretion of HCO_3^- by the ductal cells requires at least one active transport process.

  (a) HCO_3^- and H^+ are formed from the dissociation of H_2CO_3 via the same reaction that occurs in the parietal cells of the stomach [see III C 2 b (2)].

  (b) H^+ is actively transported out of the cell across its basal membrane by an Na^+–H^+ antiporter.

  (c) HCO_3^- is transported across the apical membrane of the ductal cells. The transport process responsible for HCO_3^- secretion is not yet known.

  (d) Na^+ follows the HCO_3^- into the pancreatic ducts. Most of the Na^+ flows passively between the ductal cells. However, some may be actively transported across the apical membranes.

  (e) Water flows into the ducts down the osmotic gradient established by the secretion of sodium bicarbonate ($NaHCO_3$). Flow rates as high as 1 ml/min can be established.

  (f) As the pancreatic juice flows along the ducts, Cl^- is exchanged for HCO_3^-. The higher the flow rate, the smaller the exchange. Thus, HCO_3^- concentration is highest when pancreatic secretion is greatest.

  b. Enzymes. Three major types of pancreatic enzymes are secreted by the pancreas: **amylases, lipases,** and **proteases**.

  (1) Pancreatic α-amylase is secreted in its active form. It hydrolyzes glycogen, starch, and most other complex carbohydrates, except cellulose, to form disaccharides.

  (2) Pancreatic lipases (lipase, cholesterol lipase, and phospholipase) [see IV F 3 a (1)] are secreted in their active forms. The enzymes that hydrolyze water-insoluble esters require bile salts to work. Water-soluble esters can be hydrolyzed without the action of bile salts.

  (3) Pancreatic proteases (trypsin and **the chymotrypsins)** are secreted in their inactive zymogen form (trypsinogen and the chymotrypsinogens, respectively) [see IV F 2 b (2)].

  (a) Trypsinogen is converted to trypsin by enterokinase or by trypsin itself (autocatalysis).

  (b) The chymotrypsinogens are converted to their active form by trypsin.

(4) Trypsin inhibitor is secreted by the same cells and at the same time as the pancreatic proenzymes. Trypsin inhibitor protects the pancreas from autodigestion.

3. **Control of pancreatic secretion.** Like gastric secretion, pancreatic secretion is divided into the following three phases.
   a. **Cephalic phase.** The thought, sight, smell, or taste of food produces the cephalic phase of pancreatic secretion. Both acinar and ductal cell secretions are enhanced by vagal stimulation.
      (1) Enzyme secretion by the acinar cells is stimulated by enteric neurons that release ACh.
      (2) HCO_3^- secretion by ductal cells is stimulated by enteric neurons that release a noncholinergic, nonadrenergic transmitter that is thought to be VIP.
   b. **Gastric phase.** Pancreatic secretion is enhanced during the gastric phase by distension and food breakdown products.
      (1) Distension of the antrum and corpus initiates a vago-vagal reflex resulting in a low volume of pancreatic secretion containing both HCO_3^- and enzymes. ACh is the transmitter.
      (2) Food breakdown products (primarily amino acids and peptides) can stimulate pancreatic secretions because of their ability to cause the G cells of the antrum to release gastrin. Gastrin produces a low-volume, high-enzyme pancreatic secretion.
   c. **Intestinal phase.** The major stimulants for pancreatic secretion are the hormones CCK and secretin. They are released from endocrine cells in the duodenum and jejunum during the intestinal phase of pancreatic secretion.
      (1) CCK, in addition to its effect on the gallbladder [see IV D 5 b (1)], is a potent stimulant of pancreatic enzyme secretion.
         (a) Like gastrin, CCK is found in two physiologically active forms, an octapeptide called CCK-8 and a 33-chain polypeptide, CCK-33.
         (b) The actions of CCK are potentiated by secretin. By itself, secretin has no effect on enzyme secretion.
      (2) Secretin was the first hormone ever discovered. Its primary effect is to increase HCO_3^- secretion by the pancreas.
         (a) The actions of secretin are potentiated by CCK. By itself, CCK has no effect on HCO_3^- secretion.
         (b) Because they are potentiators of each other's action, small concentrations of CCK and secretin together can produce significant amounts of pancreatic HCO_3^- and enzyme secretions, while either one alone would have little or no effect.
      (3) Control of CCK and secretin release. CCK and secretin are secreted from endocrine cells in response to the entrance of chyme into the small intestine.
         (a) Amino acids (primarily **phenylalanine**), **fatty acids,** and **monoglycerides** are the major stimuli for CCK secretion.
         (b) Low pH (< 4.5), caused by the presence of gastric acid (HCl) in the intestine, is a potent stimulus for the release of secretin.
      (4) A vago-vagal reflex, which greatly potentiates the effects of secretin and CCK, is activated during the intestinal phase of digestion.
      (5) ACh potentiates the effects of both CCK and secretin. Thus, vagal stimulation is much more potent in stimulating pancreatic secretions when CCK and secretin are present in the plasma.

D. **Biliary secretions**

1. **General features of bile**
   a. **Function.** Bile is required for the digestion and absorption of fats and for the excretion of water-insoluble substances such as cholesterol and bilirubin.
   b. **Formation.** Bile is formed by liver epithelial cells, called **hepatocytes,** and by epithelial cells lining the bile ducts, called **ductal cells**. Between 250 and 1100 ml of bile are secreted daily.
   c. **Storage.** Although it is secreted continuously, bile is stored in the gallbladder during the interdigestive period.
   d. **Release.** Bile is released into the duodenum during the digestive period only after chyme has triggered the release of CCK, which then produces contraction of the gallbladder and relaxation of the sphincter of Oddi.

2. Composition of bile
    a. Bile acids
        (1) Primary bile acids (trihydroxycholic acid and **dihydroxychenodeoxycholic acid)** are synthesized from cholesterol and converted into bile salts by the hepatocytes as follows.
            (a) Cholesterol is absorbed through microvilli lining the serosal (antiluminal) border of the hepatic epithelial cells.
            (b) The bile acids are conjugated with either taurine or glycine to form bile salts.
            (c) The bile salts are actively secreted into a canaliculus on the lateral (luminal) surface of the hepatocyte, from which they then drain into the bile duct.
            (d) Because bile salts are not lipid soluble, they remain within the intestine until reaching the ileum, where they are actively absorbed (see IV F 3 b).
        (2) Secondary bile acids are formed by deconjugation and dehydroxylation of the primary bile salts by intestinal bacteria, forming **deoxycholic acid** and **lithocholic acid.**
    b. Bile pigments
        (1) Bilirubin and **biliverdin,** the two principal bile pigments, are metabolites of hemoglobin formed in the liver and conjugated as glucuronides for excretion. They are responsible for the golden yellow color of bile.
        (2) Intestinal bacteria metabolize bilirubin further to **urobilin,** which is responsible for the brown color of stool.
        (3) If bilirubin is not secreted by the liver, it builds up in the blood and tissues, producing **jaundice** (see IV D 6 b).
    c. Phospholipids (primarily **lecithins**) are, after bile salts, the most abundant organic compound in bile.
        (1) Although the phospholipids are normally insoluble in water, they are solubilized by the bile salt micelles.
        (2) The **micelles** are able to solubilize other lipids more effectively when they are composed of bile salts and phospholipids than when they are composed of bile salts alone.
    d. Cholesterol, although present only in small amounts, is an important component of bile.
        (1) Cholesterol is essentially insoluble in water and thus must be solubilized by bile salt micelles before it can be secreted in the bile (see IV F 3 b).
        (2) Biliary secretion of cholesterol is important because it is one of the few ways in which cholesterol stores can be regulated.
    e. Electrolytes. The electrolyte composition of bile is similar to that of pancreatic juice and plasma (see IV C 2 a).

3. Enterohepatic circulation is the recirculation of bile salts from the liver to the small intestine and back again. This circulation is necessary because of the limited pool of bile salts available to help break down and absorb fat (Figure 6-7).
    a. Path of circulation. Bile salts travel from the liver to the duodenum via the common bile duct. When the bile salts reach the terminal ileum, they are reabsorbed into the portal circulation. The liver then extracts them from the portal blood and secretes them once again into the bile.
        (1) Bile salts are reabsorbed only in the terminal ileum. No reabsorption of bile salts occurs in the duodenum or jejunum.
        (2) From 90%–95% of the bile salts that enter the small intestine are actively reabsorbed from the lower ileum back into the portal circulation.
        (3) The remaining bile salts are excreted into the feces.
    b. Circulating pool. The total circulating pool of bile salts (consisting of primary and secondary bile acids) is approximately 3.6 g. Because 4–8 g of bile salts are required to digest and absorb a meal (more if the meal is high in fat), the total pool of salts must circulate twice during the digestion of each meal. Consequently, the bile salts usually circulate 6–8 times daily.
    c. Bile salt synthesis and replacement. The rate of bile salt synthesis is determined by the rate of return to the liver. The usual rate is 0.2–0.4 g/day, which replaces normal fecal losses. The maximal rate is 3–6 g/day. If fecal losses exceed this rate, the total pool size decreases.
    d. Clinical implications. Because bile salts are required for proper digestion and absorption of fats, any condition that disrupts the enterohepatic circulation (e.g., ileal resection or small intestinal diseases such as sprue or Crohn's disease) leads to a decreased bile acid pool and malabsorption of fat and fat-soluble vitamins. The clinical manifestations of such conditions

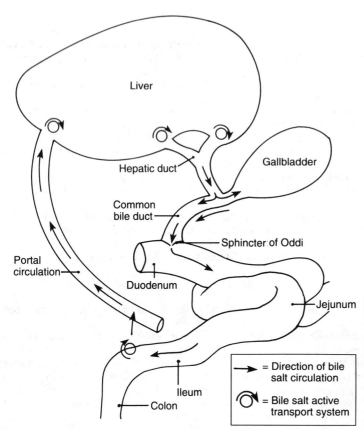

Figure 6-7. The enterohepatic circulation. Bile acids are absorbed from the terminal ileum by a Na^+-dependent active transport system and then rapidly sequestered by hepatocytes in the liver and returned to the gallbladder or duodenum. Each day, about 20% of the bile acid pool escapes the enterohepatic circulation (thus becoming lost to excretion) and must be resynthesized by the liver.

In the figure:
- Liver
- Hepatic duct
- Common bile duct
- Gallbladder
- Sphincter of Oddi
- Portal circulation
- Duodenum
- Jejunum
- Ileum
- Colon
- → = Direction of bile salt circulation
- ↻ = Bile salt active transport system

are steatorrhea and nutritional deficiency. An increase in fecal losses of bile salts results in watery diarrhea, since bile salts inhibit water and Na^+ absorption in the colon.

4. **Control of biliary secretion.** The volume of biliary secretion and the amount of bile in that secretion are regulated separately.

   a. **The bile-independent fraction of biliary secretion** refers to the amount of **fluid,** composed of electrolytes and water, that is secreted each day by the liver. Although this fluid, by definition, is secreted with the bile, its secretion is controlled separately from bile secretion.

   (1) Secretion of this fluid is controlled by the hormone **secretin.**

   (2) The fluid resembles the secretion of the pancreatic ductal cells in the following ways (see IV C 2 a, 3).

   (a) The fluid is secreted by ductal cells.

   (b) Its secretion is controlled by secretin.

   (c) It has a high concentration of HCO_3^-.

   b. **The bile-dependent fraction of biliary secretion** refers to the quantity of **bile salts** secreted by the liver.

   (1) The amount of bile salts secreted is directly related to the amount of bile reabsorbed by the hepatocytes (i.e., the more bile reabsorbed from the portal circulation, the more bile secreted by the liver).

   (a) The total amount of bile is relatively constant. Because the liver has limited synthetic capacity, there is a limit to the amount of bile that can be secreted.

   (b) Substances that enhance bile secretion are called **choleretics.** Bile salts and bile acids are the major choleretics.

   (2) Unlike the bile-independent secretion, the **synthesis and secretion of bile** by the liver is not under any direct hormonal or nervous control. However, CCK increases bile flow indirectly by increasing the release of bile from the gallbladder [see IV D 5 b (1)].

5. Gallbladder

  a. Functions. The gallbladder stores and concentrates the bile during the interdigestive period and empties its contents into the duodenum during digestion.

   (1) Storage. During the interdigestive period, the bile secreted by the liver is collected in the gallbladder. The gallbladder typically stores 20–50 ml of bile.

   (a) The bile is highly **concentrated** within the gallbladder by the reabsorption of water.

   (b) Water is reabsorbed by the osmotic gradient produced by the active reabsorption of Na^+ and HCO_3^-.

   (2) Contraction. During digestion, the gallbladder contracts, emptying its contents into the duodenum.

  b. Control

   (1) CCK is the major stimulus for gallbladder contraction and sphincter of Oddi relaxation. When chyme enters the small intestine, **fat** and **protein digestion products** directly stimulate the secretion of CCK [see IV C 3 c (3) (a)].

   (2) Vagal stimulation of the gallbladder also causes gallbladder contraction and sphincter of Oddi relaxation. Vagal stimulation occurs directly during the cephalic stage of digestion and indirectly via a vago-vagal reflex during the gastric phase of digestion.

  c. Effects of cholecystectomy. Bile, not the gallbladder, is essential to digestion. After removal of the gallbladder, bile empties slowly but continuously into the intestine, allowing digestion of fats sufficient to maintain good health and nutrition. Only high-fat meals need to be avoided.

  d. Gallstones form in an estimated 10%–30% of the population, although only a fraction, perhaps 20% of these, ever produce symptoms. In Western societies, about 85% of gallstones are composed chiefly of cholesterol; the remainder are pigment stones, composed chiefly of calcium bilirubinate.

   (1) Cholesterol and lecithin, which are both insoluble in water, are kept in solution in bile through the formation of micelles [see IV F 3 b (1)]. When the proportions of lecithin, cholesterol, and bile salts are altered, cholesterol crystalizes, leading to stone formation. Cholesterol stones are radiolucent.

   (2) Calcium bilirubinate stones can form when infection of the biliary tree leads to bacterial deconjugation of conjugated bilirubin. Unconjugated bilirubin, which is insoluble in bile, then precipitates to begin the stone-forming process. Calcium bilirubinate stones are radiopaque.

6. Bilirubin metabolism

  a. Formation of bilirubin. Bilirubin is a yellowish pigment formed as an end product of hemoglobin catabolism.

   (1) Hemoglobin is released from red blood cells when their membranes rupture at the end of their life span.

   (2) Hemoglobin is taken up by the cells of the reticuloendothelial system where the porphyrin ring and iron of the heme moiety are separated.

   (3) The porphyrin ring is then converted to bilirubin and gradually released into the plasma.

   (4) The bilirubin combines with plasma albumin and circulates to the liver where it is absorbed by the hepatocytes, conjugated with glucuronic acid, and secreted into the gallbladder along with the bile salts.

  b. Jaundice is a yellowing of the skin due to the accumulation of bilirubin within the tissues. Jaundice may result from:

   (1) Excess production of bilirubin caused by excessive red blood cell destruction (e.g., in hemolytic anemia)

   (2) Obstruction of the bile ducts or liver cells preventing the secretion of bilirubin

E. Intestinal secretion

  1. Mucus most likely serves a protective role, preventing HCl and chyme from damaging the intestinal wall. Mucus is secreted by:

  a. Brunner's glands, which are located within the duodenum

  b. Goblet cells located along the length of the intestinal epithelium and in the intestinal crypts, called the crypts of Lieberkühn

  2. Enzymes capable of breaking down small peptides and disaccharides are associated with the microvilli of the epithelial cells lining the intestine. Although these enzymes are not secreted into the intestine, they are able to digest small peptides and disaccharides during the absorptive process.

3. Water and electrolytes are secreted by all the epithelial cells of the intestine.
   a. The watery secretion provides a solvent into which the products of digestion are dissolved.
   b. If excessive amounts of fluid are produced (as happens when the enterotoxin responsible for cholera stimulates massive fluid secretion), potentially life-threatening watery diarrhea can result.

F. Digestion and absorption

1. **Carbohydrates.** The **three major carbohydrates** in the human diet are the disaccharides, **sucrose** (cane sugar) and **lactose** (milk sugar), as well as the polysaccharide starches (which may be in either the straight chain form, **amylose,** or the branched chain form, **amylopectin**). **Cellulose,** another plant polysaccharide, is present in the diet in large amounts, but no enzymes in the human digestive tract can digest it, so it is excreted unused. Dietary intake of carbohydrates is 250–800 g/day, which represents 50%–60% of the diet.
   a. Digestion. Carbohydrates must be digested into monosaccharides before being absorbed from the GI tract.
      (1) Although starch digestion, by **salivary α-amylase,** begins in the mouth, almost all carbohydrate digestion occurs within the small intestine.
      (2) Pancreatic α-amylase digests carbohydrates into a variety of oligosaccharides.
      (3) The oligosaccharides are digested into monosaccharides by brush border enzymes such as **maltase, lactase,** and **sucrase.**
      (4) The end products of carbohydrates are **fructose, glucose,** and **galactose.**
   b. Mechanisms of absorption
      (1) Glucose and **galactose** are absorbed by a common **Na$^+$-dependent active transport system**.
         (a) The carrier has two binding sites for Na$^+$ and one to which either one molecule of glucose or galactose can bind.
         (b) Because two Na$^+$ are transported down their electrochemical gradient, a large amount of energy is available for transport; thus, almost all of the glucose and galactose present in the intestine can be absorbed.
      (2) Fructose is absorbed by **facilitated transport.** Fructose absorption occurs readily because most of the fructose is rapidly converted into glucose and lactic acid within the intestinal epithelial cells, thus maintaining a high concentration gradient for diffusion.
      (3) After being absorbed into the enterocytes, the monosaccharides are transported across the basolateral membrane by facilitated diffusion. They then diffuse from the intestinal interstitium into the capillaries of the villus.
      (4) Absorption of monosaccharides is not regulated. The intestine can absorb over 5 kg of sucrose each day.
      (5) Failure to absorb carbohydrates results in diarrhea and intestinal gas.
         (a) The unabsorbed carbohydrates act as osmotic particles and draw excessive fluids into the intestine, which results in diarrhea.
         (b) The flora of the intestine and colon metabolize the unabsorbed carbohydrates producing a variety of gases [hydrogen (H_2), methane (CH_4), and CO_2], as well as a variety of intestinal irritants.
         (c) Lactose intolerance is the most common cause of carbohydrate malabsorption. It results from the inability of the goblet cells to produce lactase.
            (i) Avoidance of milk or milk products prevents the symptoms from developing.
            (ii) Lactose intolerance in adults, where it is most common, is not usually a problem. However, in infants, the diarrhea-produced dehydration can be life threatening.

2. **Proteins.** The daily dietary protein requirement for adults is 0.5–0.7 g/kg of body weight. For children 1–3 years old, it is 4 g/kg.
   a. Sources. The protein that is found in the intestines comes from two sources.
      (1) Endogenous proteins, totaling 30–40 g/day, are secretory proteins as well as the protein components of desquamated cells.
      (2) Exogenous proteins are dietary proteins, which total at least 75–100 g daily in the average American diet.
   b. Digestion. Proteins must be digested into small polypeptides and amino acids before being absorbed.
      (1) About 10%–15% of the protein entering the GI tract is digested by **gastric pepsin** secreted by chief cells. Protein digestion within the stomach is important primarily

because the protein digestion products act as secretagogues, stimulating the secretion of proteases by the pancreas.

(2) **Pancreatic proteases** play a major role in protein digestion. The proteases, such as **trypsin,** are secreted in an inactive form and must be converted into an active form within the intestine [see IV C 2 b (3)].

(a) **Enterokinase,** an enzyme secreted by the epithelial cells of the duodenum and jejunum, converts the inactive trypsinogen into trypsin.

(b) Trypsin then autocatalyzes the conversion of trypsinogen to trypsin as well as activating the other proteases.

(3) **Peptidases,** secreted by the intestinal epithelial cells, continue the digestive process begun by the pancreatic proteases, eventually converting the ingested proteins to small polypeptides and amino acids.

c. **Mechanisms of absorption**

(1) A variety of **Na^+-dependent active transport systems** have been identified for the transport of tripeptides, dipeptides, and amino acids. Polypeptides with more than three peptides are poorly absorbed.

(a) Separate transporters are present for the absorption of basic, acidic, and neutral amino acids. At least two different polypeptide transporters exist.

(b) Tripeptides and dipeptides are absorbed in greater quantities than amino acids.

(2) Once inside the enterocytes, intercellular peptidases digest some of the polypeptides to amino acids.

(3) Amino acids and the remaining polypeptides are transported across the basolateral membrane of the enterocytes by facilitated or simple diffusion. They then enter the capillaries of the villus by simple diffusion.

(4) Almost all of the ingested protein is absorbed by the intestine. Any protein that appears in the stool derives from the bacteria within the colon or from cellular debris.

(5) **Malabsorption** of amino acids due to lack of adequate transporters (e.g., the malabsorption of neutral amino acids that occurs in **Hartnup disease**) is relatively rare. Inadequate absorption of proteins due to lack of trypsin is a common consequence of pancreatic diseases.

3. **Fats. Daily dietary fat intake** varies widely from 25–160 g.

a. **Digestion.** Although the serous glands of the tongue secrete lingual lipase, very little, if any, lipid digestion occurs in the mouth or stomach. Unlike carbohydrates and proteins, lipids are absorbed from the GI tract by **passive diffusion.** However, before the lipids can be absorbed, they must first be made soluble in water. **Bile salts** are required for the solubilization of lipids.

(1) **Pancreatic lipases.** The pancreas secretes three different lipases [see IV C 2 b (2)].

(a) **Pancreatic lipase** is a fairly specific lipase that cleaves fatty acids from the 1 and 1′ positions of triglycerides, leaving a 2-monoglyceride.

(b) **Cholesterol esterase** cleaves the fatty acid from cholesterol esters, leaving free cholesterol.

(c) **Phospholipase A_2** cleaves the fatty acids from phospholipids such as phosphatidylcholine.

(2) **Emulsification of lipids.** Lipids must be broken down into small droplets (less than 1 μm in diameter) or **emulsified** into fat globules by bile acids and lecithin (a component of bile) before being digested.

(3) Fat digestion by the pancreatic lipases occurs very rapidly after emulsification because of the large surface-to-volume ratio of the small globules.

b. **Mechanism of absorption**

(1) **Micelle formation.** The emulsified products of lipid digestion (e.g., monoglycerides, cholesterol) must form **micelles** with bile salts before they can be absorbed.

(a) **Micelles** are small (about 5 nm in diameter) spherical aggregates containing some 20–30 molecules of lipids and bile salts.

(b) The **bile salts** are on the outside of the micelle. The 2-monoglycerides and lysophosphatides have their hydrophobic chains facing the interior of the micelle and their polar ends facing the surrounding water phase. The cholesterol and fat-soluble vitamins are located within the fat-soluble interior of the micelle.

(2) **Absorption of lipids and bile salts from micelles**

(a) The micelles move along the microvilli surface allowing their lipids to diffuse across the microvilli membrane and into the enterocytes.

(b) Lipids, cholesterol, and the fat-soluble vitamins are removed rapidly from the micelles once the micelles make contact with the microvilli.

(c) The rate-limiting step in lipid absorption is the migration of the micelles from the intestinal chyme to the microvilli surface.

(d) The bile salts, freed of their associated lipids, are absorbed in the terminal ileum by a Na^+-dependent active transport process.

(e) Normally, all of the ingested lipid is absorbed. Fat present in the stool is derived from the intestinal flora.

(3) Formation of chylomicrons by enterocytes

(a) Once inside the enterocytes, the digested lipids enter the **smooth endoplasmic reticulum** (ER), where they are reconstituted.

(i) 2-Monoglycerides are combined with fatty acids to produce triglycerides.

(ii) Lysophosphatides are combined with fatty acids to form phospholipids.

(iii) Cholesterol is re-esterified.

(b) The reformed lipids coalesce into **chylomicrons** (small lipid droplets about 1 nm in diameter) within the smooth ER.

(c) The chylomicrons are transported out of the cell by exocytosis. β-**lipoprotein,** which is synthesized by the enterocytes, covers the surface of the chylomicrons. In the absence of β-lipoprotein, exocytosis will not occur, and the enterocytes become engorged with lipids.

(4) Transport of lipids into circulation

(a) After exiting the cell, the chylomicrons merge into larger droplets that vary in size from 50–500 nm, depending on the amount of lipids being absorbed

(b) The large lipid droplets then diffuse into the lacteals, from which they enter the lymphatic circulation.

(c) Almost all digested lipids are totally reabsorbed by the time the chyme reaches the midjejunum, with most of the absorption occurring in the duodenum.

(5) Lipid malabsorption is much more common than carbohydrate or protein malabsorption. It usually results from one of the following two conditions.

(a) The pancreas does not secrete sufficient quantities of lipase.

(b) The liver does not secrete sufficient quantities of bile.

4. Water and electrolytes

a. Water

(1) The small intestine, in addition to absorbing most of the dietary Na^+ and water, must also absorb the 7–8 L of water and 20–30 g of Na^+ that are contained in salivary, gastric, biliary, and pancreatic secretions. Failure to reabsorb water from the intestine can lead to rapid dehydration and circulatory collapse.

(2) Water undergoes **passive, iso-osmotic reabsorption** in the small intestine.

(a) Active reabsorption of electrolytes and nutrients creates an osmotic gradient favoring the reabsorption of water.

(b) Because osmotic equilibrium is rapidly achieved, the fluid in the intestine is always isotonic to plasma.

(3) In the duodenum, the osmotic pressure created by the entering chyme causes water to flow into the intestine.

(4) In the jejunum and ileum, the reabsorption of sodium chloride (NaCl) creates an osmotic gradient favoring the reabsorption of water.

b. NaCl

(1) Na^+ reabsorption is a two-step process.

(a) First, Na^+ and Cl^- are transported from the lumen into the enterocyte.

(b) Then, they are transported across the basolateral membrane into the intestinal interstitium.

(2) Na^+ enters the enterocyte in three ways.

(a) About 30% is transported into the cell by a Na^+–glucose, Na^+–amino acid, or Na^+– (di- or tri-) peptide cotransport system.

(b) About 30% is transported into the cell by a neutral Na^+–Cl^- cotransport system.

(c) The remainder enters the cell passively down an electrochemical gradient.

(3) Once inside the enterocyte, Na^+ is transported across the basolateral membrane by a Na^+–K^+ ATPase active transport system.

(4) For the most part, Cl^- flows passively through the enterocyte down the electrochemical gradient established by the active transport of Na^+.

5. **Vitamins and minerals**
   a. **Fat-soluble vitamins (A, D, E,** and **K)** become part of the micelles formed by bile salts and are absorbed along with other lipids in the proximal intestine.
   b. **Water-soluble vitamins (C,** and the **B vitamins biotin, folic acid, nicotinic acid, B_6 or pyridoxine, B_2 or riboflavin,** and **B_1 or thiamine)** are absorbed by facilitated transport or a Na^+-dependent active transport system in the proximal small intestine.
   c. **Vitamin B_{12}** absorption is more complex than other vitamins.
      (1) In the stomach, vitamin B_{12} is bound to an **R protein,** which is a specific binding protein.
      (2) The gastric parietal cells secrete another vitamin B_{12}–binding protein called **intrinsic factor**. However, the affinity of intrinsic factor for vitamin B_{12} is less than that of R protein, so most of the B_{12} is bound to R protein in the stomach.
      (3) In the intestine, pancreatic proteases cleave vitamin B_{12} from the R protein, allowing it to bind to intrinsic factor.
      (4) The intrinsic factor–B_{12} complex binds to a receptor on ileal enterocytes.
         (a) Absorption of the vitamin B_{12} from the intrinsic factor–B_{12} complex can occur only after the complex binds to the receptor.
         (b) In the absence of intrinsic factor, minimal amounts of vitamin B_{12} can be absorbed by diffusion. Thus, if large amounts of the vitamin are ingested, enough B_{12} can be absorbed to prevent **pernicious anemia**.
   d. **Ca^{2+}** absorption within the small intestine is regulated to maintain Ca^{2+} balance. Normally, about 25%–80% of the daily intake of Ca^{2+} (1000 mg) is absorbed.
      (1) Ca^{2+} absorption occurs via a membrane-bound carrier that is activated by **vitamin D** (see also Ch 7 XIV E 1 c).
         (a) Vitamin D_3 is converted to **25-hydroxyvitamin D_3** by the liver.
         (b) The kidney converts the 25-hydroxyvitamin D_3 to **1,25-dihydroxyvitamin D_3** by a process that is regulated by parathyroid hormone.
         (c) 1,25-dihydroxyvitamin D_3 then enters the enterocyte where it induces the formation of a Ca^{2+} carrier that inserts on the luminal surface of the enterocyte.
      (2) Ca^{2+} is transported out of the cell by a Ca^{2+}–ATPase active transport system and by a Na^+–Ca^{2+} exchange system.
   e. **Iron** absorption is necessary to maintain normal iron balance. However, very little (0.75 mg for men and 1.5 mg for women) of the 15–25 mg of iron ingested each day is actually absorbed.
      (1) Iron is absorbed primarily within the **duodenum** and **jejunum**.
      (2) Iron can be absorbed either as **heme** (derived from meat) or as a **free ion**.
      (3) **The ferrous ion (Fe^{2+})** is absorbed more efficiently than the **ferric ion (Fe^{3+})**.
      (4) **Ascorbic acid (vitamin C)** promotes iron absorption by reducing Fe^{3+} to Fe^{2+} and by preventing iron from forming insoluble complexes within the chyme.
      (5) **Stomach acid** tends to break insoluble iron complexes apart and thus facilitates iron absorption.
      (6) Four separate steps are involved in the transport of iron from the intestine to the plasma.
         (a) First, the iron is transported across the apical membrane of the enterocyte by a specific iron carrier system.
         (b) Second, the iron binds to **apoferritin,** an iron-binding protein, to form **ferritin**.
         (c) In order to leave the enterocyte, the iron must dissociate from ferritin and bind to an intracellular carrier protein that shuttles it to the basolateral membrane, where it is transported out of the cell.
         (d) Upon entering the intestinal interstitium, the iron is transported to the plasma by **transferrin,** a **β-globulin**.
      (7) The **amount of iron absorbed** depends on the amount of intracellular and extracellular transport protein (transferrin) compared to the amount of ferritin.
         (a) If a large amount of transferrin is available, iron can be transported rapidly from the enterocyte to the plasma.
         (b) If little transferrin is available, most of the iron remains trapped in the enterocyte and is eventually excreted when the cells are desquamated.
         (c) When iron stores are depleted, such as after a hemorrhage, transferrin synthesis increases.

V. THE COLON, OR LARGE INTESTINE (Figure 6-8), absorbs some of the nutrients and most of the fluids passed into it from the small intestine. Under normal circumstances, all but 50–100 ml of the 1500 ml received from the small intestine is absorbed. Any nutrients or fluids that cannot be absorbed are passed into the feces.

A. Motility

   1. Function. The **contractile activity** of the large intestine serves two functions.
      a. It enhances the efficiency of water and electrolyte absorption.
      b. It promotes the excretion of the fecal material remaining in the colon.

   2. Types of movements
      a. Haustral shuttling
         (1) Bands of muscle divide the large intestine into sac-like segments called **haustrations**. Although the haustrations are present when the colon is empty, the entry of food into the colon causes an increase in colonic contractile activity.
         (2) The dynamic formation and disappearance of haustrations squeeze the chyme, moving it back and forth along the colon in a manner similar to that described for the segmentation contractions in the small intestine (see IV B 2 a).
      b. Peristalsis, here, as elsewhere in the gut, is a progressive contractile wave preceded by a wave of relaxation. Peristaltic-like segmentation contractions move the chyme very slowly (5 cm/hr) along the colon. It can take up to 48 hours for chyme to traverse the colon.
      c. Mass movements. Occasionally (three to four times daily) the chyme is swept rapidly along the colon by a peristaltic wave called a mass movement. The mass movement forces fecal material into the rectum.
      d. The **frequency** of contractions is greater in the rectum than in the sigmoid colon, causing retrograde movement of fecal material. Because of this orad movement of fecal material, the rectum is usually empty and material placed into it, such as a suppository, will be pushed up into the colon.
      e. The overall effect of the **neural input to the colon is inhibitory**. Thus, elimination of the enteric nervous system, as occurs in **Hirschsprung's disease,** leads to a large increase in colonic tone.

   3. Defecation
      a. Fecal material entering the rectum is evacuated by defecation, during which:
         (1) The smooth muscles of the distal colon and rectum contract, propelling the fecal material into the anal canal
         (2) The **internal and external anal sphincters** both relax
         (3) The abdominal and diaphragmatic muscles contract, increasing the intra-abdominal pressure and forcing the feces through the anal canal

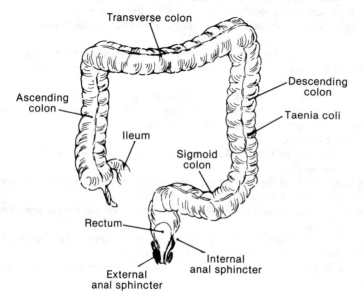

Figure 6-8. The large intestine.

   b. Defecation involves both voluntary and reflex activity.

      (1) When fecal material expands the rectum, a **rectosphincteric reflex** relaxes the anal sphincters and generates the urge to defecate.

      (2) Defecation can be prevented, however, by voluntarily contracting the external anal sphincter (which is composed of skeletal muscle innervated by the pudendal nerves).

      (3) If defecation does not occur, the internal anal sphincter closes, and the rectum relaxes to accommodate the fecal material within it.

      (4) Individuals lacking α-motoneuronal control over the external anal canal will defecate whenever the rectum is filled with fecal material.

B. Absorption, secretion, and gas production

   1. Water. The colon is unable to absorb more than 2–3 L/day. Thus, if most of the 8–10 L entering the intestine (either as ingested water or as gastric, pancreatic, or biliary secretions) is not absorbed in the small intestine, severe diarrhea can occur.

   2. Na$^+$ and Cl$^-$. The colon absorbs most of the **Na$^+$ and Cl$^-$** that escapes absorption in the small intestine.

   3. K$^+$, on the other hand, is secreted by the colon. Its concentration typically rises from its ileal concentration of 9 mEq/L to 75 mEq/L by the time the fluid reaches the end of the large intestine.

   4. Aldosterone. While the small intestine has no way to regulate Na$^+$ or K$^+$ absorption, in the colon, the hormone aldosterone controls these processes. **Aldosterone** enables the colon to absorb all of the Na$^+$ in the fecal fluid. However, in doing so, it causes significant amounts of K$^+$ to be lost from the body.

   5. Intestinal gas

     a. There are three sources of gas in the GI tract.

      (1) Swallowed air, including air released from food and carbonated beverages, enters the stomach, from which it is removed by eructation or passed into the intestines with chyme.

      (2) Gas is formed by **bacterial action** in the ileum and large intestine.

      (3) Some gases diffuse into the GI tract from the **bloodstream.**

     b. Gas in the colon differs in volume and source from gas in the small intestine.

      (1) Small intestine. The small amount of gas present is usually the result of swallowed air. This gas most likely will be passed on to the colon.

      (2) Colon

         (a) Colonic gas, or flatus, is produced in large volumes—up to 7–10 L/day.

         (b) The gas is produced chiefly through the breakdown of undigested nutrients that reach the colon.

         (c) The main components of flatus are CO_2, CH_4, H_2, and nitrogen gas (N_2). Since all of these gases except N_2 diffuse readily through the intestinal mucosa, the volume of flatus expelled is reduced to about 600 ml/day.

STUDY QUESTIONS

Directions: Each of the numbered items or incomplete statements in this section is followed by answers or by completions of the statement. Select the **one** lettered answer or completion that is **best** in each case.

1. Which of the following secretions is most dependent on vagal stimulation?

(A) Saliva
(B) HCl
(C) Pepsin
(D) Pancreatic juice
(E) Bile

2. The major stimulus for primary peristalsis in the esophagus is

(A) placing food in the esophagus
(B) swallowing
(C) regurgitation of food from the stomach
(D) closing of the upper esophageal sphincter
(E) opening of the lower esophageal sphincter

3. Which of the following will inhibit stomach contractions?

(A) ACh
(B) Motilin
(C) Gastrin
(D) Secretin
(E) Histamine

4. Gastric acid secretion increases when food enters the stomach because

(A) protein digestion products directly stimulate the parietal cells to release HCl
(B) food raises the pH of the stomach, allowing more acid to be released
(C) both
(D) neither

5. Gastric parietal cells secrete

(A) gastrin
(B) motilin
(C) CCK
(D) intrinsic factor
(E) secretin

6. Na^+-dependent transport is responsible for the absorption of all of the following EXCEPT

(A) vitamin E
(B) amino acids
(C) glucose
(D) bile salts

7. Which of the following can occur without brain stem coordination?

(A) Chewing
(B) Swallowing
(C) Primary esophageal peristalsis
(D) Vomiting
(E) Gastric emptying

8. The major stimulus for gastric acid (HCl) secretion during the cephalic stage is

(A) histamine
(B) gastrin
(C) secretin
(D) somatostatin
(E) ACh

9. The major stimulus for the release of secretin is

(A) protein digestion products
(B) histamine
(C) somatostatin
(D) HCl
(E) CCK

10. Fats are transported from intestinal cells to blood plasma primarily in the form of

(A) micelles
(B) chylomicrons
(C) triglycerides
(D) fatty acids
(E) monoglycerides

11. ACh is required for the contraction of the

(A) lower esophageal sphincter
(B) upper esophageal sphincter
(C) both
(D) neither

12. The major stimulus for receptive relaxation of the stomach is

(A) food in the stomach
(B) food in the intestine
(C) secretin
(D) CCK
(E) motilin

13. The motility pattern primarily responsible for the propulsion of chyme along the small intestine is

(A) the migrating motor complex (MMC)
(B) peristaltic waves
(C) myogenic contractions
(D) haustrations
(E) segmentation

14. Intestinal motility is increased by all of the following EXCEPT

(A) CCK
(B) secretin
(C) gastrin
(D) insulin
(E) motilin

15. Gastric acid (HCl) secretion is inhibited by

(A) somatostatin
(B) entero-oxyntin
(C) high pH
(D) amino acids
(E) ACh

16. The major factor controlling the secretion of bile salts from the liver is the amount of

(A) secretin released during a meal
(B) fat entering the small intestine
(C) bile acids produced by the liver
(D) bile reabsorbed from the intestine
(E) CCK released during a meal

17. Micelle formation is necessary for absorption of

(A) bile salts
(B) iron
(C) cholesterol
(D) alcohol
(E) B vitamins

18. All of the following stimulate CCK secretion EXCEPT

(A) amino acids
(B) fatty acids
(C) HCl
(D) bile acids

19. All of the following effects are caused by secretin EXCEPT

(A) stimulation of pancreatic HCO_3^- secretion
(B) enhancement of bile acid secretion
(C) potentiation of pancreatic enzyme secretion by CCK
(D) inhibition of gastric muscle contraction

20. A Na^+-dependent active transport system is necessary for the intestinal absorption of all of the following EXCEPT

(A) dipeptides
(B) bile salts
(C) fructose
(D) vitamin C
(E) glucose

21. Secondary bile acids are formed

(A) in the liver from cholesterol
(B) by the conjugation of bile acids with taurine or glycine
(C) both
(D) neither

Directions: Each question below contains four suggested answers of which **one or more** is correct. Choose the answer.

A if **1, 2, and 3** are correct
B if **1 and 3** are correct
C if **2 and 4** are correct
D if **4** is correct
E if **1, 2, 3, and 4** are correct

22. Removal of the duodenum will cause an increase in

(1) gastric acid (HCl) secretion
(2) pancreatic HCO_3^- secretion
(3) gastric emptying
(4) gallbladder emptying

23. Elimination of pancreatic secretions to the small intestine may cause

(1) fat malabsorption
(2) increased small bowel motility
(3) ulceration of the duodenum
(4) increased gastric emptying

Directions: Each group of items in this section consists of lettered options followed by a set of numbered items. For each item, select the **one** lettered option that is most closely associated with it. Each lettered option may be selected once, more than once, or not at all.

Questions 24–28

For each of the following GI processes, choose the stimulus that is most important for its regulation.

(A) Secretin
(B) Histamine
(C) CCK
(D) Bombesin
(E) Motilin

24. Gastric acid (HCl) secretion
25. Gastrin secretion
26. Pancreatic enzyme secretion
27. Gallbladder emptying
28. Emptying of the intestine during the interdigestive period

Questions 29–33

For each of the following functions of the GI tract, choose the area where it occurs.

(A) Fundus of the stomach
(B) Antrum of the stomach
(C) Duodenum of the intestine
(D) Ileum of the intestine
(E) Colon

29. Absorption of bile acids
30. Secretion of gastrin
31. Secretion of intrinsic factor
32. Secretion of K^+
33. Absorption of iron

ANSWERS AND EXPLANATIONS

1. The answer is A. [*II B 3 a; III C 2 c–d, 4 b (1)–(2); IV C 3, D 4 b*] Salivary flow is entirely dependent on the autonomic nervous system. Vagal stimulation produces a large volume of watery fluid, while sympathetic stimulation causes the secretion of proteins (mucus and some enzymes). Secretion of HCl, pepsin, pancreatic juice, and bile is influenced by vagal stimulation but can occur without it.

2. The answer is B. [*II D 1 c (2) (a)*] Primary esophageal peristalsis is part of the swallowing response and occurs whether or not food enters the esophagus. The intensity of the peristalsis, however, increases if food is present. If the esophagus is not emptied by primary peristalsis, the presence of food in the esophagus will initiate another peristaltic reflex, called secondary peristalsis.

3. The answer is D. [*III B 6 b (2) (b)*] Secretin has a direct inhibitory effect on the smooth muscle fibers forming the stomach wall. Acetylcholine (ACh) and motilin, and possibly gastrin, increase the force of stomach contractions. Histamine has no direct effect on stomach contractions.

4. The answer is C. [*III C 2 d (2)*] Gastric acid secretion is increased directly by protein digestion products and inhibited when the pH of the stomach is reduced. The buffering action of food promotes gastric acid secretion by keeping the pH from falling too low.

5. The answer is D. [*III C 1 c (1) (a)*] Parietal cells, located in the oxyntic glands of the orad stomach (fundus and corpus) secrete HCl and intrinsic factor. Gastrin is secreted by G cells located in the pyloric glands of the distal stomach (antrum). Secretin and cholecystokinin (CCK) are secreted by endocrine cells in the proximal intestine. Motilin is secreted from endocrine cells within the epithelium of the small intestine.

6. The answer is A. [*IV F 1 b (1), 2 c (1), 3 b (2) (d), 5 a*] Amino acids and glucose are absorbed in the proximal intestine by Na^+-dependent active transport systems. Bile salts are reabsorbed by a Na^+-dependent active transport system located within the terminal ileum. Vitamin E is a fat-soluble vitamin and is absorbed by passive diffusion along with other lipids in the proximal intestine.

7. The answer is E. [*III B 5, 6 a (1)*] Gastric emptying of solids occurs when the contractile activity of the stomach reduces the size of the particles within the food sufficiently for them to pass through the pyloric sphincter. Stomach contractions are elicited by reflexes initiated by antral distension and by gastrin. Although the strength of the contractions is reduced by vagotomy, contractions can still occur. Chewing, swallowing, primary (but not secondary) peristalsis, and vomiting all are coordinated by specific regions of the brain stem.

8. The answer is E. [*III C 2 d*] During the cephalic stage of gastric acid secretion, the sight, smell, or thought of food activates cholinergic (acetylcholine-releasing) vagal fibers, which stimulate the release of HCl from antral parietal cells.

9. The answer is D. [*IV C 3 c (3) (b)*] Secretin and cholecystokinin (CCK) are hormones released from endocrine cells located in the proximal intestine. Although the release of both hormones is stimulated by the presence of chyme in the small intestine, it is low pH resulting from HCl in the chyme that is the major stimulus for the release of secretin. Protein digestion products (e.g., amino acids, particularly phenylalanine) stimulate both secretin and CCK secretion, but are the major stimuli for CCK secretion. CCK potentiates the effects of secretin but does not affect its release. Somatostatin and histamine have no effect on CCK or secretin secretion.

10. The answer is B. [*IV F 3 b (3)*] Chylomicrons are small lipid droplets within the enterocytes that are reconstituted from the lipid digestion products formed in the intestine. The chylomicrons are transported from the enterocytes by exocytosis by the intestinal fluid surrounding the intestine. The chylomicrons are then absorbed into the intestinal lacteals, from which they enter the circulation.

11. The answer is B. [*II D 1 c (1)*] The upper esophageal sphincter is composed of striated muscle and is stimulated by cholinergic (acetylcholine-releasing) vagal fibers. The lower esophageal sphincter is composed of smooth muscle, which normally is maintained in a contracted state by a myogenic process.

12. The answer is A. [*III B 2 b*] When food distends the orad stomach (fundus and corpus), it produces a vago-vagal reflex by which noncholinergic, nonadrenergic fibers relax the stomach. About 2 L of food can be accommodated in the stomach when receptive relaxation is at its maximum.

13. The answer is E. [*IV B 2 a*] Although the major functions of segmentation are the mixing of chyme with digestive juices and exposing the products of digestion to the intestinal wall, segmentation is also responsible for pushing the chyme along the intestine. Propulsion occurs because the frequency of segmentation is higher in the more proximal intestine than it is in the distal intestine. Thus, it is more likely for the chyme to move towards the colon than it is to move towards the stomach. Peristaltic waves will move chyme along the intestine, but these are not frequent enough to propel the food into the colon. The migrating motor complex (MMC) empties the intestine of the small amount of chyme remaining in the intestine during the interdigestive period.

14. The answer is B. [*IV B 4 c (2)*] Secretin decreases intestinal and gastric contractile force. Cholecystokinin (CCK), gastrin, insulin, and motilin all increase intestinal contractions.

15. The answer is A. [*III C 2 c (2)*] Gastric acid (HCl) secretion is inhibited by low pH in the stomach and by somatostatin released from interneurons within the enteric nervous system. Acetylcholine (ACh), the hormone known as entero-oxyntin, and amino acids all stimulate gastric acid secretion.

16. The answer is D. [*IV D 4 b (1)*] The amount of bile synthesized each day is not sufficient to absorb all of the fat digested. However, because the bile salts are absorbed by the intestine and returned to the liver, they can be used over and over again. The amount of bile secreted each day is thus proportional to the amount absorbed from the intestine.

17. The answer is C. [*IV F 3 b (1), (2) (b)*] Micelles are necessary for absorption of dietary lipids such as cholesterol. Bile salts, iron, and B vitamins are absorbed by membrane-bound active transport systems. Alcohol is both water and fat soluble and so can diffuse directly across the membranes.

18. The answer is D. [*IV C 3 c (3)*] Amino acids, particularly phenylalanine, fatty acids, and monoglycerides all directly stimulate the release of cholecystokinin (CCK) from intestinal endocrine cells. HCl, which lowers the pH of chyme entering the intestine, also causes CCK secretion. Bile acids do not influence CCK secretion.

19. The answer is B. [*III B 6 b (2); IV C 3 c (2); D 4 a (1)*] Although secretin regulates the volume of biliary secretion, it neither enhances nor inhibits the secretion of bile salts. Secretin's major effect is to stimulate HCO_3^- secretion by the pancreas and the liver. In addition, it potentiates the effect of cholecystokinin (CCK) on pancreatic enzyme secretion. Secretin also diminishes gastric emptying, probably by directly reducing gastric smooth muscle contractility.

20. The answer is C. [*IV F 1 b (1)–(2), 2 c (1), 3 b (2) (d), 5 b*] Unlike glucose, which depends on a Na^+-dependent active transport system for absorption, fructose is absorbed by facilitated diffusion. Active transport for fructose is not required because the intracellular fructose concentration is maintained at a low value by intracellular enzymes that rapidly convert fructose to glucose and lactose. Bile salts, vitamin C, and dipeptides all rely on Na^+-dependent active transport systems to be absorbed.

21. The answer is D. [*IV D 2 a*] Secondary bile acids are formed in the intestine by the deconjugation and dehydroxylation of primary bile acids. Primary bile acids are synthesized from cholesterol in the liver and then conjugated with taurine or glycine to form primary bile salts.

22. The answer is B (1,3). [*III B 6 b (2), C 2 d (5) (b); IV C 3 c, D 5 b (1)*] The duodenum releases cholecystokinin (CCK) and secretin when chyme enters the small intestine. Since both of these hormones inhibit gastric emptying, removal of the duodenum will cause gastric emptying to increase. The duodenum also releases gastrin inhibitory peptide (GIP), which inhibits the release of gastrin and HCl. Thus, removing the duodenum also will enhance gastric acid (HCl) secretion. Because secretin stimulates pancreatic HCO_3^- secretion and CCK stimulates gallbladder contraction, removing the duodenum would hinder both of these activities.

23. The answer is B (1,3). [*IV C 1 b–c*] Pancreatic secretions contain lipases, which are necessary for the proper absorption of fats, and large amounts of HCO_3^-, which are necessary for the neutralization of acids entering the duodenum. In the absence of HCO_3^-, gastric acid (HCl) will produce ulcerations of the duodenum.

24–28. The answers are: 24-B, 25-D, 26-C, 27-C, 28-E. [*III B 6 c (3), C 2 c (1), 3 b (1); IV C 3 c (1), D 5 b (1)*] Gastric acid (HCl) secretion is stimulated by acetylcholine (ACh), histamine, and gastrin. It is inhibited by somatostatin. Gastrin secretion is stimulated by bombesin [also called gastrin-releasing peptide (GRP)] and inhibited by somatostatin. Cholecystokinin (CCK) and secretin both stimulate the pancreas. CCK, however, is responsible for the secretion of pancreatic enzymes. CCK is also responsible for stimulating the gallbladder to contract. Motilin, a hormone released from the small intestine, increases the strength of the migrating motor complex (MMC) and may be responsible for initiating it. The MMC, which occurs every 60–90 minutes during the interdigestive period, starts in the stomach and sweeps along the entire GI tract. It is thought to be responsible for emptying the small intestine of any chyme remaining after the completion of a meal.

29–33. The answers are: 29-D, 30-B, 31-A, 32-E, 33-C. [*III C 1 c (1) (a), (2); IV F 3 b (2) (d), 5 e (1); V B 3*] Bile acids are absorbed in the terminal ileum by a Na^+-dependent cotransport process. Gastrin is secreted by G cells contained in the pyloric glands of the distal stomach (antrum and pylorus). Intrinsic factor, necessary for the absorption of vitamin B_{12}, is secreted by parietal cells located in the proximal stomach (fundus). The colon secretes K^+ into the chyme. About 10% of the daily K^+ load is excreted by the colon; the remainder is excreted by the kidney. Iron is absorbed primarily within the duodenum and jejunum.

7
Endocrine Physiology
John Bullock

I. GENERAL FEATURES OF HORMONES

A. Definition. In the classic definition, hormones are secretory products of the ductless glands, which are released in catalytic amounts into the bloodstream and transported to specific target cells (or organs), where they elicit physiologic, morphologic, and biochemical responses. In reality, the requirement that hormones be secreted into the bloodstream is too restrictive, because they also can act locally (Figure 7-1). For example:

  1. Paracrine hormones can be conveyed over short distances by diffusion through the interstitial space, to act on neighboring cells as regulatory substances

  2. Autocrine hormones can regulate the activity of the same cells that produce them

B. Hormone-secreting tissues. Virtually all organs in the body exhibit endocrine function.

  1. The most-studied endocrine organs and examples of the hormones they produce are listed in Table 7-1.

  2. Other organs with endocrine function and the hormones they produce are:
  a. Heart: atrial natriuretic factor (ANF)
  b. Kidney: 1,25-dihydroxycholecalciferol (calcitriol)
  c. Liver: 25-hydroxycholecalciferol (calcidiol), somatomedin
  d. Pineal gland: melatonin
  e. Skin: calciferol (vitamin D_3)
  f. Gastrointestinal tract: gastrin, pancreozymin, secretin, vasoactive intestinal peptide (VIP)

C. Functions. Hormones regulate existing fundamental bodily processes but do not initiate cellular reactions de novo. In contrast to vitamins, hormones serve no nutritive role in responsive tissues and are not incorporated as a structural moiety into another molecule.

Table 7-1. Principal Endocrine Organs and Some Hormones They Produce

Organ	Examples of Hormones
Pituitary gland	Tropic hormones (e,g., adrenocorticotropic hormone, growth hormone, prolactin)
Hypothalamus	Releasing hormones (e.g., thyrotropin releasing hormone), antidiuretic hormone, oxytocin
Thyroid gland	Thyroxine, 3,5,3'-triiodothyronine
Adrenal glands	Mineralocorticoids (e.g., aldosterone), glucocorticoids (e.g., cortisol), catecholamines (e.g., epinephrine, norepinephrine)
Parathyroid glands	Parathyroid hormone
Gonads	Testosterone, estradiol
Pancreatic islets	Insulin, glucagon

1. **Regulation of biochemical reactions.** As regulators, hormones stimulate or inhibit the rate and magnitude of biochemical reactions by their control of enzymes and, thereby, cause morphologic, biochemical, and functional changes in target tissues. Although they are not used as energy sources in biochemical reactions, hormones modulate energy-producing processes and regulate the circulating levels of energy-yielding substrates (e.g., glucose, fatty acids).

2. **Regulation of bodily processes.** Hormones regulate growth, maturation, differentiation, regeneration, reproduction, pigmentation, behavior, metabolism, and chemical homeostasis. Slower processes (e.g., growth, reproduction, metabolism) require longer periods of continual hormone stimulation in contrast to rapid coordination of the body (e.g., reflex contraction of a somatic muscle), which is regulated by the nervous system.

D. Physical characteristics

1. **Chemical composition.** The three major classes of hormones are: steroids, proteins and polypeptides, and amino acid derivatives (i.e., catecholamines and thyroid hormones). No polysaccharides or nucleic acids are known to function as hormones.

2. **Plasma concentration.** Hormones usually are secreted into the circulation in extremely low concentrations.
   a. Peptide hormone concentration is between 10^{-12} mol/L and 10^{-10} mol/L.
   b. Epinephrine and norepinephrine concentrations are 2×10^{-10} mol/L and 13×10^{-10} mol/L, respectively.
   c. Steroid and thyroid hormone concentrations are 10^{-9} mol/L and 10^{-6} mol/L, respectively.

3. **Latent period** is the time interval between the application of a stimulus and a response. In contrast to a latent period of 8 msec between a neural stimulus and the contraction of a muscle, the latent period associated with hormones can be as long as seconds, minutes, hours, or days.
   a. Following the administration of **oxytocin,** milk ejection occurs in a few seconds.

A. Endocrine signaling

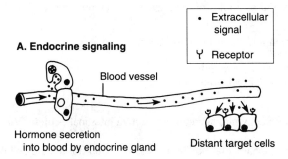

Blood vessel

Hormone secretion into blood by endocrine gland

Distant target cells

- Extracellular signal
- Ψ Receptor

B. Paracrine signaling

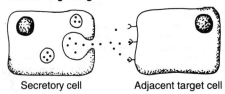

Secretory cell   Adjacent target cell

C. Autocrine signaling

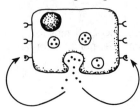

Target sites on same cell

Figure 7-1. The three methods of hormone information transfer. Cell-to-cell signaling can occur over long distances via hormone secretion into the bloodstream (*A*) or over short distances via hormone diffusion through the interstitium (*B*). A hormone also may act directly on the cell that produces it (*C*). (Adapted from Darnell J, Lodish H, Baltimore D (eds): *Molecular Cell Biology,* 2nd edition. New York, WH Freeman, 1990, p 710.)

b. The metabolic response to **thyroxine** can take as long as 3 days.

4. Postsecretory modification of hormones occurs by the proteolytic cleavage of peptide hormones or by enzymatic conversion of steroids and thyroid hormones at sites beyond the site of secretion. This peripheral conversion to more active hormonal forms occurs in the liver, kidney, fat, or bloodstream as well as in the target tissues themselves.

5. Circulating forms. Protein binding of hormones protects them against clearance by the kidneys, slows the rate of degradation by the liver, and provides a circulating reserve of hormones. Only unbound hormones pass through capillaries to produce their effects or to be degraded.

6. Hormone receptors are unique molecular groups in or on target cells that interact with hormones to initiate a characteristic response. The specificity of a hormone depends on the formation of a strong covalent bond with its hormone receptor (see II A).

7. Half-life. Most hormones are metabolized rapidly after secretion. In general, peptide hormones are short-lived in the circulation, whereas steroids have a significantly longer half-life.

8. Degradation. The interaction of hormones with their target cells is followed by intracellular degradation.
   a. Degradation of protein hormones and amines occurs after binding to membrane receptors and internalization of the hormone-receptor complex.
   b. Degradation of steroids and thyroid hormones occurs after binding of the hormone-receptor complex to the chromatin.

9. Inactivation and excretion. Only a small fraction of the circulating hormone is removed by most target tissues. Hormone inactivation occurs in the liver and kidney.
   a. Hormone degradation uses many enzymatic mechanisms such as hydrolysis, oxidation, hydroxylation, methylation, decarboxylation, sulfation, and glucuronidation.
   b. Only a small fraction ($< 1\%$) of any hormone is excreted intact in the urine or feces.

E. Chemistry (Table 7-2)

1. Proteins and polypeptides generally are water soluble and circulate unbound in plasma.
   a. Structure
      (1) The peptide and protein hormones vary greatly in size. For example, **thyrotropin releasing hormone (TRH)** is a tripeptide, whereas **human chorionic gonadotropin (HCG)** consists of 243 amino acid residues.
      (2) The molecular weights of the pituitary tropic hormones, which consist of about 200 amino acid residues, vary from 23,000 to 25,000 daltons. (The average molecular weight of an amino acid residue is 120; thus, multiplying the number of residues by 120 is a good estimate of the molecular weight of a peptide or protein.)
   b. Synthesis. Many of the protein-type hormones are synthesized on the rough endoplasmic reticulum as **prohormones** or **preprohormones**. These precursor hormones undergo post-translational cleavage by an endopeptidase within the Golgi complex prior to secretion of the biologically active hormone.
      (1) Growth hormone (GH) and **prolactin** are synthesized as prohormones.
      (2) Pro-opiomelanocortin (POMC), synthesized in the pituitary and hypothalamus, is a prohormone complex that contains peptide hormone moieties including adrenocorticotropic hormone (ACTH; corticotropin), melanotropin, lipotropin, and endorphins.
      (3) Insulin and **parathyroid hormone (PTH)** are synthesized as preprohormones, which are hydrolyzed to prohormones and then further hydrolyzed to the hormone that is secreted.
   c. Storage and secretion. Protein and polypeptide hormones are secreted by endocrine organs derived from ectoderm (pituitary gland*; tuberoinfundibular, supraoptic, and paraventricular nuclei) as well as organs derived from endoderm (pancreatic islets of Langerhans, thyroid gland, parathyroid glands). The primordial cells that give rise to the parafollicular cells (C cells) of the thyroid gland are derived from neural crest precursors.
      (1) Protein and polypeptide hormones probably are stored exclusively in subcellular membrane-bound secretory granules within the cytoplasm of endocrine cells.

---

*The anterior and posterior lobes of the pituitary gland are derivatives of buccal ectoderm and neural ectoderm, respectively.

Table 7-2. Characteristics of the Principal Classes of Hormones

Characteristic	Peptides	Steroids and Calcitriol*	Amines	
			Catecholamines	Thyroid Hormone
Solubility				
In aqueous solvents	Excellent	Limited	Good	Limited
In nonaqueous solvents	Poor	Excellent	Limited	Good
Biosynthetic pathway	Single peptide, prohormone, or preprohormone	Multiple enzymes	Multiple enzymes	Multiple enzymes
Postsecretory modifications	Very rare	Common	None	Common
Storage of preformed hormone	Often substantial	Minimal	Substantial	Substantial
Degradation products	Irreversibly inactive	Sometimes retain or regain activity	Inactive	Inactive
Plasma binding proteins	Very rare	Yes	Limited	Yes
Half-life	Short (minutes)	Long (hours)	Very short (seconds)	Very long (hours to days)
Receptors	Cell surface	Nucleus	Cell surface	Nucleus
Site of action	Plasma membrane	Nucleus	Plasma membrane	Nucleus
Mechanism of action	Stimulates production of second messenger	Stimulates production of specific mRNAs	Stimulates production of second messenger	Stimulates production of specific mRNAs

*Calcitriol = 1,25-dihydroxyvitamin D_3.

(2) These hormones are released into the blood by exocytosis, which involves fusion of the secretory granule and cell membrane followed by extrusion of the granular contents into the bloodstream.

  d. Half-life of various peptide/protein hormones is as follows:
    (1) Antidiuretic hormone (ADH) and oxytocin: < 1 minute
    (2) Insulin: 7 minutes
    (3) Prolactin: 12 minutes
    (4) ACTH: 15–25 minutes
    (5) Luteinizing hormone (LH): 15–45 minutes
    (6) Follicle-stimulating hormone (FSH): 180 minutes

2. Amino acid derivatives. Catecholamines, which are water soluble, and thyroid hormones, which are lipid soluble, circulate in the plasma bound mainly to binding globulins.

  a. Structure
    (1) Catecholamines are derived from the amino acid, tyrosine. Thyroid hormones are derived from two iodinated tyrosine residues. Hormones derived from tyrosine also are called **phenolic derivatives**.
    (2) Catecholamines and thyroid hormones both retain the aliphatic α-amino group. Introduction of a second hydroxyl group in the ortho- position on the benzene ring is characteristic of the catecholamines, whereas iodination of the benzene ring distinguishes the thyroid hormones.
    (3) Thyroid hormones are the only substances in the body that contain iodine.

  b. Synthesis. Epinephrine and norepinephrine are synthesized in the chromaffin cells, which are modified postganglionic neurons (see VII B–D). Thyroid hormones are synthesized in thyroid follicular cells (see XIII A and D).

  c. Storage and secretion. The amine hormones are secreted by endocrine tissues derived from the neural crest (adrenal medulla) and from endoderm (thyroid gland).
    (1) Catecholamines are stored in secretory granules. Secretion occurs when the membrane of the chromaffin granules fuses with plasma membranes, causing the granular contents to be extruded into the circulation.
    (2) Thyroid gland secretions are called **iodothyronines**—compounds resulting from the coupling of two iodinated tyrosine molecules. Thyroid hormones are stored outside follicular cells in the form of **thyroglobulin,** a glycoprotein precursor found in the lumen of the cells. Several weeks' supply of thyroid hormone is stored in this form. Following endocytosis and proteolysis of thyroglobulin, thyroid hormone is secreted into the bloodstream by simple diffusion.

  d. Circulation and half-life
    (1) Epinephrine and norepinephrine exist in plasma either in the free form or in conjugation with sulfate or glucuronide. Most circulating epinephrine is bound to blood proteins (mainly albumin); norepinephrine does not bind to blood proteins to any appreciable degree.
    (2) Most thyroid hormones are bound to thyroxine-binding globulin.
    (3) The half-life of various amine hormones is as follows:
      (a) Epinephrine: 10 seconds
      (b) Norepinephrine: 15 seconds
      (c) Triiodothyronine: 1 day
      (d) Thyroxine: 7 days

3. Steroid hormones are lipid soluble and circulate in the plasma bound to carrier proteins called **steroid-binding globulins**.

  a. Structure. Steroids are a group of biologically active substances, including androgens, estrogens, progesterone, glucocorticoids, and mineralocorticoids. 25-Hydroxyvitamin D_3 and 1,25-dihydroxyvitamin D_3 are modified steroids called **secosteroids**.
    (1) Steroids consist of three cyclohexyl rings and one cyclopentyl ring combined into a single structure. They are derived from the cyclopentanoperhydrophenanthrene nucleus consisting of a fully hydrogenated phenanthrene (rings A, B, and C), to which is attached a hydrogenated cyclopentane ring (D-ring). This fully saturated, four-ring structure consisting of 17 carbon atoms is the hypothetical parent compound, gonane or sterane.
    (2) The secosteroids, which are vitamin hormones, lack a B-ring and, therefore, consist of two cyclohexyl rings (rings A and C) and one cyclopentyl ring (D-ring).

b. Synthesis and secretion. Steroids are synthesized and secreted by the endocrine organs derived from mesoderm (adrenal cortex, testis, ovary). The placenta also synthesizes and secretes steroids.

(1) Steroids are derived from cholesterol. The first, and rate-limiting, step in steroid synthesis is conversion of cholesterol to pregnenolone. Depending on the product, hydroxylations of the steroid nucleus may occur at carbons 11, 17, 18, or 21.

(2) There is little storage of steroids. Instead, steroid-producing cells store esterified cholesterol in the form of lipid droplets, which serve as prohormones. Steroids are released in the circulation by simple diffusion.

c. Half-life of various steroid hormones is as follows:

(1) Aldosterone: 30 minutes

(2) Cortisol: 90–100 minutes

(3) 1,25-Dihydroxyvitamin D_3: 15 hours

(4) 25-Hydroxyvitamin D_3: 15 days

II. TYPES OF HORMONE ACTION

A. Hormone-receptor interaction. Hormones produce their effects by first combining with specific cell components called receptors. Only cells with receptors for a specific hormone respond. Hormones may be found on the target cell membrane (**external receptors**) or within the cytoplasm or nucleus (**internal receptors**).

1. Mechanisms (Figure 7-2). Hormones and receptors interact in the following ways to affect intracellular metabolism.

a. Polypeptide hormones and catecholamines bind irreversibly to **fixed receptors** on the outer surface of target cells, which initiates the transduction of the hormone signal across the membrane. Transduction occurs after activation of adenyl cyclase (see II B 1).

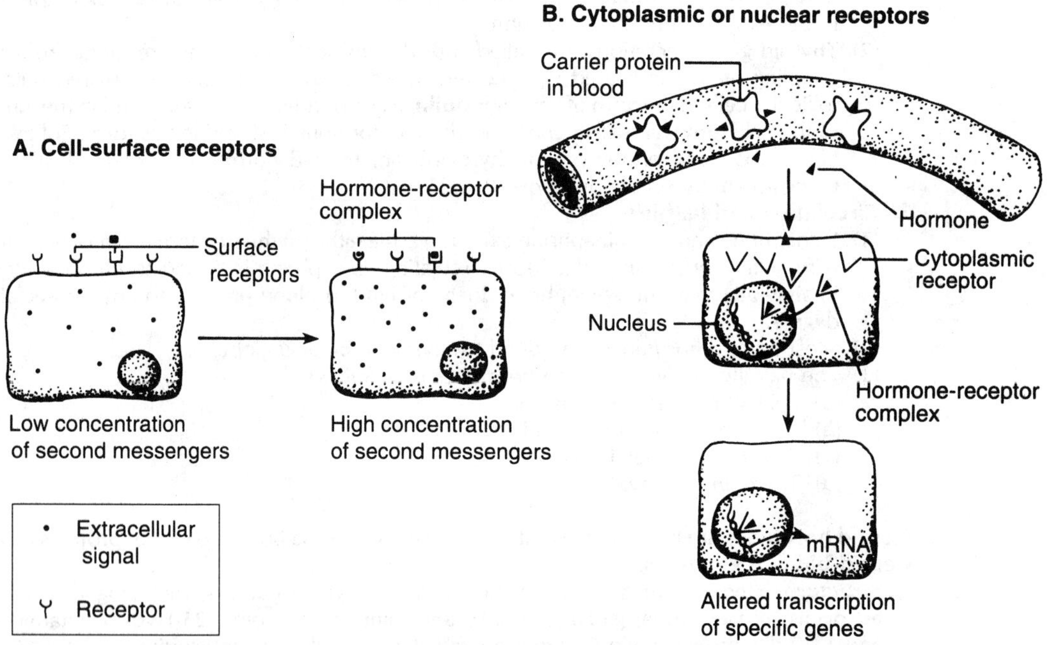

Figure 7-2. Mechanisms of hormone-receptor interaction. (*A*) Some hormones (e.g., polypeptide hormones, catecholamines) bind to specific receptors on the surface of target cells, triggering an increase or decrease in the concentration of cyclic adenosine 3',5'-monophosphate (cAMP) or some other second messenger (e.g., calcium, 1,2-diacylglycerol). (*B*) Other hormones (e.g., steroids, thyroid hormones) are transported by carrier proteins in the blood and, once dissociated from the carrier, enter the target cell and bind to specific receptors in the cytoplasm or nucleus. These hormone-receptor complexes then act on nuclear DNA to alter transcription of specific genes. (Adapted from Darnell J, Lodish H, Baltimore D (eds): *Molecular Cell Biology,* 2nd edition. New York, WH Freeman, 1990, p 711.)

(1) Approximately 10^4–10^5 receptors exist on the surface of a polypeptide hormone target cell.

(2) Polypeptide hormones that are known to enter cells include insulin, prolactin, PTH, and gonadotropins.

    (a) HCG remains bound to its receptor when it enters ovarian cells.

    (b) Since receptors may enter cells, it is possible that receptors (not hormones) are the biologically active component. For example, insulin antibody causes the insulin receptor to change its configuration, allowing the receptors to exert insulin-like effects on cells.

b. Steroids enter target cells and bind to specific **mobile receptors** in the cytoplasm. The hormone-receptor complex then binds to specific DNA sequences to initiate transcription and translation (see II B 2). Approximately 3000 to 10^4 intracellular receptors exist per steroid target cell.

c. Thyroid hormones enter target cells and combine with **nuclear receptors**. The hormone-receptor complex then binds to specific DNA sequences to initiate transcription and translation (see II B 2).

2. Effects

a. Increased membrane permeability. Hormonal control of membrane permeability is well documented for ions and for metabolites such as glucose and amino acids.

(1) The increased entry of glucose into cells, which is mediated by a special transport system, is the major method by which insulin controls glucose utilization by muscle.

(2) Other peptide hormones that change membrane permeability include GH, ACTH, calcitonin, thyroid-stimulating hormone (TSH), and thyroxine.

(3) Steroids that alter membrane permeability include mineralocorticoids, glucocorticoids, estrogens, and androgens.

b. Changes in receptor number

(1) Hormone-sensitive cells respond to high concentrations of certain hormones by reducing the number of cell surface receptors, thus decreasing sensitivity to the circulating hormone. For example, elevated ambient insulin concentrations cause a loss or inactivation of insulin receptors in liver cells, fat cells, and white blood cells.

(2) Catecholamines exert their effects via plasma membrane receptors, and thyroid hormones have receptors in the target cell nucleus. Excess thyroid hormone leads to an increased number of catecholamine receptors in the myocardium of experimental animals. This may explain the catecholamine-like effects noted in hyperthyroid patients who produce normal amounts of catecholamines.

    (a) Hyperthyroid patients have tachycardia and palpitations, effects observed with excessive catecholamines (pheochromocytoma). These cardiac symptoms can be ameliorated by the administration of a β-blocker such as propranolol.

    (b) The increase in cardiac β-adrenergic receptors in hyperthyroidism makes the heart more responsive to catecholamines.

(3) Angiotensin II decreases the number of its receptors on adrenocortical cells and increases the number of its receptors in vascular smooth muscle.

B. Mechanisms of information transfer: distant signals.

For hormones that communicate via the bloodstream, two mechanisms for hormone-receptor coupling operate.

1. Cyclic AMP-mediated hormone activity (see Figure 7-2A) involves cyclic adenosine 3′,5′-monophosphate (cAMP) as a second messenger that changes enzyme activities. cAMP is the intracellular nucleotide messenger that mediates the effects of certain hormones on subcellular processes, leading to a variety of physiologic responses.

a. Most polypeptide hormones, many biogenic amines, and some prostaglandins activate their target cells by stimulating the synthesis of cAMP, which causes enzyme phosphorylation, a reaction usually associated with enzyme activation. The first messenger is the hormone that binds to the membrane receptor and leads to **activation of membrane-bound adenyl cyclase**. This enzyme converts adenosine triphosphate (ATP) to cAMP in the presence of magnesium ion (Mg^{2+}).

b. cAMP exerts biologic activity via the **phosphorylation of cAMP-dependent protein kinases**.

(1) Kinases are a family of enzymes that phosphorylate their substrate. Protein kinases, a subgroup of the kinases, can be soluble or membrane-bound. Protein kinases transfer a phosphate group from ATP to the hydroxyl group of the substrate. Only the protein kinases that are regulated by cAMP are called cAMP-dependent protein kinases.

(2) The phosphorylation of a cAMP-dependent protein kinase can lead to the activation (e.g., via phosphorylase kinase, phosphorylase, and triglyceride lipase) or the inactivation (e.g., via glycogen synthetase and pyruvate dehydrogenase) of the substrate.

(3) The effects of the cAMP-dependent protein kinases on their substrates are reversed by a group of enzymes called **phosphoprotein phosphatases,** which remove the phosphate group from the protein enzyme by hydrolysis.

(4) The enzyme that inactivates cAMP is called **cyclic nucleotide phosphodiesterase**. Most of the phosphodiesterases are cytosolic enzymes.

c. An **example of cAMP-mediated hormone activity** is the hormone regulation of glycogen metabolism. Glucagon and epinephrine stimulate glycogenolysis and also inhibit glycogen synthesis.* Insulin has the opposite effect on both processes.

(1) The enzymes that promote glycogenolysis are active in their phospho- form and inactive in their dephospho- form. The reverse is true for glycogen synthetase, which is the major enzyme in glycogen synthesis.

(2) In muscle, a cAMP-dependent protein kinase is activated either by catecholamines (via their β-adrenergic receptors) or by glucagon (via its receptor in the liver).

(3) This activated protein kinase phosphorylates another protein kinase, phosphorylase kinase, thereby activating it.

(4) The activated phosphorylase kinase phosphorylates the enzyme, phosphorylase, which initiates glycogen breakdown. Phosphorylase kinase also phosphorylates glycogen synthetase, thereby inactivating it.

2. **Transcription and translation effects** (see Figure 7-2B). Steroids and thyroid hormones modulate transcription in specific areas of the nuclear chromatin by interacting with DNA molecules in the chromatin to cause **enzyme induction**. Steroid-receptor complexes are translocated through the cytosol and enter the nucleus. Thyroid hormones enter the nucleus in the free state and then combine with nuclear receptors.

a. Chromatin consists of DNA, histone proteins, and nonhistone (acidic) proteins. The specificity of nuclear binding is a property of a particular acidic protein.

b. As a result of the interaction of steroid and thyroid hormones with the chromatin, transcription is stimulated and specific messenger RNA (mRNA) synthesis increases.

c. The specific mRNAs enter the cytoplasm, where they direct the synthesis (translation) of specific proteins. These proteins may be enzymes, structural proteins, receptor proteins, or secretory proteins.

C. **Mechanisms of information transfer: local signals.** Some cell-to-cell signaling occurs over very small distances.

1. **Paracrine communication** (see Figure 7-1B) involves local diffusion of a peptide or other regulatory molecule to its target cell through the interstitium. Target cells are in the vicinity of the paracrine cell that releases the messenger. Although paracrine substances may diffuse into the blood, it is not necessary that they do.

a. Somatostatin is a paracrine substance whose secretion is stimulated by glucose, glucagon, and gut hormones. Somatostatin is enzymatically degraded in the blood, a process that protects distant cells from the paracrine substance, should it enter the blood.

b. Signal transmission also occurs through the release of peptides into the lumen of the gastrointestinal tract, where these peptides interact with endocrine cells to cause the release of a second endocrine or paracrine messenger. Gastrin, somatostatin, and substance P are released into the gut lumen following nerve stimulation. Possibly, these substances are released as precursor molecules that are activated by digestive enzymes. Gastrin also is secreted into the bloodstream.

2. **Neurocrine communication** involves the release of chemical messengers from nerve terminals. Neurocrine substances may reach their target cells via one of three routes.

a. The **neurotransmitter can be released directly** into the intercellular space, cross the synaptic junction, and inhibit or activate the postsynaptic cell.

---

*The glycogenolytic effect of epinephrine in human liver probably occurs via activation of a cAMP-independent phosphorylase.

(1) Neurocrine substances are inactivated by degrading enzymes and by reuptake of the substances by neurons.

(2) Examples of neurocrine substances secreted by this route are acetylcholine (ACh) and norepinephrine.

   b. A **neural signal also can be transferred via a gap junction,** which is a membrane specialization between nerve cells, between nerve terminals and endocrine cells, and between endocrine cells. Gap junctions allow the movement of small molecules and electric signals from one cell to another, creating a functional **syncytium.**

   c. The third potential route for the transmission of a neural signal is identical to the classic neurosecretory mechanism, which involves the **release of a peptide or neurohormone from a neurosecretory neuron** into the blood followed by the interaction of this neurohormone with specific receptors on distant target cells (see III C–D). Examples of such neurocrine substances are oxytocin and ADH. The effector sites of neurohormones are not always endocrine cells.

3. Autocrine communication (see Figure 7-1C). **Autacoid** is a term used to designate a compound that is synthesized at, or close to, its site of action. This is in contrast to the circulating hormones, which act on tissues distant from their site of synthesis. Prostaglandins are autacoids.

III. BASIC CONCEPTS OF ENDOCRINE CONTROL

A. Homeostasis and steady state. A major function of the endocrine system is to maintain the homeostasis of the internal environment. This condition of relative constancy in the concentration of dissolved substances, in temperature, and in pH is a basic requirement for the normal function of cells.

1. The concept of **homeostasis** as a constancy of physiologic variables must be modified, because many regulated organismic processes are not constant but conform to a persistent endogenous or exogenous **rhythm.** For example, humans demonstrate a **circadian pattern** in the levels of plasma 17-hydroxycorticosteroids. Such 24-hour cycles are not solely a response to fluctuating environmental stimuli but also are due to internal endogenous oscillators whose phases are influenced by environmental stimuli.

   a. In humans, certain corticosteroids (e.g., cortisol) have a rhythmic pattern of secretion, with secretory rates highest early in the morning and lowest late at night. Accordingly, plasma cortisol concentration is at a peak between 6 A.M. and 8 A.M. and at a nadir between midnight and 2 A.M.

   (1) This circadian rhythm persists but shifts to correspond with a change in sleeping habit (e.g., during illness, night work, changes in longitude, and total bed rest or confinement).

   (2) For this reason, treatment of patients with exogenous corticosteroids is on an alternate-day dosage regimen, whereby the entire dose is given in the morning of every other day. This dosage schedule simulates the normal adrenocortical secretory rhythm.

   b. The rhythmic pattern of corticosteroid secretion occurs in isolated adrenal glands and even in single adrenocortical cells.

2. The term **steady state** indicates that a function or a system is unvarying with time, but that the system is not in true equilibrium. The system is said to be in a **dynamic equilibrium,** because matter and energy flow into the system at a rate equal to that at which matter and energy flow out of the system.

B. Hypothalamic-hypophysial axes and feedback control. The hypothalamus has **neural control** over hormone secretion by the posterior lobe of the pituitary gland. The secretory activity of the anterior lobe is controlled by **hypothalamic hormones,** which are secreted into the **hypothalamic-hypophysial portal system** (the hypophysial portal system). Only those hypothalamic hormones that regulate the anterior pituitary are hypophysiotropic hormones.

1. Hypophysial portal system (Figure 7-3). The median eminence has a poorly developed blood-brain barrier, and there is relatively little arterial blood perfusing the cells of the anterior lobe. The blood supply of the anterior lobe is derived from branches of the internal carotid arteries (mainly the superior hypophysial artery).

   a. The posterior lobe derives its blood from a capillary plexus emanating from the inferior hypophysial artery. This capillary plexus drains into the dural sinus. The neural tissue of the upper infundibular stem (neural stalk) and of the median eminence is supplied largely by

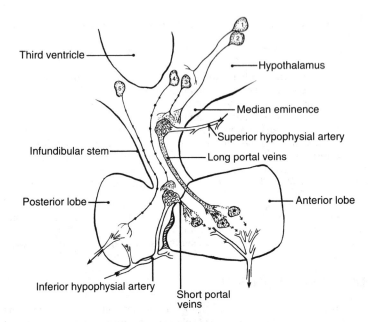

Third ventricle

Hypothalamus

Median eminence

Superior hypophysial artery

Infundibular stem

Long portal veins

Posterior lobe

Anterior lobe

Inferior hypophysial artery

Short portal veins

Figure 7-3. Anatomic relationship of the hypothalamus and pituitary gland, showing the hypophysial portal system and neurons involved in control of the pituitary gland. Neuron five (*5*) represents the peptidergic neurons of the supraopticohypophysial and paraventriculohypophysial tracts. Neurons four (*4*) and three (*3*) are the peptidergic neurons of the tuberohypophysial tract. Neuron one (*1*) and neuron two (*2*) are monoaminergic neurons. (Adapted from Gay VL: The hypothalamus: physiology and clinical use of releasing factors. *Fertil Steril* 23:51, 1972.)

branches of the superior hypophysial artery. The median eminence is the specialized area of the hypothalamus located beneath the inferior portion of the third ventricle. It is a release center for hypophysiotropic hormones.

   b. The **primary capillary plexus,** which emanates from the superior hypophysial artery, forms a set of long portal veins that carry blood downward into the anterior lobe.

      (1) The portal veins, which give rise to the **secondary capillary plexus,** constitute about 90% of the blood supply to the cells of the anterior lobe. The secondary capillary plexus drains into the dural sinus.

      (2) The anterior lobe receives its remaining blood from the short portal veins, which originate in the capillary plexus of the inferior hypophysial artery at the base of the infundibular stem.

  2. Feedback control is an important mechanism regulating hormone synthesis and secretion (Figure 7-4).

   a. Hypothalamic-pituitary-target gland model. The paradigm for feedback control is the interaction of the pituitary gland with target endocrine tissues (thyroid gland, adrenal cortex, gonads). Unbound circulating hormones produced by target endocrine organs inhibit the hypothalamic-pituitary system, causing a decrease in the secretion of pituitary tropic hormones, which, in turn, control the secretion by the endocrine target glands. Virtually all hormone secretions are controlled by some type of feedback control.

   b. Negative feedback control occurs on three levels.

      (1) Long-loop feedback. Peripheral gland hormones and substrates arising from tissue metabolism can exert long-loop feedback control on both the hypothalamus and the anterior lobe of the pituitary gland. Long-loop feedback usually is negative but occasionally can be positive and is particularly important in the control of thyroidal, adrenocortical, and gonadal secretions.

      (2) Short-loop feedback. Negative feedback also can be exerted by the anterior pituitary tropic hormones on the synthesis or release of the hypothalamic releasing or inhibiting hormones, which collectively are called **hypophysiotropic hormones**.

      (3) Ultrashort-loop feedback. Evidence suggests that the hypophysiotropic hormones may inhibit their own synthesis and secretion via a control system referred to as ultrashort-loop feedback.

C. Neurosecretory neurons (see Figure 7-3). Neural control of the pituitary gland is exerted through neurohumoral secretions that arise from specialized neurosecretory neurons (peptidergic neurons) and are carried by the bloodstream to a target site.

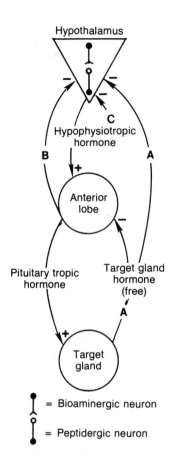

Figure 7-4. The three levels of feedback mechanisms for controlling hormone secretion: long-loop feedback (*A*), short-loop feedback (*B*), and ultrashort-loop feedback (*C*). *Plus signs* indicate stimulation, and *minus signs* indicate negative feedback.

1. Structure and function

   a. Neurosecretory neurons are glandular, unmyelinated neurosecretory cells with two functions.

   (1) They function as typical neurons, in that they conduct action potentials.

   (2) They also function as endocrine glands, in that they synthesize and release neurohormones either directly into the general circulation (as in the case of the neurosecretory neurons of the pars nervosa) or into a portal system (as in the case of the hypophysiotropic neurons, which release their neurohormones into the primary plexus of the hypophysial portal system).

   b. A neurosecretory cell system consists of axons that terminate directly on or near blood vessels. This differentiates these cells from typical neurons, which release neurotransmitters at localized synaptic regions. The functional complex of a neurosecretory neuron together with a blood vessel (hemocoele) is called a **neurohemal organ**.

2. Classification.

Neurosecretory neurons in humans are restricted to the hypothalamus, where they occur as two distinct populations of cells that secrete neurohormones (Tables 7-3, 7-4).

   a. The **magnocellular neurosecretory system** refers to the neurosecretory neurons of the supraoptic and paraventricular nuclei, which together form the supraopticohypophysial tract. Magnocellular neurons synthesize and secrete the neurohormones ADH and oxytocin.

   b. The **parvicellular** (also called parvocellular) **neurosecretory system** refers to the neurosecretory neurons of the tuberoinfundibular tract. These neurons of the medial basal hypothalamus have axons that terminate directly on the capillaries of the portal vessels in the median eminence, and they form a final common pathway for neuroendocrine function.

   (1) The parvicellular neurosecretory neurons mainly are peptidergic. An important exception, however, is the dopaminergic neurosecretory neurons that form and secrete **prolactin-inhibiting factor (PIF)**.

   (2) The secretory products of the parvicellular neurosecretory neurons are called hypophysiotropic hormones.

Table 7-3. Neuroendocrine Transducer Systems

Neuroendocrine System	Hormone
Magnocellular neurosecretory neurons	
Posterior lobe	
Supraoptic	Antidiuretic hormone
Paraventricular	Oxytocin
Parvicellular neurosecretory neurons	
Median eminence	Hypophysiotropic hormones
Preganglionic fibers	
Adrenal medulla	Epinephrine*
Postganglionic fibers	
Pineal gland	Melatonin[†]
Juxtaglomerular apparatus	Renin[†]

After Martin JB, et al: Neuroendocrine transducers and neurosecretion. In *Clinical Endocrinology*. Philadelphia, FA Davis, 1977, p 4.
*Acetylcholine is the neurotransmitter preceding epinephrine release.
[†]Norepinephrine precedes the release of both melatonin and renin.

3. **Neural control.** Neural information is transmitted to the parvicellular and magnocellular neuro-secretory cells by **monoaminergic neurons**. Most of the cell bodies of the monoaminergic neurons are located in the mesencephalon and lower brain stem.
    a. The monoaminergic neurons that innervate the parvicellular neurons produce and secrete biogenic amines, which modulate the hypothalamic release of the hypophysiotropic hormones.
    b. The function of the magnocellular neurosecretory neurons is controlled by cholinergic and noradrenergic neurotransmitters.
        (1) ACh stimulates the release of ADH and oxytocin.
        (2) Norepinephrine inhibits the secretion of ADH and oxytocin.
    c. Since the secretion of the parvicellular and magnocellular peptidergic neurons is regulated by biogenic amines, the neurosecretory neurons can correctly be viewed as **neuroeffector** cells.

D. **Neuroendocrine transducers** are endocrine glands that convert neural signals into hormonal signals. Neural control of endocrine tissues occurs in three ways (Figure 7-5).

1. **Direct innervation by autonomic secretomotor neurons**
    a. **Pancreatic islets of Langerhans**
        (1) The islets of Langerhans have a postganglionic parasympathetic innervation. Increased vagal activity to the beta cells stimulates insulin release only during periods of elevated blood sugar.

Table 7-4. Characteristic Features of the Neurosecretory Control Systems

Feature	Parvicellular System	Magnocellular System
Neural input	Norepinephrine, dopamine, serotonin	Acetylcholine
Nuclei	Arcuate nucleus	Supraoptic and paraventricular nuclei
Tract	Tuberoinfundibular	Supraopticohypophysial
Terminus	Median eminence, upper infundibular stem	Pars nervosa (infundibular process)
Neurohormones	Neuropeptides (hypophysiotropic hormones), polypeptides	Neuropeptides (arginine-ADH, oxytocin), nonapeptides
Type of endocrine neuron	Peptidergic	Peptidergic
Stimuli	Monoamines	Acetylcholine
Vascular elements	Hypophysial portal system	Capillary bed (systemic)

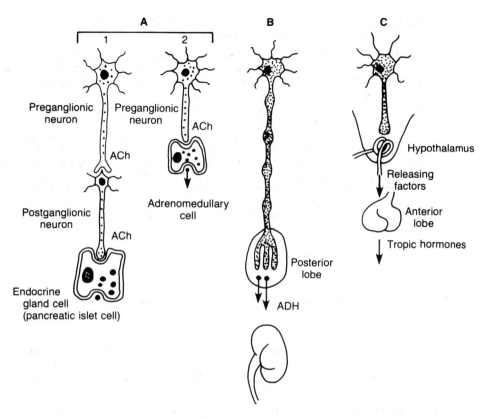

Figure 7-5. The three types of neuroendocrine transducers. (*A*) Secretomotor neurons control endocrine glands by direct innervation via autonomic fibers; the adrenal medulla is the only autonomic neuroeffector that is innervated by preganglionic neurons. *A1* = the parasympathetic nervous system; *A2* = the sympathoadrenomedullary axis; *ACh* = acetylcholine. (*B*) Magnocellular neurosecretory neurons control the posterior lobe of the pituitary gland, and (*C*) parvicellular neurosecretory neurons control the anterior lobe. (Reprinted from Martin JB, et al: Neuroendocrine transducers and neurosecretion. In *Neuroendocrinology.* Philadelphia, FA Davis, 1977, p 5.)

    (2) The islets of Langerhans also have a postganglionic sympathetic innervation. When the sympathetic nerves to the beta cells are stimulated or when norepinephrine or epinephrine is infused, the predominant effect is inhibition of insulin secretion.
  b. Pineal gland. This endocrine structure of the diencephalon is classified as a periventricular organ because it borders on the third ventricle.
    (1) The pinealocytes are innervated by the postganglionic (adrenergic) sympathetic fibers, which originate in the superior cervical ganglia of the sympathetic chain.
    (2) When the neurotransmitter norepinephrine is released by the autonomic fibers, it stimulates the synthesis and release of melatonin and other indoleamine hormones.
    (3) The pineal gland, like other periventricular organs (e.g., the median eminence), has a poorly developed blood-brain barrier.
  c. Juxtaglomerular cells. These granular cells of the juxtaglomerular apparatus receive a postganglionic input, which, when stimulated, leads to the release of the proteolytic enzyme renin. Renin can be classified, according to the neuroendocrine transduction concept, as a hormone.
  d. Adrenal medulla. This endocrine structure is composed of chromaffin cells and is innervated by preganglionic (cholinergic) sympathetic fibers, which, when stimulated, cause the release of the adrenomedullary hormones epinephrine and norepinephrine. The release of ACh at the synapses causes the secretion of these catecholamines.

2. Magnocellular neurosecretory regulation of the posterior lobe (see Table 7-4)
  a. Depolarization of the magnocellular neurosecretory cells by ACh released at synapses on the cell bodies of these neurons causes the release of ADH and oxytocin. The axons of these neurons terminate directly on the blood vessels of the posterior lobe.

  b. The neural input to the cell bodies of the magnocellular neurons is cholinergic, and the hormonal output consists of peptide hormones.

  3. Parvicellular neurosecretory regulation of the anterior lobe (see Table 7-4)

  a. The anterior lobe lacks a direct nerve supply, but the pituitary gland does possess an innervation. The neurons in the anterior lobe are exclusively postganglionic sympathetic, which are **vasomotor fibers** and not secretomotor fibers.

   (1) The hypothalamic regulation of the anterior lobe is achieved through the tuberohypophysial neurons of the medial basal hypothalamus. These peptidergic neurons synthesize and secrete specific hypophysiotropic hormones, which enter the hypophysial portal system and stimulate or inhibit the secretion of anterior pituitary hormones.

   (2) The arcuate nucleus (nucleus infundibularis) is the main site of origin of the fine unmyelinated axons of the tuberoinfundibular pathway; however, tuberohypophysial neurons exist throughout the hypophysiotropic area, including the ventromedial nuclei and the periventricular area. The arcuate nucleus serves as the final neural link in the neurovascular connection between the hypothalamus and the anterior lobe.

  b. The cells of the intermediate lobe, unlike those of the anterior lobe, receive a direct bioaminergic secretomotor supply from the hypothalamus and are not perfused directly by the hypophysial portal system.

IV. ENDOCRINE CELLS OF THE GASTROENTEROPANCREATIC (GEP) SYSTEM

  A. Neuropeptides of the GEP system

  1. The endocrine cells of the gut produce peptides that also are found in a variety of other tissues, including the pancreas, pituitary gland, and central and peripheral nerves. The same peptide might function as both a gut hormone and a neurotransmitter. The pancreatic hormones insulin, glucagon, and somatostatin are considered to be part of the GEP system.

  2. The cells of the GEP system share with certain neural tissues the capacity of **a**mine **p**recursor **u**ptake and **d**ecarboxylation and, therefore, are called **APUD cells**. APUD cells are located in the gastrointestinal tract and pancreatic islets.

  a. The endocrine cells of the gastrointestinal tract are found singly or in small groups and are dispersed among the other epithelial cells of the mucosa.

  b. The GEP cells usually are situated at or near the base of the intestinal glands.

  B. APUD cells represent a group of neurons and endocrine cells, which take up amino acids and modify them into amines and peptides. APUD cells synthesize and secrete all of the hormones of the body except the steroids.

  1. Most APUD cells are of ectodermal origin. Those in the gut and pancreas, however, originate from endoderm. Some APUD cells, such as those of the adrenal medulla and neurons of sympathetic ganglia, are derived from the neuroectodermal component called the **neural crest**.

  2. Many APUD endocrine cells contain dopamine, serotonin, and histamine as well as polypeptide hormones, and these substances tend to be released together.

  a. In some cases, enough hormone is secreted to enter the plasma and be transported to other tissues in the body and, therefore, an **endocrine** function is performed.

  b. In other instances, little hormone is bound to distant receptors (due to too few receptors, too little available hormone to bind them, or both) and, therefore, a **paracrine** action results.

V. ENDOGENOUS OPIOID PEPTIDES

  A. Terminology

  1. Opioid denotes opiate-like in terms of a functional similarity with a chemical dissimilarity.

  a. Opioid peptides refer to endogenous or synthetic compounds that have a spectrum of pharmacologic activity similar to that of morphine.

  b. Since the opioid peptides bind with morphine receptors in the central nervous system (CNS), they are called **morphinomimetic peptides**.

2. The two chemical groups of opioid peptides are called **endorphins** and **enkephalins**.
   a. The term endorphin originally was used to designate all opiate-like compounds occurring in the brain. This term now is restricted to the specific endogenous morphine-like substances, α-, β-, and γ-endorphins.
   b. The enkephalins include met-enkephalin, leu-enkephalin, dynorphin, and α-neo-endorphin.

B. Endorphins

1. **Chemistry.** Endorphins are structural derivatives of β-lipotropic hormone (β-LPH). The lipotropin-related peptides (β-LPH and β-endorphin) and ACTH have a common glycoprotein precursor molecule (molecular weight of about 31,000 daltons), which is **proopiomelanocortin (POMC)**.
   a. POMC is activated by proteolytic cleavage to yield 10 individual peptides that have been **classified into four groups:**
      (1) ACTH and corticotropin-like intermediate lobe peptide (CLIP)
      (2) Lipotropic hormones (lipotropins), represented by β- and γ-LPH
      (3) Melanocyte-stimulating hormones (MSHs), represented by α-, β-, and γ-MSH
      (4) Opioid peptides or endorphins, represented by α-, β- and γ-endorphins
   b. POMC consists of **three major chemical moieties**.
      (1) The N-terminal fragment (16 K) is cleaved to form γ-MSH.
      (2) ACTH consists of the fragment containing amino acid residues 1–39.
         (a) ACTH is cleaved to produce α-MSH, which consists of the fragment containing amino acid residues 1–13. In humans, α-MSH normally is not synthesized in significant quantities.
         (b) ACTH also is cleaved into CLIP, which consists of the fragment containing amino acid residues 18–39.
      (3) β-LPH consists of the fragment containing amino acid residues 1–91. β-LPH is the precursor of the endogenous opioids, γ-LPH and β-endorphin.
         (a) γ-LPH consists of the β-LPH fragment containing amino acid residues 1–58.
            (i) In nonhuman species, γ-LPH is converted by proteolytic cleavage to β-MSH, which is the fragment containing amino acid residues 41–58.
            (ii) β-MSH and α-MSH are not formed in humans in significant quantities because the intermediate lobe is vestigial in adult humans. β-MSH does not exist as such in normal human plasma.*
         (b) β-Endorphin consists of the β-LPH fragment containing amino acid residues 61–91.
            (i) β-Endorphin is cleaved into γ-endorphin, which is the fragment containing amino acid residues 61–77.
            (ii) γ-Endorphin forms α-endorphin, which is the fragment containing amino acid residues 61–76.

2. **Distribution.** The endogenous opioids (endorphins) do not cross the blood-brain barrier.
   a. **POMC** is synthesized in the anterior and intermediate lobes of the pituitary gland, in the hypothalamus and other areas of the brain, and in several peripheral tissues including the placenta, gastrointestinal tract, and lung.
      (1) In the human **pituitary gland,** POMC exists principally in the anterior lobe, where it is synthesized by basophils. POMC is the precursor of ACTH and β-LPH, which are secreted together from the basophils.
         (a) ACTH and β-LPH occur within the same pituitary cell and possibly within the same secretory granule.
         (b) The intermediate lobe does not produce ACTH and β-LPH as final secretory products.
      (2) In the **brain,** the concentrations of ACTH and β-LPH are much lower than in the anterior lobe, and all neural cells or fibers that contain ACTH also contain β-LPH. The highest extrapituitary concentrations of ACTH, α-MSH, β-LPH, γ-LPH, and β-endorphin exist in the hypothalamus followed by the limbic system.

---

*β-MSH has been isolated from the human pituitary gland, but the MSH activity in human plasma probably is due largely to the MSH-related peptides, β-LPH and γ-LPH.

b. β-Endorphin

(1) β-Endorphin is the principal opioid peptide in the **pituitary gland,** and it exists in the highest concentration in the intermediate lobe of experimental animals. In the human pituitary, β-endorphin generally is confined to the cells of the anterior lobe. In response to acute stress, the pituitary gland secretes concomitantly ACTH and β-endorphin.

(2) **Other sites**

(a) β-Endorphin has been found in the human pancreas, placenta, semen, and in the male reproductive tract.

(b) Immunoassayable β-endorphin also exists in the plasma and cerebrospinal fluid (CSF).

c. CNS endorphins. The endorphin system exists in the CNS and is characterized by long fiber projection systems from the arcuate region of the hypothalamus to the periventricular region.

(1) Only in the arcuate area do ACTH, β-LPH, and the endorphins occur within **cell bodies** of neurons. Therefore, endorphin cell bodies exist only in the ventral hypothalamus.

(2) Outside the arcuate area, the endorphins are found within the **fibers** of neurons, which project mainly to the mesencephalic periaqueductal gray area. Other areas of projection include the periventricular thalamus, medial amygdala, locus ceruleus, and the zona incerta.

(3) Brain β-endorphin content is unaltered by hypophysectomy, suggesting that the pituitary gland is not the source of CNS endorphins.

3. Physiologic effects

a. Opiate-receptor binding. Endorphins bind to opiate receptors in the brain to cause analgesia, sedation, respiratory depression, and miosis (pupillary contraction).

(1) The intraventricular administration of β-endorphin in humans relieves intractable pain.

(a) It is believed that this analgesic effect involves the pituitary release of endorphins because the effect is abolished with hypophysectomy.

(b) The characteristic effect of opiates in humans is less a specific blunting of pain than it is an induced state of indifference or emotional detachment from the experience of suffering.

(2) The intracerebral administration of endorphins produces a profound sedation and immobilization (catatonia).

(3) Endorphins play a central role in controlling affective states and may be involved in controlling the drive for food, water, and sex, all of which are associated with the limbic nervous system.

b. Stimulation of the hypothalamus. The intracisternal or intracerebral injection of β-endorphin elicits a hypophysiotropic effect via the stimulation of the hypothalamus.

(1) This hypothalamic effect on pituitary function causes the secretion of GH, prolactin, ACTH, and ADH.

(2) In addition, there is a diminished pituitary secretion of TSH and the gonadotropic hormones, LH and FSH.

(3) β-Endorphin in the gastrointestinal tract stimulates the secretion of glucagon and insulin but inhibits the secretion of somatostatin.

c. β-LPH is a more potent stimulator of **aldosterone synthesis** than is angiotensin II, and β-LPH is thought to be the stimulator of the zona glomerulosa in primary aldosteronism.

d. β-LPH, γ-LPH, and MSH have a lipolytic effect resulting in **fat mobilization;** however, these compounds do not have significant regulatory effects on fat metabolism in humans.

(1) β-MSH has been isolated from the human pituitary gland.

(2) MSH activity in human plasma is due largely to the MSH-related peptides, β-LPH and γ-LPH.

e. β-LPH, γ-LPH, and ACTH have a **weak MSH activity,** which explains the hyperpigmentation associated with increased ACTH secretion.

C. Enkephalins

1. Chemistry

a. Enkephalins, represented by met-enkephalin and leu-enkephalin, are pentapeptide opioids, which are not breakdown products of POMC. The precursor substance for met- and leu-enkephalin is **proenkephalin A.**

    b. Proenkephalin B contains the sequences of α-neo-endorphin, dynorphins A and B, and leu-enkephalin. Although α-neo-endorphin and dynorphin contain the leu-enkephalin sequence, neither is considered a precursor for leu-enkephalin.

2. **Distribution**
   a. **The pituitary gland** is virtually devoid of enkephalins.
      (1) Of the pituitary enkephalins, including dynorphin (amino acid residues 1–13), the greatest concentrations exist in the pars nervosa.
      (2) Brain enkephalins are not depleted following hypophysectomy, suggesting a neural origin for these peptides.
   b. **CNS.** Enkephalins usually are localized in neuronal processes and terminals. The enkephalin-containing neurons have the widest distribution throughout the CNS and are characterized by short axon projections. Dynorphin also is distributed widely in the CNS, with the highest concentrations in the hypothalamus, medulla-pons, midbrain, and spinal cord.
      (1) The highest concentrations of met-enkephalin are in the **basal ganglia** (globus pallidus) and substantia nigra.*
      (2) In the **spinal cord,** enkephalins exist in highest concentration in the dorsal gray matter, which corresponds to the synapses of primary sensory nerve endings. Enkephalins also are found in other areas of the spinal cord, notably the substantia gelatinosa, which is known to be involved in the transmission of pain impulses.
      (3) Many **limbic structures** have relatively high enkephalin levels, including the lateral septal nucleus, the interstitial nucleus of the striae terminalis, the prefornical area, and the central amygdala.
   c. **Other sites**
      (1) **Enkephalin-like material** has been demonstrated in the CSF and has been isolated from the adrenal medulla, where it exists in high concentrations both in axon terminals of the splanchnic nerve and in adrenomedullary chromaffin cells.
         (a) Met-enkephalin in the circulation originates from the adrenal medulla.
         (b) High plasma concentrations of met-enkephalin are found in patients with pheochromocytoma.
      (2) **Enkephalin immunoreactivity** has been reported in the human gastrointestinal tract (myenteric plexus and mucosa), gallbladder, and pancreas.
      (3) **Ectopic production of enkephalins** has been reported in carcinoid tumors of the lung and thymus.

3. **Physiologic effects.** Enkephalins are most appropriately classified as neurotransmitters or neuromodulators. A **neuromodulator** is a substance that modifies (positively or negatively) the action of a neurotransmitter at a presynaptic site (by modulating the release of a neurotransmitter) or at a postsynaptic site (by modulating the neurotransmitter action).
   a. **Spinal cord.** The enkephalins in the dorsal gray matter of the spinal cord function to suppress substance P–containing nerve endings and provide an analgesic (antinociceptive) role. Electric stimulation of the periaqueductal gray area of the spinal cord of humans, which causes analgesia, is associated with an increase in the concentration of enkephalins in the CSF.
      (1) Opioids are believed to modulate pain impulses by acting to reduce sensitivity to pain peripherally as well as centrally by activating opiate receptors in the periaqueductal gray area.
      (2) Acupuncture is associated with the release of enkephalins into the CSF.
   b. **Medulla.** The vagal nuclear localization of enkephalins corresponds to the emetic actions and antitussive properties of morphine.
   c. **Brain stem.** Morphine and enkephalins inhibit the firing of the noradrenergic neurons of the locus ceruleus, which is the origin of the ascending noradrenergic fibers.
      (1) The localization of enkephalins in the locus ceruleus may account for the euphoria-producing actions of morphine.
      (2) The amygdala is considered the prime site for morphine-generated euphoria.
      (3) The enkephalin tracts and opiate receptors in the limbic system may explain the euphoric effects of opiates.

---

*The substantia nigra is classified by some neuroanatomists as a component of the basal ganglia.

(4) Respiratory depression, which accounts for the lethal effects of opiates, may involve receptors in the nucleus solitarius of the brain stem, which regulates visceral reflexes including respiration.

VI. PITUITARY GLAND

A. Embryology. The pituitary gland is in close anatomic relation to the hypothalamus. This relationship has both embryologic and functional significance.

1. The anterior lobe of the pituitary gland, the **adenohypophysis,** is derived from the primitive gut by an upward extension (Rathke's pouch) of the epithelium of the primitive mouth cavity (stomodeum). The adenohypophysis is a derivative of buccal ectoderm.

2. The neural or posterior lobe, the **neurohypophysis,** develops as a downward evagination of the neural tube at the base of the hypothalamus (infundibulum) and, therefore, represents a true extension of the brain. Neuroregulation of this structure is achieved by direct neural connections. The neurohypophysis is a derivative of neural ectoderm.

B. Morphology

1. **Gross anatomy** (see Figure 7-3)
   a. General structure
   (1) The pituitary gland lies in a bony walled cavity, the **sella turcica,** in the sphenoid bone at the base of the skull.
   (2) The **dura mater** completely lines the sella turcica and nearly surrounds the gland.
   (3) The **pituitary (hypophysial) stalk** and its blood vessels reach the main body of the gland through the diaphragma sellae. The pituitary stalk consists of the infundibular stem and the adenohypophysial tissue that is continuous with the infundibular stem.
   b. Adenohypophysial structure. The anterior lobe has three components.
   (1) The **pars distalis** represents the bulk of the anterior lobe in humans and receives most of its blood supply from the superior (anterior) hypophysial artery, which gives rise to the hypophysial portal system.
   (2) The **pars intermedia** lies between the pars distalis and the neural lobe and is a vestigial structure in humans. It is relatively avascular and is considered almost nonexistent in humans.
   (3) The **pars tuberalis** is an elongated collection of secretory cells, which superficially envelops the infundibular stem and extends upward as far as the basal hypothalamus. It is the most vascular portion of the anterior lobe.
   c. Neurohypophysial structure
   (1) Components
   (a) The **median eminence,** located beneath the third ventricle, is a small, highly vascular protrusion of the dome-shaped base of the hypothalamus, which is designated grossly as the tuber cinereum. The floor of the third ventricle is designated as the infundibulum because it resembles a funnel.
   (b) The **infundibular stem** (neural stalk) of the posterior lobe arises in the median eminence.
   (c) The **pars nervosa** retains its neural connection with the ventral diencephalon.
   (2) Dominant features of the posterior lobe are the **neurosecretory neurons** that form the **magnocellular neurosecretory system**. These unmyelinated nerve tracts arise from the supraoptic and paraventricular nuclei within the ventral diencephalon and descend through the infundibulum and neural stalk to terminate in the posterior lobe. The posterior lobe is a storage site for hormones and, therefore, is not correctly termed an endocrine gland.

2. **Histology**
   a. Cells. Two major cell types are found in equal numbers in the anterior lobe.
   (1) Chromophils (granular secretory cells) exist in two forms.
   (a) Acidophils (eosinophils) account for about 80% of the chromophils and are the cellular source of prolactin and GH.
   (b) Basophils comprise about 20% of the chromophils and are the source of TSH, ACTH, LH, FSH, and β-LPH.

(2) **Chromophobes** (agranular cells) are not precursors of the chromophils and are now known to have an active secretory function. Most of these cells likely are degranulated secretory cells

b. **Neurons** in the anterior lobe are almost exclusively postganglionic sympathetic fibers that innervate blood vessels.

c. **Nerve fibers** of the neurohypophysial system terminate mostly in the pars nervosa. Interspersed between these neurosecretory fibers are numerous glial cells called **pituicytes,** whose function, other than structural support, remains unknown.

3. **Vascular supply** (see also III B 1 and Figure 7-3). A basic tenet of the neurovascular hypothesis is that the concentration of the hypophysiotropic hormones is greater in hypophysial portal blood than at any other site in the vasculature.

a. In humans, the capillaries at the base of the hypothalamus are formed directly from branches of the superior hypophysial arteries, which arise from the internal carotid arteries. There are few vascular anastomoses between the hypothalamic artery and the superior hypophysial artery. The crucial regulatory connection between the hypothalamus and the anterior lobe is via the hypophysial portal vessels.

b. The intermediate lobe is not perfused directly by the hypophysial portal system but is regulated by bioaminergic secretomotor fibers originating in the hypothalamus.

c. The blood supply to the posterior lobe is largely separate from that of the anterior lobe. The blood supply to the median eminence is greater than that to the entire pituitary gland.

C. **Hormones of the posterior lobe: ADH and oxytocin** (see Table 7-4). The physiologic aspects of ADH are described in section VIII of Chapter 4. The physiologic aspects of oxytocin are described below.

1. **Synthesis and storage**

a. Like ADH, oxytocin is a nonapeptide* that is synthesized within the cell bodies of the peptidergic neurons of the magnocellular neurosecretory system.

b. This polypeptide is synthesized mainly in the paraventricular nuclei of the hypothalamus and, like ADH, is stored in the posterior lobe of the pituitary gland.

2. **Stimuli for release.** Oxytocin secretion is brought about by stimulation of cholinergic nerve fibers.

a. Stimulation of the tactile receptors in the areolar region of the female breast during suckling activates somesthetic neural pathways, which transmit this signal to the hypothalamus. This leads to the reflex secretion of oxytocin into the bloodstream and to milk release following a latent period of 30–60 seconds. This reflex is called the **milk let-down** or **milk ejection reflex.**

(1) Oxytocin causes milk release in lactating women by contraction of the myoepithelial cells, which cover the stromal surface of the epithelium of the alveoli, ducts, and cisternae of the mammary gland.

(2) Oxytocin secretion can be conditioned so that the physical stimulation of the nipple no longer is required. Thus, lactating women can experience milk release in response to the sight and sound of a baby.

(3) While oxytocin aids in the process, its presence is not absolutely required for successful nursing in humans.

b. Oxytocin secretion also can occur in response to genital tract stimulation, such as that which occurs during coitus and parturition.

c. Oxytocin is produced in men and also is released during genital tract stimulation. The role of this neurohypophysial hormone in men is unknown.

3. **Inhibition of release**

a. Milk let-down can be inhibited by emotional stress and psychic factors such as fright.

b. Excitation of adrenergic fibers to the hypothalamus inhibits peptide release. Activation of the sympathetic neurons with the concomitant release of norepinephrine and epinephrine inhibits oxytocin secretion.

c. Ethanol inhibits endogenous oxytocin release, resulting in reduced myometrial contractility.

d. Enkephalins also inhibit oxytocin release.

---

*If the two cysteine residues are counted together as a single cystine residue, ADH and oxytocin are correctly classified as octapeptides.

4. Physiologic effects

    a. Oxytocin stimulates contraction of the smooth muscle (myoepithelium) of the lactating mammary gland.

    b. It also stimulates contraction of the smooth muscle of the uterus (myometrium).

        (1) The sensitivity of the myometrium to exogenous oxytocin during pregnancy increases as pregnancy advances.

        (2) Oxytocin plays a role in labor and has been shown to be a useful therapeutic agent in the induction of labor.

D. Hormones of the anterior lobe. The principal hormones of the anterior lobe of the pituitary gland can be classified conceptually into two groups: hormones that stimulate other endocrine glands to secrete hormones and hormones that have a direct effect on nonendocrine target tissues. Only the latter group is described here, using GH and prolactin as examples.

    1. GH also is known as human growth hormone (HGH), somatotropic hormone (STH), somatotropin, and somatocrinin.

        a. Synthesis, chemistry, and general characteristics

          (1) GH is synthesized by the acidophils of the anterior lobe and is stored in very large amounts in the human pituitary gland. GH represents approximately 4%–10% of the wet weight of the pituitary gland, which is equivalent to 5–15 mg.

          (2) GH is a single, unbranched polypeptide chain containing 191 amino acid residues. It has been synthesized in bacteria using recombinant DNA techniques.

          (3) Humans exhibit a **species specificity** for GH, and only human and monkey GH preparations have biologic activity in humans.

          (4) Like all other pituitary hormones, GH is secreted episodically in periods of 20–30 minutes. The large diurnal fluctuations represent integrations of many small secretory episodes. A regular nocturnal peak in GH secretion occurs 1–2 hours after the onset of deep sleep, which correlates with stage 3 or stage 4 slow-wave sleep.

          (5) The plasma GH concentration in the growing child is not significantly higher than that in the adult whose growth has ceased.

        b. Control of secretion. The release of GH is primarily under the control of two hypophysiotropic hormones.

          (1) Stimuli for release. Somatotropin releasing factor (SRF) is the putative releasing hormone for GH. SRF has not been identified or chemically synthesized. However, several pharmacologic, physiologic, and psychic agents are known to stimulate GH release.

            (a) Release of GH is mediated by monoaminergic and serotoninergic pathways; thus, α-adrenergic, dopaminergic, and serotoninergic agonists as well as β-adrenergic antagonists all stimulate GH release in humans.

            (b) Bromocriptine (a dopamine agonist), enkephalins, endorphins (β-endorphin), and opiates stimulate GH secretion.

            (c) Insulin-induced hypoglycemia is a potent stimulus of GH secretion as are pharmacologic doses of glucagon and vasopressin.

            (d) Physiologic stimuli include hypoglycemia, increased plasma concentrations of amino acids (arginine, leucine, lysine, tryptophan, and 5-hydroxytryptophan), and decreased free fatty acid concentrations. In addition, estrogens stimulate GH synthesis and secretion.

            (e) GH secretion is stimulated by moderate to vigorous exercise, emotional stress, and stress due to fever, surgery, anesthesia, trauma, pyrogen administration, and repeated venipuncture. Fasting leads to elevated GH secretion after 2 or 3 days.

          (2) Inhibitors of secretion

            (a) GH can inhibit its own secretion via a short-feedback loop mechanism that operates between the anterior lobe and the median eminence. Somatotropin-inhibiting hormone (SIH; somatostatin) inhibits the synthesis and release of GH.

              (i) Somatostatin is a tetradecapeptide (14 amino acid residues) that has been chemically synthesized.

              (ii) It is a product of the parvicellular neurosecretory neurons that terminate in the median eminence and produce hypophysiotropic hormones.

              (iii) Somatostatin also is found in other parts of the brain, in the gastrointestinal tract, and in the delta cells of the pancreatic islets.

              (iv) In addition to inhibiting GH secretion, somatostatin blocks the secretion of insulin, glucagon, and gastrin and inhibits the intestinal absorption of glucose. These effects of somatostatin produce a state of hypoglycemia.

(b) The secretion of GH in response to the aforementioned stimuli often is blunted in obese individuals.

(c) Glucocorticoids decrease GH secretion, but their predominant effect is the interference with the metabolic actions of GH.

(d) A decline in GH secretion is observed in late pregnancy, despite the presence of high estrogen levels.

  (i) Impaired glucose tolerance is common, and clinical diabetes occurs frequently, despite the above normal insulin secretion in response to a glucose load during pregnancy.

  (ii) Pregnancy regularly antagonizes the action of insulin and increases the pancreatic secretory capacity of both normal and diabetic individuals.

  (iii) The development of gestational diabetes probably is due to a greater degree of insulin antagonism caused by normal plasma concentrations of human placental lactogen (HPL).

c. Physiologic and metabolic effects

  (1) Stimulation of growth of bone, cartilage, and connective tissue

   (a) The effects of GH on skeletal growth are mediated by a family of polypeptides called **somatomedins** [also termed insulin-like growth factors (IGF)], which are synthesized mainly in the liver. (Thyroid hormone and insulin also are necessary for normal osteogenesis.)

   (b) Somatomedin may be produced in nonhepatic tissue as well, in that somatomedin activity has been found in the serum, kidney, and muscle tissue.

    (i) Receptors for somatomedin exist in chondrocytes, hepatocytes, adipocytes, and muscle cells.

    (ii) Somatomedin has insulin-like effects on tissues, including lipogenesis in adipose tissue, increased glucose oxidation in fat, and increased glucose and amino acid transport by muscle.

   (c) GH, through somatomedin, stimulates proliferation of chondrocytes and the appearance of osteoblasts. The increase in the thickness of the epiphysial (cartilaginous) end-plate accounts for the increase in linear skeletal growth.

   (d) After epiphysial fusion, bone length can no longer be increased by GH, but bone thickening can occur through periosteal growth. It is this growth that accounts for the changes seen in hypersomatotropism (acromegaly).

   (e) These reactions are the biochemical correlates of protein synthesis in general body growth and also account for the hyperplasia and hypertrophy associated with increased tissue mass.

  (2) Protein metabolism

   (a) GH has predominantly anabolic effects on skeletal and cardiac muscle, where it stimulates the synthesis of protein, RNA, and DNA.

   (b) GH reduces circulating levels of amino acids and urea (i.e., it promotes nitrogen retention), which accounts for the term **positive nitrogen balance**. Urinary urea concentration also is decreased.

   (c) GH promotes amino acid transport and incorporation into proteins.

  (3) Fat metabolism

   (a) GH has an overall catabolic effect on adipose tissue. It stimulates the mobilization of fatty acids from adipose tissue, leading to a decreased triglyceride content of fatty tissue and increased plasma levels of free fatty acids and glycerol.

   (b) GH increases hepatic oxidation of fatty acids to the ketone bodies, acetoacetate and β-hydroxybutyrate.

  (4) Carbohydrate metabolism. GH is a diabetogenic hormone. Because of its anti-insulin effect, GH has the tendency to cause hyperglycemia.

   (a) GH can produce an insulin-resistant diabetes mellitus primarily because of its lipolytic effect.

    (i) Free fatty acids can antagonize the effect of insulin to promote glucose uptake by skeletal muscle and adipose tissue.

    (ii) Free fatty acids can stimulate gluconeogenesis.

    (iii) Excess acetyl coenzyme A (acetyl-CoA) production favors gluconeogenesis, because pyruvate carboxylase requires acetyl-CoA to form oxaloacetate from pyruvate. Oxaloacetate is the rate-limiting factor in gluconeogenesis.

    (iv) Acetyl-CoA also inhibits glycolysis by inhibition of pyruvate kinase.

      (v) Free fatty acids stimulate hepatic glucose synthesis mainly via the stimulation of fructose diphosphatase. At the same time, pyruvate kinase and phosphofructokinase both are inhibited by free fatty acids, which block glycolysis and favor gluconeogenesis. Citrate also blocks glycolysis at the phosphofructokinase step.

    (b) GH induces an elevation in basal plasma insulin levels.

    (c) Because of its anti-insulin effect, GH inhibits glucose transport in adipose tissue. Since adipose tissue requires glucose for triglyceride synthesis, GH antagonizes insulin-stimulated lipogenesis.

  (5) Mineral metabolism. GH promotes renal reabsorption of Ca^{2+}, phosphate, and Na^{+}.

d. Endocrinopathies

  (1) Disorders associated with increased GH

    (a) Growth retardation can occur when GH levels are increased and somatomedin levels are depressed (e.g., in kwashiokor). (In the African pygmy, who is resistant to the action of GH, both GH and somatomedin levels are normal. This condition is due to a decrease in GH receptors.)

    (b) Overproduction of GH during adolescence results in **giantism,** which is characterized by excessive growth of the long bones. Patients may grow to heights of as much as 8 feet.

    (c) Excessive GH secretion during adulthood, after the epiphysial plates of long bones have fused, causes growth in those areas where cartilage persists. This leads to **acromegaly,** a condition characterized by coarse facial features, underbite, prominent brow, enlarged hands and feet, and soft tissue hypertrophy (e.g., cardiomegaly, hepatosplenomegaly).

  (2) Disorders associated with decreased GH

    (a) Decreased GH secretion in immature persons leads to stunted growth, or **dwarfism,** which is accompanied by sexual immaturity, hypothyroidism, and adrenal insufficiency.

    (b) GH deficiency may be part of an overall lack of anterior pituitary hormones (**panhypopituitarism**) or due to an isolated genetic deficiency. Selective GH deficiency is rare in adults; clinical manifestations may include impaired hair growth and a tendency toward fasting hypoglycemia.

  (3) Treatment

    (a) The treatment of choice for hypersomatotropism is selective surgical extirpation of the pituitary adenoma without damage to other pituitary functions. Bromocriptine is effective in suppressing, but not normalizing, GH levels in most acromegalic patients. This substance tends to stimulate GH secretion in normal individuals.

    (b) Disorders associated with GH deficiency can be treated with HGH.

2. Prolactin also is known as lactogenic hormone, mammotropic hormone, and galactopoietic hormone.

  a. Synthesis, chemistry, and general characteristics

    (1) Prolactin is synthesized in the pituitary acidophils.

    (2) Human prolactin is a single peptide chain containing 198 amino acid residues.

    (3) It does not regulate the function of a secondary endocrine gland in humans.

  b. Control of secretion. Two hypothalamic neurosecretory substances have been implicated in prolactin secretion.

    (1) Stimuli for release. Prolactin releasing factor (PRF) is the putative releasing hormone for prolactin. PRF has not been identified or chemically synthesized; however, one of these releasing factors is TRH, which causes the release of TSH.

      (a) Prolactin secretion increases about 1 hour after the onset of sleep, and this increase continues throughout the sleep period. The nocturnal peak occurs later than that for GH.

      (b) Prolactin secretion is enhanced by exercise and by stresses such as surgery under general anesthesia, myocardial infarction, and repeated venipuncture.

      (c) Plasma prolactin levels begin to increase by the eighth week of pregnancy and usually reach peak concentrations by the thirty-eighth week.

      (d) Nursing and breast stimulation are known to stimulate prolactin release. Oxytocin does not stimulate prolactin secretion.

      (e) Serum prolactin levels are elevated in those patients with primary hypothyroidism who are believed to have high TRH levels in the hypophysial portal circulation.

      (f) Dopamine antagonists (phenothiazine and tranquilizers), adrenergic blockers, and serotonin agonists stimulate prolactin secretion.

      (g) Pituitary stalk section and lesions that interfere with the portal circulation to the pituitary gland also cause prolactin release.

    (2) **Inhibitors of release.** Normally, the control of prolactin secretion is a constant tonic inhibition via prolactin inhibiting factor (PIF).

      (a) Dopamine is secreted into the hypophysial portal vessels and may be the PIF. It is physiologically the most important PIF.

      (b) Serotonin antagonists and dopamine agonists (bromocriptine) block the secretion of prolactin. Bromocriptine administered during the postpartum period reduces prolactin secretion to nonlactating levels and terminates lactation.

c. Physiologic effects

    (1) Prolactin does not have an important role in maintaining the secretory function of the corpus luteum and, therefore, is not a gonadotropic hormone in women.

    (2) Prolactin plays an important role in the development of the mammary gland and in milk synthesis.

      (a) During pregnancy, the mammary duct gives rise to lobules of alveoli, which are the secretory structures of this tissue. This differentiation requires prolactin, estrogens, and progestogens. Once the lobuloalveolar system is developed, the role of prolactin and corticosteroids in milk production, although essential, becomes minimal. GH and thyroid hormone enhance milk secretion.

      (b) Immediately following pregnancy, prolactin stimulates galactosyltransferase activity, leading to the synthesis of lactose.

      (c) In women, high serum levels of prolactin are associated with suppressed LH secretion and anovulation, which account for an absence of menses (amenorrhea) during postpartum lactation.

        (i) With continued nursing, FSH levels rise, but LH levels remain low.

        (ii) In the early postpartum period, both the FSH and LH levels are low.

d. Endocrinopathies

    (1) **Hyperprolactinemia** is not a rare condition but frequently is undiagnosed because galactorrhea occurs in only about 30% of cases.

      (a) In women, elevated serum prolactin manifests as infertility and amenorrhea.

      (b) In men, hyperprolactinemia is a cause of impotence and decreased libido.

    (2) **Treatment** of prolactin hypersecretion includes administration of bromocriptine, which lowers prolactin levels and usually restores normal gonadal function.

VII. ADRENAL MEDULLA

A. Embryology

1. The neural crest gives rise to neuroblasts, which eventually give rise to the autonomic postganglionic neurons, the adrenal medulla, and the spinal ganglia.

2. The adrenal medulla consists of chromaffin cells (pheochromocytes), which are neuroectodermal derivatives and the functional analogs of the sympathetic postganglionic fibers of the autonomic nervous system.

3. In early fetal life, the adrenal medulla contains only norepinephrine.

B. Morphology

1. Gross anatomy

a. The adrenal medulla represents essentially an enlarged and specialized sympathetic ganglion and is called a **neuroendocrine transducer** because a neural signal to this organ evokes hormonal secretion.

b. The adrenal medulla is the only autonomic neuroeffector organ without a two-neuron motor innervation. It is innervated by long sympathetic preganglionic, cholinergic neurons that form synaptic connections with the chromaffin cells.

c. Small clumps of chromaffin cells also can be found outside the adrenal medulla, along the aorta and the chain of sympathetic ganglia.

2. Histology
  a. Cells. There are two types of adrenomedullary chromaffin cells. Individual cells contain either **norepinephrine** or **epinephrine,** which is stored largely in subcellular particles called chromaffin granules. These granules are osmiophilic, electron-dense, membrane-bound secretory vesicles.
  (1) Approximately 80% of the chromaffin granules in the human adrenal medulla synthesize epinephrine (adrenalin). The remaining 20% synthesize norepinephrine (noradrenalin).
  (2) The chromaffin granules contain catecholamines, protein, lipids, and adenine nucleotides (mainly ATP).
      (a) One of the proteins localized in the particulate fraction is the enzyme, dopamine-β-hydroxylase.
      (b) Soluble acidic proteins found in the granules are called **chromogranins**.
  b. Neurons
  (1) The preganglionic sympathetic fibers that innervate the adrenal medulla traverse the splanchnic nerve, which contains myelinated (type B) secretomotor fibers emanating mainly from lower thoracic segments (T5 to T9) of the ipsilateral intermediolateral gray column of the spinal cord.
  (2) The cell bodies of the chromaffin cells do not have axons.

3. Vascular supply (Figure 7-6)
  a. Arterial blood to the adrenal gland reaches the outer capsule from branches of the renal and phrenic arteries, with a less important arterial input directly from the aorta. The adrenal medulla is perfused by blood vessels in two ways.
  (1) A type of **portal circulation** exists in the adrenal gland where the cortex and medulla are in juxtaposition. From the capillary plexus on the outer adrenal capsule most of the blood enters venous sinuses, which drain into and supply the medullary tissue. Thus, most of the blood perfusing the adrenal medulla is derived from the portal system and is partly deoxygenated.
  (2) There also exists a direct arterial blood supply to the medulla via the **medullary arteries,** which traverse the cortex.
  b. Venous blood drains via a single central vein, composed almost entirely of bundles of longitudinal smooth muscle fibers, which passes along the longitudinal axis of the gland.

C. Adrenomedullary hormones: monoamines

  1. The adrenal medulla synthesizes and secretes biogenic amines. These dihydroxylated phenolic amines, or **catecholamines,** are epinephrine and norepinephrine. Most of the met-enkephalin in the circulation also originates in the adrenal medulla. Enkephalins are pentapeptides that function as neurotransmitters or neuromodulators, which normally are localized in neuronal processes and terminals (see V C).
  a. Epinephrine is produced almost exclusively in the adrenal medulla, with smaller amounts synthesized in the brain. Essentially all circulating epinephrine is derived from the adrenal medulla.
  b. Norepinephrine is widely distributed in neural tissues, including the adrenal medulla, sympathetic postganglionic fibers, and CNS. In the brain, the concentration of norepinephrine

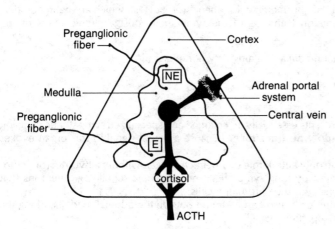

Figure 7-6. The adrenal portal vascular system constitutes a functional connection between the cortex and medulla and has a high cortisol concentration. *NE* = norepinephrine; *E* =epinephrine; *ACTH* = adrenocorticotropic hormone. (Adapted from Pohorecky LA, Wurtman RJ: Adrenocortical control of epinephrine synthesis. *Pharmacol Rev* 23(1):1–35, 1971.)

is the highest in the hypothalamus. The norepinephrine content of a tissue reflects the extent or density of its sympathetic innervation. Norepinephrine has been demonstrated in almost all tissues except the placenta, which is devoid of nerve fibers.

2. Bilaterally adrenalectomized human patients excrete practically no epinephrine in the urine. Urinary levels of norepinephrine remain within normal limits, however, indicating that the norepinephrine comes from extra-adrenal sources (i.e., the terminals of the postganglionic sympathetic fibers and the brain).

D. Control of catecholamine synthesis

1. The biosynthetic pathway originates with **L-tyrosine,** which is derived from the diet or from the hepatic hydroxylation of L-phenylalanine. Tyrosine is hydroxylated in the cytoplasm by tyrosine hydroxylase to L-dopa (3,4-dihydroxyphenylalanine).

2. Dopa is converted in the cytosol to **dopamine** (3,4-dihydroxyphenylethylamine) by a nonspecific aromatic L-amino acid decarboxylase.

3. Dopamine enters the chromaffin granule, where it is converted to **L-norepinephrine** by dopamine-β-hydroxylase, which exists exclusively in the granule.
   a. Norepinephrine is the end product in approximately 20% of chromaffin cells.
   b. In about 80% of chromaffin cells, norepinephrine diffuses back into the chromaffin cytoplasm. There, it is N-methylated by phenylethanolamine-N-methyltransferase (PNMT) using S-adenosylmethionine as a methyl donor to form L-epinephrine.
      (1) PNMT is selectively localized in the adrenal medulla, the only site where it exists in significant concentrations.
      (2) PNMT activity is induced by very high local concentrations of glucocorticoids, which are found only in the adrenal portal blood draining the adrenal cortex.

E. Control of catecholamine secretion

1. **General considerations**
   a. **ACh** provides the major physiologic stimulus for the secretion of the adrenomedullary hormones. In addition, angiotensin II, histamine, and bradykinin stimulate catecholamine secretion.
      (1) Catecholamine release is stimulated by ACh from the preganglionic sympathetic nerve endings innervating the chromaffin cells.
      (2) The final common effector pathway activating the adrenal medulla is the cholinergic preganglionic fibers in the greater splanchnic nerve.
      (3) ACh causes the depolarization of the chromaffin cells followed by the release of catecholamines by **exocytosis.** Ca^{2+} influx secondary to membrane depolarization is the central event in **stimulus-secretion coupling.**
   b. **Cortisol and ACTH.** Because catecholamine synthesis is dependent on cortisol, the functional integrity of the adrenal medulla indirectly depends on a functional pituitary gland for ACTH secretion and a functional median eminence for corticotropin releasing hormone (CRH) secretion. CRH is a hypophysiotropic hormone produced by the parvicellular nuclei of the ventral diencephalon.

2. **Physiologic and psychological stimuli for release.** The adrenal medulla constitutes the neuroeffector of the sympathoadrenomedullary axis that is activated during states of emergency. This response to stress is called the **fight-or-flight reaction.** Among the conditions in which the sympathetic nervous system is activated are fear, anxiety, pain, trauma, hemorrhage and fluid loss, asphyxia and hypoxia, changes in blood pH, extreme cold or heat, severe exercise, hypoglycemia, and hypotension. During hypoglycemia, the adrenal medulla is activated selectively. In humans, epinephrine and norepinephrine appear to be released independently by specific stimuli.
   a. Anger and active aggressive states or situations are associated with increased norepinephrine secretion.
   b. States of anxiety are associated with increased epinephrine secretion. In addition, epinephrine release is increased by tense but passive emotional displays or threatening situations of an unpredictable nature.
   c. Angiotensin II potentiates the release of catecholamines
   d. Plasma concentrations of epinephrine vary according to physiologic or pathologic state as follows:

(1) Basal level: 25–50 pg/ml (6×10^{-10} mol/L)
(2) Hypoglycemia: 230 pg/ml
(3) Diabetic ketoacidosis: 500 pg/ml
(4) Severe hypoglycemia: 1500 pg/ml

3. **Regulation of adrenergic receptors.** A reciprocal relationship exists between catecholamine concentration and the number and function of adrenergic receptors.
   a. A sustained decrease in catecholamine secretion is associated with an increased number of adrenergic receptors in target cells and an increased responsiveness to catecholamines. Conversely, a chronic increase in catecholamine secretion is associated with a decreased number of adrenergic receptors in target cells and a decreased responsiveness to catecholamines.
   b. This relationship may account in part for the phenomenon of **denervation hypersensitivity,** which is observed in sympathetic neuroeffectors following autonomic fiber denervation.

F. **Metabolism and inactivation of circulating catecholamines**

1. **General considerations.** The plasma half-life of epinephrine and norepinephrine is 10 seconds and 15 seconds, respectively. The biologic effects of circulating catecholamines are terminated rapidly by both nonenzymatic and enzymatic mechanisms.
   a. **Neuronal uptake.** Sympathetic nerve endings have the capacity to take up amines actively from the circulation. This uptake of circulating catecholamines leads to nonenzymatic inactivation by intraneuronal storage and to enzymatic inactivation by a mitochondrial enzyme called **monoamine oxidase (MAO).**
   b. **Extraneuronal uptake.** The formation of catecholamine metabolites locally in innervated tissues and systemically in the liver, kidney, lung, and gut implies catecholamine uptake by a variety of cells.
   c. **Inactivation.** Circulating epinephrine and norepinephrine are metabolized predominantly in the liver and kidney.

2. **Metabolic pathways for catecholamine inactivation**
   a. **MAO** is found in very high concentrations in the mitochondria of the liver, kidney, stomach, and intestine. MAO catalyzes the oxidative deamination of a number of biogenic amines, including the intraneuronal and circulating catecholamines.
      (1) The combined actions of MAO and **aldehyde oxidase** on epinephrine and norepinephrine produce 3,4-dihydroxymandelic acid by oxidative deamination.
      (2) The combined actions of MAO and aldehyde oxidase on the meta-O-methylated metabolites of epinephrine and norepinephrine (metanephrine and normetanephrine, respectively) produce 3-methoxy-4-hydroxymandelic acid (vanillylmandelic acid or VMA) by oxidative deamination.
   b. **Catechol-O-methyltransferase (COMT)** is found in the soluble fraction of tissue homogenates with the highest levels in liver and kidney. COMT is considered mainly as an extraneuronal enzyme, but it is also found in postsynaptic membranes. COMT metabolizes circulating catecholamines in the kidney and liver and metabolizes locally released norepinephrine in the effector tissue.
      (1) COMT requires S-adenosylmethionine as a methyl donor.
      (2) The action of COMT produces normetanephrine from norepinephrine, metanephrine from epinephrine, and VMA from 3,4-dihydroxymandelic acid by 3-O-methylation.

3. **Significance of catecholamine metabolites**
   a. Only 2%–3% of the catecholamines are excreted directly into the urine, mostly in conjugation with sulfuric or glucuronic acid. Most of the catecholamines produced daily are excreted as the deaminated metabolites, VMA and 3-methoxy-4-hydroxyphenylglycol (MOPG). Only a small fraction is excreted unchanged or as metanephrines.
   b. Under normal circumstances, epinephrine accounts for a very small proportion of urinary VMA and MOPG. Because the majority is derived from norepinephrine, urinary VMA and MOPG reflect the activity of the nerve terminals of the sympathetic nervous system rather than that of the adrenal medulla.
   c. The excretion of unchanged epinephrine or plasma epinephrine provides a better index of the physiologic activity of the sympathoadrenomedullary system than does the excretion of catecholamine metabolites, since the latter reflects, to a considerable extent, norepinephrine that is metabolized within nerve endings and the brain and never released at adrenergic synapses in the active form.

G. Physiologic actions of catecholamines (Table 7-5). The effects of adrenomedullary stimulation and sympathetic nerve stimulation generally are similar. In some tissues, however, epinephrine and norepinephrine produce different effects due to the existence of two types of adrenergic receptors, **alpha** (α) and **beta** (β) receptors, which have different sensitivities for the various catecholamines and, therefore, produce different responses. **Epinephrine** is the single most active endogenous amine on both α and β receptors.

   1. The α-**adrenergic receptors** are sensitive to both epinephrine and norepinephrine. These receptors are associated with most of the excitatory functions of the body and with at least one inhibitory function (i.e., inhibition of intestinal motility).

   2. The β-**adrenergic receptors** respond to epinephrine and, in general, are relatively insensitive to norepinephrine. These receptors are associated with most of the inhibitory functions of the body and with one important excitatory function (i.e., excitation of the myocardium).

H. Biochemical effects of catecholamines. Norepinephrine has little direct effect on carbohydrate metabolism; however, both norepinephrine and epinephrine can inhibit glucose-induced secretion of insulin from the beta cells of the pancreatic islets of Langerhans.

Table 7-5. Some Physiologic Effects of Catecholamines and Types of Adrenergic Receptors

Effector Organ	Receptor Type	Response
Eye		
Radial muscle	α	Contraction (mydriasis)
Ciliary muscle	β	Relaxation for far vision
Heart		
Sinoatrial node	β	Increase in heart rate (increase in rate of diastolic depolarization and decrease in duration of phase 4 of sinoatrial nodal action potential)
Atrioventricular node	β	Increase in conduction velocity and shortening of functional refractory period
Atria	β	Increase in contractility
Ventricles	β	Increase in contractility and irritability
Blood vessels	α	Constriction (arterioles and veins)
	β	Dilation (predominates in skeletal muscle)*
Bronchial muscle	β	Relaxation (bronchodilation)
Gastrointestinal tract		
Stomach	β	Decrease in motility
Intestine	α, β	Decrease in motility
Sphincters	α	Contraction
Urinary bladder		
Detrusor muscle	β	Relaxation
Trigone and sphincter	α	Contraction
Skin		
Pilomotor muscles	α	Piloerection
Sweat glands	α	Selective stimulation (adrenergic sweating)*
Uterus	α	Contraction
	β	Relaxation
Liver	α	Glycogenolysis
Muscle	β	Glycogenolysis
Pancreatic islets	α	Inhibition of insulin secretion
	β	Stimulation of insulin secretion

Adapted from Morgan HE: Function of the adrenal glands. In *Best and Taylor's Physiological Basis of Medical Practice*, 10th edition. Edited by Brobeck JR. Baltimore, Williams & Wilkins, 1979.
*Mediated via sympathetic cholinergic fibers.

1. **Carbohydrate metabolism.** Because hepatic stores of glycogen are limited (about 100 g) and decrease only transiently after epinephrine activation, lactate derived from muscle glycogen (300 g) is the major precursor for hepatic **gluconeogenesis,** the process that sustains hepatic glucose formation and secretion. Gluconeogenesis mainly accounts for the hyperglycemic action of epinephrine in normal physiologic states. In pathologic states (e.g., pheochromocytoma), the diabetogenic action of catecholamines is caused by the inhibition of insulin secretion and the gluconeogenic effect of these hormones (usually norepinephrine), which are secreted in excessive amounts. Since propranolol attenuates hyperlactacidemia and hyperglycemia, this implies that epinephrine-induced glycogenolysis in muscle and in the liver is mediated by the β and α receptors, respectively (see Table 7-5).

   a. **Glycogenolysis in the liver**
      (1) Epinephrine stimulates glycogenolysis in the liver via the Ca^{2+}-activated glycogen phosphorylase and the inhibition of glycogen synthetase. Glucose-6-phosphatase, found mainly in the liver and in lesser amounts in the kidney, forms free glucose, which increases blood glucose.
      (2) Since glucagon stimulates and insulin suppresses hepatic glycogenolysis, the effects of epinephrine on insulin secretion (suppression) and glucagon secretion (stimulation) reinforce the breakdown of glycogen and the increase in hepatic glucose secretion.
      (3) Epinephrine also increases the hepatic production of glucose from lactate, amino acids, and glycerol, all of which are gluconeogenic substances.

   b. **Glycogenolysis in muscle**
      (1) Epinephrine stimulates glycogenolysis in muscle by a β-adrenergic receptor mechanism involving the stimulation of adenyl cyclase and cAMP-induced stimulation of glycogen phosphorylase. Concomitantly, glycogen synthetase activity is reduced.
      (2) Muscle lacks glucose-6-phosphatase, and epinephrine-induced glycogenolysis in muscle **does not directly increase blood glucose**. The glucose-6-phosphate is metabolized to lactate or pyruvate, which is converted to glucose by the liver.
      (3) The ultimate physiologic effect of epinephrine-stimulated glycogenolysis in muscle is increased hepatic glucose secretion (hyperglycemia) via the hepatic conversion of muscle lactate to glucose.

   c. **Hyperglycemic effects of epinephrine** on the liver are important only in conjunction with the effects of epinephrine on glucagon and insulin secretion together with its glycogenolytic effect on muscle in acute emergency situations. Epinephrine in physiologic concentrations does not have a direct glycogenolytic effect in the liver.
      (1) Much higher amounts of epinephrine than glucagon are required to cause hyperglycemia. However, epinephrine has a more pronounced hyperglycemic effect than glucagon for the following important reasons.
         (a) Epinephrine inhibits insulin secretion, while glucagon stimulates insulin secretion; therefore, the hyperglycemic effect of glucagon is attenuated.
         (b) Epinephrine stimulates glycogenolysis in muscle, thereby providing lactate for hepatic gluconeogenesis.
         (c) Epinephrine stimulates glucagon secretion, which amplifies its hyperglycemic effect.
         (d) Epinephrine stimulates ACTH secretion, which then stimulates cortisol secretion. Cortisol also is a potent gluconeogenic hormone via the hepatic conversion of alanine to glucose.
         (e) Circulating catecholamines inhibit muscle glucose uptake, which is in contrast to the effect of glucagon.
      (2) Epinephrine indirectly inhibits insulin-mediated facilitated diffusion of glucose by muscle and adipose tissue via its blockade of insulin secretion.
      (3) Catecholamines also directly inhibit glucose uptake by the suppression of glucose transporter proteins in the cell membranes of skeletal and cardiac muscle cells and adipocytes.

2. **Fat metabolism.** In humans, the major site of lipogenesis from glucose is the liver.
   a. A man of average size has fat stores that contain about 15 kg of triglyceride, some of which can be mobilized as free fatty acids.
   b. Epinephrine stimulates lipolysis by activating triglyceride lipase, which is called the intracellular **hormone-sensitive lipase**. The activation of this enzyme is via the β-adrenergic receptor (i.e., cAMP).

    c. Mobilization of free fatty acids from stores in adipose tissue supplies a substrate for ketogenesis in the liver. Acetoacetate and β-hydroxybutyrate are transported from the liver to the peripheral tissues, where they are quantitatively important as energy sources.

      (1) Cardiac muscle and the renal cortex use fatty acids and acetoacetate in preference to glucose, whereas resting skeletal muscle uses fatty acids as the major source of energy.

      (2) During extreme conditions, such as starvation and diabetes, the brain adapts to the use of ketoacids. Ketoacids also are oxidized by skeletal muscle during starvation.

3. Gluconeogenesis refers to the formation of glucose from noncarbohydrate sources. Gluconeogenesis occurs in the liver and the kidney.

    a. Gluconeogenic substances include pyruvate, lactate, glycerol, odd-chain fatty acids, and amino acids. However, the major source of endogenous glucose production is protein, with a smaller fraction available from the glycerol contained in fat. All of the constituent amino acids in protein tissue, with the exception of leucine, can be converted to glucose.

    b. The conversion of even-chain fatty acids is not possible in the mammalian liver because of the absence of the enzymes necessary for the de novo synthesis of the four-carbon dicarboxylic acids from acetyl-CoA.

I. Endocrinopathies

1. Hyposecretion of catecholamines, such as occurs during tuberculosis and malignant destruction of the adrenal glands or following adrenalectomy, probably produces no symptoms or other clinical features.

    a. Catecholamine production from the sympathetic nerve endings appears to satisfy the normal biologic requirements, because the adrenal medulla is not necessary for life.

    b. The functional integrity of the adrenal medulla can be determined experimentally by the administration of 2-deoxy-D-glucose. This nonmetabolizable carbohydrate induces intracellular glycopenia and extracellular hyperglycemia.

    c. Spontaneous deficiency of epinephrine is unknown as a disease state, and adrenalectomized patients do not require epinephrine replacement therapy.

2. Hypersecretion of catecholamines from chromaffin cell tumors (pheochromocytomas) produces demonstrable clinical features.

    a. Pheochromocytoma patients have sustained or paroxysmal hypertension.

    b. The hypersecretion of catecholamines is associated with severe headache, sweating (cold or **adrenergic sweating**), palpitations, chest pain, extreme anxiety with a sense of impending death, pallor of the skin caused by vasoconstriction, and blurred vision.

    c. Most pheochromocytomas contain predominantly norepinephrine, and most affected patients secrete predominantly norepinephrine into the bloodstream. Evidence of epinephrine hypersecretion increases the likelihood that the tumor origin is in the adrenal medulla. However, an extra-adrenal site should not be excluded.

      (1) If epinephrine is secreted primarily, the heart rate is increased.

      (2) If norepinephrine is the predominant hormone, the pulse rate decreases reflexly in response to marked hypertension.

    d. Urinary excretion of catecholamines, metanephrines, and VMA is increased.

3. Clinical tests

    a. The **adrenolytic test** involves the administration of an α-blocker (phentolamine) to observe the effect of systemic blood pressure. A dramatic fall in blood pressure is pathognomonic of pheochromocytoma because this procedure is without significant effect in normotensive subjects.

    b. The **provocative test** involves the administration of histamine, glucagon, or tyramine, which will cause a rise in blood pressure.

4. Treatment requires surgical removal of the tumor.

VIII. ADRENAL CORTEX. This section describes the physiologic and biochemical aspects of **glucocorticoids,** the most important of which is **cortisol** (also known as **hydrocortisone**). The biologic characteristics of **mineralocorticoids** (specifically, aldosterone) are described in section IX of Chapter 4.

A. Embryology. Morphologically and physiologically, the fetal adrenal gland differs strikingly from that of the adult. However, through all stages of life, the function of the adrenal cortex is dependent on ACTH.

1. The adrenal cortex is a mesodermal derivative. It is axiomatic that all endocrine glands derived from mesoderm synthesize and secrete steroid hormones.

2. The outer **neocortex,** which is the progenitor of the adult adrenal cortex, comprises about 15% of the total volume of this organ, and the inner **fetal zone** (inner zone or **fetal cortex**) constitutes about 85%.

3. The adrenal gland is larger at birth than it is during adulthood. This is due to the fact that the fetal zone undergoes rapid involution during the first few months of extrauterine life and completely disappears by 3–12 months postpartum.

4. Near term, the fetal cortices of the fetal adrenal glands secrete 100–200 mg of steroids daily in the form of sulfoconjugates, the principal one of which is the biologically weak androgen, **dehydroepiandrosterone (DHEA) sulfate**. This 17-ketosteroid is an androgen containing 19 carbon atoms.

B. Morphology

1. Gross anatomy
   a. The adrenal glands are paired structures situated above the kidneys.
   b. Normally, each gland weighs about 5 g, of which the cortex constitutes approximately 80%.

2. Histology and function. The adrenal cortex consists of three distinct layers or **zones** of cells.
   a. The outermost layer, the **zona glomerulosa,** is the site of **aldosterone** and **corticosterone** synthesis. Aldosterone is the principal mineralocorticoid of the human adrenal cortex.
   b. The wider, middle zone is the **zona fasciculata,** and the innermost layer is the **zona reticularis**. The two inner zones of the adrenal cortex should be considered a functional unit, where mainly **cortisol** (and some corticosterone) and **DHEA** are synthesized.

C. Adrenocortical hormones: corticosteroids (Tables 7-6, 7-7)

1. Secretion
   a. The human adrenal cortex secretes two glucocorticoids (cortisol and corticosterone*), one mineralocorticoid (aldosterone), biosynthetic precursors of three end products (progesterone, 11-deoxycorticosterone, and 11-deoxycortisol), and androgenic substances (DHEA and its sulfate ester).
   b. The normal human adrenal cortex does not secrete physiologically effective amounts of testosterone or estrogenic substances.

2. Transport
   a. Under physiologic circumstances, about 75% of the plasma cortisol is bound to **cortisol-binding globulin** (CBG; transcortin), which is an α-globulin. (In addition to binding cortisol,

Table 7-6. Average 8:00 A.M. Plasma Concentration and Secretion Rate of Corticosteroids in Adults

Corticosteroid	Plasma Concentration (μg/dl)	Secretion Rate (mg/day)
Cortisol	13	15
Corticosterone	1	3
11-Deoxycortisol	0.16	0.40
Deoxycorticosterone	0.07	0.20
Aldosterone	0.009	0.15
18-Hydroxycorticosterone	0.009	0.10
Dehydroepiandrosterone (DHEA)	0.5	15
DHEA sulfate	115	15

Reprinted from Genuth SM: The adrenal glands. In *Physiology*. Edited by Berne RM, Levy MN. St. Louis, CV Mosby, 1983, p 1046.

*At physiologic concentrations, corticosterone has glucocorticoid activity.

Table 7-7. Blood Production Rates of Adrenal Androgens

Androgenic Steroid	Plasma Concentration (μg/dl)	Blood Production Rate (mg/day)
Testosterone		
Men	0.8	7.0
Women*	0.034	0.34
Androstenedione		
Men	0.06	1.4
Women	0.14	3.4

Reprinted from Mulrow PJ: The adrenals. In *Physiology and Biophysics.* Edited by Ruch TC, Patton HD. Philadelphia, WB Saunders, 1973, p 229.
*In women, about 50% of the blood testosterone is derived from androstenedione.

transcortin has a high binding affinity for progesterone, deoxycorticosterone, corticosterone, and some synthetic analogs.) About 15% of the plasma cortisol is bound to plasma **albumin,** and about 10% is **unbound** and represents the physiologically active steroid.
   b. The 90% of the plasma cortisol bound to plasma protein represents the metabolically inactive pool, which serves as a reservoir for free hormone.

3. **Corticosteroidogenesis** (Figure 7-7)
   a. Uptake of cholesterol. Free cholesterol is the preferred precursor of the corticosteroids, although the adrenal cortex can form cholesterol from acetyl-CoA. Most stored adrenal cholesterol is esterified with fatty acids; it is the cholesterol ester content that is reduced following ACTH stimulation of the fasiculata-reticularis complex.
   b. Side-chain cleavage of cholesterol. The rate-limiting step in corticosteroidogenesis is the mitochondrial conversion of cholesterol to pregnenolone by 20,22-desmolase.
   c. Pregnenolone: the common precursor of all steroid hormones. Conversion of pregnenolone to progesterone requires two enzymes—3β-hydroxysteroid dehydrogenase and Δ5-isomerase—which are found in the endoplasmic reticulum (microsomes).
   d. Hydroxylation reactions follow sequentially after the formation of pregnenolone and progesterone.
     (1) The hydroxylation reactions in the biosynthesis of aldosterone occur sequentially at the C-21, C-11, and C-18 positions.

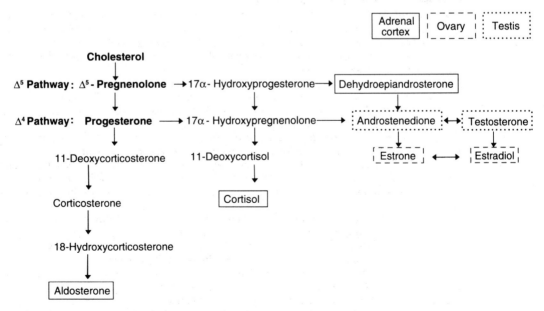

Figure 7-7. Summary of steroidogenesis in the adrenal cortex, testis, and ovary. Notice that there is no interconversion between mineralocorticoids and glucocorticoids.

(2) The hydroxylation reactions in the biosynthesis of cortisol occur sequentially at the C-17, C-21, and C-11 positions.

e. Sex steroid pathways. Both pregnenolone and progesterone are substrates for the microsomal enzyme, 17α-hydroxylase, which is found not only in the adrenal cortex but also in the testis and ovary.

(1) Pregnenolone is converted to 17α-hydroxypregnenolone, which can either continue along the Δ^5-pathway to the synthesis of androgens or enter the glucocorticoid pathway.

(2) Progesterone is converted to 17α-hydroxyprogesterone, which can either proceed along the Δ^4-pathway to the synthesis of androgens or enter the mineralocorticoid pathway.

(3) The zona glomerulosa lacks 17α-hydroxylase and does not have the capacity to synthesize 17α-hydroxyprogesterone. For this reason, the zona glomerulosa cannot synthesize cortisol.

    (a) The conversion of pregnenolone to DHEA is in the Δ^5-pathway.

    (b) The conversion of progesterone to androstenedione is in the Δ^4-pathway.

f. Glucocorticoid/mineralocorticoid pathways. Both progesterone and 17α-hydroxyprogesterone are substrates for the microsomal enzyme, 21-hydroxylase.

(1) Progesterone is converted to 11-deoxycorticosterone, which is in the mineralocorticoid pathway.

(2) 17α-Hydroxyprogesterone is converted to 11-deoxycortisol, which is in the glucocorticoid pathway.

(3) 11-Deoxycorticosterone and 11-deoxycortisol are not precursors for the synthesis of each other.

g. Diverging pathways. Both 11-deoxycorticosterone and 11-deoxycortisol are acted upon by mitochondrial 11β-hydroxylase.

(1) 11-Deoxycorticosterone forms corticosterone in the mineralocorticoid pathway.

(2) 11-Deoxycortisol is converted to cortisol, the principal glucocorticoid of the human adrenal cortex.

(3) Corticosterone differs from cortisol only in that the former lacks a 17α-hydroxyl group. Although this structural difference is small, it accounts for a very large difference in the biologic activity of these two hormones.

h. Aldosterone synthesis. Some of the corticosterone serves as the substrate for another mitochondrial enzyme, 18-hydroxylase.

(1) This reaction leads to the formation of 18-hydroxycorticosterone.

(2) 18-Hydroxycorticosterone is converted to aldosterone by the mitochondrial enzyme, 18-hydroxysteroid dehydrogenase, which is found only in the zona glomerulosa.

i. Synthesis and secretion of sex steroids

(1) **Androgens.** The adrenal cortex secretes four androgenic hormones: androstenedione, testosterone, DHEA, and DHEA sulfate. Quantitatively, the most important sex steroids produced by the human adrenal cortex are DHEA and DHEA sulfate. Except for testosterone, adrenocortical androgens are relatively weak and serve as precursors for hepatic conversion to testosterone.

    (a) DHEA is derived from 17α-hydroxypregnenolone by the action of 17,20-desmolase.

    (b) DHEA is mainly conjugated with sulfuric acid and, as such, is bound to plasma protein. While in this bound form, DHEA is not readily excreted but circulates in higher concentrations than any other steroid.

    (c) DHEA is the principal precursor of urinary 17-ketosteroids; however, the most abundant urinary 17-ketosteroids are androsterone and etiocholanolone.

    (d) DHEA and DHEA sulfate have androgenic activity by virtue of their peripheral conversion to testosterone. DHEA sulfate is active as a minor precursor of other 19-carbon steroids formed in the gonads and placenta, which are sites of sulfatase activity.

        (i) Normal excretion rates for 17-ketosteroids are 5–14 mg/day in women and 8–20 mg/day in men.

        (ii) Normally, adrenocortical precursors represent the bulk of the urinary 17-ketosteroid pool, with a smaller contribution from the gonads.

        (iii) The main androgen secreted by premenopausal women is androstenedione, about 60% of which is of adrenocortical origin.

(2) **Estrogens.** The human adrenal cortex can synthesize minute amounts of estrogen. The adrenal cortex makes its major contribution to the body's estrogen (estrone) pool indirectly by supplying androstenedione together with DHEA and its sulfate as substrates

for conversion to estrogens by subcutaneous fat, hair follicles, mammary adipose tissue, and other tissues.

4. Metabolism of corticosteroids
   a. General considerations
   (1) Corticosteroid inactivation occurs by:
      (a) Enzymatic reduction of the Δ^{4-5} double bond in the A-ring to form **dihydrocortisol**
      (b) Enzymatic reduction of the ketonic oxygen substituent at the C-3 position to form **tetrahydrocortisol**
      (c) Conjugation with glucuronic acid to form a water-soluble metabolite that is readily excreted by the kidney
   (2) The major urinary metabolite of cortisol is **tetrahydrocortisol glucuronide.**
   b. Hepatic conversion. The liver is the major extra-adrenal site of corticosteroid metabolism.
   (1) Cortisol can be enzymatically converted to cortisone by 11β-hydroxysteroid dehydrogenase and excreted as **tetrahydrocortisone glucuronide.**
   (2) The ketonic oxygen substituent on the C-20 position of cortisol and other steroids can be enzymatically converted to a hydroxyl group. These steroids can be subjected to A-ring reduction, conjugated, and excreted as glucuronides.
      (a) Cortisol is converted to **cortol glucuronide.**
      (b) Cortisone is converted to **cortolone glucuronide.**
      (c) Progesterone forms **pregnanediol glucuronide.**
      (d) 17α-Hydroxyprogesterone forms **pregnanetriol glucuronide.**
   (3) Steroids that contain a 17α-hydroxyl group are ketogenic. Cortisol can be enzymatically converted to a 17-ketosteroid by 17,20-desmolase. About 5% of cortisol appears in the urine as a 17-ketosteroid.
   (4) Aldosterone also can undergo A-ring reduction and conjugation to form **tetrahydroaldosterone glucuronide.**
   c. Conversion in other extra-adrenal tissues. Muscle, skin, fibroblasts, intestine, and lymphocytes also can carry out oxidation-reduction reactions at the C-3, C-11, C-17, and C-20 positions of the corticosteroid molecule.

D. Control of adrenocortical function (see Figure 7-4). Physiologic control of the rate of cortisol secretion occurs via a double negative feedback loop, a mechanism that is characteristic of most neuroendocrine control systems.

   1. Regulation of ACTH secretion
      a. The parvicellular peptidergic neurons of the hypothalamus release **CRH,** which stimulates the secretion of ACTH from the anterior lobe of the pituitary gland via the hypophysial portal system. CRH is a polypeptide consisting of 41 amino acid residues.
      b. CRH secretion from the tuberoinfundibular neurons is regulated by neurotransmitters secreted by the monoaminergic neurons that innervate the peptidergic neurons.
         (1) Hypothalamic secretion of CRH is stimulated by cholinergic neurons. Serotonin also is a stimulatory signal to CRH neurons.
         (2) Adrenergic neuron activity inhibits release of CRH. **Gamma-aminobutyric acid (GABA)** also is a known inhibitor of CRH secretion.

   2. Negative feedback action of corticosteroids
      a. Of the endogenous corticosteroids, only cortisol has ACTH-suppressing activity. The synthetic glucocorticoid, **dexamethasone,** is a potent inhibitor of ACTH secretion and, therefore, of endogenous glucocorticoid secretion.
      b. The negative feedback of cortisol is exerted at the level of the pituitary gland and the ventral diencephalon.
         (1) If free cortisol levels are supraphysiologic, ACTH secretion is suppressed and the adrenal cortex ceases its secretory activity and undergoes **disuse atrophy.**
         (2) Conversely, if plasma free cortisol levels are subnormal, the anterior lobe is released from inhibition by cortisol, ACTH secretion rises, and the adrenal cortex secretes more cortisol and becomes hypertrophic.

   3. Hypophysial-adrenocortical rhythm. Normally, blood ACTH levels are higher in the morning than in the evening. This accounts for the diurnal rhythms in cortisol secretion, plasma cortisol concentration, and 17-hydroxycorticosteroid excretion (see III A 1).

4. Hypophysial-adrenocortical response to stress. The normal hypothalamic-hypophysial-adrenocortical control system can be overridden by a variety of challenges, which collectively are referred to as **stress.**

    a. Among the stresses shown to induce increased activity in this system are severe trauma, pyrogens, hypoglycemia, histamine injection, electroconvulsive shock, acute anxiety, burns, hemorrhage, exercise, infections, chemical intoxication, pain, surgery, psychological stress, and cold exposure.

    b. In stress conditions, ACTH secretion is stimulated despite the fact that systemic levels of cortisol are much higher than those required to inhibit ACTH secretion completely in unstressed conditions.

E. Physiologic effects of glucocorticoids. Of the naturally occurring steroids, only cortisol, cortisone, corticosterone, and 11-dehydrocorticosterone have appreciable glucocorticoid activity. Full recovery from hypothalamic-hypophysial-adrenocortical suppression may require as long as 1 year following cessation of all steroid therapy.

    1. Anti-inflammatory effects. Glucocorticoids inhibit inflammatory and allergic reactions in several ways.

        a. They stabilize lysosomal membranes, thereby inhibiting the release of proteolytic enzymes.

        b. They decrease capillary permeability, thereby inhibiting diapedesis of leukocytes.

        c. Glucocorticoids reduce the number of circulating lymphocytes, monocytes, eosinophils, and basophils.

            (1) The decreased number of these formed elements in blood is due primarily to a redistribution of the cells from the vascular compartment into the lymphoid tissue (e.g., spleen, lymph nodes, bone marrow). Cellular lysis is not a major mechanism for decreasing the number of these cells in the human circulation.

            (2) The decrease in circulating basophils accounts for the fall in blood histamine levels and the abrogation of the allergic response.

                (a) The migration of inflammatory cells from capillaries is decreased.

                (b) Glucocorticoids lessen the formation of edema and, thereby, reduce the swelling of inflammatory tissue.

            (3) Glucocorticoids cause an increase in the number of circulating neutrophils due to the accelerated release from bone marrow and a reduced migration from the circulation. Steroids also inhibit the ability of neutrophils to marginate to the vessel wall.

        d. Glucocorticoids cause involution of the lymph nodes, thymus, and spleen, which leads to decreased antibody production.

            (1) This lymphocytopenic effect aids in the prevention or reduction of the immune response by an organ transplant recipient.

            (2) Since recipients pretreated with large doses of glucocorticoids are susceptible to intercurrent infections, antibiotics are a necessary adjunct to the steroid therapy.

            (3) Antibody production is not suppressed in humans at conventional steroid doses; however, chronic administration of high doses of glucocorticoids leads to an impairment of host defense mechanisms.

        e. Glucocorticoids lead to an increase in the total blood count because of the increased numbers of neutrophils, erythrocytes, and platelets. The increase in circulating erythrocytes (polycythemia) is due to the stimulation of hematopoiesis.

    2. Renal effects

        a. Glucocorticoids restore glomerular filtration rate and renal plasma flow to normal following adrenalectomy. Mineralocorticoids do not have these effects.

        b. Glucocorticoids facilitate free-water excretion (clearance) and uric acid excretion.

    3. Gastric effects. Cortisol increases gastric flow and gastric acid secretion, while it decreases gastric mucosal cell proliferation. The latter two effects lead to peptic ulceration following chronic cortisol treatment.

    4. Psychoneural effects of glucocorticoids have been noted following chronic high-dose therapy with these steroids. Patients may become initially euphoric and then psychotic, paranoid, and depressed.

    5. Antigrowth effects

        a. Large doses of cortisol have been shown to antagonize the effect of active vitamin D metabolites on the absorption of Ca^{2+} from the gut, to inhibit mitosis of fibroblasts, and to

cause degradation of collagen. All of these effects lead to osteoporosis, which is a reduction in bone mass per unit volume with a normal ratio of mineral-to-organic matrix.
  b. Glucocorticoids delay wound healing because of the reduction of fibroblast proliferation. Connective tissue is reduced in quantity and strength.
  c. Chronic supraphysiologic doses of glucocorticoids suppress GH secretion and inhibit somatic growth.
  d. Although glucocorticoids increase the ability of muscle to perform work, large doses lead to muscle atrophy and muscular weakness.

6. Vascular effect. Cortisol in pharmacologic doses enhances the vasopressor effect of norepinephrine. In the absence of cortisol, the vasopressor action of catecholamines is diminished, and hypotension ensues. Thus, corticosteroids have a role in the maintenance of normal arterial systemic blood pressure and volume through their support of vascular responsiveness to vasoactive substances. (Cortisol enhances catecholamine synthesis via its activation of the epinephrine-forming enzyme, PNMT.)

7. Stress adaptation. Glucocorticoids allow mammals to adapt to various stresses in order to maintain homeostasis.
  a. Resistance to stress is not increased by the administration of glucocorticoids.
  b. Stress is associated with the activation of the hypothalamic-hypophysial-adrenal axis.

F. Metabolic effects of glucocorticoids

1. Carbohydrate metabolism. Cortisol is a carbohydrate-sparing hormone and, therefore, exerts an anti-insulin effect, which leads to hyperglycemia and insulin-resistance.
  a. Glucocorticoids maintain blood glucose and the glycogen content of the liver, kidney, and muscle by promoting the conversion of amino acids to carbohydrates and the storage of carbohydrate as glycogen.
  b. Cortisol is hyperglycemic principally because of its gluconeogenic activity, which is related to its protein catabolic effect on extrahepatic tissues, especially muscle.
      (1) The proteolytic effect of glucocorticoids results in the mobilization of amino acids from muscle and in an increase in plasma amino acid concentration.
      (2) Alanine is quantitatively the major gluconeogenic amino acid precursor in the liver. Like acetyl-CoA, alanine inhibits pyruvate kinase activity.
  c. Cortisol also exerts an anti-insulin effect by blocking glucose transport in muscle and adipose tissue. This effect accounts for the phenomenon of glucose intolerance or an eventual "steroid diabetes."
  d. Glucocorticoids augment the activity of key gluconeogenic enzymes by the induction of hepatic enzyme synthesis.
      (1) The gluconeogenic pathway has three steps that differ from those in the glycolytic pathways as a result of their thermodynamic irreversibility. These enzymes and their substrates are as follows.
          (a) Glucose-6-phosphate → glucose (glucose-6-phosphatase)
          (b) Fructose 1,6-diphosphate → fructose-6-phosphate (fructose 1,6-diphosphatase)
          (c) Conversion of pyruvate to phosphoenolpyruvate requires two steps:
              (i) Pyruvate → oxaloacetate (pyruvate carboxylase)
              (ii) Oxaloacetate → phosphoenolpyruvate (phosphoenolpyruvate carboxykinase)
      (2) Glucocorticoids are associated with activation of glycogen synthetase by glucose-6-phosphate and of pyruvate carboxylase by acetyl-CoA. They also are associated with indirect inhibition of glycolysis by free fatty acids, resulting in increased glucogenesis and glycogenesis.
  e. Cortisol also indirectly inhibits the activities of glycolytic enzymes, which accounts for its anti-insulin effect. Enzymes that are blocked by the effect of glucocorticoids include:
      (1) Glucokinase
      (2) Phosphofructokinase
      (3) Pyruvate kinase
  f. Glucocorticoids mobilize fatty acids from adipose tissue to the liver, where metabolism of free fatty acids may lead to products that inhibit glycolytic enzymes and favor gluconeogenesis.
      (1) The glycerol released from the fat cell with the fatty acids also serves as a secondary substrate for gluconeogenesis.

(2) The cortisol-inhibited glycolysis in peripheral tissue probably is indirectly blocked via the inhibition of the key glycolytic enzyme, phosphofructokinase, by the elevated concentration of plasma free fatty acids.

   g. Glucocorticoids are associated with compensatory hyperinsulinemia following hyperglycemia.

2. Protein metabolism. The most important gluconeogenic substrates are amino acids derived from proteolysis in skeletal muscle.

   a. Cortisol enhances the release of amino acids from proteins in skeletal muscle and other extrahepatic tissues, including the protein matrix of bone.

   (1) The amino acids released, especially the glucogenic amino acid **alanine,** are transported to the liver and converted to glucose.

   (2) Increased glucose production by cortisol via gluconeogenesis is associated with increased urea production via the conversion of amino nitrogen to urea. This effect accounts for the increased urinary nitrogen excretion.

   (3) The proteolysis in skeletal muscle brings about a negative nitrogen balance.

   b. The amino acids taken up by the liver are used not only to form glucose or glycogen but also to build new protein. This protein anabolic effect at the level of the liver is an important exception to the overall protein catabolic effect of cortisol.

   c. Equally important is the ability of glucocorticoids to inhibit the de novo synthesis of protein, probably at the translational level. This is called an **antianabolic effect** of cortisol.

3. Fat metabolism. Glucocorticoids are lipolytic hormones. Their lipolytic effect is in part due to the potentiation of the lipolytic actions of other hormones, such as GH, catecholamines, glucagon, and thyroid hormone.

   a. Glucocorticoids favor the mobilization of fatty acids from adipose tissue to the liver, where the metabolism of fatty acids inhibits glycolytic enzymes and promotes gluconeogenesis. As a result of increased fatty acid oxidation, glucocorticoids may lead to ketosis, especially in the context of diabetes mellitus.

   (1) The major site of stimulation is the gluconeogenic enzyme, fructose 1,6-diphosphatase, which is activated by fatty acids.

   (2) At the same time, pyruvate kinase and phosphofructokinase are inhibited by fatty acids. Thus, glycolysis is inhibited while gluconeogenesis proceeds.

   b. Glucocorticoids also indirectly stimulate lipolysis by blocking peripheral glucose uptake and utilization. They inhibit re-esterification of fatty acids within adipocytes by inhibiting the use of glucose.

   c. Fatty acid synthesis is inhibited in the liver by cortisol, an effect not observed in adipose tissue. The overall effect of glucocorticoids on fat is to induce a redistribution of fat together with an increase in total body fat (i.e., truncal obesity). The increase in body weight is not due to a growth effect that is the accretion of protein. (Recall that cortisol is an antigrowth hormone.)

   (1) There is a characteristic centripetal distribution of fat (i.e., an accumulation of fat in the central axis of the body).

   (a) The deposition of fat in the facies is called "moonface."

   (b) The deposition of fat in the suprascapular region is referred to as "buffalo hump."

   (c) Excessive fat distribution leads to a pendulous abdomen.

   (2) Glucocorticoid-induced obesity reflects increased food intake rather than a change in the rate of lipid metabolism.

   d. Chronic excessive amounts of cortisol lead to hyperlipidemia and hypercholesterolemia. Glucocorticoids increase appetite and, thereby, play a role in obesity.

IX. TESTIS

A. Embryology

1. Internal genitalia (Figure 7-8). The internal reproductive tract in males and females is derived from one of two pairs of genital ducts. The **wolffian (mesonephric) ducts** give rise to the male internal genitalia, and the **müllerian ducts** give rise to the female internal genitalia.

   a. In the male, the internal genitalia (also called **accessory sex organs**) consist of the following ducts and glands for the conveyance of sperm:

   (1) Seminiferous tubules

INDIFFERENT STAGE

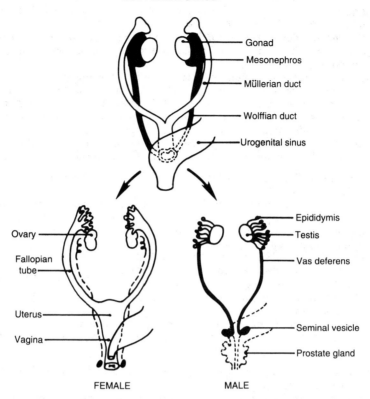

Figure 7-8. Sex differentiation of the gonad and the internal genitalia. Up to 6 weeks gestation, the gonad is a bipotential structure (*indifferent stage*), and the urogenital tract in both sexes consists of two pairs of genital ducts (i.e., the wolffian and müllerian ducts) and a mesonephros, all of which terminate in the urogenital sinus. In the female, the gonad develops into an ovary and the müllerian ducts become organized into the fallopian tubes, uterus, and upper vagina, while the wolffian ducts remain vestigial. In the male, a testis develops and the wolffian ducts differentiate into the epididymis, vas deferens, and seminal vesicle, while the müllerian ducts regress. (Reprinted from Wilson JD: Embryology of the genital tract. In *Campbell's Urology,* 4th edition. Edited by Harrison JH, et al. Philadelphia, WB Saunders, 1977, p 1473.)

(2) Rete testis*
(3) Ductuli efferentes
(4) Epididymis
(5) Vas deferens
(6) Ejaculatory duct
(7) Seminal vesicles
(8) Prostate gland
(9) Bulbourethral glands (Cowper's glands)
  b. The wolffian ducts are the excretory ducts of the mesonephric kidney and are attached to the primordial gonad; the müllerian duct is formed from the wolffian duct and is not contiguous with the primitive gonad.
  c. At 9–10 weeks gestation, following the appearance of the **Leydig (interstitial) cells** in the fetal testis, the two wolffian ducts begin to differentiate and give rise to the following male internal genitalia:
   (1) Epididymis
   (2) Vas deferens
   (3) Seminal vesicles
   (4) Ejaculatory duct

---

*The duct system distal to the rete testis is known as the excurrent (excretory) duct system.

d. It is the primary function of **androgen** to induce the formation of the male accessory sex organs during fetal life. Specifically, testicular androgen is required for the differentiation of the male genital (wolffian) duct system.

e. Genetic sex determines gonadal sex, and gonadal sex determines phenotypic sex. The control over the formation of the male phenotype requires the action of the following three hormones.

    (1) **Müllerian duct inhibiting factor** is produced by the Sertoli cells of the fetal testis and induces the regression of the müllerian duct system. Müllerian duct inhibiting factor functions as a paracrine secretion that diffuses to the paired müllerian ducts.

    (2) **Testosterone** is secreted by the fetal testis and promotes growth and differentiation of the wolffian duct system and the accessory sex organs.

    (3) Testosterone also is the precursor for **dihydrotestosterone,** which is necessary for the growth and differentiation of the male external genitalia (see IX A 2 b).

2. External genitalia. The external genitalia also begin to differentiate between 9 and 10 weeks gestation.

    a. In contrast to the internal genitalia, the external genitalia and urethra in both sexes develop from common anlagen, which are the urogenital sinus, the genital sinus, the genital tubercle, the genital swelling, and the genital (urethral) folds. The anlagen of the male external genitalia give rise to the following structures.

        (1) The urogenital sinus forms the male urethra and prostate gland.

        (2) The genital tubercle grows into the glans penis.

        (3) The genital swelling forms the scrotum.

        (4) The urethral folds enlarge to form the outer two-thirds of the penile urethra and corpora spongiosa.

    b. The growth and development of the male external genitalia require dihydrotestosterone, which is formed from the conversion of fetal testicular testosterone within the urogenital sinus and lower urogenital tract.

3. Testis (see Figure 7-8). The primitive gonads develop midabdominally in association with the mesonephric ridges.

    a. Through 5 weeks gestation, the gonads are undifferentiated and consist of a medulla, a cortex, and primordial germ cells. These germ cells are embedded in a layer of cortical epithelium surrounding a core of medullary mesenchymal tissue.

    b. At 6 weeks, the seminiferous tubules begin to form from the medulla. The cortical region of the primitive gonad undergoes regression. (The female gonad develops from the cortical region.) The primordial germ cells (gonocytes) arise outside the gonads and migrate from the endodermal yolk sac epithelium to the urogenital ridge. During migration, the gonocytes undergo continuous mitosis.

    c. At 7 weeks gestation, the **Sertoli (sustentacular) cells** begin to form and secrete the **H-Y (histocompatibility-Y) antigen,** which is under control of the Y chromosome. The Sertoli cells are derivatives of the mesenchymal tissue of the urogenital ridge.

    d. At 8–9 weeks of fetal life, the Leydig cells are formed and secrete testosterone in response to **chorionic gonadotropin,** which is secreted by the placenta. The Leydig cells also are mesenchymal derivatives, which appear in the connective tissue surrounding the seminiferous cords.

    e. At 9 weeks gestation, definitive testes are present.

    f. At 7–9 months gestation, the testes normally descend through the inguinal rings into the scrotum.

B. Morphology

1. Gross anatomy

    a. The adult testis is an ovoid gland, which is approximately 4–5 cm long and 2.5–3.0 cm wide and ranges in weight from 10 to 45 g.

    b. The testes normally are situated in the scrotum, where they are maintained at a temperature that is about 2° C lower than normal body temperature. The lower temperature is necessary for normal spermatogenesis.

    c. Each gland is surrounded by a fibrous connective tissue membrane called the **tunica albuginea,** which, in turn, is surrounded by a serous membrane called the **tunica vaginalis.**

    d. By puberty, the testes usually have developed sufficiently to perform the functions of spermatogenesis and steroidogenesis. Generally, puberty begins between the ages of 12 and

14 years. In the United States, 95% of normal boys show signs of puberty by the age of 16 years.

2. **Functional histology of testicular parenchyma.** The testicular parenchyma consists mainly of the seminiferous tubules and Leydig cells, which comprise about 90% and 10% of the testicular volume, respectively. These two compartments are separated by boundary tissue, which includes a basement membrane.
   a. **Tubular tissue.** The seminiferous tubules have a total collective length of approximately 800 m (a half mile)!
      (1) The spermatogenic tubules are organized into coiled loops, each of which begins and terminates in a single duct called the **tubulus rectus**.
      (2) The tubuli recti join to form the **rete testis** and eventually drain via the **ductuli efferentes** into the **epididymis**. The epididymis is the primary storage and final maturation site for spermatozoa.
      (3) From the epididymis, the spermatozoa are transmitted via the **vas deferens** and **ejaculatory duct** into the penile urethra.
   b. **Tubular cells.** The basic cellular components of the tubules are the germinal cells and the nongerminal (Sertoli) cells. The spermatogonia and the Sertoli cells are the only tubular cells that lie on the tubular membrane called the **basal lamina**. The other germinal cells (spermatocytes) are found between the Sertoli cells.
      (1) Spermatogonia are nonmotile stem cells that divide by mitosis to form two cellular pools: a pool of additional stem cells, which undergo continual renewal by mitosis, and a pool of **type A spermatogonia,** which enter the maturation process called **spermatogenesis**. Each spermatogonial cell (type A) gives rise to 64 sperm cells.
         (a) Spermatogenesis requires about 64 days in man, and the transport of sperm cells through the epididymis to the ejaculatory duct requires an additional 12–21 days. There are three phases of spermatogenesis.
            (i) The type A spermatogonia divide by mitosis to form **primary spermatocytes,** which are **diploid** cells.
            (ii) The primary spermatocytes continue the gametogenic process, which leads to the formation of **secondary spermatocytes** and then, through two meiotic divisions, to the formation of **spermatids.** Spermatids are **haploid** cells.
            (iii) The third phase of spermatogenesis produces mature **spermatozoa** via the metamorphosis of spermatids. This process, called **spermiogenesis,** is characterized by the absence of cell division.
         (b) Because testosterone and FSH act directly on the seminiferous epithelium, these hormones are required for spermatogenesis.
      (2) Sertoli cells are nonmotile and, in the mature testis, nonproliferating tubular cells that lie on the basal lamina.
         (a) Structural features. The Sertoli cells extend through the entire thickness of the germinal epithelium from the basement membrane to the lumen. The tight junctions between the bases of the Sertoli cells serve two functions.
            (i) They divide the seminiferous tubular epithelium into two functional pools: a basal compartment containing the spermatogonia and an adluminal compartment containing the spermatogonia and spermatids.
            (ii) They form an effective permeability barrier within the seminiferous epithelium, which is defined in man as the blood-testis barrier that limits the transport of many substances from the blood to the seminiferous tubular lumen. This barrier maintains germ cells in an immunologically privileged location, since mature sperm cells are very immunogenic when introduced into the systemic circulation.
         (b) Functions
            (i) Sertoli cells provide mechanical support for the maturing gametes.
            (ii) They also have a role in spermatogenesis, which may be attributable to their high glycogen concentration being a potential energy source.
            (iii) In the fetal testis, the Sertoli cells secrete H-Y antigen, which is a product of the testis-organizing genes and is the cell surface glycoprotein responsible for the induction of testicular organogenesis. H-Y antigen has been found in all cell membranes from normal XY males except the cell membranes of immature germ cells.
            (iv) The Sertoli cells secrete müllerian duct inhibiting factor, the glycoprotein that causes regression of the müllerian duct system.

     (v) With the onset of fetal testicular differentiation induced by H-Y antigen and the incorporation of primitive germ cells into the seminiferous tubules, the Sertoli cells secrete a meiosis-inhibiting factor, which suppresses germ cell proliferation and differentiation beyond the spermatogonial stage.

     (vi) Sertoli cells participate in the release of spermatids by enveloping the residual lobules of spermatid cytoplasm. Thus, the residual bodies are not cast off into the lumen but are retained within the epithelium throughout the spermiation process.

     (vii) Sertoli cells secrete a watery, solute-rich (K^+ and HCO_3^-) fluid into the seminiferous lumen. These cells actively pump ions into the intercellular spaces to create a standing osmotic gradient that moves water from the Sertoli cell base to the free surface of the lumen. This fluid movement provides a driving force for conveying sperm from the testis to the epididymis, where most of this isosmotic fluid is reabsorbed.

     (viii) Sertoli cells synthesize estradiol from androgenic precursors.

     (ix) Sertoli cells also produce and secrete an **androgen-binding protein** and **inhibin** (see IX D 1 a).

    c. Leydig cells are located between the seminiferous tubules and produce androgenic steroids. These cells also are called **interstitial cells**.

     (1) Leydig cells are extensive at birth but virtually disappear within the first 6 months of postnatal life. Their reappearance marks the onset of puberty.

     (2) At puberty, the fibroblast-like cells of the testis serve as stem cells that differentiate into Leydig cells.

C. Hormones of the testis: steroids

  1. Secretion and transport

    a. Testosterone is the major hormone produced by the Leydig cells of the testis. Like all naturally occurring androgens, testosterone consists of 19 carbon atoms.

     (1) Testosterone is not stored in the testis. Cholesterol esters, the major precursor for testosterone biosynthesis, are stored in the lipid droplets in the Leydig cells.

     (2) A normal man secretes 4–9 mg of testosterone daily. More than 97% of secreted testosterone is bound to plasma proteins; 68% is bound to albumin, and 30% is bound to testosterone-binding globulin (also called sex hormone–binding globulin, because it binds estradiol as well). A very small percentage of the plasma testosterone is unbound.

    b. Androstenedione also is secreted by the testis at a rate of about 2.5 mg/day and is an important steroid precursor for blood estrogens in men.

     (1) Many nonendocrine tissues (e.g., brain, skin, fat, liver) have the cytochrome P-450-dependent aromatase, which converts androgens to estrogens.

     (2) Major portions of blood estradiol and estrone in normal men are derived from blood testosterone and androstenedione, respectively.

     (3) In addition, the Sertoli and Leydig cells of the testis secrete small amounts of estradiol.

    c. Dihydrotestosterone is synthesized by the testis, probably due to the action of **5α-reductase** from the Sertoli cells on testosterone secreted by the Leydig cells.

     (1) Only 20% of plasma dihydrotestosterone is synthesized in the testis. The remainder is derived from the peripheral conversion of testosterone, which serves as a prohormone in the skin and male reproductive tract (prostate gland and seminal vesicles).

     (2) Dihydrotestosterone has more than twice the biologic activity of testosterone.

  2. Testicular steroidogenesis (see Figure 7-7)

    a. The key step in steroidogenesis is the conversion of cholesterol to pregnenolone. This reaction is catalyzed by 20,22-desmolase, the rate-limiting enzyme for steroid biosynthesis in all steroid-producing tissues.

     (1) LH* activates 20,22-desmolase and, therefore, is the pituitary gonadotropic hormone that regulates testosterone synthesis by the Leydig cells.

     (2) In the developing male fetus, the stimulus for testosterone synthesis is chorionic gonadotropin, which is the placental hormone secreted in the highest amounts during the first trimester.

    b. The Leydig cells contain 17α-hydroxylase, which hydroxylates pregnenolone at position 17.

---

*In the male, LH is known as **interstitial cell–stimulating hormone (ICSH)**.

(1) Some pregnenolone may be converted via 3-hydroxysteroid dehydrogenase/Δ^5-isomerase to progesterone, which can be hydroxylated at the C-17 position prior to its conversion to androstenedione along the Δ^4-pathway.

(2) In the human testis, the Δ^5-pathway is the preferential pathway for the synthesis of testosterone, which is a Δ^4-steroid.

   c. Androstenedione is the common final precursor in the synthesis of testosterone.

3. Metabolism

   a. Dihydrotestosterone formation. Testosterone can serve as a prohormone and be metabolized by 5α-reductase to the more active androgen, dihydrotestosterone. The activity of 5α-reductase is high in the skin, prostate gland, seminal vesicles, epididymis, and liver. This enzyme also is present in testicular tissue.

   (1) Androgen target tissues are thought to be the principal sites of dihydrotestosterone formation. The nuclear membrane and microsomes of androgen-sensitive tissues are the physiologically important sites for conversion of testosterone to dihydrotestosterone. Therefore, dihydrotestosterone not only is secreted by the testis into the circulation but also is synthesized from testosterone that has entered the cells of androgen-dependent tissues.

   (2) The formation of dihydrotestosterone is an irreversible reduction reaction; therefore, dihydrotestosterone cannot serve as a proestrogen.

   (3) The anlagen of the prostate gland and external genitalia can form dihydrotestosterone prior to the onset of virilization. Dihydrotestosterone is formed from testosterone prior to the secretion of significant amounts of testosterone from the testis. The wolffian duct derivatives can form dihydrotestosterone after the onset of androgen secretion and after differentiation of the male genital system is far advanced.

   (4) Dihydrotestosterone can be metabolized further to 17-ketosteroids and polar derivatives found in the urine.

   b. Estradiol and estrone formation. Testosterone and androstenedione can be converted to estradiol and estrone, respectively, by the action of aromatase. Thus, the estrogens in the male are derived from direct secretion by the testes and from peripheral conversion of circulating androstenedione and testosterone.

   (1) Aromatization of circulating androgens is the major pathway for estrogen formation in the male.

   (2) Aromatases are membrane-bound enzymes found in the brain, skin, liver, mammary tissues, and most significantly, the adipose tissue.

   c. 17-Ketosteroid formation. Testosterone can be metabolized to less active metabolites that are conjugated in the liver and excreted into the urine as 17-ketosteroids.

   (1) Androsterone and **etiocholanolone** are the major urinary metabolites of testosterone. **Testosterone glucuronide** and **5α-androstanediol glucuronides** are among the other androgenic metabolites measured in urine.

   (a) Testosterone glucuronide originates mainly in the liver from testosterone, androstenedione, and dihydrotestosterone. The measurement of plasma testosterone is the mainstay for assessing Leydig cell function.

   (b) 5α-Androstanediol glucuronides arise from the testosterone metabolites in both the liver and the skin. Because an increased 5α-reductase activity in the skin and other extrahepatic tissues is produced by increased androgen secretion, the measurement of urinary 5α-androstanediol glucuronides has been recommended as an index of clinical androgenicity.

   (2) The **excretory rate** for urinary 17-ketosteroids in normal men is 15–20 mg/day.

   (a) Of this amount, 20%–40% are of testicular origin. The remainder are adrenocortical secretions, the major one of which is **DHEA**.

   (b) Because the urinary 17-ketosteroid pool reflects mainly adrenocortical activity, a measurement of the 17-ketosteroid excretion is not a good index of testicular function.

D. Hormonal control of testicular function (see Figure 7-2)

1. Hypothalamic-hypophysial-seminiferous tubular axis

   a. LH and FSH are required for spermatogenesis. Since the effects of LH are mediated by testosterone, testosterone and FSH are two hormones that act directly on Sertoli cells to promote gametogenesis.

(1) Exogenous testosterone alone does not promote spermatogenesis in men lacking Leydig cells. Spermatogenesis requires that a high concentration of testosterone be produced locally by LH action on Leydig cells.

(2) Sertoli cells synthesize **androgen-binding protein** by an FSH-dependent process. This protein binds testosterone and dihydrotestosterone and functions as a local androgenic pool to support spermatogenesis.

(3) Sertoli cells also synthesize **inhibin** in response to FSH secretion. This protein inhibits FSH secretion by direct negative feedback on the pituitary gland. Inhibin is not known to suppress secretion of FSH releasing hormone (FSH-RH), a decapeptide produced by the parvicellular peptidergic neurons. The selective rise in plasma FSH levels in individuals with damaged seminiferous tubules is due to a reduced secretion of inhibin.

b. Since normal spermatogenesis occurs in men with a 5α-reductase deficiency, dihydrotestosterone is not required for normal sperm development.

c. Plasma physiologic levels of androgens have little effect on the inhibition of FSH secretion.

d. Testosterone administration has little effect on FSH secretion, and very large doses are required to suppress FSH in the male.

2. Hypothalamic-hypophysial-Leydig cell axis

a. The rate of testosterone synthesis and secretion by Leydig cells is stimulated primarily by LH. The secretion of testosterone, in turn, inhibits LH secretion. It is unbound testosterone that suppresses LH secretion. The likely major target of negative feedback is the hypothalamus.

b. Both testosterone and estradiol can inhibit LH secretion; however, since dihydrotestosterone can also suppress LH, androgen conversion to estrogen is not a prerequisite for this inhibitory action on the hypothalamus and pituitary gland.

c. Some neurosecretory neurons of the hypothalamus secrete a releasing hormone into the hypophysial portal system, which then conveys it to the anterior lobe of the pituitary gland.

(1) This decapeptide, called **gonadotropin-releasing hormone (Gn-RH),** stimulates the pituitary basophils.

(2) Gn-RH also is called LH releasing hormone (LH-RH) and FSH-RH because it elicits the secretion of both LH and FSH.

E. Physiologic effects of androgens (Table 7-8)

1. Reproductive function. Androgens are essential for the control of spermatogenesis, the maintenance of the secondary sex characteristics, and the functional competence of the accessory sex organs.

a. The accessory sex organs consist of excretory ducts and glands that transmit spermatozoa and that secrete seminal fluid necessary for the survival and motility of spermatozoa after ejaculation. The major accessory sex organs are the prostate gland and seminal vesicles.

b. The secondary sex characteristics are the physiologic characteristics of masculinity (e.g., growth of facial hair, recession of hair at the temples, enlargement of the larynx, thickening of the vocal cords).

c. Seminal plasma, the fluid in which spermatozoa normally are ejaculated, originates almost entirely from the prostate gland and seminal vesicles.

(1) The volume of the human ejaculate (semen) is 2–5 ml, most of which is contributed by the seminal vesicles.

(a) The prostate gland is the origin of citric acid, acid phosphatase, zinc, and spermine.

Table 7-8. Major Actions of Androgenic Hormones

Life Stage	Testosterone	Dihydrotestosterone
Fetal period	Development of epididymis, vas deferens, and seminal vesicles	Development of penis, penile urethra, scrotum, and prostate gland
Puberty	Growth of penis, seminal vesicles, musculature, skeleton, and larynx	Growth of scrotum, prostate gland, pubic hair, and sebaceous glands
Adulthood	Spermatogenesis	Prostatic secretions

Reprinted from Genuth SM: The reproductive glands. In *Physiology,* 2nd edition. Edited by Berne RM, Levy MN. St. Louis, CV Mosby, 1988, p 999.

(b) The seminal vesicles are the source of prostaglandins, fructose, ascorbic acid, and phosphorylcholine. Metabolism of fructose provides energy for sperm motility.

(2) Sperm represents less than 10% of the ejaculate volume. The ejaculate contains 100–300 million spermatozoa!

    (a) Euspermia is defined as greater than 20 million sperm/ml of ejaculate.

    (b) Oligospermia is defined as 5–20 million sperm/ml of ejaculate.

    (c) Azoospermia is defined as 5 million sperm/ml of ejaculate.

2. **Biologic effects.** Androgens stimulate cell division as well as tissue growth and maturation and are classified as protein anabolic hormones. (This anabolic effect in muscle is referred to as the **myotropic effect** of androgens.) Only testosterone and dihydrotestosterone have significant biologic activity. The 17-ketosteroids, androstenedione, DHEA, and etiocholanolone, are weak androgens but important metabolites of testosterone.

  a. In the adolescent, androgens produce linear growth, muscular development, and retention of nitrogen, potassium, and phosphorus. Testosterone also accelerates epiphysial fusion of the long bones. The skeletal development during puberty, particularly of the shoulder girdle, is pronounced.

  b. Testosterone stimulates differentiation of the wolffian duct system into the epididymis, vas deferens, and seminal vesicles; dihydrotestosterone stimulates organogenesis of the urogenital sinus and tubercle into the prostate gland, penis, urethra, and scrotum.

  c. Androgens produce a low-pitched voice; stimulate growth of chest, axillary, and facial hair; and cause temporal hair recession.

  d. Androgens are responsible for libido.

  e. Testosterone is a requisite for normal spermatogenesis.

X. OVARY

A. Embryology

1. **Internal genitalia** (see Figure 7-8). The primordia of both male and female genital ducts, which are derived from the mesonephros, are present in the fetus at 7 weeks gestation. The paired **müllerian ducts** form parallel to the paired **wolffian ducts**. In the female fetus, the upper ends of the müllerian ducts are the anlagen of the fallopian tubes (oviducts), whereas the lower ends join to form the uterus, cervix, and upper end of the vagina.

  a. The uterus and fallopian tubes, which develop from the müllerian ducts, do not require the presence of an ovary.

  b. In the absence of a fetal testis, müllerian duct inhibiting factor and testosterone are not secreted by the fetal Sertoli cells and Leydig cells, respectively. Moreover, dihydrotestosterone is not formed from testosterone. Without the presence of these three hormones, at 10–11 weeks gestation, the müllerian ducts begin to differentiate and the wolffian ducts undergo regression.

  c. This process of female genital duct development is completed at 18–20 weeks gestation.

2. **External genitalia**

  a. The external genitalia of both sexes begin to differentiate at 9–10 weeks gestation. In the female, this process proceeds without any known hormonal influence.

  b. The external genitalia and urethra in both sexes develop from common anlagen, which are the urogenital sinus and the genital tubercle, folds, and swelling.

    (1) The urogenital sinus gives rise to the lower portion of the vagina and to the urethra.

    (2) The genital tubercle is the origin of the clitoris.

    (3) The genital swelling is the primordium of the labia majora.

    (4) The genital folds develop into the labia minora.

3. **Ovary.** In the absence of the H-Y antigen, the gonadal primordium develops into an ovary, provided that germ cells are present. Ovarian development occurs several weeks later than does testicular differentiation.

  a. At 8 weeks, when the testicular secretion of testosterone begins and before the ovarian differentiation is completed, the fetal "ovary" has the capacity to synthesize **estradiol.**

    (1) It is unlikely that the fetal ovary contributes significantly to the circulating estrogens in the fetus. The fetus is exposed to estrogen (**estriol**) of placental origin. The site of estradiol synthesis by the primordial ovary is not known.

    (2) The ovary has no role in sex differentiation of the female genital tract.

b. Also at about 8 weeks gestation, the cortex of the primitive gonad undergoes active mitosis, and epithelial cells infiltrate the gonad as a syncytium of tubules and cords. Primordial germ cells are carried along with this inward migration.
  (1) Proliferation of the cortex ceases at about 6 months.
  (2) The **rete ovarii** secretes a meiosis-inducing factor.
  (3) The germ cells begin to proliferate to form **oogonia**. This mitotic process is maximal between 8 and 20 weeks, after which it diminishes, ceases, and never is resumed.
c. Unlike the fetal testis, the fetal ovary begins gametogenesis. **Oogenesis,** the formation of primary oocytes from oogonia, begins at 15 weeks and reaches a peak between 20 and 28 weeks gestation.
  (1) The **primary oocytes** enter a prolonged prophase (diplotene stage) of the first meiotic division and remain in this state until ovulation first occurs between 10 and 45 years later!
  (2) The **diploid primary oocytes** become enveloped by a single layer of flat granulosa cells and in this form are called **primordial follicles**. The formation of primordial follicles reaches a peak between 20 and 25 weeks.
d. Also between 20 and 25 weeks, the gonad has acquired the morphologic appearance of an ovary. During this period, the plasma concentration of pituitary FSH reaches a peak and the first **primary follicles** appear [see X E 3 a (1)].

B. Oogenesis

1. Fetal oogenesis. The period of oogonial proliferation results in a peak population of about 6–7 million germ cells in the two ovaries at 5 months gestation. Included in this group of cells are oogonia, oocytes in various stages of prophase, and degenerating germ cells. This total number of germ cells decreases to 2 million at term. The number of primordial follicles present in the ovary at birth rapidly diminishes thereafter.

2. Postnatal oogenesis. By 6 months postpartum, all of the oogonia have been converted to primary oocytes. By the onset of puberty, the number of primary oocytes has decreased to about 400,000.

3. Prepubertal oogenesis. Between birth and puberty, the primary oocyte is surrounded by the zona pellucida and six to nine layers of granulosa cells. These follicles are in varying stages of development.

4. Pubertal oogenesis. In contrast to the male who produces spermatogonia and primary spermatocytes continuously throughout life, the female cannot form oogonia beyond 28 weeks gestation and must function with a declining pool of oocytes.
a. Meiosis in the female results in the formation of one viable oocyte (**ootid**). In contrast, each primary spermatogonium in the male ultimately gives rise to 64 spermatozoa.
b. Oogenesis in the female begins in utero in response to meiosis-stimulating factor, while in the male spermatogenesis is arrested at the spermatogonial stage in response to meiosis-inhibiting factor.
c. Just prior to ovulation, the first polar body is extruded from the primary oocyte, which completes the first meiotic division, and forms a secondary oocyte.
  (1) This haploid cell immediately begins the second meiotic division but remains in metaphase.
  (2) Extrusion of the second polar body (**polocyte**) does not occur until the mature ovum (ootid) is fertilized by a sperm cell. Fertilization normally occurs in the ampulla of the fallopian tube.

C. Morphology

1. Gross anatomy
a. The ovaries are ovoid glands with a combined weight of 10–20 g during the reproductive years.
b. The ovaries are anchored to the **broad ligament** by the **mesovarium**.

2. Functional histology of the ovary and uterus
a. Ovary
  (1) Structural divisions include the **cortex** (which is lined by the germinal epithelium and contains all of the oocytes), the **inner medulla,** and the **hilus** (i.e., the point where the ovary attaches to the mesentery).

(2) Functional subunits include the **follicle** and **oocyte** (each consisting of **theca cells** and **granulosa cells**) and the **corpus luteum.**

(3) Gametogenesis in the female denotes **folliculogenesis,** which leads to the formation of a mature ovum.

(a) During the preovulatory phase, the functional unit of the ovary is the follicle.

(b) During the postovulatory phase, the functional unit of the ovary is the corpus luteum.

(4) Steroidogenesis in the ovary is the synthesis and secretion of **estradiol** and **progesterone**. Although steroidogenesis occurs in three morphologic units (i.e., the follicle, corpus luteum, and stroma), only the follicle and corpus luteum are major steroid-producing units.

(a) Theca interna cells produce androstenedione and testosterone, which diffuse into the granulosa cells.

(b) Granulosa cells synthesize estradiol and estrone from androgenic precursors produced by the theca interna cells.

b. Uterus

(1) Layers. The uterus consists of two major tissue layers.

(a) The outer layer, the **myometrium,** is a thick layer of smooth muscle.

(b) The inner layer of the uterus is the **endometrium**. At the height of its development, during the luteal phase, the endometrium is approximately 5 mm thick. On the basis of blood supply, the endometrium can be divided into two major layers.

(i) The **stratum basalis** is the abluminal (deeper) layer of the endometrium. This layer functions as the regenerative layer in the growth and differentiation of endometrial tissue that is sloughed during menses.

(ii) The **stratum functionalis** is the adluminal (superficial) layer of the endometrium, which is shed during menses.

(2) Endometrial blood supply

(a) The stratum basalis receives its vascular supply from the **basal (straight) arteries,** which arise from the uterine radial arteries.

(b) The stratum functionalis is perfused by the **spiral (coiled) arteries,** which also emanate from the radial arteries.

(c) Thus, the arcuate arteries give off radial arteries, and the radial arteries bifurcate to form the basal and coiled arteries.

D. Hormones of the ovary: steroids. Ovarian hormones include two phenolic steroids—**estradiol** (C-18) and **estrone** (C-18)—and the progestogen, **progesterone** (C-21).

1. Secretion and transport (Tables 7-9, 7-10)

a. Estrogens. Over 70% of circulating estrogens are bound to sex steroid–binding globulin, and 25% are bound to plasma albumin.

(1) Estradiol is the principal and biologically most active estrogen secreted by the ovary. Ovarian estradiol accounts for more than 90% of the circulating estradiol.

(2) Estrone, a weak ovarian estrogen, also is formed by the peripheral conversion of **androstenedione.**

(a) In premenopausal women, most of the circulating estrone is derived from estradiol by conversion via 17-hydroxysteroid dehydrogenase.

(b) In postmenopausal women, estrone is the dominant plasma estrogen and is derived via the prohormone pathway. Specifically, estrone is derived from the conversion of adrenocortical androstenedione in peripheral tissues (mainly liver). In obese women, there is a significant peripheral conversion of androstenedione to estrone by adipose tissue. This extraovarian synthesis of estrogen is implicated in the higher incidence of endometrial carcinoma in obese women.

(3) Estriol, the weakest of all of the naturally occurring estrogens, is synthesized by the placenta and the liver but not the ovary. In nongravid women, estriol is formed in the liver as a conversion product of estradiol and estrone.

b. Progesterone is not bound to sex hormone–binding globulin. The progesterones are bound primarily to CBG (transcortin) and albumin.

Table 7-9. Types of Steroids and their Systemic Concentrations and Rates of Synthesis in Women

Steroid	Plasma Concentration (ng/dl)	Production Rate (μg/day)
Estrogens (C-18)		
Estradiol		
Early follicular phase	6	80
Late follicular phase	50	700
Middle luteal phase	20	300
Estrone		
Early follicular phase	5	100
Late follicular phase	20	500
Middle luteal phase	10	250
Progestogens (C-21)		
17-Hydroxyprogesterone		
Early follicular phase	30	600
Late follicular phase	200	4000
Middle luteal phase	200	4000
Progesterone		
Follicular phase	100	2000
Luteal phase	1000	25,000
Testosterone	40	250
Dihydrotestosterone	20	50
Androstenedione	150	3000
Dehydroepiandrosterone	500	8000

Reprinted from Lipsett MB: Steroid hormones. In *Reproductive Endocrinology: Physiology, Pathophysiology and Management.* Edited by Yen SSC, Jaffe RB. Philadelphia, WB Saunders, 1978, p 84.

Table 7-10. Serum FSH and LH Concentrations During the Life Cycle of the Normal Female*

Stage of Life	FSH[†]	LH[†]
Prepubertal period (5–11 years)	4.5	3.9
Puberty (11–13 years)	6.8	8.2
Reproductive period		
Follicular phase	8.3	12.8
Midcycle	19.3	83.5
Luteal phase	6.9	11.6
Postmenopausal period	96.0	66.0

Adapted from Ontjes DA, Walton J, Ney RL: The anterior pituitary gland. In *Metabolic Control of Disease,* 8th edition. Edited by Bondy PK, Rosenberg LE. Philadelphia, WB Saunders, 1980, p 1192.
*Mean values without standard deviations.
[†]Concentrations expressed in milli-International units.

2. **Ovarian steroidogenesis** (see Figure 7-7). The two pathways for biosynthesis of ovarian steroids have in common the conversion of cholesterol to pregnenolone, a reaction stimulated by LH and FSH via 20,22-desmolase.
   a. One pathway proceeds by way of the Δ^5-pathway, which involves the synthesis of 17α-hydroxypregnenolone and DHEA via 17α-hydroxylase and 17,20-lyase, respectively.
      (1) DHEA, a 17-ketosteroid,* is converted to another androgenic 17-ketosteroid,* androstenedione, by 3β-hydroxysteroid dehydrogenase and Δ^5-reductase.

---

*These 17-ketosteroids are steroids that consist of 19 carbon atoms.

(2) Androstenedione can be reduced by 17-hydroxysteroid dehydrogenase to testosterone, which is a reversible reaction.

(3) Testosterone is a precursor of estradiol via an aromatase reaction.

b. The other pathway proceeds via the conversion of pregnenolone to progesterone by 3β-hydroxysteroid dehydrogenase and Δ^5-isomerase. Progesterone is the initial compound in the Δ^4-pathway.

(1) Progesterone is converted to 17α-hydroxyprogesterone by 17α-hydroxylase.

(2) 17α-Hydroxyprogesterone is another precursor for androstenedione via another 17, 20-lyase step.

(3) Androstenedione and testosterone are interconvertible with the enzyme 17-hydroxysteroid dehydrogenase.

(4) Androstenedione and testosterone are converted to estrone and estradiol, respectively, by the action of aromatase. These two 18-carbon steroids (estrogens) also are interconvertible with 17-hydroxysteroid dehydrogenase.

3. Metabolism of ovarian steroids. The liver is the major site of steroid metabolism.

a. Catabolism of progestogens

(1) Progesterone is converted to **pregnanediol**.

(2) 17α-Hydroxyprogesterone is catabolized to **pregnanetriol**.

b. Catabolism of estrogens

(1) Large quantities of both estradiol and estrone are hydroxylated (primarily in the liver) at the C-16 position to form **estriol**.

(2) Another major catabolic route for estrogens is hydroxylation at the C-2 and C-4 positions, which yields the **catecholestrogens**.

c. Estrogens are excreted in the urine in the form of soluble conjugates.

(1) Estriol, catecholestradiol, and catecholestrone are excreted primarily as glucuronidates.

(2) Estrone is excreted primarily as a sulfate conjugate.

E. Ovarian function

1. Menstruation

a. Menarche refers to the onset of menstruation, which normally occurs between the ages of 12 and 14 years. Prior to menarche, minimal amounts of estrogen are produced by the peripheral conversion of androgens.

b. Menstrual cycle

(1) Duration. Although a duration of 25–30 days is considered typical, a cycle length of 28 days is the exception rather than the rule in adult women. In early adolescence, the cycle is characterized by irregular menses and anovulation.

(2) Temporal reference points (Figure 7-9). The menstrual cycle conventionally begins with the first day of menstruation, when the endometrial lining is shed along with blood and uterine secretions. Days of the menstrual cycle are measured using two different reference points.

(a) One system designates the first day of menses as **day 0** and the last day as **day 28**.

(b) The other system designates the day of the LH peak (ovulation) as **day 0,** with preovulatory days indicated with a **minus sign** and postovulatory days indicated with a **plus sign**.

c. Menopause refers to the cessation of menses, which typically occurs at about age 50 following a gradual decrease in frequency. Menopause is due to a primary hypogonadism (i.e., cessation of ovarian steroid secretion) and is associated with an increased gonadotropin (predominantly FSH) secretion.

2. Ovarian cycle

a. Preovulatory phase. This phase is marked by follicular growth and maturation and by endometrial proliferation. The preovulatory phase generally lasts 8–9 days but can be quite variable (10–25 days). During this phase, the stratum basalis regenerates a stratum functionalis, and, by the end of this phase, one follicle (rarely more) has reached the final stage of growth.

(1) A **primary follicle** begins as an oocyte surrounded by a single layer of cuboidal epithelial cells called granulosa cells. The primary follicle becomes multilaminar by the mitosis of the granulosa cells, which occurs with maturation.

(2) Upon cavitation of the granulosa, an **antrum** is formed, which is filled with **liquor folliculi** secreted by the granulosa cells. The developing follicle now is called the **secondary follicle** (vesicular follicle or, more commonly, graafian follicle).

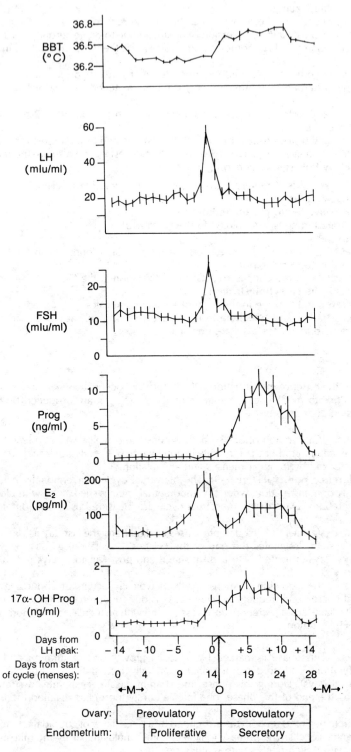

Days from LH peak: −14 −10 −5 0 +5 +10 +14

Days from start of cycle (menses): 0 4 9 14 19 24 28

←M→ O ←M→

Ovary:	Preovulatory	Postovulatory
Endometrium:	Proliferative	Secretory

Figure 7-9. Hormonal (ovarian and pituitary), uterine (endometrial), and basal body temperature (*BBT*) correlates of the normal menstrual cycle. Mean plasma concentrations (± SEM) of LH, FSH, progesterone (*Prog*), estradiol (*E₂*) (day +1) and 17α-hydroxyprogesterone (*17α-OHProg*) are shown as a function of time. Ovulation occurs on day 15 (day +1) following the LH surge, which occurs at midcycle on day 14 (day 0). *M* = menses, *O* = ovulation. (Adapted from Thorneycroft IA, et al: The relation of serum 17-hydroxyprogesterone and estradiol 17-β levels during the human menstrual cycle. *Am J Obstet Gynecol* 111:947–951, 1971.)

(a) The granulosa cells secrete a protective shell, the **zona pellucida,** which surrounds the oocyte.

(b) The stroma gives rise to a bilaminar **theca,** which surrounds the granulosa cells but is separated from them by a basal lamina (**lamina propria**).

(i) The **theca interna** is a well-vascularized layer consisting of steroid-secreting cells that lie on the basal lamina. The blood vessels do not penetrate this membrane and, therefore, the granulosa cells are avascular until after ovulation.

(ii) The **theca externa** is peripheral to the theca and is composed mainly of fibrous connective tissue. It is less vascular than the theca interna.

(3) Just prior to ovulation, the primary oocyte of the secondary follicle completes the first meiotic division, which began prior to birth, and forms a secondary oocyte with a haploid nucleus and the first polar body. The second meiotic division takes place in the ampulla of the oviduct and occurs only if fertilization occurs.

b. Ovulation. The secondary oocyte is released from the secondary follicle by a process called **ovulation,** which usually occurs on **day 15** of the average cycle. (In reference to the LH peak, this midcycle event occurs on **day 1,** that is, one day following the LH surge.)

c. Postovulatory phase. The next 13–14 days constitute the postovulatory phase, during which the endometrium is prepared for the possible implantation of the fertilized ovum, which is a blastocyst when it arrives in the uterine cavity. This phase is relatively constant in duration. As a result, the day of ovulation can be estimated by subtracting 14 days from the total length of the menstrual cycle.

(1) If fertilization does not occur, implantation also does not occur because the hormonal maintenance of the endometrial growth and differentiation is withdrawn.

(2) Without conception, ischemia and necrosis of the luminal endometrium result after 14 days, and the ensuing menses marks the beginning of another menstrual cycle.

(3) If conception occurs, the functional lifespan of the corpus luteum is extended, and it continues to secrete estradiol and progesterone at increasing rates during the first 6–8 weeks gestation.

F. Neuroendocrine control of ovarian function (see Figure 7-4). The ovarian cycle is associated with the secretion of ovarian steroids (estradiol and progesterone), which, in turn, are regulated by FSH and LH from the pituitary gland. A single hypothalamic hormone, Gn-RH, differentially regulates the secretion of FSH and LH.

1. Ovarian cycle

a. Preovulatory phase. Under the influence of FSH and LH, the primary follicle begins to develop. The combined effects of LH on the theca* cells to produce androgenic proestrogens and of FSH on the granulosa cells to aromatize these androgens to estrogens (estradiol) result in a slowly increasing blood estradiol concentration. During the preovulatory phase, the dominant gonadotropic hormone is FSH, and the dominant steroid is estradiol. Therefore, the preovulatory phase of the ovarian cycle also is referred to as the **follicular phase** and the **estrogenic phase**.

(1) Plasma estradiol concentration reaches a peak about 24 hours prior to the surge in LH secretion, or about 48 hours prior to ovulation. The peak in plasma estradiol concentration occurs on day 13 (day −1).[†]

(a) As the plasma estradiol concentration increases, it exerts a negative feedback on the hypothalamic-hypophysial complex, resulting in a gradual decline in FSH. LH levels rise slightly through the follicular phase.

(b) Inhibin secretion by the granulosa cells also exerts a negative feedback effect on FSH secretion.

(2) The peak in plasma estradiol concentration exerts a positive feedback effect on the hypothalamic-hypophysial axis, causing a reflex release of Gn-RH and a concomitant surge in pituitary LH secretion 24 hours later, on day 14 (day 0).

(a) A lesser increase in plasma FSH secretion occurs on day 14 (day 0).

(b) Increased plasma estradiol levels inhibit FSH secretion both directly and indirectly at the level of the pituitary gland and ventral diencephalon, respectively.

---

*For the remainder of this chapter, the term theca denotes theca interna.

[†]Also throughout this chapter, the cycle days are numbered with reference to the onset of menses and with reference to the LH peak (in parentheses).

(3) The follicular phase usually lasts 14 days, but any variability in the duration of the menstrual cycle usually is attributable to variability in the length of the follicular phase. It should be noted that the follicular phase begins with the first day of menses, while the proliferative phase of the endometrium begins with the last day of menses.

(4) There is a fall in the plasma estradiol level following the estradiol peak. Note in Figure 7-9 that this fall in plasma estradiol precedes ovulation.

(5) During the preovulatory phase, the plasma progesterone concentration remains low.

    (a) The bulk of this progesterone is derived from the peripheral conversion of adrenal progestogens; however, large amounts of progesterone exist within the follicular antrum.

    (b) The principal progestin secreted by the granulosa cells during the late follicular phase is 17α-hydroxyprogesterone.

b. Ovulatory phase. Ovulation occurs on day 15 (day +1) in response to the surge of LH secretion that occurred 24 hours earlier.

(1) Ovulation refers to the extrusion of a haploid secondary oocyte into the peritoneal cavity. The oocyte enters the oviduct (fallopian tube) where fertilization occurs.

(2) The rupture of the secondary follicle by the midcycle surge of LH leads to the formation of a new endocrine tissue (i.e., the corpus luteum), which involves proliferation, vascularization, and luteinization of the theca and granulosa cells. LH, therefore, is called the **luteotropic hormone of the menstrual cycle.**

c. Postovulatory phase. LH also maintains the functional status of the corpus luteum during the postovulatory phase. The hormone secreted in the greatest amounts by the corpus luteum during this phase is progesterone. For these reasons, the postovulatory phase also is called the **luteal phase** and the **progestational phase**.

(1) The decline in estrogen secretion prior to ovulation and at the time of ovulation removes the positive feedback effect of estradiol on gonadotropin secretion.

(2) Both FSH and LH levels fall after their midcycle peaks but remain sufficiently high to stimulate the newly formed lutein-theca cells and lutein-granulosa cells to secrete estradiol, estrone, and progesterone.

(3) About 6 days after ovulation, on day 21 (day +7), the plasma concentrations of progesterone and 17α-hydroxyprogesterone peak coincidently with the second peak of plasma estradiol concentration. Note in Figure 7-9 that this second peak is lower and broader than the estradiol peak that occurs during the preovulatory phase.

(4) The plasma concentrations of progesterone during the entire ovarian cycle are higher than those of estradiol. It is imperative to note that the units of concentration for progesterone (ng/ml) are 1000 times greater than for estradiol (pg/ml).

(5) The effect of the raised plasma estradiol and progesterone levels is a negative feedback on FSH and LH, respectively. Progesterone acts as an antiestrogen at this time because it inhibits LH secretion when the second estradiol peak occurs.

(6) LH levels continue to decline during the luteal phase, while FSH levels begin to rise progressively during the late luteal phase.

(7) The total amount of estradiol secreted during the follicular phase is comparable to that secreted during the luteal phase. This can be appreciated by comparing the areas under the estradiol curve during these two phases, using day 15 (day +1) as the dividing line between the follicular and luteal phases.

(8) Unless conception occurs and is followed by implantation of the blastocyst, the corpus luteum undergoes involution following the reduction in gonadotropin secretion.

    (a) The declines in estradiol and progesterone secretion remove the negative feedback effect on the hypothalamic-hypophysial complex.

    (b) The corpus luteum regresses after about 12 days of steroid hormone secretion.

(9) Progesterone is associated with a 0.2–0.5° C rise in basal body temperature, which occurs immediately following ovulation and which persists during most of the luteal phase (see Figure 7-9).

    (a) The basal body temperature dips during the follicular phase.

    (b) This temperature increment is used clinically as an index of ovulation.

2. Endometrial cycle

a. Hormonal effects on the myometrium. Estradiol and progesterone are antagonistic with respect to their effects on the myometrium: estrogens promote uterine motility, and progestogens inhibit myometrial contractility.

b. Hormonal effects on the endometrium. Estradiol and progesterone are synergistic with respect to their effects on the endometrium during the proliferative and secretory phases.

   (1) The **proliferative (preovulatory) phase** of the menstrual cycle refers to the endometrial changes that occur in response to estradiol.

   (a) Estrogens stimulate mitosis of the stratum basalis, also correctly termed the stratum basale, which regenerates the stratum functionalis.

   (b) Estrogens stimulate angiogenesis (neovascularization) in the stratum functionalis as well as stimulate the growth of secretory glands. The blood vessels become the spiral arteries that perfuse the stratum functionalis, also correctly termed the stratum functionale. The glands contain glycogen but are nonsecretory at this time.

   (c) The cervical epithelium secretes a watery mucus in response to estrogen stimulation.

   (2) The **secretory (postovulatory) phase** is characterized by secretion of large amounts of both progesterone and estradiol by the corpus luteum. The endometrium during this secretory phase is hyperemic and has a "lace curtain" or "swiss cheese" appearance.

   (a) Progesterone promotes differentiation of the endometrium, including elongation and coiling of the mucous glands (which secrete a thick viscous fluid containing glycogen) and spiraling of the blood vessels.

   (b) Unless fertilization occurs, hormone secretion by the hypothalamic-hypophysial complex and ovarian steroid secretion decline on about day 25 (day +11).

   (i) Menses, beginning on day 0 (day −14) of the following cycle, starts with vasoconstriction of the spiral arteries, which causes ischemia and necrosis.

   (ii) The necrotic tissue releases vasodilator substances, causing vasodilation. The necrotic walls of the spiral arteries rupture, causing hemorrhage and shedding of cells over a period of 4–5 days.

G. Physiologic effects of ovarian steroids

1. Estrogens have important protein anabolic effects. Estrogens are responsible for the growth and development of the fallopian tubes, uterus, vagina, and external genitalia as well as the maintenance of these organs in adulthood. These steroids also promote cellular proliferation in the mucosal linings of these structures.

   a. Endometrium. Estrogens stimulate the regeneration of the stratum functionalis during the proliferative phase of the endometrial cycle.

   (1) The water content and blood flow to the endometrium are increased markedly.

   (2) The spiral arteries of the stratum functionalis are especially sensitive to estrogens and grow rapidly under their influence.

   b. Myometrium. Estrogens increase the amount of contractile proteins (i.e., actin and myosin) in the myometrium and, thereby, increase spontaneous muscular contractions. Estrogens also sensitize the myometrium to the action of oxytocin, which promotes uterine contractility.

   c. Cervix. Under the influence of estrogens, the uterine cervix secretes an abundance of thin, watery mucus.

   (1) This fluid can be drawn into very long threads when placed between two glass slides. This is a clinical index of estrogen activity called **spinnbarkheit**.

   (2) Cervical mucus also demonstrates the phenomenon of crystallization when it is dried on a glass slide. The characteristic **ferning pattern** is due to the accumulation of sodium chloride. This phenomenon also is used diagnostically as an index of endogenous estrogen secretion.

   d. Breast. Estrogens promote the development of the tubular duct system of the mammary gland. Estrogens are synergistic with progesterone in stimulating the growth of the lobuloalveolar portions of this gland.

   e. Bone. Estrogens, like androgens, exert a dual effect on skeletal growth in that they cause an increase in osteoblastic activity, which results in a growth spurt at puberty.

   (1) Estrogens hasten bone maturation and promote the closure of the epiphysial plates in the long bones more effectively than does testosterone. Therefore, the female skeleton usually is shorter than the male skeleton.

   (2) Estrogens are responsible for the oval or roundish shape of the female pelvic inlet. This inlet in the male is spade-shaped.

   (3) Estrogens, to a lesser degree than testosterone, promote the deposition of bone matrix by causing Ca^{2+} and HPO_4^{2-} retention. In large amounts, estrogens also promote retention of Na^+ and water.

    f. Liver. Estrogens stimulate the hepatic synthesis of the transport globulins, including thyroxine-binding globulin and transcortin.

      (1) This results in increased plasma concentrations of thyroxine and cortisol but unchanged amounts of free thyroxine.

      (2) Pregnant women often are in a state of mild hyperadrenocorticism because the elevated placental progesterone competes with cortisol for binding sites on transcortin, thus increasing plasma free cortisol.

2. Progesterone

    a. Endometrium. The endometrium, which proliferates under the influence of estrogens, becomes a secretory structure under the influence of progesterone.

      (1) The endometrial glands become elongated and coiled and secrete a glycogen-rich fluid.

      (2) Progesterone accounts for the differentiation of the stratum functionalis.

    b. Cervix. Under the influence of progesterone, the mucus secreted by the cervical glands is reduced in volume and becomes thick and viscid. This consistency of cervical mucus together with the absence of "ferning" provide presumptive evidence that ovulation and luteinization have occurred.

    c. Myometrium. Progesterone decreases the frequency and amplitude of myometrial contractions.

    d. Breast. This steroid also promotes lobuloalveolar growth in the mammary gland.

    e. Kidney. Progesterone promotes renal excretion of Na^+.

XI. ENDOCRINE PLACENTA

A. Placenta formation

    1. Timetable of early placental function (days measured from ovulation). The gestational period, measured from the time of conception (ovulation) to parturition, is 38 weeks (266 days).

    a. Day 0: Fertilization occurs in the distal portion of the oviduct, or **ampulla**.

      (1) Fertilization triggers the final stage of the second meiotic division of the oocyte. The second polar body is extruded from the oocyte, and a haploid number of chromosomes are present in the female pronucleus.

      (2) The lifespan of an unfertilized ovum is less than 20 hours following ovulation; sperm cells are viable for about 24 hours after ejaculation.

    b. Day +3 or +4: The morula enters the uterine cavity.

    c. Day +5 or +6: The morula forms a cavity, the **blastocoele,** which is transformed into a **blastocyst.**

    d. Day +7: The blastocyst is implanted into the endometrium. By the end of 1 week, a primitive uteroplacental circulation begins to develop.

    e. Day +21: The placenta is fully functional.

    2. Early placental formation

    a. At the time of implantation, or **nidation,** the blastocyst consists of two cellular masses (Figure 7-10).

      (1) The inner cell mass, the **embryoblast,** will form the embryo and, eventually, the fetus.

      (2) The outer rim of cells, the **trophoblast,** forms the attachment to the endometrium and gives rise to the fetal membranes.

    b. The endometrium, under the influence of progesterone secreted by the corpus luteum, is transformed into a **decidua,** which is the maternal portion of the placenta that surrounds the conceptus.

    3. Placental development. The trophoblast, which is entirely fetal in origin, develops into the placenta. The trophoblast forms two cell layers (see Figure 7-10).

    a. The inner layer forms the **cytotrophoblast,** which is on the fetal side of the blastocyst and is the progenitor of the syncytiotrophoblast.

    b. The outer layer forms the **syncytiotrophoblast**.

B. Hormones of the placenta (Figures 7-11, 7-12, 7-13). The fetus, placenta, and mother are interdependent and constitute a functional unit called the **feto-placento-maternal unit**. The pregnant woman at or near term produces 15–20 mg/day of estradiol, 50–100 mg/day of estriol, 250–300 mg/day of progesterone, 1–2 mg/day of aldosterone, and 3–8 mg/day of deoxycorticosterone (DOC). By itself, the placenta is an incomplete steroid-producing organ.

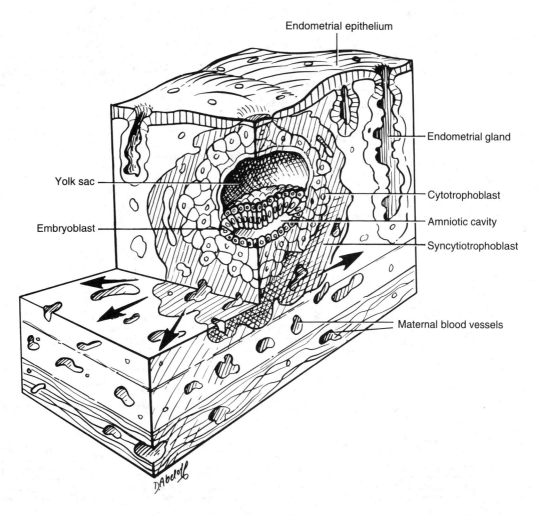

Figure 7-10. Structures formed by an implanting conceptus within the uterine endometrium. (Reprinted from Johnson KE: *Human Developmental Anatomy.* Baltimore, Williams & Wilkins, 1988, p 52.)

1. **HCG** is a polypeptide containing 236 amino acid residues, making it the largest active peptide hormone produced in humans.
    a. **Synthesis and secretion.** The syncytiotrophoblast is the source of HCG, which is secreted soon after fertilization.
    b. **Plasma concentration**
        (1) HCG reaches a plasma peak between 60 and 90 days gestation. This peak is 200 times greater than the LH peak at the height of the ovulatory surge.
        (2) HCG is detectable in maternal blood as early as 6–8 days after conception, which forms the basis for the immunologic pregnancy test.
        (3) HCG measurement in maternal blood is a useful index of the functional status of the trophoblast.

2. **Progesterone** (see Figures 7-11, 7-12, 7-13)
    a. **Synthesis and secretion**
        (1) Placental progesterone is derived mainly from maternal cholesterol; the fetus does not contribute significantly to placental progesterone formation.
        (2) This C-21 steroid is synthesized by the trophoblast. Most (85%) of the progesterone formed in the trophoblast is secreted into the maternal compartment.
    b. **Plasma concentration** of placental progesterone rises steadily throughout gestation, reaching a maximal plateau at 36–40 weeks. There is no significant drop in plasma progesterone concentration prior to labor.

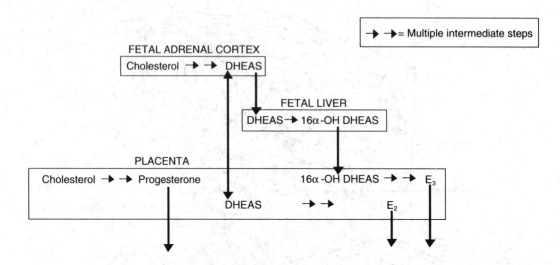

Figure 7-11. Pathways of steroid hormone biosynthesis in the feto-placental unit. *DHEAS* = dehydroepiandrosterone sulfate; *16α-OH DHEAS* = 16α-hydroxydehydroepiandrosterone sulfate; E_2 = estradiol; E_3 = estriol. (Adapted from Wilson JD, Foster DW: *Williams Testbook of Endocrinology,* 7th edition. Philadelphia, WB Saunders, 1985, p 423.)

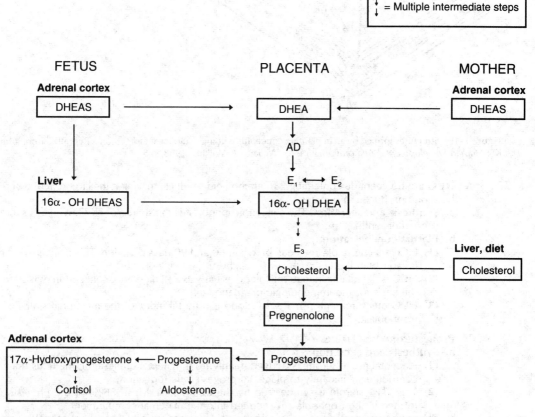

Figure 7-12. The primary pathways of estrogen and progesterone synthesis in the human feto-materno-placental unit. *DHEA(S)* = dehydroepiandrosterone (sulfate); *16α-OH DHEA(S)* = 16α-hydroxydehydroepiandrosterone (sulfate); *AD* = androstenedione; E_1 = estrone; E_2 = estradiol; E_3 = estriol.

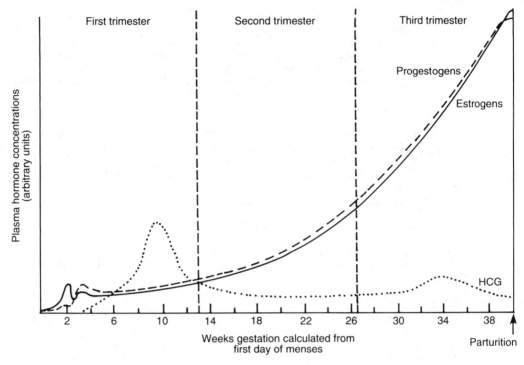

Figure 7-13. The patterns of maternal plasma hormone concentrations that occur during pregnancy. *HCG* = human chorionic gonadotropin. (Adapted from Laycock J, Wise P: *Essentials of Endocrinology,* 2nd edition. New York, Oxford University Press, 1983, p 155.)

> **(1)** Both 17α-hydroxyprogesterone and progesterone from the corpus luteum reach a peak 3–4 weeks postconception in response to HCG secretion.
>
> **(2)** At 6–8 weeks postconception, progesterone reaches a nadir, while 17α-hydroxyprogesterone continues to decline. The 17α-hydroxyprogesterone levels reflect corpus luteal secretion. The secondary rise in plasma progesterone reflects placental (trophoblast) secretion.
>
> **(3)** An intact materno-placental circulation will produce essentially normal progesterone levels, even in the event of fetal death.

  c. Metabolism. The principal urinary metabolite of progesterone is pregnanediol, which is an estimate of placental function.

  d. Conversion to fetal corticoids

> **(1)** The placenta produces pregnenolone from maternal cholesterol, but it lacks the enzymes necessary for androgen synthesis (i.e., 17α-hydroxylase and 17,20-desmolase).
>
> **(2)** Pregnenolone synthesized by the placenta is oxidized to progesterone by 3β-hydroxysteroid dehydrogenase/Δ^5-isomerase.
>
> **(3)** Placental progesterone circulates to the fetal adrenal cortex, where it is hydroxylated at positions C-17, C-21, and C-11 to form aldosterone and cortisol. Thus, in early pregnancy, the fetus requires placental progesterone to synthesize corticoids, because the fetal zone of the adrenal cortex has a relative block in the 3β-hydroxysteroid dehydrogenase/Δ^5-isomerase system.
>
> **(4)** Beyond 10 weeks gestation, the fetal adrenal cortex no longer depends on placental progesterone for synthesis of aldosterone and cortisol.

3. Estrogens (see Figures 7-11, 7-12, 7-13). Quantitatively, estriol is the major estrogen of human pregnancy, with smaller amounts of estradiol and estrone produced.

  a. Synthesis and secretion

> **(1)** These C-18 steroids are synthesized in the placental trophoblasts.
>
> **(2)** Estriol is produced primarily from androgenic precursors formed in the fetal zone of the adrenal cortex and the liver. The principal adrenal steroid is **DHEA,** which is a ketoste-

roid. DHEA is sulfoconjugated (sulfurylated) by sulfokinase* in the fetal adrenal to **DHEAS**.

   (a) DHEAS is the predominant precursor for estradiol and estrone synthesis after DHEAS is deconjugated by placental sulfatase. These two estrogens contribute to the estriol pool by conversion in the maternal liver.

   (b) DHEAS is converted in the fetal liver to 16α-hydroxydehydroepiandrosterone sulfate by 16α-hydroxylase. This enzyme is not found in the placenta.

   b. Plasma concentration. Like progesterone, the plasma estriol concentration rises steadily throughout gestation, reaching a maximal plateau at 36–40 weeks. The secretory curve for estriol parallels that for progesterone and correlates well with the fetal growth curve.

   (1) Plasma estriol concentrations reflect the functional status of the feto-placental unit. Falling levels indicate impending fetal death.

   (2) Even with fetal death, plasma progesterone levels can remain within normal limits.

C. Physiologic effects of placental hormones

1. **HCG** is classified as an anterior pituitary-like hormone with biologic actions that mimic those of LH (i.e., it can be used to induce ovulation).

   a. HCG is a second luteotropic hormone, in that it maintains the function of the corpus luteum until the feto-placental unit is autonomous in terms of hormone synthesis (about 6–7 weeks postconception).

   b. HCG converts the corpus luteum of menstruation into the corpus luteum of pregnancy, thereby extending the functional lifespan of the corpus luteum.

   c. HCG stimulates the corpus luteum of early pregnancy to secrete 17α-hydroxyprogesterone and lesser amounts of progesterone, which reach a peak 3–4 weeks postconception. Blood 17α-hydroxyprogesterone level is an excellent indicator of corpus luteal function during early pregnancy, because the placenta lacks significant 17α-hydroxylase activity.

   d. HCG stimulates the fetal testis to secrete testosterone at a time prior to fetal pituitary LH secretion.

   e. HCG may serve as a tropic agent for the fetal zone of the adrenal cortex, which secretes DHEA.

2. **Progesterone**

   a. Progesterone inhibits uterine motility by hyperpolarization of the uterine myometrium.

   b. It converts the secretory endometrium of the luteal phase of the menstrual cycle to the decidua during pregnancy. Progesterone maintains the decidua.

   c. Synergistic action of progesterone and estrogen is required to induce development of the lobuloalveolar compartment, which prepares the breasts for lactation. Progesterone acts primarily on the lobuloalveolar compartment.

   d. Progesterone has an immunosuppressive role in protecting the fetus.

   e. Progesterone contributes to the growth and development of the fetus (e.g., by acting as a precursor for corticoid synthesis by the fetal adrenal cortex).

   f. Progesterone promotes renal excretion of Na^+, which antagonizes the effects of increased aldosterone levels found in pregnancy.

3. **Estrogens.** The estrogenic effects of pregnancy are primarily due to estradiol, the most potent of the estrogens.

   a. Estrogens mediate the growth and development of the maternal reproductive organs.

   (1) The gravid uterus increases 18-fold (about 1700%) in weight, beginning as a 60 g organ in the nongravid state.

   (2) During the gestational period, the uterus lengthens from 7 cm to 30 cm.

   (3) Uterine volume at term is 500–1000 times greater than that before pregnancy.

   (4) The increase in uterine size during pregnancy occurs by stretching and hypertrophy of the myometrium.

   b. Estrogens stimulate hepatic synthesis of thyroxine-binding globulin, steroid hormone–binding globulin, and angiotensinogen as well as renal renin secretion. The latter two effects lead to increased angiotensin II synthesis.

   c. Estrogens stimulate development of the lactiferous ductal system in the mammary gland.

---

*Fetal sulfokinase conjugates metabolites of pregnenolone, progesterone derivatives, and the androgens (DHEA) found in high levels in the feto-placental unit.

d. Just before term, the estrogen-to-progesterone ratio increases and the uterus is dominated by estrogen.

XII. ENDOCRINE PANCREAS

A. **Histology and function of the islets of Langerhans** (Figure 7-14). The endocrine pancreas consists of **islet of Langerhans,** which form less than 2% of the pancreatic tissue.

1. **Cells.** Four cell types have been identified.
   a. **Alpha cells** make up about 25% of the islet cells and are the source of glucagon, which consists of 29 amino acid residues.
   b. **Beta cells** constitute about 60% of the islet cells and are associated with insulin synthesis. This polypeptide consists of 51 amino acid residues.
   c. **Delta cells** comprise about 10% of the islet cells and are the source of somatostatin, which is a tetradecapeptide.
   d. **Pancreatic polypeptide cells** comprise approximately 5% of the islet cells and synthesize a polypeptide that contains 36 amino acid residues.

2. **Neurotransmitters and epinephrine.** Unmyelinated postganglionic sympathetic and parasympathetic nerve fibers terminate close to the three cell types (alpha, beta, and delta cells) and modulate pancreatic endocrine function via the secretion of neurotransmitters.
   a. **ACh** secretion causes insulin release only when glucose levels are elevated. ACh appears to inhibit somatostatin release.
   b. **Norepinephrine** secretion due to sympathetic stimulation via activation of the α-receptor leads to inhibition of insulin release. Norepinephrine stimulates somatostatin release.
   c. **Epinephrine.** Despite the dual α- and β-adrenergic receptor system in beta cells, the α-adrenergic action of epinephrine predominates, so that insulin secretion in inhibited. (Insulin release is mediated by a β-adrenergic receptor.)

3. **Control of secretions.** The alpha, beta, and delta cells constitute a functional syncytium, which forms a paracrine control system for the coordinated secretion of pancreatic polypeptides (see Figure 7-14).
   a. Insulin inhibits alpha cell (glucagon) secretion, which increases peripheral glucose uptake and opposes glucagon-mediated glucose production.
   b. Glucagon stimulates beta cell (insulin) secretion and delta cell (somatostatin) secretion, which increases hepatic glucose production and opposes hepatic glucose storage.
   c. Somatostatin inhibits alpha cell (glucagon) and beta cell (insulin) secretion, which produces hypoglycemia and inhibition of intestinal glucose absorption. The lowering of blood glucose levels by somatostatin in diabetic patients probably is due both to inhibition of glucagon secretion and to reduced intestinal absorption of glucose.
   d. Pancreatic polypeptide inhibits insulin and somatostatin secretion via a direct pancreatic effect.

B. **Biosynthetic organization of the beta cell**

1. **Human proinsulin** is a single-chain polypeptide of 86 amino acid residues, with a molecular weight of approximately 9000 daltons.
   a. Intracellular proteolytic cleavage of proinsulin forms insulin and C-peptide.

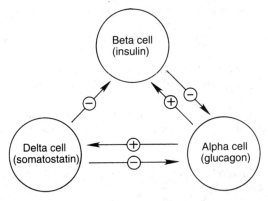

Figure 7-14. The paracrine system of the pancreatic islet cells. The pattern of islet cell hormone secretion represents an integrated response by all of the islet cells to humoral, neural, and paracrine regulation. *Plus signs* indicate stimulation, and *minus signs* indicate inhibition. (Reprinted from Tepperman J: Endocrine function of the pancreas. In *Metabolic and Endocrine Physiology*, 4th edition. Chicago, Year Book, 1980, p 233.)

    b. The conversion of proinsulin to insulin is not fully completed, and about 5% of the secretory product of the beta cell is proinsulin.

    c. The biologic activity of proinsulin is about 5%–10% that of insulin.

    d. The plasma half-life of proinsulin is 15 minutes.

2. C-peptide is the connecting peptide remaining after cleavage of proinsulin to insulin. In humans, it consists of 31 amino acid residues.

    a. Beta cell secretory products consist of equimolar amounts of insulin and C-peptide; therefore, circulating C-peptide concentrations reflect beta cell activity.

    b. The normal fasting concentration of C-peptide in peripheral blood is approximately 1.0–3.5 ng/ml.

    c. C-peptide has no detectable biologic activity.

    d. The plasma half-life of C-peptide is 30 minutes.

3. Insulin is stored in the beta cell granules as a crystalline hexamer complex with two atoms of zinc per hexamer. In plasma, insulin is transported as a monomer.

    a. Insulin has a molecular weight of about 6000 daltons and contains 51 amino acid residues.

    b. Insulin secretion requires the presence of extracellular Ca^{2+}. Inside the beta cell, Ca^{2+} binds to a Ca^{2+}-binding protein called **calmodulin**.

    c. The plasma half-life of insulin is 5 minutes.

C. Control of insulin secretion

1. Carbohydrates

    a. Monosaccharides that can be metabolized (e.g., hexose, triose) are more potent stimuli of insulin secretion than carbohydrates that cannot be metabolized (e.g., mannose, 2-deoxy-D-glucose).

    b. The principal stimulus for insulin release is glucose. As the blood glucose level rises above 4.5 mmol/L (80 mg/dl), it stimulates the release and synthesis of insulin.

    c. Substances that inhibit glucose metabolism (e.g., 2-deoxy-D-glucose, D-mannoheptulose) interfere with insulin secretion.

    d. The reduction of glucose to sorbitol may contribute to insulin secretion.

    e. Glucose also stimulates somatostatin release.

2. Gastrointestinal hormones

    a. The plasma concentration of insulin is higher after oral administration of glucose than after it has been administered intravenously, even though the arterial blood glucose concentration remains lower. This augmented release of insulin following an oral glucose dose is due to the secretion of gastrointestinal hormones, including:

        (1) Gastric inhibitory peptide (GIP), which appears to be the principal gastrointestinal potentiator of insulin release

        (2) Gastrin

        (3) Secretin

        (4) CCK

    b. Gastrointestinal hormones also augment somatostatin release.

3. Amino acids

    a. Amino acids vary in their ability to stimulate beta cells. Among the essential amino acids, in decreasing order of effectiveness, are arginine, lysine, and phenylalanine.

    b. The stimulation of insulin secretion by oral administration of amino acids exceeds that of intravenously administered amino acids. Protein-stimulated secretion of CCK, gastrin, or both may mediate this effect.

    c. The analogs of leucine and arginine that cannot be metabolized also stimulate insulin secretion.

4. Fatty acids and ketone bodies are not known to have an important role in the regulation of insulin secretion in humans. The ingestion of medium-chain triglycerides causes a small increment in insulin levels.

5. Islet hormones. Glucagon stimulates insulin secretion and somatostatin inhibits insulin secretion. Somatostatin inhibits gastrin and secretin secretion, glucose absorption, and gastrointestinal motility (Table 7-11).

6. Other hormones

    a. GH induces an elevation in basal insulin levels, which precedes a change in blood glucose levels, suggesting a direct beta-cytotropic effect.

Table 7-11. Physiologic Effects of Somatostatin

Site	Action
Anterior pituitary	Inhibits secretion of growth hormone and thyrotropin
Pancreas	Inhibits secretion of insulin, glucagon, and pancreatic polypeptide
Gastrointestinal tract	Inhibits secretion of gut hormones (gastrin, secretin, VIP, cholecystokinin), gastric acid, and pepsin; decreases blood flow, motility, and carbohydrate absorption; increases water and electrolyte absorption

VIP = vasoactive intestinal peptide.

    b. Hyperinsulinemia also has been observed with exogenous and endogenous increments of corticosteroids, estrogens, progestogens, and PTH. Since blood glucose concentrations are not reduced with these hormones, it is inferred that these hormones have an anti-insulin effect.

  7. Obesity. Hyperinsulinemia is observed in obese patients. An increase in body weight in the absence of a disproportionate increase in body fat does not affect insulin levels.

  8. Ions. Both K^+ and Ca^{2+} are necessary for normal insulin and glucagon responses to glucose. Therefore, hypokalemia leads to glucose intolerance.

  9. Cyclic nucleotides. cAMP is a releaser of insulin.

D. Physiologic actions of insulin (Figure 7-15). Target cells with insulin receptors have been demonstrated in liver, muscle, adipose tissue, lymphocytes, monocytes, and granulocytes.

  1. Carbohydrate metabolism
    a. Liver
      (1) The liver is freely permeable to glucose, and glucose transport can occur without insulin by simple diffusion.
        (a) Insulin acts on the liver to promote glucose uptake and to inhibit enzymatic processes involved in glucose production and release (glycogenolysis).
        (b) Because the hepatocyte is permeable to glucose, uptake of glucose in the liver is not rate-limiting.
      (2) A control point in glucose metabolism occurs when metabolism is initiated by the phosphorylation of glucose to glucose-6-phosphate, which is catalyzed by hexokinase and glucokinase.

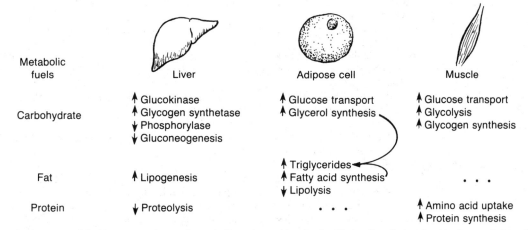

Figure 7-15. The major target sites and metabolic actions of insulin. Insulin is primarily involved in the regulation of metabolic processes, the principal manifestation of which is the control of plasma glucose concentration. (Reprinted from Felig P: Pathophysiology of diabetes mellitus. *Med Clin North Am* 55:821–834, 1971.)

       (a) Hexokinase is saturated at normal plasma glucose concentrations and is not regulated by insulin.

       (b) Glucokinase is only half-saturated at blood glucose concentrations between 90 and 100 mg/dl (5–10 mmol/L). Thus, the activity of this enzyme is insulin- and glucose-dependent.

    (3) The phosphorylation of fructose-6-phosphate by phosphofructokinase is enhanced by insulin. A decrease in phosphofructokinase activity favors the reversal of glycolysis.

    (4) Insulin diminishes hepatic glucose output by activating glycogen synthetase and by inhibiting gluconeogenesis. The key intermediary reaction in gluconeogenesis is between pyruvate and phosphoenolpyruvate, which requires the enzymes pyruvate carboxylase and phosphoenolpyruvate carboxykinase. The latter enzyme is inhibited in the presence of glucose and insulin.

  b. Muscle. The **insulin-dependent facilitated diffusion mechanism** for glucose is found in skeletal and cardiac muscle.

    (1) Glucose transport across muscle cell membranes requires insulin.

    (2) Insulin activates glycogen synthetase and phosphofructokinase, which cause glycogen synthesis and glucose utilization, respectively.

    (3) It should be emphasized that glucose uptake in exercising muscle is not dependent on *increased* insulin secretion. In resting muscle, glucose is a relatively unimportant fuel, with the oxidation of fatty acids supplying most of the energy.

  c. Adipose tissue. The **insulin-dependent facilitated diffusion mechanism** for glucose is found also in adipose tissue.

    (1) Insulin acts primarily to stimulate glucose transport.

    (2) It activates glycogen synthetase and phosphofructokinase.

    (3) The major end products of glucose metabolism in fat cells are fatty acids and α-glycerophosphate. The fat cell depends on glucose as a precursor of α-glycerophosphate, which is important in fat storage because it esterifies with fatty acids to form triglycerides.

2. Fat metabolism. Insulin is a lipogenic as well as an antilipolytic hormone.

  a. Liver

    (1) When insulin and carbohydrate are available, the human liver is quantitatively a more important site of fat synthesis than is adipose tissue.

    (2) In the absence of insulin, the liver does not actively synthesize fatty acids, but it is capable of esterifying fatty acids with **glycerol,** which is phosphorylated by glycerokinase.

       (a) Glycerol must be phosphorylated before it can be used in the synthesis of fat.

       (b) In the absence of glycolytic breakdown of glucose to α-glycerophosphate, glycerokinase permits the esterification of fatty acids.

    (3) In the absence of insulin, there is an increase in fat oxidation and the production of ketone bodies. Insulin exerts a potent antiketogenic effect.

    (4) Insulin promotes the synthesis and release of **lipotropin lipase,** which is an extracellular enzyme that hydrolyzes both chylomicron and very-low-density lipoprotein (VLDL) triglyceride.

       (a) Lipoprotein lipase catalyzes the hydrolysis of circulating lipoprotein triglyceride to fatty acids and glycerol.

       (b) Lipoprotein lipase is the key enzyme in the removal of lipoprotein triglyceride and thereby is important in the formation of both light and heavy lipoprotein (LDL and HDL).

       (c) Lipoprotein lipase is active at the luminal surface of the capillary endothelial cell and under normal conditions is completely absent from the circulation.

       (d) The highest lipoprotein lipase activity is found in the heart, but its distribution includes adipose tissue, lactating mammary gland (and milk), lung, skeletal muscle, aorta, corpus luteum, brain, and placenta.

  b. Adipose tissue

    (1) Insulin deficiency also decreases the formation of fatty acids in adipose tissue.

    (2) The major effect of insulin-stimulated glucose uptake in human fat cells is to provide α-glycerophosphate for esterification of free fatty acids. The absence of α-glycerophosphate formation from glycolysis during insulin deficiency prevents the esterification of free fatty acids, which are constantly released from triglycerides in the adipocytes.

    (3) The lipolytic effect in the absence of insulin is due to an increase in the hormone-sensitive lipase known as triglyceride lipase, the activity of which is normally inhibited by insulin.

3. **Amino acid and protein metabolism.** Insulin is an important protein anabolic hormone, and it is necessary for the assimilation of a protein meal. The protein anabolic effect of insulin is not dependent on increased glucose transport.

   a. In diabetic patients, the muscle uptake of amino acids is reduced and elevated postprandial blood levels are observed.

   b. During severe insulin deficiency, **hyperaminoacidemia** involving branched-chain amino acids (i.e., valine, leucine, and isoleucine) is present.

   c. Insulin increases uptake of most amino acids into muscle and increases the incorporation of amino acids into protein.

   d. Insulin increases body protein stores by four mechanisms:

      (1) Increased tissue uptake of amino acids

      (2) Increased protein synthesis

      (3) Decreased protein catabolism

      (4) Decreased oxidation of amino acids

4. **Electrolyte metabolism**

   a. Insulin lowers serum K^+ concentration. This hypokalemic action of insulin is due to stimulation of K^+ uptake by muscle and hepatic tissue.

   b. Diabetic patients have a proclivity toward developing hyperkalemia in the absence of acidosis.

   c. Insulin has an antinatriuretic effect.

5. **Membrane polarization**

   a. Insulin decreases membrane permeability to both Na^+ and K^+, but it decreases Na^+ permeability to a greater extent, causing hyperpolarization of mammalian muscle.

   b. The membrane hyperpolarization produced by insulin is the cause of the net shift of K^+ from the extracellular to the intracellular space, and not the result of the shift.

6. **Integration of insulin action: a summary** (Figure 7-16)

   a. Insulin is a very effective hypoglycemic hormone for two major reasons.

      (1) It promotes both hepatic and muscle glycogen deposition.

      (2) It enhances glucose utilization (glycolysis).

   b. In terms of glucose transport, the major insulin-independent tissues are brain, erythrocytes, liver, and epithelial cells of the kidney and intestine.

   c. Insulin is the primary anabolic hormone in the body for the following reasons.

      (1) Inhibition of hepatic gluconeogenesis decreases the hepatic requirement for amino acids.

      (2) The protein anabolic effect of insulin reduces the output of amino acids from muscle, thereby decreasing the availability of glucogenic amino acids for gluconeogenesis.

      (3) Glucose uptake by muscle is stimulated, providing an energy source to spare fatty acids, the release of which is inhibited by the antilipolytic action of insulin.

      (4) Fat accumulation is enhanced by increased hepatic lipogenesis.

      (5) The antilipolytic action of insulin (inhibition of hepatic oxidation of fatty acids) is due to the formation of α-glycerophosphate from glucose in the fat cell.

      (6) The antilipolytic action of insulin at the level of the adipose cell reinforces the insulin-mediated inhibition of hepatic ketogenesis and gluconeogenesis by depriving the liver of precursor substrates for ketogenesis and an energy source (fatty acids) and cofactors (acetyl-CoA) necessary for gluconeogenesis.

E. **Control of glucagon secretion**

   1. **Metabolic fuels**

      a. Hypoglycemia stimulates and hyperglycemia inhibits glucagon secretion.

      b. Amino acids (e.g., arginine, alanine) also are stimuli for glucagon release.

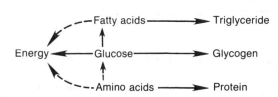

Figure 7-16. Insulin exerts integrated and synergistic actions in the promotion of the storage of body fuels. It enhances the storage of fat and protein, and it promotes both the storage and utilization of carbohydrates. *Solid arrows* denote stimulation, and *dashed arrows* indicate inhibition. (Reprinted from Felig P: Disorders of carbohydrate metabolism. In *Metabolic Control and Disease,* 8th edition. Edited by Bondy PK and Rosenberg LE. Philadelphia, WB Saunders, 1980, p 294.)

    c. Decreasing circulatory levels of fatty acids are associated with glucagon release.

  2. Gastrointestinal hormones
    a. CCK, gastrin, secretin, and GIP stimulate glucagon secretion.
    b. The potentiation of glucagon secretion by the ingestion of a protein meal is probably mediated via CCK secretion.

  3. Fatty acids inhibit glucagon release.

F. Physiologic actions of glucagon. The major site of action of glucagon is the liver.

  1. Carbohydrate metabolism
    a. Glucagon has a hyperglycemic action, resulting primarily from stimulation of hepatic glycogenolysis. It should be emphasized, however, that the hyperglycemic effect of glucagon does not involve inhibition of the peripheral utilization of glucose.
    b. Glucagon is an important gluconeogenic hormone.
    c. The hyperglycemic action of epinephrine is amplified by its stimulation of glucagon secretion and its inhibition of insulin secretion.
    d. Suppression of glucagon secretion by glucose is not essential for normal glucose tolerance as long as insulin is available.

  2. Fat metabolism
    a. Glucagon is a lipolytic hormone *because* of its activation of glucagon-sensitive lipase (triglyceride lipase) in adipose tissue by cAMP.
    b. Glucagon causes an elevation in the plasma level of fatty acids and glycerol.
      (1) Glycerol is utilized as a glyconeogenic substrate in the liver.
      (2) The oxidation of fatty acids as an energy substrate accounts for the glucose-sparing effect of glucagon.
      (3) Glucagon is essential for the ketogenesis brought about by the oxidation of fatty acids. In the absence of insulin, glucagon can accelerate ketogenesis, which leads to metabolic acidosis.

  3. Protein metabolism
    a. Glucagon has a net proteolytic effect in the liver.
    b. This peptide is gluconeogenic, an effect that leads to increased amino acid oxidation and urea formation.
    c. In addition to its protein catabolic effect, glucagon has an antianabolic effect—inhibition of protein synthesis.

XIII. THYROID GLAND

A. Histology and function

  1. Thyroid components. The functional unit of the thyroid gland is the **follicle (acinus)** surrounded by a rich capillary plexus.
    a. The follicular (acinar) **epithelium** consists of a single layer of cuboidal cells.
      (1) The cell height of the follicular epithelium varies with the degree of stimulation of TSH.
      (2) The glandular epithelium varies with the degree of stimulation, becoming columnar when active and flat when inactive.
    b. The **lumen** of the follicle is filled with a clear, amber proteinaceous fluid called **colloid,** which is the major constituent of the thyroid mass.
    c. **Microvilli** extend into the colloid from the apical (adluminal) border, which is the site of the iodination reaction. The initial phase of thyroid hormone secretion (i.e., resorption of the colloid by endocytosis) also occurs at the apical border.
    d. The **parafollicular (C) cells,** which secrete **calcitonin,** do not border on the follicular lumen.

  2. Thyroid functions
    a. Hormone secretion. The thyroid gland secretes two hormones, which are **iodothyronines** and, therefore, derivatives of the amino acid **tyrosine.**
      (1) The major secretory product of the thyroid gland is **3,5,3′,5′-tetraiodothyronine (thyroxine),** which is abbreviated as T_4 to denote the four iodide atoms. The other thyroid

hormone is **3,5,3′-triiodothyronine,** which is abbreviated as T_3. T_3 is secreted in small amounts.

   (2) Only these two thyronines have biologic activity.
   (a) The **molar activity** ratio of T_3 to T_4 is 3–5:1
   (b) The **secretory ratio** of T_4 to T_3 is 10–20:1
   (c) The **plasma concentration ratio** of free T_4 to free T_3 is 2:1. Most of the T_3 in the plasma is derived from monodeiodination of T_4 by the action of monodeiodinase (5′-deiodinase) found in peripheral tissue.
   (3) **Reverse 3,3′,5′-triiodothyronine (rT$_3$)** is a biologically inactive thyronine also formed by peripheral conversion catalyzed by 5-deiodinase.
   b. **Related functions.** The thyroid cell performs two parallel functions in the synthesis of thyroid hormone.*
   (1) It synthesizes a protein substrate called **thyroglobulin.**
   (a) This glycoprotein serves as a matrix in which thyroid hormone is formed.
   (b) Thyroglobulin also is the storage form of thyroid hormone.
   (c) Each thyroglobulin molecule contains approximately 120 tyrosyl residues.
   (2) The thyroid cell accumulates inorganic iodide from the plasma.

B. Distribution of thyroid iodide

1. Iodide intake
   a. In the United States, the daily dietary iodine intake is about 500 μg.
   b. About 1 mg of iodide is required per week (or 150 μg/day) to maintain euthyroidism.
   c. The thyroid gland stores enough thyroid hormone to maintain a euthyroid state for 3 months without hormone synthesis.

2. Thyroid iodide
   a. The thyroid gland contains 5–7 mg of iodide.
   (1) Of the total thyroid iodide, 95% is in the extracellular space (i.e., stored in the colloid as thyroglobulin).
   (a) Two-thirds of the total iodide content in the colloid is in the form of biologically inactive **iodotyrosines.**
   (b) One-third of the colloid iodide content is in the form of biologically active **thyronines** (i.e., T_4 and T_3).
   (c) The molar storage ratio of T_4 to T_3 is 9:1, and the molar storage ratio of iodotyrosines to iodothyronines is 2:1.
   (2) The remaining 5% of the total thyroid iodide is in the intracellular space of the follicular epithelium.
   b. The thyroid gland contains the body's largest iodide pool.

C. Hormone transport: extracellular binding proteins. Extracellular binding proteins for thyroid hormone are in the plasma, while the storage form of thyroid hormone (thyroglobulin) is in the follicular lumen (colloid).

1. Thyroxine-binding proteins. Virtually all (99.95%) of T_4 is bound to plasma proteins, leaving about 0.05% unbound (free). This portion of unbound thyroxine represents the biologically active hormone. Thyroxine is mainly associated with two of the three binding proteins.
   a. Thyroxine-binding globulin (TBG) binds about 75% of the plasma T_4. In normal individuals, less than half of the available binding sites on TBG are saturated with T_4.
   b. Thyroxine-binding prealbumin (TBPA) binds about 15%–20% of the circulating T_4.
   c. About 9% of the T_4 is bound to albumin.

2. Triiodothyronine-binding proteins
   a. Almost all (99.5%) of T_3 is transported bound to TBG.
   b. Very little T_3 is bound to albumin, and practically none is bound to TBPA.
   c. About 0.5% of the T_3 is unbound. The lower affinity of T_3 for the plasma binding proteins and, thus, the higher concentration of unbound T_3 contributes to the greater biologic activity of T_3.

---

*The term "thyroid hormone" denotes thyroxine (T_4) and triiodothyronine (T_3).

D. Biosynthesis and release of thyroid hormone. The thyroid gland accumulates or "traps" iodide by an active transport mechanism that operates against a concentration and an electric gradient. The normal thyroid iodide-to-plasma iodide concentration ratio is 25–40:1.

  1. Synthesis. All of the biosynthetic steps are stimulated by TSH.

    a. Iodide uptake (Figure 7-17)

      (1) Active iodide uptake occurs at the basal membrane of the follicular cell and is not an essential step in thyroid hormone synthesis.

      (2) Iodide diffuses along an electric gradient into the lumen, where the luminal iodide-to-follicular cell iodide concentration ratio is 5:1.

      (3) Radioactive iodide uptake by the thyroid gland is a useful therapeutic index of the functional status of the thyroid gland. A 24-hour uptake normally ranges between 10% and 35% of the administered dose.

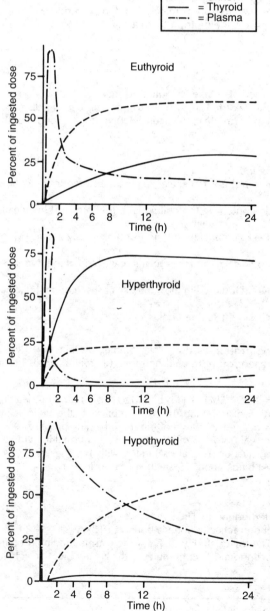

Figure 7-17. Radioactive iodine uptake by individuals on a relatively low-iodine diet. Percentages are plotted against time after an oral dose of radioactive iodine for plasma, urine, and the thyroid gland. In hyperthyroidism, plasma radioactivity falls rapidly, then rises again as a result of release of labeled T_4 and T_3 from the thyroid gland. (Adapted from Ingbar SH, Woeber KA: *Textbook of Endocrinology,* 4th edition. Edited by Williams RH. Philadelphia, WB Saunders, 1968, p 144.)

    b. Oxidation of iodide is mediated by a peroxidase and forms active iodide, which may be in the form of iodinium ion (I^+), a free radical of iodine (IO_3^-), or iodine (I_2).

    c. Iodination of active iodide denotes the addition of iodide to the tyrosyl residues of thyroglobulin. The substrate for iodination is thyroglobulin.

        (1) Iodination leads to the formation of the iodotyrosines within the preformed thyroglobulin molecule rather than in free amino acids that are then incorporated into protein. The hormonally inactive substrates formed by the iodination of thyroglobulin are called **mono-** and **diiodotyrosine**.

        (2) Iodination occurs by the catalytic action of thyroid peroxidase.

    d. Coupling (condensation) of iodotyrosines occurs and forms biologically active thyronines (T_3 and T_4).

        (1) The iodothyronines are formed at the apical border of the follicular cell and are held in peptide linkage with thyroglobulin (as are the tyrosines).

        (2) T_4 synthesis requires the fusion of two diiodotyrosine molecules, and T_3 synthesis requires the condensation of a monoiodotyrosine molecule with a diiodotyrosine molecule.

        (3) Thyroid peroxidase also mediates the coupling reaction.

2. Release

    a. Secretion begins with endocytosis of the colloid at the apical border. This brings colloid "droplets" into contact with protease-containing lysosomes.

    b. The release of hormones involves the following reactions:

        (1) Hydrolysis of thyroglobulin by the thyroid protease and by peptidases, which liberate free amino acids

        (2) Secretion of iodothyronines into the blood and deiodination of the iodotyrosines, which form a second iodide pool that can be recycled into hormone synthesis

E. Metabolism and excretion of thyroid hormone

    1. T_4 is the iodothyronine found in highest concentration in plasma and is the only one that arises solely by direct secretion from the thyroid gland.

    2. Most of the T_3 present in plasma is derived from the peripheral conversion of T_4 by monodeiodination via 5'-deiodinase.

        a. The extrathyroid deiodination of T_4 accounts for over 80% of the circulating T_3.

        b. The liver and kidney deiodinate T_4 to form T_3.

    3. Thyroid hormone is metabolized by deiodination, deamination, and by conjugation with glucuronic acid. The conjugate then is secreted via the bile duct into the intestine.

    4. In normal individuals, T_4 and T_3 are excreted mainly in the feces, with a small amount appearing in the urine.

F. Control of thyroid function (see Figure 7-4)

    1. Hypothalamic-hypophysial-thyroid axis. TSH secretion is influenced by four factors: TRH secretion from the median eminence, the blood level of unbound T_4, the blood level of unbound T_3 generated by the peripheral conversion of T_4 to T_3, and the peripheral conversion of T_4 to T_3 within the pituitary gland.

        a. TRH is a tripeptide synthesized by the parvicellular peptidergic neurons in the hypothalamus.

        (1) TRH is transported to the median eminence, where it is stored. From there, TRH is released into the hypophysial portal system and is carried to the anterior lobe of the pituitary gland.

        (2) TRH stimulates some of the basophils (**thyrotrophs**) to secrete TSH.

        b. TSH stimulates the thyroid follicle to secrete thyroid hormone, most of which is bound to plasma protein carriers.

        (1) TSH stimulates the series of chemical reactions that lead to the synthesis of the iodothyronines.

        (2) It is the circulating free T_3 and T_4 concentrations that influence (regulate) TSH release by exerting a negative feedback effect at the level of the anterior lobe and probably at the level of the hypothalamus. Both free T_3 and free T_4 are effective inhibitors of TSH secretion when their plasma concentrations are increased.

        (3) The pituitary gland also converts T_4 to T_3 by monodeiodinase, and this intrapituitary T_3 plays a major role in the negative feedback that occurs at the pituitary level.

    c. Other regulators
      (1) Estrogens enhance TSH secretion.
      (2) Large doses of iodide inhibit thyroid hormone release and, thereby, cause decreases in serum T_4 and T_3 concentrations and an increase in TSH secretion.
      (3) Somatostatin inhibits TSH secretion and the response to TRH.
      (4) Dihydroxyphenylethylamine (dopamine), dopa, and bromocriptine decrease the basal secretion of TSH.

  2. Thyroid autoregulation
    a. Thyroid function also is regulated by an intrinsic control system that maintains the constancy of thyroid hormone stores.
      (1) The high concentrations of intrathyroidal **inorganic** iodide lead to the inhibition of thyroid release.
      (2) High concentrations of **organic** iodide (thyroid hormone) lead to a decrease in iodide uptake.
    b. Both of these effects reduce the fluctuation in thyroid hormone secretion when an acute change occurs in the availability of a requisite substrate (e.g., iodide).

  3. Goiter
    a. Any enlargement of the thyroid gland is called a **goiter,** and antithyroid substances that cause thyroid enlargement are called **goitrogens**.
      (1) Goitrogens are substances that block the synthesis of thyroid hormone.
      (2) A goiter does not define the functional state of the thyroid gland.
    b. If the goitrogen reduces thyroid hormone synthesis to subnormal levels, TSH secretion is enhanced.
    c. Goitrogens lead to the increased synthesis of endogenous TSH, which is responsible for the formation of a hypertropic thyroid gland (goiter).
    d. Goitrogenic agents include:
      (1) Perchlorate, thiocyanate, and pertechnetate, which are monovalent anions that block iodide trapping
      (2) Thionamides (propylthiouracil and methimazole), which block the coupling of iodotyrosines
      (3) Iodide deficiency
      (4) Excess iodide

G. Physiologic effects of thyroid hormone

  1. Thyroid hormone increases the **basal metabolic rate (BMR)** of most cells in the body. The normal BMR for adult euthyroid males is 35–40 kcal/m² body surface/hr. Normal BMR is 6%–10% lower in euthyroid females.
    a. Exceptions to this effect occur in the gonads, brain, lymph nodes, thymus, lung, spleen, dermis, and some accessory sex organs.
    b. A correlate of the increase in BMR is an increase in the size and the number of mitochondria together with an increase in the enzymes that regulate oxidative phosphorylation.
    c. The increase in BMR also is associated with an increase in Na^+-K^+-ATPase (Na^+ pump) activity. The fluxes of Na^+ (efflux) and K^+ (influx) are estimated to require 10%–30% of the total energy consumed by cells.
    d. The increase in BMR accounts for the thermogenic effect of thyroid hormone.

  2. Thyroid hormone is essential for normal **bone growth and maturation** as well as for the maturation of neurologic tissue, especially the brain.
    a. In hypothyroidism, there is a marked decrease in the myelination and arborization of neurons in the brain.
    b. If hypothyroidism is untreated, mental retardation occurs.

  3. Thyroid hormone is necessary for normal **lactation**.

H. Metabolic effects of thyroid hormone

  1. Carbohydrate metabolism
    a. In physiologic amounts, thyroid hormone potentiates the action of insulin and promotes glycogenesis and glucose utilization.
    b. In pharmacologic amounts, thyroid hormone is a hyperglycemic agent.

(1) Thyroid hormone potentiates the glycogenolytic effect of epinephrine, causing glycogen depletion.

(2) Thyroid hormone is gluconeogenic in that it increases the availability of precursors (lactate and glycerol).

(3) In large doses, thyroid hormone promotes intestinal glucose absorption.

2. Protein metabolism

   a. In physiologic amounts, thyroid hormone has a potent protein anabolic effect.

   b. In large doses, thyroid hormone has a protein catabolic effect.

3. Fat metabolism

   a. Thyroid hormone stimulates all aspects of lipid metabolism, including synthesis, mobilization, and utilization. On a net basis, the lipolytic effect is greater than the lipogenic effect.

   b. There is a general inverse relationship between thyroid hormone levels and plasma lipids.

   (1) Elevated thyroid hormone levels are associated with decreases in blood triglycerides, phospholipids, and cholesterol.

   (2) High levels of thyroid hormone also are associated with increases in plasma free fatty acids and glycerol.

4. Vitamin metabolism. The metabolism of fat-soluble vitamins is affected by thyroid hormone. For example, thyroid hormone is required for the synthesis of vitamin A from carotene and the conversion of vitamin A to retinene.

   a. In hypothyroid states, the serum carotene is elevated, and the skin becomes yellow.

   b. This skin condition differs from that observed in jaundice in that the sclera of the eye is not yellow.

XIV. PARATHYROID HORMONE, CALCITONIN, AND VITAMIN D

A. Role of Ca^{2+} in physiologic processes

1. Hemostasis. Ca^{2+} is necessary for the activation of clotting enzymes in plasma.

2. Ca^{2+} controls **membrane excitation,** and Ca^{2+} influx occurs during the excitatory process of nerve and muscle.

   a. Excitable membranes contain specific Ca^{2+} channels.

   b. Ca^{2+} entry does not require an active transport process, since the concentration gradient across the membrane is larger for Ca^{2+} than for any other ion.

   (1) Ca^{2+} concentration ($[Ca^{2+}]$) in the intracellular fluid (ICF) is about 10^{-7} mol/ L.

   (2) $[Ca^{2+}]$ in the extracellular fluid (ECF) is about 10^{-3} mol/L (the actual value is 2.5×10^{-3} mol/L).

   (3) The $[Ca^{2+}]$ gradient from outside to inside the cell is on the order of 10,000 to 1!

3. Ca^{2+} is bound to cell surfaces and has a role in the **stabilization of the membrane** and **intercellular adhesion.**

4. Ca^{2+} is necessary for **muscle contraction** (excitation-contraction coupling).

5. Ca^{2+} is essential in all **excitation-secretion processes,** such as the release of hormone by endocrine cells and the release of other products by exocrine cells. It is also essential for **neurotransmitter release.**

6. Ca^{2+} is necessary for the production of **milk** and the formation of **bone** and **teeth.**

B. Ca^{2+} distribution

1. Skeletal storage. More than 99% of the total body Ca^{2+} is stored in the skeleton.

   a. The skeleton of a 70-kg individual contains about 1000 g of Ca^{2+} compared to about 1 g in the extracellular pool.

   b. The skeleton also serves as a storage depot for phosphorus and contains about 80% of the total body phosphorus.

   c. Bone serves as a third-line defense in acid-base regulation by virtue of its CO_3^{2-}, HCO_3^-, and PO_4^{3-} content.

2. Plasma. The plasma concentration of total (ionized and nonionized) Ca^{2+} is about 10 mg/dl, which is equivalent to 5 mEq/L or 2.5 mmol/L.

    a. Ca^{2+} is present in the plasma as:
      (1) Ionized or free (45%)
      (2) Complexed with HPO_4^{2-}, HCO_3^-, or citrate ion (10%)
      (3) Bound to protein [primarily to albumin] (45%)
    b. The sum of the ionized and complexed Ca^{2+} constitutes the diffusible fraction (55%) of Ca^{2+}. The protein-bound form constitutes the nondiffusible fraction (45%).
    c. PTH, calcitonin, and vitamin D regulate the serum-ionized Ca^{2+} concentration.

  3. Ca^{2+} **pools.** Total body Ca^{2+} can be conceptualized as two major "pools."
    a. The larger Ca^{2+} pool, which contains over 99% of the total Ca^{2+}, consists of stable (mature) bone. This represents the Ca^{2+} pool that is not readily exchangeable, and it is not available for rapid mobilization.
    b. The smaller Ca^{2+} pool, which contains less than 1% of the total body Ca^{2+}, consists of labile (young) bone. This Ca^{2+} pool is readily exchangeable because it is in physicochemical equilibrium with the ECF. The pool consists of calcium phosphate salts and provides an immediate reserve for sudden changes in blood $[Ca^{2+}]$.

C. Bone chemistry

  1. Bone Ca^{2+} is found in the form of **hydroxyapatite crystals**. The empirical chemical formula for this substance is $Ca_{10}(PO_4)_6(OH)_2$ or $[(Ca_3PO_4)_2]_3 \cdot Ca(OH)_2$. Fluoride ion can replace the OH^- group and form **fluoroapatite,** or $[(Ca_3PO_4)_2]_3 \cdot CaF_2$.
    a. The calcium:phosphorus ratio in bone is about 1.7:1.
    b. A large surface area is provided by the microcrystalline structure of bone; it is estimated to be 100 acres in humans!

  2. Dry, fat-free bone consists of two-thirds mineral (inorganic) and one-third organic matrix.
    a. Over 90% of the organic matrix is collagen.
    b. The inorganic crystalline structure of bone imparts to it an elastic modulus similar to that of concrete.

D. Bone development. Bone is both a tissue and an organ. It consists of cells and an extracellular matrix containing organic and inorganic components. In its early development, bone exists as **osteoid,** an organic, unmineralized matrix surrounding the bone cells that deposited it.

  1. Mesoderm is the embryonic germ layer that gives rise to cartilage, bone, and muscle.

  2. Neural crest cells also can differentiate as bone cells, including:
    a. Cells that deposit dentin in teeth
    b. Cartilage and bone of the head

E. Ca^{2+} regulation: overview. Ca^{2+} regulation involves three tissues (bone, intestine, and kidney), three hormones (PTH, calcitonin, and activated vitamin D_3), and three cell types (osteoblasts, osteocytes, and osteoclasts).

  1. Hormonal control of Ca^{2+} metabolism. The pituitary gland does not play a major role in regulating the cells that produce PTH, calcitonin, and activated vitamin D_3.
    a. PTH is a polypeptide containing 84 amino acid residues. PTH is secreted by chief cells of the four parathyroid glands.
      (1) PTH is the **hypercalcemic hormone** of the body; it exerts its effects on the bone, intestine, and kidney.
      (2) PTH regulates only the plasma $[Ca^{2+}]$. An inverse linear relationship exists between plasma $[Ca^{2+}]$ and PTH secretion (Figure 7-18).
        (a) When plasma $[Ca^{2+}]$ falls, PTH secretion increases.
        (b) As plasma $[Ca^{2+}]$ increases, PTH secretion decreases.
    b. Calcitonin is a 32 amino acid residue polypeptide secreted by the parafollicular (C) cells of the thyroid gland.
      (1) Calcitonin is the **hypocalcemic hormone** of the body; it exerts a biologic effect on the bone, intestine, and kidney.
      (2) A positive linear relationship exists between plasma $[Ca^{2+}]$ and calcitonin secretion (see Figure 7-18).
        (a) As plasma $[Ca^{2+}]$ increases, calcitonin secretion increases.
        (b) When plasma $[Ca^{2+}]$ decreases, calcitonin secretion decreases.
      (3) Calcitonin release is stimulated by pentagastrin.

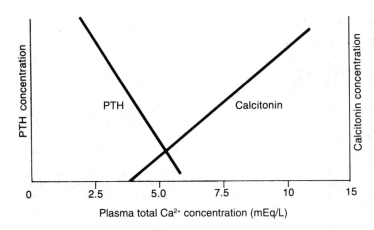

Figure 7-18. Relationship between plasma total Ca^{2+} concentration and blood levels of PTH and calcitonin. (Reprinted from Tepperman J: Hormonal regulation of calcium homeostasis. In *Metabolic and Endocrine Physiology*, 4th edition. Chicago, Year Book, 1980, p 297.)

   c. Vitamin D₃ (cholecalciferol) is a secosteroid containing 27 carbon atoms, which makes it the largest steroid hormone. The term "vitamin D" denotes both vitamins D_2 and D_3. In humans, the storage, transport, metabolism, and potency of vitamin D_2 and vitamin D_3 are identical.

     (1) Active metabolites. Only the active metabolites of vitamin D exert biologic activity.

        (a) Calcidiol (25-hydroxyvitamin D_3; 25-hydroxycholecalciferol) is the major blood form of vitamin D. This active metabolite is two to five times more effective than vitamin D_3 in preventing rickets.

        (b) Calcitriol (1,25-dihydroxyvitamin D_3; 1,25-dihydroxycholecalciferol) is another active metabolite of vitamin D. On a molar basis, it is 100 times more potent than calcidiol.

     (2) Synthesis of active vitamin D₃

        (a) In the epidermis, the previtamin 7-dehydrocholesterol is transformed into the lipid-soluble vitamin D_3, by nonenzymatic photoactivation upon exposure to the sun's ultraviolet rays. Photoactivation converts the 4-ring sterol into a 3-ring sterol.

           (i) Exposure of the skin to sunlight for 15–20 min/day or the irradiation of food has an antirachitic effect.

           (ii) The plant sterol, **ergosterol,** is transformed into vitamin D_2 (ergocalciferol) by irradiation and has been the main source of vitamin D that is added to foods (e.g., milk).

        (b) In the liver, vitamin D_3 is converted to calcidiol by 25-hydroxylase.

        (c) In cells of the proximal convoluted tubule, calcidiol is converted to calcitriol by the action of 1α-hydroxylase. The activity of this enzyme is enhanced by PTH.

2. Cellular aspects of bone metabolism (Figure 7-19)

   a. Osteoblasts are highly differentiated cells that are nonmitotic in their differentiated state. They are the **bone-forming cells** and are located on the bone-forming surface.

     (1) Osteoblasts **synthesize and secrete collagen.**

     (2) They contain abundant alkaline phosphatase activity.

     (3) They are derived from bone marrow mesenchyme.

   b. Osteocytes are osteoblasts that have become **buried in bone matrix** and are the most numerous of the bone cells in mature bone.

     (1) Each cell is surrounded by its own lacuna, but an extensive canalicular system connects osteocytes and surface osteoblasts, forming a functional syncytium.

     (2) In the osteocytic form, these cells no longer synthesize collagen.

     (3) Osteocytes have an **osteolytic activity,** which is stimulated by PTH. The "osteocytic osteolysis" in the bone matrix provides for the rapid movement of Ca^{2+} from bone into the ECF space.

   c. Osteoclasts are large, multinucleated cells containing numerous lysosomes. They mediate **bone resorption** at bone surfaces.

     (1) These cells contain acid phosphatase.

     (2) Osteoclasts are stimulated by PTH and form significant amounts of lactic and hyaluronic acids.

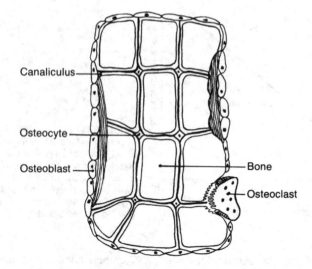

Figure 7-19. Anatomic relationships among the three types of bone cells. The canalicular system provides the structure for a functional syncytium between the osteocytes and osteoblasts. These intercellular connections between these two cell types are disrupted at sites where osteoblasts are found. (Reprinted from Avioli LV, Raisz LG: Bone metabolism and disease. In *Metabolic Control and Disease*, 8th edition. Philadelphia, WB Saunders, 1980, p 1715.)

(3) Osteoclasts might cause bone dissolution via an increased local concentration of H^+, which solubilizes bone mineral and increases the activity of enzymes that degrade matrix.

(4) They are derived from circulating monocytes.

F. Physiologic actions of PTH (Figure 7-20)

1. Osseous tissue

a. PTH action on bone is increased mobilization of Ca^{2+} and phosphate (i.e., bone dissolution) from the **nonreadily exchangeable Ca^{2+}** pool.

(1) The long-term effects of PTH on bone (i.e., the release of Ca^{2+} from bone) may be related to its effects on bone remodeling, which involves bone resorption and accretion.

(2) PTH also is known to stimulate bone synthesis.

(3) The effects of PTH on osteogenesis can be both anabolic and catabolic in terms of collagen metabolism.

b. PTH has three important effects on bone that account for its overall osteolytic activity.

(1) It stimulates osteoclastic and osteocytic activity.

(2) It stimulates the fusion of progenitor cells to form the multinucleated osteoclastic cells.

(3) It causes a transient suppression of osteoblastic activity.

c. Bone forms from cartilage, which serves as a template for cortical bone by periosteal apposition and for trabecular bone by endochondral ossification.

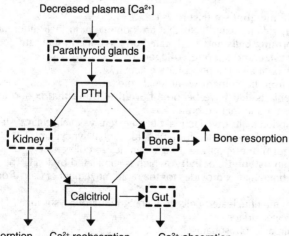

Figure 7-20. Physiologic actions of PTH and calcitriol on major target organs.

(1) Some bones, particularly those in the skull, are formed without a cartilage anlage by intramembranous bone formation.

(2) In adults, hematopoietic tissue is more abundant in trabecular bone.

   d. cAMP is a mediator of bone resorption since PTH stimulates adenyl cyclase in bone cells.
   e. With the dissolution of stable bone, **hydroxyproline** is excreted in the urine. This forms the basis for assessing collagen metabolism and thereby the relative rate of bone resorption.

2. **Intestinal tissue**
   a. Ca^{2+} and phosphate are absorbed in the intestine by both active and passive transport, but most of the intestinal absorption of Ca^{2+} occurs via facilitated diffusion.
   b. PTH alone does not directly affect the intestinal absorption of Ca^{2+}. PTH and calcitriol act synergistically to absorb Ca^{2+} and phosphate.
      (1) Intestinal absorption of Ca^{2+} does reflect parathyroid status, in that hypoparathyroid states are associated with low absorption and hyperparathyroid states are associated with high absorption.
      (2) The increased intestinal absorption of Ca^{2+} promoted by PTH is mediated indirectly through the increased synthesis of calcitriol.
      (3) Calcitriol acts on the intestine to promote the transport of Ca^{2+} and phosphate.

3. **Renal tissue**
   a. PTH increases the renal threshold for Ca^{2+} by promoting the active reabsorption of Ca^{2+} by the distal nephron. PTH inhibits the proximal tubular reabsorption of Ca^{2+}. Thus, PTH increases the tubular maximum (Tm) for Ca^{2+}.
   b. PTH inhibits phosphate reabsorption in the proximal tubules [i.e., it lowers the renal threshold for HPO_4^-, which leads to a phosphate diuresis (phosphaturia)]. Thus, PTH decreases the Tm for phosphate.
   c. Both increased PTH secretion and phosphate depletion stimulate the formation of calcitriol via the activation of 1α-hydroxylase.
   d. The phosphaturic effect of PTH may be mediated by cAMP, because PTH activates adenyl cyclase in the renal cortex.
   e. PTH also increases the urinary excretion of Na^+, K^+, and HCO_3^- and decreases the excretion of NH_4^+ and H^+. These effects account for the metabolic acidosis that occurs in hyperparathyroid states.

4. **Summary.** The unopposed effects of PTH on bone, intestine, and kidney include:
   a. Hypercalcemia
   b. Hypophosphatemia
   c. Hypocalciuria, initially due to increased Ca^{2+} reabsorption (however, in chronic hyperparathyroid states, the hypercalcemia exceeds the renal threshold for Ca^{2+}, and **hypercalciuria** is observed)
   d. Hyperphosphaturia

G. **Physiologic actions of calcitonin**

1. **Osseous tissue**
   a. Calcitonin **inhibits osteoclastic activity,** and the antihypercalcemic effect of calcitonin is due principally to the direct inhibition of bone resorption. This effect is not dependent on a functioning kidney, intestine, or parathyroid gland.
   b. Calcitonin also diminishes the osteolytic activity of osteoclasts and osteocytes.
   c. Calcitonin activity is associated with an increase in alkaline phosphatase synthesis from the osteoblasts.

2. **Intestinal tissue**
   a. Calcitonin inhibits gastric motility and gastrin secretion; however, it stimulates intestinal secretion.
   b. Calcitonin inhibits the intestinal (jejunal) absorption of Ca^{2+} and PO_4^{3-}.

3. **Renal tissue**
   a. Calcitonin promotes the urinary excretion of phosphate, Ca^{2+}, and Na^+.
   b. It also inhibits renal 1α-hydroxylase activity, which leads to a decrease in the synthesis of calcitriol.

H. **Physiologic actions of the biologically active calciferols.** The actions of calcitriol raise the plasma Ca^{2+} in concert with PTH. This vitamin hormone acts directly on the bone, small intestine, and kidney.

1. **Osseous tissue**
   a. Calcitriol, together with PTH, increases the mobilization of Ca^{2+} and phosphate from the bone.
      (1) Paradoxically, by raising serum Ca^{2+} and HPO_4^{2-} levels, it also fosters bone deposition.
      (2) This effect of calcitriol on Ca^{2+} and HPO_4^{2-} mobilization is not observed with vitamin D_3 (cholecalciferol) or vitamin D_2 (ergocalciferol).
   b. Ca^{2+}-binding protein in bone can be activated by PTH and calcitriol.
   c. The antirachitic action of vitamin D has traditionally been measured by an increase in bone formation following vitamin D administration to vitamin D–depleted rats.
      (1) It appears that the direct effect of the calciferols on bone is resorption.
      (2) The antirachitic influence of the calciferols appears to be due to an indirect effect on bone through the direct stimulating effect of the calciferols on the intestinal absorption of Ca^{2+} and phosphate.
   d. In summary, calcitriol acts synergistically with PTH to cause bone dissolution through the proliferation of osteoclasts. In short, increased osteoclastic activity by PTH requires calcitriol.

2. **Intestinal tissue**
   a. Calcitriol is the principal factor in increased Ca^{2+} absorption in intestinal tissue.
      (1) Calcitriol is the mediator hormone for the intestinal actions of PTH and calcitonin.
      (2) In turn, the action of calcitriol is mediated by the induction of a Ca^{2+}-binding protein.
   b. Calcitriol causes a lesser increase in intestinal HPO_4^{2-} absorption.

3. **Renal tissue**
   a. Calcitriol promotes the distal tubular reabsorption of Ca^{2+}.
   b. It promotes proximal tubular reabsorption of HPO_4^{2-}.
   c. The renal effects of PTH and calcitriol are similar in that both promote Ca^{2+} reabsorption; however, the renal effects of these two hormones on phosphate reabsorption are different.
      (1) PTH promotes phosphate diuresis.
      (2) Calcitriol promotes phosphate reabsorption; however, pharmacologic doses of calcitriol do have phosphaturic effects.

STUDY QUESTIONS

Directions: Each of the numbered items or incomplete statements in this section is followed by answers or by completions of the statement. Select the **one** lettered answer or completion that is **best** in each case.

1. Which of the following peptides is synthesized by neurosecretory neurons?

(A) Epinephrine
(B) Norepinephrine
(C) Somatomedin
(D) Somatostatin
(E) Somatotropin

2. Which of the following adrenomedullary enzymes is correctly paired with its substrate?

(A) Phenylethanolamine-N-methyltransferase/epinephrine
(B) Phenylalanine hydroxylase/tyrosine
(C) Dopa decarboxylase/phenylalanine
(D) Dopamine β-hydroxylase/dihydroxyphenylethylamine
(E) Tyrosine hydroxylase/norepinephrine

3. When is the second meiotic division of the developing ovarian follicle completed?

(A) At puberty
(B) Just prior to ovulation
(C) During the follicular phase of the menstrual cycle
(D) Just after conception
(E) During fetal development

4. In terms of serum concentration, the major postmenopausal steroid hormone and pituitary tropic hormone are

(A) estradiol and FSH
(B) estradiol and LH
(C) estrone and FSH
(D) estrone and LH
(E) estriol and LH

5. A decrease in cortisol secretion would lead to

(A) increased storage of glycogen in the liver
(B) decreased ACTH secretion
(C) decreased adrenomedullary synthesis of epinephrine
(D) increased plasma glucose concentration
(E) increased hepatic protein synthesis

6. Glucose transport occurs by insulin-dependent facilitated diffusion in which of the following tissues?

(A) Cardiac muscle
(B) Intestinal epithelium
(C) Renal epithelium
(D) Brain

7. Because hepatic glycogen stores are limited and decrease only temporarily after epinephrine secretion, muscle glycogenolysis is the major mechanism for providing gluconeogenic precursors for hepatic glucogenesis. This gluconeogenic substance derived from muscle is

(A) lactate
(B) acetyl-CoA
(C) glucose
(D) glycerol
(E) alanine

8. Progesterone secretion during the second and third trimesters of pregnancy is a measure of the functional status of the

(A) materno-placental unit
(B) corpus luteum
(C) fetal liver
(D) fetal adrenal gland
(E) maternal ovary

9. The most biologically active iodothyronine secreted by the thyroid follicles is

(A) T_3
(B) T_4
(C) rT_3
(D) thyroglobulin
(E) triiodothyroacetic acid

10. The primary site of 1,25-dihydroxycholecalciferol formation from its immediate precursor is the

(A) bone
(B) liver
(C) skin
(D) nephron
(E) bloodstream

11. Which of the following endocrine organs is larger at birth than in adulthood?

(A) Hypophysis
(B) Thyroid gland
(C) Adrenal gland
(D) Parathyroid glands
(E) Endocrine pancreas

12. The major steroid hormone secreted by the inner zone of the fetal adrenal cortex is

(A) cortisol
(B) DHEA
(C) progesterone
(D) estriol
(E) corticosterone

13. Prostaglandins found in the seminal fluid are secretory products of the

(A) prostate gland
(B) Sertoli cells
(C) seminal vesicles
(D) Leydig cells
(E) epididymis

14. Which of the following is a secretory product of the lutein-granulosa cells in nongravid women?

(A) Androstenedione
(B) Pregnenolone
(C) Pregnanediol
(D) Estriol
(E) Estrone

15. The cells that contribute most to testicular volume are the

(A) interstitial cells
(B) tubular cells
(C) spermatocytes
(D) connective tissue cells

16. An increase in plasma PTH level would lead to an increase in which of the following?

(A) The number of active osteoblasts
(B) Plasma inorganic phosphate concentration
(C) Renal synthesis of calcitriol
(D) Collagen synthesis
(E) Renal proximal tubular reabsorption of Ca^{2+}

17. Active vitamin D_3 (calcitriol) and PTH have many similar effects. Which of the following physiologic effects is specific only for calcitriol?

(A) Increased renal phosphate reabsorption
(B) Increased renal Ca^{2+} reabsorption
(C) Increased intestinal Ca^{2+} absorption
(D) Increased plasma $[Ca^{2+}]$
(E) Decreased plasma $[HPO_4^{2-}]$

18. All of the following statements regarding testosterone are true EXCEPT

(A) it is produced by the fetal testis
(B) it inhibits LH secretion from the pituitary gland
(C) it is a proestrogen
(D) it is inactivated after conversion to dihydrotestosterone
(E) it accelerates epiphysial closure of the long bones

19. All of the following are biochemical effects of cortisol EXCEPT

(A) hepatic lipogenesis
(B) hepatic gluconeogenesis
(C) muscle proteolysis
(D) hepatic protein anabolism
(E) hepatic glycogenesis

20. Abnormally high glucocorticoid levels would be associated with an increase in all of the following activities in the liver EXCEPT

(A) gluconeogenesis
(B) glycogenesis
(C) glycogenolysis
(D) glucose production
(E) protein synthesis

21. Characteristics of estriol include all of the following EXCEPT

(A) production by the liver
(B) production by the placenta
(C) secretion by theca interna cells of the ovary
(D) quantitatively the major urinary metabolite of the estrogens
(E) the least biologically active of the endogenous estrogens

22. Thyroid peroxidase is required for all of the following steps in thyroid hormone synthesis EXCEPT

(A) iodide uptake
(B) oxidation of iodide
(C) iodination of active iodide
(D) coupling of iodotyrosines
(E) synthesis of iodothyronines

23. Oxytocin is a neurosecretory hormone released from the pars nervosa. Oxytocin secretion promotes all of the following actions EXCEPT

(A) myometrial contraction
(B) lactogenesis
(C) milk ejection
(D) myoepithelial cell contraction

24. Activation of the sympathetic nervous system would lead to all of the following responses EXCEPT

(A) inhibition of peristalsis
(B) contraction of the radial ocular muscle
(C) renin secretion
(D) insulin secretion
(E) vasodilation in skeletal muscle

25. Insulin exerts all of the following effects EXCEPT

(A) hyperpolarization of skeletal muscle cells
(B) promotion of lipogenesis
(C) stimulation of glycogen synthase activity
(D) increase in secondary active transport of glucose into muscle cells
(E) increase in glucose transport in adipocytes

26. Progesterone serves several important functions during pregnancy. All of the following physiologic or biochemical effects require progesterone EXCEPT

(A) stimulation of myometrial contraction
(B) promotion of differentiation and growth of the lactiferous ducts
(C) inhibition of renal Na^+ excretion
(D) formation of cortisol by the fetal adrenal cortex in early pregnancy
(E) formation of aldosterone by the fetal adrenal cortex in early pregnancy

27. All of the following are substrates for MAO EXCEPT

(A) norepinephrine
(B) VMA
(C) epinephrine
(D) metanephrine
(E) normetanephrine

28. Estriol synthesis during gestation requires all of the following organs EXCEPT the

(A) fetal pituitary gland
(B) fetal liver
(C) neocortex of the fetal adrenal gland
(D) placenta
(E) trophoblast

29. All of the following are neuropeptide hormones EXCEPT

(A) ADH
(B) β-endorphin
(C) oxytocin
(D) somatomedin
(E) TRH

30. All of the following are stimuli for GH release EXCEPT

(A) bromocriptine
(B) hypoglycemia
(C) stress
(D) obesity
(E) vigorous exercise

31. True statements about β-endorphin include all of the following EXCEPT

(A) it reacts with the same receptors that bind morphine
(B) it is synthesized by pituitary basophils
(C) it is synthesized from corticotropin
(D) it is a proteolytic cleavage product of POMC

32. A 24-year-old woman has regular menstrual cycles of 21–23 days. Ovulation can be expected to occur between cycle days

(A) 7 and 9
(B) 10 and 12
(C) 13 and 15
(D) 16 and 18
(E) 19 and 21

33. A 17-year-old patient with a normal female phenotype is referred to an endocrinology clinic because of sparse pubic and axillary hair and amenorrhea. Chromosomal analysis reveals a male genotype. Further evaluation reveals intra-abdominal testes and circulating testosterone and estrogen concentrations that are characteristic of a normal man. The physician concludes that the patient has testicular feminization syndrome, a genetic end-organ insensitivity to androgen due to the absence of androgen receptors. Which of the following findings would be consistent with this syndrome?

(A) Normal wolffian duct development
(B) Normal müllerian duct development
(C) Regression of the internal genitalia
(D) Beard growth following androgen treatment
(E) Normal fertility

Questions 34–37

The graph below shows plasma steroid hormone levels as a function of time during a normal ovarian cycle in a 22-year-old woman. The woman becomes pregnant during this cycle.

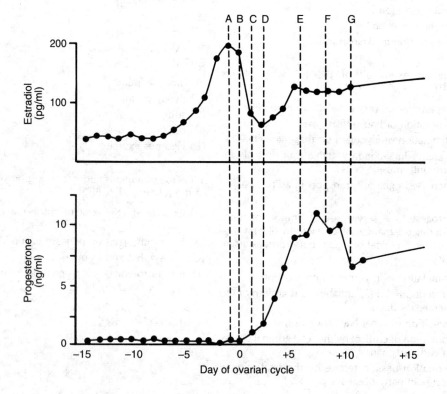

Adapted from Goodman HM: *Basic Medical Endocrinology.* New York, Raven, 1988, p 306.

34. Ovulation is indicated by which of the following lettered points on the graph?

(A) A
(B) B
(C) C
(D) D
(E) E

35. The LH peak is indicated by which of the following lettered points on the graph?

(A) A
(B) B
(C) C
(D) D
(E) E

36. Implantation of the blastocyst corresponds to which of the following points on the graph?

(A) C
(B) D
(C) E
(D) F
(E) G

37. The gradual increase in plasma concentrations of estradiol and progesterone beginning at point G indicate steroid secretion by the

(A) placenta
(B) corpus luteum
(C) adrenal cortex
(D) theca interna
(E) ovarian follicle

Directions: Each question below contains four suggested answers of which **one or more** is correct. Choose the answer.

A if **1, 2, and 3** are correct
B if **1 and 3** are correct
C if **2 and 4** are correct
D if **4** is correct
E if **1, 2, 3, and 4** are correct

38. Hormones that may cause negative nitrogen balance include

(1) glucagon
(2) thyroid hormone (excess)
(3) cortisol
(4) GH

39. Hormones that participate in maintenance of the corpus luteum include

(1) LH
(2) FSH
(3) HCG
(4) progesterone

40. Epinephrine is a potent hyperglycemic agent due to its ability to

(1) stimulate glucagon secretion
(2) stimulate ACTH secretion
(3) stimulate hepatic and muscle glycogenolysis
(4) inhibit insulin secretion

41. Administration of exogenous thyroid hormone would likely lead to

(1) negative feedback inhibition of TSH secretion
(2) decreased secretion of T_3
(3) decreased iodide uptake by the thyroid gland
(4) increased O_2 consumption by the brain

Directions: Each group of items in this section consists of lettered options followed by a set of numbered items. For each item, select the **one** lettered option that is most closely associated with it. Each lettered option may be selected once, more than once, or not at all.

Questions 42–45

Match each statement concerning plasma catecholamines with the specific catecholamine(s).

(A) Epinephrine
(B) Norepinephrine
(C) Both
(D) Neither

42. Decreases markedly following bilateral adrenalectomy
43. Mainly derived from tissues other than the adrenal medulla
44. Enzymatically inactivated by COMT
45. Elevated in most pheochromocytoma patients

Questions 46–49

Match each hormone classification with the hormone(s) to which it applies.

(A) Somatostatin
(B) Insulin
(C) Both
(D) Neither

46. Hypoglycemic hormone
47. Polypeptide hormone
48. Neurosecretory hormone
49. Hypophysiotropic hormone

Questions 50–55

Match each physiologic effect with the appropriate adrenergic receptor(s).

(A) Alpha receptor
(B) Beta receptor
(C) Both
(D) Neither

50. Increased insulin secretion
51. Increased pupillary diameter
52. Bronchodilation
53. Increased renin secretion
54. Decreased peristalsis
55. Relaxation of the detrusor muscle

ANSWERS AND EXPLANATIONS

1. The answer is D. [*VI D 1 b (2) (a) (ii)*] Somatostatin is a hypophysiotropic hormone produced by the parvicellular neurosecretory neurons that terminate in the median eminence. Somatostatin inhibits secretion of growth hormone (GH; somatotropin), a pituitary tropic hormone. The catecholamines, norepinephrine and epinephrine are mainly products of the sympathetic postganglionic neurons and the adrenal medulla, respectively. Somatomedin is a hepatic hormone.

2. The answer is D. [*VII D*] The substrate for dopamine β-hydroxylase is dihydroxyphenylethylamine (dopamine), which is converted to norepinephrine within the cytoplasmic granules of the adrenomedullary chromaffin cell or within the vesicle of the postganglionic sympathetic neuron. The other substrates and their enzymes are: epinephrine or norepinephrine/catechol-O-methyltransferase (COMT) or monoamine oxidase (MAO), tyrosine/tyrosine hydroxylase, phenylalanine/phenylalanine hydroxylase, and dihydroxyphenylalanine (dopa)/dopa decarboxylase.

3. The answer is D. [*X B 4 c (2)*] The second meiotic division of the oocyte is not completed until just after fertilization. From the third month of embryonic life, the primary oocytes are arrested in the diplotene phase of the early prophase of meiosis. Completion of this first division occurs just prior to ovulation by the extrusion of the first polar body, resulting in the formation of a secondary oocyte with a haploid number of chromosomes (22 autosomes and 1 sex chromosome). The onset of ovulatory menstrual cycles may lag several months behind menses, which usually occurs between the ages of 12 and 16. The second meiotic division occurs just after conception, yielding a fertilized ovum (zygote) and a second polar body.

4. The answer is C. [*X D 1 a (2) (b); Table 7-10*] In terms of serum concentration, the major postmenopausal steroid and pituitary tropic hormones are estrone and follicle-stimulating hormone (FSH). Menopause, defined as the physiologic cessation of menses, results from a loss in the cyclic ovarian function caused by the failure of the ovary to respond to gonadotropins. The decline in ovarian function causes an increase in pituitary tropic hormones, with a striking increase in plasma FSH concentration relative to the increase in plasma luteinizing hormone (LH) level. Some estrogens continue to be produced in postmenopausal women by the extraovarian conversion of androstenedione of adrenal origin to estrone.

5. The answer is C. [*VII D 3 b (2)*] Since cortisol activates the epinephrine-forming enzyme, phenylethanolamine-N-methyltransferase, a decrease in cortisol secretion would lead to decreased adrenomedullary synthesis of epinephrine. Another effect of cortisol is to elevate blood glucose through gluconeogenesis and inhibition of peripheral glucose transport and utilization. Cortisol's proteolytic effect in extrahepatic tissues leads to increased transport of amino acids to the liver, where they are used for gluconeogenesis, glycogenesis, and protein synthesis. Hypocortisolism, therefore, would result in decrements in plasma glucose concentration, hepatic glycogen, and hepatic protein. A decline in free blood cortisol levels would also lead to an increase in adrenocorticotropic hormone (ACTH; corticotropin) secretion due to the reduced negative feedback effect of cortisol on the hypothalamic-pituitary complex.

6. The answer is A. [*XII D 1 b*] Glucose is transported by facilitated diffusion in erythrocytes, skeletal and cardiac muscle, and adipocytes. Facilitated diffusion for glucose transport requires a protein carrier molecule in the cell membrane. The insulin-dependent facilitated diffusion mechanism for glucose is found in skeletal and cardiac muscle and in adipose tissue. Some tissues obtain their glucose requirements by insulin-independent mechanisms. For example, CNS cells and hepatocytes transport glucose by simple diffusion, whereas erythrocytes transport glucose by insulin-independent facilitated diffusion. In renal and gastrointestinal epithelium, glucose transport occurs by a secondary active transport system that is not insulin-dependent but is Na^+-dependent. Since Na^+ and glucose are transported in the same directions, this active transport system is called a symport system.

7. The answer is A. [*VII H 1 b (2)*] After an overnight fast, about 75% of the hepatic glucose output is derived from glycogen and about 25% from gluconeogenesis. If fasting is prolonged or combined with exercise, the hepatic glycogen stores are depleted much more rapidly and the percentage contribution from gluconeogenesis increases. The gluconeogenic substrates are lactate, glycerol, and amino acids. The lactate is derived from incomplete oxidation of glucose by the action of epinephrine on muscle, which comprises about 45% of the body mass. Thus, net hepatic glucose synthesis and secretion requires lactate production from muscle glycogenolysis. The major gluconeogenic source of endogenous glucose production by the action of cortisol is alanine, with a smaller fraction available from the glycerol released from triglycerides hydrolyzed in adipose tissue. Acetyl-CoA is not a gluconeogenic substance in mammalian liver. Thus, epinephrine causes hyperglycemia because it stimulates hepatic glycogenolysis and gluconeogenesis and muscle glycogenolysis while it inhibits insulin secretion.

8. The answer is A. [*XI B 2 (b) (3)*] The major source of progesterone during the second and third trimesters of pregnancy is the trophoblast of the placenta. Maternal cholesterol is the principal source of precursor substrate for the biosynthesis of progesterone. Thus, progesterone levels during pregnancy are an index of the functional status of the materno-placental unit. Plasma progesterone levels during pregnancy are of limited value as an indicator of placental function alone because of the variability of normal levels.

9. The answer is A. [*XIII C 2*] Triiodothyronine (T_3) is the most biologically active iodothyronine secreted by the thyroid follicles. In the secretory process, thyroglobulin (the major storage form of thyroid hormone) is degraded to free amino acids, including tetraiodothyronine (T_4), T_3, monoiodotyrosine, and diiodotyrosine. Of these, only T_4 and T_3 are released into the bloodstream, in a ratio of about 20:1. T_3 has three to five times the biologic activity of T_4 and is considered the most biologically active form of thyroid hormone. T_3 also can be formed from T_4 by the action of 5'-deiodinase in peripheral tissues, especially the liver and kidney. The other circulating iodothyronines, such as reverse triiodothyronine (rT_3), tetraiodothyroacetic acid, and triiodothyroacetic acid have much less biologic activity than T_3 and T_4.

10. The answer is D. [*XIV E 1 c (3)*] Active vitamin D_3 (1,25-dihydroxycholecalciferol; calcitriol) is formed in the proximal convoluted tubule by conversion (via the action of 1α-hydroxylase) from its immediate precursor, 25-hydroxycholecalciferol (calcidiol). Vitamin D activation begins in the skin, where the previtamin, 7-dehydrocholesterol, is photoactivated by sunlight to lipid-soluble vitamin D_3. Vitamin D_3 then is converted in the liver to calcidiol, the major circulating form of vitamin D_3.

11. The answer is C. [*VIII A 3*] In the fetus, the adrenal glands are much larger relative to body size than in the adult. In absolute size, they are almost as large at term as the fetal kidneys and are as large as the adult adrenal glands. The fetal zone, or inner zone, of the fetal adrenal cortex undergoes complete involution 4–12 weeks postpartum. The outer zone, or neocortex, undergoes further differentiation into the adult adrenal cortex.

12. The answer is B. [*IX B 3 a (2)*] The chief product of the fetal zone of the adrenal cortex is the weak androgen dehydroepiandrosterone (DHEA), which is secreted as the inactive sulfate ester. DHEA sulfate is converted in the fetal liver to 16α-hydroxydehydroepiandrosterone sulfate, which, upon aromatization in the placenta, becomes estriol (estriol is synthesized only by the placenta and adult liver). After the first trimester, the inner zone of the fetal adrenal cortex also can secrete cortisol and aldosterone. Progesterone and corticosterone are synthesized, but not secreted, by the fetal adrenal cortex.

13. The answer is C. [*IX E 1 c (1) (b)*] Prostaglandins found in seminal fluid are secretory products of the seminal vesicles. Chemical analysis of seminal fluid provides an indirect measure of testicular function, because the male accessory sex organs (i.e., the seminal vesicles and prostate gland) are androgen-dependent. The seminal vesicles secrete fructose, prostaglandins, and ascorbate into the seminal plasma; the prostate gland secretes acid phosphatase and citrate into the seminal fluid. Originally, the prostate was believed to be the source of the prostaglandins.

14. The answer is E. [*X F 1 c (2)*] The secretory products of the ovary are estradiol and estrone, which are produced by the granulosa cells of the unruptured follicle and the lutein-granulosa cells of the corpus luteum. Luteinization of the granulosa cells, which depends on LH, involves the appearance of lipid droplets in the cytoplasm. The estrogenic precursors produced by the theca interna (and lutein-theca interna) are androstenedione and testosterone, which are converted by the granulosa (and lutein-granulosa) cells into estrone and estradiol, respectively. The lutein-granulosa cells also produce and secrete progesterone. Thus, androstenedione and pregnenolone are produced, but not secreted, by the ovary. Estriol is not an ovarian product but is synthesized in the liver from estradiol and estrone and in the placenta by the conversion of imported androgens. Pregnanediol is the urinary metabolite of progesterone and is formed in the liver.

15. The answer is B. [*IX B 2*] Ninety percent of the volume of the testis is composed of tubular tissue, with the remaining 10% consisting mainly of nontubular, or interstitial, cells (also called Leydig cells). The seminiferous tubular epithelium contains three cell types: spermatogonia, spermatocytes, and Sertoli cells. Spermatogonia and spermatocytes are germinal cells, and the Sertoli cells are nongerminal cells.

16. The answer is C. [*XIV F 3 c*] The major renal effect of parathyroid hormone (PTH) is stimulation of proximal tubular 1α-hydroxylase, an enzyme that converts calcidiol to calcitriol. PTH affects renal Ca^{2+} reabsorption in two ways: it reduces Ca^{2+} reabsorption in the proximal tubule and increases Ca^{2+} reabsorption in the distal tubule. The net effect is an increase in tubular Ca^{2+} reabsorption. This renal effect of PTH can be lessened by PTH's effect on bone. PTH increases plasma $[Ca^{2+}]$ by promoting bone resorption, which causes an increase in the renal filtered load of Ca^{2+} and, thus, hypercalciuria. Also, PTH inhibits the proximal reabsorption of phosphate, thereby favoring phosphate excretion. PTH acts on bone by dissolving the nonreadily exchangeable calcium phosphate "pool" known as stable bone. PTH activates the osteoclasts—cells that cause osteolysis by their high content of lysosomal enzymes. PTH also stimulates osteocytes, which are bone-bound osteoblasts that mediate osteocytic osteolysis.

17. The answer is A. [*XIV H 3 c*] PTH and active vitamin D_3 (calcitriol) have several similar effects, particularly in Ca^{2+} regulation. However, these two hormones exert different effects on renal phosphate handling: calcitriol promotes phosphate reabsorption, whereas PTH promotes phosphate diuresis. PTH and calcitriol both promote bone resorption, which leads to an increase in serum $[Ca^{2+}]$. Calcitriol promotes intestinal absorption of Ca^{2+} and phosphate. Although PTH has no direct intestinal effect on Ca^{2+} or phosphate absorption, it does stimulate renal 1α-hydroxylase activity and subsequent calcitriol formation; thus, PTH indirectly promotes intestinal absorption of Ca^{2+} and phosphate. In the kidney, PTH promotes distal tubular Ca^{2+} reabsorption, inhibits proximal phosphate reabsorption, and activates proximal 1α-hydroxylase. Calcitriol enhances renal tubular reabsorption of both Ca^{2+} and phosphate. At pharmacologic doses, calcitriol mimics the effects of PTH (i.e., it promotes phosphate excretion).

18. The answer is D. [*IX C 1 c*] In skin and target cells of the male reproductive tract, testosterone may be reduced to the more potent androgen, dihydrotestosterone, in a one-way reaction by the enzyme 5α-reductase. Testosterone also is metabolized to estradiol by the action of aromatase in various tissues, including brain, breast, and adipose tissue.

During fetal organization, the testis is stimulated by placental human chorionic gonadotropin (HCG) to produce testosterone. The characteristic adolescent "growth spurt" in the male results from an interplay of testosterone and GH, which promote growth of the vertebrae, shoulder girdle, and long bones. This growth is self-limited, as androgens also accelerate epiphysial closure. In adulthood, testosterone is secreted from the testis in response to LH secretion from the pituitary gland; testosterone acts as a negative feedback regulator of LH secretion and, thus, its own secretion.

19. The answer is A. [*VIII F 3 a–c*] The overall metabolic effects of cortisol are the release of amino acids from muscle and both the storage and release of glucose and fatty acids. Cortisol inhibits fatty acid synthesis in the liver, and it increases blood glycerol and fatty acid concentrations in concert with other hormones (norepinephrine, epinephrine, glucagon) to increase lipolysis in adipose tissues. Glycerol is an excellent index of lipolysis, because, unlike free fatty acids, it is not reused by adipocytes in the resynthesis of triglycerides. Rather, glycerol is used by the liver as a gluconeogenic substrate. Cortisol also promotes synthesis of gluconeogenic enzymes needed to convert the amino acids released from muscle to carbohydrate synthesis in the liver. Glucocorticoids promote proteolysis and inhibit protein synthesis in most tissues except the liver.

20. The answer is C. [*VIII F 1 a*] Glucocorticoids are potent hyperglycemic hormones that, at high levels, lead to glucose intolerance. Glucocorticoids stimulate the liver to produce glucose and glycogen due, in part, to the increased synthesis of key gluconeogenic enzymes. They also promote muscle proteolysis, resulting in the release of amino acids, which (aside from leucine) are gluconeogenic. The synthesis of hepatic gluconeogenic enzymes accounts for the protein anabolic effect of glucocorticoids in liver. Hepatic glycogenesis occurs via the activation of glycogen synthase by glucocorticoids. The increased hepatic glucose production and secretion involves gluconeogenesis, not glycogen degradation (glycogenolysis).

21. The answer is C. [*X D 1 a (3); XI B 3*] Estriol, the least biologically active natural estrogen, is produced only in the liver and placenta. It is not an ovarian product but represents the predominant urinary end product of estrogen metabolism. In pregnancy, the levels of urinary estriol increase 1000-fold, because estriol is formed by the trophoblast of the placenta. The liver converts estrone and estradiol of ovarian origin into estriol, whereas the trophoblast converts 16α-hydroxydehydroepiandrosterone (16α-OH DHEA) sulfate of fetal adrenal origin into estriol.

22. The answer is A. [*XIII D 1*] The biosynthetic pathway for thyroid hormone has four steps: active uptake of inorganic iodide, oxidation of iodide to active iodide, iodination of the tyrosyl residues on the thyroglobulin molecule, and coupling of iodotyrosines to form iodothyronines. Only the first step does not require thyroid peroxidase. Secretion depends on a lysosomal protease.

23. The answer is B. [*VI C 2, 4, D 2 c*] Prolactin, not oxytocin, causes milk synthesis (lactogenesis); oxytocin evokes milk secretion (ejection). Oxytocin is an octapeptide produced mainly in the paraventricular nucleus of the ventral diencephalon. Since it is synthesized in the magnocellular neurosecretory neurons, oxytocin is classified as a neuropeptide, or neurosecretory hormone. The unmyelinated axons of the paraventricular nucleus contribute to the supraoptico-hypophysial tract, which terminates in the pars nervosa. The pars nervosa is a release center and not the site of oxytocin synthesis. Oxytocin has two important smooth muscle effects: It causes milk ejection via contraction of the myoepithelial cells of the mammary gland, and it promotes myometrial (uterine) contraction.

24. The answer is D. [*VII H 1 c (1) (a)*] Activation of the sympathetic nervous system involves stimulation of both α- and β-adrenergic receptors by both norepinephrine and epinephrine. The α-adrenergic response of the pancreatic islet cells is dominant during sympathetic stimulation; thus, insulin secretion (a response mediated by β-adrenergic receptors) is suppressed. Other responses that are mediated by β-adrenergic receptors include renin secretion, vasodilation, and inhibition of intestinal motility. Responses mediated by α-adrenergic receptors include inhibition of intestinal motility, contraction of the radial eye muscle (mydriasis), and vasoconstriction. Vasodilation in skeletal muscle is caused by epinephrine and by stimulation of sympathetic cholinergic nerves.

25. The answer is D. [*XII D 1 b*] Insulin decreases cell membrane permeability to both Na^+ and K^+, but it decreases Na^+ permeability more, causing hyperpolarization of skeletal and cardiac muscle cells and adipocytes but not liver or pancreatic cells. Insulin also promotes lipogenesis, protein anabolism, glycogenesis, and glycolysis. Insulin promotes glucose uptake by muscle and adipose tissue but not by the liver, and it promotes storage and utilization of glucose in all three tissues. Glucose storage by muscle and liver is achieved by formation of glycogen via an increase in glycogen synthase activity. Glucose transport across skeletal and cardiac muscle cells and adipocytes occurs by insulin-dependent facilitated diffusion.

26. The answer is A. [*XI B 2 d, C 2*] By 8 weeks gestation, maternal cholesterol can be converted to progesterone by the placental trophoblast, which becomes the major producer of progesterone. Progesterone is necessary for maintenance of the decidual cells of the endometrium, inhibition of myometrial contraction by the hyperpolarization of uterine smooth muscle cells, and formation of a dense, viscous mucus that seals off the uterine cavity. Progesterone, in synergy with estrogen, promotes the growth and branching of the lobuloalveolar ductal system of the mammary gland. In early pregnancy, placental progesterone serves as a precursor for the synthesis of cortisol and aldosterone by the fetal zone of the adrenal cortex under the stimulation of HCG. Beyond 10 weeks gestation, the outer zone of the fetal adrenal cortex (neocortex) can produce these two corticoids under the stimulation of ACTH. Progesterone also exhibits a natriuretic effect by antagonizing the action of aldosterone on the renal tubule.

27. The answer is B. [*VII F 2 a (1), (2)*] Vanillylmandelic acid (VMA) is the major urinary end product of catecholamine metabolism. Monoamine oxidase (MAO) is a mitochondrial enzyme that catalyzes the oxidative deamination of the catecholamines—dopamine, norepinephrine, and epinephrine. MAO also inactivates the indolamine, serotonin (5-hydroxytryptamine). Substrates for MAO also include normetanephrine and metanephrine.

28. The answer is C. [*XI B 3 a*] The neocortex of the fetal adrenal gland does not participate in estriol synthesis during gestation. Estriol is a product of the combined activities of the fetal pituitary gland, the fetal zone of the adrenal cortex, the fetal liver, and the trophoblast of the placenta. The fetal zone of the adrenal gland is the inner zone that persists only during gestation; the neocortex is the outer zone of the fetal adrenal gland, which is the progenitor of the adult adrenal cortex. In the first trimester, the fetal zone is primarily stimulated by HCG, with ACTH (from the fetal pituitary) gaining prominence thereafter. The fetal zone synthesizes DHEA and its sulfate (DHEAS), which are transported to the placenta after 16α-hydroxylation by the fetal liver. It is 16α-OH DHEAS that is the precursor of placental estriol.

29. The answer is D. [*VI D 1 c (1) (a)*] The protein anabolic effects of GH are mediated through a GH-dependent peptide known as somatomedin, which is synthesized in the liver. Important biologic effects of somatomedin include mitogenesis of chondrocytes and bone cells, stimulation of lipogenesis and muscle glycogenesis, and GH-like actions in cartilage (e.g., sulfation of chondroitin). ADH and oxytocin are neuropeptides synthesized mainly in the neurosecretory neurons of the supraoptic and paraventricular nuclei, respectively. Thyrotropin releasing hormone (TRH) is produced in the neurosecretory neurons of the arcuate nucleus. Endorphins are secretory products of basophils in the adenohypophysis and of brain neurons. Thus, endorphins also can be classified as neuropeptides.

30. The answer is D. [*VI D 1 b (2) (b)*] Secretion of GH in response to various stimuli often is blunted in obesity. Stimuli for GH secretion include norepinephrine, bromocriptine, hypoglycemia, stress, vigorous exercises, insulin, and arginine. The long-term metabolic effects of GH include hyperglycemia due to gluconeogenesis, glycogenolysis, and lipolysis, all of which explain its anti-insulin effects. GH has a fat catabolic effect, which increases plasma free fatty acid levels. It should be noted that these free fatty acids are not substrates for gluconeogenesis.

31. The answer is C. [*V B 1, 2*] The polypeptide known as β-endorphin is formed from β-lipotropin (not from ACTH), which is a cleavage product of the prohormone pro-opiomelanocortin (POMC). POMC and its derivatives are synthesized in the basophils of the pars distalis of the anterior lobe of the pituitary gland. Endorphins, β-lipotropin, and ACTH (corticotropin) also are found in the brain. ACTH can be cleaved to form α-melanocyte-stimulating hormone (α-MSH) and a corticotropin-like peptide called CLIP. β-Endorphin binds with morphine receptors and produces morphine-like responses, such as respiratory depression, analgesia, and miosis.

32. The answer is A. [*X E 3 c*] In this woman with a menstrual cycle of 21–23 days, ovulation would be expected to occur between day 7 and day 9, counting from the first day of menses. The normal menstrual cycle has an average duration of 28 days. The luteal, or postovulatory, phase corresponds to the secretory phase of the endometrium or the progestational phase of the corpus luteum. This phase of the menstrual cycle usually is very constant, lasting about 14 days, while the duration of the follicular phase can be highly variable. Because of the constancy of the postovulatory phase, the time of ovulation can be estimated by subtracting 14 days from the cycle length.

33. The answer is C. [*IX A 1 e (1), (2)*] Patients with testicular feminization have a female phenotype but a male genotype. They have intra-abdominal testes that produce testosterone and estrogen concentrations that are characteristic of normal men, but their tissues are totally unresponsive to androgens due to a lack of androgen receptors and they do not produce sperm. The external genitalia are female because the primordial tissue develops in the female pattern unless stimulated by androgen. The androgen resistance syndromes are disorders in which müllerian duct regression and testosterone synthesis are normal. Male development of the embryo requires müllerian duct inhibiting factor (to prevent the development of female internal genitalia), testosterone (to mediate wolffian duct development), and dihydrotestosterone (to stimulate development of the prostate and male external genitalia). This patient lacks female internal genitalia (because the müllerian duct inhibiting factor exerts its normal action) and lacks male wolffian duct development (because testosterone does not exert its normal stimulatory action). Since both duct systems regress, neither male nor female internal genitalia develop. Secondary sex characteristics, including breast development, appear at puberty in response to the unopposed action of estrogen formed extragonadally from testosterone. This patient has male pseudohermaphroditism, in that the female phenotype in this genetic male is not due to the presence of both testes and ovaries.

34–37. The answers are: 34-C, 35-B, 36-D, 37-B. [*X E 3 b, c (3), F 1 b; XI A 1 d; Figure 7-9*] In a normal ovarian cycle, the plasma concentration of estradiol peaks (*point A*), triggering a sudden discharge of pituitary gonadotropins (LH and FSH) 12–24 hours later. This preovulatory gonadotropin surge (*point B*) is more pronounced for LH than for FSH. Ovulation (*point C*) occurs 16–20 hours after the LH peak and marks the formation of the corpus luteum. LH maintains the functional and morphologic integrity of the corpus luteum for 14 days.

The ovum can be fertilized between 6 and 20 hours following ovulation (*point D*). If fertilization occurs, a morula is formed, which enters the uterine cavity on the third to fourth day postovulation (*point E*). The morula is converted to the blastocyst during the fifth to sixth day postovulation, which is implanted in the endometrial wall of the uterus on the seventh day (*point F*).

The corpus luteum is the principal source of estrogen (estradiol) and progesterone during the first 6–8 weeks gestation. The principal gonadotropic hormone during this period is HCG, which is secreted by the syncytiotrophoblast of the placenta. This anterior pituitary–like hormone has mainly LH activity and is the luteotropic hormone of pregnancy. The trophoblast takes over as the major source of progesterone and estrogen (estriol) secretion by 8 weeks gestation. HCG also stimulates the fetal testis to produce testosterone and the fetal adrenal gland to produce corticoids in early pregnancy.

38. The answer is A (1, 2, 3). [*VI D 1 c (2) (b)*] GH, like insulin and normal levels of thyroid hormone, is a potent protein anabolic hormone (i.e., it promotes nitrogen retention and, thus, a positive nitrogen balance). Negative nitrogen balance results when nitrogen losses exceed nitrogen intake, as occurs with deficient dietary protein or in stressful conditions (e.g., tissue trauma, disease states, burns, surgery). Utilization of amino acids in hepatic gluconeogenesis results in negative nitrogen balance. Glucocorticoids (e.g., cortisol) promote muscle proteolysis and inhibit protein synthesis, leading to increased BUN and enhanced nitrogen excretion. Alanine is the predominant amino acid released by muscle. Glucagon is a catabolic hormone for protein in muscle, and it also is an important gluconeogenic hormone. At normal levels, thyroid hormone stimulates protein synthesis and degradation. The positive effect of thyroid hormone on body growth is derived largely from stimulation of protein synthesis. In excess amounts, thyroid hormone causes accelerated protein catabolism, leading to increased nitrogen excretion.

39. The answer is B (1, 3). [*X F 1 b (2); XI C 1*] Hormones that maintain the functional and morphologic integrity of the corpus luteum (called luteotropic hormones) include LH and HCG. LH is the luteotropic hormone during the luteal phase of the menstrual cycle; it converts the ovulated follicle into a functioning corpus luteum. HCG is the luteotropic hormone of pregnancy; it stimulates progesterone secretion during the first 6–8 weeks gestation, and it converts the corpus luteum of menstruation to the corpus luteum of pregnancy. FSH and progesterone are not luteotropic hormones.

40. The answer is E (all). [*VII H 1 c*] Epinephrine is a potent hyperglycemic hormone due mainly to its effects on the liver and pancreas. In the liver, it promotes glycogenolysis and gluconeogenesis, and the glucose-6-phosphate formed by glycogenolysis is hydrolyzed to glucose. Epinephrine-induced glycogenolysis in muscle leads to formation of lactic acid, which is converted to glucose in the liver, further elevating blood glucose. In the pancreas, epinephrine stimulates glucagon secretion and inhibits insulin secretion, both hyperglycemic effects. Epinephrine also stimulates ACTH secretion, which, in turn, leads to secretion of cortisol, which is a major hyperglycemic hormone.

41. The answer is A (1, 2, 3). [*XIII D 1, F 1 b (2), G 1*] Normally, thyroid hormone (T_3 and T_4) is regulated by thyroid-stimulating hormone (TSH), which, in turn, is controlled by TRH. The circulating levels of T_3 and T_4 also influence TSH release by exerting negative feedback control at the adenohypophysial and, to a lesser extent, hypothalamic levels. TSH influences the structure and function of thyroid follicular cells and regulates all phases of thyroid hormone synthesis, storage, and secretion. Administration of exogenous thyroid hormone leads to disuse atrophy of the thyroid gland via TSH suppression. Thus, iodide trapping and endogenous thyroid hormone secretion are reduced when TSH is suppressed by exogenous thyroid hormone. In most cells of the body, thyroid hormone stimulates the cell membrane enzyme Na^+-K^+-ATPase, thereby increasing O_2 consumption. This effect is not exerted on cells of the brain, lymph nodes, gonads, lungs, spleen, and dermis.

42–45. The answers are: 42-A, 43-B, 44-C, 45-B. [*VII C 2, F 2 b, I 2 c*] The adrenal medulla is a modified sympathetic ganglion consisting of chromaffin cells (pheochromocytes) that are the functional analogs of the postganglionic neurons. About 80% of the adrenomedullary secretion is epinephrine, and 20% is norepinephrine. The plasma concentration ratio of epinephrine to norepinephrine normally is 1:4, the inverse of the secretory ratio. All circulating epinephrine originates in the adrenal medulla; therefore, bilaterally adrenalectomized patients have very low plasma epinephrine levels. In contrast, plasma norepinephrine levels in these patients remain within normal limits, which demonstrates that the major source of circulating norepinephrine is not the adrenal medulla.

Most circulating norepinephrine is derived from the postganglionic sympathetic neurons. Norepinephrine has been demonstrated in almost all tissues except the placenta, which lacks nerve fibers. The norepinephrine content of a tissue or organ serves as an index of the density of its sympathetic autonomic innervation.

Catecholamines are catabolized by two enzymes: catecholamine-O-methyltransferase (COMT) and monoamine oxidase (MAO). Norepinephrine and epinephrine are metabolized extracellularly by COMT to normetanephrine and metanephrine, respectively. COMT is localized within the cytosol of sympathetic effector cells.

Tumors may arise within the sympathoadrenomedullary system. Adrenal chromaffin cell tumors, known as pheochromocytomas, usually arise within the adrenal medulla, are benign, and secrete primarily excessive amounts of norepinephrine. However, pheochromocytoma patients may have signs of both norepinephrine and epinephrine hypersecretion. Most patients have paroxysms of hypertension, tachycardia, sweating, tremor, palpitations, and nervousness. Most lose weight and are hyperglycemic because of the catecholamine-induced inhibition of insulin secretion.

46–49. The answers are: 46-C, 47-C, 48-A, 49-A. [*VI D 1 b (2) (a); XII A 1 b, c, 3 c, 6 a*] Somatostatin and insulin both are hypoglycemic hormones. Somatostatin decreases blood glucose levels by decreasing hepatic glucose production. It also inhibits the secretion of GH, glucagon, and insulin and decreases intestinal glucose absorption. Insulin exerts its hypoglycemic action by promoting glucose uptake by muscle and adipose tissue, by increasing storage and utilization of glucose (mainly by liver, muscle, and adipose tissue), and by suppressing hepatic production and release of glucose.

Somatostatin and insulin are polypeptides, consisting of 14 and 51 amino acid residues, respectively. Insulin is synthesized in the pancreatic beta cells; somatostatin is synthesized in the pancreatic delta cells as well as many other sites, including the hypothalamus, cerebrum, thymus, thyroid, gastric and intestinal epithelium, skin, and heart. Insulin inhibits glucagon secretion, and somatostatin inhibits secretion of GH, TSH, insulin, glucagon, pancreatic polypeptide, gut hormones, gastric acid, and pepsin.

Somatostatin is synthesized in the arcuate nucleus of the tuberoinfundibular neurosecretory neurons that terminate in the median eminence. Somatostatin of hypothalamic origin functions as a neurosecretory hormone. It also is a hypophysiotropic hormone that inhibits GH and thyrotropin secretion following its release into the hypophysial portal system. Insulin is neither a neurosecretory hormone nor a hypophysiotropic hormone.

50–55. The answers are: 50-B, 51-A, 52-B, 53-B, 54-C, 55-B. [*VII G; Table 7-5*] Insulin secretion is elicited by stimulation of the β-adrenergic receptor on the surface of the pancreatic beta cell.

In order to increase pupillary diameter, the radial eye muscle must contract. Contraction of this intrinsic ocular muscle occurs by stimulation of the α-adrenergic receptor. The radial ocular muscle is innervated only by postganglionic sympathetic nerves. A decrease in pupillary diameter occurs by stimulation of the parasympathetic innervation of the pupillary sphincter.

Bronchodilation is caused by stimulation of β-adrenergic receptors. There are no α-adrenergic receptors in bronchial smooth muscle. Bronchoconstriction occurs by stimulation of the parasympathetic innervation of the bronchi.

Renin is synthesized by the juxtaglomerular (JG) cells of the afferent arteriole, which receive postganglionic sympathetic innervation. Thus, these autonomic nerves secrete norepinephrine, which stimulates β-adrenergic receptors on the surface of the JG cells.

Intestinal smooth muscle contains both α- and β-adrenergic receptors. Intestinal motility is inhibited by stimulation of either receptor type. Sphincters in the gastrointestinal tract contain only α-adrenergic receptors, which, when stimulated, lead to sphincter contraction.

The detrusor muscle of the urinary bladder, in contrast to the trigone and sphincter, contains only β-adrenergic receptors, which, when stimulated, evoke relaxation of this smooth muscle. Stimulation of the α-adrenergic receptors of the bladder produces contraction of these smooth muscles. Thus, sympathetic stimulation of the bladder blocks micturition.

Comprehensive
Exam

Introduction

One of the least attractive aspects of pursuing an education is the necessity of being examined on what has been learned. Instructors do not like to prepare tests, and students do not like to take them.

However, students are required to take many examinations during their learning careers, and little if any time is spent acquainting them with the positive aspects of tests and with systematic and successful methods for approaching them. Students perceive tests as punitive and sometimes feel that they are merely opportunities for the instructor to discover what the student has forgotten or has never learned. Students need to view tests as opportunities to display their knowledge and to use them as tools for developing prescriptions for further study and learning.

A brief history and discussion of the National Board of Medical Examiners (NBME) examinations (i.e., Parts I, II, and III and FLEX) are presented here, along with ideas concerning psychological preparation for the examinations. Also presented are general considerations and test-taking tips, as well as ways to use practice exams as educational tools. (The literature provided by the various examination boards contains detailed information concerning the construction and scoring of specific exams.)

National Board of Medical Examiners Examinations

Before the various NBME exams were developed, each state attempted to license physicians through its own procedures. Differences between the quality and testing procedures of the various state examinations resulted in the refusal of some states to recognize the licensure of physicians licensed in other states. This made it difficult for physicians to move freely from one state to another and produced an uneven quality of medical care in the United States.

To remedy this situation, the various state medical boards decided they would be better served if an outside agency prepared standard exams to be given in all states, allowing each state to meet its own needs and have a common standard by which to judge the educational preparation of individuals applying for licensure.

One misconception concerning these outside agencies is that they are licensing authorities. This is not the case; they are examination boards only. The individual states retain the power to grant and revoke licenses. The examination boards are charged with designing and scoring valid and reliable tests. They are primarily concerned with providing the states with feedback on how examinees have performed and with making suggestions about the interpretation and usefulness of scores. The states use this information as partial fulfillment of qualifications upon which they grant licenses.

Students should remember that these exams are administered nationwide and, although the general medical information is the same, educational methodologies and faculty areas of

The author of this introduction, Michael J. O'Donnell, holds the positions of Assistant Professor of Psychiatry and Director of Biomedical Communications at the University of New Mexico School of Medicine, Albuquerque, New Mexico.

expertise differ from institution to institution. It is unrealistic to expect that students will know all the material presented in the exams; they may face questions on the exams in areas that were only superficially covered in their classes. The testing authorities recognize this situation, and their scoring procedures take it into account.

The Exams

The first exam was given in 1916. It was a combination of written, oral, and laboratory tests, and it was administered over a 5-day period. Admission to the exam required proof of completion of medical education and 1 year of internship.

In 1922, the examination was changed to a new format and was divided into three parts. Part I, a 3-day essay exam, was given in the basic sciences after 2 years of medical school. Part II, a 2-day exam, was administered shortly before or after graduation, and Part III was taken at the end of the first postgraduate year. To pass both Part I and Part II, a score equaling 75% of the total points available in each was required.

In 1954, after a 3-year extensive study, the NBME adopted the multiple-choice format. To pass, a statistically computed score of 75 was required, which allowed comparison of test results from year to year. In 1971, this method was changed to one that held the mean constant at a computed score of 500, with a predetermined deviation from the mean to ascertain a passing or failing score. The 1971 changes permitted more sophisticated analysis of test results and allowed schools to compare among individual students within their respective institutions as well as among students nationwide. Feedback to students regarding performance included the reporting of pass or failure along with scores in each of the areas tested.

During the 1980s, the ever-changing field of medicine made it necessary for the NBME to examine once again its evaluation strategies. It was found necessary to develop questions in multidisciplinary areas such as gerontology, health promotion, immunology, and cell and molecular biology. In addition, it was decided that questions should test higher cognitive levels and reasoning skills.

To meet the new goals, many changes have been made in both the form and content of the examination. These changes include reduction in the number of questions to approximately 800 in Part I and Part II to allow students more time on each question, with total testing time reduced on Part I from 13 to 12 hours and on Part II from 12.5 to 12 hours. The basic science disciplines are no longer allotted the same number of questions, which permits flexible weighing of the exam areas. Reporting of scores to schools includes total scores for individuals and group mean scores for separate discipline areas. Only pass/fail designations and total scores are reported to examinees. There is no longer a provision for the reporting of individual subscores to either the examinees or medical schools. Finally, the question format used in the new exams, now referred to as Comprehensive (Comp) I and II, is predominately multiple-choice, best-answer.

The New Format

New questions, designed specifically for Comp I, are constructed in an effort to test the student's grasp of the sciences basic to medicine in an integrated fashion— the questions are designed to be interdisciplinary. Many of these items are presented as vignettes, or case studies, followed by a series of multiple-choice, best-answer questions.

The scoring of this exam is altered. Whereas in the past the exams were scored on a normal curve, the new exam has a predetermined standard, which must be met in order to pass. The exam no longer concentrates on the trivial; therefore, it has been concluded that there is a common base of information that all medical students should know in order to pass. It is anticipated that a major shift in the pass/fail rate for the nation is unlikely. In the past, the average student could only expect to feel comfortable with half the test and eventually would complete approximately 67% of the questions correctly, to achieve a mean score of 500. Although with the standard setting method it is likely that the mean score will change and become higher, it is unlikely that the pass/fail rates will differ significantly from those in the past. During the first testing in 1991, there will not be differential weighing of questions. However, in the future, the NBME will be researching methods of weighing questions based on both the time it takes to answer questions vis-à-vis their difficulty and the perceived importance of the information. In addition, the NBME is attempting to design a method of delivering feedback to the student that will have considerable importance in discovering weaknesses and pinpointing areas for further study in the event that a retake is necessary.

Since many of the proposed changes will be implemented for the first time in June 1991, specific information regarding actual standards, question emphasis, pass/fail rates, and so forth were unavailable at the time of publication. The publisher will update this section as information becomes available and as we attempt to follow the evolution and changes that occur in the area of physician evaluation.

Materials Needed for Test Preparation

In preparation for a test, many students collect far too much study material only to find that they simply do not have the time to go through all of it. They are defeated before they begin because either they leave areas unstudied, or they race through the material so quickly that they cannot benefit from the activity.

It is generally more efficient for the student to use materials already at hand; that is, class notes, one good outline to cover or strengthen areas not locally stressed and to quickly review the whole topic, and one good text as a reference for looking up complex material needing further explanation.

Also, many students attempt to memorize far too much information, rather than learning and understanding less material and then relying on that learned information to determine the answers to questions at the time of the examination. Relying too heavily on memorized material causes anxiety, and the more anxious students become during a test, the less learned knowledge they are likely to use.

Positive Attitude

A positive attitude and a realistic approach are essential to successful test taking. If concentration is placed on the negative aspects of tests or on the potential for failure, anxiety increases and performance decreases. A negative attitude generally develops if the student concentrates on "I must pass" rather than on "I can pass." "What if I fail?" becomes the major factor motivating the student to **run from failure rather than toward success**. This results from placing too much emphasis on scores rather than understanding that scores have only slight relevance to future professional performance.

The score received is only one aspect of test performance. Test performance also indicates the student's ability to use information during evaluation procedures and reveals how this ability might be used in the future. For example, when a patient enters the physician's office with a problem, the physician begins by asking questions, searching for clues, and seeking diagnostic information. Hypotheses are then developed, which will include several potential causes for the problem. Weighing the probabilities, the physician will begin to discard those hypotheses with the least likelihood of being correct. Good differential diagnosis involves the ability to deal with uncertainty, to reduce potential causes to the smallest number, and to use all learned information in arriving at a conclusion.

The same thought process can and should be used in testing situations. It might be termed **paper-and-pencil differential diagnosis**. In each question with five alternatives, of which one is correct, there are four alternatives that are incorrect. If deductive reasoning is used, as in solving a clinical problem, the choices can be viewed as having possibilities of being correct. The elimination of wrong choices increases the odds that a student will be able to recognize the correct choice. Even if the correct choice does not become evident, the probability of guessing correctly increases. Just as differential diagnosis in a clinical setting can result in a correct diagnosis, eliminating choices on a test can result in choosing the correct answer.

Answering questions based on what is incorrect is difficult for many students since they have had nearly 20 years experience taking tests with the implied assertion that knowledge can be displayed only by knowing what is correct. It must be remembered, however, that students can display knowledge by knowing something is wrong, just as they can display it by knowing something is right. **Students should begin to think in the present as they expect themselves to think in the future.**

Paper-and-Pencil Differential Diagnosis

The technique used to arrive at the answer to the following question is an example of the paper-and-pencil differential diagnosis approach.

A recently diagnosed case of hypothyroidism in a 45-year-old man may result in which of the following conditions?

(A) Thyrotoxicosis

(B) Cretinism

(C) Myxedema

(D) Graves' disease

(E) Hashimoto's thyroiditis

It is presumed that all of the choices presented in the question are plausible and partially correct. If the student begins by breaking the question into parts and trying to discover what the question is attempting to measure, it will be possible to answer the question correctly by using more than memorized charts concerning thyroid problems.

- The question may be testing if the student knows the difference between "hypo" and "hyper" conditions.
- The answer choices may include thyroid problems that are not "hypothyroid" problems.
- It is possible that one or more of the choices are "hypo" but are not "thyroid" problems, that they are some other endocrine problems.
- "Recently diagnosed in a 45-year-old man" indicates that the correct answer is not a congenital childhood problem.
- "May result in" as opposed to "resulting from" suggests that the choices might include a problem that **causes** hypothyroidism rather than **results from** hypothyroidism, as stated.

By applying this kind of reasoning, the student can see that choice **A,** thyroid toxicosis, which is a disorder resulting from an overactive thyroid gland ("hyper"), must be eliminated. Another piece of knowledge, that is, Graves' disease is thyroid toxicosis, eliminates choice **D.** Choice **B,** cretinism, is indeed hypothyroidism, but is a childhood disorder. Therefore, **B** is eliminated. Choice **E** is an inflammation of the thyroid gland—here the clue is the suffix "itis." The reasoning is that thyroiditis, being an inflammation, may **cause** a thyroid problem, perhaps even a hypothyroid problem, but there is no reason for the reverse to be true. Myxedema, choice **C,** is the only choice left and the obvious correct answer.

Preparing for Board Examinations

1. Study for yourself. Although some of the material may seem irrelevant, the more you learn now, the less you will have to learn later. Also, do not let the fear of the test rob you of an important part of your education. If you study to learn, the task is less distasteful than studying solely to pass a test.

2. Review all areas. You should not be selective by studying perceived weak areas and ignoring perceived strong areas. This is probably the last time you will have the time and the motivation to review **all** of the basic sciences.

3. Attempt to understand, not just memorize, the material. Ask yourself: To whom does the material apply? Where does it apply? When does it apply? Understanding the connections among these points allows for longer retention and aids in those situations when guessing strategies may be needed.

4. Try to anticipate questions that might appear on the test. Ask yourself how you might construct a question on a specific topic.

5. Give yourself a couple days of rest before the test. Studying up to the last moment will increase your anxiety and cause potential confusion.

Taking Board Examinations

1. In the case of NBME exams, be sure to **pace yourself** to use the time optimally. Each booklet is designed to take 2 hours. You should use all of your allotted time; if you finish too early, you probably did so by moving too quickly through the test.

2. Read each question and all the alternatives carefully before you begin to make decisions. Remember the questions contain clues, as do the answer choices. As a physician, you would not make a clinical decision without a complete examination of all the data: the same holds true for answering test questions.

3. Read the directions for each question set carefully. You would be amazed at how many students make mistakes in tests simply because they have not paid close attention to the directions.

4. It is not advisable to leave blanks with the intention of coming back to answer the questions later. Because of the way Board examinations are constructed, you probably will not pick up any new information that will help you when you come back, and the chances of getting numerically off on your answer sheet are greater than your chances of benefiting by skipping around. If you feel that you must come back to a question, mark the best choice and place a note in the margin. Generally speaking, it is best not to change answers once you have made a decision. Your intuitive reaction and first response are correct more often than changes made out of frustration or anxiety. **Never turn in an answer sheet with blanks.** Scores are based on the number that you get correct; you are not penalized for incorrect choices.

5. Do not try to answer the questions on a stimulus-response basis. It generally will not work. Use all of your learned knowledge.

6. Do not let anxiety destroy your confidence. If you have prepared conscientiously, you know enough to pass. Use all that you have learned.

7. Do not try to determine how well you are doing as you proceed. You will not be able to make an objective assessment, and your anxiety will increase.

8. Do not expect a feeling of mastery or anything close to what you are accustomed to. Remember, this is a nationally administered exam, not a mastery test.

9. Do not become frustrated or angry about what appear to be bad or difficult questions. You simply do not know the answers; you cannot know everything.

Specific Test-Taking Strategies

Read the entire question carefully, regardless of format. Test questions have multiple parts. Concentrate on picking out the pertinent key words that might help you begin to problem-solve. Words such as "always," "never," "mostly," "primarily," and so forth play significant roles. In all types of questions, distractors with terms such as "always" or "never" most often are incorrect. Adjectives and adverbs can completely change the meaning of questions—pay close attention to them. Also, medical prefixes and suffixes (e.g., "hypo-," "hyper-," "-ectomy," "-itis") are sometimes at the root of the question. The knowledge and application of everyday English grammar often is the key to dissecting questions.

Multiple-Choice Questions

Read the question and the choices carefully to become familiar with the data as given. Remember, in multiple-choice questions there is one correct answer and there are four distractors, or incorrect answers. (Distractors are plausible and possibly correct or they would not be called distractors.) They are generally correct for part of the question but not for the entire question. Dissecting the question into parts aids in discerning these distractors.

If the correct answer is not immediately evident, begin eliminating the distractors. (Many students feel that they must always start at option A and make a decision before they move to B, thus forcing decisions they are not ready to make.) Your first decisions should be made on those choices you feel the most confident about.

Compare the choices to each part of the question. **To be wrong,** a choice needs to be **incorrect for only part** of the question. **To be correct,** it must be **totally** correct. If you believe a choice is partially incorrect, tentatively eliminate that choice. Make notes next to the choices regarding tentative decisions. One method is to place a minus sign next to the choices you are certain are incorrect and a plus sign next to those that potentially are correct. Finally, place a zero next to any choice you do not understand or need to come back to for further inspection. Do not feel that you must make final decisions until you have examined all choices carefully.

When you have eliminated as many choices as you can, decide which of those that are left has the highest probability of being correct. Remember to use paper-and-pencil differential diagnosis. Above all, be honest with yourself. If you do not know the answer, eliminate as many choices as possible and choose reasonably.

Vignette-Based Questions

Vignette-based questions are nothing more than normal multiple-choice questions that use the same case, or grouped information, for setting the problem. The NBME has been re-searching question types that would test the student's grasp of the integrated medical basic sciences in a more cognitively complex fashion than can be accomplished with traditional testing formats. These questions allow the testing of information that is more medically relevant than memorized terminology.

It is important to realize that several questions, although grouped together and referring to one situation or vignette, are independent questions; that is, they are able to stand alone. Your inability to answer one question in a group should have no bearing on your ability to answer other questions in that group.

These are multiple-choice questions, and just as with single best-answer questions, you should use the paper-and-pencil differential diagnosis, as was described earlier.

Single Best-Answer–Matching Sets

Single best-answer–matching sets consist of a list of words or statements followed by several numbered items or statements. Be sure to pay attention to whether the choices can be used more than once, only once, or not at all. Consider each choice individually and carefully. Begin with those with which you are the most familiar. It is important always to break the statements and words into parts, as with all other question formats. **If a choice is only partially correct, then it is incorrect.**

Guessing

Nothing takes the place of a firm knowledge base, but with little information to work with, even after playing paper-and-pencil differential diagnosis, you may find it necessary to guess at the correct answer. A few simple rules can help increase your guessing accuracy. Always guess consistently if you have no idea what is correct; that is, after eliminating all that you can, make the choice that agrees with your intuition or choose the option closest to the top of the list that has not been eliminated as a potential answer.

When guessing at questions that present with choices in numerical form, you will often find the choices listed in ascending or descending order. It is generally not wise to guess the first or last alternative, since these are usually extreme values and are most likely incorrect.

Using the Comprehensive Exam to Learn

All too often, students do not take full advantage of practice exams. There is a tendency to complete the exam, score it, look up the correct answers to those questions missed, and then forget the entire thing.

In fact, great educational benefits can be derived if students would spend more time using practice tests as learning tools. As mentioned earlier, incorrect choices in test questions are plausible and partially correct or they would not fulfill their purpose as distractors. This means that it is just as beneficial to look up the incorrect choices as the correct choices to discover specifically why they are incorrect. In this way, it is possible to learn better test-taking skills as the subtlety of question construction is uncovered.

Additionally, it is advisable to go back and attempt to restructure each question to see if all the choices can be made correct by modifying the question. By doing this, four times as much will be learned. By all means, look up the right answer and explanation. Then, focus on each of the other choices and ask yourself under what conditions they might be correct. For example, the entire thrust of the sample question concerning hypothyroidism could be altered by changing the first few words to read:

"Hyperthyroidism recently discovered in..."
"Hypothyroidism prenatally occurring in..."
"Hypothyroidism resulting from..."

This question can be used to learn and understand thyroid problems in general, not only to memorize answers to specific questions.

In the practice exams that follow, every effort has been made to simulate the types of questions and the degree of question difficulty in the NBME Part I Comprehensive exam. While taking these exams, the student should attempt to create the testing conditions that might be experienced during actual testing situations. Approximately 1 minute should be allowed for each question, and the entire test should be finished before it is scored.

Summary

Ideally, examinations are designed to determine how much information students have learned and how that information is used in the successful completion of the examination. Students will be successful if these suggestions are followed:

- Develop a positive attitude and maintain that attitude.
- Be realistic in determining the amount of material you attempt to master and in the score you hope to attain.
- Read the directions for each type of question and the questions themselves closely and follow the directions carefully.
- Guess intelligently and consistently when guessing strategies must be used.
- Bring the paper-and-pencil differential diagnosis approach to each question in the examination.
- Use the test as an opportunity to display your knowledge and as a tool for developing prescriptions for further study and learning.

National Board examinations are not easy. They may be almost impossible for those who have unrealistic expectations or for those who allow misinformation concerning the exams to produce anxiety out of proportion to the task at hand. They are manageable if they are approached with a positive attitude and with consistent use of all the information that has been learned.

Michael J. O'Donnell

QUESTIONS

Directions: Each of the numbered items or incomplete statements in this section is followed by answers or by completions of the statement. Select the **one** lettered answer or completion that is **best** in each case.

1. At the end of a normal expiration, a young woman has an interpleural pressure of –5 cm H_2O. Without expiring further, she closes off her nose and mouth and performs a Valsalva maneuver against the closed airway. If the airway pressure is +20 mm Hg during the expiratory maneuver, the interpleural pressure would be

(A) –5 cm H_2O
(B) 5 cm H_2O
(C) 10 cm H_2O
(D) 15 cm H_2O
(E) 20 cm H_2O

2. Which of the following is characteristic of myopic eyes?

(A) They are too short for the refractive power of their lenses
(B) They have a focal plane behind the retina
(C) They have a far point that is greater than normal
(D) They require a diverging lens to correct the optical defects
(E) They have a greater than normal power of accommodation

3. Assuming a total body fluid volume of 40 L, 15 L of which are ECF with an osmolarity of 300 mOsm/L, what will be the equilibrium ICF osmolarity if 500 ml of a 0.15 mol/L NaCl solution are infused intravenously?

(A) 164 mOsm/L
(B) 277 mOsm/L
(C) 286 mOsm/L
(D) 300 mOsm/L
(E) 324 mOsm/L

4. The following data were obtained from a 55-year-old male patient during cardiac catheterization: O_2 consumption = 210 ml/min, O_2 content of right ventricular blood = 11 ml/dl, O_2 content of brachial artery blood = 18 ml/dl, and heart rate = 75 bpm. These data indicate that which of the following was true during the catheterization of this patient?

(A) The tissues received 29 ml O_2/dl blood flow
(B) Cardiac output was about 1470 ml/min
(C) Pulmonary venous O_2 content was about 145 ml/dl blood
(D) Right ventricular stroke volume averaged about 40 ml
(E) Cardiac output was dangerously high

5. The regression of the corpus luteum at the end of the postovulatory phase is caused by

(A) a decrease in FSH secretion
(B) a decrease in LH secretion
(C) an increase in HCG secretion
(D) a reduced capacity of the corpus luteum to synthesize steroids
(E) ovarian failure

6. The rate of gastric emptying is controlled primarily by reflexes that occur

(A) during chewing
(B) during swallowing
(C) when chyme enters the stomach
(D) when chyme enters the intestine
(E) during the interdigestive period

7. Spironolactone, which is an aldosterone antagonist, is injected into the renal artery of a laboratory animal. What are the effects on Na^+ and K^+ excretion, assuming that there is no change in GFR or RBF by this drug?

	Na^+	K^+
(A)	↑	↑
(B)	↓	↓
(C)	↑	↓
(D)	↓	↑
(E)	↑	unchanged

8. Which of the following conditions would cause an increase in aortic systolic pressure and a decrease in aortic pulse pressure?

(A) Increased heart rate
(B) Increased arterial compliance
(C) Decreased peripheral resistance
(D) Increased stroke volume
(E) Increased elastic modulus

9. A 32-year-old male electrician consults an internist and complains of episodes of palpitations and sweating that occur when he climbs a ladder at work. He feels "washed out" after these attacks, which he ascribes to "nerves." Subsequent examination by the physician leads to a diagnosis of pheochromocytoma. Body fluid analysis of this patient is most likely to reveal a low plasma concentration of

(A) free fatty acids
(B) insulin
(C) fasting glucose
(D) lactate
(E) pyruvate

10. The following data are collected from a 40-year-old male patient:

airway resistance = 27 cm H_2O/L/sec (normal = 2–3 cm H_2O/L/sec)
diffusing capacity of the lungs = 10 ml/min/mm Hg (normal = 30 ml/min/mm Hg)
arterial pH = 7.32
minute ventilation = 6 L/min

These data indicate that this patient most likely has

(A) an arterial PCO_2 less than 40 mm Hg
(B) an arterial PO_2 less than 85 mm Hg
(C) a mean pulmonary artery pressure less than 15 mm Hg
(D) a total work of breathing less than normal
(E) an arterial $[HCO_3{}^-]/[H_2CO_3]$ ratio greater than 20:1

11. Elimination of the terminal ileum will result in malabsorption of

(A) vitamin B_{12}
(B) fats
(C) both
(D) neither

12. Acetazolamide is administered to a glaucoma patient. Given that this drug inhibits carbonic anhydrase in the renal proximal tubule, which of the following substances will be excreted at a lower rate?

(A) Na^+
(B) H_2O
(C) $HCO_3{}^-$
(D) $NH_4{}^+$
(E) K^+

13. A decrease in the total osmotic pressure of arterial blood would lead to an increase in urine volume by

(A) increasing the hydrostatic pressure inside the glomerulus
(B) increasing the permeability of the glomerular capillaries to water
(C) inhibiting ADH secretion
(D) stimulating the secretion of aldosterone
(E) directly inhibiting the reabsorption of water by the collecting ducts

14. During moderate exercise, a patient has a cardiac index of 6.5 L/min/m², a hemoglobin concentration of 12 g/dl, a venous PO_2 of 30 mm Hg, and a venous O_2 saturation of 50%. Assuming 100% hemoglobin saturation in arterial blood, what is this patient's O_2 consumption?

(A) 150 ml/min/m²
(B) 275 ml/min/m²
(C) 520 ml/min/m²
(D) 790 ml/min/m²
(E) 1030 ml/min/m²

15. A red blood cell is placed in a solution. The cell initially shrinks and then returns to its original volume. The red cell's reaction indicates that the solution is

(A) hyperosmotic and hypertonic
(B) hypo-osmotic and hypertonic
(C) hyperosmotic and isotonic
(D) hyperosmotic and hypotonic
(E) hypo-osmotic and isotonic

16. A 32-year-old woman is admitted to the hospital with suspected partially compensated respiratory acidosis. Which of the following sets of laboratory data would confirm this suspicion?

	$[HCO_3^-]$ (mEq/L)	P_{CO_2} (mm Hg)	pH
(A)	17	19	7.9
(B)	31	80	7.22
(C)	9.8	30	7.14
(D)	24	45	7.5
(E)	20	25	7.5

17. Following a massive hemorrhage during delivery, a 34-year-old woman experiences a failure to lactate and to menstruate. Which of the following is most likely to be associated with this clinical picture?

(A) Elevated prolactin secretion

(B) Excessive urinary Na^+ excretion

(C) Excessive water excretion

(D) Increased sensitivity to insulin

(E) Elevated gonadotropin secretion

18. If the mean electrical axis of a ventricular depolarization is $-30°$, the resultant EKG should reveal the largest positive QRS complex in lead

(A) I

(B) II

(C) aVR

(D) aVL

(E) aVF

19. Normally, most of the H^+ is excreted by the kidneys in the form of

(A) HCO_3^-

(B) phosphate ion

(C) NH_4^+

(D) titratable acid

(E) β-hydroxybutyrate ion

20. A delirious 5-year-old boy is brought to the emergency room. His parents report that their son was in good health until 2 hours beforehand, when his mental state suddenly began to deteriorate. They suspect he may have swallowed a bottle of aspirin, because they discovered an empty bottle before leaving for the hospital. Laboratory evaluation reveals the following arterial blood data: $[H^+]$ = 18 nmol/L, $[HCO_3^-]$ = 13 mmol/L, and P_{CO_2} = 10 mm Hg. This patient's history and blood data are most likely associated with

(A) respiratory alkalosis with partial renal compensation

(B) metabolic alkalosis with partial respiratory compensation

(C) a reduced $[HCO_3^-]$/dissolved CO_2 ratio

(D) a greater than normal CO_2 content

21. If acidosis and hypokalemia result from loss of fluid from the GI tract, the fluid was most likely drained from the

(A) stomach

(B) intestine

(C) gallbladder

(D) pancreas

(E) colon

22. Tapping the patella tendon initiates a reflex contraction of the quadriceps muscle. When the contractile force of the quadriceps muscle reaches its peak, there is an increase in the activity of

(A) the Ia afferent fibers innervating the quadriceps muscle

(B) the Ib afferent fibers innervating the quadriceps muscle

(C) both

(D) neither

23. When the ACh receptors on the pacemaker cells of the heart are activated, there is an increase in the membrane conductance to

(A) K^+

(B) Na^+

(C) Ca^{2+}

(D) Cl^-

(E) K^+ and Na^+

24. Normal O_2 delivery to the tissues would be cut in half by a 50% decrease in the normal value of

(A) arterial Po_2
(B) minute ventilation
(C) hemoglobin concentration
(D) inspired Po_2

25. A 45-year-old woman with severe vomiting caused by pyloric obstruction would be expected to show

(A) hyperchloremia
(B) an increase in plasma $[HCO_3^-]$
(C) an increase in alveolar ventilation
(D) acid urine
(E) a decrease in arterial Pco_2

26. In the formation of HCl by the parietal cells, the transport of H^+ across the parietal cell is coupled with the transport of

(A) K^+
(B) Cl^-
(C) HCO_3^-
(D) H^+
(E) Na^+

27. Primary hyperkalemic periodic paralysis is an inherited disease characterized by alterations in K^+ homeostasis. The K^+ defect leads to intermittent periods of muscle weakness due to failure of action potential propagation. This failure is due to

(A) inactivation of Na^+ channels
(B) membrane hyperpolarization
(C) both
(D) neither

28. The normal sequence of phases of the menstrual cycle is

(A) menses, preovulatory, ovulatory, estrogenic
(B) preovulatory, ovulatory, progestational, menses
(C) ovulatory, progestational, menses, luteal
(D) progestational, menses, follicular, preovulatory
(E) menses, follicular, ovulatory, estrogenic

29. Gastric digestion is most important for which of the following substances?

(A) Fats
(B) Carbohydrates
(C) Proteins
(D) Vitamins
(E) Minerals

30. Resistance to blood flow through the kidney can be determined by

(A) measuring the clearance of PAH
(B) measuring the hydrostatic pressure difference between the renal artery and renal vein
(C) measuring the RBF
(D) dividing the arteriovenous hydrostatic pressure difference by the RBF
(E) dividing the RBF by the arteriovenous hydrostatic pressure difference

31. The following blood data are collected from a 27-year-old male patient: pH = 7.50, $[HCO_3^-]$ = 38 mmol/L, and Po_2 = 80 mm Hg. Given these findings, what is the expected Pco_2 for this patient?

(A) 30 mm Hg
(B) 40 mm Hg
(C) 50 mm Hg
(D) 60 mm Hg
(E) 70 mm Hg

32. Electrically excitable channels are required for

(A) the propagation of the action potential in nerve fibers
(B) the release of synaptic transmitter from presynaptic nerve terminals
(C) both
(D) neither

33. A 57-year-old male patient who is breathing air at sea level has a respiratory exchange ratio of 1. Arterial blood gas analysis of this patient reveals the following: Po_2 = 85 mm Hg, Pco_2 = 30 mm Hg, and pH = 7.52. This patient's blood data indicate which of the following?

(A) His alveolar-to-arterial Po_2 difference exceeds 20 mm Hg
(B) His plasma $[HCO_3^-]$ is increased
(C) He has been hypoventilating
(D) He has metabolic alkalosis
(E) He has chronic obstructive lung disease

34. Given the following data: glomerular capillary hydrostatic pressure = 47 mm Hg, glomerular capillary colloid osmotic pressure = 28 mm Hg, Bowman's space hydrostatic pressure = 10 mm Hg, and Bowman's space oncotic pressure = 0 mm Hg, what is the glomerular filtration pressure?

(A) 2 mm Hg

(B) 4 mm Hg

(C) 6 mm Hg

(D) 9 mm Hg

(E) 10 mm Hg

35. Following the intravenous administration of 1 L of a 150 mmol NaCl solution into a patient with a blood loss, there will be

(A) a decrease in the plasma Na^+ concentration

(B) an increase in the osmolarity of the ICF compartment

(C) an increase in the volume of the ICF compartment

(D) edema

(E) a decrease in the colloid osmotic pressure of the plasma

36. A 45-year-old male patient is studied and found to have a respiratory rate of 15 breaths/min, a tidal volume of 0.5 L, and a dead space of 200 ml. The patient is asked to increase his respiratory rate to 30 breaths/min, and his tidal volume is measured at 350 ml. Assuming no change in dead space, which of the following is true regarding alveolar Pco_2?

(A) Pco_2 will increase because of the decreased ventilation

(B) Pco_2 will decrease because of the increased ventilation

(C) Pco_2 will not change because it is not affected by respiration

(D) Pco_2 will not change because alveolar ventilation remains constant

(E) The arterial blood pH will decrease because of the increased ventilation

37. Cortisol affects the biosynthesis of epinephrine in the adrenal medulla by

(A) augmenting the conversion of dopamine to epinephrine

(B) activating the epinephrine-forming enzyme

(C) inhibiting the methylation of norepinephrine

(D) activating catechol-O-methyltransferase

(E) increasing the release of acetylcholine

38. A positive QRS complex in leads aVR and aVF indicates

(A) no axis deviation

(B) right axis deviation

(C) left axis deviation

(D) left ventricular hypertrophy

(E) a mean electrical axis between 0° and 90°

39. The following arterial blood data are collected from a 42-year-old female patient: $[H^+]$ = 49 nEq/L, Pco_2 = 30 mm Hg, and Po_2 = 95 mm Hg. Given these findings, what is the expected arterial $[HCO_3^-]$ for this patient?

(A) 13.2 mEq/L

(B) 14.7 mEq/L

(C) 15.8 mEq/L

(D) 16.5 mEq/L

(E) 17.1 mEq/L

40. Which of the following statements best characterizes the transpulmonary pressure at the base of the lung of a person who is standing?

(A) It is independent of lung volume

(B) It is equal to the transpulmonary pressure at the apex of the lung

(C) It may be negative if the lung is at residual volume

(D) It causes the basal alveoli to be more dilated than the apical alveoli

(E) It causes the bronchioles at the lung base to be more dilated than those at the apex

41. Which of the following reflexes is most dependent on a vago-vagal reflex?

(A) Chewing

(B) Swallowing

(C) Receptive relaxation

(D) Gastric emptying

(E) Intestinal segmentation

42. A 23-year-old woman with diabetes mellitus is admitted to the hospital. She is dehydrated and hyperpneic. Laboratory examination reveals high urinary concentrations of acetoacetic acid and glucose, blood pH of 7.39, plasma $[HCO_3^-]$ of 19.0 mmol/L, and plasma P_{CO_2} of 33 mm Hg. These data are most suggestive of

(A) metabolic alkalosis with complete renal compensation
(B) metabolic acidosis with complete respiratory compensation
(C) respiratory acidosis with partial renal compensation
(D) respiratory alkalosis with partial renal compensation

43. Epinephrine inhibits glucose uptake by muscle and adipose tissue. This inhibitory effect is attributed to

(A) glucagon secretion
(B) thyroid hormone secretion
(C) inhibition of insulin secretion
(D) inhibition of GH secretion
(E) inhibition of cortisol secretion

44. The pars nervosa is derived from which of the following embryonic tissues?

(A) Oral ectoderm
(B) Neural ectoderm
(C) Neural crest
(D) Endoderm
(E) Mesoderm

45. A 32-year-old man can generate an inspiratory pressure of –50 mm Hg intermittently for several minutes. How deep can this man lie underwater while breathing through a tube, if the tube offers no significant resistance to air flow?

(A) 37 mm
(B) 37 cm
(C) 68 mm
(D) 68 cm
(E) 74 cm

46. Which of the following conditions is most likely to cause acidosis with marked dehydration?

(A) Severe diarrhea
(B) Severe, persistent vomiting
(C) Excessive sweating
(D) Drinking sodium lactate solution
(E) Complete water deprivation for 24 hours

47. During the first 6–8 weeks of pregnancy, progesterone is secreted mainly by the

(A) maternal adrenal glands
(B) maternal theca interna
(C) corpus luteum
(D) fetal adrenal gland
(E) decidua

48. Which of the following factors best explains an increase in the filtration fraction?

(A) Increased ureteral pressure
(B) Increased efferent arteriolar resistance
(C) Increased plasma protein concentration
(D) Decreased glomerular capillary hydrostatic pressure
(E) Decreased glomerular filtration area

49. The greatest amount of fat absorption occurs in the

(A) stomach
(B) duodenum
(C) ileum
(D) colon
(E) rectum

50. A 27-year-old, anxious man is examined in the emergency room, after which the following arterial blood data are obtained:

Blood Chemistry	Blood Gas
$[Na^+]$ = 140 mEq/L	pH = 7.6
$[K^+]$ = 4 mEq/L	P_{CO_2} = 20 mm Hg
$[HCO_3^-]$ = 19 mEq/L	P_{O_2} = 98 mm Hg
$[Cl^-]$ = 109 mEq/L	

From the above data, the most likely diagnosis is

(A) hypoxia
(B) metabolic alkalosis
(C) metabolic acidosis with respiratory compensation
(D) obstructive lung disease
(E) respiratory alkalosis

51. Epinephrine-forming enzyme activity is increased directly by

(A) hydrocortisone
(B) acetylcholine
(C) norepinephrine
(D) ACTH
(E) 11-deoxycortisol

52. Two patients are studied and the following data are collected:

Patient	Respiratory Rate (breaths/min)	Tidal Volume (ml)	Dead Space (ml)
A	20	200	150
B	10	400	150

Which of the following statements about these two patients is true?

(A) The alveolar ventilation in patient A is greater than in patient B
(B) The alveolar ventilation in patient B is greater than in patient A
(C) The alveolar ventilation in both patients is equal
(D) The dead space ventilation in both patients is equal

53. A semicomatose 19-year-old woman is brought to the emergency room with dry skin, hyperventilation, hypotension, and a rapid pulse rate. The following blood data are obtained:

pH = 7.14
$[Na^+]$ = 140 mEq/L
$[K^+]$ = 4.5 mEq/L
$[Cl^-]$ = 82 mEq/L
$[HCO_3^-]$ = 11 mEq/L
P_{CO_2} = 30 mm Hg
[glucose] = 180 mg/dl

From the above history and laboratory data, the most likely diagnosis is

(A) metabolic alkalosis
(B) metabolic acidosis
(C) respiratory alkalosis
(D) respiratory acidosis
(E) hypoglycemia

54. Which of the following hormones is correctly paired with its effect on renal electrolyte reabsorption?

(A) Calcitriol / increased HPO_4^{2-} reabsorption
(B) Calcitonin / increased Ca^{2+} reabsorption
(C) Aldosterone / increased K^+ reabsorption
(D) Progesterone / increased Na^+ reabsorption
(E) Calcitonin / increased HPO_4^{2-} reabsorption

55. During the process of vitamin B_{12} absorption, almost all of the ingested vitamin B_{12}

(A) binds to intrinsic factor in the stomach
(B) is absorbed in the stomach
(C) both
(D) neither

56. When the arterial P_{O_2} drops from 100 mm Hg to 27 mm Hg with an arterial pH of 7.4 and a P_{CO_2} of 40 mm Hg, O_2 content in the blood decreases by about

(A) 10%
(B) 25%
(C) 33%
(D) 50%
(E) 75%

57. A newborn genotypic male is found to have an adrenogenital syndrome due to a 17α-hydroxylase defect. Which of the following biochemical reactions in the biosynthesis of gonadal hormones is decelerated in this case of congenital adrenal hyperplasia?

(A) Pregnenolone → 17α-hydroxypregnenolone
(B) Cholesterol → pregnenolone
(C) Progesterone → corticosterone
(D) Deoxycorticosterone → corticosterone
(E) 17α-Hydroxypregnenolone → 17α-hydroxyprogesterone

58. β-Adrenergic receptors mediate all of the following responses EXCEPT

(A) ciliary muscle contraction
(B) increased myocardial contractility
(C) vasodilation
(D) insulin secretion
(E) decreased intestinal motility

59. All of the following substances can be converted to more biologically active forms after secretion EXCEPT

(A) angiotensinogen
(B) angiotensin I
(C) tetraiodothyronine
(D) testosterone

60. A large dose of insulin is administered intravenously to a normal 34-year-old female patient. This is likely to cause an increase in all of the following EXCEPT

(A) plasma epinephrine concentration
(B) plasma K^+ concentration
(C) ACTH secretion
(D) GH secretion
(E) glucagon secretion

61. Findings consistent with metabolic acidosis include all of the following EXCEPT

(A) increased excretion of titrable acid
(B) increased excretion of NH_4^+
(C) decreased excretion of $NaHCO_3$
(D) decreased respiratory rate
(E) a negative base excess

62. All of the following are the renal responses to severe hemorrhage EXCEPT

(A) increased Na^+ reabsorption by the distal nephron
(B) decreased GFR
(C) increased permeability of the collecting duct to water
(D) decreased filtration fraction
(E) increased renin secretion

63. Untreated type I diabetes mellitus is associated with all of the following biochemical changes EXCEPT

(A) positive nitrogen balance
(B) ketonemia
(C) ketonuria
(D) low plasma C-peptide concentration
(E) glycosuria

64. All of the following would increase alveolar Po_2 EXCEPT

(A) breathing gas containing 24% O_2
(B) breathing gas containing 28% O_2
(C) ascending a mountain
(D) hyperventilating

65. Hypophysectomy results in the functional decline of many endocrine organs. All of the following changes are likely to occur after removal of the pituitary gland EXCEPT

(A) atrophy of the thyroid gland
(B) dwarfism if performed during early adolescence
(C) deficiency of aldosterone
(D) cessation of menstrual cycles
(E) impaired testosterone secretion

66. All of the following are substrates for COMT EXCEPT

(A) epinephrine
(B) norepinephrine
(C) normetanephrine
(D) dihydroxymandelic acid

67. Stimuli for aldosterone secretion include all of the following EXCEPT

(A) angiotensin II
(B) hyperkalemia
(C) hypovolemia
(D) corticotropin
(E) atrial natriuretic factor

68. Cutting the vagus nerve has a major effect on all of the following EXCEPT

(A) the rate of liquid emptying from the stomach
(B) receptive relaxation
(C) migrating motor complex
(D) primary esophageal peristalsis

69. GH secretion is increased by all of the following factors EXCEPT

(A) insulin administration
(B) arginine administration
(C) somatostatin administration
(D) onset of sleep
(E) exercise

70. All of the following substances are required for thyroxine biosynthesis EXCEPT

(A) active iodide
(B) diiodotyrosine
(C) monoiodotyrosine
(D) thyroglobulin
(E) thyroid peroxidase

71. Hormones with lipolytic activity include all of the following EXCEPT

(A) glucagon
(B) epinephrine
(C) insulin
(D) cortisol
(E) GH

72. Expansion of the antrum causes an increase in all of the following EXCEPT

(A) secretion of gastrin
(B) secretion of pancreatic enzymes
(C) secretion of gastric acid (HCl)
(D) gastric motility
(E) receptive relaxation

73. The neural crest gives rise to all of the following tissues and cells EXCEPT

(A) Schwann cells
(B) cartilage and bone of the skull
(C) the neural lobe of the pituitary gland
(D) melanocytes
(E) the adrenal medulla

74. In hyperaldosteronemia, all of the following conditions are likely to be observed EXCEPT

(A) decreased hematocrit
(B) fall in plasma oncotic pressure
(C) increased ECF volume
(D) hyperkalemia
(E) metabolic alkalosis

Questions 75–76

The following data were obtained from intracellular recordings of three cardiac cells (*bpm* = beats per minute).

	Cell A	Cell B	Cell C
Resting membrane potential	–60 mV	–60 mV	–80 mV
Spontaneous depolarization	yes	yes	no
Intrinsic rate of depolarization	80 bpm	45 bpm	none

75. Cell A most likely is located in

(A) the SA node
(B) atrial muscle
(C) the AV node
(D) Purkinje fibers
(E) ventricular muscle

76. Pacemaker activity is exhibited by

(A) cell A
(B) cell B
(C) cell C
(D) cells A and B
(E) cells A, B, and C

Questions 77–80

A patient presents with crushing chest pain, shortness of breath, and marked anxiety. A preliminary diagnosis of acute myocardial infarction is made. Physical examination reveals evidence of pulmonary edema, cardiomegaly, peripheral edema, and pulmonary hypertension. Arterial blood gas analysis, on room air, reveals the following: P_{CO_2} = 40 mm Hg, P_{O_2} = 60 mm Hg, pH = 7.32, and $[HCO_3^-]$ = 20 mEq/L.

77. This patient's blood data are most indicative of

(A) a diffusion abnormality
(B) hypoventilation
(C) a ventilation:perfusion abnormality
(D) left-to-right cardiac shunt

78. The patient is admitted to the CCU, sedated, and given 40% O_2 by respirator, which is set to deliver a tidal volume of 0.6 L at a rate of 16 breaths/min. An inspiratory pressure of 20 mm Hg is required to deliver the tidal volume. If the patient's predicted dead space is 150 ml, then his calculated alveolar ventilation is approximately

(A) 5 L/min
(B) 7 L/min
(C) 9 L/min
(D) 12 L/min
(E) 16 L/min

79. While ventilation continues with 40% O_2, a Swan-Ganz catheter is inserted into the patient's pulmonary artery. A repeat arterial blood analysis reveals a P_{O_2} of 120 mm Hg, a P_{CO_2} of 43 mm Hg, and pH of 7.31. These findings indicate that

(A) the pulmonary edema has cleared
(B) gas exchange is completely normal
(C) hemoglobin concentration has increased
(D) the alveolar-to-arterial P_{O_2} difference is increased

80. This patient's respiratory compliance is

(A) 0.03 L/mm Hg
(B) 0.1 L/mm Hg
(C) 9.0 L/mm Hg
(D) 12.0 mm Hg/L
(E) 26.7 mm Hg/L

Questions 81–82

Lung compliance in a 32-year-old female patient is studied. Data collected under control and experimental conditions are listed in the following table.

	Respiratory Rate (breaths/min)	Tidal Volume (ml)	Change in Interpleural Pressure during Inspiration (cm H_2O)
Control	15	600	4
Experimental	25	600	10

81. This patient's lung compliance during control and experimental conditions was

(A) unchanged

(B) 40 ml/breath and 24 ml/breath, respectively

(C) 150 ml/cm H_2O and 60 ml/cm H_2O, respectively

(D) 150 cm H_2O/ml and 60 cm H_2O/ml, respectively

82. This patient can be characterized as having frequency-dependent compliance, which indicates

(A) abnormal surfactant function

(B) obstructive lung disease

(C) restrictive lung disease

(D) pulmonary vascular disease

Questions 83–87

The diagram below represents an experimental record obtained from an anesthetized dog.

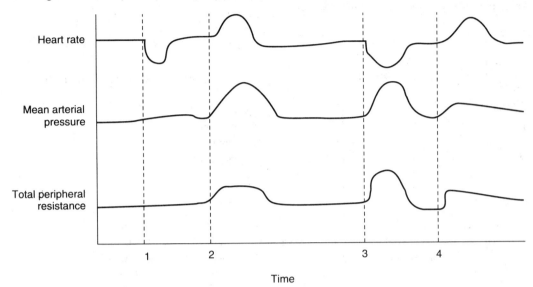

Time

83. The experimental intervention at time 1 most likely represents

(A) electrical stimulation of the lumbar sympathetic nerve roots
(B) electrical stimulation of the superior cervical ganglion (cardiac sympathetic nerves)
(C) administration of a β-adrenergic blocking drug
(D) stimulation of the right vagus nerve
(E) administration of a cholinergic blocking drug

84. The experimental intervention at time 2 most likely represents

(A) electrical stimulation of the sacral sympathetic nerve roots
(B) electrical stimulation of the superior cervical ganglion (cardiac sympathetic nerves)
(C) administration of a sympathetic blocking drug
(D) stimulation of the right vagus nerve
(E) administration of a cholinergic blocking drug

85. A drug was administered at time 3. This drug is most likely classified as

(A) a cholinergic blocking agent
(B) an α-adrenergic agonist
(C) a β-adrenergic agonist
(D) a β-adrenergic blocking agent
(E) an α-adrenergic blocking agent

86. The decrease in heart rate at time 3 was most likely caused by

(A) the direct effect of the drug on the SA node
(B) the direct effect of the drug on the ventricular muscle
(C) the occurrence of ventricular extrasystoles
(D) a reflex effect mediated by the chemoreceptors
(E) a reflex effect mediated by the baroreceptors

87. A nerve was stimulated at time 4. This nerve most likely was the

(A) cardiac end of a cut vagus nerve
(B) superior cervical ganglion of the sympathetic chain
(C) central end of a cut carotid sinus nerve
(D) peripheral end of a cut carotid sinus nerve
(E) postganglionic fibers of the lumbar sympathetic ganglia

Questions 88–93

The following six questions refer to the ventricular pressure-volume (PV) relationships depicted below. Loop ABEH represents a normal PV loop.

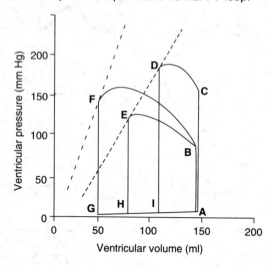

88. Loop ACDI shows the effect of

(A) increased heart rate
(B) increased contractility
(C) increased end-diastolic volume
(D) increased afterload
(E) decreased preload

89. Loop ABFG shows the effect of

(A) increased heart rate
(B) increased contractility
(C) increased end-diastolic volume
(D) increased afterload
(E) decreased preload

90. In loop ABEH, mitral valve closure occurs at

(A) point A
(B) point B
(C) point E
(D) point H
(E) point I

91. In loop ABEH, ventricular filling is indicated by

(A) segment AB
(B) segment BE
(C) segment EH
(D) segment HA

92. The slope of the line from point G to point A represents

(A) diastolic ventricular compliance
(B) systolic ventricular compliance
(C) diastolic ventricular elastance
(D) systolic arterial elastance
(E) diastolic arterial compliance

93. The volume of blood in the ventricles at points D, E, and F is referred to as ventricular

(A) stroke volume
(B) end-diastolic volume
(C) end-systolic volume
(D) diastolic reserve volume
(E) residual volume

Questions 94–97

The data below represent the volume of distribution of tritiated water, inulin, and Evan's blue dye in a 60-kg man after allowing time for equilibration.

Space	Volume (L)
Tritiated water	35
Inulin	8
Evan's blue	3

94. With a hematocrit ratio of 0.40, the man's blood volume is

(A) 2 L
(B) 3 L
(C) 4 L
(D) 5 L
(E) 6 L

95. The interstitial fluid volume is

(A) 3 L
(B) 5 L
(C) 8 L
(D) 11 L
(E) 24 L

96. The ECF volume is

(A) 2 L
(B) 4 L
(C) 6 L
(D) 8 L
(E) 10 L

97. Assuming that the total body water constitutes 70% of the lean body mass, the amount of body fat in this man is approximately

(A) 5 kg
(B) 10 kg
(C) 15 kg
(D) 20 kg
(E) 25 kg

Questions 98–100

A normal 55-year-old man who lives at an altitude of 11,500 feet is seen for an annual physical examination. Findings on examination include a hemoglobin concentration of 18 g/dl, an arterial Po_2 of 27 mm Hg, and an arterial pH of 7.40.

98. Based on the above findings, this man's arterial O_2 content (ml/dl) would be approximately

(A) 8
(B) 12
(C) 15
(D) 18
(E) 24

99. This man's pulmonary artery pressure is likely to be

(A) normal
(B) lower than normal due to inhibition of chemoreceptors caused by the low Pco_2
(C) higher than normal due to increased cardiac output
(D) higher than normal due to hypoxic pulmonary vasoconstriction

100. The hemoglobin saturation in this man's venous blood is likely to be

(A) normal
(B) less than normal due to decreased arterial hemoglobin saturation
(C) greater than normal due to increased cardiac output
(D) greater than normal due to increased hemoglobin concentration

101. The curves below represent the clearances of various substances as a function of their plasma concentrations. *Curve* E represents the clearance curve for glucose. Following the administration of a substance (phlorizin) that blocks epithelial transport of glucose, the clearance curve for glucose would resemble which of the following curves?

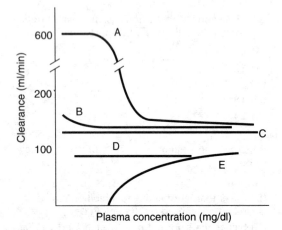

Adapted from Bauman JW, Chinard FP: *Renal Function: Physiological and Medical Aspects.* St. Louis, Mosby, 1975, p 43.

(A) A
(B) B
(C) C
(D) D
(E) E

Questions 102–103

The left ventricular and aortic pressure tracings below were recorded during cardiac catheterization of a 62-year-old patient who complains of chest pain and dizziness on exertion.

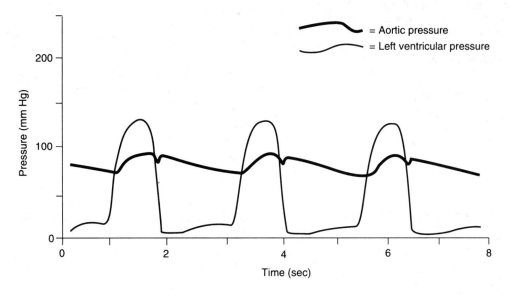

102. The left ventricular and aortic pressure tracings indicate that this patient has

(A) pulmonary stenosis
(B) aortic stenosis
(C) mitral stenosis
(D) aortic insufficiency
(E) mitral insufficiency

103. The most likely physical sign of this patient's condition is

(A) a systolic murmur
(B) a diastolic murmur
(C) a presystolic murmur
(D) a mid-diastolic murmur

Questions 104–106

A male patient undergoes lung volume studies using the helium dilution method. The initial fraction of helium in the spirometer is 0.05, and the helium fraction after equilibration with the lungs is 0.03. The volume of gas in the spirometer is kept constant at 4 L during the procedure by the addition of O_2. According to a spirogram, this patient's VC is 5 L and his ERV is 2 L.

104. What is this patient's FRC?

(A) 1.0 L
(B) 1.7 L
(C) 2.7 L
(D) 3.0 L
(E) 5.0 L

105. What is this patient's RV?

(A) 0.7 L
(B) 1.0 L
(C) 1.7 L
(D) 2.7 L
(E) 3.0 L

106. What is this patient's TLC?

(A) 1.7 L
(B) 2.7 L
(C) 3.0 L
(D) 5.0 L
(E) 5.7 L

Questions 107–108

The following two questions refer to the pressure-volume loop below.

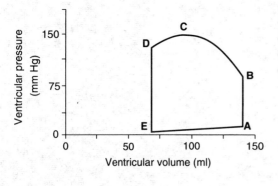

107. Mitral valve opening is represented by which Point on the curve?

(A) Point A
(B) Point B
(C) Point C
(D) Point D
(E) Point E

108. The period of isovolumic ventricular contraction is represented by which segment of the curve?

(A) A to B
(B) B to C
(C) C to D
(D) D to E
(E) E to A

109. The graph below shows the diurnal variation in the plasma concentration of which of the following hormones?

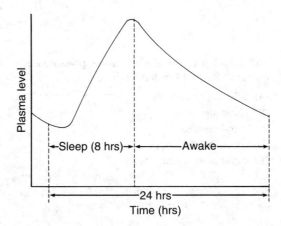

(A) Thyroxine
(B) Insulin
(C) Testosterone
(D) Cortisol
(E) Estradiol

Questions 110–112

The following three questions refer to the diagram below.

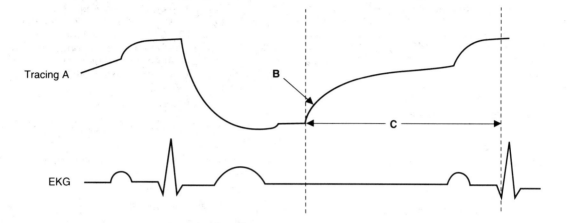

110. What does *tracing A* represent?

(A) A ventricular pressure pulse
(B) A ventricular volume curve
(C) An atrial pressure pulse
(D) An atrial volume pulse
(E) An aortic pressure pulse

111. *Point B* occurs in what phase of the cardiac cycle?

(A) Atrial contraction
(B) Isovolumic contraction
(C) Rapid ejection
(D) Reduced ejection
(E) Rapid filling

112. The duration of *interval C* is decreased by

(A) vagus stimulation
(B) baroreceptor stimulation
(C) arteriolar constriction
(D) increasing heart rate
(E) decreasing venous pressure

Questions 113–115

A 50-year-old male patient who has smoked two packs of cigarettes a day for 35 years complains of shortness of breath, chronic cough, and production of yellowish, foul-smelling sputum. He has clubbing of his finger nails, and the nail beds and lips are noted to be cyanotic. Arterial blood gas analysis of this patient reveals the following: $PO_2 = 55$ mm Hg, $PCO_2 = 56$ mm Hg, $HCO_3^- = 35$ mEq/L, and pH = 7.4.

113. These blood gas data are most consistent with

(A) acute respiratory failure
(B) inadequate alveolar ventilation
(C) anatomic shunt
(D) anemia
(E) carbon monoxide poisoning

114. Which of the following pathophysiologic phenomena most likely initiated this patient's syndrome?

(A) Decreased alveolar ventilation followed by increased work of breathing
(B) Bronchial narrowing due to inflammation and edema followed by pulmonary hypertension
(C) Bronchial narrowing due to inflammation and edema followed by increased work of breathing
(D) Hypercapnia followed by pulmonary hypertension
(E) Pulmonary hypertension followed by hypercapnia

115. The acid-base status of this patient's blood is best categorized as

(A) respiratory acidosis, uncompensated
(B) metabolic acidosis, compensated
(C) respiratory acidosis, compensated
(D) lactic acidosis secondary to hypoxia
(E) respiratory alkalosis

Questions 116–118

The following arterial blood data are obtained from a patient who is cyanotic at rest: $PO_2 = 60$ mm Hg, hemoglobin saturation = 85%, $PCO_2 = 40$ mm Hg, pH = 7.39, and hemoglobin concentration = 18 g/dl. Arterial PO_2 rises to 295 mm Hg after the patient breathes 100% O_2 at sea level for 20 minutes. Cardiac catheterization reveals normal pressures and a PO_2 of 40 mm Hg in the right atrium, right ventricle, and pulmonary artery while the patient breathes 100% O_2.

116. The most likely cause of hypoxia in this patient is

(A) a ventilation:perfusion abnormality (physiologic shunt)
(B) a right-to-left anatomic shunt
(C) a left-to-right anatomic shunt
(D) a diffusion defect
(E) hypoventilation

117. The fraction of the cardiac output that represents shunted blood is

(A) 0.2
(B) 0.3
(C) 0.4
(D) 0.5
(E) 0.6

118. The most useful data for locating the site of the shunt would be

(A) PO_2 in the left ventricle and left atrium
(B) airway resistance
(C) right ventricular and pulmonary arterial pressures
(D) left ventricular pressures
(E) compliance of the lungs

Questions 119–121

A lightly anesthetized male patient who is breathing spontaneously has a cardiac output of 6 L/min, a heart rate of 75 beats/min, an O_2 consumption of 240 ml/min, and a mixed venous P_{O_2} of 40 mm Hg. He is ventilated for 10 minutes at his normal tidal volume but at twice the normal frequency with a gas mixture that is 20% O_2 and 80% N_2. The airway pressure at the end of inspiration is 5 cm H_2O. On cessation of artificial ventilation, the patient does not breathe for 1 minute.

119. The most important factor responsible for the temporary apnea in this patient is reduced activity of the

(A) peripheral chemoreceptors due to the high P_{O_2}
(B) peripheral chemoreceptors due to the low P_{CO_2}
(C) pulmonary stretch receptors that inhibit inspiration
(D) medullary chemoreceptors due to the low P_{CO_2}
(E) medullary chemoreceptors due to the high P_{O_2}

120. During the artificial ventilation, the mixed venous P_{O_2} in this patient would be

(A) equal to 40 mm Hg because the metabolic rate was unchanged
(B) greater than 40 mm Hg because the hyperventilation would cause vasoconstriction and less gas exchange with the tissues
(C) greater than 40 mm Hg because of the significant increase in P_{50} of hemoglobin.
(D) less than 40 mm Hg because of the hypoxia that would occur from breathing this gas mixture
(E) less than 40 mm Hg because of a decrease in cardiac output

121. During the period of spontaneous breathing, this patient had an arteriovenous O_2 difference of

(A) 4 ml/dl
(B) 5 ml/dl
(C) 6 ml/dl
(D) 40 ml/dl
(E) 80 ml/dl

Directions: Each group of items in this section consists of lettered options followed by a set of numbered items. For each item, select the **one** lettered option that is most closely associated with it. Each lettered option may be selected once, more than once, or not at all.

Questions 122–126

Match each disorder with its associated acid-base state.

(A) Metabolic acidosis
(B) Respiratory acidosis
(C) Both
(D) Neither

122. Hypoventilation
123. Increased [total CO_2]
124. Decreased [total CO_2]
125. Decreased P_{CO_2}
126. Increased [HCO_3^-]/S × P_{CO_2} ratio

Questions 127–132

Match each characteristic with the appropriate substance.

(A) Aldosterone
(B) ADH
(C) Atrial natriuretic factor
(D) Renin
(E) Angiotensin I

127. An octapeptide
128. Regulates plasma [Na$^+$]
129. Controls body Na$^+$ content
130. Secreted by the zona glomerulosa of the adrenal cortex
131. Substrate for dipeptidyl carboxypeptidase
132. Produced by modified smooth muscle cells

Questions 133–136

Each of the mechanisms described below is used to alter the force of contraction in certain types of muscle. For each mechanism, select the muscle type or types that alter their contractile force in the manner described.

(A) Cardiac muscle
(B) Skeletal muscle
(C) Smooth and cardiac muscle
(D) Smooth and skeletal muscle
(E) Smooth, cardiac, and skeletal muscle

133. Alteration in the amount of Ca^{2+} released from the sarcoplasmic reticulum
134. Recruitment of additional muscle fibers
135. Summation of contractions
136. Alteration in preload

Questions 137–140

For each patient described below, select the set of arterial blood values that coincides with that patient's acid-base disorder.

Patient	P$_{CO_2}$ (mm Hg)	[HCO$_3^-$] (mmol/L)	pH	[H$^+$] (nmol/L)
(A)	25	15.0	7.40	40.0
(B)	30	16.0	7.31	49.0
(C)	40	25.0	7.41	38.9
(D)	45	35.0	7.51	30.9
(E)	55	27.5	7.37	42.7

137. Patient with metabolic alkalosis and partial respiratory compensation
138. Patient with metabolic acidosis and partial respiratory compensation
139. Patient with highest CO$_2$ content
140. Patient with highest alveolar ventilation

Questions 141–143

For each of the following nutrients, choose the substance that is most important for its absorption from the intestine.

(A) Intrinsic factor
(B) Ferritin
(C) Bile salts
(D) Vitamin D
(E) Trypsin

141. Iron
142. Ca^{2+}
143. Cholesterol

Questions 144–147

Match each site of secretory activity described below with the appropriate lettered region of the nephron.

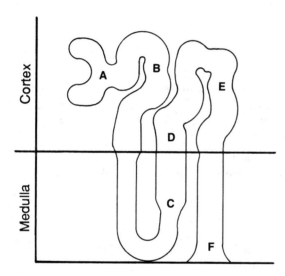

144. Primary site of H^+ secretion
145. Site of K^+ secretion
146. Primary site of NH_4^+ secretion
147. Site of PAH secretion

Questions 148–150

Match each hormone imbalance with its effect on plasma ion concentrations.

(A) Low plasma inorganic $[HPO_4^{2-}]$
(B) Low plasma $[Ca^{2+}]$
(C) Both
(D) Neither

148. Excessive PTH secretion
149. Deficient PTH secretion
150. Vitamin D intoxication

Questions 151–154

Match each motor deficit listed below with the component of the motor control system that is affected.

(A) Motor cortex
(B) Subthalamic nucleus
(C) Posterior parietal lobe
(D) Vestibular nucleus

151. Spontaneous movements
152. Spasticity
153. Apraxia
154. Nystagmus

Questions 155–159

Match each person described below with the set of blood data that best coincides with that person's condition.

	Arterial P_{O_2} (mm Hg)	Arterial P_{CO_2} (mm Hg)	O_2 Content (ml/dl)	Arterial pH
(A)	98	40	arterial = 10	7.40
(B)	50	65	arterial = 15	7.32
(C)	50	40	arterial = 16	7.38
(D)	105	35	venous = 18	7.45
(E)	50	32	arterial = 10	7.45

155. A normal 40-year-old man who has been mountain climbing for two days
156. A 30-year-old anemic woman
157. A 73-year-old man who is hypoventilating
158. A 45-year-old woman who is hyperventilating
159. A 56-year-old man with moderately severe obstructive lung disease

Questions 160–164

The following information was obtained from a healthy 24-year-old man who was studied in a renal laboratory:

Inulin Concentration (mg/ml)	Glucose Concentration	Urine Flow Rate	Hematocrit Ratio
Urine = 150 Renal arterial plasma = 1.50 Renal venous plasma = 1.20	Urine = 1 mg/ml Plasma = 90 mg/dl	1.2 ml/min	0.40

Using the above data, match each of the following measurements of renal function with the appropriate lettered value.

(A) 0.20
(B) 108 mg/min
(C) 120 ml/min
(D) 600 ml/min
(E) 1000 ml/min

160. Glomerular filtration rate
161. Renal blood flow
162. Filtration fraction
163. Renal plasma flow
164. Filtered load of glucose

Questions 165–167

Match each cause of an acid-base disturbance with the characteristic set of body fluid changes.

	Plasma pH	Plasma [HCO$_3^-$] (mEq/L)	Urine pH
(A)	7.27	37	acid
(B)	7.31	16	acid
(C)	7.40	15	alkaline
(D)	7.40	24	acid
(E)	7.55	22	alkaline

165. Hyperventilation
166. Chronic respiratory tract obstruction
167. Diabetic ketoacidosis

Questions 168–171

Match each visual defect described below with the condition that causes it.

(A) Cataracts
(B) Astigmatism
(C) Presbyopia
(D) Myopia
(E) Hyperopia

168. A progressive decrease in the power of accommodation

169. A progressive loss of lens transparency

170. An optical defect that requires a diverging lens to see distant objects clearly

171. An optical defect that requires continuous accommodation to see distant objects clearly

Questions 172–175

Match each of the clinical conditions listed below with the digestive system malfunction most likely to cause the condition.

(A) Blockage of the bile duct
(B) Malabsorption of vitamin B$_{12}$
(C) Malabsorption of water
(D) Malabsorption of carbohydrates

172. Circulatory collapse
173. Malnutrition and steatorrhea
174. Jaundice
175. Anemia

ANSWERS AND EXPLANATIONS

1. The answer is D. [*Ch 3 II D 1*] The interpleural pressure during the expiratory maneuver would be 15 cm H_2O. Before the Valsalva maneuver, the transmural pressure across the lungs (i.e., transpulmonary pressure) is equal to the alveolar pressure minus interpleural pressure, or $0 - (-5) = 5$ cm H_2O. Since lung volume remained constant, the transmural pressure must also remain constant during the Valsalva maneuver, so that when the alveolar pressure increases to 20 mm Hg, the interpleural pressure must equal 15 in order to maintain the difference of 5 cm H_2O between the inside and outside of the lungs.

2. The answer is D. [*Ch 1 VIII A 5 a*] In myopia, the focal plane of the unaccommodated eye is in front of the retina because the axial length of the eye is too long for the refractive power of the eye. To be seen clearly without accommodation, an object must be placed close to the eye (i.e., the far point is less than normal). However, if a diverging lens is placed in front of the eye, distant objects can be brought into focus on the retina.

3. The answer is D. [*Ch 4 II A 1–3; Figures 4-3, 4-4; Table 4-1*] Assuming complete dissociation, this 0.15 mol/L solution of sodium chloride (NaCl) is equivalent to a solution of 300 mOsm/L (0.15 mol/L of Na^+ + 0.15 mol/L of Cl^-). Because this NaCl solution is osmotically balanced with body fluids, there is no shift of water between the major fluid compartments. Thus, the extracellular fluid (ECF) volume increases with no change in the osmolar concentration of the ECF or intracellular fluid (ICF) compartment.

4. The answer is D. [*Ch 2 VIII D 1*] The data obtained during cardiac catheterization of this patient can be used to calculate the cardiac output ($\dot{Q}$) using the Fick equation, which states that

$$\dot{Q} = \frac{O_2 \text{ consumption}}{\text{arterial } O_2 \text{ content} - \text{mixed venous } O_2 \text{ content}}$$

Mixed venous blood (from the right ventricle or pulmonary artery) must be used to calculate cardiac output. For this calculation, as in all others, the units must be consistent. So, using milliliters, the cardiac output is calculated as

$$\dot{Q} = \frac{210 \text{ ml/min}}{0.18 \text{ ml } O_2/\text{ml blood} - 0.11 \text{ ml } O_2/\text{ml blood}}$$

$$= 210/0.07 = 3000 \text{ ml/min}$$

The normal cardiac output in a man is approximately 5 L/min (5000 ml/min); thus, a cardiac output of 3000 ml/min is *not* dangerously high. The amount of O_2 transferred to the tissues is given by the difference between arterial and venous O_2 content and, thus, is 7 ml O_2/dl blood, not 29 ml. The arterial O_2 content is equal to or only slightly less than the pulmonary venous O_2 content (due to thebesian venous flow into the left ventricle), meaning that the pulmonary venous O_2 content would not be 145 ml/dl. Finally, stroke volume equals cardiac output/heart rate; thus, the stroke volume in this patient is 3000/75 = 40 ml, which is less than the normal stroke volume of 70 ml and the likely cause of the subnormal cardiac output in this patient.

5. The answer is D. [*Ch 7 X F 1 c (8)*] During the postovulatory phase of the menstrual cycle, the thickened, secretory endometrium depends on the continued presence of estradiol and progesterone. If fertilization does not occur, the corpus luteum remains functional for 13–14 days and then undergoes regression (luteolysis). After the regressing corpus luteum loses its ability to produce adequate amounts of estradiol and progesterone, the innermost layer (adluminal, or stratum functionale, layer) of the endometrium becomes ischemic, degenerates, becomes necrotic, and is sloughed into the uterine cavity. The loss of proliferated endometrium (stratum functionale) is accompanied by bleeding known as menstruation (menses). The discontinuation of oral contraceptives after 3 weeks also permits menstruation.

6. The answer is D. [*Ch 6 III B 6*] Gastric motility is controlled primarily by enterogastric reflexes that are elicited when chyme enters the small intestine. The reflexes are both neural and hormonal and act to inhibit gastric contractions. They prevent food from entering the intestine too rapidly. Although distension of the antrum will elicit neuronal and hormonal excitatory reflexes, these reflexes are overcome by the inhibitory enterogastric reflexes.

7. The answer is C. [*Ch 4 IX F 1*] Spironolactone is a competitive aldosterone antagonist that interferes with the aldosterone-stimulated Na^+ reabsorption in the distal tubular cell and in the collecting duct. Inhibition of Na^+ reabsorption is associated with a marked decrease in urinary excretion of K^+ and H^+.

Spironolactone is effective as a diuretic in normal subjects or in patients on a low-Na$^+$ diet, but not in adrenalectomized patients.

8. The answer is A. [*Ch 2 VII C 1*] Of the conditions listed, only an increase in heart rate would cause an increase in aortic systolic pressure and a decrease in aortic pulse pressure. This can be verified using a derivation of Hook's law, which states that arterial pressure is determined by the interaction of mean arterial volume (V), pulse pressure (dP), arterial uptake during systole (dV), and the elasticity of the arterial system (a constant represented by E), or $E = dP \times V/dV$. An increase in heart rate increases cardiac output and raises V, which is proportional to mean aortic pressure. If stroke volume remains constant, then dV will not change significantly; so that dP must decrease to maintain a valid equation. An increase in arterial compliance would decrease the value of E (compliance is the reciprocal of elastance), which would lower systolic pressure. A decrease in peripheral resistance would decrease systolic pressure and increase pulse pressure. An increase in either stroke volume or the elastic modulus would increase systolic pressure but also increase pulse pressure.

9. The answer is B. [*Ch 7 VII H 1 c (1), I 2*] Pheochromocytoma is an adrenomedullary tumor characterized by the hypersecretion of catecholamines, usually norepinephrine. Catecholamines block insulin release (hypoinsulinemia) and increase both gluconeogenesis and fatty acid mobilization, which account for glucose intolerance, fasting hyperglycemia, and glycosuria. The breakdown of muscle glycogen leads to elevated plasma pyruvate or lactate levels. The metabolic features of pheochromocytoma resemble hyperthyroidism and include tremor, weight loss, heat intolerance, and increased basal metabolic rate. The cardinal sign is hypertension, which may be persistent or paroxysmal. Patients usually complain of attacks that may be precipitated by emotion or physical exercise, which consist of pounding headaches, sweating, pallor, pain or "tightness" in the chest, apprehension, paresthesia, nausea, and vomiting. In this patient, the attacks were provoked by mechanical pressure that was exerted on the tumor by changes in body position.

10. The answer is B. [*Ch 3 II F 1; IX C 1 c*] This patient has obstructive lung disease—probably emphysema—as indicated by the high airway resistance and the decreased diffusion capacity of the lungs. The minute ventilation is decreased probably because of the high respiratory work load, which would increase arterial PCO_2. The pulmonary artery pressure would be increased due to the destruction of pulmonary capillaries secondary to the emphysema and to the pulmonary vasoconstriction caused by the hypoxia and hypercapnia. Arterial pH would tend to be decreased due to the hypercapnia, and the [HCO_3^-]/[H_2CO_3] ratio would be decreased in accordance with the Henderson-Hasselbalch equation [see Ch 5 II D].

11. The answer is C. [*Ch 6 IV D 3 a (1), F 5 c*] Vitamin B$_{12}$ is absorbed in the terminal ileum in association with intrinsic factor. Bile salts are also absorbed in the terminal ileum. If bile salts cannot be absorbed and recirculated, lipid absorption will be compromised because the liver is unable to synthesize enough bile to absorb all the fats ingested during a meal.

12. The answer is D. [*Ch 4 VI C 2 b, 4; Figure 5-3*] The primary effect of carbonic anhydrase inhibitors such as acetazolamide is to inhibit both H$^+$ secretion and NaHCO$_3$ reabsorption, making the urine alkaline. NH$_4^+$ excretion is reduced as a result of the diminished H$^+$ secretion. Carbonic anhydrase inhibitors restrict H$^+$ secretion by inhibiting the intracellular hydration of CO$_2$, a primary source of intracellular H$^+$. The decline in H$^+$ secretion inhibits the Na$^+$-H$^+$ exchange at the luminal membrane of the proximal tubule, which is the primary site of NaHCO$_3$ reabsorption. HCO$_3^-$ reabsorption is also inhibited because only a limited amount of HCO$_3^-$ can be reabsorbed by the distal segment. The elevated intraluminal HCO$_3^-$ augments Na$^+$ and K$^+$ excretion and results in NaHCO$_3$ diuresis. Carbonic anhydrase inhibitors also block the dehydration of H$_2$CO$_3$ formed in the tubular lumen. Chronic doses of such drugs can lead to hyperchloremic acidosis (metabolic acidosis).

13. The answer is C. [*Ch 4 VIII B 1–2, C 1*] A decrease in plasma osmolality (osmolarity) leads to a decrease in antidiuretic hormone (ADH) secretion. This results in an increase in free-water clearance and diuresis. These osmoreceptors are found in the vicinity of the supraoptic nucleus of the hypothalamus. ADH augments the water permeability of the cortical collecting duct and the water and urea permeabilities of the medullary collecting duct. It increases renal water reabsorption, resulting in the excretion of a small volume of hypertonic urine. The major stimuli for ADH are an increase in the plasma osmolality and a decrease in the effective circulating blood volume.

14. The answer is C. [*Ch 2 VIII D 1*] This patient's O_2 consumption is 520 ml/min/m². O_2 consumption can be determined using the Fick principle, an important concept that has wide applicability in physiology and medicine. To solve for O_2 consumption ($\dot{V}O_2$), the Fick principle is expressed as

$$\dot{V}O_2 = \dot{Q} \times (CaO_2 - C\bar{v}O_2)$$

where $\dot{Q}$ = blood flow, and $CaO_2 - C\bar{v}O_2$ = the arteriovenous O_2 difference. In this problem, $\dot{Q}$ is given, and, since arterial hemoglobin saturation is assumed to be 100%, CaO_2 equals O_2 capacity, which equals hemoglobin $\times$ 1.34, or 16 ml/dl (160 ml/L). Using these values in the above formula gives

$$\dot{V}O_2 = 6.5 \times (160 - 80)$$
$$= 520 \text{ ml/min}$$

It is important to use similar units for blood flow and O_2 content. Also, converting O_2 content to ml/L before calculating is recommended.

15. The answer is C. [*Ch 1 I C 3 d–f*] The initial shrinkage of the red blood cell results from a higher osmotic pressure in the solution than within the cell. However, since the red blood cell eventually returns to its initial volume, the solution must be isotonic. The particles producing the higher osmotic pressure in the solution have reflection coefficients of less than 1 and, thus, are able to diffuse across the membrane. Although these particles initially cause water to flow out of the cell, they eventually reach diffusional equilibrium across the cell membrane and, thus, are unable to keep the water from returning to the cell.

16. The answer is B. [*Ch 5 VIII D 3, E 1; Figures 5-13, 5-14*] In respiratory acidosis, there is a primary increase in plasma PCO_2. The renal compensation for respiratory acidosis is increased HCO_3^- reabsorption, which increases the plasma $[HCO_3^-]$. With partial compensation, the pH would not return to normal. Only one set of data (*B*) indicates hypercapnia with a decrease in pH. Note that the pH and the $[HCO_3^-]/S \times PCO_2$ ratio are below normal in this patient, which is consistent with acidosis.

17. The answer is D. [*Ch 7 VI D 1 c (4) (b)*] This woman's inability to lactate and to menstruate following parturition stems from pituitary failure related to the massive bleeding she experienced during delivery. Most likely, this woman has postpartum pituitary necrosis (Sheehan's syndrome) due to ischemia of the hypophysial portal system, which supplies 90% of the blood to the anterior pituitary gland. The resultant pituitary failure may be complete (panhypopituitarism) or partial. In this case, the acidophils responsible for production of prolactin, follicle-stimulating hormone (FSH), and luteinizing hormone (LH) were affected, resulting in this patient's inability to lactate and to menstruate. This patient also would exhibit hypoglycemia due to low plasma growth hormone (GH) levels and to low plasma adrenocorticotropic hormone (ACTH) levels, which would lead to decreased secretion of cortisol (a potent hyperglycemic hormone) and decreased synthesis of epinephrine (a hyperglycemic hormone that depends on cortisol). The decreased levels of these hyperglycemic hormones would cause insulin sensitivity.

Since aldosterone secretion does not depend mainly on ACTH secretion, this patient should be able to regulate Na^+ balance. She also should have normal ADH activity, since the supraoptic nuclei and pars nervosa are not perfused by the hypophysial portal system. Thus, this patient should demonstrate normal water balance.

18. The answer is D. [*Ch 2 III A 5, B 3 d*] The largest deflection in the EKG occurs when the electrical vector is parallel to the lead axis. Lead aVL lies at −30° and, therefore, should exhibit the largest positive deflection of any lead.

19. The answer is C. [*Ch 4 VI C 4 c; Table 4-6*] The kidney excretes H^+ as NH_4^+ and titratable acid. Titratable acid exists mainly in the form of $H_2PO_4^-$. The titratable acid is excreted mainly as NaH_2PO_4, which is also called acid phosphate, monosodium phosphate, or monobasic phosphate. The normal kidney excretes almost twice as much acid combined with NH_3 than it excretes titratable acid. The rate of NH_4^+ excretion increases during metabolic acidosis.

20. The answer is A. [*Ch 5 VIII D 4, E 2; Figure 5-14*] The variable that shows the greatest degree of change in this patient is arterial PCO_2, which is decreased by 75% compared to a 46% decrease in $[HCO_3^-]$. These findings are consistent with respiratory alkalosis, which, in this case, is due to hyperventilation brought on by salicylate toxicity. The resultant hypocapnia reduces tubular H^+ secretion, so that

HCO_3^- reabsorption is attenuated, causing a compensatory loss of urinary HCO_3^-. The arterial pH of this patient can be calculated using the Henderson-Hasselbalch equation as

$$pH = pK + \log \frac{[HCO_3^-]}{S \times P_{CO_2}}$$

$$= 6.1 + \log \frac{13 \text{ mmol/L}}{0.3 \text{ mmol/L}}$$

$$= 6.1 + \log 43.3$$

$$= 6.1 + 1.6$$

$$= 7.7$$

Note that the increased ratio of $[HCO_3^-]/S \times P_{CO_2}$ is consistent with alkalotic states. This condition must be differentiated from metabolic alkalosis, which is associated with increases in both arterial $[HCO_3^-]$ and P_{CO_2}. Metabolic acidosis is associated with decreases in $[HCO_3^-]$, P_{CO_2}, and pH.

21. The answer is E. [*Ch 6 V B 1, 3*] Although acidosis can result from loss of fluid from both the intestine and the colon, only the colon secretes K^+. Thus, excessive fluid loss from the colon will result in both acidosis and hypokalemia.

22. The answer is B. [*Ch 1 XI B 1, 2*] The Ib afferent fibers convey information from the Golgi tendon organs, which are stimulated by an increase in muscle tension. Thus, when the muscle contracts in response to a stretch reflex, Ib afferent activity increases. Ia afferent activity decreases during the contraction because its receptor, the muscle spindle, is unloaded (i.e., becomes slack) during contraction.

23. The answer is A. [*Ch 1 III B 2 b; Ch 2 II C 1 b, d*] The pacemaker cells of the heart contain muscarinic receptors, which are activated by acetylcholine (ACh). When the ACh receptor on these cells is activated, K^+ conductance is increased, which causes the membrane to hyperpolarize and slows pacemaker activity.

24. The answer is C. [*Ch 3 IV C 3*] O_2 delivery to the tissues is the product of arterial O_2 content and cardiac output, which normally equals 1 L/min. Of the factors listed, only a 50% reduction in normal hemoglobin concentration would reduce O_2 delivery by half. This would produce a proportional change in O_2 content and an equivalent change in the O_2 delivery. Due to the nonlinear relationship between P_{O_2} and hemoglobin saturation, changing the P_{O_2}, ventilation rate, or inspired P_{O_2} would have a relatively small effect on O_2 delivery.

25. The answer is B. [*Ch 5 VIII D 2; Figure 5-13; Table 5-8*] The severe loss of gastric fluid in this patient would produce metabolic alkalosis due to the loss of HCl (a noncarbonic acid). She also would exhibit hypovolemia due to the fluid loss. The development of alkalosis is sensed by the chemoreceptors controlling ventilation, resulting in hypoventilation, and an increase in the arterial P_{CO_2} (hypercapnia), which reduces the pH toward normal. The kidneys would be expected to excrete the excess HCO_3^-, raising the urinary pH.

26. The answer is A. [*Ch 6 III C 2 b*] The formation of HCl by the parietal cells is a two-step process. First, Cl^- is transported into the parietal cell canaliculi. The negative potential developed by the flow of Cl^- allows K^+ to flow into the canaliculi. The K^+ is then actively transported out of the canaliculi in exchange for H^+. Since both K^+ and H^+ are transported against their concentration gradients, an active transport system (H^+-K^+ ATPase) is required.

27. The answer is A. [*Ch 1 II C 2, 4*] During attacks of hyperkalemic periodic paralysis, large amounts of K^+ are released from skeletal muscle fibers, causing a rise in extracellular K^+ concentration. When the extracellular K^+ concentration increases, the membrane potential depolarizes, causing inactivation of the Na^+ channels.

28. The answer is B. [*Ch 7 X E 3 a–c, F 1; Figure 7-9*] The menstrual cycle consists of four phases. These phases (with synonymous terms) are: menses (4–5 days), the preovulatory phase (also called follicular, estrogenic, or proliferative phase; 10–12 days), the ovulatory phase, and the postovulatory phase (also called luteal, progestational, or secretory phase; 14 days).

29. The answer is C. [*Ch 6 III C 4, E; IV E 2 b (1)*] A significant amount of protein digestion occurs in the stomach because of the action of HCl, which begins to break proteins apart, and pepsin, a protease enzyme secreted by gastric chief cells. These protein digestion products then act as secretagogues that stimulate the secretion of pancreatic proteases.

30. The answer is D. [*Ch 4 V C 1 b; Table 4-5*] The simplest flow (F) equation that summarizes the relationship between hydrostatic pressure gradient (ΔP) and resistance (R) is

$$F = \frac{\Delta P}{R}$$

The pressure gradient is the hydrostatic pressure difference between the renal artery and renal vein. The resistance to blood flow can be determined by dividing this hydrostatic pressure gradient by the renal blood flow (RBF) as

$$R = \frac{\Delta P}{F}$$
$$= \frac{\Delta P}{RBF}$$

31. The answer is C. [*Ch 5 II D 3 b (1), E 2 b*] From the data given, this patient's arterial P_{CO_2} is determined to be 50 mm Hg (normal = 40 mm Hg). The Henderson equation is used to estimate P_{CO_2}, but $[H^+]$ must be determined first using the Henderson-Hasselbalch equation. Given a pH of 7.5, $[H^+]$ is calculated as

$$[H^+] = \text{antilog } (9 - pH)$$
$$= \text{antilog } (9 - 7.5)$$
$$= \text{antilog } 1.5 = 31.6 \text{ nmol/L}$$

Substituting to solve for P_{CO_2},

$$P_{CO_2} = \frac{[H^+] [HCO_3^-]}{24}$$
$$= \frac{(31.6 \text{ nmol/L}) (38 \text{ mmol/L})}{24}$$
$$= 50 \text{ mm Hg}$$

This patient's increased $[HCO_3^-]$ indicates a metabolic alkalosis. The elevated $[HCO_3^-]$ suppresses respiratory drive, leading to compensatory elevation of P_{CO_2}. Note that unit analysis cannot be used in this equation.

32. The answer is C. [*Ch 1 II C 5; III A 3*] The flow of Ca^{2+} into the presynaptic nerve terminal stimulates the release of neurotransmitter from synaptic vesicles by exocytosis. Ca^{2+} enters the nerve terminal down its electrochemical gradient through channels opened by membrane depolarization. Propagation of an action potential is accomplished by generating new action potentials along the nerve fiber. The generation of action potentials is dependent on electrically excitable Na^+ and K^+ channels.

33. The answer is A. [*Ch 3 III C 3; IX C 1; Table 3-2*] This patient has an alveolar-to-arterial P_{O_2} difference of 35 mm Hg. To calculate this, the alveolar P_{O_2} must be determined using the alveolar gas equation. Given an arterial P_{CO_2} of 30 mm Hg, the alveolar P_{O_2} is determined as

$$\text{alveolar } P_{O_2} = (760 - 47) \times .21 - 30$$
$$= 150 - 30 = 120 \text{ mm Hg}$$

Thus, the alveolar-to-arterial P_{O_2} difference is $120 - 85 = 35$ mm Hg. This patient's low arterial P_{CO_2} and high pH indicate that he has respiratory alkalosis; thus, he has been hyperventilating and does not have metabolic alkalosis. Respiratory alkalosis causes a decrease in plasma $[HCO_3^-]$ due to the slope of the blood buffer line. Lastly, chronic obstructive lung disease leads to hypoxia and CO_2 retention, not hyperventilation.

34. The answer is D. [*Ch 4 V A 1 a*] The oncotic and hydrostatic pressures within the capillary and within the interstitium contribute to the regulation of fluid exchange between the plasma and the interstitial fluid (or Bowman's space). To determine whether there is net reabsorption or filtration, it is necessary to compare the magnitude and direction of these two different pressures as

Outward Forces (mm Hg)			Inward Forces (mm Hg)	
Capillary hydrostatic pressure	= 47		Bowman's space hydrostatic pressure	= 10
Bowman's space oncotic pressure	= 0		Capillary oncotic pressure	= 28
	47			38

Because the outwardly directed forces exceed the inwardly directed forces, there will be a net filtration force of 9 mm Hg. Note that the oncotic pressure in Bowman's capsule is assigned a value of zero because the fluid in this region is an ultrafiltrate of plasma.

35. The answer is E. [*Ch 4 II C 3 a; Figure 4-4; Table 4-2*] As a result of the hemorrhage, this patient has undergone the loss of water and electrolytes in isotonic concentration leading to dehydration (reduced extracellular fluid volume). In this patient there will be no net water movement across the cell membranes because both NaCl and water are administered as 1 L of isotonic (isosmotic) NaCl. Since the administered NaCl will remain initially in the extracellular space, the only effects will be a 1 L increase in the ECF volume and a dilution of the plasma protein concentration, the latter effect accounting for the decrease in the plasma colloid osmotic pressure.

A 0.9% NaCl solution is isotonic because it maintains the normal red blood cell volume. Furthermore, this 0.9% NaCl solution contains dissociated Na^+ and Cl^-, each in a concentration of 150 mEq/L. Because NaCl is an electrolyte, this same solution is equivalent to a concentration of 300 mOsm/L, assuming 100% dissociation of NaCl. Thus, a 150 mmol/L NaCl solution is isotonic and isosmotic to human plasma. Edema is ruled out in this patient because it represents the retention of both water and NaCl as an isotonic solution, resulting in isotonic overhydration. The infusion of a large volume of isotonic saline at a rate that exceeds urinary excretion results in edema.

36. The answer is D. [*Ch 3 III B 3 b*] The alveolar PCO_2 is directly proportional to the rate at which CO_2 is produced by metabolism and inversely proportional to alveolar ventilation. The increased minute ventilation would not alter the metabolic rate significantly, so any changes in PCO_2 must be related to alveolar ventilation. Alveolar ventilation (V_A) is the difference between minute ventilation and dead space ventilation. During control conditions, alveolar ventilation is $(0.5 L \times 15) - (0.2 L \times 15) = 4.5 L/min$, which is unchanged by the alteration in respiratory pattern [$30 \times (0.35 L - 0.2 L)$]. An increase in alveolar ventilation would cause a decrease in PCO_2, which would lead to respiratory alkalosis (i.e., increased arterial pH).

37. The answer is B. [*Ch 7 VII D 3 b (2)*] Norepinephrine can be converted to epinephrine by methylation with the epinephrine-forming enzyme, phenylethanolamine-N-methyltransferase (PNMT), which is highly localized in the cytosol of the adrenomedullary chromaffin cells. The methyl group donor in this reaction is S-adenosylmethionine. Very high local concentrations of cortisol from the adrenal cortex reach the adrenomedullary chromaffin cells via the adrenal portal system. PNMT is inducible by glucocorticoids. The secretion of adrenomedullary catecholamines is stimulated by ACh from the preganglionic nerve endings, which innervate the chromaffin cells. Thus, the adrenal medulla is a functional extension of the nervous system. The synthesis of epinephrine in the adrenal medulla depends on PNMT, cortisol, corticotropin (ACTH), and corticotropin-releasing hormone.

38. The answer is B. [*Ch 2 III C 2 a, b*] A positive EKG complex recorded from a unipolar electrode indicates that the electrical vector is directed toward the electrode. Positive QRS complexes in aVR and aVF indicate that the vector must lie in the right axis deviation quadrant (i.e., between $+120°$ and $+180°$). No axis deviation or a mean electrical axis between $0°$ and $90°$ indicates that the vector would lie in the normal quadrant, whereas left axis deviation or left ventricular hypertrophy would have the vector in the left axis deviation quadrant (i.e., between $-30°$ and $-90°$).

39. The answer is B. [*Ch 5 II E 2 b*] From the data given, this patient's arterial $[HCO_3^-]$ is determined to be 14.7 mmol/L (mEq/L). Arterial $[HCO_3^-]$ is easily estimated using the Henderson equation, which is stated as

$$[HCO_3^-] = 24 \frac{PCO_2}{[H^+]}$$

where $[HCO_3^-]$ is expressed in mmol/L, $[H^+]$ in nmol/L, and PCO_2 in mm Hg.

Substituting,
$$[HCO_3^-] = 24 \frac{30}{49}$$

$$= 14.7 \text{ mmol/L}$$

The acid-base disturbance in this case is an almost completely compensated metabolic acidosis (pH 7.32).

40. The answer is C. [*Ch 3 VI A 1*] The lung is a passive structure whose transmural pressure varies as a function of volume and phase of respiration. At residual volume, the transpulmonary pressure may be negative at the base of the lung, since interpleural pressure could be positive due to the hydrostatic effects of lung weight and the decreased retractile force at low lung volume. The transpulmonary pressure always is greater at the apex than at the base when standing because of the hydrostatic effects of gravity. Thus, the airways and alveoli at the apex of the lung always are more distended than those at the base of the lung of a person who is standing.

41. The answer is C. [*Ch 6 III B 2*] Receptive relaxation occurs when food enters the stomach. Although a small amount of receptive relaxation occurs as part of the esophageal reflex, the relaxation of the orad stomach necessary to accommodate the food entering it during a meal is brought about by a vago-vagal reflex initiated by the presence of food in the stomach. Chewing is entirely dependent on a nonvagal reflex involving stretch receptors and motor efferents in the jaw muscles. Swallowing is coordinated by a swallowing center in the brain stem and does not involve vagal reflexes. Although gastric emptying is modified to some extent by vago-vagal reflexes responding to gastric distension and the presence of chyme in the intestine, local reflexes and hormones are primarily responsible for regulating the rate of gastric emptying. Intestinal contractions are controlled almost entirely by local reflexes and the presence of circulating hormones.

42. The answer is B. [*Ch 5 VII B 2 C (2); VIII D 1, E 3; Figure 5-14*] This diabetic patient has metabolic acidosis due to ketoacidosis. The ketoacidosis is attributable mainly to the formation of β-hydroxybutyric acid from the partial oxidation of fatty acids. The high concentration of ketoacids in the form of anions is responsible for the increased anion gap; however, it is H^+ retention, not anion accumulation, that is responsible for the acidosis. Osmotic diuresis accounts for this patient's dehydration, which is not only water loss but also increased renal excretion of Na^+, K^+, Cl^-, and glucose. Na^+ and K^+ also are lost when they are excreted in association with the excess quantities of organic anions. In this case, the decline in $[HCO_3^-]$ is the primary change, and the decreased Pco_2 (i.e., hyperventilation) is the compensatory response. Complete compensation is evidenced by the normal pH of 7.39 and the normal ratio of $[HCO_3^-]/S \times Pco_2$. It is possible to have an acidosis with a normal blood pH, or $[H^+]$, because secondary changes diminish the extent of the acid-base imbalance.

43. The answer is C. [*Ch 7 VII H 1 c (2)*] Epinephrine is a potent hyperglycemic agent for several reasons. It stimulates α-adrenergic receptors on the pancreatic beta cell, inhibiting insulin secretion and, therefore, subsequent facilitated transport of glucose by muscle and adipose tissue. Epinephrine also promotes hepatic and muscle glycogenolysis by activating cyclic adenosine 3',5'-monophosphate (cAMP)-dependent phosphorylase; glycogenolysis in muscle leads to an increase in the plasma level of lactate, which provides the liver with an important gluconeogenic substrate. The lipolytic effect of epinephrine mobilizes free fatty acids, which enhances gluconeogenesis. Additionally, catecholamines directly inhibit peripheral glucose uptake, partly due to the suppression of glucose transporters.

44. The answer is B. [*Ch 7 VI A 2*] The pars nervosa is a functional component of the neurohypophysis (also called the neural lobe of the pituitary gland), which is derived from neural ectoderm. The neurohypophysis consists of three anatomic components: the median eminence, the infundibular stem, and the pars nervosa. Like the median eminence, the pars nervosa does not synthesize hormones but, rather, functions as a release center for hormones. The anterior lobe of the pituitary gland, or adenohypophysis, is derived from oral ectoderm. Endocrine tissues that arise from endoderm include the pancreas, thyroid gland, and parathyroid glands. Those derived from mesoderm include the adrenal cortex and gonads. The adrenal medulla originates from the neural crest, as do the parafollicular cells of the thyroid gland.

45. The answer is D. [*Ch 3 II D 3*] The man can lie under 68 cm of water. To calculate this depth requires converting the pressure given in mm Hg to a pressure expressed in cm H_2O, or 50 mm Hg $\times$ 1.36 cm H_2O/mm Hg = 68 cm H_2O. This man's alveolar pressure is equal to atmospheric pressure, while the pressure outside his chest wall is 68 cm H_2O higher. To expand his lungs, this man must generate a pressure slightly more negative than –68 cm H_2O. Under these conditions, the man is undergoing negative pressure breathing, which, if prolonged, can lead to pulmonary edema.

46. The answer is A. [*Ch 4 II C 2 a (2); Figure 4-4; Ch 5 VIII D 1 a*] Severe diarrhea causes volume depletion and electrolyte loss, which defines a state of dehydration. It is important to remember that volume depletion refers to effective circulating blood volume. Since intestinal secretions are rich in K^+ and HCO_3^-, diarrhea causes K^+ depletion and HCO_3^- loss, which lead to hypokalemia and metabolic acidosis, respectively. Vomiting, excessive sweating, and water deprivation can lead to volume depletion but not acidosis. In fact, vomiting causes loss of high concentrations of gastric H^+ and Cl^-, which leads to metabolic alkalosis and hypochloremia. Daily sweat production can exceed 10 L in subjects exercising in a hot climate. Severe sweating is associated with a significant loss of K^+, which may contribute to heat stroke.

47. The answer is C. [*Ch 7 XI B 2 b (2)*] The corpus luteum is the initial source of plasma progesterone and 17α-hydroxyprogesterone, which peak at 3–4 weeks postconception. At 6–8 weeks postconception, the progesterone reaches its lowest point, while the 17α-hydroxyprogesterone continues to decline. Since the placenta cannot synthesize 17α-hydroxyprogesterone, the secondary rise in progesterone reflects placental (trophoblast) function, and the 17α-hydroxyprogesterone curve is a correlate of corpus luteal function of pregnancy. The theca interna and adrenal glands do not secrete progesterone. The decidua is not the source of any hormones but is a specialized region of the endometrium that develops into the maternal component of the placenta.

48. The answer is B. [*Ch 4 V C 3 b; Table 4-5*] The filtration fraction is that fraction of the plasma flowing through the kidneys that is filtered into Bowman's capsule. Normally about one-fifth of the plasma entering the 2 million glomerular capillaries is filtered. Filtration fraction is the ratio of glomerular filtration rate (GFR) [125 ml/min] to renal plasma flow (RPF) [625 ml/min]. Both GFR and RPF are regulated in parallel at the afferent arteriole (e.g., constriction decreases both) and inversely at the efferent arteriole (e.g., constriction augments GFR and reduces RPF). As a result, only changes in efferent arteriolar resistance, not those in afferent arteriolar resistance, affect the ratio of GFR to the RPF. Since fluid movement across the glomerulus is governed by Starling's forces, it is proportional to the permeability and surface area of the filtering membrane and to the balance between the hydrostatic and oncotic forces. Ureteral obstruction results in an increase in the hydrostatic pressure in Bowman's space, reducing the hydrostatic pressure gradient and, therefore, the GFR and the filtration fraction. An increase in the plasma oncotic pressure contributes to a decrease in GFR as do decreases in the glomerular capillary hydrostatic pressure and the glomerular filtration area.

49. The answer is B. [*Ch 6 IV F 3 b (4) (c)*] Although fats can be absorbed all along the intestine, from the point at which the bile duct enters the duodenum until the bile salts are absorbed from the terminal ileum, almost all of the digested lipids are absorbed by the time the chyme reaches the midjejunum. Little, if any, lipid absorption occurs in the ileum.

50. The answer is E. [*Ch 5 VIII D 4, E 2; Figures 5-13, 5-14*] The most likely diagnosis is respiratory alkalosis due to anxiety-induced hyperventilation. The key determinants of the cause and compensation of the acid-base disorder are the arterial Pco_2 and $[HCO_3^-]$. The greater reduction in Pco_2 than in $[HCO_3^-]$ indicates that the primary disturbance is respiratory alkalosis. That both the arterial $[HCO_3^-]$ and $[H^+]$ change in the same direction further supports the diagnosis of a respiratory acid-base imbalance. Both obstructive and restrictive lung diseases are common causes of respiratory acidosis.

51. The answer is A. [*Ch 7 VII D 3 b (2)*] Hydrocortisone (cortisol) directly increases epinephrine-forming enzyme activity. Epinephrine is synthesized in the adrenal medulla and certain brain neurons from norepinephrine by the action of the enzyme PNMT. The adrenal cortex and adrenal medulla are related both anatomically and functionally. Venous blood from the sinusoids of the cortex enters the adrenal portal system and perfuses the adrenal medulla before entering the systemic circulation. Therefore, the chromaffin cells are exposed to a high concentration of cortisol. ACTH indirectly activates the epinephrine-forming enzyme, because it stimulates the secretion of cortisol.

52. The answer is B. [*Ch 3 III B 3*] The alveolar ventilation in *patient B* is greater than in *patient A*. Alveolar ventilation is minute ventilation minus dead space ventilation or respiratory rate times tidal volume minus dead space. Thus, *patient B* has an alveolar ventilation of 10 × (400 – 150) = 2500 ml/

min, whereas *patient A* has an alveolar ventilation of $20 \times (200 - 150) = 1000$ ml/min. Dead space ventilation in *patient A* is 3000 ml/min, whereas in *patient B* it is 1500 ml/min.

53. The answer is B. [*Ch 5 VII B 2 c; VIII D 1, E 3; Figures 5-13, 5-14*] The decreases in arterial pH, [HCO_3^-], and Pco_2 are consistent with metabolic acidosis. There also is a decreased [HCO_3^-]/S $\times$ Pco_2 ratio (i.e., 12.2) and a widened anion gap (i.e., 47 mEq/L). The primary disturbance is the marked reduction in [HCO_3^-], and the compensatory response is hyperventilation (as indicated by the hypocapnia). That the [HCO_3^-] decreases and the [H^+] increases is evidence of metabolic acid-base imbalance. The hyperglycemia and dehydration (as evidenced by dry skin) support a diagnosis of diabetes mellitus.

54. The answer is A. [*Ch 7 XIV H 3 b*] Calcitriol promotes renal tubular reabsorption of both Ca^{2+} and HPO_4^{2-}, leading to hypercalcemia and hyperphosphatemia. Calcitonin has the opposite effects, promoting the excretion of both electrolytes and, thus, leading to hypocalcemia and hypophosphatemia. Aldosterone favors increased reabsorption of Na^+ and increased excretion of K^+, leading to hypokalemia and alkalemia; the [Na^+] usually remains within normal limits because of a commensurate increase in water reabsorption. Progesterone has an anti-aldosterone–like effect, in that it promotes Na^+ excretion.

55. The answer is D. [*Ch 6 IV F 5 c*] Although intrinsic factor is secreted by parietal cells in the stomach, it is not able to bind vitamin B_{12} in the stomach because the vitamin is bound to another protein, called protein R. When the vitamin reaches the intestine, the R protein is removed, and intrinsic factor is able to bind to the vitamin B_{12}. The B_{12}–intrinsic factor complex is absorbed in the terminal ileum.

56. The answer is D. [*Ch 3 IV C 1 a*] A drop in arterial Po_2 from 100 mm Hg to 27 mm Hg would decrease the O_2 content of the blood by about 50%. The key to answering this question is to recognize that 27 mm Hg is the normal value for the P_{50}, which is the Po_2 at a hemoglobin saturation of 50%. Since hemoglobin is nearly 100% saturated at a Po_2 of 100 mm Hg, then O_2 content must decrease by 50%.

57. The answer is A. [*Ch 7 VIII C 3 e (1)–(3); Figure 7-7*] A deficiency of 17α-hydroxylase leads to reduction in the 17α-hydroxylation of pregnenolone and progesterone, resulting in hypogonadism and elevated blood gonadotropin levels. This enzyme deficiency also leads to increased production of 11-deoxycorticosterone (11-DOC). This mineralocorticoid causes Na^+ retention, extracellular volume expansion, and hypertension—effects that suppress renin and aldosterone secretion. The 17α-hydroxylase defect also affects the gonads, preventing testicular and adrenal androgen synthesis in males and ovarian estrogen synthesis in females and, thus, resulting in a female phenotype regardless of genotypic sex. These patients require not only cortisol to suppress ACTH secretion but also sex steroid treatment consistent with the genotypic sex.

58. The answer is A. [*Ch 7 VII G; Table 7-5*] The ciliary muscle is innervated by both postganglionic sympathetic and postganglionic parasympathetic fibers. Only β-adrenergic, not α-adrenergic, receptors are found in this smooth muscle. β-Adrenergic stimulation causes relaxation of the ciliary muscle, which, in turn, increases the tension on the lens, causing the lens to become thinner and adapted for far vision. The parasympathetic innervation of the ciliary muscle causes it to contract, which, in turn, decreases the tension on the lens, causing the lens to become thick and adapted for near vision (accommodation). The heart contains only β-adrenergic receptors, whereas vascular smooth muscle, pancreatic beta cells, and intestinal smooth muscle contain both α- and β-adrenergic receptors. Stimulation of either type of receptor in the intestine causes inhibition of peristalsis.

59. The answer is A. [*Ch 4 IX D 2 e; Ch 7 IX C 3 a; XIII A 2 a, E 2*] Angiotensin I, tetraiodothyronine (T_4), and testosterone all can be considered prohormones. Angiotensinogen is a liver-derived globulin that has no effect on aldosterone secretion. Angiotensinogen is converted by renin into angiotensin I, an inactive decapeptide that is rapidly converted into the active octapeptide, angiotensin II, by angiotensin-converting enzyme (ACE), found mainly in pulmonary endothelial cells. Most of the circulating triiodothyronine (T_3) is formed extrathyroidally from T_4 by 5'-deiodinase, an enzyme found mainly in the liver and kidney. On a molar basis, T_3 has 3–5 times the bioactivity of T_4. Testosterone can be reduced to the more potent androgen dihydrotestosterone (DHT) in an irreversible reaction catalyzed by 5α-reductase, an enzyme found in the prostate, seminal vesicles, epididymis, and skin. DHT has twice the bioactivity of testosterone.

60. The answer is B. [*Ch 7 XII D 4 a, 6 a, b*] Insulin is stimulatory to K^+ uptake by cells, and high concentrations of exogenous insulin cause extracellular hypokalemia. This hypokalemic action of insulin is due to the increased K^+ uptake by muscle and liver. Although the primary stimulus to the alpha cell is hypoglycemia, hypoglycemia-induced catecholamine release undoubtedly plays a role. Hypoglycemia also is a potent stimulus for the release of GH (somatotropin) and ACTH (corticotropin). Among the stimuli for ACTH release are pain, anxiety, pyrogens, and hypoglycemia.

61. The answer is D. [*Ch 5 VIII D 1 b*] Metabolic acidosis is associated with a low arterial pH, a reduced plasma [HCO_3^-], and a compensatory increase in alveolar ventilation and renal excretion of H^+. Renal excretion of titrable acid is limited by the amount of filtered buffers (e.g., phosphate buffer) and by the ability of the kidney to lower urinary pH. Thus, the net acid excretion by the kidney is largely controlled by factors that regulate the urinary secretion of NH_3 and the urinary excretion of NH_4^+. There is a concomitant, but limited, increase in titratable acid. Base excess is negative in metabolic acidosis.

62. The answer is D. [*Ch 4 V C 3 b; IX D 2 a (1) (a), c (1) (a), (b)*] Since hemorrhage decreases both GFR and RPF commensurately, there is no change in the filtration fraction, which is the ratio of GFR to RPF. Severe hemorrhage results in a state of hypovolemia, which, in turn, is a major stimulus for the secretions of both aldosterone and ADH. Bodily responses to hypovolemia take precedence over the regulation of solute concentration. The decrease in blood volume initiates a neurocirculatory reflex arc composed of afferent and efferent limbs. The afferent limb is activated by a reduction in plasma volume. The reduced plasma volume leads to a decrease in central blood volume, decreased atrial and pulmonary venous pressures, and a resultant reduction in the activity of low-pressure baroreceptors in these regions. There also is a concomitant reduction in the firing rate of the high-pressure baroreceptors in the carotid sinus and aortic arch. The decreased firing rates of these receptors cause an increase in ADH secretion. Another afferent input originates in the low-pressure baroreceptors in the kidney. Stimulation of the juxtaglomerular (JG) cells by hypovolemia evokes the secretion of renin, which forms angiotensin I, a decapeptide converted to angiotensin II by ACE in the pulmonary circulation. Angiotensin II is also a stimulator of aldosterone secretion. Thus, aldosterone enhances Na^+ reabsorption, and ADH enhances water reabsorption by increasing the permeability of the collecting duct to water.

63. The answer is A. [*Ch 7 XII D 3*] Untreated diabetes mellitus is associated with negative nitrogen balance. The disease reflects a state of severe insulin deficiency combined with a decreased concentration of plasma C-peptide. Since insulin and C-peptide are secreted in equimolar amounts, C-peptide directly reflects pancreatic beta cell secretory activity and, therefore, endogenous insulin secretion.

The most characteristic feature of untreated diabetes mellitus is fasting hyperglycemia due, in part, to the lack of insulin and, in larger part, to the unopposed action of the counter-regulatory (insulin-antagonizing) hormones (glucagon, cortisol, epinephrine, GH). These four hormones cause the liver to change from a glucose-utilizing to a glucose-producing organ, leading to hyperglycemia and glycosuria. The hormones also increase the rate of lipolysis, which leads to a fatty acid–induced decrease in glucose uptake. Cortisol and glucagon promote muscle proteolysis, leading to hyperaminoacidemia, hyperaminoaciduria, elevated blood urea nitrogen (BUN), and negative nitrogen balance. This hormonal imbalance switches liver fatty acid metabolism from oxidation, reesterification, or both to ketogenesis. The sum of these effects is an increase in blood fatty acids, amino acids, and ketones due to the gradual loss of muscle and adipose tissue. The ketosis leads to ketonemia and, eventually, ketonuria.

64. The answer is C. [*Ch 3 III A, B 3 a*] Ascending a mountain results in a decrease in PO_2 due to the reduction in the barometric pressure. Air contains about 21% O_2; therefore, administering supplemental O_2 would increase the PO_2. Hyperventilation would reduce PCO_2, resulting in a higher PO_2, according to the alveolar gas equation.

65. The answer is C. [*Ch 4 IX D 1 a (2) (a); Ch 7 VI D 1 d (2)*] Aldosterone secretion is not significantly depressed following hypophysectomy and patients can regulate Na^+ and K^+ balance. However, hypophysectomy without hormone replacement therapy is incompatible with human life. The pituitary gland is essential for cellular differentiation, somatic growth, adaptation to stress, and reproduction. If the pituitary gland is removed from young, rapidly growing animals, dwarfism occurs due to the absence of GH. Atrophy of the gonads, thyroid gland, and adrenal cortex also occurs in the absence of pituitary tropic hormones.

66. The answer is C. [*Ch 7 VII F 2 b (2)*] Catecholamine-O-methyltransferase (COMT) catalyzes the metabolic inactivation of dopamine, norepinephrine, and epinephrine. Normetanephrine is a deaminated catecholamine produced by the action of COMT on norepinephrine. Dihydroxymandelic acid also is a substrate for COMT. COMT is found in the soluble fraction of cells, especially the liver and kidney, and it is more important in the metabolic degradation of circulating catecholamines.

67. The answer is E. [*Ch 4 IX D 2 e; X C 1 d*] Atrial natriuretic factor (ANF) is secreted by the atrial myocytes in response to increased blood volume. ANF prevents angiotensin formation by inhibiting renin release. The cardiac hormones inhibit angiotensin II and ACTH—two stimuli for aldosterone secretion. Angiotensin II is the primary stimulus for aldosterone secretion. Angiotensin II synthesis is regulated by renin release from the JG cells. The principal stimuli for renin secretion are decreased perfusion pressure in the afferent arterioles, decreased [Na^+] in the macula densa area of the nephron, and norepinephrine secretion from the sympathetic neurons innervating the JG cells. All actions of renin are mediated through the generation of angiotensin II. Other stimuli for aldosterone include corticotropin (ACTH), a high plasma [K^+], and a low plasma [Na^+]. The stimulatory effect of ACTH on aldosterone secretion is powerful but short-lived and is not a major factor in the control of aldosterone production.

68. The answer is C. [*Ch 6 II D 1 c (3) (a); III B 2, 6 a (1); IV B 2 c*] The migrating motor complex (MMC) is initiated by motilin (a hormone released from the endocrine cells in the small intestine) and is not affected by vagotomy. Cutting the vagus nerve prevents receptive relaxation from occurring and thus increases the rate of liquid emptying from the stomach. Primary esophageal peristalsis is coordinated by the swallowing center and thus is prevented by vagotomy.

69. The answer is C. [*Ch 7 VI D 1 b (2) (a)*] Somatostatin is the GH-inhibiting hormone produced in the arcuate nucleus of the tuberoinfundibular neural tract. Somatostatin is released from the median eminence and enters the hypothalamic-hypophysial portal system that perfuses the adenohypophysis. This neuropeptide also is produced in duodenum and the pancreatic delta cells. The plasma GH level is elevated in response to any form of stress as well as to exercise and to deep sleep (stages III and IV). Both insulin- and arginine-induced hypoglycemia are potent stimuli for GH secretion and can be used as provocative tests of pituitary GH reserve.

70. The answer is C. [*Ch 7 XIII D 1 a–d*] Thyroid hormone biosynthesis occurs in four steps: (1) active uptake of inorganic iodide (I^-), (2) oxidation of inorganic iodide to active iodide (I^+, known as iodinium), (3) formation of iodotyrosines by the iodination of tyrosine residues within the matrix of thyroglobulin, and (4) coupling (condensation) of iodotyrosines to form iodothyronines. All four steps require TSH; steps 2, 3, and 4 are catalyzed by the membrane-bound enzyme thyroid peroxidase. Diiodotyrosine and monoiodotyrosine have no biologic activity; however, the iodothyronines—T_3 and T_4 (thyroxine)—are biologically active. Therefore, thyroxine synthesis requires TSH, active iodide, thyroglobulin, thyroid peroxidase, and the coupling of two molecules of diiodotyrosine.

71. The answer is C. [*Ch 7 XII D 2*] Insulin is a lipogenic as well as antilipolytic hormone. It decreases the activity of the intracellular lipase, triglyceride lipase ("hormone-sensitive lipase"), by inhibiting the formation of cAMP. Insulin activates the extracellular lipase, lipoprotein lipase, which is responsible for the hydrolysis of plasma lipoproteins. The hydrolysis of triglycerides in the adipocyte delivers long-chain fatty acids and glycerol into the circulation. Lipolysis in adipocytes is mediated by triglyceride lipase. Catecholamines play a primary role in lipolysis, whereas glucagon, GH, cortisol, and ACTH have lesser lipolytic effects. Cortisol and T_3 modulate the sensitivity of adipocytes to the lipolytic effects of catecholamines.

72. The answer is E. [*Ch 6 III B 2 b, 6 a (1), b (1), C 1 b (2) (b), 2 c (1); IV C 3 b*] Receptive relaxation occurs when the proximal stomach (fundus and corpus) is stretched by the presence of food. Distension of the antrum causes gastrin to be released from G cells. Gastrin enhances gastric acid (HCl) and pancreatic enzyme secretion as well as gastric motility. In addition, expansion of the antrum initiates a vago-vagal reflex that enhances antral contractions.

73. The answer is C. [*Ch 7 VI A 2*] The neural lobe of the pituitary gland, or neurohypophysis, is derived from the neural tube (neural ectoderm). The neural crest gives rise to a wide variety of cells, including neurons with perikarya outside the CNS, parafollicular cells of the thyroid gland, fibroblasts, dentin-producing cells, vascular smooth muscle cells, melanocytes, and Schwann cells. Also derived from the neural crest are the cartilage and bone of the skull and the adrenal medulla.

74. The answer is D. [*Ch 4 IX B 1–3, C 1 a, b, E 1*] Aldosterone acts on the connecting tubule and collecting tubules to increase the reabsorption of Na^+ and the secretion of K^+ and H^+. Thus, excess amounts of this hormone lead to hypokalemia (kaliuresis), alkalemia (alkalosis), and hypertension. By promoting Na^+ retention, excess aldosterone secretion leads to an increase in blood volume via an increase in plasma volume. This effect, in turn, brings about a decline in hematocrit and in plasma oncotic pressure.

amounts of this hormone lead to hypokalemia (kaliuresis), alkalemia (alkalosis), and hypertension. By promoting Na^+ retention, excess aldosterone secretion leads to an increase in blood volume via an increase in plasma volume. This effect, in turn, brings about a decline in hematocrit and in plasma oncotic pressure.

75–76. The answers are: 75-A, 76-D. [*Ch 2 II B 3 a, c (5) (b)*] Sinoatrial (SA) cells are characterized by relatively low membrane potentials, spontaneous depolarization (which indicates pacemaker activity), and a relatively fast rate of depolarization. Cell A meets all of these criteria and, thus, is most likely derived from the SA node.

Pacemaker activity is synonymous with spontaneous depolarization, which is a decrease in the diastolic membrane potential. The decrease in membrane potential is caused by the decline in K^+ conductance; less K^+ leaves the cell during diastole while Na^+ continues to leak into the cell. Thus, membrane potential declines until threshold is reached, at which time an all-or-none action potential is generated. Pacemaker cells lack the fast Na^+ channels, so the rate of membrane depolarization during phase 0 of the cardiac action potential is slow.

77–80. The answers are: 77-C, 78-B, 79-D, 80-A. [*Ch 3 II D 1 c; III B 3; VII B; Table 3-2; Ch 5 VIII D 1*] This patient's blood data are most consistent with a ventilation:perfusion abnormality. Although the Pco_2 is normal, the Po_2 is decreased, as are pH and $[HCO_3^-]$. These latter two findings indicate a metabolic acidosis that is most likely due to accumulation of lactic acid caused by the decreased Po_2 and cardiac output. Ventilation:perfusion abnormalities are the most common cause of clinical hypoxia. A variety of factors affecting the lungs and cardiovascular system may cause these abnormalities.

Given a predicted dead space of 150 ml, this patient's calculated alveolar ventilation is 7.2 L/min. Alveolar ventilation is the volume/min that is effective in gas exchange, which is calculated as follows: (tidal volume − dead space) × respiratory rate. Thus, in this patient, the alveolar ventilation equals (0.6 − 0.15) × 16 = 7.2 L/min.

This patient's Po_2 value while breathing 40% O_2 (120 mm Hg) indicates an increase in the alveolar-to-arterial Po_2 difference. Applying the alveolar gas equation, the alveolar Po_2 is calculated as

$$\text{alveolar } Po_2 = (760 - 47) \times 0.40 - \frac{43}{0.8}$$

$$= 230 \text{ mm Hg}$$

Thus, there is a difference of 110 mm Hg in alveolar-to-arterial Po_2 in this patient (230 − 120 = 110). Normally, the alveolar-to-arterial Po_2 is 5–10 mm Hg. The large difference in this patient is due to a physiologic shunt secondary to the ventilation:perfusion abnormality. Changes in the hemoglobin concentration do not affect the arterial Po_2, which depends solely on the amount of O_2 that is in physical solution in the blood.

This patient's respiratory compliance is 0.03 L/mm Hg. The compliance of the respiratory system is calculated as the change in volume (i.e., tidal volume) divided by the change in distending pressure. Since the respirator is delivering intermittent positive pressure (i.e., 20 mm Hg in order to distend the lungs to accept the delivered volume), the pressure in the lungs at the end of expiration is zero (i.e., atmospheric). Thus, compliance of the respiratory system is 0.6 L/20 mm Hg = 0.03 L/mm Hg. Normal respiratory compliance is about 0.1 L/mm Hg. The decreased compliance ("stiff lungs") in this patient could be due to the pulmonary edema.

81–82. The answers are: 81-C, 82-B. [*Ch 3 II C 4 b, D 2 c; VI B 1*] Lung compliance is calculated as the change in volume per unit change in distending pressure. Since alveolar pressure is zero at the beginning and end of inspiration, the transmural (distending) pressure for the lung is zero minus the interpleural pressure. Only the change in interpleural pressure between the beginning and end of inspiration is given. This difference divided into the tidal volume gives the lung compliance during dynamic conditions, or 150 and 60 ml/cm H_2O. Note that compliance is expressed as volume/pressure.

The change in this patient's lung compliance when she alters her respiratory rate is termed frequency-dependent compliance (FDC). FDC occurs in the presence of high airway resistance, which causes some acini not to fill completely at rapid rates of respiration, due to long time constants. Thus, FDC indicates the presence of high airway resistance, which is synonymous with obstructive lung disease.

83–87. The answers are: 83-D, 84-B, 85-B, 86-E, 87-B. [*Ch 2 IX B 1–2, 4 a, b*] From the record, it is obvious that the major effect at time 1 was a slowing of the heart rate. Cardiac output (Q) must have remained relatively constant, since there is little change in arterial pressure (P) and total peripheral resistance (TPR) [Q = P/TPR]. Both electrical stimulation of the superior cervical ganglion and administration of a cholinergic blocking drug would cause an increased heart rate, whereas electrical stimulation of the lumbar sympathetic nerve roots would result in a significant rise in resistance in the lower portion of the body. The administration of a β-adrenergic blocking drug would reduce heart rate as well as cardiac contractility, so that cardiac output would be significantly reduced, which it is not. Vagal stimulation can cause marked cardiac slowing, but ventricular end-diastolic volume can increase, so that stroke volume increases sufficiently to compensate for the reduced heart rate.

At time 2, there is obviously a rise in heart rate, arterial pressure, and total peripheral resistance. The only factor that could cause this sequence of changes is stimulation of the sympathetic system, which includes the heart.

An α-adrenergic agonist increases peripheral resistance, which raises arterial blood pressure. This results in baroreceptor stimulation, causing a reflex slowing of the heart rate. An α-adrenergic blocking drug has the opposite effect. Cholinergic blocking agents and β-adrenergic agonists increase heart rate. A β-adrenergic blocking agent would reduce heart rate and contractility, resulting in a decreased cardiac output and a reduced arterial pressure.

All adrenergic agonists would cause an increase in heart rate as a direct effect on the SA node, isoproterenol having the greatest effect and norepinephrine the least. A drug such as epinephrine, which has both α- and β-adrenergic properties, can increase contractility and peripheral resistance so that a large increase in arterial pressure occurs, which can then affect the baroreceptors and result in a reflex slowing of cardiac rate. The decrease in heart rate at time 3 was not caused by a reflex effect mediated by the chemoreceptors, because the chemoreceptors, when activated, would lead to a rise in heart rate as well as an increase in contractility. Ventricular extrasystoles have no effect on peripheral resistance.

Arterial pressure, peripheral resistance, and heart rate all rise after time 4, indicating a primary effect on the heart, since the increased pressure would otherwise result in a reflexly produced decreased heart rate and vasodilation. The only nerves likely to induce these effects are the superior cervical ganglia, which are the sympathetic innervation to the heart. Postganglionic fibers of the lumbar sympathetic ganglia would not affect the heart directly. Stimulating the peripheral end of a cut carotid sinus nerve would have no effect on the cardiovascular system, whereas stimulating the central end of this nerve would cause cardiac slowing and vasodilation rather than the results presented here. Vagal stimulation would cause cardiac slowing.

88–93. The answers are: 88-D, 89-B, 90-A, 91-D, 92-C, 93-C. [*Ch 2 V B 1; VI A 3 b, d, 4 b, c; Figure 2-19; Ch 3 II D, E*] The maximal pressure-volume (PV) point for *loop ACDI* lies on the same line as normal, which indicates that there is no change in contractility. The increased ventricular pressure reflects an increase in arterial resistance and results in the increased pressure against which the ventricle must work. The end-diastolic volume is indicated by *point A*; all three loops start at the same end-diastolic volume. Preload is a synonym for end-diastolic volume, so this choice is not correct.

Loop ABFG shows the effect of increased contractility. An increase in contractility causes the maximal PV relationship to move to the left and assume a steeper slope. Thus, stroke volume is increased as a result of these changes, and there is a smaller volume of blood that remains in the ventricles at the end of systole.

Mitral valve closure occurs at the onset of systole (*point A*) and represents the beginning of the isovolumic contraction period. During this interval, as the name implies, the ventricular volume remains constant, and ventricular pressure rises until it exceeds the pressure in the aorta.

Ventricular filling occurs during diastole (indicated by *segment HA*) and is the increase in volume that occurs in preparation for the next contraction. During steady state conditions, the ventricular filling will be equal to the stroke volume that is ejected during the next systole. Changes in venous return, duration of diastole, afterload, and ventricular contractility all can influence the ventricular filling and the stroke volume in a complex fashion. The ventricular PV curve provides an excellent model for understanding these relationships.

The slope of the line from *point G* to *point A* has the units of mm Hg/ml, which is compatible with elastance. Compliance is the reciprocal of elastance and has units of volume/pressure. The line from *point G* to *point A* occurs during diastole, and the slope is a measure of the diastolic elastance of the ventricles.

Stroke volume is represented as the width of each PV loop (i.e., *segments AI, AH,* and *AG*). The end-diastolic volume occurs at *point A* for the three loops. The diastolic reserve volume is the volume of blood from the end-diastolic point to the maximal volume that the ventricle can hold—this volume is not shown on the graph. The residual volume represents a minimal volume of blood that lies between the trabeculae carneae and the papillary muscles. This volume is a part of the end-systolic volume and can never be ejected from the ventricles.

94–97. The answers are: 94-D, 95-B, 96-D, 97-B. [*Ch 4 II A 1 b, 2 a, B 1 a, 2 a–c; Figures 4-3, 4-4*] The man's plasma volume is equal to the Evan's blue space, which is 3 L. Since his plasma volume represents 60% of his blood volume, the total blood volume is calculated by dividing the Evan's blue space by 0.6, or

$$\text{blood volume (L)} = \frac{\text{plasma volume (L)}}{(1-\text{hematocrit})}$$

$$\text{blood volume} = \frac{3 \text{ L}}{0.6} = 5 \text{ L}$$

The interstitial fluid volume (ISFV) cannot be calculated directly by the dilution principle because there is no substance that is confined to the interstitial fluid space. Since the interstitial fluid space constitutes part of the ECF volume, it can be determined by subtracting the plasma volume from the ECF. The ECF volume of this man is equal to the inulin space. Therefore, the ISFV is

$$\text{ISFV} = \text{inulin space} - \text{Evan's blue space}$$
$$= 8 \text{ L} - 3 \text{ L}$$
$$= 5 \text{ L}$$

The ECF volume is determined directly by the dilution of such substances as inulin, mannitol, sucrose, thiosulfate, radiosodium, and radiochloride. Thus, the volume of the ECF is equal to the inulin space, which is given as 8 L.

The mathematical relationship between lean body mass (LBM) and total body water (TBW) is

$$\text{TBW (L)} = 0.7 \text{ LBM (kg)}$$

Therefore,

$$\text{LBM (kg)} = \frac{\text{TBW (L)}}{0.7}$$

Since the total body water is given as the tritiated water space, the calculation is

$$\text{LBM (kg)} = \frac{35 \text{ L}}{0.7} = 50 \text{ L}$$

But 50 L of water weighs 50 kg. Subtracting the 50 kg LBM from the 60 kg body weight leaves 10 kg for the weight of the body fat.

98–100. The answers are: 98-B, 99-D, 100-B. [*Ch 3 IV C 1 a, 3; V B 2; IX C, D 2 b; Figure 3-31*] This man's arterial O_2 content is 12 ml/dl. To calculate this, it is important to recognize that the normal P_{50} is 27 mm Hg, meaning that this man's arterial hemoglobin saturation must be 50%. One reason that this man's hemoglobin saturation is low is his exposure to a low barometric pressure (PB) at an altitude of 11,500 feet. The PB at this altitude is about 490 mm Hg, providing an alveolar P_{O_2} of about 50 mm Hg. The O_2 capacity is calculated from the hemoglobin concentration as follows: 18 g/dl × 1.34 ml/g = 24.12 ml/dl. Disregarding the small amount of O_2 that is dissolved in plasma, the arterial O_2 content would be O_2 capacity × saturation, or 24.12 × 0.5 = 12 ml/dl.

This man's pulmonary artery pressure is likely to be higher than normal due to hypoxic pulmonary vasoconstriction. At high altitudes there would be an increased resistance to flow through the pulmonary vasculature. This increased resistance is the result of smooth muscle contraction in the walls of the pulmonary blood vessels in response to an unknown mediator brought on by hypoxia. The cardiac output in people who live for long periods at high altitudes is in the normal range, although it increases upon initial exposure to such altitudes. Chemoreceptors have no significant effect on the pulmonary circulation.

Venous hemoglobin saturation is decreased at high altitudes, because the hemoglobin starts in the lungs somewhat unsaturated due to the low P_{O_2}. Residents at high altitudes have an increased hemoglobin concentration as a compensation for the low P_{O_2} in the tissues. If hemoglobin concentration rises sufficiently, the O_2 content may be in the normal range, but the hemoglobin saturation would be reduced secondary to the low P_{O_2}. Although cardiac output is normal in people acclimated to high altitude, the venous hemoglobin saturation *is* above normal in people with increased cardiac output. This is due to less O_2 being removed from each unit of blood when O_2 delivery is increased by the higher cardiac output.

101. The answer is C. [*Ch 4 IV B; VI B 2; Figures 4-7, 4-11C*] The Na^+-dependent reabsorption of glucose is blocked by phlorizin, a phenolic glycoside. Thus, this competitive inhibitor virtually blocks the secondary active transport of glucose by the proximal tubule, which leads to glycosuria. Therefore, following phlorizin administration, the clearance of glucose becomes equal to the clearance of inulin. *Curve C* depicts the clearance curve for a filtered substance that is neither reabsorbed nor secreted, such as inulin.

102–103. The answers are: 102-B, 103-A. [*Ch 2 VI B 1*] The gradient that occurs between the ventricular and aortic systolic pressures is diagnostic of aortic stenosis. The normal aortic valve provides a negligible resistance, and the aortic pressure is nearly identical to the ventricular pressure during the phase of rapid ventricular ejection. A similar picture is seen if right ventricular and pulmonary pressures are measured in the presence of pulmonary valve stenosis, but the pressures are proportionately reduced because of the low resistance of the pulmonary circulation.

Semilunar valve stenosis represents an impediment to the ejection of blood from the ventricle and results in an ejection-type murmur during systole. An ejection murmur is diamond-shaped (i.e., it is a crescendo-decrescendo sound that has maximal intensity in midsystole, when the pressure gradient is largest).

104–106. The answers are: 104-C, 105-A, 106-E. [*Ch 3 II H 1 d (1), 2 a*] This patient's functional residual capacity (FRC) is 2.7 L. Since the volume of gas in the spirometer is kept constant, the degree of dilution produced by the lungs after equilibration must be determined by calculating the ratio of F_1/F_2, which equals $0.05 \div 0.03 = 1.67$. Thus, the volume of the spirometer plus the lung volume is 1.67 times the volume of the spirometer, or $1.67 \times 4 = 6.7$ L. Subtracting the volume of the spirometer leaves the lung volume at the start of the test (i.e., FRC), or $6.7 - 4 = 2.7$ L.

This patient's residual volume (RV) is 0.7 L. Since FRC is the sum of RV and expiratory reserve volume (ERV), RV is determined as: $RV = FRC - ERV$, or $2.7 - 2.0 = 0.7$ L.

This patient's total lung capacity (TLC) is 5.7 L. TLC is the sum of vital capacity (VC) and RV, or $5 + 0.7 = 5.7$ L.

107–108. The answers are: 107-E, 108-A. [*Ch 2 Figure 2-19*] Opening of the mitral valve signals the onset of ventricular filling. This occurs when ventricular pressure and volume are at their lowest point (i.e., *point E* on the pressure-volume loop).

Isovolumic contraction is the first period of ventricular systole. It begins (*point A*) when ventricular pressure rises above atrial pressure and the mitral valve closes. Isovolumic contraction ends (*point B*) when the semilunar valve opens and ventricular ejection begins. The onset of ejection is signaled by the point where ventricular volume begins to decrease.

109. The answer is D. [*Ch 7 III A 1*] In humans, there is a diurnal variation in the secretory patterns of ACTH and cortisol (hydrocortisone) and in the excretion of 17-hydroxycorticoids. In individuals who sleep regularly from about 11:00 P.M. to 7:30 A.M., the peak of this circadian rhythm occurs between 6 A.M. and 8 A.M., while the nadir is observed between midnight and 2 A.M. Thus, the peak plasma level of glucocorticoid is entrained to the activity cycle with maximal ACTH-cortisol secretion appearing about 1 hour after awakening. Changes in sleep periods or in longitude cause phase shifts in the ACTH-cortisol secretory pattern; however, the rhythm itself persists. There is an abrogation of the pituitary-adrenocortical rhythm in patients with hypercortisolism.

110–112. The answers are: 110-B, 111-E, 112-D. [*Ch 2 VI A 4 b, c, 6*] *Tracing A* is a ventricular volume curve. The ventricular volume curve is distinctive because of the decline in left ventricular volume during systole and the rise during diastole. The timing of systole and diastole can be identified from the curve's relationship with the EKG tracing.

Rapid filling begins with the opening of the AV valves early in diastole. Diastole normally begins shortly after the end of the T wave and is indicated by the occurrence of the second heart sound (not shown).

Interval C should be readily recognized as diastole. The duration of diastole is largely determined by heart rate, which alters the length of the cardiac cycle. The period of systole does shorten with an increase in heart rate, but diastole is affected much more. At very high heart rates, stroke volume is reduced because of such an abbreviation of diastole that ventricular filling time becomes inadequate. Thus, it is important to terminate ventricular tachycardia because of the reduced cardiac output that results from the rapid rate.

113–115. The answers are: 113-B, 114-C, 115-C. [*Ch 3 II G 1; III B 2 a (3), 3 b; IX C 1 a; Ch 5 VII D 1 a; VIII D 3, E 1; Figure 5-14*] The increased P_{CO_2} in this patient indicates that alveolar ventilation is inadequate. Minute ventilation may be normal or increased while alveolar ventilation is reduced, because the total dead space increases secondary to ventilation:perfusion abnormalities caused by the pulmonary disease. (It is important to remember that total dead space equals alveolar dead space plus anatomic dead space.) The normal arterial pH in this patient indicates that renal compensation has occurred. Renal compensation may require 1–2 weeks, meaning that the abnormality is chronic, ruling out acute respiratory failure. Anemia and carbon monoxide poisoning are not viable alternatives, since arterial P_{O_2} is normal in these two conditions. An anatomic shunt causes hypoxia and typically results in a lowered P_{CO_2} due to the increased ventilation that results from this type of hypoxia.

Smoking causes inflammation and edema in the airways (bronchitis), which can lead to infection of the airways and consequent mucus production. The resultant airway narrowing increases airway resistance and the work of breathing. If respiratory work increases sufficiently, there will be respiratory muscle fatigue and a decrease in alveolar ventilation, which results in hypercapnia and hypoxia. These latter events cause pulmonary artery vasoconstriction and pulmonary hypertension.

This patient has respiratory acidosis, by definition, due to the hypercapnia. Respiratory acidosis is compensated by renal retention of HCO_3^- (normal = 24 mEq/L). Metabolic or lactic acidosis is not present because of the increased HCO_3^- levels. Respiratory alkalosis is produced by a decrease in P_{CO_2}.

116–118. The answers are: 116-B, 117-E, 118-A. [*Ch 3 VII B 2, 3; IX C 1; Table 3-5*] A right-to-left anatomic shunt causes mixed venous blood to flow from the right to the left ventricle without being oxygenated, leading to hypoxia and an increased alveolar-arterial P_{O_2} difference. Ventilation:perfusion imbalance also leads to hypoxia but is not the cause in this patient. With ventilation:perfusion imbalance, arterial P_{O_2} should exceed 500 mm Hg when the patient breathes 100% O_2 at sea level, and alveolar P_{O_2} should be 673 mm Hg (which can be calculated using the alveolar gas equation). A left-to-right anatomic shunt does not cause hypoxia, since oxygenated blood (from the left side of the circulation) enters the right ventricle or the pulmonary artery. Administration of 100% O_2 will completely correct the hypoxia due to diffusion abnormalities or hypoventilation.

The fraction of the cardiac output that represents shunted blood can be calculated using the shunt equation, which states that

$$\frac{Qs}{Qt} = \frac{(Ci_{O_2} - Ca_{O_2})}{(Ci_{O_2} - C\bar{v}_{O_2})}$$

where Qs = the shunted blood flow, Qt = cardiac output, Ci_{O_2} = pulmonary capillary O_2 content, and Ca_{O_2} and $C\bar{v}_{O_2}$ = arterial and venous O_2 content, respectively. The calculation usually is sufficiently accurate only if hemoglobin O_2 is used and the dissolved O_2 is disregarded. In the pulmonary capillaries, hemoglobin should be 100% saturated while the patient is breathing O_2 so that

$$Ci_{O_2} = O_2 \text{ capacity}$$
$$= 18 \text{ g hemoglobin/dl} \times 1.34 \text{ ml } O_2/g$$
$$= 24.1 \text{ ml/dl}$$

Arterial O_2 content can be determined as

$$Ca_{O_2} = Ci_{O_2} \times \text{hemoglobin saturation}$$
$$= 24.1 \text{ ml/dl} \times 0.85$$
$$= 20.5 \text{ ml/dl}$$

Venous O_2 content can be determined if it is remembered that the normal venous P_{O_2} also is 40 mm Hg, which is equivalent to 75% hemoglobin saturation, giving

$$C\bar{v}_{O_2} = 24.1 \times 0.75$$
$$= 18.1 \text{ ml/dl}$$

Thus,

$$\frac{Qs}{Qt} = \frac{24.1 - 20.5}{24.1 - 18.1}$$
$$= \frac{3.6}{6.0} = 0.6$$

To determine the site of a right-to-left shunt, the P_{O_2} must be measured in the left heart rather than the right (i.e., the unsaturated venous blood is used to locate the point where P_{O_2} drops). Thus, if there were an interatrial septal defect with right-to-left flow, the P_{O_2} in the left atrium would be lower than the P_{O_2} in the pulmonary veins. A ventricular septal defect would show a drop in P_{O_2} in the left ventricle compared to that in the left atrium. If the P_{O_2} were low and equal in the pulmonary veins, left atrium, and left ventricle, then the shunt would necessarily be within the lungs, which could indicate the presence of an arteriovenous anastomosis.

119–121. The answers are: 119-D, 120-E, 121-A. [*Ch 2 VIII A, D 1; Ch 3 III B 3 b, C 3 d; VIII A 3; IX C 2*] The major drive for respiration comes from the medullary chemoreceptors, which respond to local H^+ concentration, which, in turn, depends on the P_{CO_2} in the surrounding tissues. During the period of increased ventilation, P_{CO_2} in the body is reduced, thus eliminating the respiratory drive until CO_2 again reaches a threshold value at the end of apnea. The medullary chemoreceptors do not respond to changes in P_{O_2} and, thus, are not the cause of the apnea.

During positive pressure ventilation, the intrathoracic pressure increases, which reduces the pressure gradient between the peripheral tissues and the right heart. Thus, venous return and cardiac output decline during positive pressure ventilation, and the tissues remove more O_2 from each unit of blood as the blood flows through systemic capillaries. Consequently, during positive pressure ventilation venous blood contains less O_2 than normal and the mixed venous P_{O_2} decreases. The mixed venous P_{O_2} would not be less than 40 mm Hg because of the hypoxia that would occur from breathing this gas mixture, which contains only 0.9% less O_2 than air; the hyperventilation would more than make up for the slight reduction in P_{O_2}. Hyperventilation causes a decrease in P_{CO_2} and an increased pH, both of which would reduce the P_{50} of hemoglobin, not increase it.

The arteriovenous O_2 content difference is calculated by dividing the O_2 consumption by the cardiac output: 240 ml/min ÷ 6 L/min = 40 ml O_2/L of blood. Since the measurements in the question are given in ml/dl, the correct answer is 4 ml/dl.

122–126. The answers are: 122-B, 123-B, 124-A, 125-A, 126-D. [*Ch 5 VII A 1–5, D; VII D 1, 3*] Hypoventilation, or decreased alveolar ventilation, is the cause of respiratory acidosis. In metabolic acidosis, the increased $[H^+]$ in the arterial blood is a stimulus for hyperventilation and a decline in P_{CO_2}.

The main components of total CO_2 content ($[$total $CO_2]$) are $[HCO_3^-]$ and dissolved CO_2; an increase in either factor will increase the CO_2 content. In respiratory acidosis, $[$total $CO_2]$ increases via an increase in P_{CO_2}, and in metabolic alkalosis, $[$total $CO_2]$ increases via an increase in $[HCO_3^-]$. The compensatory responses to respiratory acidosis and metabolic alkalosis increase the CO_2 content further by elevating the $[HCO_3^-]$ and $S \times P_{CO_2}$, respectively.

$[$Total $CO_2]$ is decreased when either $[HCO_3^-]$ or dissolved CO_2 is decreased. Thus, in respiratory alkalosis, $[$total $CO_2]$ decreases due to a decline in arterial P_{CO_2}, and in metabolic acidosis, $[$total $CO_2]$ decreases due to a decline in $[HCO_3^-]$. The renal compensation to respiratory alkalosis is increased $[HCO_3^-]$ excretion, which lowers $[$total $CO_2]$ further. Similarly, the respiratory compensation to metabolic acidosis is hyperventilation, which also reduces the CO_2 content further.

The reduced $[HCO_3^-]$ and increased $[H^+]$ of metabolic acidosis lead to a compensatory hyperventilation, which results in decreased arterial P_{CO_2} (hypocapnia). Respiratory acidosis is due to hypoventilation, which results in increased arterial P_{CO_2} (hypercapnia).

The normal $[HCO_3^-]/S \times P_{CO_2}$ ratio is 20:1. This ratio is decreased in acidotic states, by either the increased P_{CO_2} of respiratory acidosis or the decreased $[HCO_3^-]$ of metabolic acidosis. This ratio is increased in alkalotic conditions, by either the decreased P_{CO_2} of respiratory alkalosis or the elevated $[HCO_3^-]$ of metabolic alkalosis.

127–132. The answers are: 127-B, 128-B, 129-A, 130-A, 131-E, 132-D. [*Ch 4 VIII A 2, C 1; IX A 1, B 1, D 2 a (1), e (3) (a)*] ADH, also known as arginine vasopressin, is an octapeptide synthesized mainly in the supraoptic nucleus of the ventral diencephalon. ADH also can be classified as a nonapeptide, if the single cystine moiety is counted as two cysteine residues. ADH is stored in the pars nervosa, from which it is secreted.

ADH regulates plasma osmolality, which normally is about 300 mOsm/kg. It promotes water reabsorption mainly from the tubular fluid in the renal collecting ducts by increasing the water permeability of these cells. ADH allows humans to elaborate a small volume, hypertonic urine in order to conserve water. Since ADH promotes free-water reabsorption, it determines the plasma $[Na^+]$.

Aldosterone, the most potent endogenous mineralocorticoid, acts primarily on the renal collecting ducts to promote Na^+ reabsorption and K^+ and H^+ excretion. It has similar effects on sweat, salivary, and intestinal glands. Thus, aldosterone controls the Na^+ content of the body. In turn, the Na^+ content determines the volume of the various fluid compartments. Aldosterone also increases Na^+ reabsorption by the connecting segment of the nephron. Aldosterone is synthesized in and secreted from the outermost layer of the adrenal cortex, called the zona glomerulosa.

Dipeptidyl carboxypeptidase, also known as ACE or kininase II, is located mainly on the endothelial surface of pulmonary capillaries. ACE catalyzes the conversion of angiotensin I (an inactive decapeptide) to angiotensin II (an active octapeptide). In addition, ACE simultaneously inactivates the nonapeptide, bradykinin. Thus, ACE leads to the increased formation of angiotensin II (a vasoconstrictor) and the decreased formation of bradykinin (a vasodilator). Angiotensin II functions as a vasoconstrictor and aldosterone-inhibiting hormone.

Renin is a proteolytic enzyme produced by the JG cells of the afferent arteriole. These cells are modified smooth muscle cells that have acquired secretory function and, thus, are described as myoepithelial cells. JG cells function as low-pressure baroreceptors that are stimulated by a decrease in renal perfusion pressure caused by a decrease in systemic blood volume or pressure.

133–136. The answers are: 133-C, 134-D, 135-D, 136-E. [*Ch 1 IV A 3, B 3, C 3*] The amount of Ca^{2+} released from the sarcoplasmic reticulum of the smooth and cardiac muscle cells is normally varied to control contractile force. In skeletal muscle, maximal amounts of Ca^{2+} enter the cell with each contraction. Skeletal and smooth muscles are able to recruit additional fibers when more force is required. The heart must contract in a coordinated fashion so that all of its muscle fibers are recruited at the same time. In addition, the heart must relax between contractions; thus, it cannot use repetitive stimulation as a mechanism for increased force production. Smooth, cardiac, and skeletal muscles all are able to influence the force of contraction by varying the initial length (preload) of their sarcomeres.

137–140. The answers are: 137-D, 138-B, 139-D, 140-A. [*Ch 5 VII A 3, 5 c, D 1 b, 2; VIII D 1, 2, 4, E 2, F*] The blood data for *patient D* include high pH ($\downarrow[H^+]$), high $[HCO_3^-]$, and a compensatory increase in PCO_2—findings that coincide with partially compensated metabolic alkalosis. That $[H^+]$ and $[HCO_3^-]$ change in opposite directions indicates a primary metabolic disturbance, and the pH of 7.5 indicates alkalosis. The increment in PCO_2 shows that partial respiratory compensation has occurred.

The blood data for *patient B* include low pH ($\uparrow[H^+]$), low $[HCO_3^-]$, low CO_2 content, and low PCO_2—findings that coincide with a compensated metabolic acidosis. When metabolic acidosis is compensated by a respiratory alkalosis, the pH tends to return to normal, the CO_2 content drops further, and the PCO_2 decreases.

Patient D has the highest CO_2 content (i.e., $[HCO_3^-] + S \times PCO_2$). This patient has partially compensated metabolic alkalosis indicated by a 42% increase in $[HCO_3^-]$ with only a 13% compensatory increase in PCO_2. The CO_2 content is equal to 36.35 mmol/L (35 mmol/L + 1.35 mmol/L).

The blood data for *patient A* include a low pH ($\uparrow[H^+]$), PCO_2, and a compensatory decrease in $[HCO_3^-]$—findings that coincide with respiratory alkalosis, a condition caused by high alveolar ventilation. Note that the change in the alternate variable ($[HCO_3^-]$) is in the same direction as the change in the primary variable (PCO_2). In this patient, the pH of 7.4 together with hypocapnia and decreased $[HCO_3^-]$ indicates full compensation for a respiratory alkalosis (i.e., a $[HCO_3^-]/S \times PCO_2$ ratio of 20:1) and thus the highest alveolar ventilation.

141–143. The answers are: 141-B, 142-D, 143-C. [*Ch 6 IV F 3 b, 5 d–e*] Iron is absorbed from the duodenum and proximal jejunum by a membrane-bound carrier protein on the luminal surface of the enterocyte. Once inside the cell, iron combines with an iron-binding protein to form a complex called ferritin. Before being extruded from the serosal surface of the cell, the iron dissociates from ferritin.

Ca^{2+} is absorbed from the duodenum by a membrane-bound carrier on the luminal surface of the enterocyte that is formed in response to the presence of vitamin D. Once inside the enterocyte, the Ca^{2+} is extruded from the serosal surface of the cell by an active transport system.

Cholesterol must be dissolved in micelles before it can be absorbed. Micelles are small spherical globules formed from bile salts. The polar, water-soluble end of the bile salt faces outward, and the lipid-soluble tail portion of the bile salt faces inward. Cholesterol dissolves in the interior of the micelle, and, when the micelle makes contact with the intestinal membrane, the cholesterol diffuses from the micelle into the enterocyte.

144–147. The answers are: 144-B, 145-E, 146-B, 147-B. [*Ch 4 VI C 3, 4 a, c; Figure 4-9; Table 4-4B*] Most H^+ secretion by the nephron occurs in the proximal tubule by Na^+-H^+ exchange. This active H^+ efflux is linked through a countertransport mechanism to Na^+ influx across the luminal membrane. Most of this secreted H^+ is not excreted but is reabsorbed in the form of H_2O. Most important, any secreted H^+ that combined with HCO_3^- in the lumen forms H_2CO_3, which is dehydrated into CO_2 and H_2O, both of which are reabsorbed. Thus, the secreted H^+ that combines with HCO_3^- does not contribute to the urinary excretion of acid.

K^+ secretion occurs mainly in the cortical collecting tubule. The excreted K^+ is derived mainly from K^+ secretion in this region of the nephron. Secretion of K^+ involves active pumping across the peritubular membrane followed by passive diffusion across the luminal membrane into the tubular lumen.

The cells of the proximal tubule are the major site of ammonia production. Addition of NH_4^+ to the tubular fluid occurs primarily in the proximal tubule. Thus, the accumulation of ammonia in the lumen involves both nonionic diffusion of NH_3 and transport of NH_4^+.

The proximal tubule is the major site for the active secretion of para-aminohippuric acid (PAH) in its anionic form.

148–150. The answers are: 148-A, 149-B, 150-D. [*Ch 7 XIV F 3, H*] Parathyroid hormone (PTH) is the hypercalcemic hormone of the body. Excessive amounts of PTH would produce hypercalcemia and hypophosphatemia together with hypercalciuria and hyperphosphaturia. The major regulator of PTH synthesis and secretion is serum ionized calcium concentration ($[Ca^{2+}]$). PTH maintains the normal plasma total calcium concentration at about 5 mEq/L by interacting with kidney, bone, and intestine. PTH stimulates bone resorption, and it decreases the tubular maximum (Tm) for phosphate by decreasing proximal tubular reabsorption of phosphate, resulting in phosphate diuresis. When PTH secretion is high, the fraction of filtered phosphate that is reabsorbed may fall from the normal 80%–95% to 5%–20%. PTH increases the Tm for Ca^{2+} by increasing distal tubular reabsorption of Ca^{2+}. Although PTH stimulates renal Ca^{2+} reabsorption, the urinary Ca^{2+} is greater than normal in states of excess PTH because of the increased filtered load of Ca^{2+}.

There is an inverse relationship between PTH secretion and plasma $[Ca^{2+}]$: when plasma $[Ca^{2+}]$ falls PTH increases, and when plasma $[Ca^{2+}]$ rises PTH secretion is suppressed. Thus, a decrease in PTH secretion is caused by an increase in plasma $[Ca^{2+}]$. This increase in $[Ca^{2+}]$ is associated with a decrease in $[HPO_4^{2-}]$. Inorganic phosphate has no direct influence on PTH secretion; rather, it is the phosphate-induced decrease in plasma $[Ca^{2+}]$ that stimulates PTH secretion. Conversely, a decrease in plasma inorganic $[HPO_4^{2-}]$ increases plasma $[Ca^{2+}]$ and indirectly inhibits PTH secretion.

Overall, the physiologic action of active vitamin D_3 (calcitriol) is to increase extracellular $[Ca^{2+}]$ and $[HPO_4^{2-}]$. The effects of calcitriol are exerted primarily on the intestine and bone and, to a lesser extent, the kidney. Calcitriol increases renal tubular reabsorption of Ca^{2+} and phosphate. It also is the principal mediator of PTH-induced intestinal Ca^{2+} and phosphate absorption. Thus, vitamin D_3 toxicity leads to hypercalcemia directly by bone resorption and renal Ca^{2+} reabsorption and indirectly by mediating PTH-induced intestinal Ca^{2+} absorption. This hypercalcemia suppresses PTH secretion, leading to increased plasma $[HPO_4^{2-}]$ and increased renal phosphate reabsorption. This is in contradistinction to hyperparathyroidism, which causes an increase in $[Ca^{2+}]$, a decrease in $[HPO_4^{2-}]$, and an increase in plasma PTH level. Thus, hyperparathyroidism must be distinguished from other causes of hypercalcemia.

151–154. The answers are: 151-B, 152-A, 153-C, 154-D. [*Ch 1 XII A 1, C 1 b (2), 2 b, c, D 2 b*] Lesions to the subthalamic nucleus produce ballismus, a movement disorder characterized by wild, flinging movements of the arms and legs on the side of the body opposite to the lesion. Lesions within the motor cortex interrupt axons that are inhibitory to brain stem nuclei, which are, in turn, excitatory to alpha motoneurons innervating antigravity muscles. Loss of these inhibitory fibers (release from inhibition) results in the overactivity of the antigravity muscles that characterizes spasticity. Lesions to the posterior parietal lobe result in apraxia, which is the inability to carry out motor tasks voluntarily, even though the muscle strength and coordination that is required to carry out the movement is intact. Lesions to the vestibular nucleus damage the circuitry required for the vestibular ocular reflex, resulting in nystagmus.

155–159. The answers are: 155-E, 156-A, 157-B, 158-D, 159-C. [*Ch 3 III B 3 a; VII D 3; VIII B 1; IX C 1, 3, D 1 a*] While mountain climbing, the arterial Po_2 would be reduced, with arterial Pco_2 reduced by reflex hyperventilation. Since this is acute exposure to high altitude, arterial pH would be increased; not enough time has passed for the kidneys to correct the blood pH. Only one set of blood data (*E*) coincides with these conditions.

Although arterial Po_2 is normal in anemia, the arterial O_2 content is reduced in proportion to the decrease in hemoglobin concentration. Since the chemoreceptors are not stimulated, ventilation does not change and, thus, arterial Pco_2 and pH are normal. Only one set of blood data (*A*) coincides with these conditions.

Hypoventilation is synonymous with hypercapnia and, thus, with an increase in arterial Pco_2. Only one set of blood data (*B*) reflects this change.

During hyperventilation, arterial Pco_2 decreases and arterial Po_2 increases. Only one set of blood data (*D*) coincides with these conditions.

Patients with chronic obstructive lung disease typically have a reduced arterial Po_2 due to the ventilation:perfusion abnormality that is present. In early stages of the disease, these patients have a normal-to-low Pco_2 and an arterial pH in the normal range, since the kidneys can maintain the correct HCO_3^-/H_2CO_3 ratio. Only one set of blood data (*C*) coincides with these conditions.

160–164. The answers are 160-C, 161-E, 162-A, 163-D, 164-B. [*Ch 4 IV B 1, C 1 a; V C; VI A 3, B 3 c (2); Figures 4-7, 4-10, 4-11; Table 4-5*] The use of inulin clearance (C_{in}) to measure glomerular filtration rate (GFR) is valid because all of the filtered inulin is excreted in the urine without being reabsorbed or secreted by the renal tubules. Thus, GFR is equal to C_{in} as

$$GFR = C_{in} = \frac{U_{in} \times \dot{V}}{P_{in}}$$

where U_{in} and P_{in} = the urinary and plasma inulin concentrations, respectively, and $\dot{V}$ = the urinary volume/minute. Substituting,

$$GFR = C_{in} = \frac{150 \text{ mg/ml} \times 1.2 \text{ ml/min}}{1.5 \text{ mg/ml}}$$

$$= \frac{180 \text{ mg/min}}{1.5 \text{ mg/ml}}$$

$$= 120 \text{ ml/min}$$

Note that with a unit analysis, both mg/ml terms cancel out.

Renal plasma flow (RPF) is calculated from the clearance of para-aminohippuric acid (PAH); however, there are no PAH data provided for this man. Thus, it is necessary to determine RPF from the filtration fraction (FF), defined as the ratio of GFR to RPF, or

$$FF = \frac{GFR}{RPF}$$

FF can be calculated from the arterial and venous inulin concentrations as

$$\text{fraction of inulin filtered} = \frac{A_{in} - V_{in}}{A_{in}}$$

where A_{in} and V_{in} are the arterial and venous inulin concentrations, respectively. Substituting,

$$FF = \frac{1.50 \text{ mg/ml} - 1.20 \text{ mg/ml}}{1.50 \text{ mg/ml}}$$

$$= \frac{0.3}{1.50} = 0.20$$

With FF and GFR determined, the RPF is easily determined as

$$FF = \frac{GFR}{RPF}; \quad RPF = \frac{GFR}{FF} = \frac{120}{0.2} \text{ ml/min} = 600 \text{ ml/min}$$

Because FF = the ratio of GFR to RPF, FF also can be calculated using the extraction of a substance such as inulin to determine RPF. If the urinary concentration of inulin is 150 mg/ml and the urine flow rate is 1.2 ml/min, the urinary excretion rate of inulin is 180 mg/min. If, when the excretion rate was measured, the inulin concentration was 1.50 mg/ml in renal arterial plasma and 1.20 mg/ml in renal venous plasma, each milliliter of plasma traversing the kidneys must have contributed 0.30 mg to the 180 mg that was excreted. Hence, the RPF must have been

$$\frac{180 \text{ mg/min}}{0.3 \text{ mg/ml}} = 600 \text{ ml/min}$$

Thus, with the RPF determined, FF can be calculated as

$$FF = \frac{GFR}{RPF} = \frac{120 \text{ ml/min}}{600 \text{ ml/min}} = 0.2$$

Once RPF is determined, renal blood flow (RBF) can also be discerned. RBF is calculated by dividing RPF by the term (1 − hematocrit), or

$$RBF = \frac{RPF}{1 - \text{hematocrit}} = \frac{600 \text{ ml/min}}{1 - 0.40} = \frac{600 \text{ ml/min}}{0.60} = 1000 \text{ ml/min}$$

The quantity (or amount) of the substance filtered per unit time is termed the filtered load (FL), or amount filtered. It is equal to the product of the GFR (or C_{in}) and the plasma concentration of that substance, or

$$FL = GFR \times P_G$$
$$= 120 \text{ ml/min} \times 0.9 \text{ mg/ml}$$
$$= 108 \text{ mg/min}$$

Note that the ml terms cancel out with a unit analysis and that it is necessary to express the plasma glucose concentration (P_G) in mg/ml and not in the unit of mg/dl given in the original data.

165–167. The answers are: 165-C, 166-A, 167-B. [*Ch 5 VII A 1, 2, D; VIII D 1, 3, 4; Figure 5-14*] Hyperventilation is defined as alveolar ventilation in excess of the body's need for CO_2 elimination, which results in decreased arterial P_{CO_2}—the underlying factor in respiratory alkalosis. Although P_{CO_2} is not given, it can be calculated using the mathematical relationship: $[HCO_3^-]/S \times P_{CO_2} = 20/1$. Applying this equation, it is clear that the values in *set C* indicate a decrease in P_{CO_2}, as

$$\frac{15}{S \times P_{CO_2}} = \frac{20}{1}, \text{ or } S \times P_{CO_2} = 0.75 \text{ mmol/L}$$

Since

$$S \times P_{CO_2} = 0.75 \text{ mmol/L}$$

and

$$S = 0.03 \text{ mmol/L/mm Hg,}$$

then

$$P_{CO_2} = \frac{0.75 \text{ mmol/L}}{0.03 \text{ mmol/L/mm Hg}} = 25 \text{ mm Hg}$$

A P_{CO_2} of 25 mm Hg is significantly lower than normal (40 mm Hg). In acute respiratory alkalosis, there is a decrease in urinary H^+ excretion and an increase in urinary HCO_3^- excretion. In this case, there has been complete renal compensation, as evidenced by the return of the $[HCO_3^-]/S \times P_{CO_2}$ ratio to normal (20:1).

In chronic respiratory tract obstruction (e.g., due to tracheal stenosis, foreign body, or tumor), alveolar ventilation is insufficient to excrete CO_2 at a rate required by the body, which leads to increased arterial P_{CO_2}—the underlying factor in respiratory acidosis. The kidney increases H^+ secretion, resulting in the addition of HCO_3^- to the ECF. The values in *set A* coincide with these changes. In this case, there is partial compensation of the respiratory acidosis as evidenced by the increase in $[HCO_3^-]$. In chronic respiratory acidosis, the respiratory centers become less sensitive to hypercapnia and acidosis and rely on the associated hypoxemia as the primary drive to ventilation. Correction of the low P_{O_2} by the administration of O_2 will diminish respiratory drive, resulting in hypoventilation, a further increase in P_{CO_2}, and possibly CO_2 narcosis. For this reason, O_2 must be given with extreme caution to patients with chronic hypercapnia.

Metabolic acidosis exhibits the characteristics of increased $[H^+]$ (decreased pH), a reduced $[HCO_3^-]$, and a compensatory hyperventilation resulting in hypocapnia. The kidney responds to the increased H^+ load by increasing the secretion and excretion of NH_4^+ and the excretion of titratable acid ($H_2PO_4^-$). The values in *set B* coincide with these changes.

168–171. The answers are: 168-C, 169-A, 170-D, 171-E. [*Ch 1 VIII A 2 b, 5*] Presbyopia (impairment of vision due to old age) is caused by a decrease in the elasticity of the lens. As a result, the eyes are unable to accommodate for near vision. Another condition associated with aging is cataracts, in which the lens becomes progressively less transparent. Myopia is caused by an overall refractive power that is too great for the axial length of the eyeball. It causes distant objects to be focused in front of the retina and can be corrected by a diverging lens. In hyperopia, the overall refractive power is too low for the axial length of the eyeball, and so the eyes must continuously accommodate to see distant objects clearly.

172–175. The answers are: 172-C, 173-A, 174-A, 175-B. [*Ch 6 IV D 3 c–d, 6 a (4), b, F 1 b (5), 4 a (1), 5 c (4) (b)*] If water is not properly absorbed in the small intestine, rapid dehydration and consequent circulatory collapse will result.

Any blockage of the bile duct prevents bile from being secreted and thus prevents the proper absorption of fat. Malabsorption of fat results in malnutrition and steatorrhea. In addition, when bile is not secreted, bilirubin cannot be secreted, and the resulting accumulation of bilirubin in the soft tissues leads to jaundice.

Vitamin B_{12} is required for the normal production of red blood cells (erythropoiesis). Thus, failure to absorb vitamin B_{12}, caused either by lack of intrinsic factor or by intestinal disease, will lead to pernicious anemia.

Carbohydrate malabsorption usually results in watery diarrhea and intestinal gas.

Index